Inquire

Interact

Inspire

Invent

This iScience Interactive Student Textbook Belongs to:

Name

Teacher/Class

Where am I located?

The dot on the map shows where my school is.

FLORIDA

COURSE 1

iSCIENCE

Glencoe

McGraw Hill Education

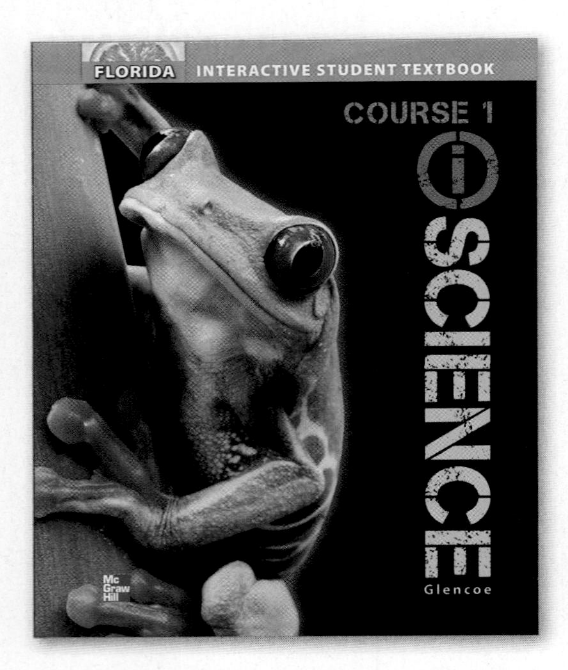

Red-Eyed Tree Frog, *Agalychnis callidryas*
This tiny frog is arboreal—it lives in rain-forest trees. Its diet includes different types of insects. Females generally are about 6 cm–7.5 cm in length. Males are slightly smaller. It can be found in Central America, southern Mexico, and northern Colombia.

The McGraw·Hill Companies

Education

Send all inquiries to:
McGraw-Hill Education
8787 Orion Place
Columbus, OH 43240-4027

ISBN: 978-0-07-660220-9
MHID: 0-07-660220-6

Printed in the United States of America.

2 3 4 5 6 7 8 9 10 QDB 15 14 13 12 11 10

Get Connected to ConnectED

connectED.mcgraw-hill.com

Your online portal to everything you need!

- One-Stop Shop, One Personalized Password
- Easy Intuitive Navigation
- Resources, Resources, Resources

For Students

Leave your books at school. Now you can go online and interact with your StudentWorks™ Plus digital Student Edition from any place, any time!

For Teachers

ConnectED is your one-stop online center for everything you need to teach, including: digital eTeacherEdition, lesson planning and scheduling tools, pacing, and assessment.

For Parents

Get homework help, help your student prepare for testing, and review science topics.

Log on today and get ConnectED!

connectED.mcgraw-hill.com

Username: flscistudent
Password: science2012

Username and password allows you to experience all of the helpful features of ConnectED. For details about how to get your personal password for the life of the adoption, contact your Glencoe sales representative.

Florida Teacher Advisory Board

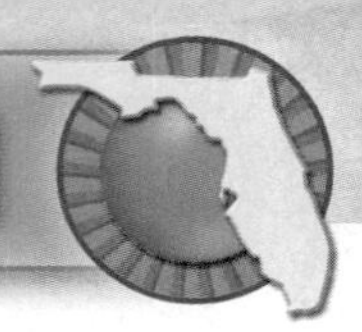

The Florida Teacher Advisory Board provided valuable input in the development of the © 2012 Florida student textbooks.

Ray Amil
Union Park Middle School
Orlando, FL

Maria Swain Kearns
Venice Middle School
Venice, FL

Ivette M. Acevedo Santiago, MEd
Resource Teacher
Lake Nona High School
Orlando, FL

Christy Bowman
Montford Middle School
Tallahassee, FL

Susan Leeds
Department Chair
Howard Middle School
Orlando, FL

Rachel Cassandra Scott
Bair Middle School
Sunrise, FL

Authors and Contributors

Authors

American Museum of Natural History
New York, NY

Michelle Anderson, MS
Lecturer
The Ohio State University
Columbus, OH

Juli Berwald, PhD
Science Writer
Austin, TX

John F. Bolzan, PhD
Science Writer
Columbus, OH

Rachel Clark, MS
Science Writer
Moscow, ID

Patricia Craig, MS
Science Writer
Bozeman, MT

Randall Frost, PhD
Science Writer
Pleasanton, CA

Lisa S. Gardiner, PhD
Science Writer
Denver, CO

Jennifer Gonya, PhD
The Ohio State University
Columbus, OH

Mary Ann Grobbel, MD
Science Writer
Grand Rapids, MI

Whitney Crispen Hagins, MA, MAT
Biology Teacher
Lexington High School
Lexington, MA

Carole Holmberg, BS
Planetarium Director
Calusa Nature Center and Planetarium, Inc.
Fort Myers, FL

Tina C. Hopper
Science Writer
Rockwall, TX

Jonathan D. W. Kahl, PhD
Professor of Atmospheric Science
University of Wisconsin-Milwaukee
Milwaukee, WI

Nanette Kalis
Science Writer
Athens, OH

S. Page Keeley, MEd
Maine Mathematics and Science Alliance
Augusta, ME

Cindy Klevickis, PhD
Professor of Integrated Science and Technology
James Madison University
Harrisonburg, VA

Kimberly Fekany Lee, PhD
Science Writer
La Grange, IL

Michael Manga, PhD
Professor
University of California, Berkeley
Berkeley, CA

Devi Ried Mathieu
Science Writer
Sebastopol, CA

Elizabeth A. Nagy-Shadman, PhD
Geology Professor
Pasadena City College
Pasadena, CA

William D. Rogers, DA
Professor of Biology
Ball State University
Muncie, IN

Donna L. Ross, PhD
Associate Professor
San Diego State University
San Diego, CA

Marion B. Sewer, PhD
Assistant Professor
School of Biology
Georgia Institute of Technology
Atlanta, GA

Julia Meyer Sheets, PhD
Lecturer
School of Earth Sciences
The Ohio State University
Columbus, OH

Michael J. Singer, PhD
Professor of Soil Science
Department of Land, Air and Water Resources
University of California
Davis, CA

Karen S. Sottosanti, MA
Science Writer
Pickerington, Ohio

Paul K. Strode, PhD
I.B. Biology Teacher
Fairview High School
Boulder, CO

Jan M. Vermilye, PhD
Research Geologist
Seismo-Tectonic Reservoir Monitoring (STRM)
Boulder, CO

Judith A. Yero, MA
Director
Teacher's Mind Resources
Hamilton, MT

Dinah Zike, MEd
Author, Consultant, Inventor of Foldables
Dinah Zike Academy; Dinah-Might Adventures, LP
San Antonio, TX

Margaret Zorn, MS
Science Writer
Yorktown, VA

Authors and Contributors

Consulting Authors

Alton L. Biggs
Biggs Educational Consulting
Commerce, TX

Ralph M. Feather, Jr., PhD
Assistant Professor
Department of Educational Studies and Secondary Education
Bloomsburg University
Bloomsburg, PA

Douglas Fisher, PhD
Professor of Teacher Education
San Diego State University
San Diego, CA

Edward P. Ortleb
Science/Safety Consultant
St. Louis, MO

Series Consultants

Science

Solomon Bililign, PhD
Professor
Department of Physics
North Carolina Agricultural and Technical State University
Greensboro, NC

John Choinski
Professor
Department of Biology
University of Central Arkansas
Conway, AR

Anastasia Chopelas, PhD
Research Professor
Department of Earth and Space Sciences
UCLA
Los Angeles, CA

David T. Crowther, PhD
Professor of Science Education
University of Nevada, Reno
Reno, NV

A. John Gatz
Professor of Zoology
Ohio Wesleyan University
Delaware, OH

Sarah Gille, PhD
Professor
University of California San Diego
La Jolla, CA

David G. Haase, PhD
Professor of Physics
North Carolina State University
Raleigh, NC

Janet S. Herman, PhD
Professor
Department of Environmental Sciences
University of Virginia
Charlottesville, VA

David T. Ho, PhD
Associate Professor
Department of Oceanography
University of Hawaii
Honolulu, HI

Ruth Howes, PhD
Professor of Physics
Marquette University
Milwaukee, WI

Jose Miguel Hurtado, Jr., PhD
Associate Professor
Department of Geological Sciences
University of Texas at El Paso
El Paso, TX

Monika Kress, PhD
Assistant Professor
San Jose State University
San Jose, CA

Mark E. Lee, PhD
Associate Chair & Assistant Professor
Department of Biology
Spelman College
Atlanta, GA

Linda Lundgren
Science writer
Lakewood, CO

Keith O. Mann, PhD
Ohio Wesleyan University
Delaware, OH

Charles W. McLaughlin, PhD
Adjunct Professor of Chemistry
Montana State University
Bozeman, MT

Katharina Pahnke, PhD
Research Professor
Department of Geology and Geophysics
University of Hawaii
Honolulu, HI

Jesús Pando, PhD
Associate Professor
DePaul University
Chicago, IL

Hay-Oak Park, PhD
Associate Professor
Department of Molecular Genetics
Ohio State University
Columbus, OH

David A. Rubin, PhD
Associate Professor of Physiology
School of Biological Sciences
Illinois State University
Normal, IL

Toni D. Sauncy
Assistant Professor of Physics
Department of Physics
Angelo State University
San Angelo, TX

Series Consultants, continued

Malathi Srivatsan, PhD
Associate Professor of Neurobiology
College of Sciences and Mathematics
Arkansas State University
Jonesboro, AR

Cheryl Wistrom, PhD
Associate Professor of Chemistry
Saint Joseph's College
Rensselaer, IN

Reading

ReLeah Cossett Lent
Author/Educational Consultant
Blue Ridge, GA

Math

Vik Hovsepian
Professor of Mathematics
Rio Hondo College
Whittier, CA

Series Reviewers

Thad Boggs
Mandarin High School
Jacksonville, FL

Catherine Butcher
Webster Junior High School
Minden, LA

Erin Darichuk
West Frederick Middle School
Frederick, MD

Joanne Hedrick Davis
Murphy High School
Murphy, NC

Anthony J. DiSipio, Jr.
Octorara Middle School
Atglen, PA

Adrienne Elder
Tulsa Public Schools
Tulsa, OK

Carolyn Elliott
Iredell-Statesville Schools
Statesville, NC

Christine M. Jacobs
Ranger Middle School
Murphy, NC

Jason O. L. Johnson
Thurmont Middle School
Thurmont, MD

Felecia Joiner
Stony Point Ninth Grade Center
Round Rock, TX

Joseph L. Kowalski, MS
Lamar Academy
McAllen, TX

Brian McClain
Amos P. Godby High School
Tallahassee, FL

Von W. Mosser
Thurmont Middle School
Thurmont, MD

Ashlea Peterson
Heritage Intermediate Grade Center
Coweta, OK

Nicole Lenihan Rhoades
Walkersville Middle School
Walkersvillle, MD

Maria A. Rozenberg
Indian Ridge Middle School
Davie, FL

Barb Seymour
Westridge Middle School
Overland Park, KS

Ginger Shirley
Our Lady of Providence Junior-Senior High School
Clarksville, IN

Curtis Smith
Elmwood Middle School
Rogers, AR

Sheila Smith
Jackson Public School
Jackson, MS

Sabra Soileau
Moss Bluff Middle School
Lake Charles, LA

Tony Spoores
Switzerland County Middle School
Vevay, IN

Nancy A. Stearns
Switzerland County Middle School
Vevay, IN

Kari Vogel
Princeton Middle School
Princeton, MN

Alison Welch
Wm. D. Slider Middle School
El Paso, TX

Linda Workman
Parkway Northeast Middle School
Creve Coeur, MO

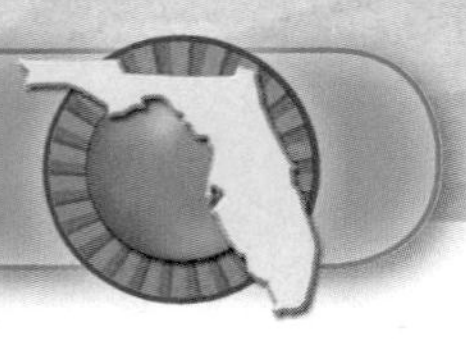

With your book!

Answer questions, record data, and interact with images directly in your book!

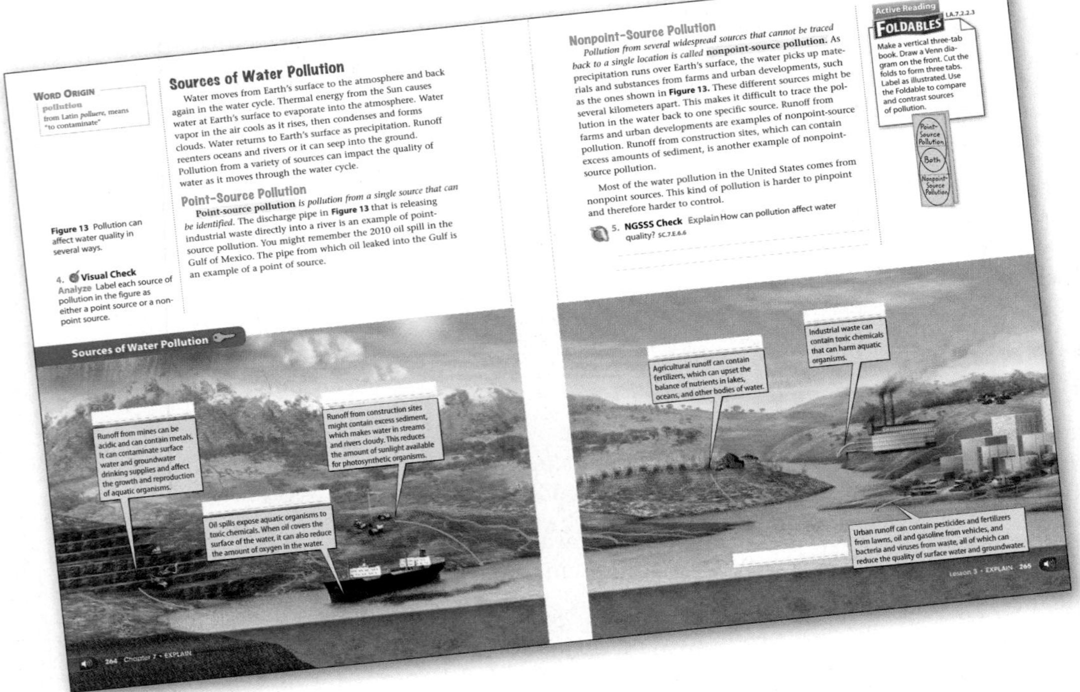

WORD ORIGIN
pollution
from Latin *polluere*, means "to contaminate"

Sources of Water Pollution

Water moves from Earth's surface to the atmosphere and back again in the water cycle. Thermal energy from the Sun causes water at Earth's surface to evaporate into the atmosphere. Water vapor in the air cools as it rises, then condenses and forms clouds. Water returns to Earth's surface as precipitation. Runoff reenters oceans and rivers or it can seep into the ground. Pollution from a variety of sources can impact the quality of water as it moves through the water cycle.

Point-Source Pollution

Point-source pollution *is pollution from a single source that can be identified.* The discharge pipe in **Figure 13** that is releasing industrial waste directly into a river is an example of point-source pollution. You might remember the 2010 oil spill in the Gulf of Mexico. The pipe from which oil leaked into the Gulf is an example of a point of source.

Figure 13 Pollution can affect water quality in several ways.

4. **Visual Check** Analyze Label each source of pollution in the figure as either a point source or a non-point source.

Nonpoint-Source Pollution

Pollution from several widespread sources that cannot be traced back to a single location is called **nonpoint-source pollution.** As precipitation runs over Earth's surface, the water picks up materials and substances from farms and urban developments, such as the ones shown in **Figure 13.** These different sources might be several kilometers apart. This makes it difficult to trace the pollution in the water back to one specific source. Runoff from farms and urban developments are examples of nonpoint-source pollution. Runoff from construction sites, which can contain excess amounts of sediment, is another example of nonpoint-source pollution.

Most of the water pollution in the United States comes from nonpoint sources. This kind of pollution is harder to pinpoint and therefore harder to control.

5. **NGSSS Check** Explain How can pollution affect water quality? SC.7.E.6.6

Active Reading FOLDABLES LA.7.2.2.3
Make a vertical three-tab book. Draw a Venn diagram on the front. Cut the folds to form three tabs. Label as illustrated. Use the Foldable to compare and contrast sources of pollution.

Online!

Log on to ConnectED for a digital version of this book that includes

- audio;
- animations;
- virtual labs.

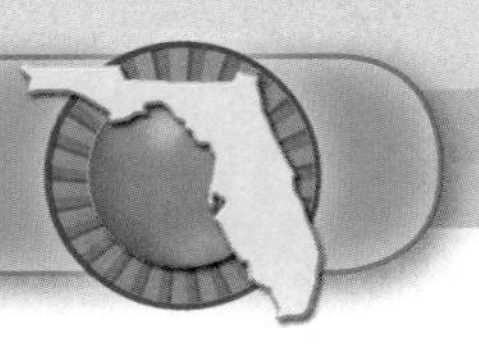

Labs, Labs, Labs

Launch Labs at the beginning of every lesson let you be the scientist! The ⓘLAB Station on ConnectED has all the labs for each chapter.

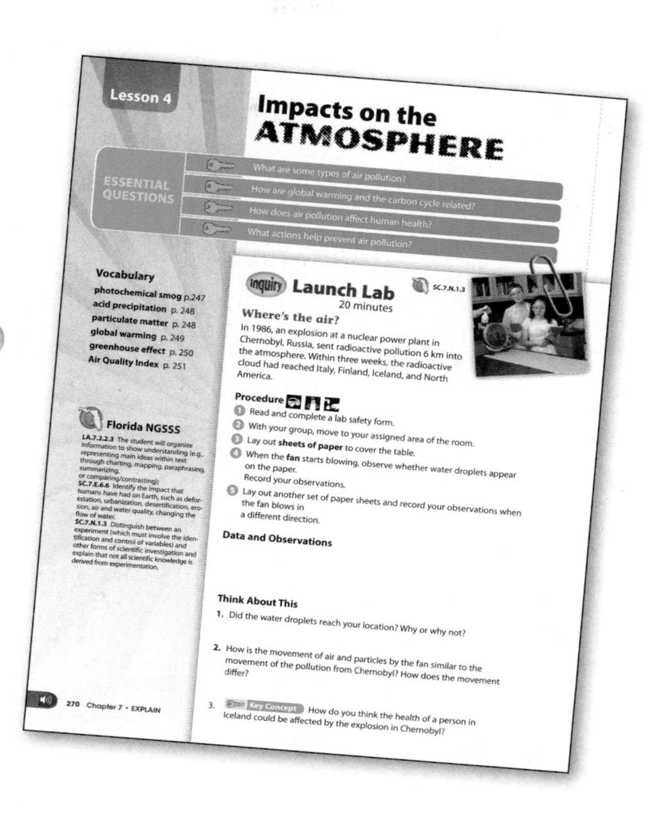

Lesson 4

Impacts on the ATMOSPHERE

ESSENTIAL QUESTIONS

- What are some types of air pollution?
- How are global warming and the carbon cycle related?
- How does air pollution affect human health?
- What actions help prevent air pollution?

Vocabulary

photochemical smog p. 247
acid precipitation p. 248
particulate matter p. 248
global warming p. 249
greenhouse effect p. 250
Air Quality Index p. 251

Florida NGSSS

LA.7.2.2.3 The student will organize information to show understanding (e.g., representing main ideas within text through charting, mapping, paraphrasing, summarizing, or comparing/contrasting);
SC.7.E.6.6 Identify the impact that humans have had on Earth, such as deforestation, urbanization, desertification, erosion, air and water quality, changing the flow of water.
SC.7.N.1.3 Distinguish between an experiment (which must involve the identification and control of variables) and other forms of scientific investigation and explain that not all scientific knowledge is derived from experimentation.

Inquiry Launch Lab SC.7.N.1.3
20 minutes

Where's the air?

In 1986, an explosion at a nuclear power plant in Chernobyl, Russia, sent radioactive pollution 6 km into the atmosphere. Within three weeks, the radioactive cloud had reached Italy, Finland, Iceland, and North America.

Procedure

1. Read and complete a lab safety form.
2. With your group, move to your assigned area of the room.
3. Lay out **sheets of paper** to cover the table.
4. When the **fan** starts blowing, observe whether water droplets appear on the paper. Record your observations.
5. Lay out another set of paper sheets and record your observations when the fan blows in a different direction.

Data and Observations

Think About This

1. Did the water droplets reach your location? Why or why not?
2. How is the movement of air and particles by the fan similar to the movement of the pollution from Chernobyl? How does the movement differ?
3. Key Concept How do you think the health of a person in Iceland could be affected by the explosion in Chernobyl?

270 Chapter 7 • EXPLAIN

Virtual Labs

Virtual Labs provide a highly interactive lab experience.

...with ⓘSCIENCE.

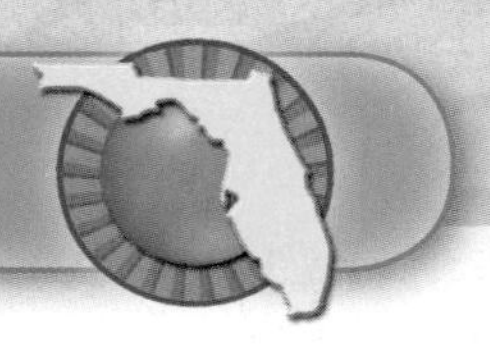

iRead iScience

Sequence Words

While you read, watch for words that show the order events happen:

- first
- next
- last
- begins
- second
- later

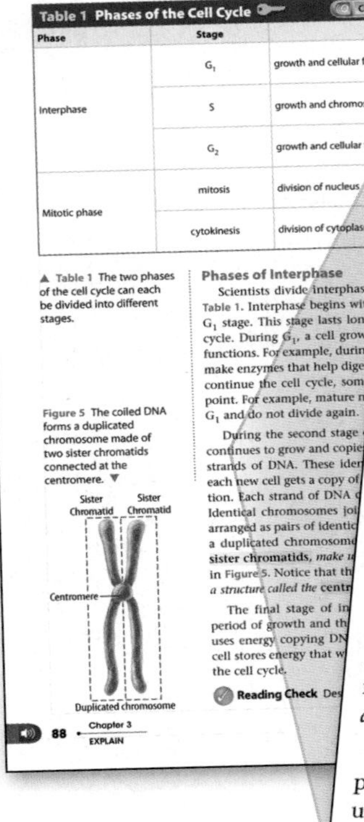

Table 1 Phases of the Cell Cycle Concepts in Motion Interactive Table

Phase	Stage	Description
Interphase	G_1	growth and cellular functions; organelle replication
	S	growth and chromosome replication
	G_2	growth and cellular functions
Mitotic phase	mitosis	division of nucleus
	cytokinesis	division of cytoplasm

▲ Table 1 The two phases of the cell cycle can each be divided into different stages.

Figure 5 The coiled DNA forms a duplicated chromosome made of two sister chromatids connected at the centromere. ▼

88 Chapter 3 EXPLAIN

Phases of Interphase

Scientists divide interphase into three stages, Table 1. Interphase begins with a period of rapid G_1 stage. This stage lasts longer than other stag cycle. During G_1, a cell grows and carries out it functions. For example, during G_1, cells that line make enzymes that help digest your food. Althou continue the cell cycle, some cells stop the cell point. For example, mature nerve cells in your br G_1 and do not divide again.

During the second stage of interphase—the S continues to grow and copies its DNA. There are strands of DNA. These identical strands of DN each new cell gets a copy of the original cell's ge tion. Each strand of DNA coils up and forms a Identical chromosomes join together. The cell's arranged as pairs of identical chromosomes. Each a duplicated chromosome. *Two identical chrom* **sister chromatids,** *make up a duplicated chromoso* in **Figure 5.** Notice that the *sister chromatids are h* *a structure called the* **centromere.**

The final stage of interphase—the G_2 stag period of growth and the final preparation for r uses energy copying DNA during the S stage. D cell stores energy that will be used during the mi the cell cycle.

✓ Reading Check

Vocabulary Help

Science terms are highlighted and reviewed to check your understanding.

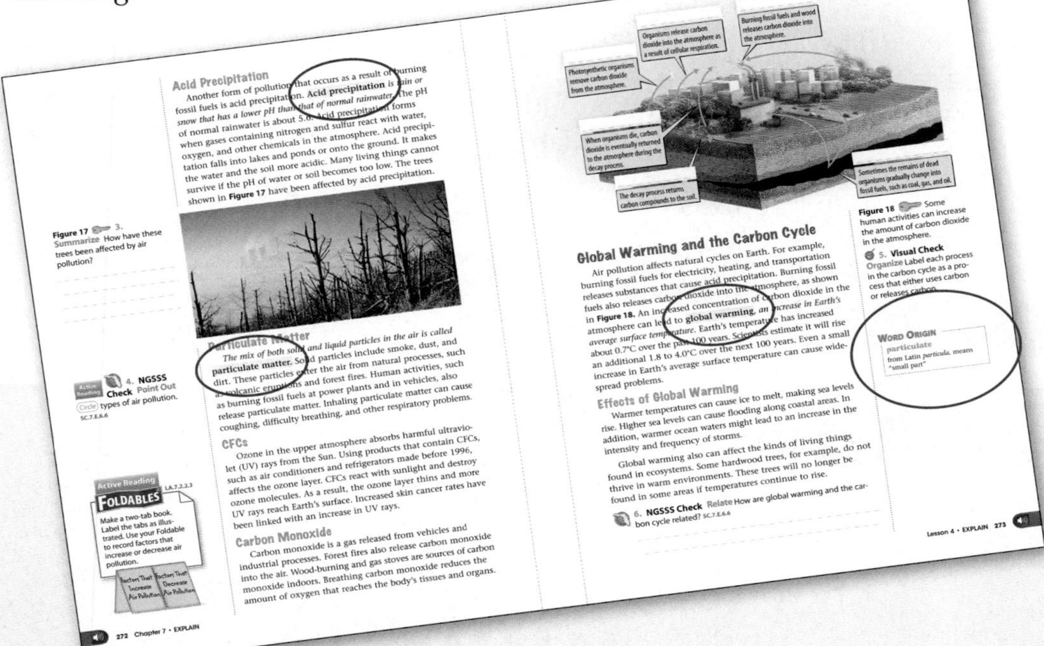

Acid Precipitation

Another form of pollution that occurs as a result of burning fossil fuels is acid precipitation. **Acid precipitation** *is rain or snow that has a lower pH than that of normal rainwater.* The pH of normal rainwater is about 5.6. Acid precipitation forms when gases containing nitrogen and sulfur react with water, oxygen, and other chemicals in the atmosphere. Acid precipitation falls into lakes and ponds or onto the ground. It makes the water and the soil more acidic. Many living things cannot survive if the pH of water or soil becomes too low. The trees shown in **Figure 17** have been affected by acid precipitation.

Figure 17 3. Summarize How have these trees been affected by air pollution?

Particulate Matter

The mix of both solid and liquid particles in the air is called **particulate matter.** Solid particles include smoke, dust, and dirt. These particles enter the air from natural processes, such as volcanic eruptions and forest fires. Human activities, such as burning fossil fuels at power plants and in vehicles, also release particulate matter. Inhaling particulate matter can cause coughing, difficulty breathing, and other respiratory problems.

Active Reading 4. NGSSS Check Point Out Circle types of air pollution. SC.7.E.6.6

CFCs

Ozone in the upper atmosphere absorbs harmful ultraviolet (UV) rays from the Sun. Using products that contain CFCs, such as air conditioners and refrigerators made before 1996, affects the ozone layer. CFCs react with sunlight and destroy ozone molecules. As a result, the ozone layer thins and more UV rays reach Earth's surface. Increased skin cancer rates have been linked with an increase in UV rays.

Active Reading FOLDABLES LA.7.2.2.3 Make a two-tab book. Label the tabs as illustrated. Use your Foldable to record factors that increase or decrease air pollution.

Factors That Increase Air Pollution | Factors That Decrease Air Pollution

Carbon Monoxide

Carbon monoxide is a gas released from vehicles and industrial processes. Forest fires also release carbon monoxide into the air. Wood-burning and gas stoves are sources of carbon monoxide indoors. Breathing carbon monoxide reduces the amount of oxygen that reaches the body's tissues and organs.

272 Chapter 7 • EXPLAIN

Figure 18 Some human activities can increase the amount of carbon dioxide in the atmosphere.

5. Visual Check Organize Label each process in the carbon cycle as a process that either uses carbon or releases carbon.

Global Warming and the Carbon Cycle

Air pollution affects natural cycles on Earth. For example, burning fossil fuels for electricity, heating, and transportation releases substances that cause acid precipitation. Burning fossil fuels also releases carbon dioxide into the atmosphere, as shown in **Figure 18.** An increased concentration of carbon dioxide in the atmosphere can lead to **global warming,** *an increase in Earth's average surface temperature.* Earth's temperature has increased about 0.7°C over the past 100 years. Scientists estimate it will rise an additional 1.8 to 4.0°C over the next 100 years. Even a small increase in Earth's average surface temperature can cause widespread problems.

WORD ORIGIN
particulate
from Latin *particula,* means "small part"

Effects of Global Warming

Warmer temperatures can cause ice to melt, making sea levels rise. Higher sea levels can cause flooding along coastal areas. In addition, warmer ocean waters might lead to an increase in the intensity and frequency of storms.

Global warming also can affect the kinds of living things found in ecosystems. Some hardwood trees, for example, do not thrive in warm environments. These trees will no longer be found in some areas if temperatures continue to rise.

6. NGSSS Check Relate How are global warming and the carbon cycle related? SC.7.E.6.6

Lesson 4 • EXPLAIN 273

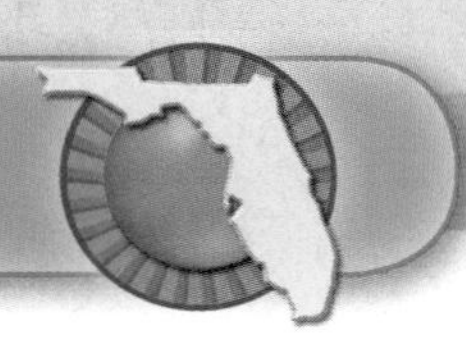

Write the answers to questions right in your book!

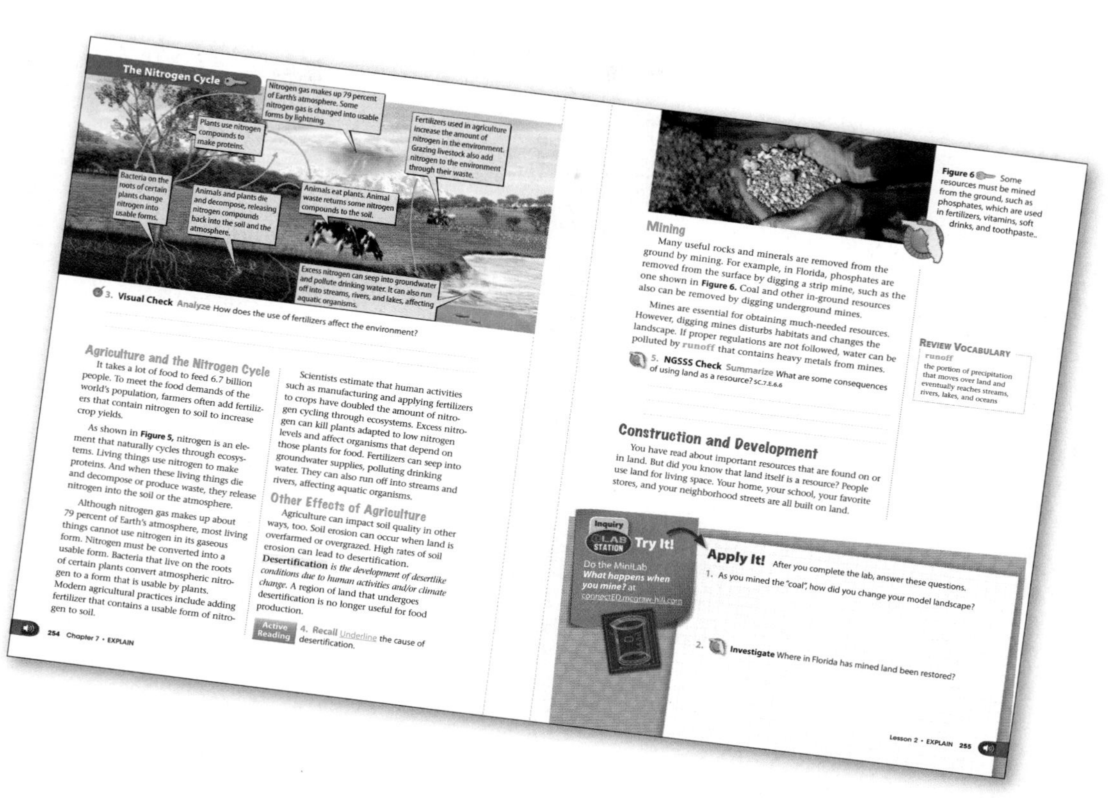

3. **Visual Check** Analyze How does the use of fertilizers affect the environment?

Agriculture and the Nitrogen Cycle

It takes a lot of food to feed 6.7 billion people. To meet the food demands of the world's population, farmers often add fertilizers that contain nitrogen to soil to increase crop yields.

As shown in **Figure 5,** nitrogen is an element that naturally cycles through ecosystems. Living things use nitrogen to make proteins. And when these living things die and decompose or produce waste, they release nitrogen into the soil or the atmosphere.

Although nitrogen gas makes up about 79 percent of Earth's atmosphere, most living things cannot use nitrogen in its gaseous form. Nitrogen must be converted into a usable form. Bacteria that live on the roots of certain plants convert atmospheric nitrogen to a form that is usable by plants. Modern agricultural practices include adding fertilizer that contains a usable form of nitrogen to soil.

Scientists estimate that human activities such as manufacturing and applying fertilizers to crops have doubled the amount of nitrogen cycling through ecosystems. Excess nitrogen can kill plants adapted to low nitrogen levels and affect organisms that depend on those plants for food. Fertilizers can seep into groundwater supplies, polluting drinking water. They can also run off into streams and rivers, affecting aquatic organisms.

Other Effects of Agriculture

Agriculture can impact soil quality in other ways, too. Soil erosion can occur when land is overfarmed or overgrazed. High rates of soil erosion can lead to desertification. **Desertification** *is the development of desertlike conditions due to human activities and/or climate change.* A region of land that undergoes desertification is no longer useful for food production.

Active Reading 4. **Recall** Underline the cause of desertification.

254 Chapter 7 • EXPLAIN

Figure 6 Some resources must be mined from the ground, such as phosphates, which are used in fertilizers, vitamins, soft drinks, and toothpaste..

Mining

Many useful rocks and minerals are removed from the ground by mining. For example, in Florida, phosphates are removed from the surface by digging a strip mine, such as the one shown in **Figure 6.** Coal and other in-ground resources also can be removed by digging underground mines.

Mines are essential for obtaining much-needed resources. However, digging mines disturbs habitats and changes the landscape. If proper regulations are not followed, water can be polluted by **runoff** that contains heavy metals from mines.

5. **NGSSS Check** Summarize What are some consequences of using land as a resource? SC.7.E.6.6

Review Vocabulary
runoff
the portion of precipitation that moves over land and eventually reaches streams, rivers, lakes, and oceans

Construction and Development

You have read about important resources that are found on or in land. But did you know that land itself is a resource? People use land for living space. Your home, your school, your favorite stores, and your neighborhood streets are all built on land.

Inquiry Lab Station **Try It!**
Do the MiniLab *What happens when you mine?* at connectED.mcgraw-hill.com

Apply It! After you complete the lab, answer these questions.

1. As you mined the "coal", how did you change your model landscape?

2. **Investigate** Where in Florida has mined land been restored?

Lesson 2 • EXPLAIN 255

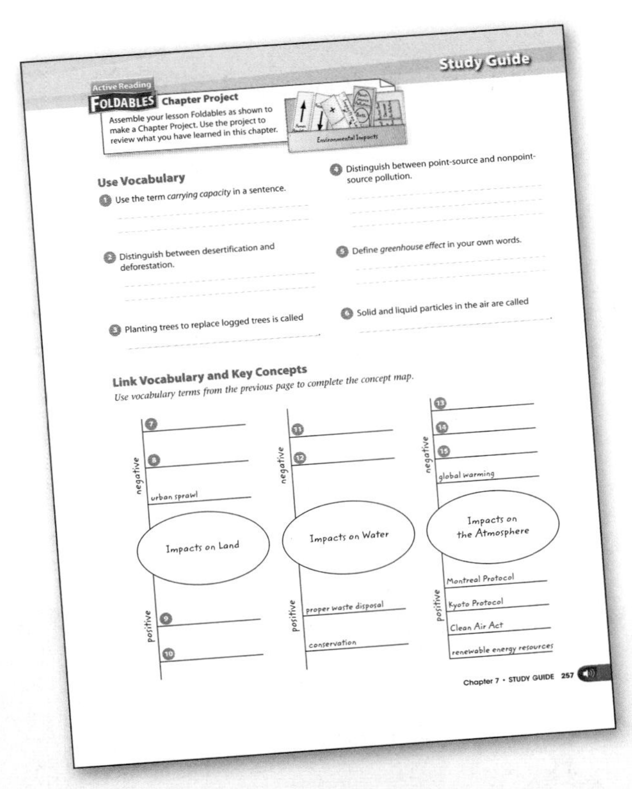

Study Guide

Active Reading **Foldables** Chapter Project

Assemble your lesson Foldables as shown to make a Chapter Project. Use the project to review what you have learned in this chapter.

Use Vocabulary

1. Use the term *carrying capacity* in a sentence.
2. Distinguish between desertification and deforestation.
3. Planting trees to replace logged trees is called
4. Distinguish between point-source and nonpoint-source pollution.
5. Define *greenhouse effect* in your own words.
6. Solid and liquid particles in the air are called

Link Vocabulary and Key Concepts

Use vocabulary terms from the previous page to complete the concept map.

Chapter 7 • STUDY GUIDE 257

Concept Map

Each chapter's **Concept Map** gives you a place to show all the science connections you've learned.

TABLE OF CONTENTS

Nature of Science:

Methods of Science

Methods of Science .... NOS 2

PAGE KEELEY SCIENCE PROBES *The Scientific Method* .... NOS 1

LESSON 1 **Understanding Science** .... NOS 4

LESSON 2 **Measurement and Scientific Tools** .... NOS 12

Science & Engineering The Design Process .... NOS 19

LESSON 3 **Case Study: The Iceman's Last Journey** .... NOS 20

Mini Benchmark Assessments .... NOS 30

Check It! ☐ Lesson 1 ☐ Lesson 2 ☐ Lesson 3

Inquiry LAB STATION

Skill Practice: What can you learn by collecting and analyzing data?

Inquiry Lab: Inferring from Indirect Evidence

Your online portal to everything you need!
Video • Audio • Review • ⓘLab Station • WebQuests • Assessment • Concepts in Motion • Personal Tutors • Virtual Labs

Here are some of the exciting digital activities you will find in this chapter!

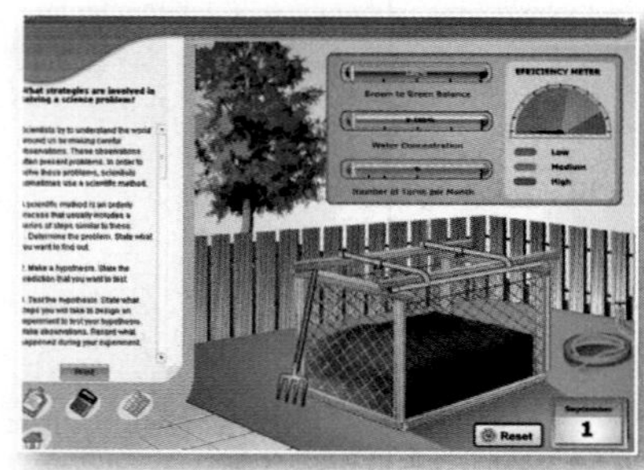

Virtual Lab: What strategies are involved in solving a science problem?

BrainPOP: Scientific Methods

Page Keeley Science Probe

Unit 1 EARTH STRUCTURES

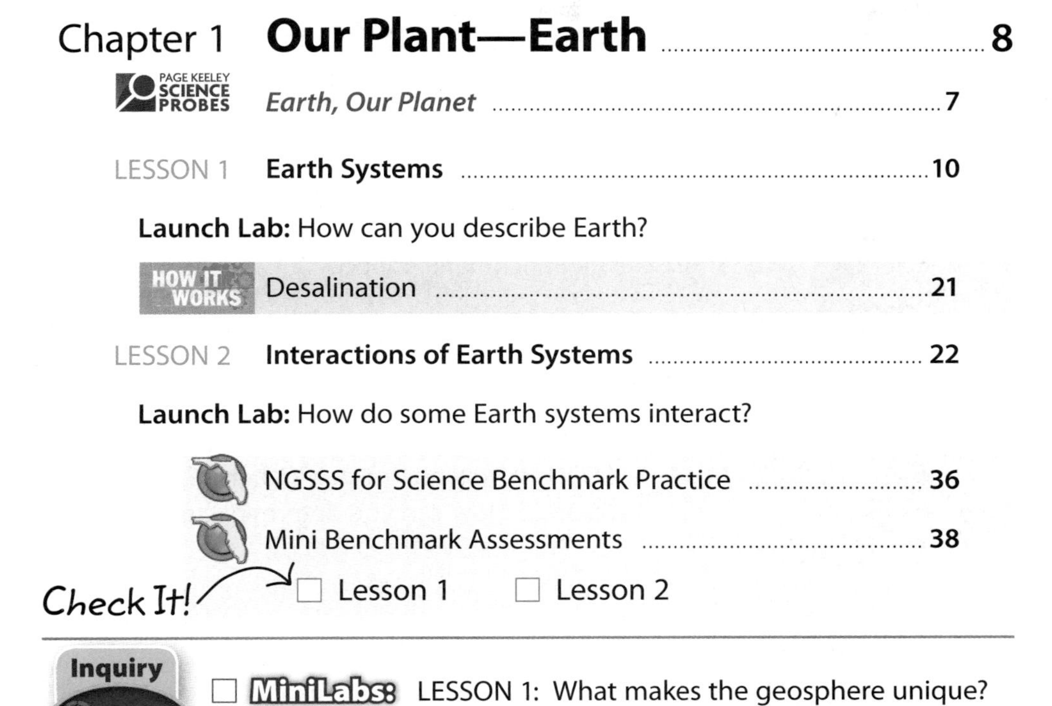

Chapter 1 Our Plant—Earth 8

PAGE KEELEY SCIENCE PROBES *Earth, Our Planet* 7

LESSON 1 **Earth Systems** 10

Launch Lab: How can you describe Earth?

HOW IT WORKS Desalination 21

LESSON 2 **Interactions of Earth Systems** 22

Launch Lab: How do some Earth systems interact?

NGSSS for Science Benchmark Practice 36

Mini Benchmark Assessments 38

Check It! ☐ Lesson 1 ☐ Lesson 2

Inquiry ⓘLAB STATION

☐ **MiniLabs:** LESSON 1: What makes the geosphere unique?
LESSON 2: How do plants contribute to the water cycle?

Try It! then Apply It!

Inquiry Lab: How do Earth's systems interact?

Your online portal to everything you need!
Video • Audio • Review • ⓘLab Station • WebQuests • Assessment • Concepts in Motion • Personal Tutors • Virtual Labs

Here are some of the exciting digital activities you will find in this chapter!

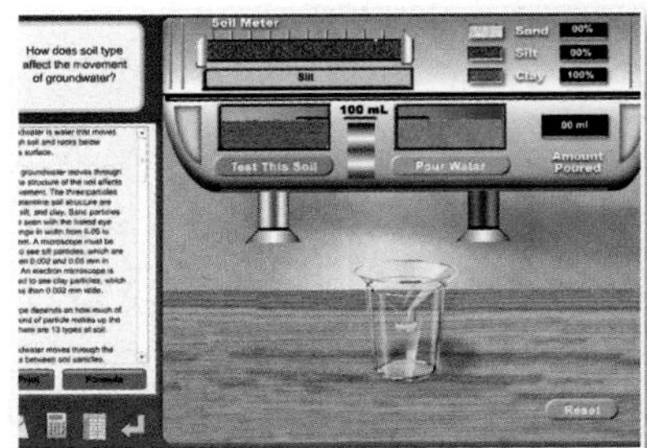

Virtual Lab: How does soil type affect the movement of groundwater?

BrainPOP: Groundwater

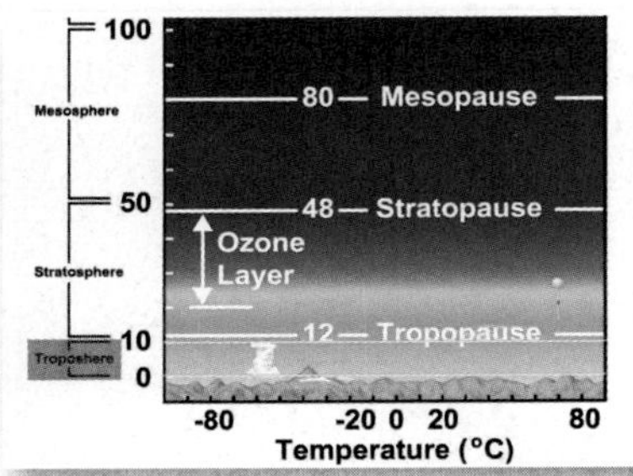

Concepts in Motion: Layers of the Atmosphere

TABLE OF CONTENTS

Chapter 2 **Weathering and Soil** 42

PAGE KEELEY SCIENCE PROBES *Which came first?* 41

LESSON 1 **Weathering** 44

Launch Lab: How can rocks be broken down?

LESSON 2 **Soil** 53

Launch Lab: What is in your soil?

NGSSS for Science Benchmark Practice 66

Mini Benchmark Assessments 68

Check It! ☐ Lesson 1 ☐ Lesson 2

☐ **MiniLabs:** LESSON 1: How are rocks weathered?
LESSON 2: How can you determine soil composition?

Try It! then Apply It!

Skill Practice: What causes weathering?

Inquiry Lab: Soil Horizons and Soil Formation

ConnectED **Your online portal to everything you need!**
Video • Audio • Review • Lab Station • WebQuests • Assessment • Concepts in Motion • Personal Tutors • Virtual Labs

Here are some of the exciting digital activities you will find in this chapter!

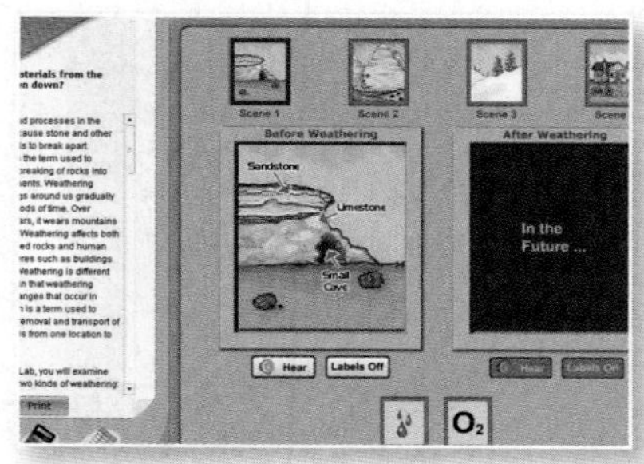

Virtual Lab: How are materials from Earth broken down?

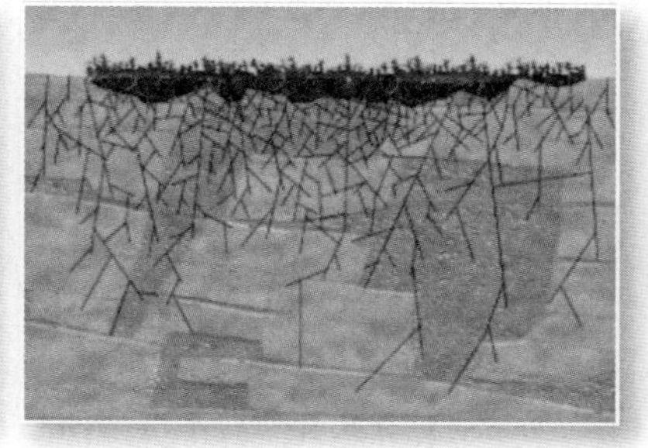

Concepts in Motion: Soil Formation

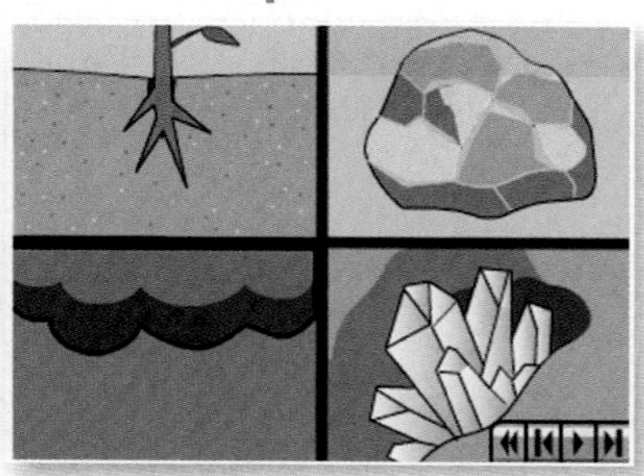

BrainPOP: Weathering

Chapter 3 Erosion and Deposition 72

PAGE KEELEY SCIENCE PROBES *What is erosion?* 71

LESSON 1 **The Erosion-Deposition Process** 74

Launch Lab: How do the shape and size of sediment differ?

FOCUS on FLORIDA Coastal Erosion in Florida 83

LESSON 2 **Landforms Shaped by Water and Wind** 84

Launch Lab: How do water and wind shape Earth?

FOCUS on FLORIDA Landforms in Florida 92

LESSON 3 **Mass Wasting and Glaciers** 93

Launch Lab: How does a moving glacier shape Earth's surface?

NGSSS for Science Benchmark Practice 104

Mini Benchmark Assessments 106

Check It! ☐ Lesson 1 ☐ Lesson 2 ☐ Lesson 3

Inquiry LAB STATION

☐ **MiniLabs:**
LESSON 1: Can weathering be measured?
LESSON 2: How do stalactites form?
LESSON 3: How does the slope of a hill affect erosion?

Try It! then Apply It!

Skill Practice: How do water erosion and deposition occur along a stream?

Inquiry Lab: Avoiding a Landslide

Your online portal to everything you need!
Video • Audio • Review • ⓘLab Station • WebQuests • Assessment • Concepts in Motion • Personal Tutors • Virtual Labs

Here are some of the exciting digital activities you will find in this chapter!

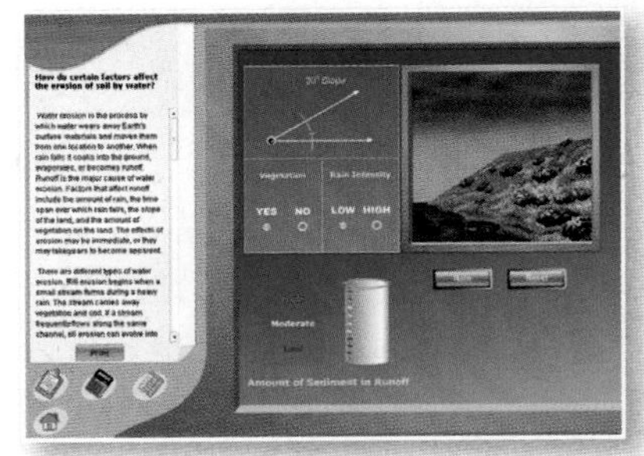

Virtual Lab: How do certain factors affect the erosion of soil by water?

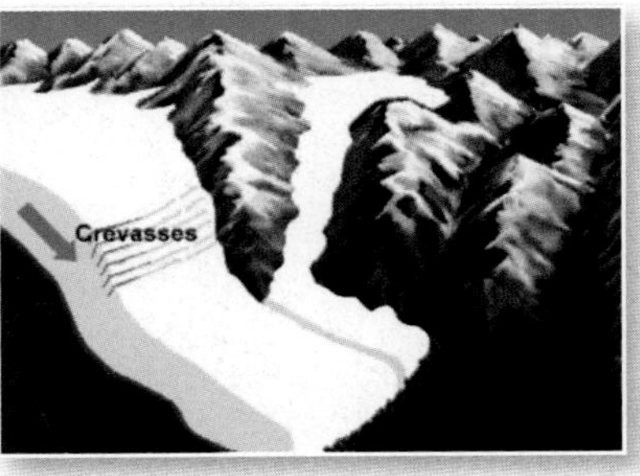

Concepts in Motion: Glacier Formation

BrainPOP: Glaciers

FLORIDA iSCIENCE COURSE 1

TABLE OF CONTENTS

Unit 2 EARTH SYSTEMS AND PATTERNS

Chapter 4 Earth's Atmosphere 116

PAGE KEELEY SCIENCE PROBES *Temperature Changes in the Atmosphere* 115

LESSON 1 **Describing Earth's Atmosphere** 118
Launch Lab: Where does air apply pressure?

CAREERS in SCIENCE A Crack in Earth's Shield 126

LESSON 2 **Energy Transfer in the Atmosphere** 127
Launch Lab: What happens to air as it warms?

LESSON 3 **Air Currents** 136
Launch Lab: Why does air move?

FOCUS on FLORIDA Salads on Mars 142

LESSON 4 **Air Quality** 143
Launch Lab: How does acid rain form?

NGSSS for Science Benchmark Practice 154

Mini Benchmark Assessments 156

Check It! ☐ Lesson 1 ☐ Lesson 2 ☐ Lesson 3 ☐ Lesson 4

Inquiry iLAB STATION

☐ **MiniLabs:** LESSON 1: Why does the furniture get dusty?
LESSON 2: Can you identify a temperature inversion?
LESSON 3: Can you model the Coriolis effect?
LESSON 4: Can being out in fresh air be harmful to your health?

Try It! then Apply It!

Skill Practice: Can you conduct, convect, and radiate?

Skill Practice: Can you model global wind patterns?

Inquiry Lab: Radiant Energy Absorption

Your online portal to everything you need!
Video • Audio • Review • iLab Station • WebQuests • Assessment • Concepts in Motion • Personal Tutors • Virtual Labs

Here are some of the exciting digital activities you will find in this chapter!

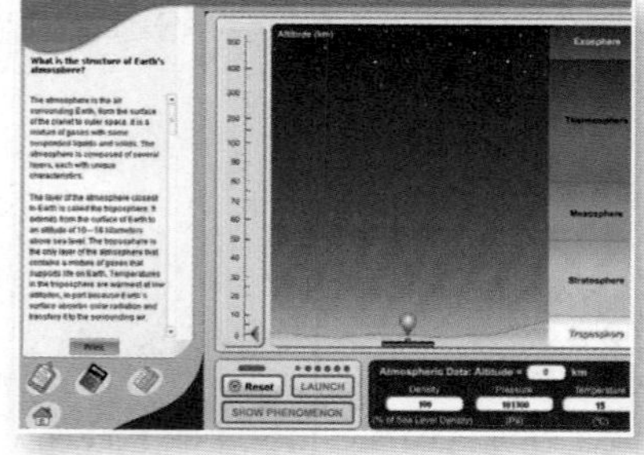

Virtual Lab: What is the structure of Earth's atmosphere?

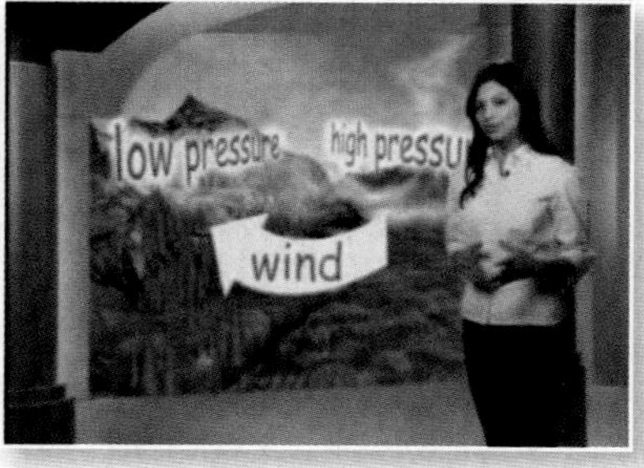

What's Science Got to do With It? Electrifying Wind

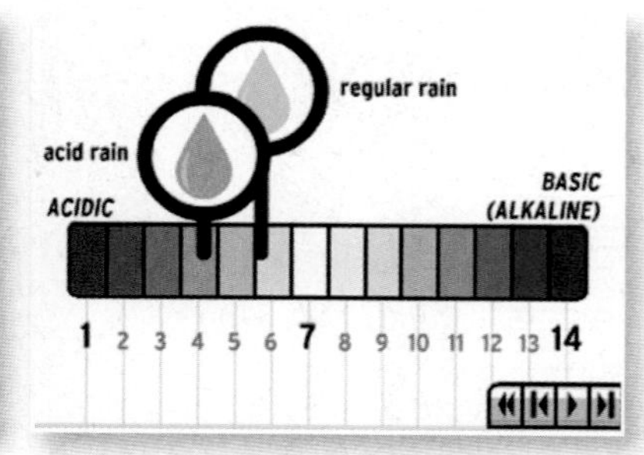

BrainPOP: Air Pollution

Chapter 5 **Weather** 162

PAGE KEELEY SCIENCE PROBES *Air Pressure Ideas* 161

LESSON 1 **Describing Weather** 164

Launch Lab: How can you model the cycling of water?

SCIENCE & SOCIETY Is there a link between hurricanes and global warming? 171

LESSON 2 **Weather Patterns** 172

Launch Lab: How can temperature affect pressure?

LESSON 3 **Weather Forecasts** 184

Launch Lab: Can you understand the weather report?

NGSSS for Science Benchmark Practice 194

Mini Benchmark Assessments 196

Check It! ☐ Lesson 1 ☐ Lesson 2 ☐ Lesson 3

☐ **MiniLabs:** LESSON 1: When will dew form?
LESSON 2: How can you observe air pressure?
LESSON 3: How is weather represented on a map?

Try It! then Apply It!

Skill Practice: Why does the weather change?

Inquiry Lab: Hurricanes and Their Effects

ConnectED **Your online portal to everything you need!**
Video • Audio • Review • Lab Station • WebQuests • Assessment • Concepts in Motion • Personal Tutors • Virtual Labs

Here are some of the exciting digital activities you will find in this chapter!

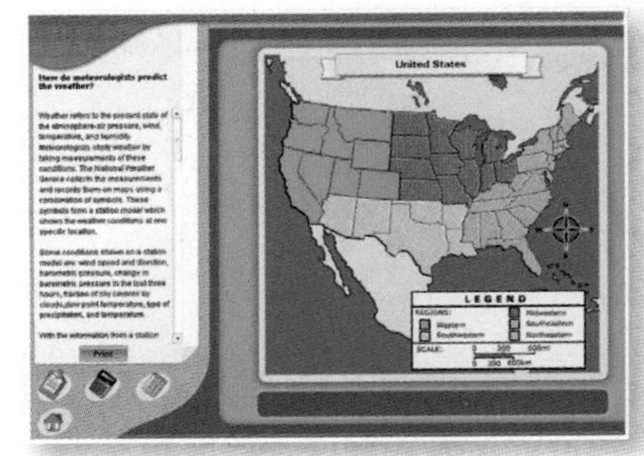

Virtual Lab: How do meteorologists predict the weather?

What's Science Got to do With It? Weather Effects

BrainPOP: Clouds

FLORIDA iSCIENCE COURSE 1

TABLE OF CONTENTS

Chapter 6 **Climate** **202**

PAGE KEELEY SCIENCE PROBES *Is Earth getting warmer?* **201**

LESSON 1 **Climates of Earth** **204**

Launch Lab: How do climates compare?

LESSON 2 **Climate Cycles** **213**

Launch Lab: How does Earth's tilted axis affect climate?

CAREERS in SCIENCE Frozen in Time **221**

LESSON 3 **Recent Climate Change** **222**

Launch Lab: What changes climates?

NGSSS for Science Benchmark Practice **234**

Mini Benchmark Assessments **236**

Check It! ☐ Lesson 1 ☐ Lesson 2 ☐ Lesson 3

Inquiry iLAB STATION

☐ **MiniLabs:** LESSON 1: What factors affect climate?
LESSON 2: How do climates vary?
LESSON 3: How do the particles in a liquid move when heated?

Try It! then Apply It!

Skill Practice: Can reflection of the Sun's rays change the climate?

Inquiry Lab: The Greenhouse Effect is a Gas!

Your online portal to everything you need!
Video • Audio • Review • iLab Station • WebQuests • Assessment • Concepts in Motion • Personal Tutors • Virtual Labs

Here are some of the exciting digital activities you will find in this chapter!

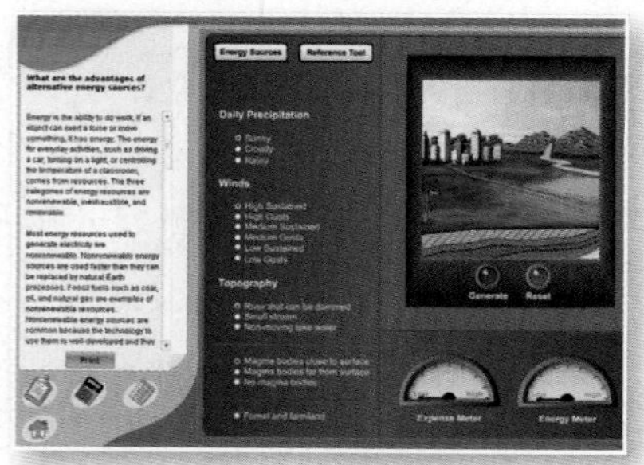
Virtual Lab: What are the advantages of alternative energy sources?

Concepts in Motion: Seasons

Page Keeley Science Probe

FLORIDA iSCIENCE COURSE 1

Unit 3 ENERGY, MOTION, AND FORCES

Chapter 7 Energy and Energy Transformations 246

PAGE KEELEY SCIENCE PROBES *What is energy?* 245

LESSON 1 **Forms of Energy** 248

Launch Lab: Can you change matter?

LESSON 2 **Energy Transformations** 256

Launch Lab: Is energy lost when it changes form?

GREEN SCIENCE Fossil Fuels and Rising CO_2 263

LESSON 3 **Thermal Energy on the Move** 264

Launch Lab: What changes climates?

FOCUS on FLORIDA Solar Energy in Florida 271

NGSSS for Science Benchmark Practice 276

Mini Benchmark Assessments 278

Check It! ☐ Lesson 1 ☐ Lesson 2 ☐ Lesson 3

Inquiry iLAB STATION

☐ **MiniLabs:** LESSON 1: Can a moving object do work?
LESSON 2: How does energy change form?
LESSON 3: How do the particles in a liquid move when heated?

Try It! then Apply It!

Skill Practice: Can you identify potential and kinetic energy?

Inquiry Lab: Pinwheel Power

ConnectED **Your online portal to everything you need!**
Video • Audio • Review • iLab Station • WebQuests • Assessment • Concepts in Motion • Personal Tutors • Virtual Labs

Here are some of the exciting digital activities you will find in this chapter!

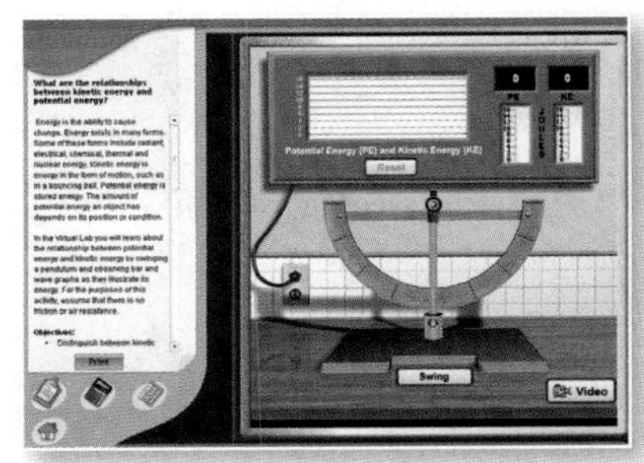

Virtual Lab: What are the relationships between kinetic energy and potential energy?

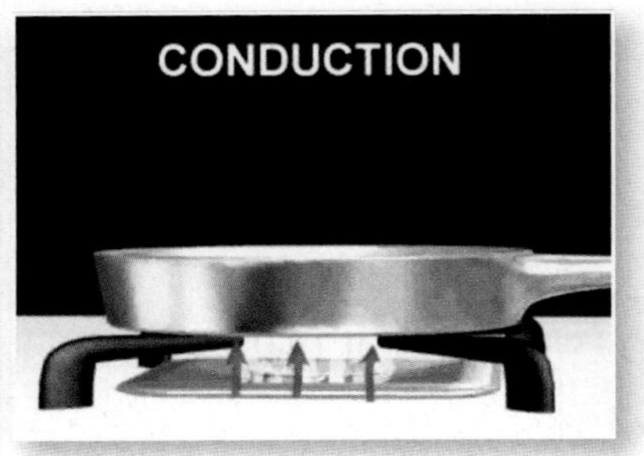

Concepts in Motion: Conduction, Convection and Radiation

BrainPOP: Forms of Energy

Chapter 8 **Motion and Forces** **284**

PAGE KEELEY SCIENCE PROBES *Beach Ball* **283**

LESSON 1 **Describing Motion** **286**

Launch Lab: Where are you right now?

LESSON 2 **Graphing Motion** **295**

Launch Lab: What does a graph show?

FOCUS on FLORIDA Using Satellites to Track the Florida Panther **302**

LESSON 3 **Forces** **303**

Launch Lab: What affects the way objects fall?

NGSSS for Science Benchmark Practice **316**

Mini Benchmark Assessments **318**

Check It! ☐ Lesson 1 ☐ Lesson 2 ☐ Lesson 3

Inquiry iLAB STATION

☐ **MiniLabs:** LESSON 1: How can you determine average speed?
LESSON 2: How can you make a speed-time graph?
LESSON 3: How do forces affect motion?

Try It! then Apply It!

Skill Practice: How can you test and describe an object's motion?

Inquiry Lab: Design a Safe Vehicle

Your online portal to everything you need!
Video • Audio • Review • ⓘLab Station • WebQuests • Assessment • Concepts in Motion • Personal Tutors • Virtual Labs

Here are some of the exciting digital activities you will find in this chapter!

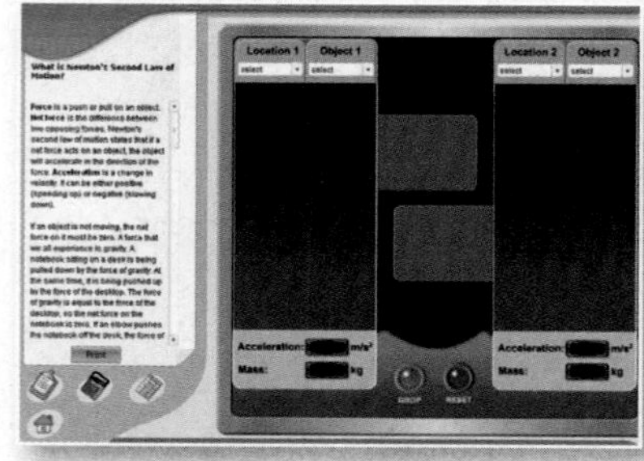

Virtual Lab: What is Newton's second law of motion?

Concepts in Motion: Distance v. Time Graph

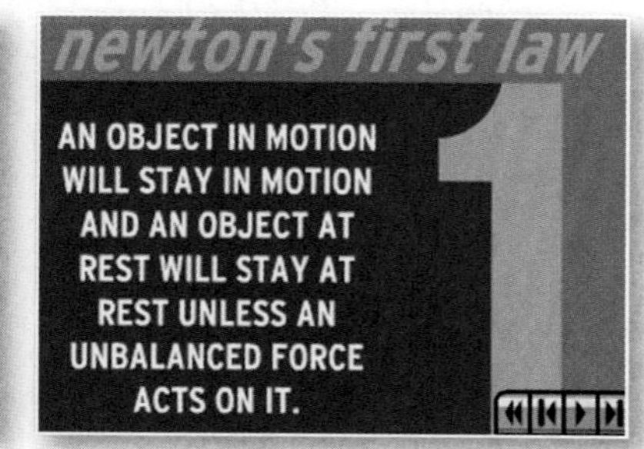

BrainPOP: Newton's Laws of Motion

FLORIDA ⓘSCIENCE COURSE 1

Unit 4 ORGANIZATION AND DEVELOPMENT OF ORGANISMS

Chapter 9 Classifying and Exploring Life 328

PAGE KEELEY SCIENCE PROBES *Classifying and Exploring Life* 327

LESSON 1 **Characteristics of Life** 330

Launch Lab: Is it alive?

CAREERS in SCIENCE The Amazing Adaptation of an Air-Breathing Catfish 339

LESSON 2 **Classifying Organisms** 340

Launch Lab: How do you identify similar items?

FOCUS on FLORIDA The Name Game 347

LESSON 3 **Exploring Life** 348

Launch Lab: Can a water drop make objects appear bigger or smaller?

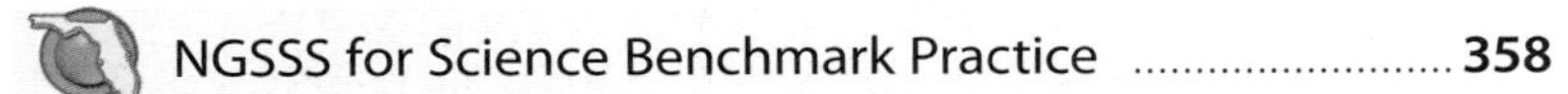

NGSSS for Science Benchmark Practice 358

Mini Benchmark Assessments 360

Check It! ☐ Lesson 1 ☐ Lesson 2 ☐ Lesson 3

☐ **MiniLabs:** LESSON 1: Did you blink?
LESSON 2: How would you name an unknown organism?
LESSON 3: How do microscopes help us compare living things?

Try It! then Apply It!

Skill Practice: How can you identify a beetle?

Inquiry Lab: Constructing a Dichotomous Key

Your online portal to everything you need!
Video • Audio • Review • ⓘLab Station • WebQuests • Assessment • Concepts in Motion • Personal Tutors • Virtual Labs

Here are some of the exciting digital activities you will find in this chapter!

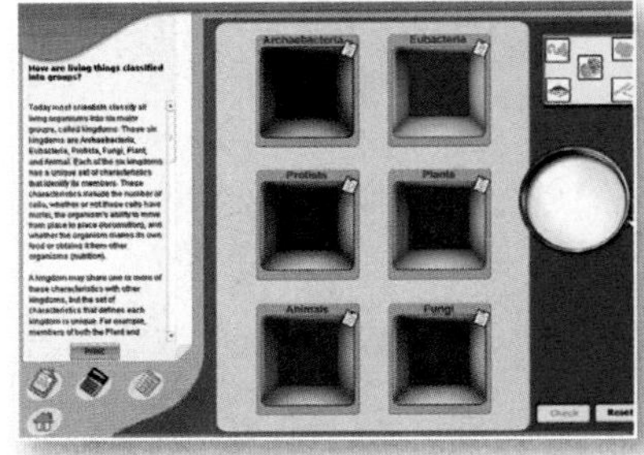

Virtual Lab: How are living things classified into groups?

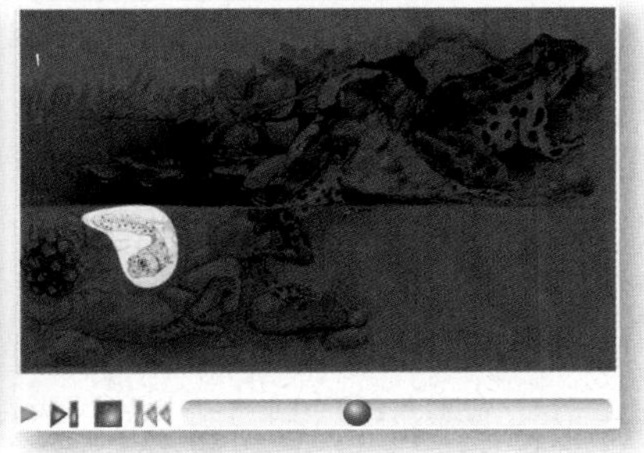

Concepts in Motion: Frog's Life Cycle

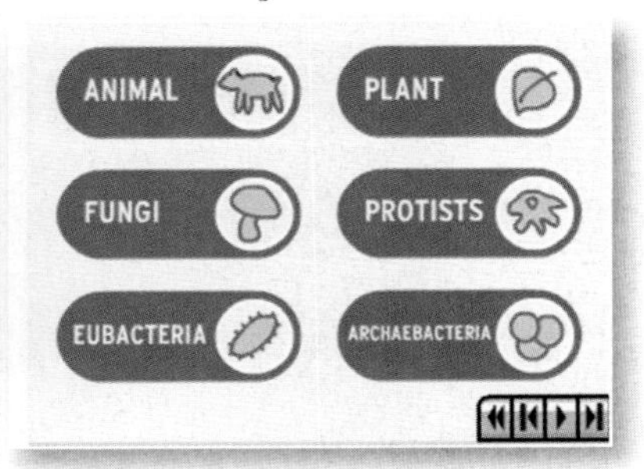

BrainPOP: Kingdoms

FLORIDA ⓘSCIENCE COURSE 1

TABLE OF CONTENTS

Chapter 10 **Cell Structure and Function** **366**

PAGE KEELEY SCIENCE PROBES *The Basic Unit of Life* **365**

LESSON 1 **Cells and Life** **368**

Launch Lab: What's in a cell?

HOW IT WORKS A Very Powerful Microscope **375**

LESSON 2 **The Cell** **376**

Launch Lab: Why do eggs have shells?

LESSON 3 **Moving Cellular Material** **386**

Launch Lab: What does the cell membrane do?

LESSON 4 **Cells and Energy** **394**

Launch Lab: What do you exhale?

NGSSS for Science Benchmark Practice **404**

Mini Benchmark Assessments **406**

Check It! ☐ Lesson 1 ☐ Lesson 2 ☐ Lesson 3 ☐ Lesson 4

Inquiry ⓘLAB STATION

☐ **MiniLabs:** LESSON 1: How can you observe DNA?
LESSON 2: How do eukaryotic and prokaryotic cells compare?
LESSON 3: How is a balloon like a cell membrane?

Try It! then Apply It!

Skill Practice: How are plant and animal cells similar and how are they different?

Skill Practice: How does an object's size affect the transport of materials?

Inquiry Lab: Photosynthesis and Light

Your online portal to everything you need!
Video • Audio • Review • ⓘLab Station • WebQuests • Assessment • Concepts in Motion • Personal Tutors • Virtual Labs

Here are some of the exciting digital activities you will find in this chapter!

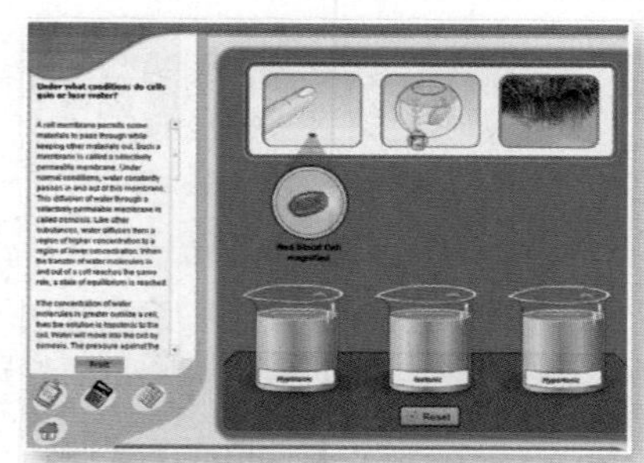

Virtual Lab: Under what conditions do cells lose and gain water?

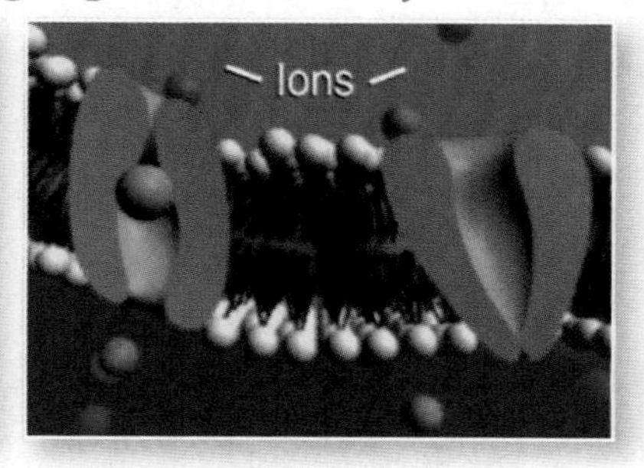

Concepts in Motion: Active Transport

BrainPOP: Cell Structures

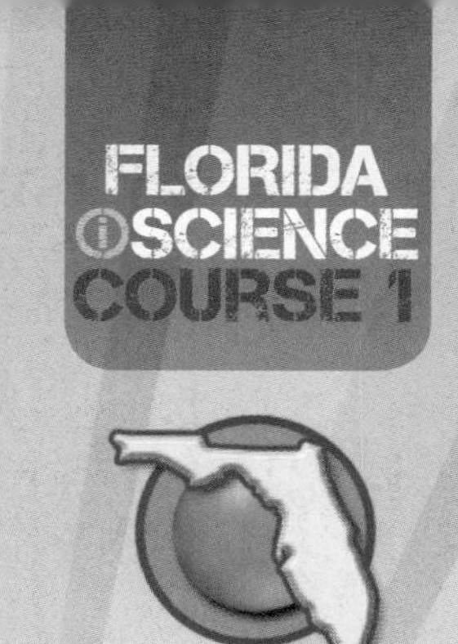

Chapter 11 From a Cell to an Organism

From a Cell to an Organism 412

PAGE KEELEY SCIENCE PROBES *Getting Bigger* 411

LESSON 1 **The Cell Cycle and Cell Division** 414

Launch Lab: Why isn't your cell like mine?

SCIENCE & SOCIETY DNA Fingerprinting 425

LESSON 2 **Levels of Organization** 426

Launch Lab: How is a system organized?

NGSSS for Science Benchmark Practice 440

Mini Benchmark Assessments 442

Check It! ☐ Lesson 1 ☐ Lesson 2

☐ **MiniLabs:** LESSON 1: How does mitosis work?
LESSON 2: How do cells work together to make an organism?

Try It! then Apply It!

Inquiry Lab: Cell Differentiation

ConnectED **Your online portal to everything you need!**
Video • Audio • Review • ⓘLab Station • WebQuests • Assessment • Concepts in Motion • Personal Tutors • Virtual Labs

Here are some of the exciting digital activities you will find in this chapter!

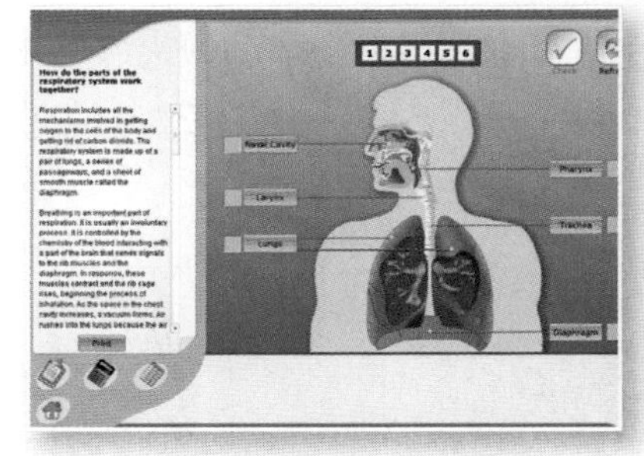

Virtual Lab: How do the parts of the respiratory system work together?

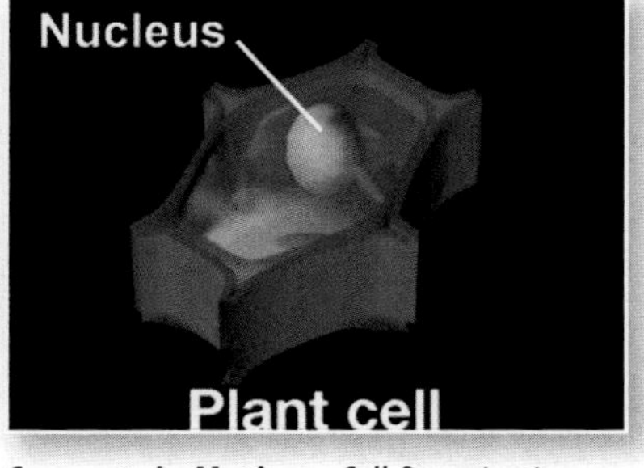

Concepts in Motion: Cell Organization

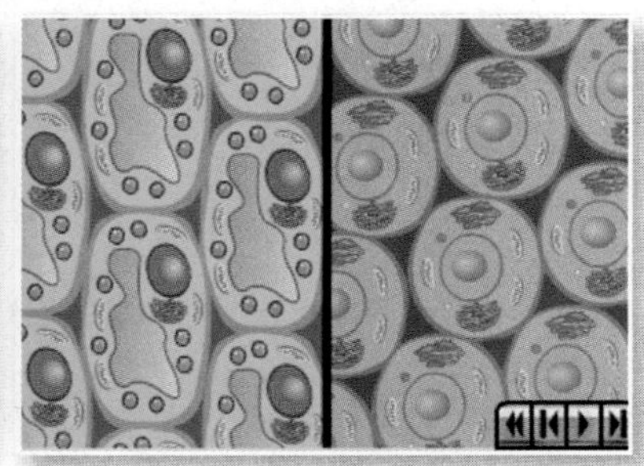

BrainPOP: Cell Specialization

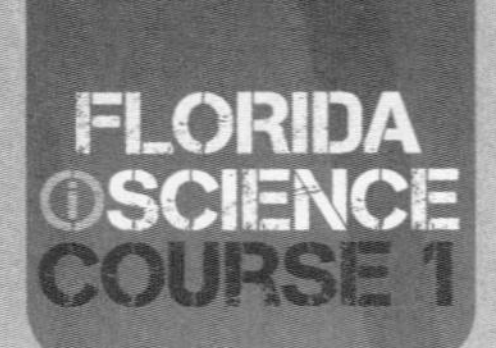

TABLE OF CONTENTS

Chapter 12 **Human Body Systems** 446

PAGE KEELEY SCIENCE PROBES *Basic Unit of Function* 445

LESSON 1 **Transport and Defense** 448

Launch Lab: Which tool can transport water quickly?

LESSON 2 **Structure, Movement, and Control** 462

Launch Lab: Why is the skeletal system so important?

SCIENCE & SOCIETY Bone Marrow Transplants 471

LESSON 3 **Reproduction and Development** 472

Launch Lab: How do the sizes of egg and sperm cells compare?

NGSSS for Science Benchmark Practice 482

Mini Benchmark Assessments 484

Check It! ☐ Lesson 1 ☐ Lesson 2 ☐ Lesson 3

Inquiry LAB STATION

☐ **MiniLabs:** LESSON 1: How much water do you lose in a day?
LESSON 2: Does your sight help you keep your balance?

Try It! then Apply It!

Skill Practice: How can you model the function of blood cells?

Inquiry Lab: Model the Body Systems

Your online portal to everything you need!
Video • Audio • Review • Lab Station • WebQuests • Assessment • Concepts in Motion • Personal Tutors • Virtual Labs

Here are some of the exciting digital activities you will find in this chapter!

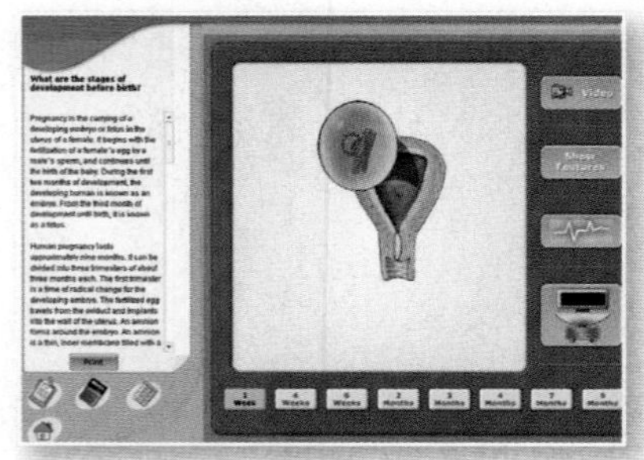

Virtual Lab: What are the stages of development before birth?

What's Science Got to do With It? Epidemic

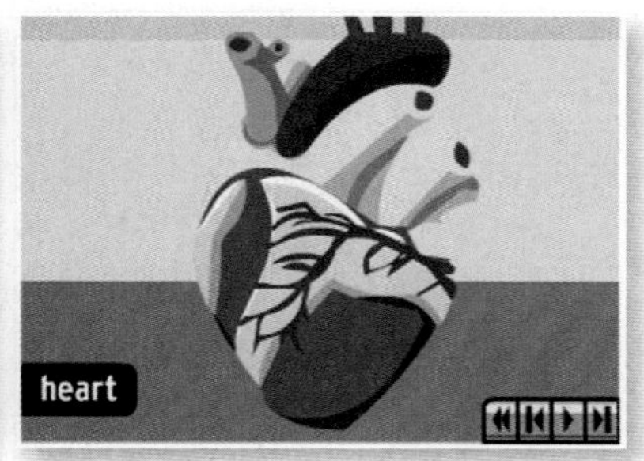

BrainPOP: Circulatory System

Chapter 13 Bacteria and Viruses 490

PAGE KEELEY SCIENCE PROBES *Is it an organism?* 489

LESSON 1 **What are bacteria?** 492

Launch Lab: How small are bacteria?

HOW IT WORKS Cooking Bacteria 499

LESSON 2 **Bacteria in Nature** 500

Launch Lab: How do bacteria affect the environment?

LESSON 3 **What are viruses?** 508

Launch Lab: How quickly do viruses replicate?

NGSSS for Science Benchmark Practice 520

Mini Benchmark Assessments 522

Check It! ☐ Lesson 1 ☐ Lesson 2 ☐ Lesson 3

Try It! then Apply It!

☐ **MiniLabs:** LESSON 1: How does a slime layer work?
LESSON 2: Can decomposition happen without oxygen?
LESSON 3: How do antibodies work?

Skill Practice: How do lab techniques affect an investigation?

Inquiry Lab: Bacterial Growth and Disinfectants

ConnectED **Your online portal to everything you need!**
Video • Audio • Review • Lab Station • WebQuests • Assessment • Concepts in Motion • Personal Tutors • Virtual Labs

Here are some of the exciting digital activities you will find in this chapter!

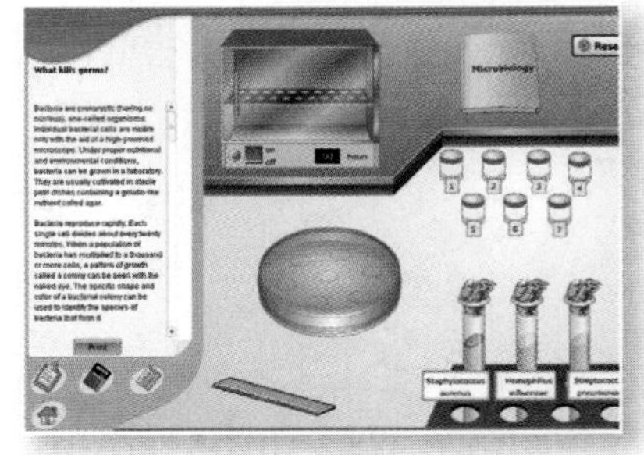

Virtual Lab: What kills germs?

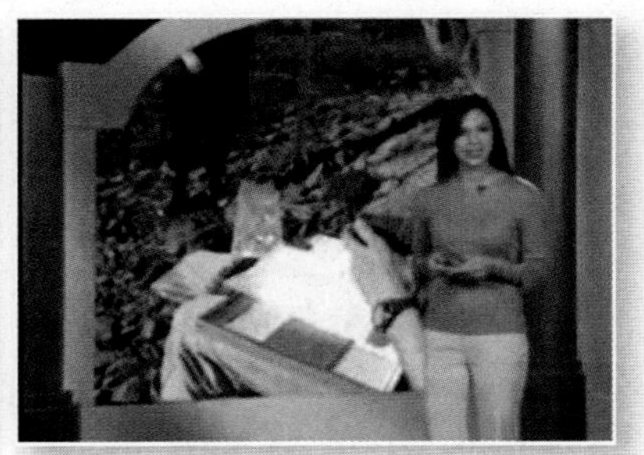

What's Science Got to do With It? Cleaning Crew

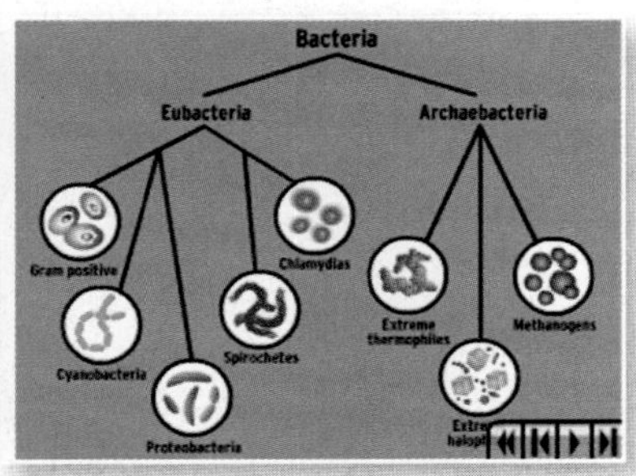

BrainPOP: Bacteria

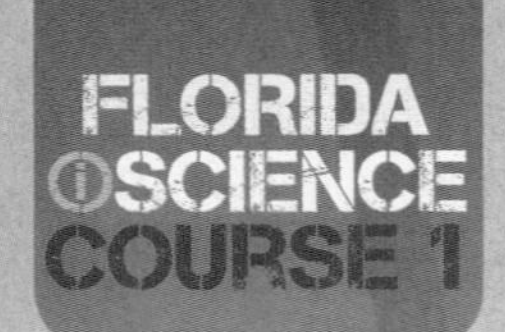

TABLE OF CONTENTS

Chapter 14 Protists and Fungi

Chapter 14 **Protists and Fungi** **528**

PAGE KEELEY SCIENCE PROBES *What are protists?* **527**

LESSON 1 **What are protists?** **530**

Launch Lab: How does a protist react to its environment?

GREEN SCIENCE The Benefits of Algae **541**

LESSON 2 **What are fungi?** **542**

Launch Lab: Is there a fungus among us?

NGSSS for Science Benchmark Practice **556**

Mini Benchmark Assessments **558**

Check It! ☐ Lesson 1 ☐ Lesson 2 ☐ Lesson 3

Inquiry LAB STATION

☐ **MiniLabs:** LESSON 1: How can you model the movement of an amoeba?
LESSON 2: What do fungal spores look like?

Try It! then Apply It!

Inquiry Lab: What does a lichen look like?

Your online portal to everything you need!
Video • Audio • Review • ⓘLab Station • WebQuests • Assessment • Concepts in Motion • Personal Tutors • Virtual Labs

Here are some of the exciting digital activities you will find in this chapter!

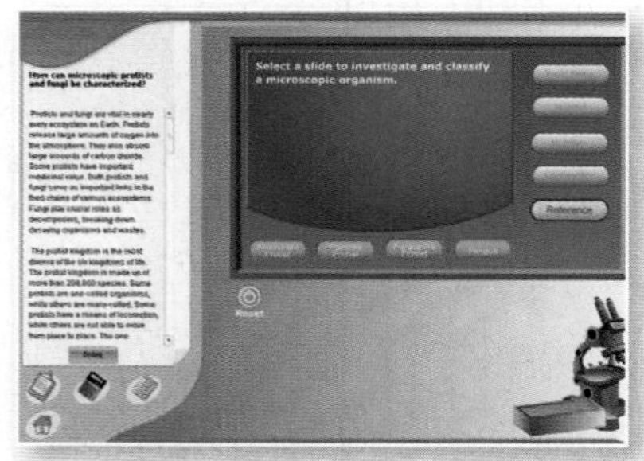

Virtual Lab: How can microscopic protists and fungi be characterized?

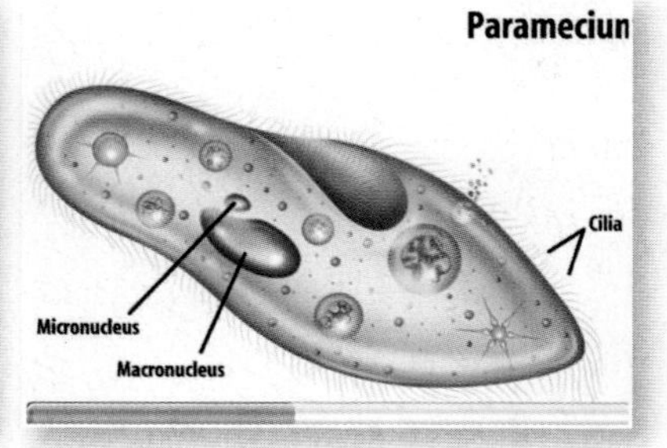

Concepts in Motion: Paramecium

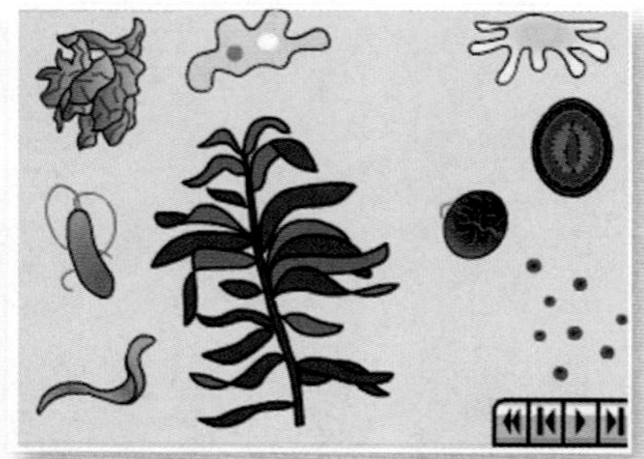

BrainPOP: Protists

The Scientific Method

Rita claims that scientists conduct scientific investigations using the scientific method. She says that even though there are many different kinds of investigations, scientists follow the same series of steps.

Do you agree or disagree with Rita? Explain your reasoning.

Methods of SCIENCE

Nature of Science

This chapter begins your study of the nature of science, but there is even more information about the nature of science in this book. Each unit begins by exploring an important topic that is fundamental to scientific study. As you read these topics, you will learn even more about the nature of science.

History	**Unit 1**
Patterns	**Unit 2**
Models	**Unit 3**
Technology	**Unit 4**

FLORIDA BIG IDEAS

1 The Practice of Science
2 The Characteristics of Scientific Knowledge
3 The Role of Theories, Laws, Hypotheses, and Models

The Big Idea

Think About It! What processes do scientists use when they perform scientific investigations?

This scientist is using pink dye to measure the speed of glacier water in the country of Greenland. Scientists are testing the hypothesis that the speed of the glacier water is increasing because amounts of meltwater, caused by climate change, are increasing.

1. What is a hypothesis?

2. What other ways do scientists test hypotheses?

3. What processes do scientists use when they perform scientific investigations?

Florida NGSSS

SC.6.N.1.1 Define a problem from the sixth grade curriculum, use appropriate reference materials to support scientific understanding, plan and carry out scientific investigation of various types, such as systematic observations or experiments, identify variables, collect and organize data, interpret data in charts, tables, and graphics, analyze information, make predictions, and defend conclusions.
SC.6.N.1.2 Explain why scientific investigations should be replicable.
SC.6.N.1.3 Explain the difference between an experiment and other types of scientific investigation, and explain the relative benefits and limitations of each.
SC.6.N.1.4 Discuss, compare, and negotiate methods used, results obtained, and explanations among groups of students conducting the same investigation.
SC.6.N.1.5 Recognize that science involves creativity, not just in designing experiments, but also in creating explanations that fit evidence.
SC.6.N.2.1 Distinguish science from other activities involving thought.
SC.6.N.2.2 Explain that scientific knowledge is durable because it is open to change as new evidence or interpretations are encountered.
SC.6.N.2.3 Recognize that scientists who make contributions to scientific knowledge come from all kinds of backgrounds and possess varied talents, interests, and goals.
SC.3.N.3.1 Recognize and explain that a scientific theory is a well-supported and widely accepted explanation of nature and is not simply a claim posed by an individual. Thus, the use of the term theory in science is very different than how it is used in everyday life.
SC.3.N.3.2 Recognize and explain that a scientific law is a description of a specific relationship under given conditions in the natural world. Thus, scientific laws are different from societal laws.
SC.3.N.3.3 Give several examples of scientific laws.
MA.6.S.6.2 Select and analyze the measures of central tendency or variability to represent, describe, analyze, and/or summarize a data set for the purposes of answering questions appropriately.

There's More Online!
Video • Audio • Review • ⓘLab Station • WebQuest • Assessment • Concepts in Motion • Multilingual eGlossary

Lesson 1

Understanding SCIENCE

ESSENTIAL QUESTIONS

- What is scientific inquiry?
- How do scientific laws and scientific theories differ?
- What is the difference between a fact and an opinion?

Vocabulary

science p. NOS 4
observation p. NOS 6
inference p. NOS 6
hypothesis p. NOS 6
prediction p. NOS 6
technology p. NOS 8
scientific theory p. NOS 9
scientific law p. NOS 9
critical thinking p. NOS 10

What is science?

Did you ever hear a bird sing and then look in nearby trees to find the singing bird? Have you ever noticed how the Moon changes from a thin crescent to a full moon each month? When you do these things, you are doing science. **Science** *is the investigation and exploration of natural events and of the new information that results from those investigations.*

For thousands of years, men and women of all countries and cultures have studied the natural world and recorded their observations. They have shared their knowledge and findings and have created a vast amount of scientific information.

People use science in their everyday lives and careers. For example, firefighters, as shown in **Figure 1,** wear clothing that has been developed and tested to withstand extreme temperatures and not catch fire. Athletes use science when they use high-performance gear or wear specially designed clothing. You use science or the results of science in almost everything you do.

Active Reading **1. Identify** List four ways you use science in your daily life.

______________ ______________

______________ ______________

Active Reading **2. Point out** Complete the statement below.

Figure 1 Firefighters' clothing, oxygen tanks, and equipment are all results of ______________.

Branches of Science

There are many different parts of the natural world. Scientists often focus their work in one branch of science or on one topic within a branch of science. There are three main branches of science—Earth science, life science, and **physical** science.

WORD ORIGIN

physical
from Latin *physica,* means "study of nature"

Active Reading **3. Analyze** Within the boxes below, list what questions you might ask each of these scientists.

Earth Science

The study of Earth, including rocks, soils, oceans, and the atmosphere, is Earth science. This Earth scientist is collecting lava samples for research.

Earth Science Questions:

-
-
-

Life Science

The study of living things is life science, or biology. These biologists are attaching a radio collar to a polar bear and marking it. The mark on a bear's back can be seen from the air. Biologists can track the bears and learn about their movements and survival techniques.

Life Science Questions:

-
-
-

Physical Science

The study of matter and energy is physical science. It includes both physics and chemistry. This chemist is preparing samples of possible new medicines.

Physical Science Questions:

-
-
-

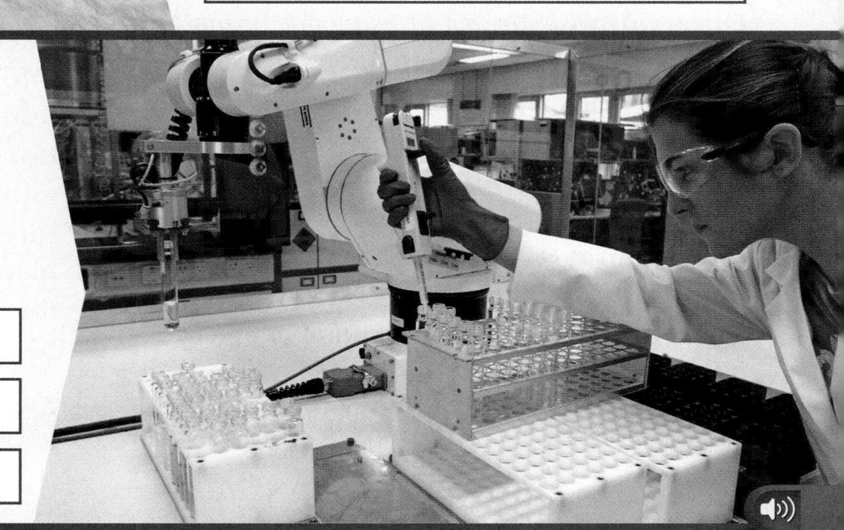

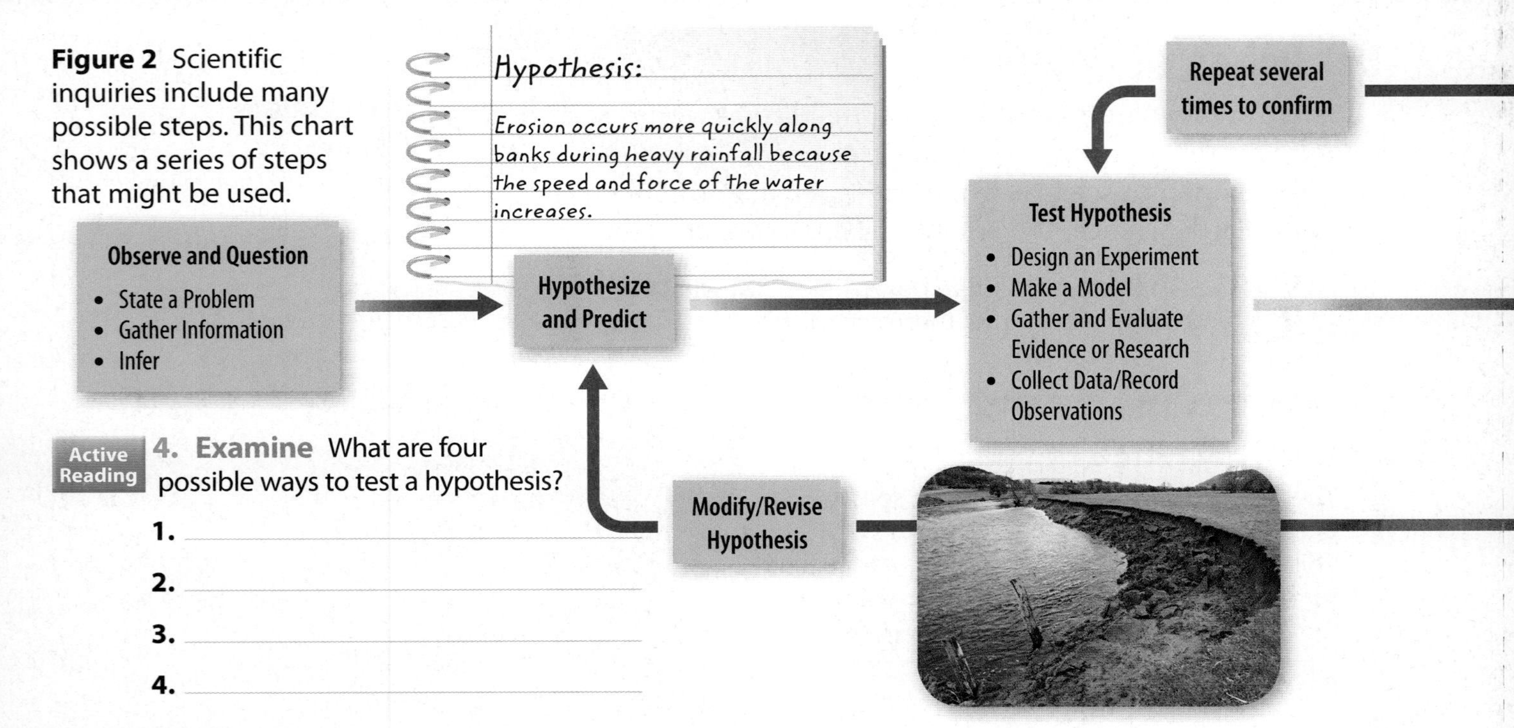

Figure 2 Scientific inquiries include many possible steps. This chart shows a series of steps that might be used.

Active Reading **4. Examine** What are four possible ways to test a hypothesis?

1. ______________________

2. ______________________

3. ______________________

4. ______________________

Scientific Inquiry

When scientists conduct scientific investigations, they use scientific inquiry. Scientific inquiry is a process that uses a set of skills to answer questions or to test ideas about the natural world. There are many kinds of scientific investigations and many ways to conduct them. The series of steps used in each investigation often varies. The flow chart in **Figure 2** shows an example of the skills used in scientific inquiry.

Active Reading **5. Define** Underline the term *scientific inquiry* and its definition.

Ask Questions

One way to begin a scientific inquiry is to observe the natural world and ask questions. **Observation** *is the act of using one or more of your senses to gather information and taking note of what occurs.* Suppose you observe that the banks of a river have eroded more this year than in the previous year, and you want to know why. You note that there was an increase in rainfall this year. After these observations, you make an inference. *An* **inference** *is a logical explanation of an observation that is drawn from prior knowledge or experience.*

You infer that the increase in rainfall caused the increase in erosion. You decide to investigate further. You develop a hypothesis and a method to test it.

Hypothesize and Predict

A **hypothesis** *is a possible explanation for an observation that can be tested by scientific investigations.* A hypothesis states an observation and provides an explanation. For example, you might make the following hypothesis: More of the riverbank eroded this year because the amount, the speed, and the force of the river water increased.

When scientists state a hypothesis, they often use it to make predictions to help test their hypothesis. *A* **prediction** *is a statement of what will happen next in a sequence of events.* Scientists make predictions based on what information they think they will find when testing their hypothesis. For example, predictions for the hypothesis above could be: If rainfall increases, then the amount, the speed, and the force of river water will increase. If the amount, the speed, and the force of river water increase, then there will be more erosion.

Active Reading **6. Define** Underline the term *hypothesis* and its definition.

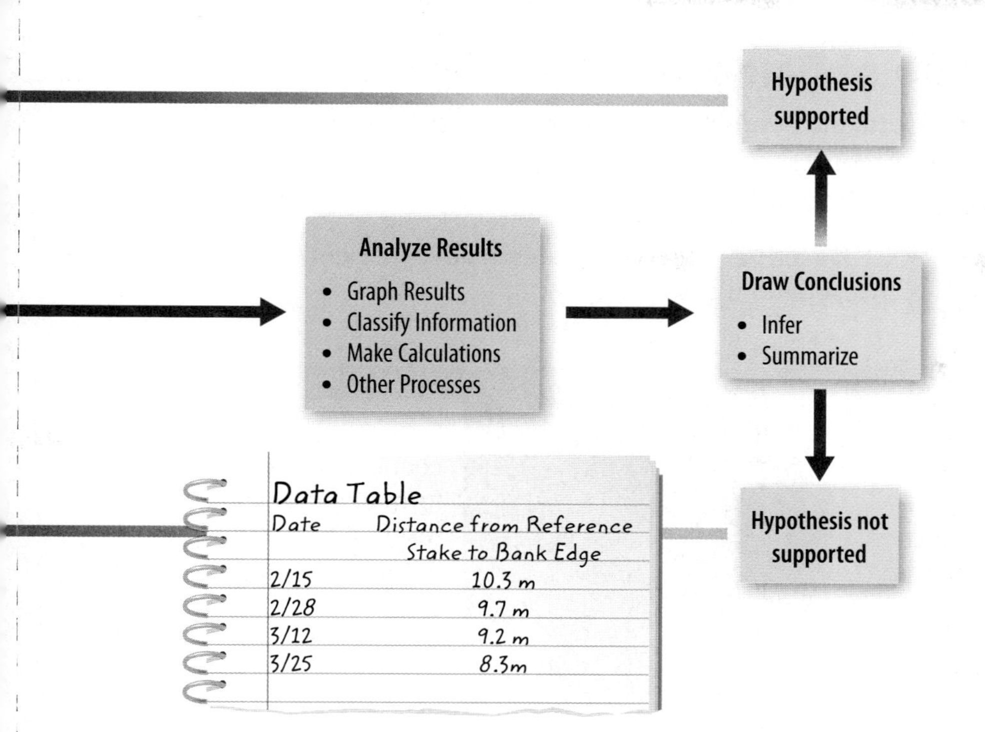

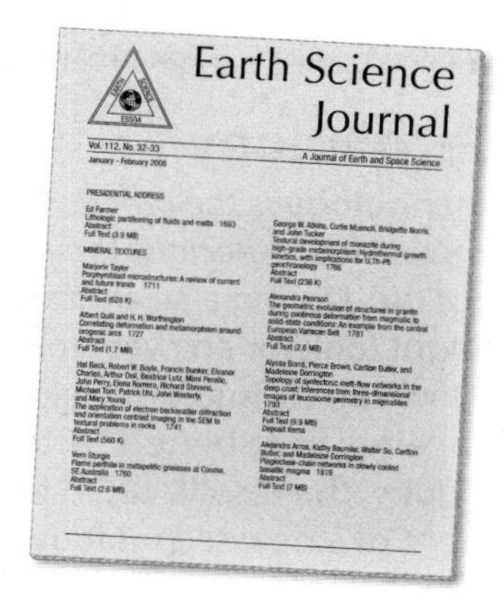

Test Hypothesis

When you test a hypothesis, you often test whether your predictions are true. If a prediction is confirmed, then it supports your hypothesis. If your prediction is not confirmed, you might need to modify your hypothesis and retest it.

There are several ways to test a hypothesis when performing a scientific investigation. Some possible ways are shown in **Figure 2.** For example, you might make a model of a riverbank in which you change the speed and the amount of water and record results and observations.

Analyze Results

Often, it is hard to see trends or relationships in data while collecting it. Data should be organized or graphed. After analyzing the data using various methods, additional inferences can be made.

Draw Conclusions

A conclusion is a summary of the information gained from testing a hypothesis. Scientists study relationships among data, make inferences from the available information, and draw conclusions based on that information.

Communicate Results

An important part of the scientific inquiry process is communicating results. Several ways to communicate results are listed in **Figure 2.** Scientists might share their information in new and different ways. Scientists communicate results of investigations to inform other scientists about their research and the conclusions of their research. Scientists might study and apply each other's conclusions to better understand their own work or to help support their own hypotheses. They may also repeat investigations to check for accuracy and for support of a scientific law or scientific theory.

Further Scientific Inquiry

Scientific inquiry is not complete with one scientific investigation. If predictions are correct and the hypothesis is supported, scientists will retest the procedure several times to ensure the conclusions are valid. If the hypothesis is not supported, it can be revised and retested.

Active Reading **7. Analyze** Why do scientists communicate their research results to other scientists?

Results of Science

The results and conclusions from an investigation can lead to many outcomes, such as the answers to a question, more information on a specific topic, or support for a hypothesis.

Active Reading **8. Describe** Discuss some results of the scientific investigations shown below.

Technology

Technology *is the practical use of scientific knowledge, especially for industrial or commercial use.* During a scientific inquiry, scientists might look for answers to questions such as, "How can technology help the hearing impaired hear better?" After much scientific inquiry, the results could be the development of a new technology, such as the cochlear implant.

Infer Discuss how the cochlear implant is a form of technology that resulted from scientific inquiry.

Assess Summarize how an astronaut's spacesuit is a form of new material.

New Materials

Space travel has unique technology challenges. Today, an astronaut's spacesuit, a result of much technological design, research, and testing, is made from a blend of 14 layers of materials. The fabric is waterproof and resistant to fire, temperature, and pressure extremes as well as small, high-speed flying objects.

Possible Explanations

Scientists often perform investigations to find explanations as to why or how something happens. NASA's *Spitzer Space Telescope,* which has aided in the understanding of star formation, shows a cloud of gas and dust with newly formed stars.

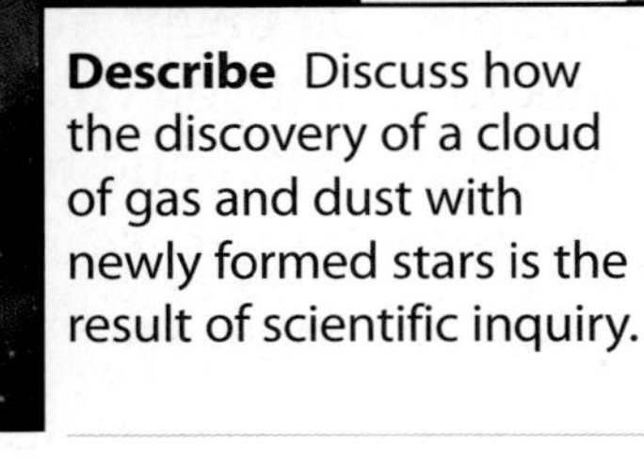

Describe Discuss how the discovery of a cloud of gas and dust with newly formed stars is the result of scientific inquiry.

Scientific Theory and Scientific Law

Another outcome of science is the development of scientific theories and laws. When a hypothesis or a group of hypotheses has been tested and supported by repeated scientific investigations, the hypothesis can become a scientific theory.

Scientific Theory

Often, the word *theory* is used to mean an idea or an opinion. However, scientists use *theory* differently. *A* **scientific theory** *is an explanation of observations or events that is based on knowledge gained from many observations and investigations.*

Scientists question theories and test them for validity. A scientific theory generally is accepted as true until it is disproved. The theory of plate tectonics explains how Earth's crust moves and how earthquakes and volcanoes occur. Another example of a scientific theory is discussed in **Figure 3.**

Figure 3 Scientists once believed Earth was the center of the solar system. In the 16th century, Nicolaus Copernicus hypothesized that Earth and the other planets actually revolve around the Sun.

Scientific Law

A social law is an agreement among people concerning a behavior. *A* **scientific law** *is a rule that describes a pattern in nature.* A scientific law only states that an event will occur under certain circumstances, unlike a scientific theory that explains why an event occurs. For example, Newton's law of gravitational force implies that if you drop an object, it will fall toward Earth. Newton's law does not explain why the object moves toward Earth when dropped, only that it will happen.

Active Reading **9. Complete** Compare and contrast scientific theory to scientific law as you fill in the blanks in the chart.

	Scientific Theory	Scientific Law
Definition	an explanation of ______ or events that is based on ______ gained from many ______ and ______	a ______ that describes a ______ in nature
Summary	attempts to explain ______ something happens	states that something ______ happen

New Information

Scientific information constantly changes as new information is discovered or as previous hypotheses are retested. New information can lead to changes in scientific theories. When new facts are revealed, a current scientific theory might be revised to include the new facts, or it might be disproved and rejected, as explained in **Figure 4.**

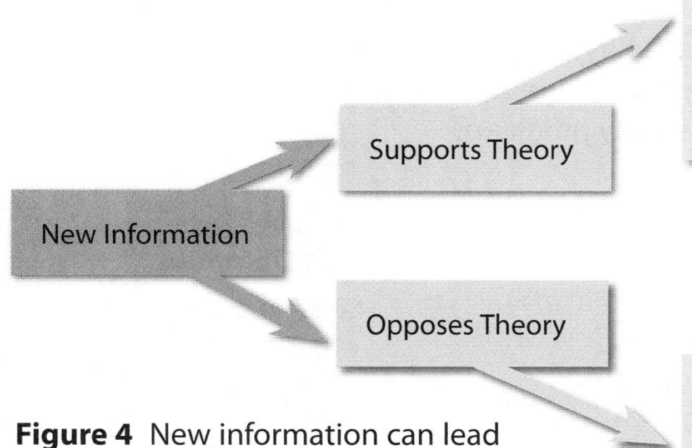

If new information supports a scientific theory, then the theory is not changed. Results might be published to show support of the theory. It might lead to scientific advancements or questions that lead to new investigations.

If new information does not support a scientific theory, the theory might be modified or rejected altogether. This can lead to new investigations with new hypotheses.

Figure 4 New information can lead to changes in scientific theories.

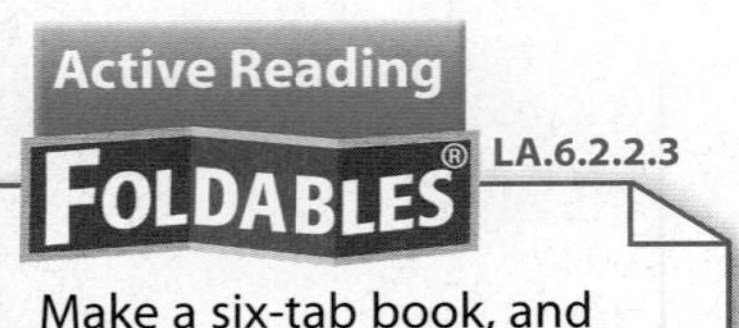

LA.6.2.2.3

Make a six-tab book, and label it as shown. Use it to organize your notes about a scientific inquiry.

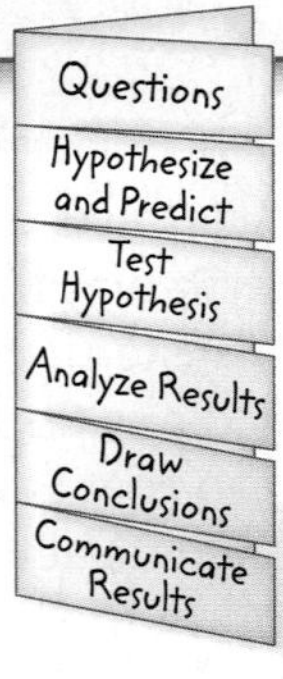

Evaluating Scientific Evidence

Did you ever read an advertisement, such as the one below, that made extraordinary claims? If so, you probably practiced **critical thinking**—*comparing what you already know with the information you are given in order to decide whether you agree with it.* To determine whether information is scientific or pseudoscience (information incorrectly represented as scientific), be skeptical and identify facts and opinions. This helps you evaluate the strengths and weaknesses of information and make informed decisions. Critical thinking is important in all decision making—from personal decisions to community, national, and international decisions.

Active Reading **10. Analyze** Underline facts in the advertisement below. Highlight opinions. Circle possible pseudoscience statements.

Learn Algebra *While You Sleep!*

Have you struggled to learn algebra? Struggle no more.

Math-er-ific's new algebra pillow is scientifically proven to transfer math skills from the pillow to your brain while you sleep. This revolutionary scientific design improved the algebra test scores of laboratory mice by 150 percent.

Dr. Tom Equation says, "I have never seen students or mice learn algebra so easily. This pillow is truly amazing."

For only $19.95, those boring hours spent studying are a thing of the past. So act fast! If you order today, you can get the algebra pillow and the equally amazing geometry pillow for only $29.95. That is a $10 savings!

Skepticism

To be skeptical is to doubt the truthfulness or accuracy of something. If someone publishes facts or if an investigation gives results that don't seem accurate, a skeptical scientist might challenge the information and test the results for accuracy.

Identifying Opinions

An opinion is a personal view, feeling, or claim about a topic. Opinions are neither true nor false.

Identifying Facts

The prices of the pillows and the savings are facts. A fact is a measurement, observation, or statement that can be strictly defined. Many scientific facts can be evaluated for their validity through investigations.

Mixing Facts and Opinions

Sometimes people mix facts and opinions. You must read carefully to determine which information is fact and which is opinion.

Science cannot answer all questions.

Some questions cannot be studied using scientific inquiry. Questions that deal with opinions, beliefs, values, and feelings cannot be answered through scientific investigation. For example, questions that cannot be answered through scientific investigation might include

- Is it ever okay to lie?
- Which food tastes best?

11. Explain Why can the answers to these questions not be determined through scientific investigations?

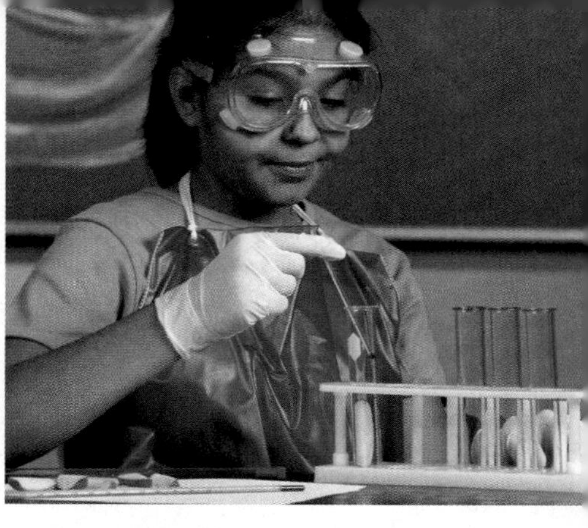

Figure 5 Always use safe lab practices when doing scientific investigations.

Safety in Science

Anyone performing scientific investigations should use safe practices, such as the student shown in **Figure 5.** Always follow your teacher's instructions. If you have questions about **potential** hazards, use of equipment, or the meaning of safety symbols, ask your teacher. Always wear protective clothing and equipment while performing scientific investigations. When using live animals in investigations, provide appropriate care and ethical treatment.

ACADEMIC VOCABULARY
potential
(adjective) possible, likely, or probable

Lesson Review 1

Use Vocabulary

1. The practical use of science, especially for industrial or commercial use, is _______________.

2. **Distinguish** between a hypothesis and a prediction.

3. **Define** *observation* in your own words.

Understand Key Concepts

4. **Define** a scientific theory and a scientific law. Give an example of each.

Scientific Theory	Scientific Law
Definition:	Definition:
Example:	Example:

5. Write an example of a fact and an example of an opinion.

Fact: _______________

Opinion: _______________

Interpret Graphics

6. **Organize** four ways to communicate results.

Communicate Results

Lesson 2

Measurement and SCIENTIFIC TOOLS

ESSENTIAL QUESTIONS

- Why is it important for scientists to use the International System of Units?
- What causes measurement uncertainty?
- What are mean, median, mode, and range?

Vocabulary

description p. NOS 12
explanation p. NOS 12
International System of Units (SI) p. NOS 12
significant digits p. NOS 14

Description and Explanation

The scientist in **Figure 6** is observing a volcano. He describes that the flowing lava is bright red with a black crust and has a temperature of 630°C. A **description** *is a spoken or written summary of observations.* There are two types of descriptions. A qualitative description, such as *bright red,* uses your senses (sight, sound, smell, touch, taste) to describe an observation. A quantitative description, such as *630°C,* uses measurements to describe an observation. Later, the scientist might explain his observations. *An* **explanation** *is an interpretation of observations.* Because the lava was bright red and 630°C, the scientist might explain that the lava is cooling.

Active Reading **1. Define** Underline the terms and definitions for the two types of description.

The International System of Units

At one time, scientists in different parts of the world used different units of measurement. Imagine the confusion when a British scientist, a French scientist, and a Japanese scientist each measured weight in different units.

In 1960, scientists adopted a new system of measurement to eliminate this confusion. *The* **International System of Units (SI)** *is the internationally accepted system for measurement.* SI uses standards of measurement, called base units, which are shown in **Table 1** on the next page. A base unit is the most common unit used in the SI system for a given measurement.

Active Reading **2. Explain** Why do scientists around the world all agree to use the SI system?

Figure 6 Scientists use descriptions and explanations when observing natural events.

Table 1 SI Base Units		
Quantity Measured	**Unit**	**Symbol**
Length	meter	m
Mass	kilogram	kg
Time	second	s
Electric current	ampere	A
Temperature	Kelvin	K
Amount of substance	mole	mol
Intensity of light	candela	cd

Table 1 Use SI units to measure the physical properties of objects.

SI Unit Prefixes

In addition to base units, SI uses prefixes to identify the size of the unit, as shown in **Table 2.** Prefixes are used to indicate a fraction or a multiple of ten. In other words, each unit is either ten times smaller than the next larger unit or ten times larger than the next smaller unit. For example, the prefix *deci–* means 1/10. A decimeter is 1/10 of a meter. The prefix *kilo–* means 1,000. A kilometer is 1,000 m.

Table 2 Prefixes are used in SI to indicate the size of the unit.

Table 2 Prefixes	
Prefix	**Meaning**
Mega– (M)	1,000,000 (10^6)
Kilo– (k)	1,000 (10^3)
Hecto– (h)	100 (10^2)
Deka– (da)	10 (10^1)
Deci– (d)	0.1 (10^{-1})
Centi– (c)	0.01 (10^{-2})
Milli– (m)	0.001 (10^{-3})
Micro– (μ)	0.000 001 (10^{-6})

Converting Between SI Units

SI is based on ten. It is easy to convert from one SI unit to another, by multiplying or dividing by a factor of ten. You also can use proportions as shown in the Math Skills activity.

Active Reading **3. Sequence** Order the SI units of length from greatest to least.

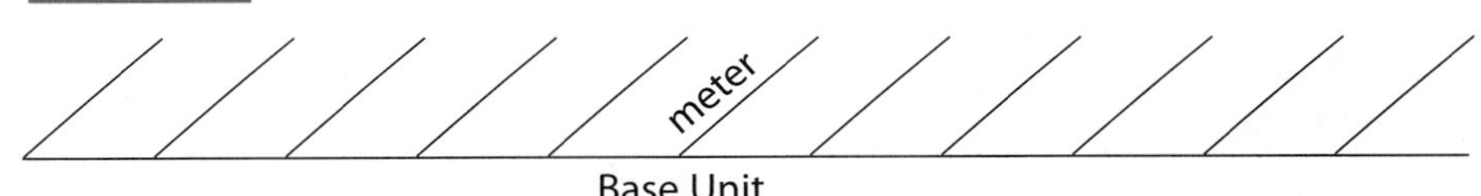

Math Skills

MA.6.A.3.6

Use Proportions

A book has a mass of **1.1 kg**. Using a proportion, find the mass of the book in grams.

1. Use the table to determine the correct relationship between the units. One kg is 1,000 times greater than 1 g. So, there are 1,000 g in 1 kg.
2. Then set up a proportion.

$$\left(\frac{x}{\mathbf{1.1\ kg}}\right) = \left(\frac{1{,}000\ g}{1\ kg}\right)$$

$$x = \left(\frac{(1{,}000\ g)(\mathbf{1.1}\ \cancel{kg})}{1\ \cancel{kg}}\right) = 1{,}100\ g$$

3. Check your units. The answer is 1,100 g.

Practice

1. Two towns are separated by 15,328 m. What is the distance in kilometers?
2. A dosage of medicine is 325 mg. What is the dosage in grams?

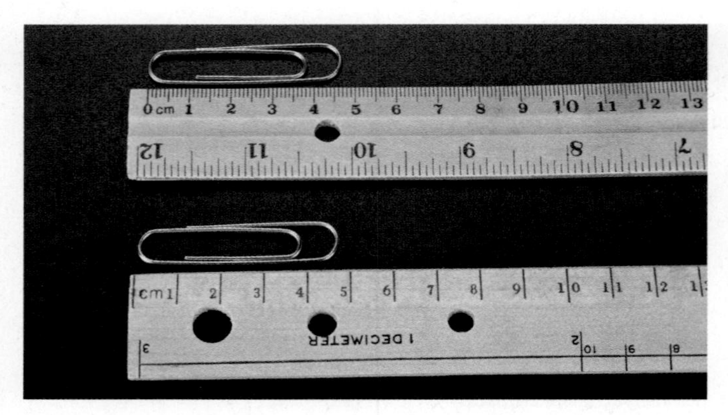

Figure 7 All measurements have some uncertainty.

Measurement and Uncertainty

Have you ever measured an object, such as a paper clip? The tools used to take measurements can limit the accuracy of the measurements. The bottom ruler in **Figure 7** is divided into centimeters. The paper clip is between 4 cm and 5 cm. You might estimate that it is 4.5 cm long. The top ruler is divided into millimeters. You can say with more certainty that the paper clip is about 4.7 cm long. This measurement is more accurate than the first.

4. Summarize What causes measurement uncertainty?

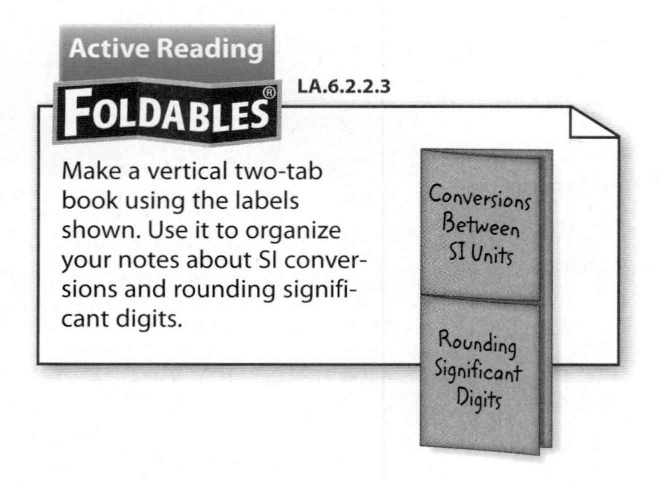

Significant Digits and Rounding

Significant digits allow scientists to record measurements with the same degree of precision. **Significant digits** *are the number of digits in a measurement that you know with a certain degree of reliability.* **Table 3** lists the rules for determining significant digits.

In order to achieve the same degree of precision as a previous measurement, it often is necessary to round a measurement to a certain number of significant digits. Suppose you have the number below.

1,348.527 g

To round to four significant digits, you need to round the 8. If the digit to the right of the 8 is less than 5, the digit being rounded (8) remains the same. If the digit to the right of the 8 is 5 or greater, the digit being rounded (8) increases by one. The rounded number is 1,349 g.

To round 1,348.527 g to two significant digits look at the third digit. 1,348.527 rounded to two significant digits would be 1,300 g. The 4 and 8 become zeros.

5. Determine Circle how to round numerals to a given number of significant digits.

Table 3 Significant Digits Rules

1. All nonzero numbers are significant.
2. Zeros between nonzero digits are significant.
3. One or more final zeros used after the decimal point are significant.
4. Zeros used solely for spacing the decimal point are NOT significant. The zeros only indicate the position of the decimal point.

* The blue numbers in the examples are the significant digits.

Number	Significant Digits	Applied Rules
1.234	4	1
0.200	3	1, 3
1,002	4	1, 2
0.001	1	1, 4
50,600	3	1, 2, 4

Mean, Median, Mode, and Range

A rain gauge measures the amount of rain that falls on a location over a period of time. A rain gauge can be used to collect data in scientific investigations, such as the data shown in **Table 4a.** Scientists often need to analyze their data to obtain information. Four values often used when analyzing numbers are median, mean, mode, and range.

Median

The median is the middle number in a data set when the data are arranged in numerical order. The rainfall data are listed in numerical order in Table 4b. If you have an even number of data items, add the two middle numbers together and divide by two to find the median.

$$\text{median} = \frac{8.18 \text{ cm} + 8.84 \text{ cm}}{2} = 8.51 \text{ cm}$$

Table 4a Rainfall Data

Month	Rainfall
January	7.11 cm
February	11.89 cm
March	9.58 cm
April	8.18 cm
May	7.11 cm
June	1.47 cm
July	18.21 cm
August	8.84 cm

Mean

The mean, or average, of a data set is the sum of the numbers in a data set divided by the number of entries in the set. To find the mean rainfall in Table 4a, add the numbers in the data set and then divide the total by the number of items in the data set (8).

$$\text{mean} = \frac{(\text{sum of numbers})}{(\text{number of items})} = \frac{72.39 \text{ cm}}{8 \text{ months}} = \frac{9.05 \text{ cm}}{\text{month}}$$

Mode

The mode of a data set is the number or item that appears most often. The number in blue in Table 4b appears twice. All other numbers only appear once.

mode = 7.11 cm

Table 4b Rainfall Data (numerical order)

Rainfall
1.47 cm
7.11 cm
7.11 cm
8.18 cm
8.84 cm
9.58 cm
11.89 cm
18.21 cm

Range

The range is the difference between the greatest number and the least number in the data set.

$$\text{range} = 18.21 \text{ cm} - 1.47 \text{ cm} = 16.74 \text{ cm}$$

Active Reading 6. **Distinguish** Underline the terms and definitions of median, mode, mean, and range.

Scientific Tools

As you engage in scientific inquiry, you will need specific tools to help you take quantitative measurements. Always follow appropriate safety procedures when using scientific tools.

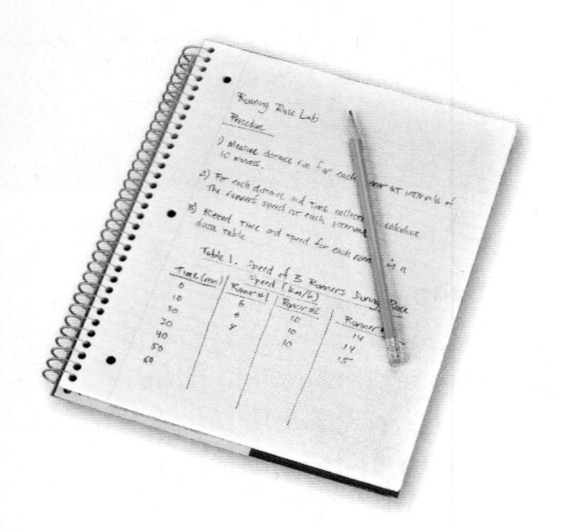

◄ Science Journal

Use a science journal to record observations, questions, hypotheses, data, and conclusions from your scientific investigations. A science journal is any notebook that you use to record notes and data while you conduct a scientific investigation. Keep it organized so you can find information easily. Write down the date whenever you record new information in the journal. Make sure you are recording your data honestly and accurately.

Rulers and Metersticks ►

Use rulers and metersticks to measure lengths and distances. The SI unit of measurement for length is the meter (m). For small objects, such as pebbles or seeds, use a metric ruler with centimeter and millimeter markings. To measure larger objects, such as the length of your classroom, use a meterstick. To measure long distances, such as the distance between cities, use an instrument that measures in kilometers. Be careful when carrying rulers and metersticks, and never point them at anyone.

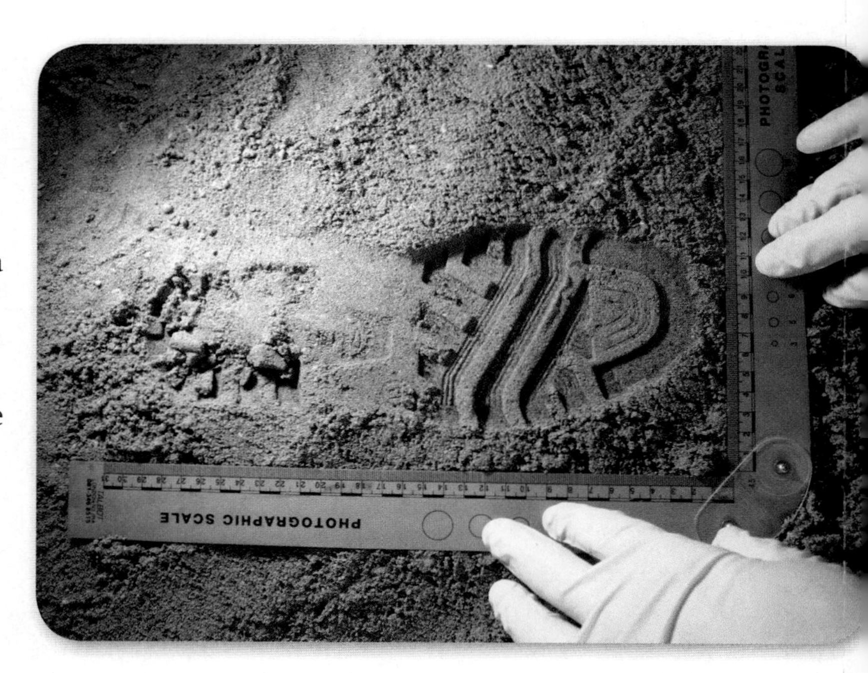

◄ Glassware

Use beakers to hold and pour liquids. The lines on a beaker are not very precise measurements. Use a graduated cylinder to measure the volume of a liquid. Volume is typically measured in liters (L) or milliliters (mL). Be careful when using glassware to avoid breaking it. Always report spills or breakage to your teacher.

Triple-Beam Balance ▶

You might use a triple-beam balance to measure the mass of an object. The mass of a light object is measured in grams. The mass of a heavy object is usually measured in kilograms. Triple-beam balances are calibrated instruments that require care when using. Follow your teacher's instructions so that you do not damage the instrument. Digital balances also might be used.

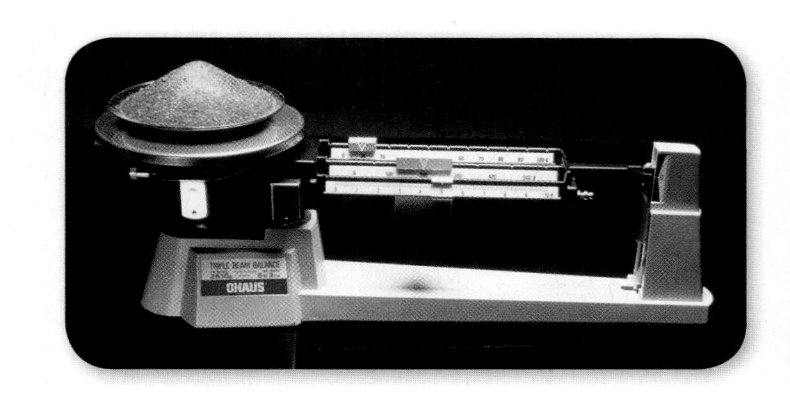

◀ Thermometer

Use a thermometer to measure the temperature of a substance. Although Kelvin is the standard SI unit for temperature, you will use a thermometer to measure temperature in degrees Celsius (°C). Place your thermometer in a secure place where it cannot roll off the table. Never use a thermometer as a stirring rod. If the thermometer breaks, notify your teacher immediately. Do not touch the broken glass or the thermometer's liquid.

Computers and the Internet

Use a computer to collect, organize, and store information about a research topic or scientific investigation. Computers are useful tools to scientists for several reasons. Scientists use computers to record and analyze data, to research new information, and to quickly share their results with others worldwide over the Internet.

7. Synthesize In the chart below, list at least one safety rule for each scientific tool previously discussed.

Rulers and Metersticks	
Glassware	
Thermometer	
Triple-Beam Balance	
Computer	

Tools Used by Earth Scientists

Binoculars

Binoculars are tools that enable people to view faraway objects more clearly. Earth scientists might to view distant landforms, animals, or even incoming weather.

Compass

A compass is an instrument that shows magnetic north. Scientists use compasses to navigate and to determine the direction of landforms or other natural objects.

Wind Vane and Anemometer

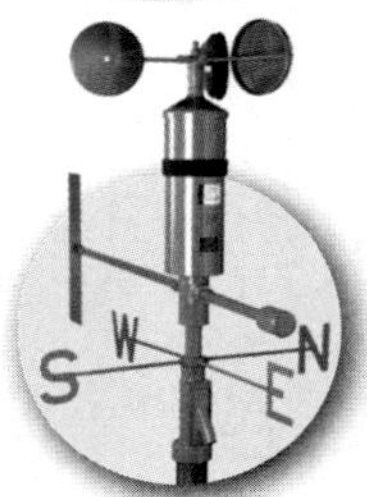

A wind vane is a device that rotates to show the direction of the wind. An anemometer, or wind-speed gauge, is used to measure the speed and the force of wind.

Streak Plate

A streak plate is a piece of hard, unglazed porcelain that helps identify minerals. A scientist might scrape a mineral across the plate. The color of the mark is the mineral's streak.

Active Reading **8. State** Give an example of a situation in which you might use each scientific tool.

Streak plate:

Compass:

Binoculars:

Wind vane or anemometer:

Lesson Review 2

Use Vocabulary

1. **Distinguish** between description and explanation.

Understand Key Concepts

2. **Give an example** of how scientific tools cause measurement uncertainty.

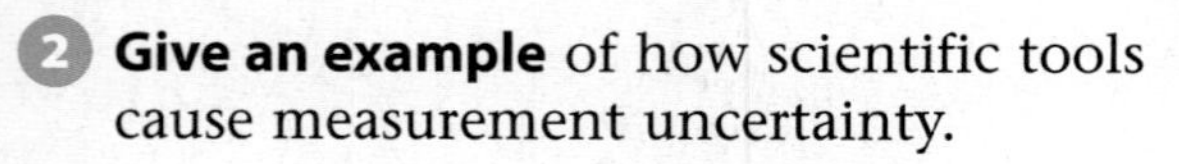

3. **Differentiate** among mean, median, mode, and range.

Interpret Graphics

4. **Determine** the number of significant digits indicated.

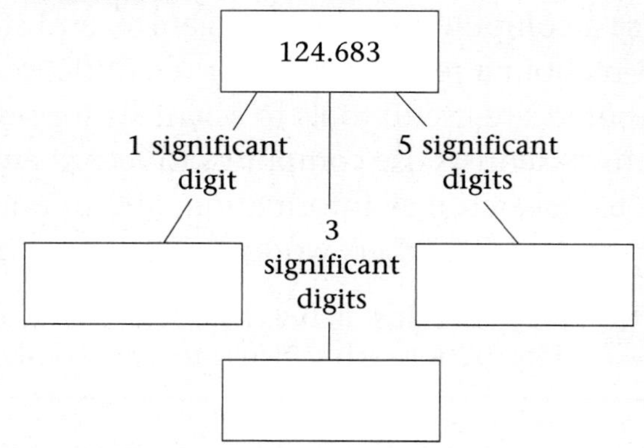

Math Skills MA.6.5.6.2

5. **Convert** 52 m to KM.

The Design Process

Science & Engineering

Create a Solution

Scientists investigate and explore natural events. Then they interpret data and share information learned from those investigations. How do engineers differ from scientists? Engineers construct and maintain the designed world. Look around you and take notice of things that do not occur in nature. Schools, roads, submarines, toys, microscopes, medical equipment, amusement park rides, computer programs, and video games all result from engineering. Science involves the practice of scientific inquiry, but engineering involves The Design Process—a set of methods used to create a solution to a problem or need.

From the first snowshoes to modern ski lifts, people have designed solutions to travel effectively in the snow. Though the term *engineer* did not exist when the first snowshoes were made, the developers of snowshoes used The Design Process just as the engineers of ski lifts do today.

The Design Process

1. Identify a Problem or Need
- Determine a problem or need
- Document all questions, research, and procedures throughout the process

2. Research and Development
- Brainstorm solutions
- Research any existing solutions that address the problem or need
- Suggest limitations

3. Construct a Prototype
- Develop possible solutions
- Estimate materials, costs, resources, and time to develop the solutions
- Select the best possible solution
- Construct a prototype

4. Test and Evaluate Solutions
- Use models to test the solution
- Use graphs, charts, and tables to evaluate results
- Analyze the solution's strengths and weaknesses

5. Communicate Results and Redesign
- Communicate the design process and results to others
- Redesign and modify the solution
- Construct the final solution

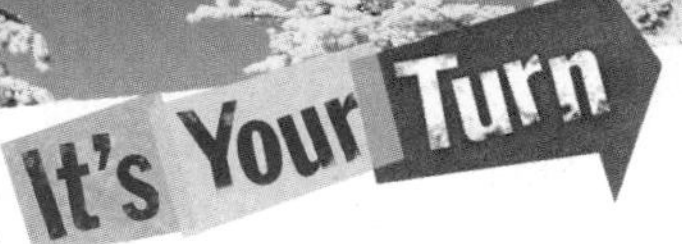

SC.6.N.1.1
SC.6.N.2.1
SC.6.N.3.4
SC.6.S.6.2

Inquiry LAB STATION

DESIGN PROCESS LAB Design a zipline ride.

Lesson 3

CASE STUDY

ESSENTIAL QUESTIONS

How are independent variables and dependent variables related?

How is scientific inquiry used in a real-life scientific investigation?

Vocabulary

variable p. NOS 21

independent variable p. NOS 21

dependent variable p. NOS 21

The Iceman's Last Journey

The Tyrolean Alps border western Austria, northern Italy, and eastern Switzerland, as shown in **Figure 8.** They are popular with tourists, hikers, mountain climbers, and skiers. In 1991, two hikers discovered the remains of a man, also shown in **Figure 8,** in a melting glacier on the border between Austria and Italy. They thought the man had died in a hiking accident. They reported their discovery to the authorities.

Initially authorities thought the man was a music professor who disappeared in 1938. However, they soon learned that the music professor was buried in a nearby town. Artifacts near the frozen corpse indicated that the man died long before 1938. The artifacts, as shown in **Figure 9,** were unusual. The man, nicknamed the Iceman, was dressed in leggings, a loincloth, and a goatskin jacket. A bearskin cap lay nearby. He wore shoes made of red deerskin with thick bearskin soles. The shoes were stuffed with grass for insulation. In addition, investigators found a copper ax, a partially constructed longbow, a quiver containing 14 arrows, a wooden backpack frame, and a dagger at the site.

Active Reading **1. Infer** How did the artifacts lead scientists to believe the Iceman died long before 1938?

Figure 8 Excavators used jackhammers to free the man's body from the ice, which caused serious damage to his hip. Part of a longbow also was found nearby.

A Controlled Experiment

Scientists and the public wanted to know the identity of the man, how he died, and when he died. Many people hypothesized about his identity, but controlled experiments were needed to unravel the mystery of who he was.

Identifying Variables and Constants

When scientists design a controlled experiment, they have to identify factors that might affect the outcome. *A* **variable** *is any factor that can have more than one value.* In controlled experiments, there are two kinds of variables. *The* **independent variable** *is the factor that you want to test. It is changed by the investigator to observe how it affects a dependent variable. The* **dependent variable** *is the factor you observe or measure during an experiment.* When the independent variable is changed, it causes the dependent variable to change.

Figure 9 These models show what the Iceman and the artifacts found with him might have looked like.

Active Reading **2. Point out** Highlight the statement that describes the relationship between the independent variable and the dependent variable.

A controlled experiment has two groups—an experimental group and a control group. The experimental group is used to study how a change in the independent variable changes the dependent variable. The control group contains the same factors as the experimental group, but the independent variable is not changed. Without a control, it is difficult to know if your observations result from the variable you are testing or from another factor.

Scientists used inquiry to investigate the mystery of the Iceman. The blue boxes in the margins point out examples of the scientific inquiry process. The notebooks in the margin identify what a scientist might have written in a journal.

Scientific investigations often begin when someone asks a question about something observed in nature.

Observation: A corpse was found buried in ice in the Tyrolean Alps.
Hypothesis: The corpse is the body of a music professor because he went missing in 1938 and has not been found.
Observation: Artifacts suggest the body is much older than the professor would have been.
Revised Hypothesis: The corpse was dead long before 1938 because artifacts appear to date before 1930.
Prediction: If the artifacts date before 1930 and belong to the corpse, then the corpse is not the professor.

An inference is a logical explanation of observations based on past experiences.

Inference: Based on its construction, the ax is at least 4,000 years old.
Prediction: If the ax is 4,000 years old, then the body is also 4,000 years old.
Test Results: Radiocarbon dating showed the man to be 5,300 years old.

After many observations, revised hypotheses, and tests, conclusions often can be made.

Conclusion: The Iceman is about 5,300 years old. He was a seasonal visitor to the high mountains. He died in autumn. When winter came, the body was buried, frozen, and preserved in the snow.

Figure 10 This ax, bow and quiver, and dagger and sheath were found with the Iceman's body.

An Early Conclusion

Konrad Spindler was a professor of archeology at the University of Innsbruck in Austria when the Iceman was discovered. Spindler estimated that the ax, shown in **Figure 10,** was at least 4,000 years old based on its construction. If the ax was that old, then the Iceman was also that old. Later, radiocarbon dating showed that the Iceman lived about 5,300 years ago.

The Iceman's body was in a mountain glacier 3,210 m above sea level. What was he doing in the snow- and ice-covered mountains? Was he hunting, shepherding his animals, or looking for metal ore?

Spindler noted that some of the wood used in the artifacts was from trees that grew at lower elevations. He concluded that the Iceman was probably a seasonal visitor to the mountains.

Spindler hypothesized that shortly before the man's death, he had driven his herds from the mountain pastures to the valleys. However, the Iceman soon returned to the mountains where he died of exposure to the cold, wintry weather.

The Iceman's body was well preserved. Spindler inferred that ice and snow covered the Iceman's body shortly after he died. Spindler concluded that the Iceman died in autumn and was quickly buried and frozen, which preserved his body and his possessions.

Active Reading **3. Identify** List two pieces of evidence and the conclusions Spindler drew based on that evidence from the Iceman and his artifacts.

Evidence	Conclusion

More Observations and Revised Hypotheses

When the Iceman's body was discovered, Klaus Oeggl was an assistant professor of botany at the University of Innsbruck. His area of study was plant life during prehistoric times in the Alps. He was invited to join the research team studying the Iceman.

Upon close examination of the Iceman and his belongings, Professor Oeggl found three plant materials—grass from the Iceman's shoe, as shown in **Figure 11,** a splinter of wood from his longbow, and a tiny fruit called a sloe berry.

Over the next year, Professor Oeggl examined bits of charcoal wrapped in maple leaves that had been found at the discovery site. Scientific examination revealed the charcoal was from the wood of eight different types of trees. All but one of the trees grew only at lower elevations than where the Iceman's body was found. Like Spindler, Professor Oeggl suspected that the Iceman had been at a lower elevation shortly before he died. From Oeggl's observations, he formed a hypothesis and made predictions.

Oeggl realized that he needed more data to support his hypothesis. He requested to examine the contents of the Iceman's digestive tract. The study would show what the Iceman had eaten just hours before his death.

Active Reading **4. Predict** Highlight ways in which the botanist predicted the Iceman's travels.

Figure 11 Professor Oeggl examined the Iceman's belongings along with the leaves and grass that were stuck to his shoe.

Scientific investigations often lead to new questions.

Observations: Plant matter near body to study—grass on shoe, splinter from longbow, sloe berry fruit, charcoal wrapped in maple leaves, wood in charcoal— 7 of 8 types of wood in charcoal grow at lower elevations

Hypothesis: Iceman had recently been at lower elevations before he died because the plants identified grow only at lower elevations.

Prediction: If the identified plants are found in the digestive tract of the corpse, then he was at lower elevations just before he died.

Question: What did the Iceman eat the day before he died?

Experiment to Test Hypothesis

The research teams provided Professor Oeggl with a tiny sample from the Iceman's digestive tract. Oeggl carefully planned his scientific inquiry. He knew that he had to work quickly to avoid the decomposition of the sample and to reduce the chances of contamination.

His plan was to divide the material from the digestive tract into four samples. Each sample would undergo several chemical tests. Then, the samples would be examined under an electron microscope to see as many details as possible.

Professor Oeggl began by adding a saline solution to the first sample. This caused it to swell, making it easier to identify particles using the microscope at low magnification. He saw particles of a wheat grain known as einkorn, a common type of wheat grown in the region during prehistoric times. He also found other edible plant material.

Oeggl noticed that the sample also contained pollen grains, as shown in the inset of **Figure 12.** To see more clearly, he used a chemical that separated unwanted substances from the pollen grains. He washed the sample with alcohol. After each wash, he examined the sample under a microscope at a high magnification. The pollen grains became more visible. More microscopic pollen grains could now be seen. Professor Oeggl identified these pollen grains as those from a hop hornbeam tree.

There is more than one way to test a hypothesis. Scientists might gather and evaluate evidence, collect data and record their observations, create a model, or design and perform an experiment. They also might perform a combination of these skills.

Test Plan:
- Divide a sample of the Iceman's digestive tract into four sections.
- Examine the pieces under microscopes.
- Gather data from observations of the pieces and record observations.

Active Reading **5. Infer** Suggest why Professor Oeggl might have divided the digestive tract sample into four smaller samples before adding chemicals and testing.

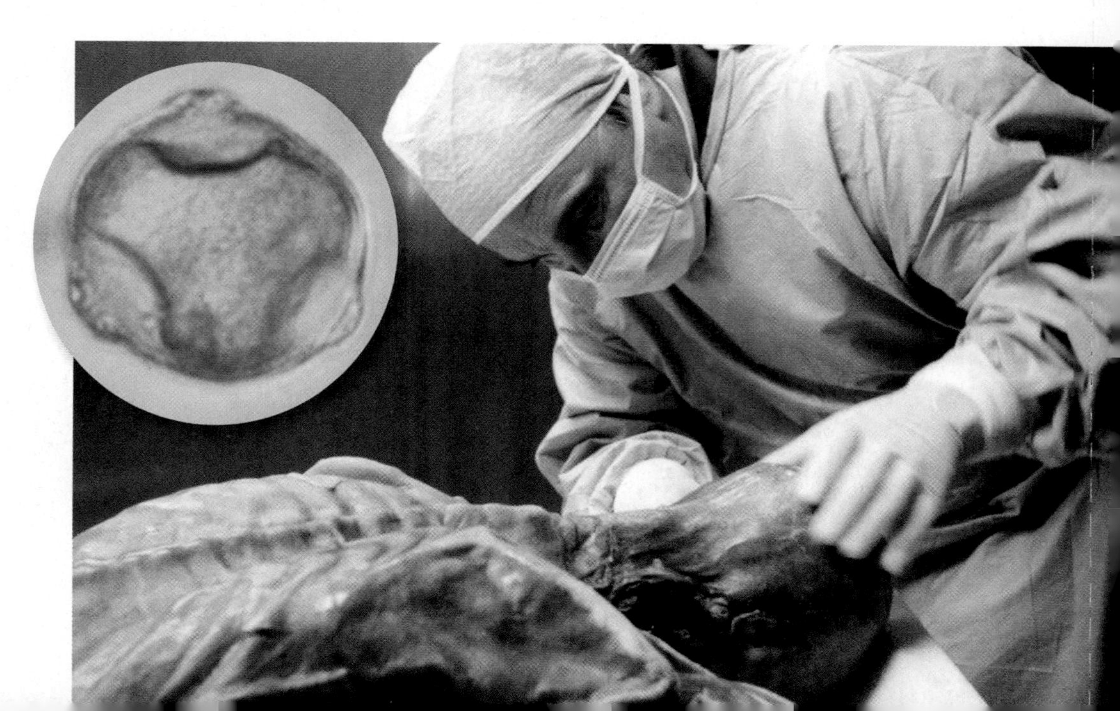

Figure 12 Eventually, Professor Oeggl identified pollen grains from hop hornbeam trees.

Analyzing Results

Oeggl observed that the hop-hornbeam pollen grains had not been digested, indicating they had been swallowed just hours before death. But, hophornbeam trees only grow in lower valleys. Oeggl was confused. How could pollen grains from low elevations be ingested within a few hours of this man dying in high mountains? Perhaps the samples from the man's digestive tract had been contaminated. Oeggl knew he needed to investigate.

Further Experimentation

Oeggl realized that the most likely source of contamination would be his own laboratory. He decided to test whether his lab equipment or saline solution contained hophornbeam pollen. He prepared two identical, sterile slides with saline solution. On one slide, he placed a sample from the Iceman's digestive tract. The slide with the sample was the experimental group. The slide without the sample was the control group.

The independent variable, the variable Oeggl changed, was the presence of the sample on the slide. The dependent variable, the variable Oeggl tested, was whether hophornbeam pollen showed up on the slides. Oeggl examined the slides carefully.

Analyzing Additional Results

The experiment showed that the control group (the slide without the digestive tract sample) contained no hophornbeam pollen grains, confirming, the pollen grains had not come from the lab equipment or solutions. Each sample from the digestive tract was re-examined. Each sample contained the same hophornbeam pollen grains. The Iceman had indeed swallowed the hophornbeam pollen grains.

Active Reading **6. Distinguish** Circle the independent variable. Underline the dependent variable.

Error is common in scientific research. Scientists are careful to document procedures and any unanticipated factors or accidents. They also are careful to document uncertainty in their measurements.

Procedure:
- Sterilize laboratory equipment.
- Prepare saline slides.
- View saline slides under electron microscope. Results: no hophornbeam pollen grains
- Add digestive tract sample to one slide.
- View this slide under electron microscope. Result: hophornbeam pollen grains present.

Controlled experiments contain two types of variables.

Dependent Variables: presence of hophornbeam pollen grains found on slide
Independent Variable: digestive tract sample on slide

Without a control group, it is difficult to determine the origin of some observations.

Control Group: sterilized slide
Experimental Group: sterilized slide with digestive tract sample

An inference is a logical explanation of an observation that is drawn from prior knowledge or experience. Inferences can lead to predictions, hypotheses, or conclusions.

Observation: The Iceman's digestive tract contains pollen grains from the hop hornbeam tree and other plants that bloom in spring.
Inference: Knowing the rate at which food and pollen decompose after swallowed, it can be inferred that the Iceman ate three times in the day and a half before he died.
Prediction: The Iceman died in the spring within hours of digesting the hop-hornbeam pollen grains.

Mapping the Iceman's Journey

The hop-hornbeam pollen grains were helpful in determining the season the Iceman died. Because the pollen grains were whole, Professor Oeggl inferred that the Iceman swallowed the pollen grains during their blooming season. The Iceman must have died between March and June.

After additional investigation, Professor Oeggl was ready to map the Iceman's final trek up the mountain. Because Oeggl knew the rate at which food travels through the digestive system, he inferred that the Iceman had eaten three times in the final day and a half of his life. From the digestive tract samples, Oeggl estimated where the Iceman was located when he ate.

First, the Iceman ingested pollen grains native to higher mountain regions. Several hours later, he swallowed pollen grains from the valley regions. Last, the Iceman swallowed more pollen grains from trees of higher mountain areas. Oeggl proposed the Iceman traveled from the southern region of the Italian Alps to the higher, northern region as shown in **Figure 13,** where he died suddenly. He did this all in a period of about 33 hours.

Active Reading **7. Summarize** What knowledge and evidence did Oeggl use to determine the route the Iceman traveled in his final days?

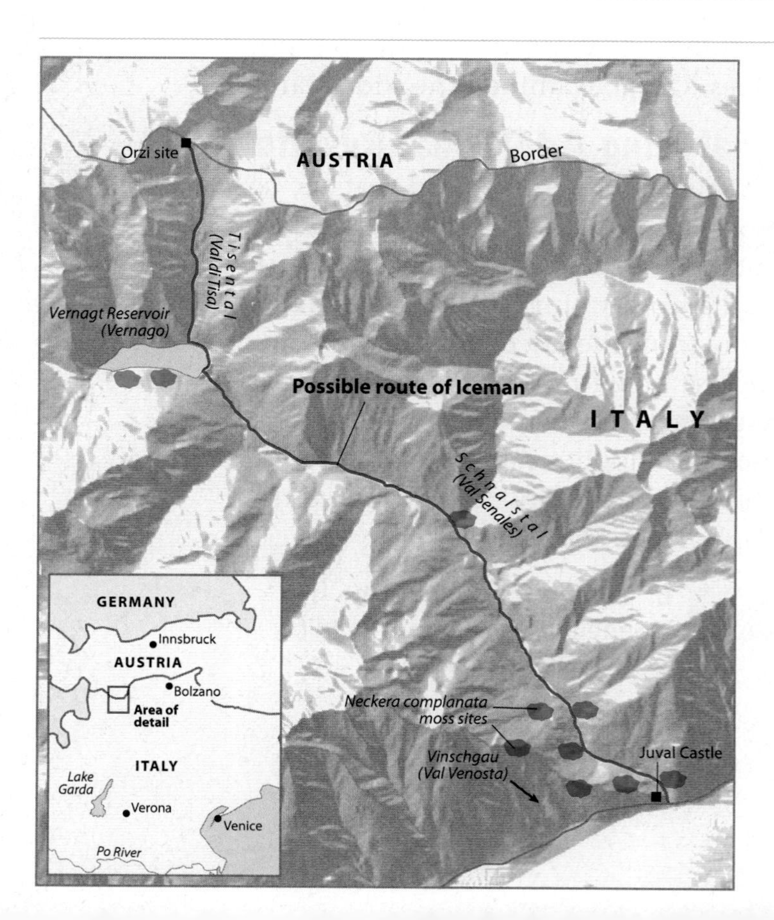

Figure 13 By examining the contents of the Iceman's digestive tract, Professor Oeggl was able to reconstruct the Iceman's last journey.

Conclusion

Researchers from around the world worked on different parts of the Iceman mystery and shared their results. Analysis of the Iceman's hair revealed his diet usually contained vegetables and meat. Examining the Iceman's one remaining fingernail, scientists determined that he had been sick three times within the last six months of his life. X-rays revealed an arrowhead under the Iceman's left shoulder. This suggested that he died from that serious injury rather than from exposure.

Finally, scientists concluded that the Iceman traveled from the high alpine region in spring to his native village in the lowland valleys. There, during a conflict, the Iceman sustained a fatal injury. He retreated back to the higher elevations, where he died. Scientists recognize their hypotheses can never be proved, only supported or not supported. However, with advances in technology, scientists are able to more thoroughly investigate mysteries of nature.

Scientific investigations may disprove early hypotheses or conclusions. However, new information can cause a hypothesis or conclusion to be revised many times.

Revised Conclusion:
In spring, the Iceman traveled from the high country to the valleys. He was involved in a violent confrontation, climbed the mountain into a region of permanent ice, and died of his wounds.

Lesson Review 3

Use Vocabulary

1. A factor that can have more than one value is a(n) ______________.

2. **Clarify** Differentiate between independent and dependent variables.

Understand Key Concepts

3. **Determine** Identify each of the components of a valid experiment in the following scenario. Underline the control group. Circle the experimental group.

   Scientists are testing a new kind of aspirin to see whether it will relieve headaches. They give one group of volunteers the aspirin. They give another group of volunteers pills that look like aspirin but are actually sugar pills.

Interpret Graphics

4. **Order** Sequence the steps of scientific inquiry that was used in one part of the case study.

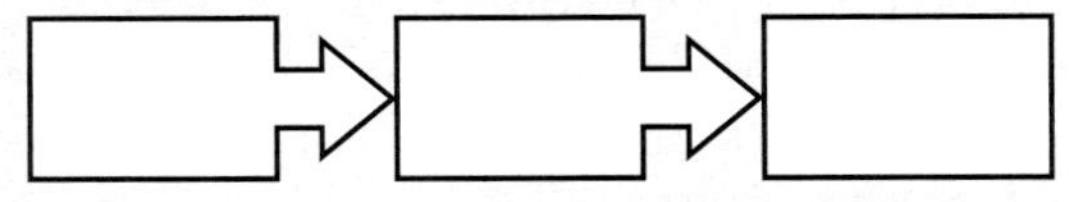

5. **Explain** the significance of the pollen found in the Iceman's digestive tract?

Critical Thinking

6. **Formulate** two questions about the Iceman. What would you want to know next?

Methods of Science

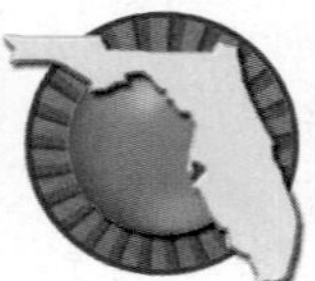

Think About It! **Scientists creatively use the process of scientific inquiry to formulate investigable questions, collect and evaluate appropriate data, and then communicate the evaluations.**

Key Concepts Summary

LESSON 1 Understanding Science

- Scientific inquiry is a process that uses a set of skills to answer questions or to test ideas about the natural world.
- A **scientific law** is a rule that describes a pattern in nature. A **scientific theory** is an explanation of things or events that is based on knowledge gained from many **observations** and investigations.
- Facts are measurements, observations, and theories that can be evaluated for validity through objective investigation. Opinions are personal views or feelings about a topic.

Vocabulary

science p. NOS 4
observation p. NOS 6
inference p. NOS 6
hypothesis p. NOS 6
prediction p. NOS 6
technology p. NOS 8
scientific theory p. NOS 9
scientific law p. NOS 9
critical thinking p. NOS 10

LESSON 2 Measurement and Scientific Tools

- Scientists worldwide agree to use the **International System of Units** because measurements are easier to compare, confirm, and repeat using standardized units.
- Measurement uncertainty occurs because no scientific tool can provide a perfect measurement.
- Mean, median, mode, and range are calculations used to evaluate sets of data.

description p. NOS 12
explanation p. NOS 12
International System of Units (SI) p. NOS 12
significant digits p. NOS 14

LESSON 3 Case Study: The Iceman's Last Journey

- The **independent variable** is the factor a scientist changes to observe how it affects a **dependent variable**. A dependent variable is the factor a scientist measures or observes during an experiment.
- Scientific inquiry is used throughout investigations as hypotheses, predictions, tests, analysis, and conclusions are developed.

variable p. NOS 21
independent variable p. NOS 21
dependent variable p. NOS 21

Use Vocabulary

Fill in the blanks with the correct vocabulary terms.

1. A ____________________ is an interpretation of observations.
2. The ____________________ are the numbers of digits in a measurement that you know with a certain degree of reliability.
3. The act of watching something and taking note of what occurs is a(n) ____________________ .
4. A ____________________ is a rule that describes a pattern in nature.

Methods of Science

Review

Understand Key Concepts

5. You have the following data set: 2, 3, 4, 4, 5, 7, and 8. Is 6 the mean, the median, the mode, or the range of the data set? **MA.6.S.6.2**
   (A) mean
   (B) median
   (C) mode
   (D) range

6. Which best describes an independent variable? **SC.6.N.3.2**
   (A) It is a factor that is not in every test.
   (B) It is a factor the investigator changes.
   (C) It is a factor you measure during a test.
   (D) It is a factor that stays the same in every test.

Critical Thinking

7. **Predict** what would happen if every scientist tried to use all the skills of scientific inquiry in the same order in every investigation. **SC.6.N.2.2**

8. **Assess** the role of measurement uncertainty in scientific investigations. **SC.6.N.2.1**

9. **Evaluate** the importance of having a control group in a scientific investigation. **SC.6.N.1.1**

Writing in Science

10. **Construct** On a separate piece of paper, write a paragraph explaining why scientists agree the International System of Units (SI) is an easier system to use when communicating with scientist from around the world. Be sure to include a topic sentence and a concluding sentence in your paragraph. **SC.6.N.2.1**

Big Idea Review

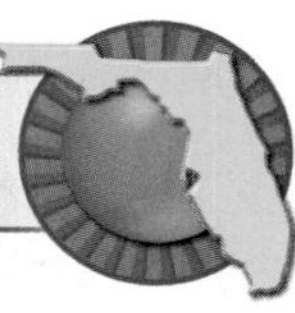

11. What process do scientists use to perform scientific investigations? List and explain three of the skills involved. **SC.6.N.2.1**

12. **Infer** the purpose of the pink dye in the scientific investigation shown in the photo. **SC.6.N.1.5**

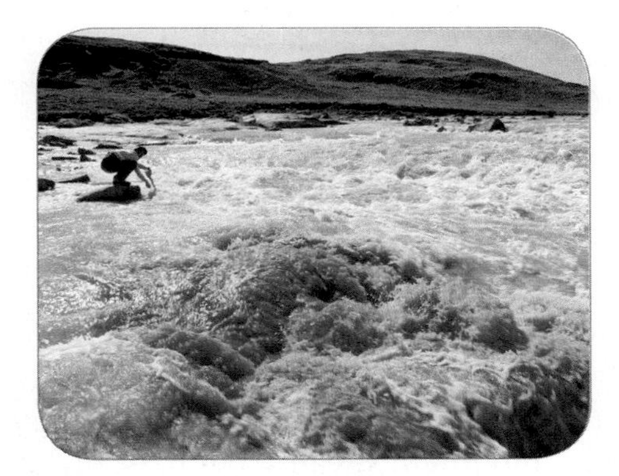

Math Skills MA.6.S.6.2

Use Numbers Convert:

13. 162.5 hg = ________ g

14. 89.7 cm = ________ mm

Name ______________________ Date ______________

Multiple Choice *Bubble the correct answer.*

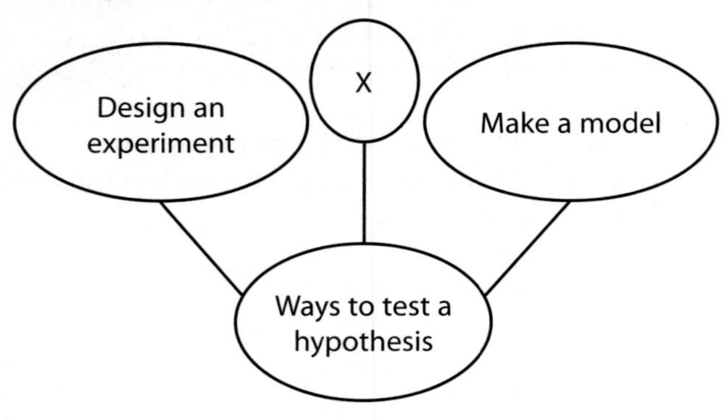

1. In the graphic organizer above, which could be placed in the circle labeled X? **SC.6.N.2.1**

- (A) Ask questions.
- (B) Share information.
- (C) Gather and evaluate evidence.
- (D) Write a science journal article.

2. Which is a prediction? **SC.6.N.2.2**

- (F) The highest snowfall amounts in the mountains generally occur in March.
- (G) Snowfall in the mountains can be a very beautiful scene.
- (H) If snowfall is above average in March, mountain streams will have higher-than-average water flow in April.
- (I) Snowfall is highest in the mountains in March because more weather systems move from the ocean to the mountains.

3. Which question would a life scientist ask? **SC.6.N.2.3**

- (A) How can molecules be used to make a new medicine?
- (B) How do earthquakes change Earth's surface?
- (C) How do sunspots affect radio transmissions on Earth?
- (D) How might climate change affect plants in the tundra?

Fact	Opinion
The video is rated PG.	It is a great movie.
This video costs $15.99.	Q

4. Which goes where Q is in the table above? **LA.6.2.2.3**

- (F) This movie is two hours long.
- (G) This movie shows conflict.
- (H) This movie has two main characters.
- (I) This movie will change your life.

Name ______________________________ Date ______________

Multiple Choice *Bubble the correct answer.*

Use the data table below to answer questions 1 and 2.

Miami High Temperatures—May 1–7

May 1	22.3°C
May 2	26.7°C
May 3	28.6°C
May 4	22.3°C
May 5	21.2°C
May 6	23.8°C
May 7	25.2°C

1. Which is the mode of the temperatures in the table? **MA.6.S.6.2**

Ⓐ 21.4
Ⓑ 22.3
Ⓒ 24.3
Ⓓ 25.2

2. Which is the range of the temperatures in the table? **MA.6.S.6.2**

Ⓕ 7.4
Ⓖ 9.3
Ⓗ 24.3
Ⓘ 25.2

3. How many significant digits are in the number 70.404753? **MA.6.S.6.2**

Ⓐ 5
Ⓑ 6
Ⓒ 7
Ⓓ 8

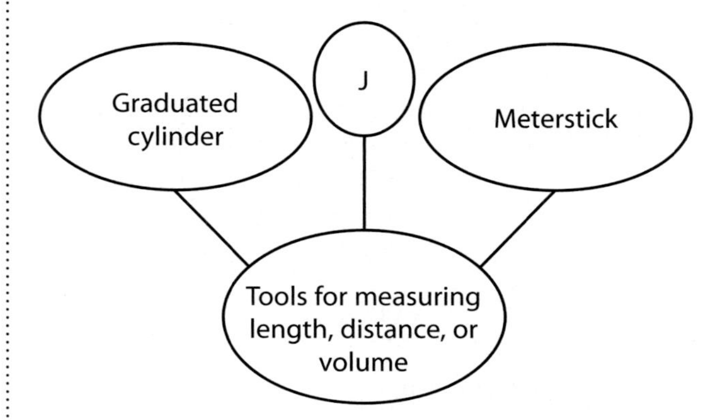

4. In the graphic organizer above, which tool could be placed in the circle labeled J? **LA.6.2.2.2**

Ⓕ compass
Ⓖ thermometer
Ⓗ metric ruler
Ⓘ streak plate

Multiple Choice *Bubble the correct answer.*

Scientific Investigation: Presence of Pollen in the Iceman Digestive System	
Component A	Saline solution plus sterilized equipment
Component B	Presence or absence of pollen
Component C	Digestive system contents plus sterilized equipment
Component D	Samples of digestive tract

1. The table above shows the variables, control group, and experimental group used in a scientific investigation of the Iceman's digestive tract. Which component is the dependent variable in this investigation? **SC.6.N.1.1**

Ⓐ Component A

Ⓑ Component B

Ⓒ Component C

Ⓓ Component D

2. Which is a hypothesis about the Iceman? **SC.6.N.1.5**

Ⓕ An axe, bow, quiver, and dagger were found with the Iceman.

Ⓖ The wood in the Iceman's tools was from low-elevation plants.

Ⓗ The Iceman was well preserved because he died in autumn and was quickly covered in snow.

Ⓘ If hop hornbeam pollen is found in the Iceman's stomach, then he was at a lower elevation before he died.

3. Which evidence helped scientists to reject the hypothesis that the Iceman was a man who went missing in 1938? **SC.6.N.2.2**

Ⓐ the artifacts surrounding the Iceman

Ⓑ the stomach contents of the Iceman

Ⓒ the preserved condition of the remains of the Iceman

Ⓓ the species of trees growing where the Iceman was found

Some Data Collected About the Iceman	
1	The Iceman carried an axe, bow, quiver, and dagger.
2	Wood used in the Iceman artifacts was from trees growing at low elevations.
3	Pollen from hop hornbeam was found in the digestive tract. This plant flowers in spring and grows at low elevations.
4	Einhorn grain and other plants were found in the digestive tract.

4. Based on the data shown above, which is an inference that could be made about the Iceman? **SC.6.N.1.5**

Ⓕ The Iceman ate plants.

Ⓖ The Iceman died of starvation.

Ⓗ The Iceman liked to take long hikes.

Ⓘ The Iceman lived in the mountains.

1950 — 2000

1960s
Geographic Information Systems (GIS) are developed. GIS displays large amounts of information and includes computer software and hardware as well as digital data and storage.

1993
The space-based Global Positioning System achieves initial operational capability.

2005
An internet mapping tool displays satellite images of Earth's surface.

Inquiry
Visit ConnectED for this unit's **STEM** activity.

Science and History

About 500,000 years ago, humans used stone to make tools, weapons, and small decorative items. About 8,000 years ago, someone might have spied a shiny rock. It was gold—thought to be the first metal discovered by humans. Gold was very different from stone. It did not break when it was struck. It could easily be shaped into useful and beautiful objects. Over time, other metals were discovered. Each metal helped humans progress from the Stone Age to the Moon, to Mars, and beyond.

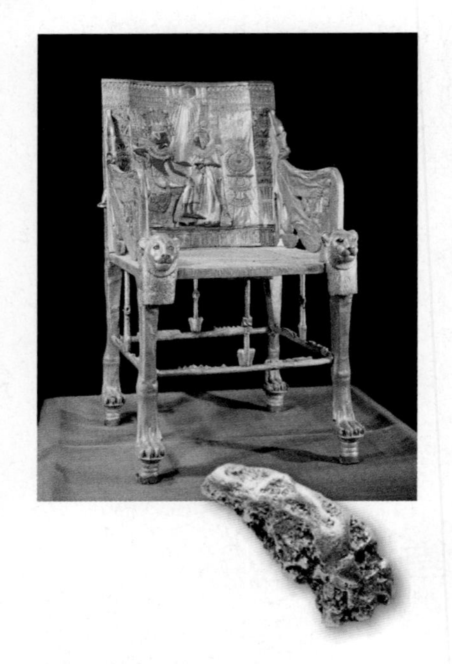

Gold

Since the time of its discovery, gold has been a symbol of wealth and power. It is used mainly in jewelry, coins, and other valuable objects. King Tut's coffin was made of pure gold. Tut's body was surrounded by the largest collection of gold objects ever discovered—chariots, statues, jewelry, and a golden throne.

Active Reading **1. Infer** Suggest two reasons gold might be a symbol of wealth and power.

Lead

Ancient Egyptians used the mineral lead sulfide, also called galena, as eye paint. About 5,500 years ago, metalworkers found that galena melts into puddles at a low temperature. Lead bends easily, and the Romans shaped it into pipes for carrying water. Over the years, the Romans realized that lead was entering the water and was toxic to humans. Despite possible danger, lead water pipes and paint containing lead were common in modern homes for decades. Finally, however, in 2004 the use of lead in home construction was banned.

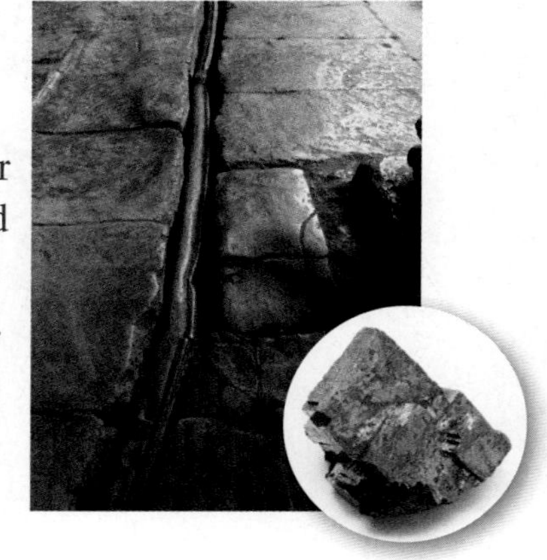

Active Reading **2. Analyze** Why was lead banned in home construction?

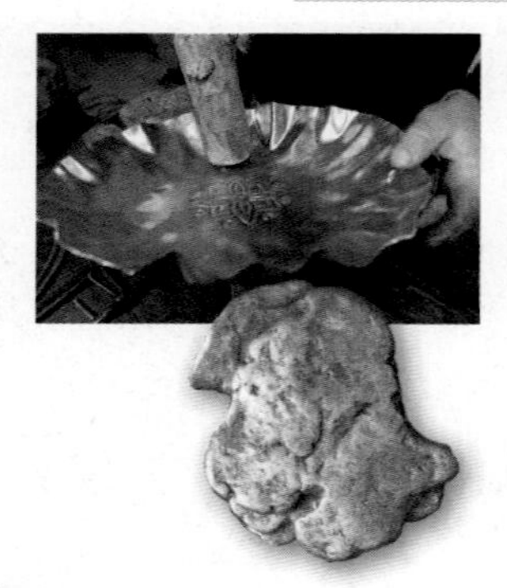

Copper

The first metal commonly traded was copper. About 5,000 years ago, Native Americans mined more than half a million tons of copper from the area that is now Michigan. Copper is stronger than gold. Back then, it was shaped into saws, axes, and other tools. Stronger saws made it easier to cut down trees. The wood from trees then could be used to build boats, which allowed trade routes to expand.

Tin and Bronze

Around 4,500 years ago, the Sumerians noticed differences in the copper they used. Some flowed more easily when it melted and was stronger after it hardened. They discovered that this harder copper contained another metal—tin. Metalworkers began combining tin and copper to produce a metal called bronze. Bronze eventually replaced copper as the most important metal to society. Bronze was strong and cheap enough to make everyday tools. It could easily be shaped into arrowheads, armor, axes, and sword blades. People admired the appearance of bronze. It continues to be used in sculptures. Bronze, along with gold and silver, is used in Olympic medals as a symbol of excellence.

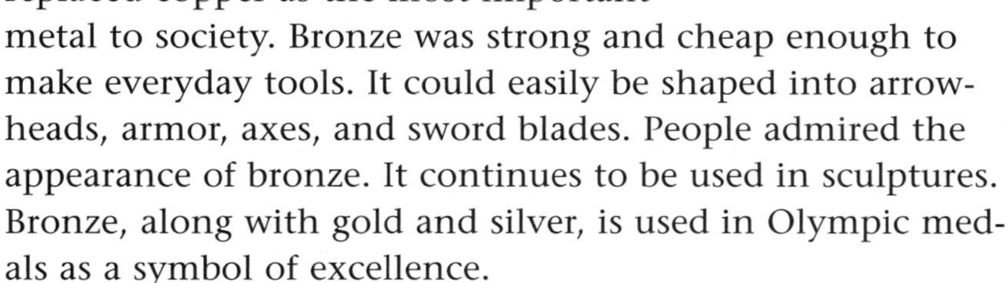

Iron and Steel

Although iron-containing rock was known centuries ago, people couldn't build fires hot enough to melt the rock and separate out the iron. As fire-building methods improved, iron use became more common. It replaced bronze for all uses except art. Iron farm tools revolutionized agriculture. Iron weapons became the choice for war. Like metals used by earlier civilizations, iron increased trade and wealth, and improved people's lives.

In the 17th century, metalworkers developed a way to mix iron with carbon. This process formed steel. Steel quickly became valued for its strength, resistance to rusting, and ease of use in welding. Besides being used in the construction of skyscrapers, bridges, and highways, steel is used to make tools, ships, vehicles, machines, and appliances.

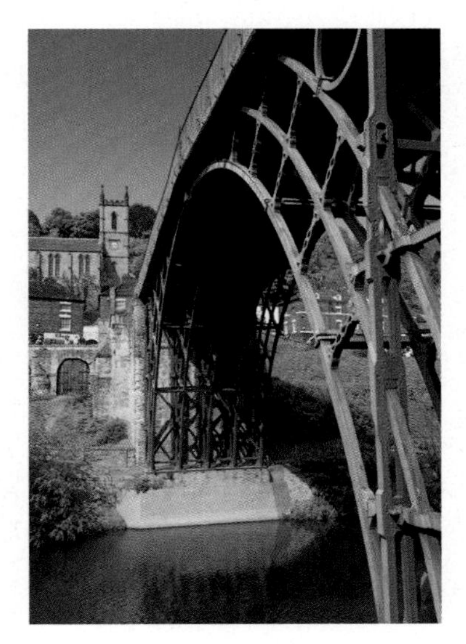

Try to imagine your world without metals. Throughout history, metals changed society as people learned to use them.

MiniLab ***How do a metal's properties affect its uses?*** at connectED.mcgraw-hill.com

Apply It! After you complete the lab, investigate and answer these questions.

1. **Research** What is the current market price of one pound of each of the metals listed?

Metal	Current Market Price
Aluminum	
Bronze	
Copper	
Gold	
Iron	
Lead	
Nickel	
Silver	
Tin	

2. **Explain** How does the cost of jewelry differ depending on the type of metal from which it is made? Support your reasoning.

Name ______________________ Date ______________

Notes

Earth, Our Planet

When we talk about our planet Earth, what parts are we describing? Circle the answer that best describes our planet, Earth.

A. Only the land parts.

B. Only land and air parts.

C. All of the land and water parts.

D. The land, water, ice, and air parts.

E. The land parts and some of the water parts.

F. The land and ice parts, and some of the air parts.

Explain your thinking about our planet Earth.

Our Planet— EARTH

FLORIDA BIG IDEAS

1 The Practice of Science
2 The Characteristics of Scientific Knowledge
3 The Role of Theories, Laws, Hypotheses, and Models
6 Earth Structures
7 Earth Systems and Patterns

Think About It!

How can you describe Earth?

From space, it's easy to see why Earth is called the blue planet. But there's more to Earth than oceans of water.

1. What other parts of Earth can you see in the photo?

2. How would you describe Earth to a friend?

Get Ready to Read

What do you think about the planet Earth?

Before you read, decide if you agree or disagree with each of these statements. As you read this chapter, see if you change your mind about any of the statements.

	AGREE	DISAGREE
1 Earth is a simple system made of rocks.	☐	☐
2 Most of Earth is covered by one large ocean.	☐	☐
3 Earth's interior is made of distinct layers.	☐	☐
4 The water cycle begins in the ocean.	☐	☐
5 Earth's air contains solids, liquids, and gases.	☐	☐
6 Rocks are made of minerals.	☐	☐

There's More Online!
Video • Audio • Review • ⓘLab Station • WebQuest • Assessment • Concepts in Motion • Multilingual eGlossary

Lesson 1

EARTH SYSTEMS

ESSENTIAL QUESTIONS

- What are the composition and the structure of the atmosphere?
- How is water distributed in the hydrosphere?
- What are Earth's systems?
- What are the composition and the structure of the geosphere?

Vocabulary

biosphere p. 12
atmosphere p. 13
hydrosphere p. 15
cryosphere p. 16
groundwater p. 16
geosphere p. 17
mineral p. 17
rock p. 18

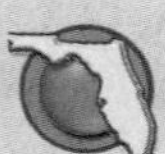

Florida NGSSS

LA.6.2.2.3 The student will organize information to show understanding (e.g., representing main ideas within text through charting, mapping, paraphrasing, summarizing, or comparing/contrasting);

LA.6.4.2.2 The student will record information (e.g., observations, notes, lists, charts, legends) related to a topic, including visual aids to organize and record information and include a list of sources used;

SC.6.E.7.4 Differentiate and show interactions among the geosphere, hydrosphere, cryosphere, atmosphere, and biosphere.

SC.6.N.1.5 Recognize that science involves creativity, not just in designing experiments, but also in creating explanations that fit evidence.

SC.6.N.1.1 Define a problem from the sixth grade curriculum, use appropriate reference materials to support scientific understanding, plan and carry out scientific investigation of various types, such as systematic observations or experiments, identify variables, collect and organize data, interpret data in charts, tables, and graphics, analyze information, make predictions, and defend conclusions.

SC.6.N.2.1 Distinguish science from other activities involving thought.

Inquiry Launch Lab

SC.6.N.1.5, SC.6.N.2.1

20 minutes

How can you describe Earth?

When you look out the window, you might see wispy white clouds, birds in the trees, and rolling hills in the distance. All these things are part of Earth. What else makes up Earth?

Procedure 

1. Read and complete a lab safety form.
2. With your partner, brainstorm a list of words that describe Earth. Limit the list to 20 words. Be creative!
3. Use **markers** to rewrite your list of words using different colors and letter shapes. Use **scissors** to cut out each word.
4. Group the words that you think relate to each other. Use a **glue stick** to fix the words to a piece of **colored paper.**

Data and Observations

Think About This

1. What words did you use to describe Earth?
2. How did your list compare to those of other students?
3. Key Concept What things do you think make up Earth?

A Hot Mix?

1. Earth is made of more than soil, minerals, and melted rock flowing out of volcanoes. What other parts of Earth do you see in the photo? How do you think these parts interact?

What is Earth?

The puffy, white clouds over your head and the hard ground under your feet are both parts of Earth. The water in the oceans and the fish that live there are also parts of Earth. The planet Earth is more than a solid ball in space. It includes air molecules that float near the boundaries of outer space and molten rock that churns deep below Earth's surface.

Earth is a complex place. People often divide complex things into smaller parts in order to study them. Scientists divide Earth into systems to help better understand the planet. The systems contain different materials and work in different ways, but they all interact. What happens in one system affects the others.

Earth's Air

The outermost Earth system is an invisible layer of gases that surrounds the planet. Even though you cannot see air, you can feel it when the wind blows. Moving air is blowing the tree in **Figure 1.**

Figure 1 Even though you cannot see air, you can see its power when it makes objects move.

Earth's Water

Below the layer of air are the systems that contain Earth's water. Like air, water can move from place to place. Some water is salty, and some is fresh. Fresh river water flows into the salty Pacific Ocean in Hawaii, as shown in **Figure 2.**

The Solid Earth

The next system is the solid part of Earth. It contains a thin layer of soil covering a rocky center. It is by far the largest Earth system. Because it is solid, materials in this system move more slowly than air or water. But they do move, and over time, landforms rise up and then wear away. It took millions of years for the canyon shown in **Figure 2** to form.

Life on Earth

The Earth system that contains all living things is the **biosphere.** Living things are found in air, water, and soil. So, the biosphere has no distinct boundaries; it is found within the other Earth systems. The living things shown in **Figure 2** are part of the biosphere. You will learn more about the biosphere when you study life science, or biology. The rest of this chapter will describe the Earth systems made of nonliving things.

Active Reading **3. Recall** Underline the reason the biosphere does not have distinct boundaries.

Figure 2 Air, water, rocks, and living things are all part of Earth.

2. Visual Check
Identify Label the Earth systems shown in the photo.

The Atmosphere

The force of Earth's gravity pulls molecules of gases into a layer surrounding the planet. *This mixture of gases forms a layer around Earth called the* **atmosphere.** The atmosphere is denser near Earth's surface and becomes thinner farther from Earth. It keeps Earth warm by trapping thermal energy from the Sun that bounces back from Earth's surface. If the atmosphere did not regulate temperature, life as it is on Earth could not exist.

WORD ORIGIN

atmosphere
from Greek *atmos-*, means "vapor"; and Greek *spharia*, means "sphere"

What makes up the atmosphere?

The atmosphere contains a mixture of nitrogen, oxygen, and smaller amounts of other gases. The graph in **Figure 3** shows the percentages of these gases. The most common gas is nitrogen, which makes up 78 percent of the atmosphere. Most of the remaining gas is oxygen.

The other gases are called trace gases because they make up only 1 percent, or a trace, of the atmosphere. Nonetheless, trace gases are important. Carbon dioxide, methane, and water vapor help regulate Earth's temperature. Note that **Figure 3** shows the percentages of gases in dry air. The atmosphere also contains water vapor. The amount of water vapor in the atmosphere generally ranges from 0 to 4 percent.

Along with gases and water vapor, the atmosphere contains small amounts of solids. Particles of dust float along with the gases and water vapor. Sometimes you can see these tiny specks as sunlight reflects off them as it shines through a window.

4. NGSSS Check
Analyze What is the composition of the atmosphere? SC.6.E.7.4

Figure 3 Dry air contains a mixture of gases. Though the atmosphere is made mainly of nitrogen and oxygen, trace gases are also important.

Active Reading **5. Calculate** What percentage of Earth's atmosphere is oxygen?

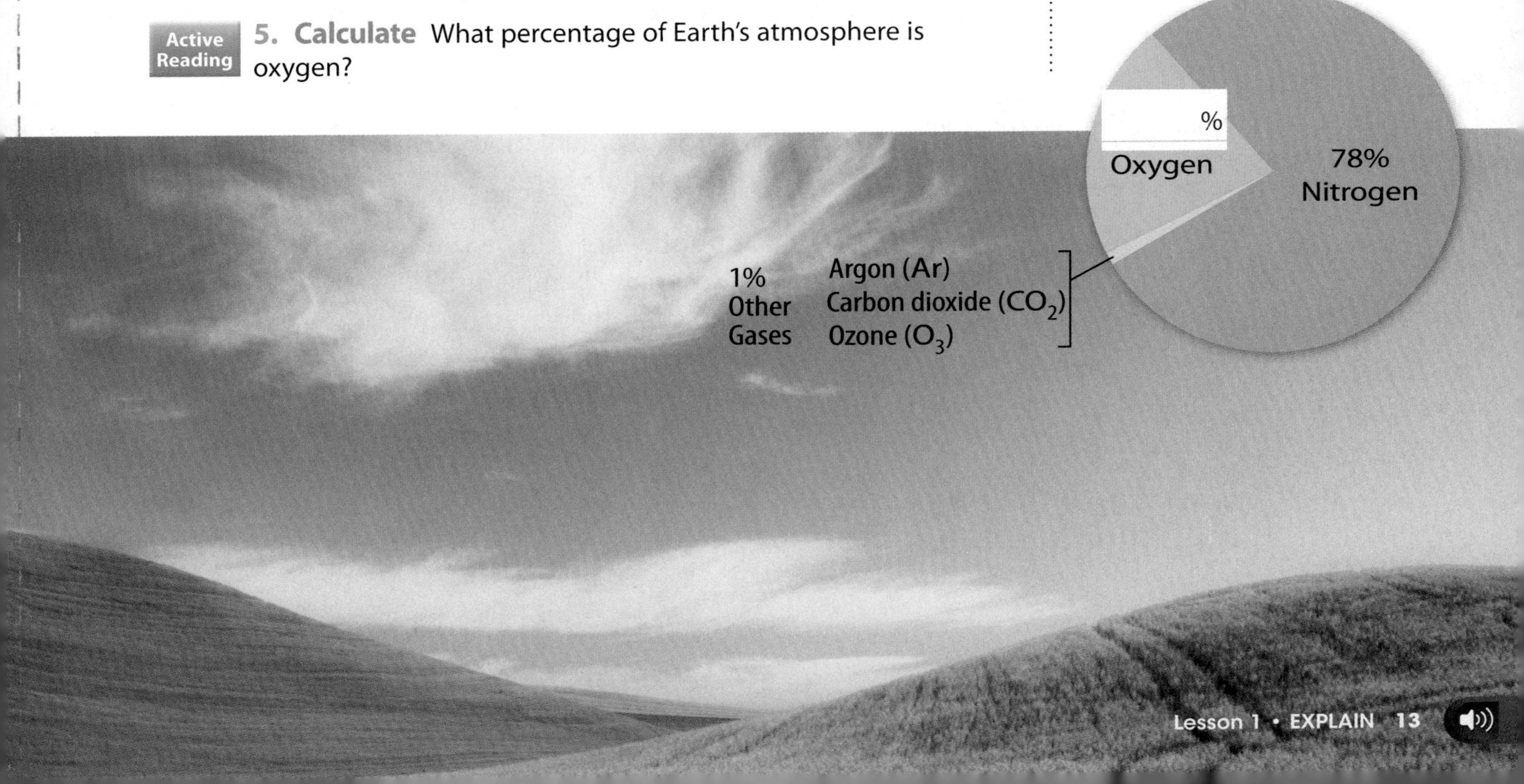

%
Oxygen
78%
Nitrogen
1% Other Gases
Argon (Ar)
Carbon dioxide (CO_2)
Ozone (O_3)

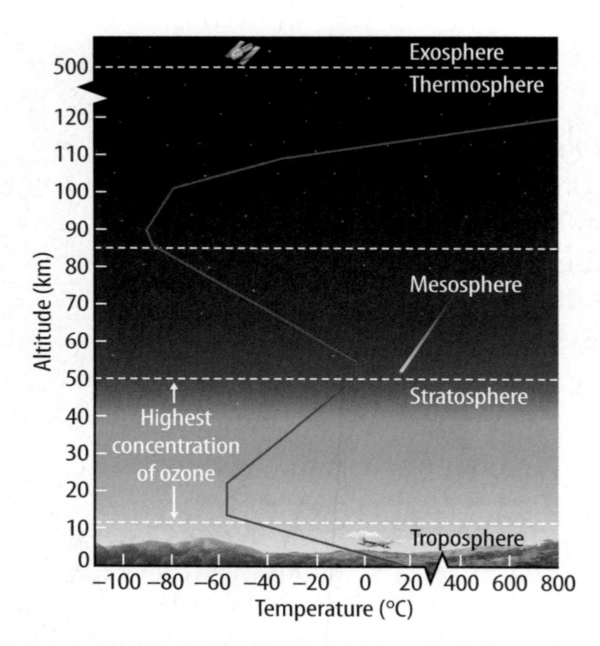

Figure 4 The atmosphere is divided into layers according to differences in temperature.

6. Visual Check Summarize How does temperature change as altitude increases?

Active Reading **7. List** What are the layers of the atmosphere?

Layers of the Atmosphere

The composition of the atmosphere does not change much over time. However, the temperature of the atmosphere does change. Thermal energy from the Sun heats the atmosphere; however, different parts of the atmosphere absorb or reflect this thermal energy in different ways. The red line in **Figure 4** shows changes in temperature as altitude increases. These temperature changes are used to distinguish layers in the atmosphere.

The Troposphere If you have ever hiked up a mountain, you might have noticed that the temperature gets cooler as you climb higher. In the bottom layer of the atmosphere, called the troposphere, temperature decreases as you move upward from Earth's surface. Gases flow and swirl in the troposphere, causing weather. Although the troposphere does not extend very far upward, it contains most of the mass in the atmosphere.

The Stratosphere Above the troposphere is the stratosphere. Unlike gases in the troposphere, gases in the stratosphere do not swirl around. They are more stable and form flat layers. Within the stratosphere is a layer of ozone, a form of oxygen. This ozone layer protects Earth's surface from harmful radiation from the Sun. It acts like a layer of sunscreen, protecting the biosphere. Because ozone absorbs solar radiation, temperatures increase in the stratosphere.

Upper Layers Above the stratosphere is the mesosphere. Temperature decreases in this layer, then increases again in the next layer, the thermosphere. The last layer of Earth's atmosphere is the exosphere. The lowest density of gas molecules is in this layer. Beyond the exosphere is outer space.

The Hydrosphere

Water is one of the most common and important substances on Earth. *The system containing all Earth's water is called the* **hydrosphere.** Most water is stored on Earth's surface, but some is located below the surface or within the atmosphere and biosphere. The hydrosphere contains more than 1.3 billion km^3 of water. The amount of water does not change. But like the gases in the atmosphere, water in the hydrosphere flows. It moves from one location to another over time. Water also changes state. It is found as a liquid, a solid, and a gas on Earth.

Active Reading **8. Point out** Circle the volume of water contained in the hydrosphere.

Active Reading

FOLDABLES® LA.6.2.2.3

Make a small, horizontal four-door book with a 1-cm center tab. Label it as shown. Use it to organize your notes on Earth systems.

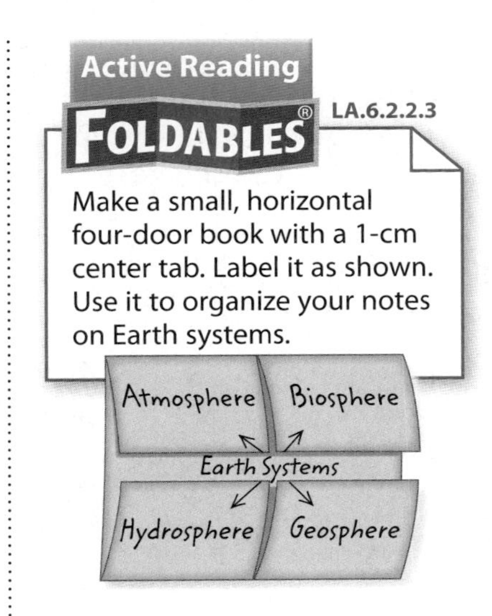

Ocean

Scientists call the natural locations where water is stored reservoirs (REH zuh vworz). The largest reservoir on Earth is the world ocean. Though the oceans have separate names, they are all connected, making one large ocean. Water flows freely throughout the world ocean. About 97 percent of Earth's water is in the ocean, as shown in **Figure 5.**

Many minerals dissolve easily in water. As water in rivers and underground reservoirs flows toward the ocean, it dissolves materials from the solid Earth. These dissolved minerals make ocean water salty. Most plants and animals that live on land, including humans, cannot use salt water. They need **freshwater** to survive.

Review Vocabulary

freshwater
water that contains less than 0.2 percent dissolved salts

Figure 5 Water in the hydrosphere is found in several different reservoirs.

Active Reading **9. Calculate** What percentage of Earth's freshwater is in icecaps and glaciers?

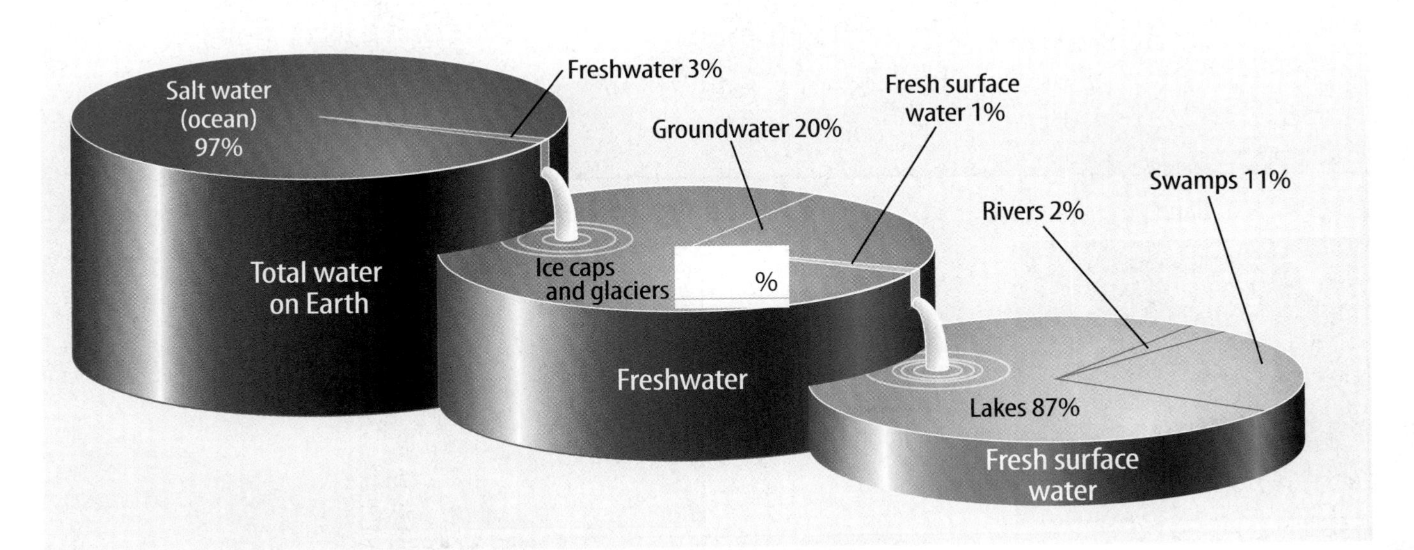

Active Reading **10. Summarize** How is water distributed in the hydrosphere?

Lakes and Rivers

Less than 1 percent of freshwater is easily accessible on Earth's surface. This small percentage of Earth's total water must meet the needs of people and other organisms that require freshwater. Rain and snow supply water to the surface reservoirs—lakes and rivers. Water in these reservoirs moves through the water cycle much faster than water frozen in glaciers and ice caps.

Groundwater

Ice, lakes, and rivers hold about 80 percent of Earth's freshwater. The remaining 20 percent is beneath the ground. Some rain and snow seep into the ground and collect in small cracks and open spaces called pores. **Groundwater** *is water that is stored in cracks and pores beneath Earth's surface.* As shown in **Figure 6,** groundwater collects in layers. Many people get their water by drilling wells down into these layers of groundwater.

The Cryosphere

Did you know that most of Earth's freshwater is frozen? The frozen portion of water on Earth's surface is called the **cryosphere.** About 79 percent of the planet's freshwater is in the cryosphere. The cryosphere consists of snow, glaciers, and icebergs. Water can be stored as ice for thousands of years before melting and becoming liquid water in other reservoirs.

Active Reading **11. Relate** How are the hydrosphere and the cryosphere related?

Figure 6 Freshwater in lakes, rivers, and glaciers is visible on Earth's surface, but large amounts of groundwater are hidden below the surface.

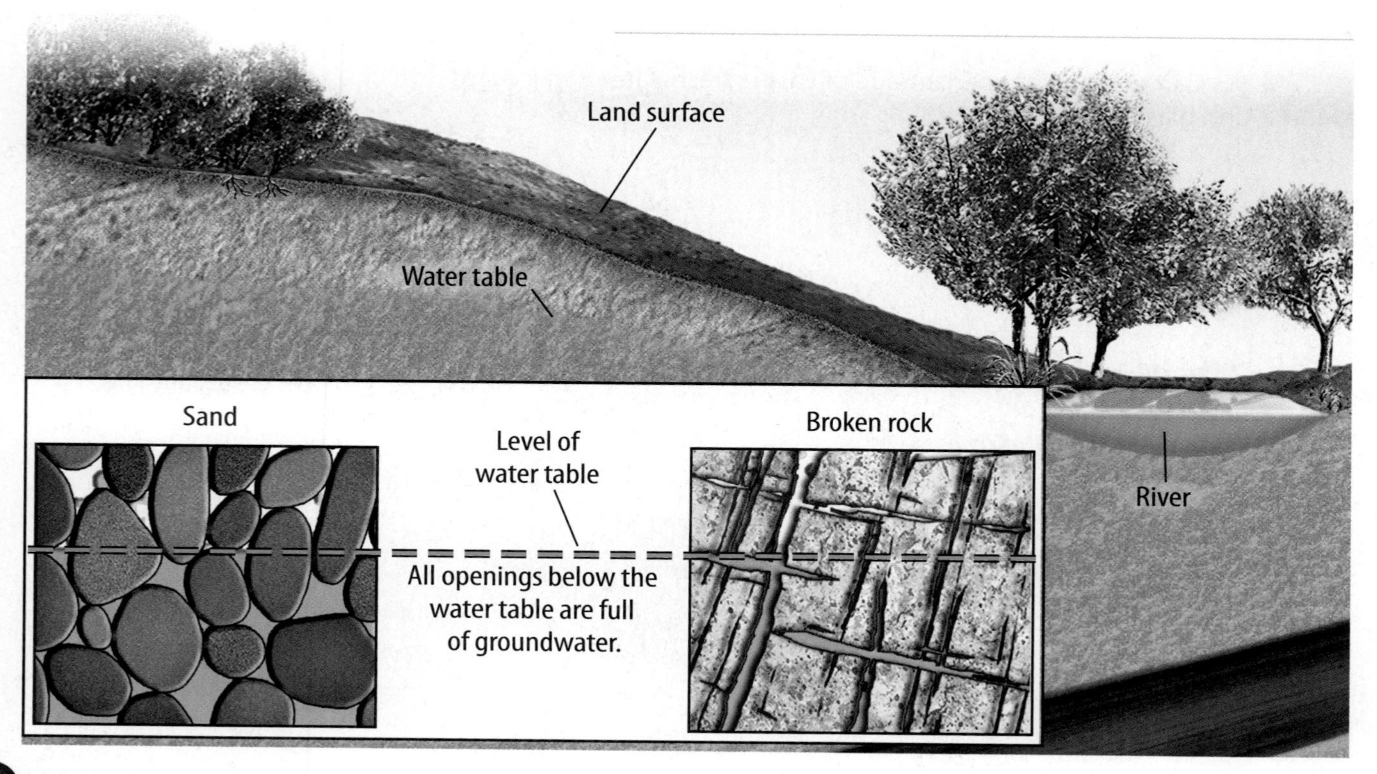

The Geosphere

The last nonliving Earth system is the geosphere. *The* **geosphere** *is the solid part of Earth.* It includes a thin layer of soil and broken rock material along with the underlying layers of rock. The rocks and soil on land and beneath the oceans are part of the geosphere.

Active Reading **12. Recall** What are Earth's systems?

Materials in the Geosphere

The geosphere is made of soil, rock, and metal. All of these materials are composed of smaller particles.

Minerals Have you ever seen a sparkling diamond ring? Diamond is a mineral that is mined and then later cut and polished. **Minerals** *are naturally occurring, inorganic solids that have crystal structures and definite chemical compositions.*

To be considered a mineral, a material must have all five characteristics listed above. For example, materials that are made by people are not minerals because they did not form naturally. Materials that were once alive are organic and cannot be minerals. A mineral must be solid, so liquids and gases are not minerals. The atoms in minerals must be arranged in an orderly, repeating pattern. Finally, each mineral has a unique composition made of specific elements.

Minerals are identified by their physical properties, which include color, streak, hardness, luster, and crystal shape. Streak is the color of a mineral's powder. Even though some minerals have different colors, the color of the streak is the same. Hardness describes how easily a mineral can be scratched. Luster describes how a mineral reflects light. Usually, you must test several properties to identify a mineral. Examples of minerals with different properties are shown in **Figure 7.**

Active Reading **13. Identify** (Circle) the physical properties used to identify minerals.

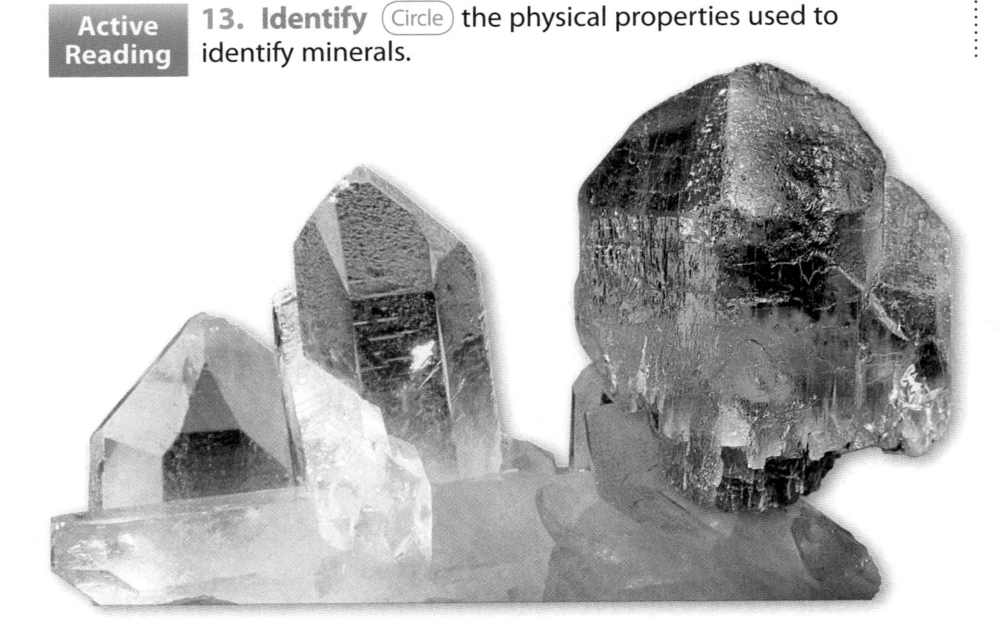

Figure 7 Minerals have different properties. The quartz shown on the left has a visible crystal structure. The olivine shown on the right has a striking color.

Figure 8 Diorite (top) is an igneous rock. Gneiss (center) is metamorphic. The conglomerate (bottom) is sedimentary.

Rocks Minerals are the building blocks of rocks. A **rock** *is a naturally occurring solid composed of minerals and sometimes other materials such as organic matter.* Scientists classify rocks according to how they formed. As shown in **Figure 8,** there are three major rock types: igneous, sedimentary, and metamorphic.

Igneous rocks form when molten material, called magma, cools and then hardens. Often the magma is found deep inside Earth, but sometimes it erupts from volcanoes and flows onto Earth's surface as lava. So, igneous rocks can form inside Earth or on Earth's surface.

Sedimentary rocks form when forces such as water, wind, and ice break down rocks into small pieces called sediment. These same forces carry and deposit the sediment in layers. The bottom layers of sediment are compressed and then cemented together by natural substances to form rocks.

Metamorphic rocks form when extreme temperatures and pressure within Earth change existing rocks into new rocks. The rocks do not melt. Instead, their compositions or their structures change.

Active Reading **14. Differentiate** How are rocks different from minerals?

MiniLab ***What makes the geosphere unique?*** at connectED.mcgraw-hill.com

Apply It! After you complete the lab, answer these questions.

1. What properties did you use to distinguish the rock sample from the mineral samples?

2. How was the rock sample related to some of the mineral samples?

Structure

Earth's internal structure is layered like the layers of a hard-cooked egg. The three basic layers of the geosphere are shown in **Figure 9.** Similar to an egg, each layer of the geosphere has a different composition.

Crust The brittle outer layer of the geosphere is much thinner than the inner layers, like the shell on a hard-cooked egg. This thin layer of rock is called the crust. The crust is found under the soil on continents and under the ocean. Oceanic crust is thinner and denser than continental crust. This is due to their different compositions. Continental crust is made of igneous, sedimentary, and metamorphic rocks. Oceanic crust is made of only igneous rock.

Mantle The middle and largest layer of the geosphere is the mantle. Like the crust, the mantle is made of rock; however, mantle rocks are hotter and denser than those in the crust. In parts of the mantle, temperatures are so high that rocks flow, a bit like partially melted plastic.

Core The center of Earth is the core. If you use a hard-cooked egg as a model of Earth, then the yolk would be the core. Unlike the crust and the mantle, the core is not made of rock. Instead, it is made mostly of the metal iron and small amounts of nickel. The core is divided into two parts. The outer core is liquid. The inner core is a dense ball of solid iron.

15. NGSSS Check Analyze What are the composition and structure of the geosphere? SC.6.E.7.4

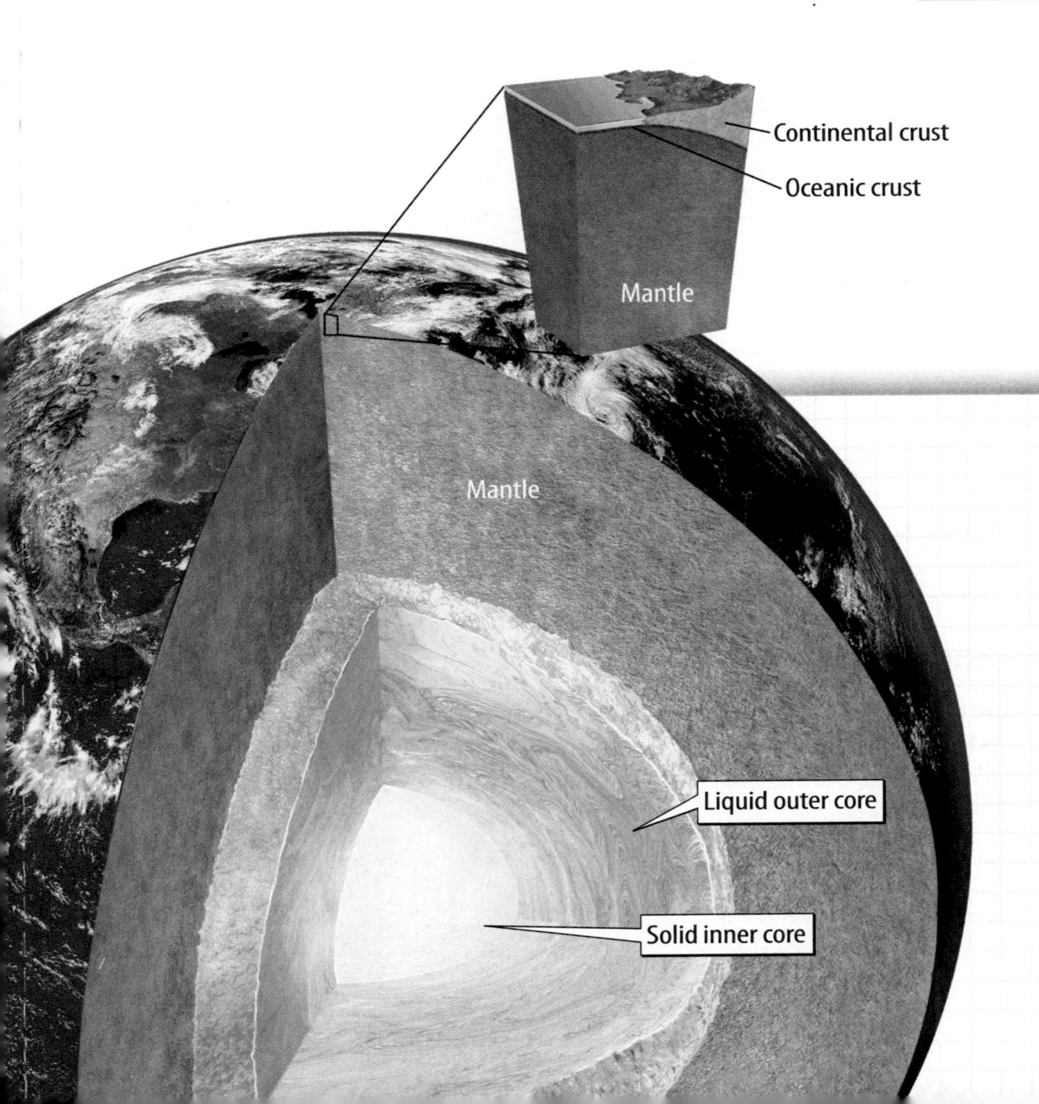

Figure 9 Earth's major layers include the crust, the mantle, and the core.

Active Reading **16. Draw** In the space below, draw an egg and relate its structure to the structure of the geosphere.

Lesson 1 Review

Visual Summary

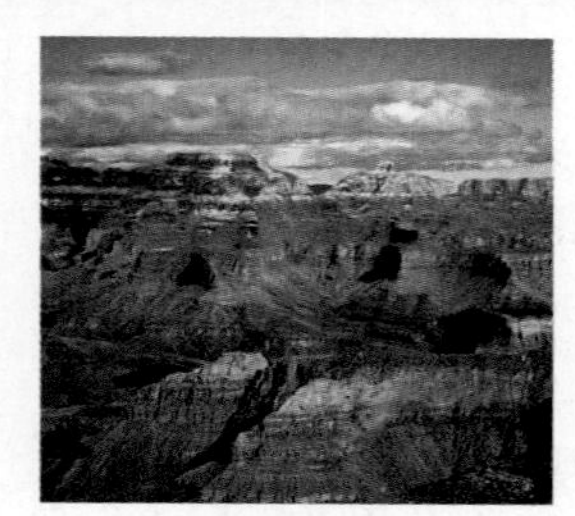

Earth is made of four interacting systems: the atmosphere, the hydrosphere, the geosphere, and the biosphere.

The atmosphere is made mainly of gases and has a layered structure.

Most water in the hydrosphere is in the world ocean.

Use Vocabulary

1. **Use the term** *atmosphere* in a sentence.

2. **Distinguish** between the geosphere and the hydrosphere.

3. **Define** *mineral* in your own words.

Understand Key Concepts

4. Which Earth system contains living things?
   - (A) atmosphere
   - (B) biosphere
   - (C) geosphere
   - (D) hydrosphere

5. **Distinguish** among Earth systems based on the states of matter found in each system.

Interpret Graphics

6. **Summarize** Copy and fill in the graphic organizer below to identify Earth systems. LA.6.2.2.3

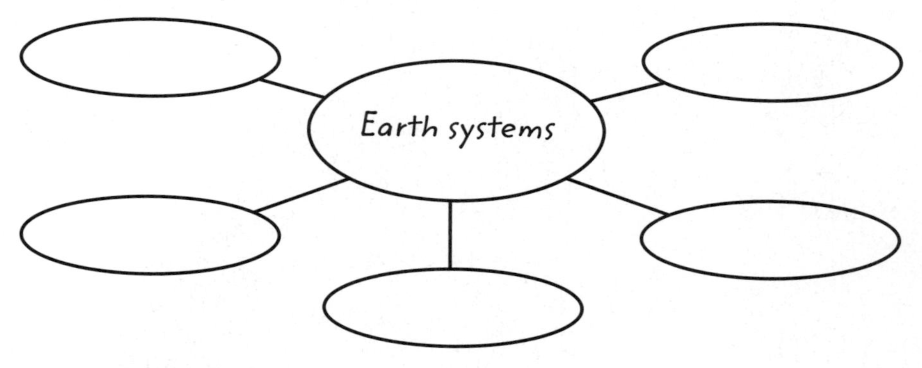

Critical Thinking

7. **Hypothesize** Earth systems interact with and affect one another. What might happen to your local hydrosphere and geosphere if conditions in the troposphere caused rain for several weeks? SC.6.E.7.4

Desalination

HOW IT WORKS

Taking the Salt out of Salt Water

Anyone who's been toppled by a big ocean wave knows salt water doesn't taste like the water we drink. People can't drink salt water. It's about 200 times more salty than freshwater. About 97 percent of Earth's water is salty. Most freshwater is frozen in glaciers and ice caps, leaving less than 1 percent of the planet's water available for 6.7 billion people.

The need for freshwater has scientists searching for efficient ways to take the salt out of salt water. One solution is a desalination plant, where dissolved salts are separated from seawater through a process called reverse osmosis. This is how it works:

Desalination plants are found all over the world, including the United States.

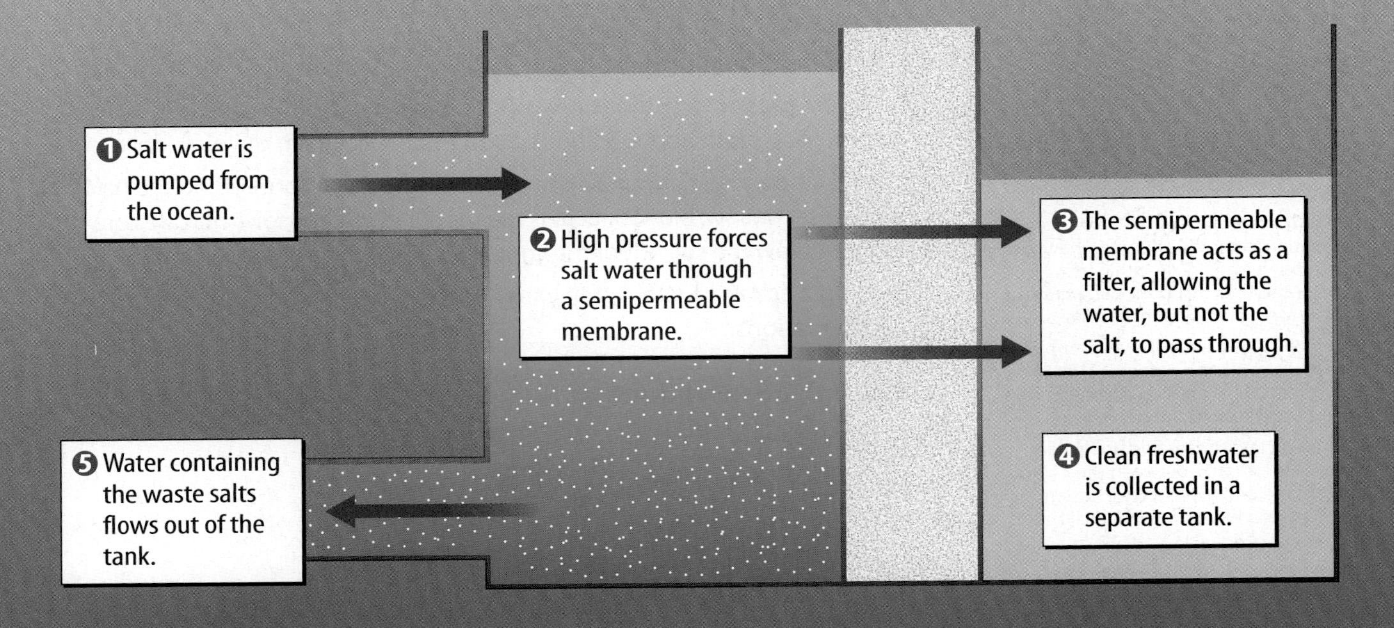

Because it takes a lot of energy to change salt water into freshwater, desalination plants are expensive to operate. But desalination is used in places such as Saudi Arabia and Japan, where millions of people have few freshwater resources.

It's Your Turn

RESEARCH What is the cost of desalinated water for households? How does it compare to the cost of water for households in your area? Present your findings to the class. LA.6.4.2.2

Lesson 2

Interactions of EARTH SYSTEMS

ESSENTIAL QUESTIONS

How does the water cycle show interactions of Earth systems?

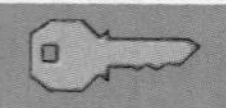
How does weather show interactions of Earth systems?

How does the rock cycle show interactions of Earth systems?

Vocabulary

water cycle p. 23
evaporation p. 24
transpiration p. 24
condensation p. 25
precipitation p. 25
weather p. 26
climate p. 26
rock cycle p. 28
uplift p. 28

Florida NGSSS

LA.6.2.2.3 The student will organize information to show understanding (e.g., representing main ideas within text through charting, mapping, paraphrasing, summarizing, or comparing/contrasting);

SC.6.E.6.1 Describe and give examples of ways in which Earth's surface is built up and torn down by physical and chemical weathering, erosion, and deposition.

SC.6.E.7.2 Investigate and apply how the cycling of water between the atmosphere and hydrosphere has an effect on weather patterns and climate.

SC.6.E.7.4 Differentiate and show interactions among the geosphere, hydrosphere, cryosphere, atmosphere, and biosphere.

SC.6.E.7.6 Differentiate between weather and climate.

SC.6.N.1.1 Define a problem from the sixth grade curriculum, use appropriate reference materials to support scientific understanding, plan and carry out scientific investigation of various types, such as systematic observations or experiments, identify variables, collect and organize data, interpret data in charts, tables, and graphics, analyze information, make predictions, and defend conclusions.

SC.6.N.3.4 Identify the role of models in the context of the sixth grade science benchmarks.

Inquiry Launch Lab

SC.6.N.1.1

20 minutes

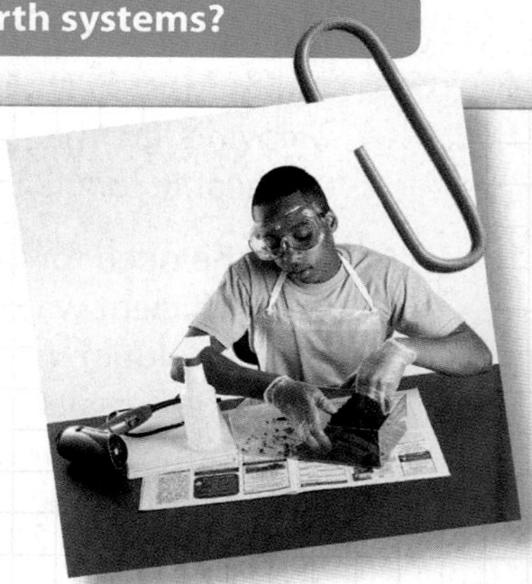

How do some Earth systems interact?

Earth's systems constantly interact with each other. In this activity, you'll model some common interactions.

Procedure

1. Read and complete a lab safety form.
2. Place a **plastic container** on a sheet of **newspaper.** In one end of the container, mold about 5 cups of **soil** into a landform of your choice.
3. Hold a **hair dryer** about 20 cm from the model landform. Using the hair dryer set on low, blow air across the model landscape for 1 min. Be careful not to blow the soil out of the container. Record your observations.
4. Using a **spray bottle,** spray water onto your landform. Record your observations.

Data and Observations

Think About This

1. How did you use the materials in this activity to model Earth's systems?

2. How could you improve your model? What changes would you make?

3. Key Concept Describe how Earth systems interacted in your model.

All Systems Go?

1. A storm is moving from over the ocean to the land. Waves are crashing against the shore. All Earth systems are affected by the storm. How do you think water in the clouds enters the atmosphere? How do you think Earth systems are interacting in this storm?

The Water Cycle

You read that the amount of water on Earth does not change. The water that you drink has been on Earth for a long time. Millions of years ago, a dinosaur might have swallowed the same water that you are drinking today. Or, that water might have raged down a river, flooding an ancient city. How does water move from place to place as time passes?

The **water cycle** *is the continuous movement of water on, above, and below Earth's surface.* The Sun provides the energy that drives the water cycle and moves water from place to place. As this occurs, water can change state to a gas or a solid and then back again to a liquid. The change of state requires either an input or an output of thermal energy. **Figure 10** illustrates how energy is absorbed during evaporation and released during condensation.

Because the water cycle is continuous, there is no beginning or end. Water continually moves through all of Earth's systems.

Active Reading **3. Recall** (Circle) the source of energy for the water cycle.

Active Reading **2. Label** Which arrow represents thermal energy being absorbed? Which one represents thermal energy being released?

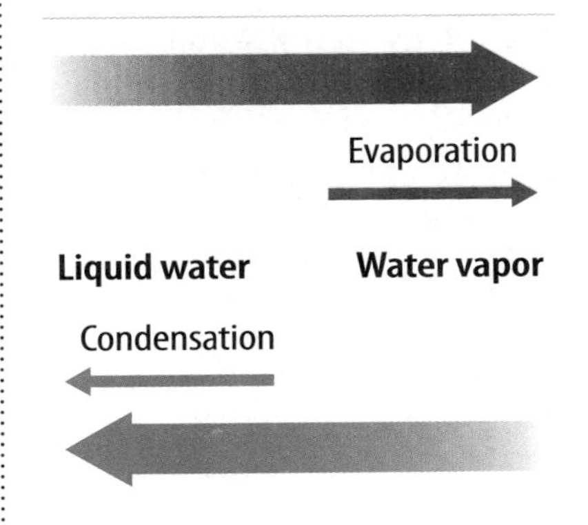

Figure 10 When water changes state from a gas to a liquid, thermal energy is released. Thermal energy is absorbed when liquid water changes into water vapor.

Active Reading **4. Compare and Contrast** How are transpiration and respiration similar? How are they different?

Evaporation

When the Sun shines on an ocean, water near the surface absorbs thermal energy and becomes warmer. As a molecule of water absorbs thermal energy, it begins to vibrate faster. When it has enough energy, it breaks away from the other water molecules in the ocean. It rises into the atmosphere as a molecule of gas called water vapor. **Evaporation** *is the process by which a liquid, such as water, changes into a gas.* Water vapor, like other gases in the atmosphere, is invisible.

Transpiration and Respiration

Oceans hold most of Earth's water, so they are major sources of water vapor. But, water also evaporates from rivers, lakes, puddles, and even soil. These sources along with oceans account for 90 percent of the water that enters the atmosphere. Most of the remaining 10 percent is produced by transpiration. **Transpiration** *is the process by which plants release water vapor through their leaves.*

Some water vapor also comes from organisms through cellular respiration. Cellular respiration takes place in many cells. Water and carbon dioxide are produced during cellular respiration. When animals breathe, they release carbon dioxide and water vapor from their lungs into the atmosphere. The blue arrows in **Figure 11** show how water vapor enters the atmosphere.

Active Reading **5. Identify** (Circle) the processes in the figure below through which water vapor enters the atmosphere.

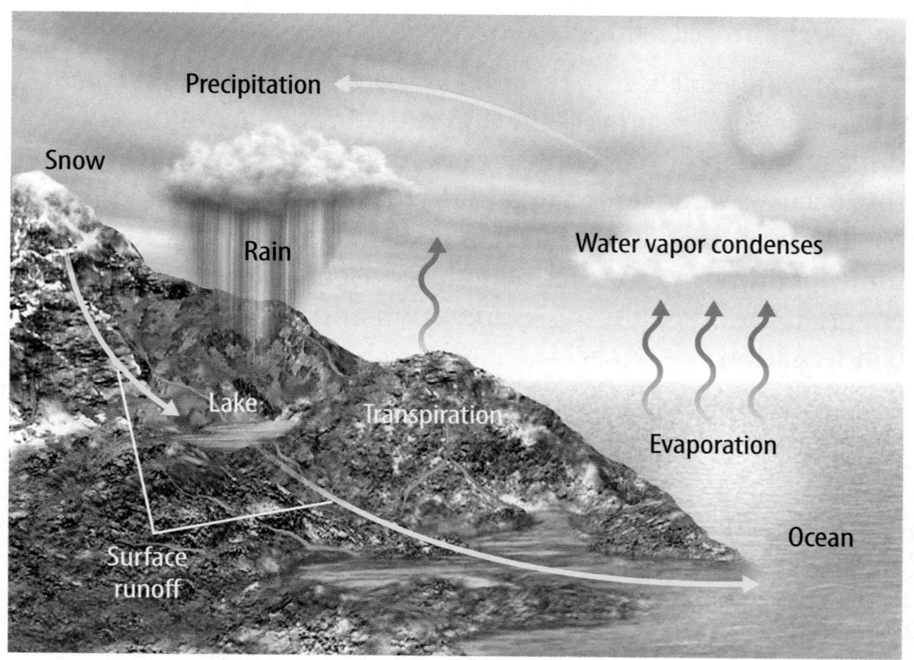

Figure 11 In the water cycle, water moves through the hydrosphere, the atmosphere, the geosphere, and the biosphere.

Condensation

Recall that temperatures in the troposphere decrease with altitude. So, as water vapor rises through the troposphere, it becomes cooler. Eventually it loses so much thermal energy that it returns to the liquid state. *The process by which a gas changes to a liquid is* **condensation.** Tiny droplets of liquid water join to form larger drops. When millions of water droplets come together, a cloud forms.

Precipitation

Eventually, drops of water in the clouds become so large and heavy that they fall to Earth's surface. *Moisture that falls from clouds to Earth's surface is* **precipitation.** Rain and snow both are forms of precipitation.

WORD ORIGIN
precipitation
from Latin *praecipitationem,* means "act or fact of falling headlong"

More than 75 percent of precipitation falls into the ocean, and the rest falls onto land. Some of this water evaporates and goes right back into the atmosphere. Some flows into lakes or rivers, and the rest seeps into soil and rocks.

In the water cycle, water continually moves between the hydrosphere, the cryosphere, the atmosphere, the biosphere, and the geosphere. As water flows across the land, it interacts with soil and rocks in the geosphere. You will learn more about these interactions when you read about the rock cycle.

6. **NGSSS Check** Synthesize How do Earth systems interact in the water cycle? SC.6.E.7.4

Inquiry LAB STATION SC.6.N.1.1 SC.6.E.7.2 SC.6.E.7.4 **Try It!**

MiniLab *How do plants contribute to the water cycle?* at connectED.mcgraw-hill.com

Apply It! After you complete the lab, answer these questions.

1. Relate the water cycle process you modeled to the way you contribute to the water cycle through respiration.

Math Skills MA.6.A.3.6

Use a Formula

The amount of water vapor in air is called vapor density. Relative humidity (RH) compares the actual vapor density in air to the amount of water vapor the air could contain at that temperature. For example, at 15°C, air can contain a maximum of **12.8 g/m³** of water vapor. If the air contains 10.0 g/m³ of water vapor, what is the RH?

1. Use the formula:

$$\text{RH} = \left(\frac{\text{actual vapor density}}{\text{maximum vapor density}}\right) \times 100$$

2. Work out the equation.

$$\text{RH} = \left(\frac{10.0\ \text{g/m}^3}{12.8\ \text{g/m}^3}\right) \times 100$$

$$\text{RH} = 0.781 \times 100 = 78.1\%$$

Practice

7. At 0°C, air can contain 4.85 g/m³ of water vapor. Assume the actual water vapor content is 0.970 g/m³. What is the RH?

Changes in the Atmosphere

The atmosphere is continually changing. These changes take place mainly within the troposphere, which contains most of the gases in the atmosphere. Some changes occur within hours or days. Others can take decades or even centuries.

Weather

Weather *is the state of the atmosphere at a certain time and place.* In most places, the weather changes to some degree every day. How do scientists describe weather and its changes?

Describing Weather Scientists use several factors to describe weather, as shown in **Figure 12.** Air temperature is a measure of the average amount of energy produced by the motion of air molecules. Air **pressure** is the force exerted by air molecules in all directions. Wind is the movement of air caused by differences in air pressure. Humidity is the amount of water vapor in a given volume of air. High humidity makes it more likely that clouds will form and precipitation will fall.

Interactions Weather is influenced by conditions in the geosphere and the hydrosphere. For example, air masses take on the characteristics of the area over which they form. So, an air mass that forms over a cool ocean will bring cool, moist air. In addition to these interactions, the hydrosphere provides much of the water for cloud formation and precipitation. Tropical waters provide the thermal energy that produces hurricanes.

8. NGSSS Check Explain How does weather show interactions of Earth systems? SC.6.E.7.4

____

____

____

Figure 12 Scientists describe weather using air temperature and pressure, wind speed and direction, and humidity.

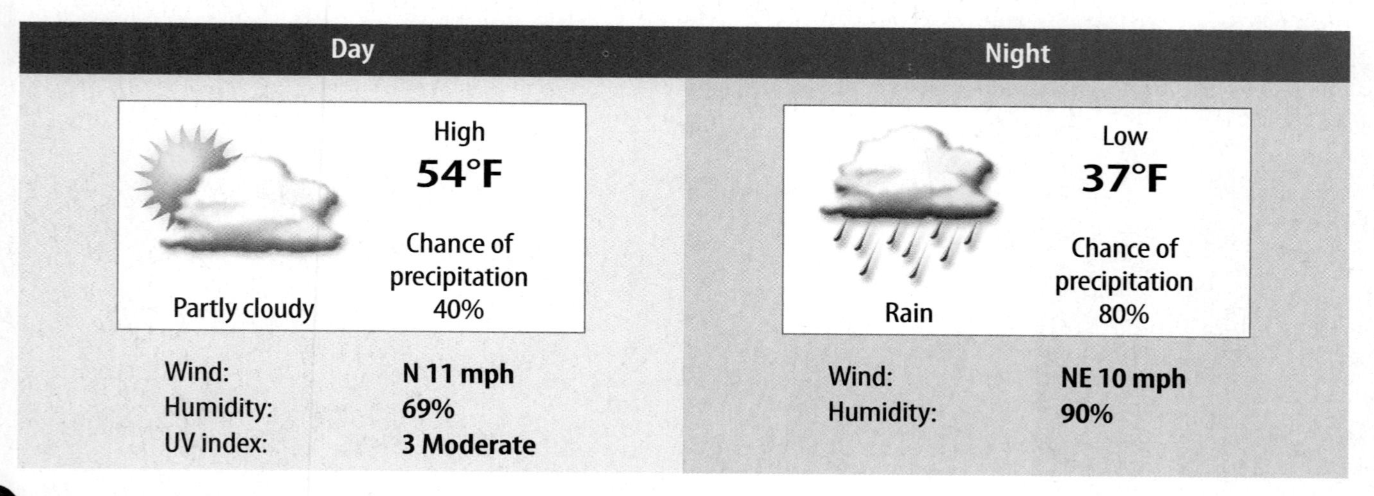

Climate

The weather in the area where you live might change each day, but weather patterns can remain nearly the same from season to season. For example, the weather might differ each day in the summer. But overall, summer is warm. These weather patterns are called climate. **Climate** *is the average weather pattern for a region over a long period of time.* Earth has many climates. Climates differ in part because of interactions between the atmosphere and other Earth systems.

Active Reading **9. Differentiate** How does weather differ from climate?

Mountains Recall that air temperature decreases with altitude. So the climate near the top of a mountain often is cooler than the climate near the mountain's base. Mountains also can affect the amount of precipitation an area receives—a phenomenon known as the rain-shadow effect. As shown in **Figure 13,** warm, wet air rises and cools as it moves up the windward side of a mountain. Clouds form and precipitation falls, giving this side of the mountain a wet climate. The air, now dry, continues to move over and down the leeward side of the mountain. This side of the mountain often has a dry climate.

Ocean Currents As wind blows over an ocean, it creates surface currents. Surface currents are like rivers in an ocean—the water flows in a predictable pattern. These currents transport the thermal energy in water from place to place. For example, the Gulf Stream carries warm waters from tropical regions to northern Europe, making the climate of northern Europe warmer than it would be without these warm waters.

Figure 13 Moist air on the windward side of mountains cools as it rises. Rain falls on this side of the mountain, resulting in a wet climate. This leaves little precipitation for the leeward side of the mountain, resulting in a dry climate.

Active Reading **10. Label** On the figure below, label which side of the mountain is wet and which side is dry.

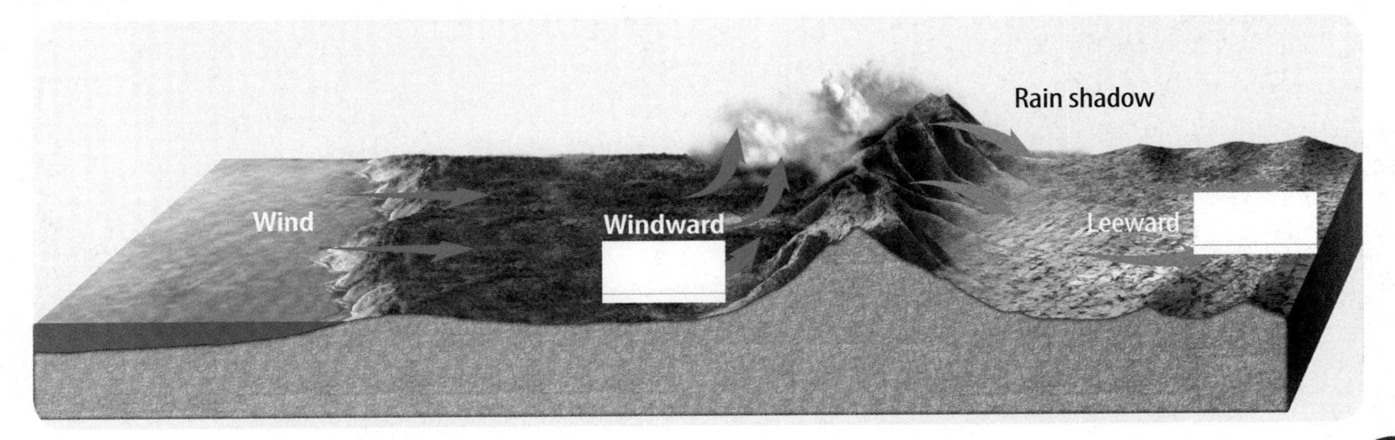

The Rock Cycle

In the water cycle, water moves throughout the hydrosphere, the cryosphere, the atmosphere, the biosphere, and the geosphere. Another natural cycle is the rock cycle. *The* **rock cycle** *is the series of processes that transport and continually change rocks into different forms.* This cycle, shown in **Figure 14,** takes place in the geosphere, but it is affected by interactions with the other Earth systems.

As rocks move through the rock cycle, they might become igneous rocks, sedimentary rocks, or metamorphic rocks. At times they might not be rocks at all. Instead, they might take the form of sediments or hot, flowing magma. Like the water cycle, the rock cycle has no beginning or end. Some processes in this cycle take place on Earth's surface, and others take place deep within the geosphere.

Cooling and Crystallization

As shown in **Figure 14,** magma is located inside the geosphere. When magma flows out onto Earth's surface, it is called lava. Mineral crystals form as magma cools below the surface or as lava cools on the surface. This crystallization changes the molten material into igneous rock.

Uplift

Even rocks formed deep within Earth can eventually be exposed at the surface. **Uplift** *is the process that moves large bodies of Earth materials to higher elevations.* Uplift is often associated with mountain building. After millions of years of uplift, rocks that formed deep below Earth's surface could move up to the surface.

Active Reading **11. Point out** (Circle) the process in the figure below that moves deep rock to Earth's surface.

The Rock Cycle

Figure 14 As rocks move slowly through the rock cycle, they change from one form to another.

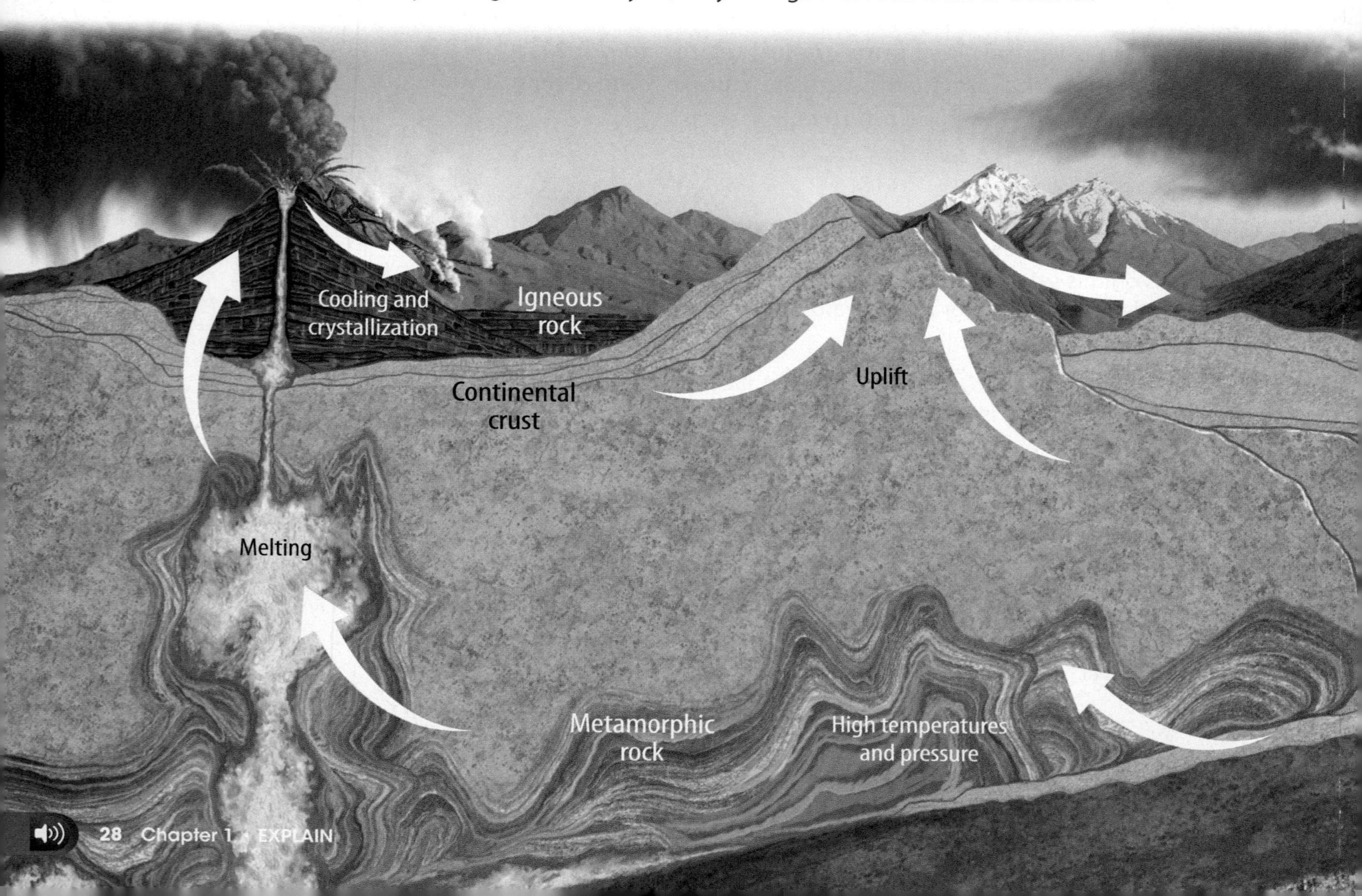

Weathering and Erosion

Rocks on Earth's surface are exposed to the atmosphere, the hydrosphere, the cryosphere, and the biosphere. Glaciers, wind, and rain, along with the activities of some organisms, break down rocks into sediment. This **process** is called weathering. In **Figure 14,** weathering is shown in the mountains, where uplift has exposed rocks. Weathering of rocks into sediments is often accompanied by erosion. Erosion occurs when the sediments are carried by agents of erosion—water, wind, or glaciers—to new locations.

Deposition

Eventually, agents of erosion lose their energy and slow down or stop. When this occurs, eroded sediments are deposited, or laid down, in new places. Deposition forms layers of sediment. Over time, more and more layers are deposited.

Compaction and Cementation

As more layers of sediment are deposited, their weight pushes down on underlying layers. The deeper layers are compacted. Minerals dissolved in surrounding water crystallize between grains of sediment and cement the sediments together. Compaction and cementation produce sedimentary rocks.

ACADEMIC VOCABULARY

process
(noun) a natural phenomenon marked by gradual changes that lead toward a particular result

Active Reading **12. State** How do weathering and erosion change rocks?

Active Reading **FOLDABLES**® LA.6.2.2.3

Make a pyramid book, and label it as shown. Use it to organize your notes on Earth-system interactions including the water cycle, the rock cycle, and weather.

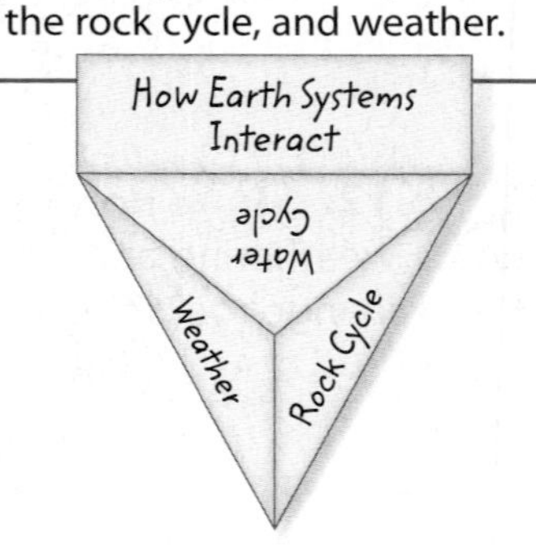

High Temperatures and Pressure

Metamorphic rocks form when rocks are subjected to high temperatures and pressure. This usually occurs far beneath Earth's surface. Igneous, sedimentary, and even metamorphic rocks can become new metamorphic rocks. Then, uplift can bring the rocks to the surface. There, the rocks are broken down and continue moving through the rock cycle.

13. NGSSS Check Analyze How do Earth systems interact in the rock cycle? SC.6.E.7.4

Most interactions between the geosphere, the hydrosphere, and the atmosphere occur on Earth's surface. The atmosphere and hydrosphere alter rocks in the geosphere, and the geosphere in turn alters the other Earth systems. For example, energy from the Sun reaches Earth. The energy is reflected by Earth's surface and heats the atmosphere. These are just a few examples of interactions among Earth's systems. As **Figure 15** shows, these systems interact and function together as one unified system—planet Earth.

Figure 15 Earth is a unified system made of interacting subsystems.

Active Reading **14. Identify** Label the Earth system represented by each photo below.

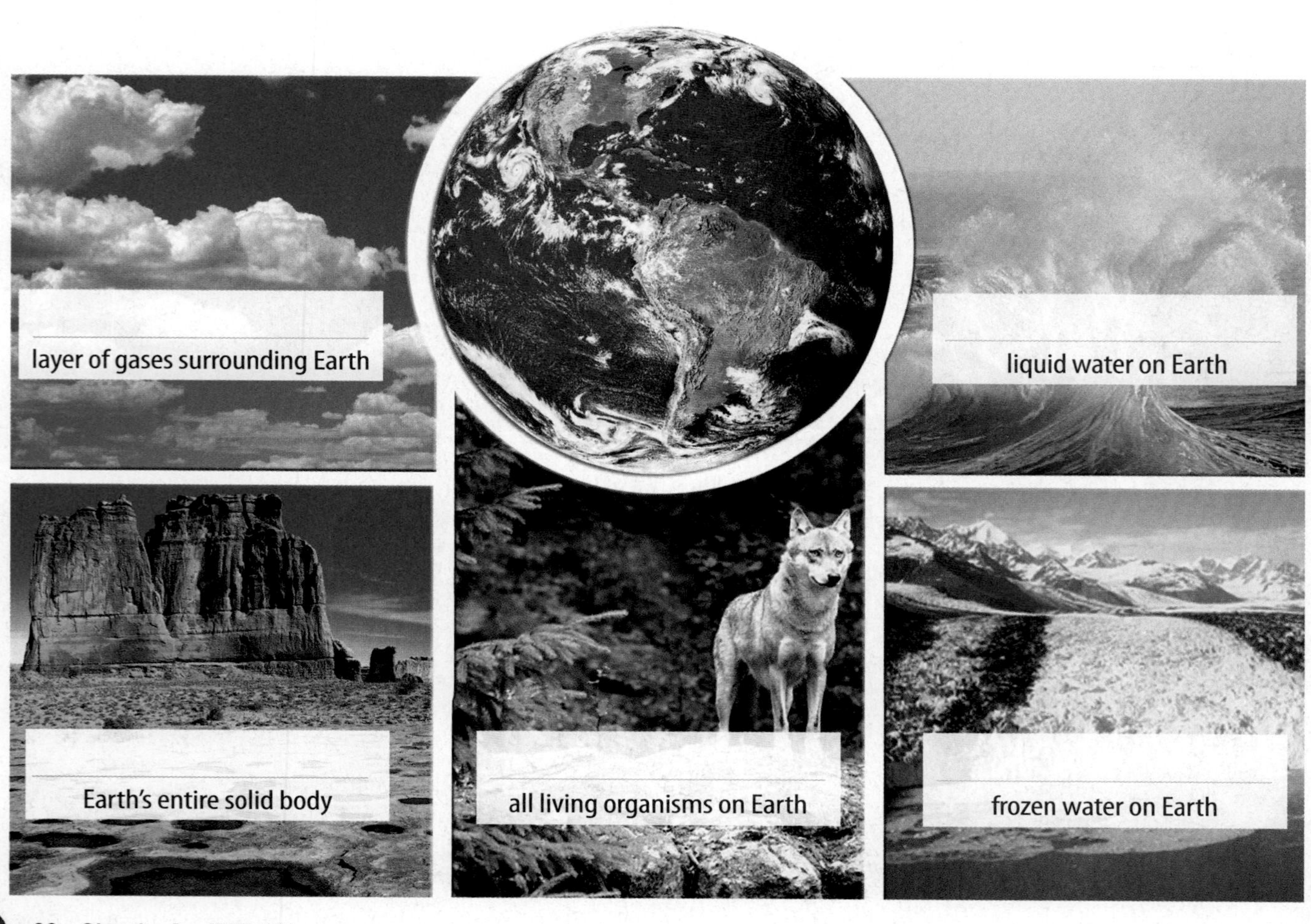

Lesson Review 2

Visual Summary

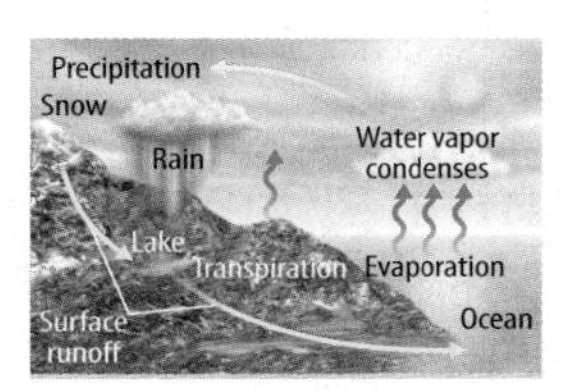

In the water cycle, water continually moves through the hydrosphere, the atmosphere, the geosphere, and the biosphere.

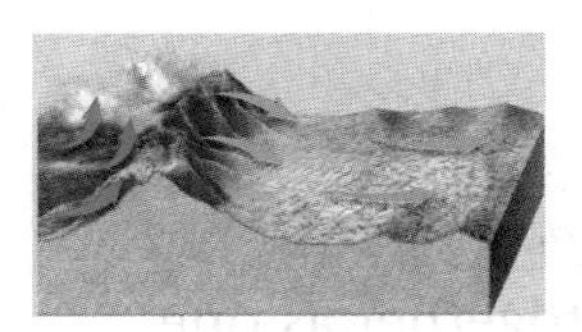

Weather and climate are influenced by interactions between the atmosphere and the other Earth systems.

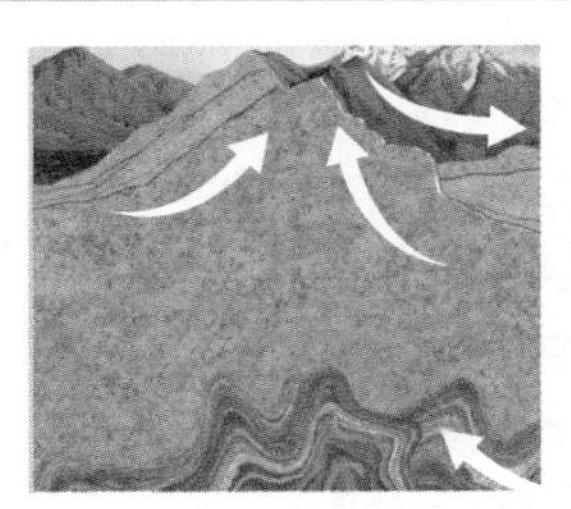

In the rock cycle, rocks continually change from one form to another.

Use Vocabulary

1. **Distinguish** between weather and climate. SC.6.E.7.6

____

2. **Define** the *water cycle* in your own words.

____

3. The process that changes liquid water to water vapor is ________.

Understand Key Concepts

4. Which is an example of an interaction between the atmosphere and the geosphere? SC.6.E.7.4
   - (A) breathing
   - (B) ocean currents
   - (C) storms
   - (D) weathering

5. **Outline** On a separate sheet of paper make an outline about the rock cycle. Include information about processes, rock types, and interactions with Earth systems. SC.6.E.7.4

Interpret Graphics

6. **Organize Information** Fill in the graphic organizer below. Identify the processes of the water cycle. LA.6.4.2.2

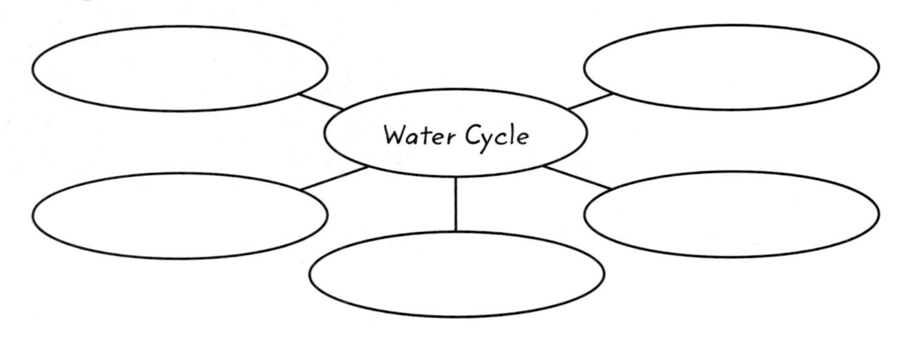

Critical Thinking

7. **Apply** Describe two Earth systems and how they interact. SC.6.E.7.4

____

____

____

Math Skills

8. Air at 20°C has a vapor density of 8.65 g/m^3. The maximum amount of vapor density at that temperature is 17.3 g/m^3. What is the relative humidity? MA.6.A.3.6

Chapter 1 Study Guide

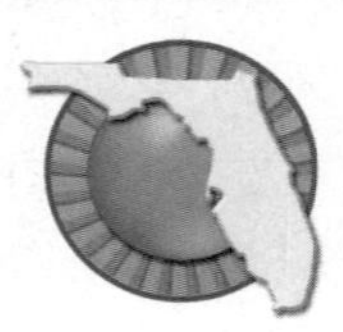

Think About It! Earth is a constantly evolving system with dynamic interactions between the biosphere, atmosphere, hydrosphere, cryosphere, and geosphere that are driven by internal and external energy processes.

Key Concepts Summary

LESSON 1 Earth Systems

- Earth is made of the **biosphere**, the **atmosphere**, the **hydrosphere**, the **cryosphere**, and the **geosphere**.
- The atmosphere has a layered structure that includes the troposphere, the stratosphere, the mesosphere, the thermosphere, and the exosphere. It is made of nitrogen, oxygen, and trace gases.
- Water is found on Earth in oceans, lakes, rivers, and as ice and **groundwater**. Small amounts of water are also found within the atmosphere and the biosphere.
- The geosphere is made of soil, metal, and **rock**. It has a layered structure that includes the crust, the mantle, and the core.

Vocabulary

biosphere p. 12
atmosphere p. 13
hydrosphere p. 15
cryosphere p. 16
groundwater p. 16
geosphere p. 17
mineral p. 17
rock p. 18

LESSON 2 Interactions of Earth Systems

- The **water cycle** shows how water moves between reservoirs of the hydrosphere, the cryosphere, the atmosphere, the geosphere, and the biosphere.
- **Weather** and **climate** are influenced by transfers of water and energy among the atmosphere, the geosphere, and the hydrosphere.
- Rocks continually change form as they move through the **rock cycle**. Processes such as weathering and erosion are examples of interactions among Earth systems.

water cycle p. 23
evaporation p. 24
transpiration p. 24
condensation p. 25
precipitation p. 25
weather p. 26
climate p. 26
rock cycle p. 28
uplift p. 28

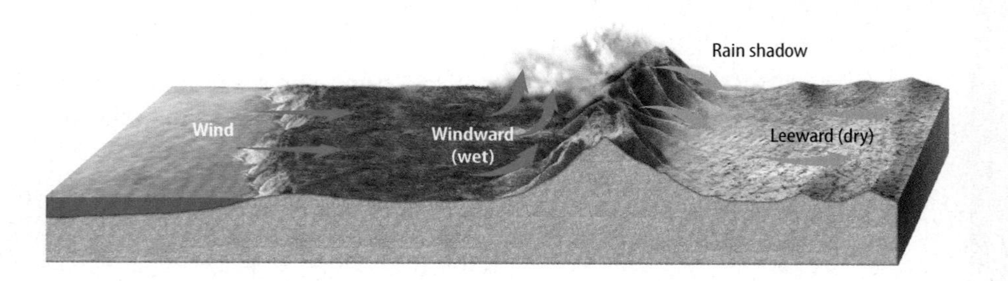

Active Reading **FOLDABLES®** **Chapter Project**

Assemble your lesson Foldables as shown to make a Chapter Project. Use the project to review what you have learned in this chapter.

Use Vocabulary

1. The Earth system containing all living things is the ______________.
2. Use the term *mineral* in a sentence.
3. Distinguish between rocks and minerals.
4. Conditions in the atmosphere at a given time and place are called ______________.
5. Define the word *uplift* in your own words.
6. Distinguish between condensation and precipitation.

Link Vocabulary and Key Concepts

Use vocabulary terms from the previous page to complete the concept map.

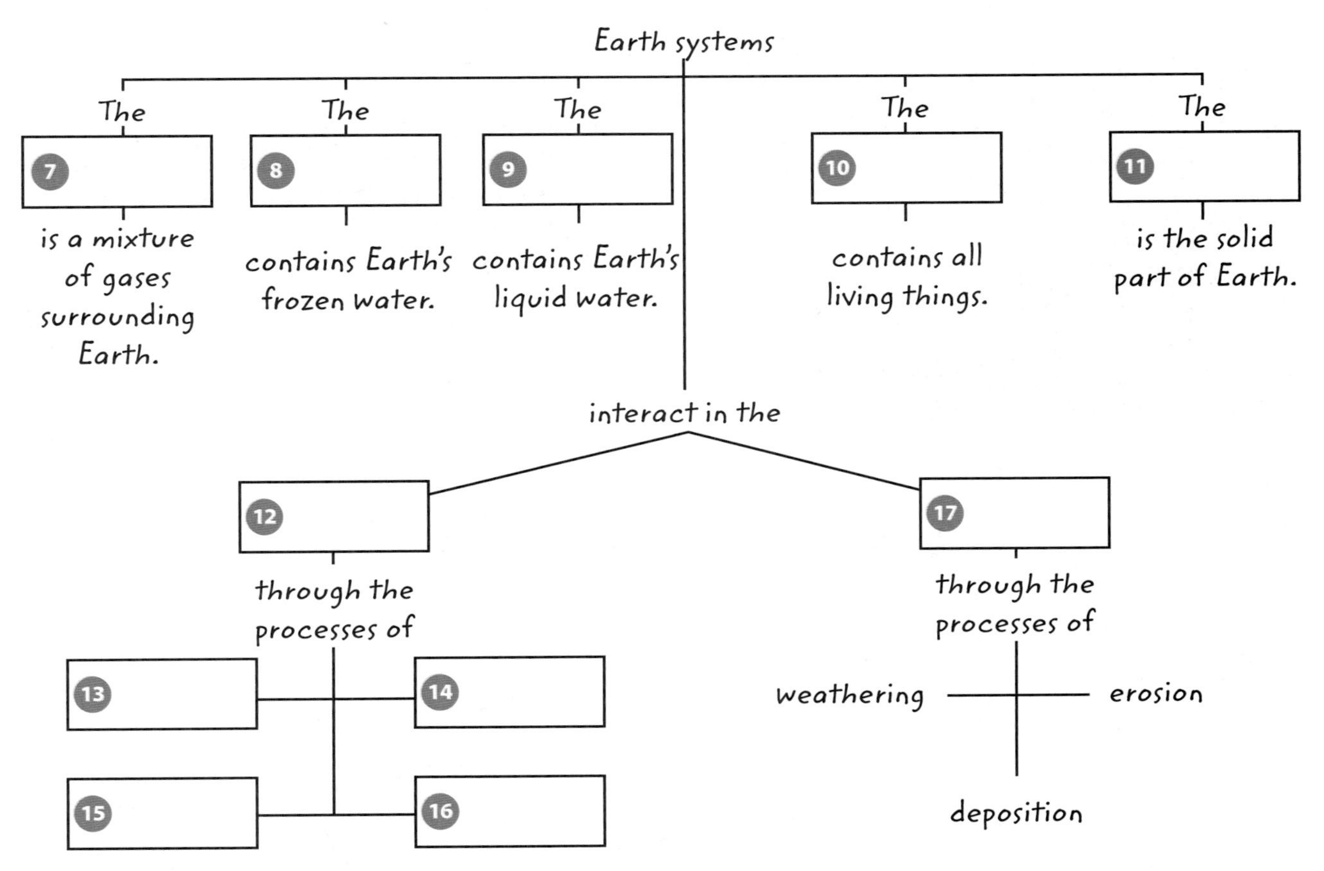

Chapter 1 Review

Fill in the correct answer choice.

Understand Key Concepts

1. Which are two characteristics of minerals? **SC.6.E.6.1**
   Ⓐ artificial and organic
   Ⓑ liquid and gas
   Ⓒ living and inorganic
   Ⓓ solid and natural

2. What are the major gases of the atmosphere? **SC.6.E.7.4**
   Ⓐ carbon dioxide and water vapor
   Ⓑ nitrogen and carbon dioxide
   Ⓒ nitrogen and oxygen
   Ⓓ oxygen and water vapor

3. Which reservoir holds the largest amount of freshwater? **SC.6.E.7.4**
   Ⓐ groundwater
   Ⓑ ice
   Ⓒ lakes
   Ⓓ rivers

4. The diagram below shows the water cycle. Which number represents precipitation? **SC.6.E.7.2**

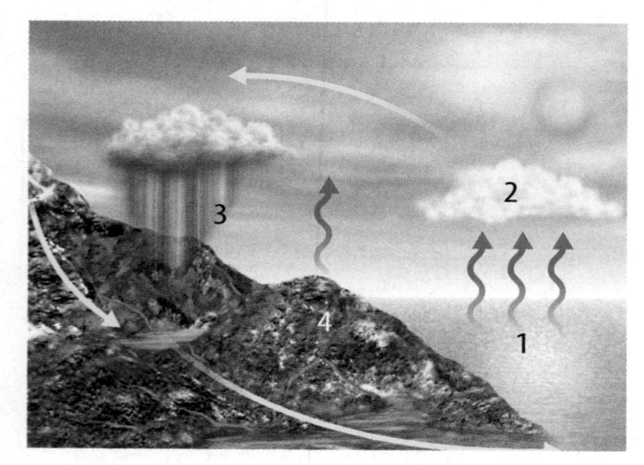

   Ⓐ 1
   Ⓑ 2
   Ⓒ 3
   Ⓓ 4

5. In which layer of the atmosphere does weather occur? **SC.6.E.7.2**
   Ⓐ hydrosphere
   Ⓑ mesosphere
   Ⓒ stratosphere
   Ⓓ troposphere

Critical Thinking

6. **Give** an example of how the water cycle impacts the rock cycle. **SC.6.E.7.4**

7. **Assess** How does the geosphere affect organisms that live in an ocean? **SC.6.E.7.4**

8. **Infer** How might the distribution of freshwater on Earth change if surface temperatures decreased? **SC.6.E.7.2**

9. **Evaluate** the relationship between weathering and erosion. How do the processes work together to change Earth's surface? How might the surface be different if only one of these processes occurred? **SC.6.E.7.4**

10 **Simplify** The diagram below shows the path of one rock through the rock cycle. What terms are missing from the diagram? Use the terms to describe how the rock changed. SC.6.E.6.1

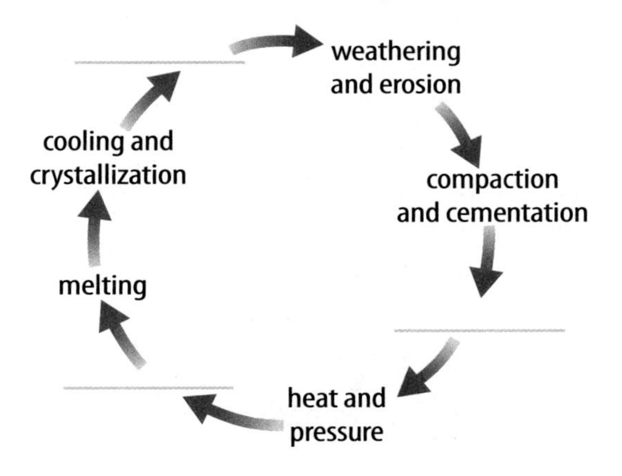

Writing in Science

11 **Create** A haiku is a poem with three lines. The lines contain five, seven, and five syllables respectively. On a separate piece of paper create a haiku that describes interactions among Earth systems. SC.6.E.7.4

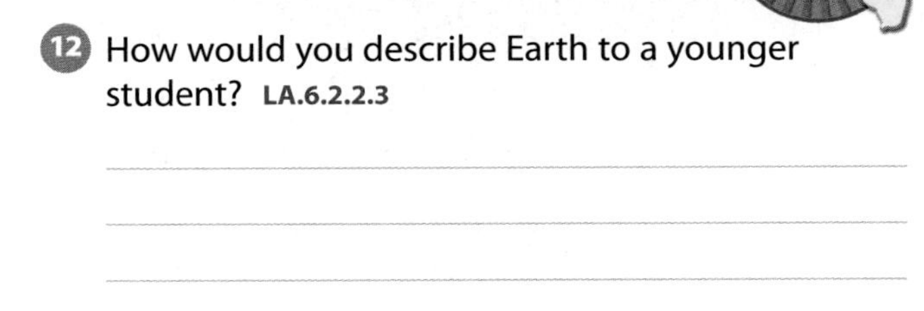

12 How would you describe Earth to a younger student? LA.6.2.2.3

Math Skills MA.6.A.3.6

Use a Formula

Use the data in the table below to answer questions 13–15.

Temperature (°C)	Maximum Vapor Density (g/m^3)
10	9.4
24	23.0
30	30.4

13 The current temperature is 24°C. The water vapor in the air has a density of 5.75 g/m^3. What is the relative humidity?

14 At a temperature of 30°C, the air contains 22.8 g/m^3 of water vapor. What is the relative humidity?

15 Based on the data in the table, what is the relationship between the temperature and the amount of water vapor air can contain?

Record your answers on the answer sheet provided by your teacher or on a sheet of paper.

Multiple Choice

1 Which term describes the state of the atmosphere at a certain time and place? **SC.6.E.7.6**

Ⓐ climate

Ⓑ weather

Ⓒ troposphere

Ⓓ ozone

2 Which is NOT a process where water vapor is released into the atmosphere? **SC.6.E.7.4**

Ⓕ evaporation

Ⓖ respiration

Ⓗ transpiration

Ⓘ precipitation

Use the diagram below to answer question 3.

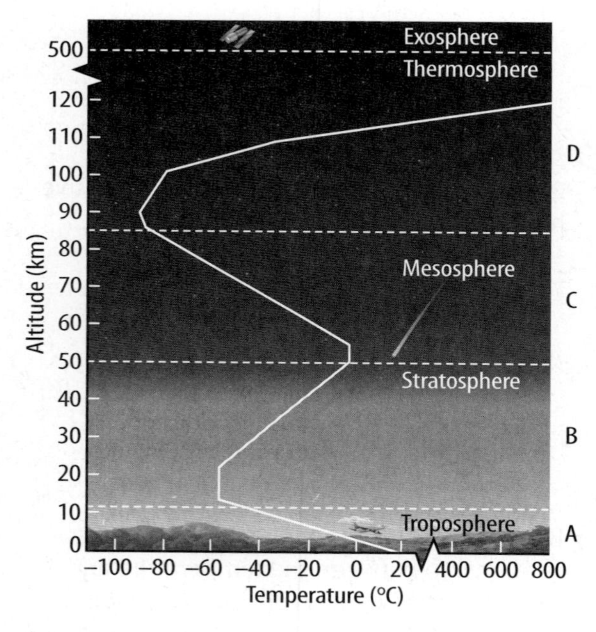

3 Earth's ozone layer absorbs solar radiation, protecting the biosphere. Which atmospheric layer includes the ozone layer? **SC.6.E.7.4**

Ⓐ A

Ⓑ B

Ⓒ C

Ⓓ D

4 Through which process does water leave the hydrosphere and enter the atmosphere? **SC.6.E.7.4**

Ⓕ condensation

Ⓖ deposition

Ⓗ evaporation

Ⓘ precipitation

5 Which Earth systems interact with the geosphere to cause erosion? **SC.6.E.7.4**

Ⓐ cryosphere, biosphere, geosphere

Ⓑ hydrosphere, biosphere, cryosphere

Ⓒ cryosphere, hydrosphere, atmosphere

Ⓓ biosphere, atmosphere, hydrosphere

Use the image below to answer question 6.

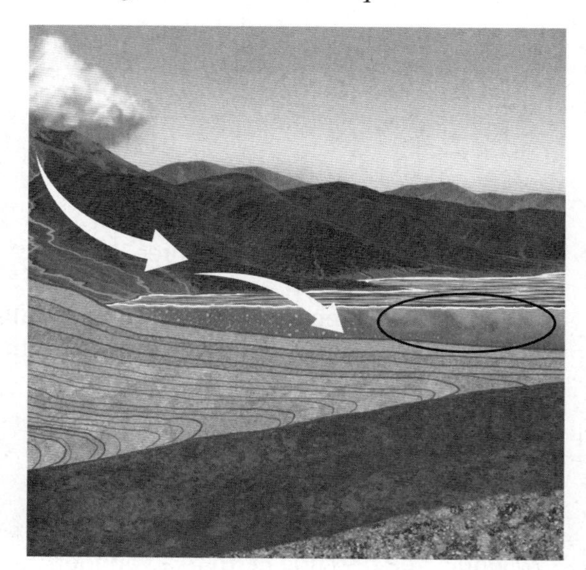

6 Erosion caused by forces in the atmosphere and hydrosphere eventually return sediment to the geosphere. Which process is occurring in the area circled? **SC.6.E.7.4**

Ⓕ condensation

Ⓖ deposition

Ⓗ precipitation

Ⓘ transpiration

7 Which process recycles water from the biosphere to the atmosphere? **SC.6.E.7.4**

Ⓐ condensation

Ⓑ deposition

Ⓒ precipitation

Ⓓ transpiration

8 Which two Earth systems influence weather? **SC.6.E.7.4, SC.6.E.7.6**

Ⓕ geosphere and hydrosphere

Ⓖ cryosphere and atmosphere

Ⓗ biosphere and geosphere

Ⓘ hydrosphere and the cryosphere

Use the diagram below to answer question 9.

9 What process is shown above by the interaction of the hydrosphere, the geosphere, and the atmosphere? **SC.6.E.7.4**

Ⓐ tectonic plate movement

Ⓑ evolution

Ⓒ the rock cycle

Ⓓ the water cycle

10 Which two systems contain all of Earth's water? **SC.6.E.7.4**

Ⓕ geosphere and hydrosphere

Ⓖ biosphere and atmosphere

Ⓗ cryosphere and hydrosphere

Ⓘ geosphere and biosphere

Use the figure below to answer question 11.

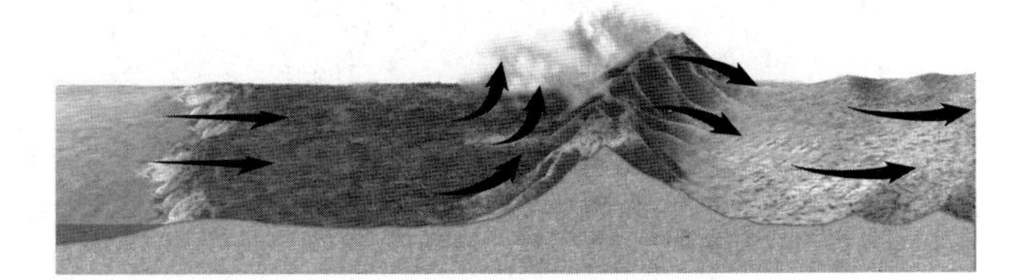

11 What phenomenon describes how mountains can affect the amount of precipitation an area receives? **SC.6.E.7.4**

Ⓐ rain shadow effect

Ⓑ climate

Ⓒ weather

Ⓓ Gulf Stream

12 Which process occurs when water returns from the atmosphere to the geosphere? **SC.6.E.7.4**

Ⓕ condensation

Ⓖ precipitation

Ⓗ transpiration

Ⓘ respiration

Need Extra Help?

If You Missed Question...	1	2	3	4	5	6	7	8	9	10	11	12
Go to Lesson...	2	2	1	2	2	2	2	2	1	2	2	2

Multiple Choice *Bubble the correct answer.*

Use the image below to answer questions 1 and 2.

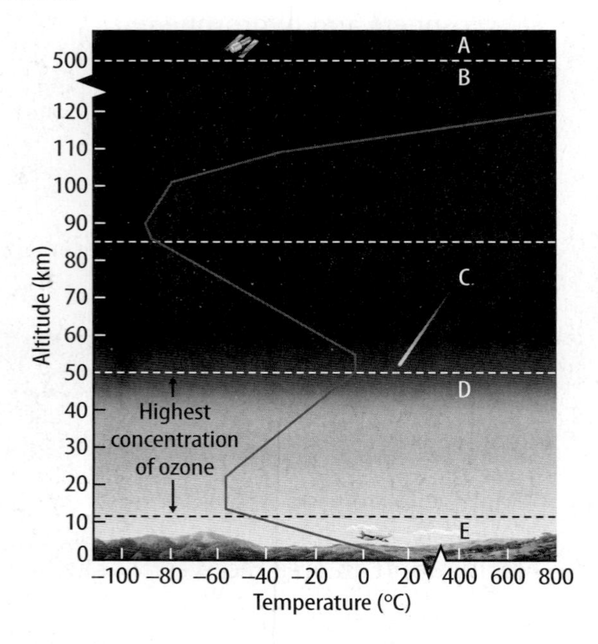

1. Which layer contains most of the mass in the atmosphere? **SC.6.E.7.4**

- (A) Layer B
- (B) Layer C
- (C) Layer D
- (D) Layer E

2. Which statement is true? **SC.6.E.7.4**

- (F) Layer B blocks solar radiation.
- (G) Layer D contains weather.
- (H) Temperatures are lowest in layer B.
- (I) Temperatures decrease with altitude in layer D.

3. The Earth system that is made of soil, rock, and metal is called the **SC.6.E.7.4**

- (A) atmosphere.
- (B) biosphere.
- (C) geosphere.
- (D) hydrosphere.

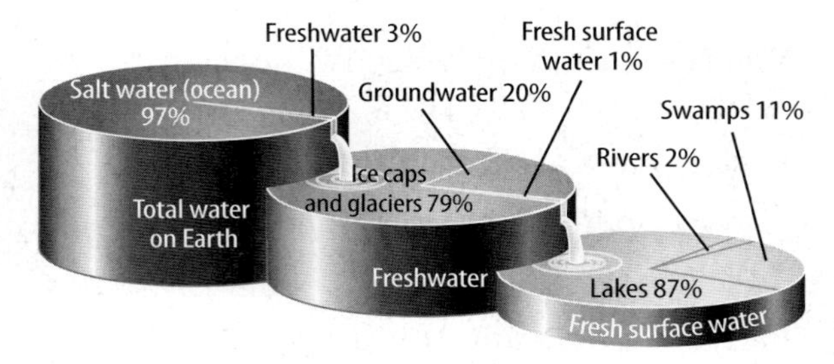

4. In the image above, which contains the largest percentage of freshwater? **SC.6.E.7.4**

- (F) ice
- (G) oceans
- (H) rivers
- (I) swamps

Name ______________________________ Date ______________

Multiple Choice *Bubble the correct answer.*

1. In the image above, what is the climate on the leeward side of the Sierra Nevada range? **SC.6.E.7.6**

- (A) cold
- (B) dry
- (C) hot
- (D) wet

2. Sheila sees that part of a limestone building in town has been worn away. This building was affected by the process of **SC.6.E.6.1**

- (F) crystallization.
- (G) deposition.
- (H) uplifting.
- (I) weathering.

3. Where do most changes in the atmosphere take place? **SC.6.E.7.4**

- (A) geosphere
- (B) hydrosphere
- (C) stratosphere
- (D) troposphere

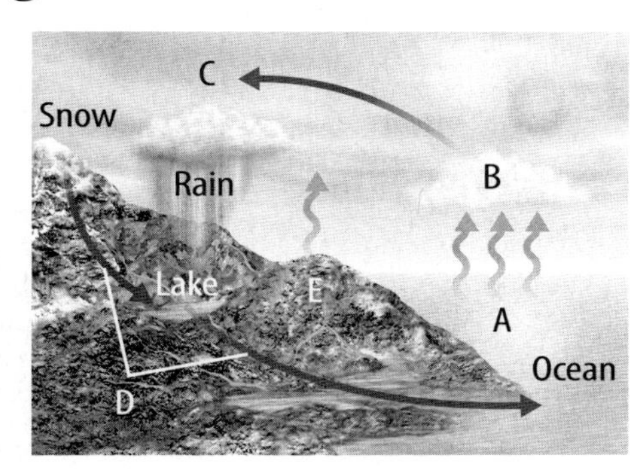

4. Thermal energy is lost during which process shown in the diagram? **SC.6.E.7.4**

- (F) Process A
- (G) Process B
- (H) Process C
- (I) Process D

Name ______________________________ Date ______________

Notes

Which came first?

Two friends were wondering which came first—soil or rock. This is what they said:

Jenna: I think soil came first. Eventually it combines, hardens, and forms rock.

Arnie: I think rock came first. Eventually the rock wears down and forms soil.

Who do you most agree with? ______________ Explain why you agree.

Weathering AND SOIL

The Big Idea

FLORIDA BIG IDEAS

1 The Practice of Science

2 The Characteristics of Scientific Knowledge

6 Earth Structures

Think About It!

What natural processes break down rocks and begin soil formation?

Natural weathering processes create dust by breaking down rock into tiny pieces. These tiny pieces of rock make up a large part of soil. Sometimes they are so small that they are easily blown by the wind.

1. How do you think rock breaks into tiny pieces of dust?

2. What natural processes break down rocks and begin soil formation?

Get Ready to Read

What do you think about weathering and soil?

Before you read, decide if you agree or disagree with each of these statements. As you read this chapter, see if you change your mind about any of the statements.

	AGREE	DISAGREE
1 Any two rocks weather at the same rate.	☐	☐
2 Humans are the main cause of weathering.	☐	☐
3 Plants can break rocks into smaller pieces.	☐	☐
4 Air and water are present in soil.	☐	☐
5 Soil that is 1,000 years old is young soil.	☐	☐
6 Soil is the same in all locations.	☐	☐

There's More Online!
Video • Audio • Review • ⓘLab Station • WebQuest • Assessment • Concepts in Motion • Multilingual eGlossary

Lesson 1

WEATHERING

ESSENTIAL QUESTIONS

 How does weathering break down or change rock?

 How do mechanical processes break rocks into smaller pieces?

 How do chemical processes change rocks?

Vocabulary

weathering p. 45
mechanical weathering p. 46
chemical weathering p. 48
oxidation p. 49

Florida NGSSS

LA.6.2.2.3 The student will organize information to show understanding (e.g., representing main ideas within text through charting, mapping, paraphrasing, summarizing, or comparing/contrasting);

MA.6.A.3.6 Construct and analyze tables, graphs, and equations to describe linear functions and other simple relations using both common language and algebraic notation.

SC.6.E.6.1 Describe and give examples of ways in which Earth's surface is built up and torn down by physical and chemical weathering, erosion, and deposition.

SC.6.N.1.1 Define a problem from the sixth grade curriculum, use appropriate reference materials to support scientific understanding, plan and carry out scientific investigation of various types, such as systematic observations or experiments, identify variables, collect and organize data, interpret data in charts, tables, and graphics, analyze information, make predictions, and defend conclusions.

Launch Lab

SC.6.N.1.1

10 minutes

How can rocks be broken down?

Have you ever looked at the rocks in a stream? What makes some rocks look different from other rocks?

Procedure

1. Read and complete a lab safety form.
2. Obtain 12 pieces of **candy-coated chocolate candies.** Put four of them in a **plastic cup.** Place the rest into a **container with a lid.**
3. Fasten the lid tightly. Shake the container vigorously 300 times.
4. Remove about half of the pieces. Place them in another plastic cup.
5. Replace the lid, and shake the container 300 more times. Remove the remaining "rocks," and place them in another cup.

Think About This

1. Compare and contrast the "rocks" in each cup.

2. Key Concept What do you think caused your "rocks" to change?

What carved this rock?

1. Rocks carved like this can be along ocean shores and rivers, in deserts, and even underground. What carved them? What do they have in common?

Weathering and Its Effects

Everything around you changes over time. Brightly painted walls and signs slowly fade. Shiny cars become rusty. Things made of wood dry out and change color. These changes are some examples of weathering. *The physical and chemical processes that change objects on Earth's surface over time are called* **weathering**.

Weathering also changes Earth's surface. Earth's surface today is different from what it was in the past and what it will be in the future. Weathering processes break, wear, abrade, and chemically alter rocks and rock surfaces. Weathering can produce strangely shaped rocks like those shown above as well.

Over thousands of years, weathering can break rock into smaller and smaller pieces. These pieces are known as sediment. Different sediments are given names based on their size. Boulders are the largest, and clay is microscopic. Weathering also can change the chemical makeup of a rock. Often, chemical changes can make a rock easier to break down.

2. **NGSSS Check** **Summarize** How does weathering break down or change rock? SC.6.E.6.1

SCIENCE USE V. COMMON USE

weather

Science Use to change from the action of the environment

Common Use the state of the atmosphere

Active Reading

FOLDABLES® LA.6.2.2.3

Make a two-tab book and label it as shown. Use it to organize your notes about how mechanical and chemical weathering affect rocks.

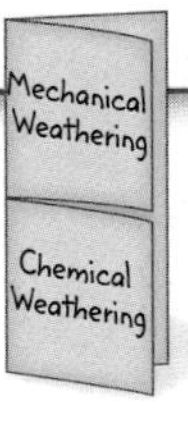

Math Skills MA.6.A.3.6

Use Geometry

The area (A) of a rectangular surface is the product of its length and its width.

$$A = \ell \times w$$

Area has square units, such as square centimeters (cm^2).

The surface area (SA) of a regular solid is the sum of the areas of all of its sides.

Practice

A rock sample is a cube and measures 3 cm on each side.

3. What is the surface area of the rock?

4. If you break the sample into two equal parts, what is the total surface area now?

Mechanical Weathering

When physical processes naturally break rocks into smaller pieces, **mechanical weathering** occurs. Mechanical weathering is also called physical weathering. The chemical makeup of a rock is not changed by mechanical weathering. For example, if a piece of granite undergoes mechanical weathering, the smaller pieces that result are still granite.

Examples of Mechanical Weathering

An example of mechanical weathering is when the intense temperature of a forest fire causes nearby rocks to expand and crack. Other causes of mechanical weathering are described in **Table 1** on the next page.

5. **NGSSS Check** **Summarize** Underline the result of a rock undergoing mechanical weathering. SC.6.E.6.1

Surface Area

As shown in **Figure 1,** when something is broken into smaller pieces, the total surface area increases. Surface area is the amount of space on the outside of an object. The rate of weathering depends on a rock's surface area that is exposed to the environment.

Sand and clay are both results of mechanical weathering. If you pour water on sand, some of the water sticks to the surface. Suppose you pour the same amount of water on an equal volume of clay. Clay particles are only about one-hundredth the size of sand. The greater total surface area of clay particles means more water sticks to its surfaces, along with any substances the water contains. The increased surface area means that weathering has a greater effect on soil with smaller particles. It also increases the rate of chemical weathering.

Active Reading 6. **Recall** Highlight why the surface area of a rock is important.

Surface Area

Figure 1 The surface area of an object is all of the area on its exposed surfaces.

Surface area of cube = 6 equal squares
Surface area = 6 squares × 64 cm^2/square
Surface area = 384 cm^2

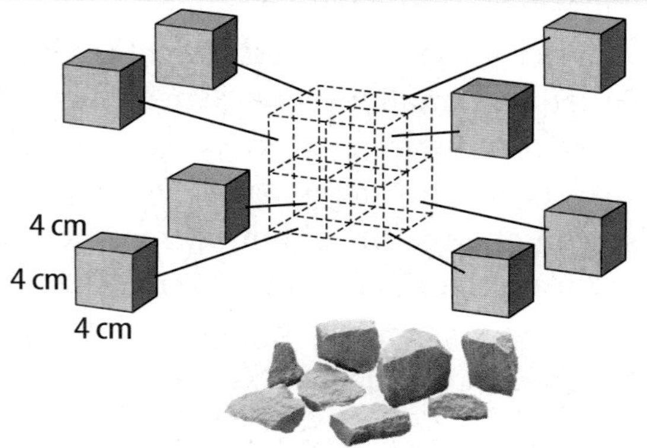

Surface area of 8 cubes = 48 equal squares
Surface area = 48 squares × 16 cm^2/square
Surface area = 768 cm^2

Table 1 Causes of Mechanical Weathering

Ice Wedging One of the most effective weathering processes is ice wedging—also called frost wedging. Water enters cracks in rocks. When the temperature reaches 0°C, the water freezes. Water expands as it freezes, and the expansion widens the crack. As shown in the photo, repeated freezing and thawing can break apart rocks.	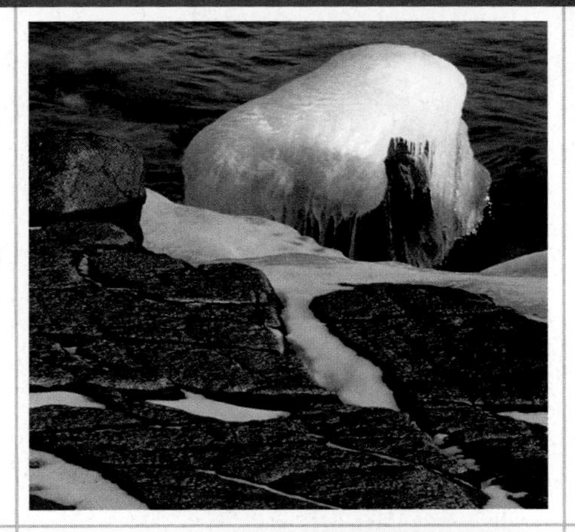	Active Reading **7. Identify** Highlight How does ice break apart rocks?
Abrasion Another effective mechanical weathering process is abrasion—the grinding away of rock by friction or impact. For example, a strong current in a stream can carry loose fragments of rock downstream. The rock fragments tumble and grind against one another. Eventually, the fragments grind themselves into smaller and smaller pieces. Glaciers, wind, and waves along ocean or lake shores can also cause abrasion.		Active Reading **8. Confirm** Circle four things that cause abrasion.
Plants Plants can cause weathering by crumbling rocks. Imagine a plant growing into a crack in a rock. Roots absorb minerals from the rock, making it weaker. As the plant grows, its stem and roots not only get longer, they also get wider. The growing plant pushes on the sides of the crack. Over time, the rock breaks.		Active Reading **9.** Underline What causes rocks to become weaker?
Animals Animals that live in soil create holes in the soil where water enters and causes weathering. Animals burrowing through loose rock can also help break down rocks as they dig.	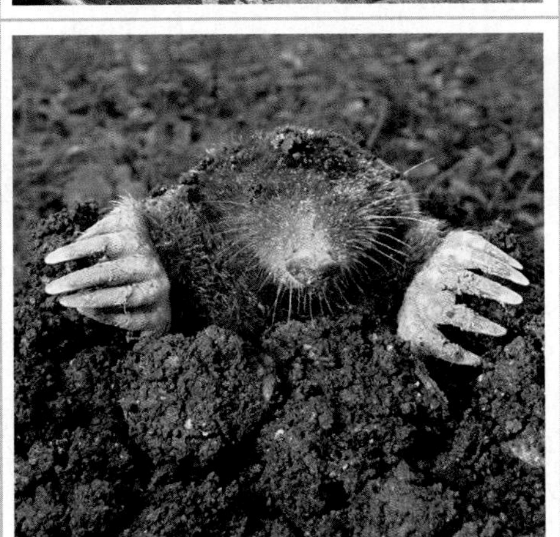	Active Reading **10. Name** List two organisms that live in soil and create holes in soil.

Figure 2 These granite obelisks were carved in a dry climate. Then one was moved to a different, wetter climate.

11. Visual Check
Explain What is the evidence that chemical weathering occurred?

Egypt

New York

Chemical Weathering

Figure 2 shows how chemical weathering can affect some rock. Both obelisks were carved in Egypt about 3,500 years ago. One was moved to New York City in the 1800s. There it has been exposed to more agents of chemical weathering. **Chemical weathering** *changes the materials that are part of a rock into new materials.* If a piece of granite weathers chemically, the composition and size of the granite changes.

12. Compare How does chemical weathering differ from mechanical weathering?

Water and Chemical Weathering

Water is important in chemical weathering because most substances dissolve in water. The minerals that make up most rocks dissolve very slowly in water. Sometimes the amount that dissolves over several years is so small that it seems as though the mineral does not dissolve at all.

For a rock, the process of dissolving happens when minerals in the rock break into smaller parts in solution. For example, table salt is the mineral sodium chloride. When table salt dissolves in water, it breaks into smaller sodium ions and chlorine ions.

Dissolving by Acids

Acids increase the rate of chemical weathering more than rain or water does. The action of acids attracts atoms away from rock minerals and dissolves them in the acid.

Scientists use pH, which is a property of solutions, to learn if a solution is acidic, basic, or neutral. They rate the pH of a solution on a scale from 0 to 14, where 7 is neutral. The pH of an acid is between 0 and 7. Vinegar has a pH of 2 to 3, so it is an acid.

Normal rain is slightly acidic, around 5.6, because carbon dioxide in the air forms a weak acid when it reacts with rain. This means rain can dissolve rocks, as it did to the obelisk in **Figure 2.**

Acid-forming chemicals enter the air from natural sources such as volcanoes. Pollutants in the air also react with rain and make it more acidic. For example, when coal burns, sulfur oxides enter the atmosphere. When these oxides dissolve in rain, they ionize the water to produce acid rain. Acid rain has a pH of 4.5 or less. It can cause more chemical weathering than normal rain causes.

13. Summarize How can pollutants create acid rain?

Oxidation

Another process that causes chemical weathering is called oxidation. **Oxidation** *combines the element oxygen with other elements or molecules.* Most of the oxygen needed for oxidation comes from the air.

The addition of oxygen to a substance produces an oxide. Iron oxide is a common oxide of Earth materials. Useful ores, such as bauxite and hematite, are oxides of aluminum and iron, respectively.

Do all parts of an iron-containing rock oxidize at the same rate? The outside of the rock has the most contact with oxygen in the air. Therefore, this outer part oxidizes the most. When rocks that contain iron oxidize, a layer of red iron oxide forms on the gray, outside surface, as shown in **Figure 3.**

Figure 3 The red outer layer of this rock is created by oxidation. The oxidized minerals in the outer layer are different from the minerals in the center of the rock.

14. NGSSS Check Recall How does chemical weathering change rock? SC.6.E.6.1

15. Summarize Analyze oxidation *using the graphic organizer.*

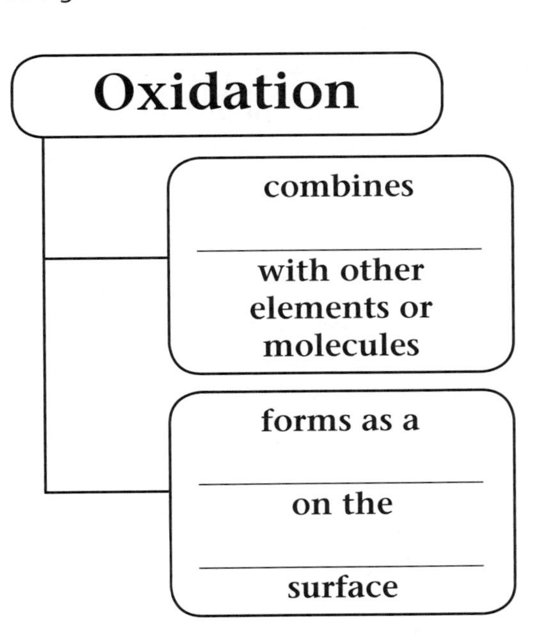

Apply It!

After you complete the lab, answer the questions below.

1. Why might statues exposed to rain show effects of weathering?

2. Why are pollutants, such as sulfur oxide, more likely than normal rain to cause chemical weathering?

Figure 4 The NIST wall was constructed of rock from almost every state and several foreign countries. It has been exposed to continuous weathering since 1948.

16. Visual Check
Identify (Circle) the rocks that have been weathered.

ACADEMIC VOCABULARY
environment
the physical, chemical, and biotic factors acting in a community

Active Reading **17. Explain** Why is weathering slow in cold, dry places?

What affects weathering rates?

You saw in **Figure 2** that similar rocks can weather at different rates. What causes this difference?

The **environment** in which weathering occurs helps determine the rate of weathering. Both types of weathering depend on water and temperature. Mechanical weathering occurs fastest in locations that have frequent temperature changes. This type of weathering requires cycles of either wetting and drying or freezing and thawing. Chemical weathering is fastest in warm, wet places. As a result, weathering often occurs fastest in the regions near the equator.

The type of rock also affects the rate of weathering. The National Institute of Standards and Technology (NIST) constructed the wall shown in **Figure 4** to observe how different rocks weather under the same conditions.

Rocks can be made of one mineral or many minerals. The most easily weathered mineral determines the rate at which the entire rock weathers. For example, rocks containing minerals with low hardness undergo mechanical weathering more easily. This increases the surface area of the rock. Because more surface area is exposed, these rocks more easily undergo chemical weathering. The size and number of holes in a rock also affect the rate at which a rock weathers.

Lesson 1 Review

Visual Summary

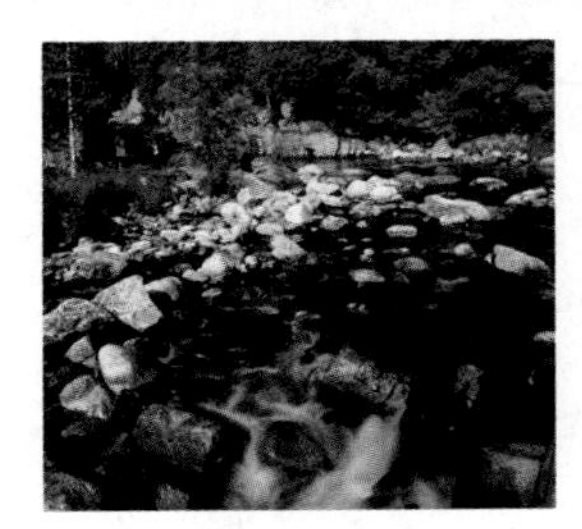
Weathering is the physical and chemical processes that change things over time.

Mechanical, or physical, weathering does not change the identity of the materials that make up rocks. It breaks rocks into smaller pieces.

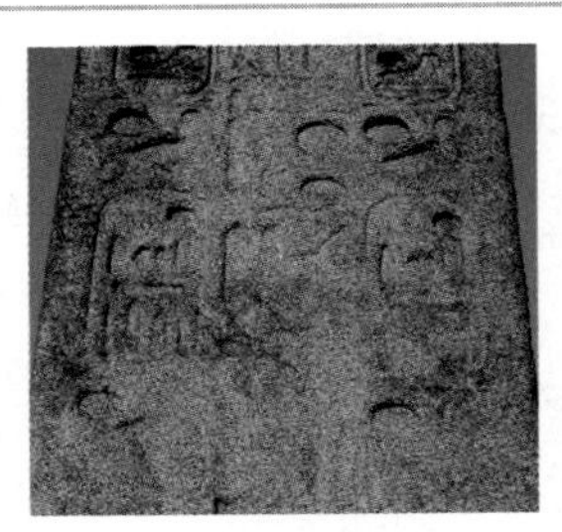
Chemical weathering is the process that changes the minerals in rock into different materials. Oxidation is a type of chemical weathering, as is reaction with an acid.

Use Vocabulary

1. The chemical and physical processes that change things over time are called ______________. SC.6.E.6.1

2. **Define** *mechanical weathering* in your own words. SC.6.E.6.1

3. **Use the term** *oxidation* in a sentence. SC.6.E.6.1

Understand Key Concepts

4. **Summarize** How does weathering change rocks and minerals? SC.6.E.6.1

Interpret Graphics

5. **Compare and contrast** types of weathering by completing this table. SC.6.E.6.1

Weathering	Alike	Different
Chemical and Physical		

Critical Thinking

6. **Explain** how rates of chemical weathering change as temperature increases. SC.6.E.6.1

Math Skills MA.6.A.3.6

7. A block of stone measures 15 cm × 15 cm × 20 cm. What is the total surface area of the stone? Hint: A block has six sides.

Sequence the process of mechanical weathering.

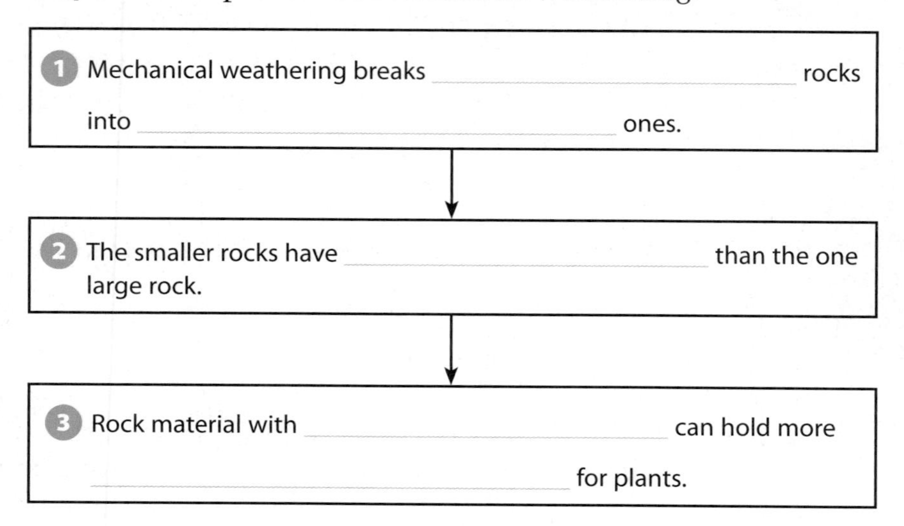

Organize information about chemical weathering.

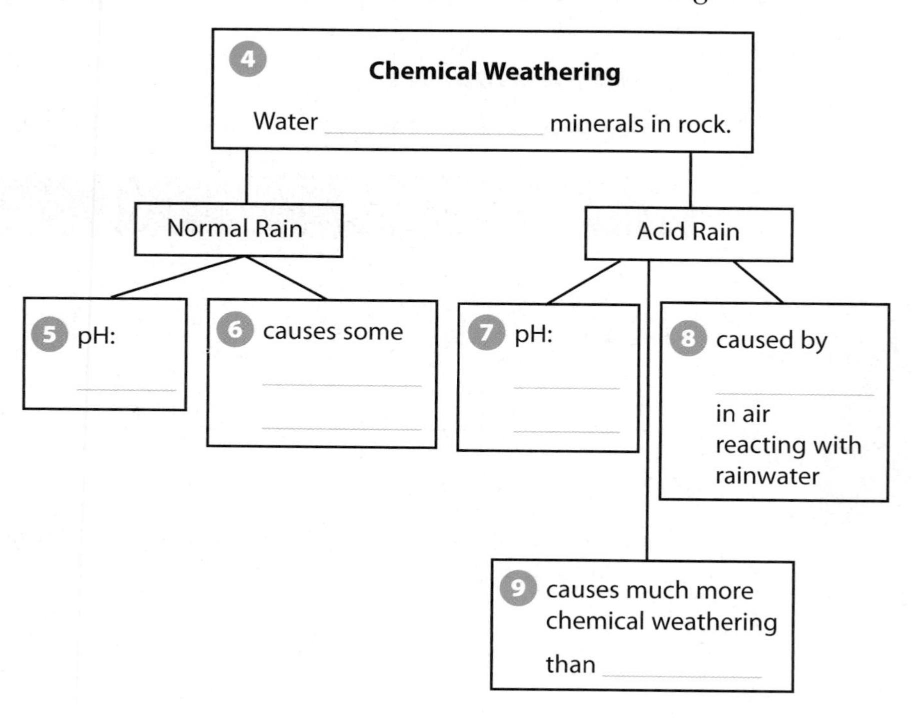

10 **Identify** pH ranges. Color the pH range for normal rain blue, and color the pH range for acid rain red.

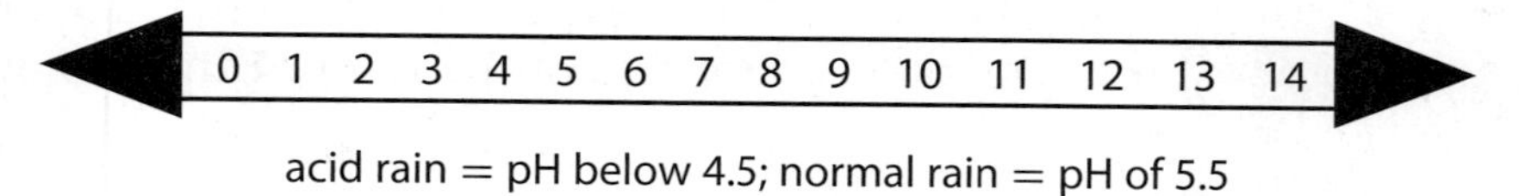

acid rain = pH below 4.5; normal rain = pH of 5.5

Lesson 2

SOIL

ESSENTIAL QUESTIONS

- How is soil created?
- What are soil horizons?
- Which soil properties can be observed and measured?
- How are soils and soil conditions related to life?

Vocabulary

soil p. 54
organic matter p. 54
pore p. 54
decomposition p. 55
parent material p. 56
climate p. 56
topography p. 57
biota p. 57
horizon p. 58

Florida NGSSS

LA.6.2.2.3 The student will organize information to show understanding (e.g., representing main ideas within text through charting, mapping, paraphrasing, summarizing, or comparing/contrasting);

SC.6.E.6.1 Describe and give examples of ways in which Earth's surface is built up and torn down by physical and chemical weathering, erosion, and deposition.

SC.6.N.1.1 Define a problem from the sixth grade curriculum, use appropriate reference materials to support scientific understanding, plan and carry out scientific investigation of various types, such as systematic observations or experiments, identify variables, collect and organize data, interpret data in charts, tables, and graphics, analyze information, make predictions, and defend conclusions.

SC.6.N.1.5 Recognize that science involves creativity, not just in designing experiments, but also in creating explanations that fit evidence.

SC.6.N.2.1 Distinguish science from other activities involving thought.

SC.6.N.1.5

Inquiry Launch Lab

10 minutes

What is in your soil?

Soils are different in different places. Suppose you look at the soil along a riverbank. Is this soil like the soil in a field? Are either of these soils like the soil near your home? What is in the soil where you live?

Procedure

1. Read and complete a lab safety form.
2. Place about a cup of **local soil** in a **jar** that has a **lid.** Add a few drops of **liquid detergent.**
3. Add **water** to the jar until it is almost full. Firmly attach the lid.
4. Shake for 1 minute and place it on your desk.
5. Observe the contents of the jar after 2 minutes and again after 5 minutes.

Data and Observations

Think About This

1. How many different layers did your sample form?

2. Key Concept From your observations, what do you think makes up each layer?

Inquiry Why do you think the soil is so red?

1. Soils have different colors because of what they contain. This soil contains iron, which makes it red. Do you think the soil is red underground, too? What color is your soil?

What is soil?

A soil scientist might think of soil as the "active skin of Earth." Soil is full of life, and life on Earth depends on soil.

If you were to dig into soil, what would you find? About half the volume of soil is solid materials. The other half is liquids and gases. **Soil** *is a mixture of weathered rock, sediment, decayed organic matter, water, and air.*

As you read in Lesson 1, weathering gradually breaks rocks into smaller and smaller fragments. These fragments, however, do not become good soil until plants and animals live in them. Plants and animals add organic matter to the rock fragments. **Organic matter** *is the remains of something that was once alive.*

Water and air are present in varying amounts in the small holes and spaces in soil. *These small holes and spaces in soil are called* **pores**. Pores are important because they can allow water to flow into and through soil. Pores can vary greatly in size depending on the particles that make up the soil. **Figure 5** shows three particles commonly found in soil—sand, silt, and clay. As particle size increases, pore size also increases. The pores between clay particles are smaller than the pores between sand particles.

WORD ORIGIN

pore
from Greek *poros,* means "passage"

Active Reading **2. Define** What is in a pore?

The Organic Part of Soil

Recall that the solid part of soil that was once part of an organism is called organic matter. Pieces of leaves, dead insects, and waste products of animals that are in the soil are examples of organic matter.

How does organic matter form? Soil is home to many organisms, from roots of plants to tiny bacteria. Over time, roots die, and leaves and twigs fall to the ground. Organisms living in the soil decompose these materials for food. **Decomposition** *is the process of changing once-living material into dark-colored organic matter.* In the end, something that was once recognizable as a pine needle becomes organic matter.

Active Reading **3. Recall** Underline how decomposition is related to organic matter.

Organic matter gives soil important properties. Dark soil absorbs sunlight, and organic matter holds water and provides plant nutrients. Organic material holds minerals together in clusters. This helps keep soil pores open for the movement of water and air in soil.

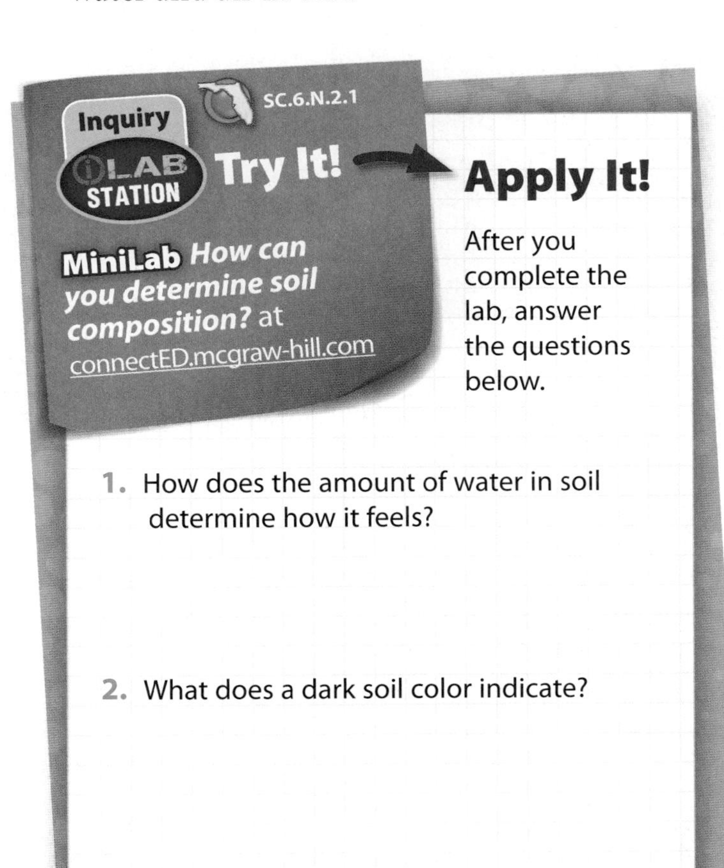

The Inorganic Part of Soil

The term *inorganic* describes materials that have never been alive. Mechanical and chemical weathering of rocks into fragments forms inorganic matter in soil. Soil scientists classify the soil fragments according to their sizes. Rock fragments can be boulders, cobbles, gravel, sand, silt, or clay. **Figure 5** shows a magnified image of the three smallest sizes of soil particles. Between large particles are large pores, which affect soil properties such as drainage and water storage.

Sand feels rough.	**Silt** feels smooth.	**Clay** feels sticky.

Figure 5 Inorganic matter contributes different properties to soil. Large pores occur between large particles, which drain rapidly; small particle pores retain more water in the soil.

4. Identify Summarize organic and inorganic parts of soil. Describe and give three examples of each part.

Soil	
Inorganic	**Organic**
Description:	Description:
Examples: **1.** ______ **2.** ______ **3.** ______	Examples: **1.** ______ **2.** ______ **3.** ______

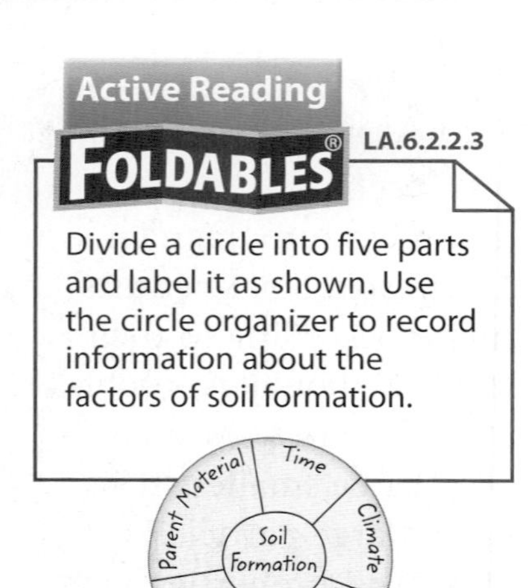

Formation of Soil

Why is the soil near your school different from the soil along a riverbank or soil in a desert? The many kinds of soils that form depend on five factors, called the factors of soil formation. The five factors are parent material, climate, topography, biota, and time.

Parent Material

The starting material of soil is **parent material**. It is made of the rock or **sediment** that weathers and forms the soil, as shown in **Figure 6.** Soil can develop from rock that weathered in the same place where the rock first formed. This rock is known as bedrock. Soil also can develop from weathered pieces of rock that were carried by wind or water from another location. The particle size and the type of parent material can determine the properties of the soil that develops.

Active Reading **5. Summarize** What is the role of parent material in creating soil?

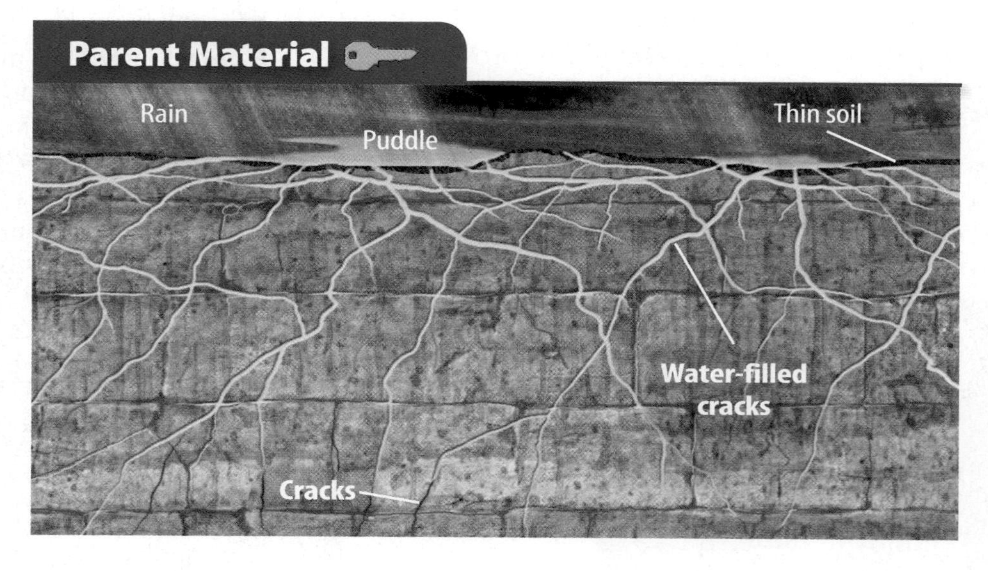

Figure 6 Parent material is broken down by mechanical and chemical weathering.

REVIEW VOCABULARY

sediment
rock material that has been broken down or dissolved in water

Climate

The average weather of an area is its **climate**. How can you describe the climate where you live? The amount of precipitation and the daily and average annual temperatures are some measures of climate. If the parent material is in a warm, wet climate, soil formation can be rapid. Large amounts of rain can speed up weathering as the drops contact the surface of the rock. Warm temperatures also speed up weathering by increasing the rate of chemical changes. Weathering rates also increase in locations where freezing and thawing occur.

Active Reading **6. Recall** Underline why soils form rapidly in warm, moist climates.

Topography

Is the land where you live flat or hilly? If it is hilly, are the hills steep or gentle? **Topography** *is the shape and steepness of the landscape.* The topography of an area determines what happens to water that reaches the soil surface. For example, in flat landscapes, most of the water enters the soil. Water speeds weathering. In steep landscapes, much of the water runs downhill. Water running downhill can carry soil with it, leaving some slopes bare of soil. **Figure 7** shows that broken rock and sediments collect at the bottom of a steep slope. There, they undergo further weathering.

7. Define What is topography?

Biota

Soil is home to a large number and variety of organisms. They range from the smallest bacteria to small rodents. *All of the organisms that live in a region are called* **biota** (bi OH tuh). Biota in the soil help speed up the process of soil formation. Some soil biota form passages for water to move through. Most soil organisms are involved in the decomposition of materials that form organic matter. As **Figure 8** shows, rock and soil are affected by organism activity.

8. Recall Highlight the sentences that explaib how biota aids in soil formation.

Time

As time passes, weathering is constantly acting on rock and sediment. Therefore, soil formation is a constant but slow process. A 90-year-old person is considered old, but soil is still young after a thousand years. It is difficult to see all the soil-producing changes in one human lifetime.

As **Figure 8** shows, mature soils develop layers as new soil forms on top of older soil. Each layer has different characteristics as organic matter is added or as water carries elements and nutrients downward.

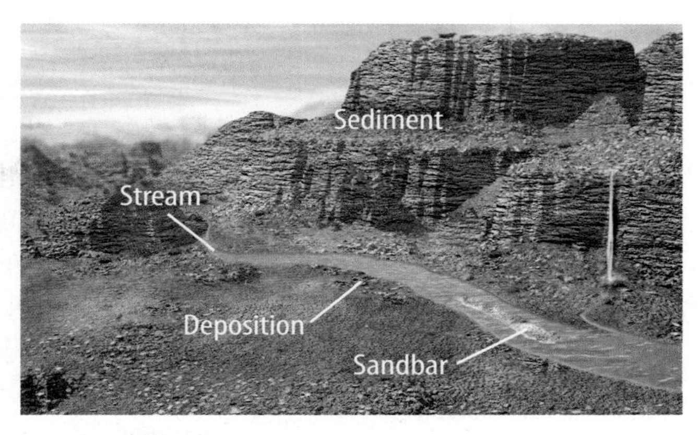

Figure 7 Broken rock and sediment collects at the bottom of steep slopes. Sediment is redistributed by streams and moving water as sandbars and shoreline deposition.

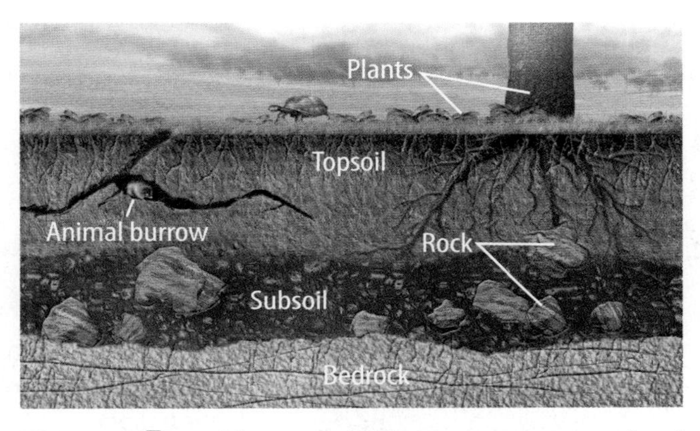

Figure 8 Mature soils form over thousands of years as plants, animals, and other processes break down the bedrock and subsoil.

9. Draw Use the space below to draw a picture of an organism that lives in the soil around your home. Be sure to label the parts it uses to form passages in soil.

WORD ORIGIN

horizon
from Latin *horizontem,* means "bounding circle"

10. Visual Check
Explain One horizon contains a lot of clay, and another horizon is dark. Of these two horizons, which is on top? Explain your answer.

Horizons

You know that soil is more than what you see when you look at the ground. If you dig into the soil, you see that it is different as you dig deeper. You might see dark soil on or near the surface. The soil you see deeper down is lighter in color and probably contains larger pieces of rock. Soil might be loosely packed on the surface, but deeper soil is more tightly packed.

Soil has layers, called horizons. **Horizons** *are layers of soil formed from the movement of the products of weathering.* Each horizon has characteristics based on the type of materials it contains. The three horizons common to most soils are identified as A-horizon, B-horizon, and C-horizon, as shown in **Figure 9.** Each horizon can appear quite different depending on where the soil forms. The top, organic layer is called the O-horizon, and the unweathered, bedrock layer is the R-horizon.

Active Reading **11. Explain** What are soil horizons?

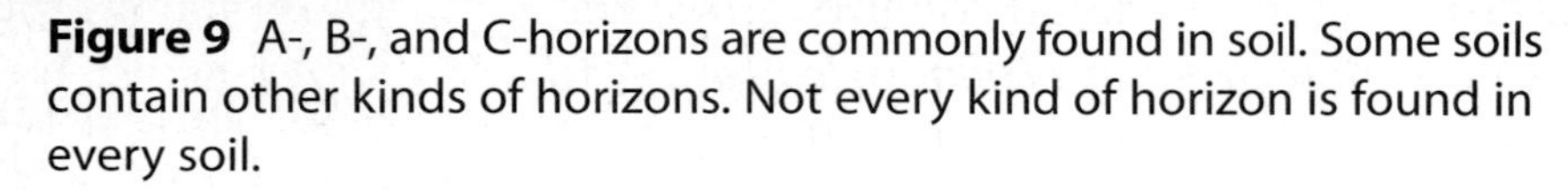

Figure 9 A-, B-, and C-horizons are commonly found in soil. Some soils contain other kinds of horizons. Not every kind of horizon is found in every soil.

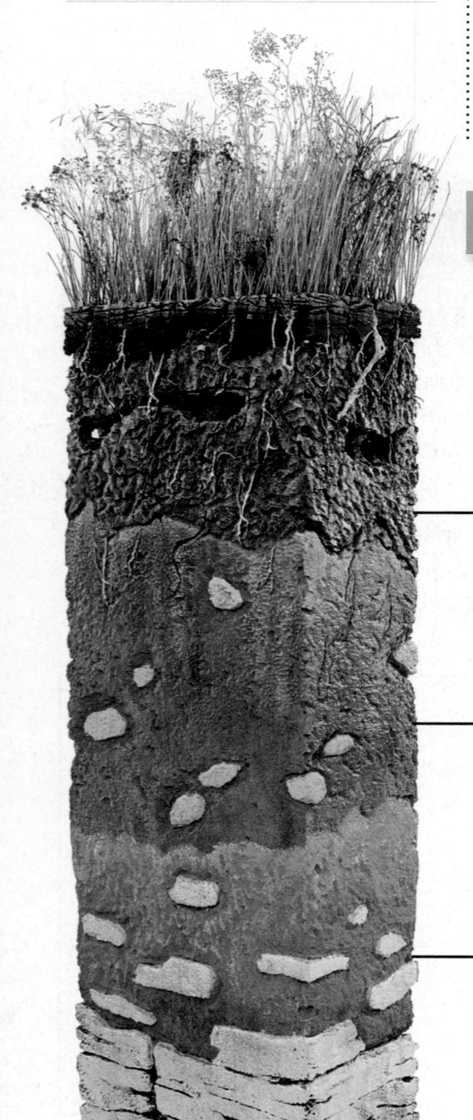

A-horizon
The A-horizon is the part of the soil that you are the most likely to see when you dig a shallow hole in the soil with your fingers. Organic matter from the decay of roots and the action of soil organisms often makes this horizon excellent for plant growth. Because the A-horizon contains most of the organic matter in the soil, it is usually darker than other horizons.

B-horizon
When water from rain or snow seeps through pores in the A-horizon, it carries clay particles. The clay is then deposited below the upper layer, forming a B-horizon. Other materials also accumulate in B-horizons.

C-horizon
The layer of weathered parent material is called the C-horizon. Parent material can be rock or sediments.

Soil Properties and Uses

Soil horizons in different locations have different properties. Recall that properties are characteristics used to describe something. Several soil properties are listed and described in **Table 2.** The properties of a soil determine the best use of that soil. For example, soil that is young, deep, and has few horizons is good for plant growth.

Observing and Measuring Soil Properties

Some properties of soil can be determined just by observation. The amount of sand, silt, and clay in a soil can be estimated by feeling the soil. The types of horizons also provide information about the soil. The color of a soil is easily observed and shows how much organic matter it contains.

Many soil properties can be measured more accurately in a laboratory. Laboratory measurements can determine exactly what is in each sample of soil. Measuring nutrient content and soil pH to determine the suitability for farming or gardening requires careful laboratory analysis.

Soil Properties That Support Life

Plants depend on the nutrients that come from organic matter and the weathering of rocks. Plant growers can observe how well plants grow in the soil to get information about soil nutrients. Crop plants depend less on weathering for nutrients because farmers usually use fertilizers that add nutrients to the soil.

It takes thousands of years to form soil from parent material. Soil that is damaged or misused is slow to replenish its nutrients. The restoration can take many human lifetimes.

Active Reading **13. Relate** How are soil nutrients related to life?

Table 2 Soil Properties

Color	Soil can be described based on the color, such as how yellow, brown, or red it is; how light or dark it is; and how intense the color is.
Texture	The texture of soil ranges from boulder-sized pieces to very fine clay.
Structure	Soil structure describes the shape of soil clumps and how the particles are held together. Structure can look grainy, blocky, or prism shaped.
Consistency	The hardness or softness of a soil is the measure of its consistency. Consistency varies with moisture. For example, some soils have a soft, slippery consistency when they are moist.
Infiltration	Infiltration describes how fast water enters a soil.
Soil moisture	The amount of water in soil pores is its moisture content. Soil scientists determine weight loss by drying samples in an oven at 100°C. The weight difference is the amount of moisture in the soil.
pH	Most soils have a pH between 5.5 and 8.2. Soils can be more acidic in humid environments.
Fertility	Soil fertility is the measure of the ability of a soil to support plant growth. Soil fertility includes the amount of certain elements that are essential for good plant growth.
Temperature	On the ground surface, soil temperature changes with daily cycles and the weather. Soil temperature in lower layers changes less.

Table 2 Soil properties such as fertility and pH are very important when determining if soil can support life.

12. Show In **Table 2,** circle the soil properties that can be observed and underline the soil properties that can be measured.

Soil Types of North America

Figure 10 The properties of soil are different in different climates. There are 12 major soil types in the world. North America contains almost every type of soil.

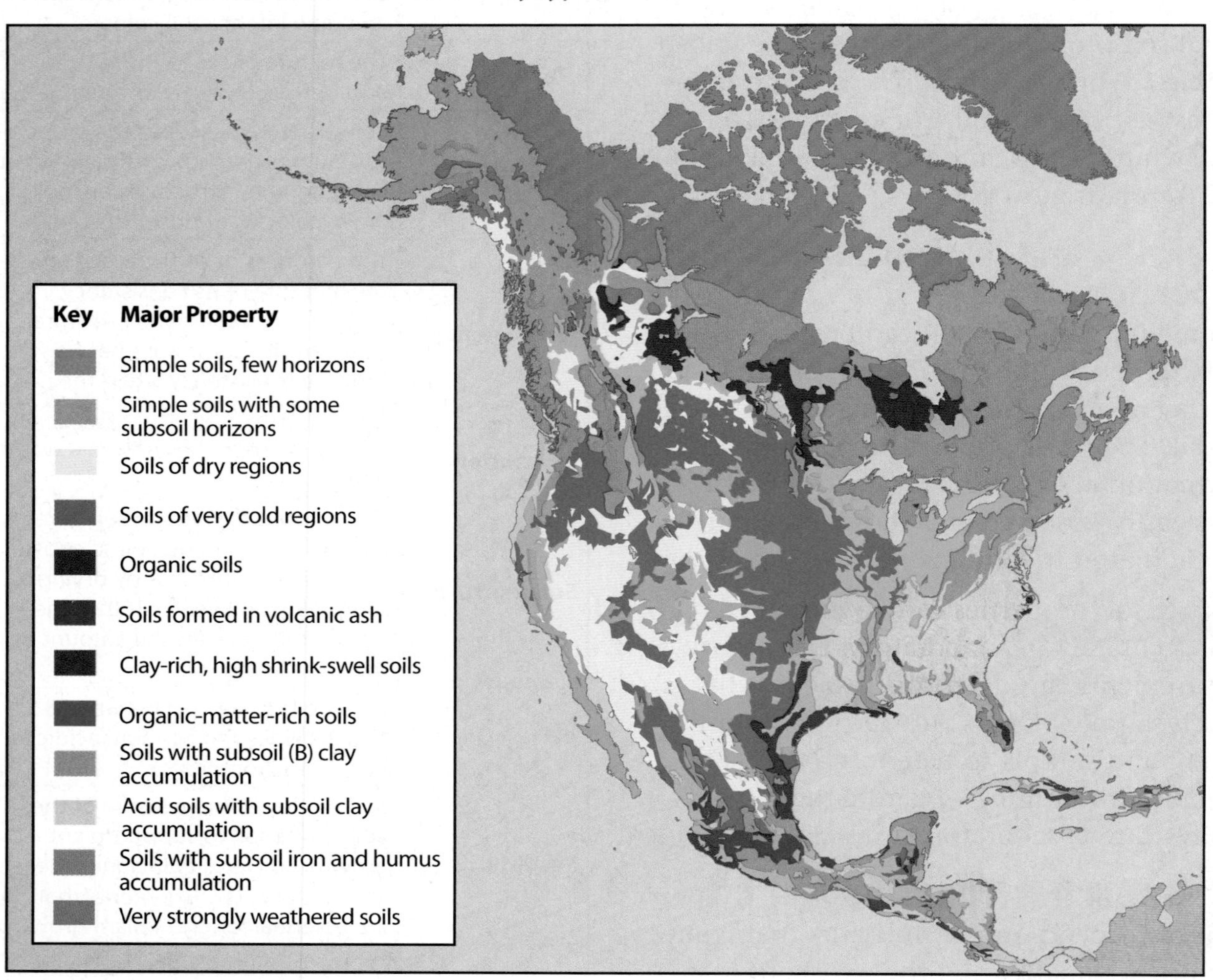

14. Visual Check
Identify Name all of the soil types found in Florida.

Soil Types and Locations

Recall that the type of soil formed depends partly on climate. Can you see how the soil types shown in **Figure 10** depend on the climate where they form? For example, in northern parts of Canada and Alaska and along mountain ranges, some soils stay frozen throughout the year. These soils are very simple and have few horizons. In the mid-latitudes, you can see a wide variety of soil types and depths. Farther toward the warm and wet climate of the tropics, soils are deeply weathered. Soils formed near volcanoes, such as those in Alaska and California, are acidic and have fine ash particles from volcanic activity.

Active Reading **15. Explain** Why aren't soils the same everywhere?

Lesson Review 2

Visual Summary

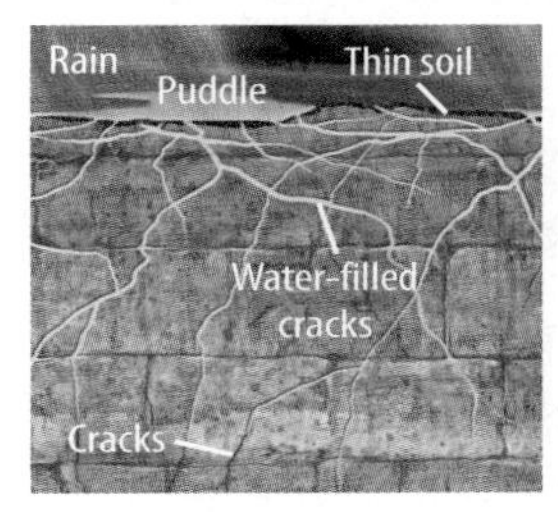

The inorganic matter in soil is made up of weathered parent material.

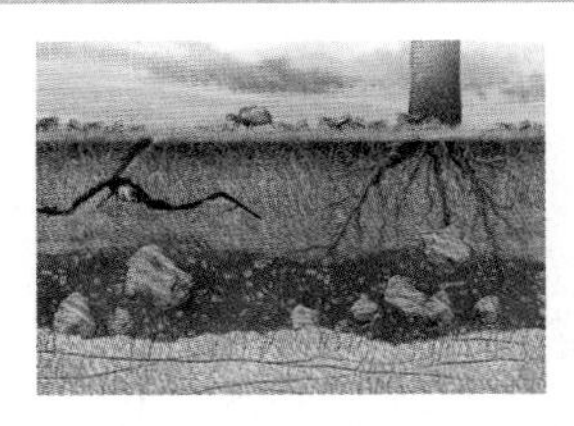

The five factors that contribute to soil formation are parent material, topography, climate, biota, and time.

Soil contains horizons, which are layers formed from the movement of the products of weathering.

Inquiry Lab ***How can you determine soil composition?*** at connectED.mcgraw-hill.com

Use Vocabulary

1. **Use the term** *decomposition* correctly in a sentence.

2. **Explain** how a leaf is organic matter.

3. **Define** *biota* in your own words.

Understand Key Concepts 

4. What is in the C-horizon?
   - (A) bedrock
   - (B) clay
   - (C) weathered stone
   - (D) organic material

5. **Contrast** rocks and soil. List three differences.

Interpret Graphics

6. **Sequence** Fill in the graphic organizer below. Starting with parent material, list steps that lead to the formation of an A-horizon. LA.6.2.2.3

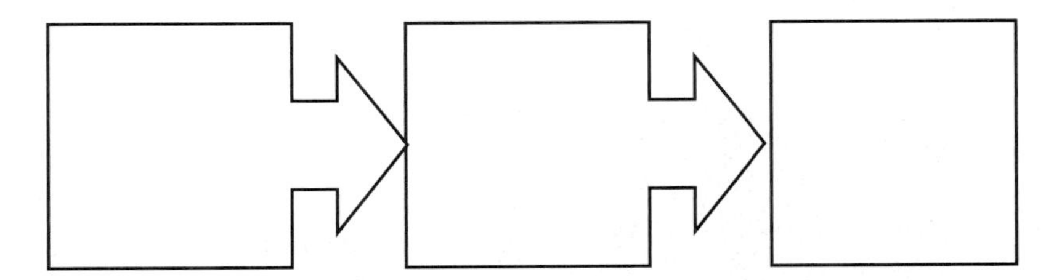

Critical Thinking

7. **Explain** What three things does soil provide for plants?

8. **Apply** Describe the soil-forming factors around your school.

Chapter 2 Study Guide

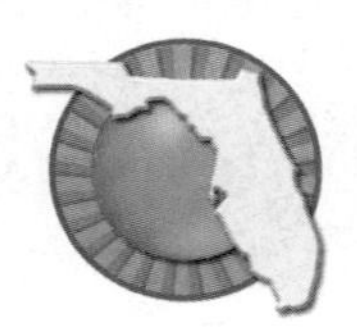

Think About It! Mechanical and chemical weathering are destructive forces that break down rocks, which begins the formation of soil.

Key Concepts Summary

LESSON 1 Weathering

- **Weathering** acts mechanically and chemically and breaks down rocks.
- Through the action of Earth processes such as freezing and thawing, **mechanical weathering** breaks rocks into smaller pieces.
- **Chemical weathering** by water and acids changes the materials in rocks into new materials.

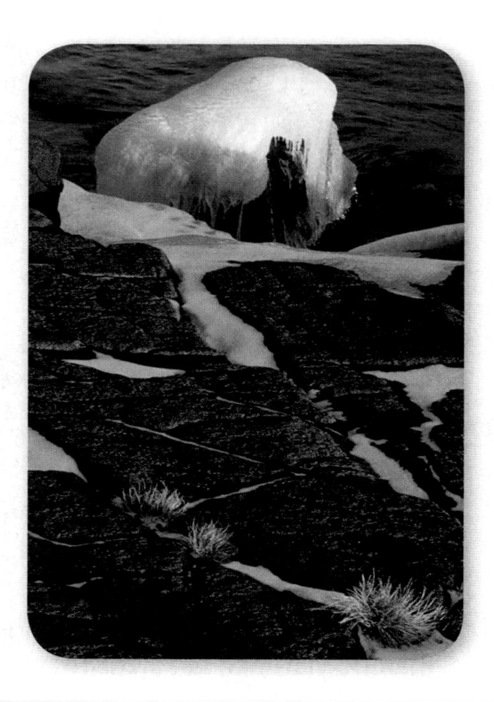

Vocabulary

weathering p. 45
mechanical weathering p. 46
chemical weathering p. 48
oxidation p. 49

LESSON 2 Soil

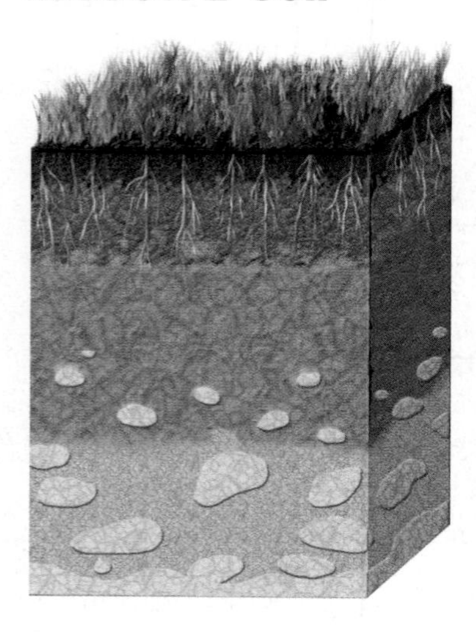

- Five factors—**parent material**, **climate, topography, biota**, and time—affect the formation of soil.
- **Horizons** are soil layers formed from the movement of the various products of weathering.
- Soil can be characterized by properties such as the amount of **organic matter** and inorganic matter.
- Plants depend on certain characteristics of soil, such as organic matter and amount of weathering.

soil p. 54
organic matter p. 54
pore p. 54
decomposition p. 55
parent material p. 56
climate p. 56
topography p. 57
biota p. 57
horizon p. 58

Active Reading FOLDABLES® **Chapter Project**

Assemble your lesson Foldables as shown to make a Chapter Project. Use the project to review what you have learned in this chapter.

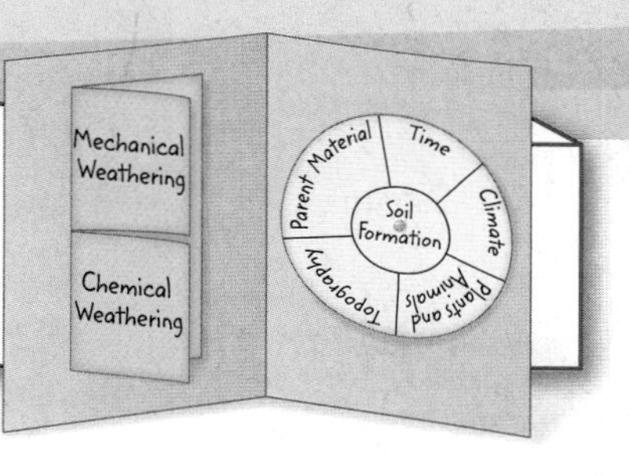

Use Vocabulary

1. When rock undergoes ______________, the product is smaller pieces of the same kind of rock.
2. Rock fragments and other materials combine to form ______________.
3. The part of soil that comes from plants and animals is ______________.
4. An important soil-forming factor that includes trees and microorganisms is ______________.
5. Oxygen combines with other elements or compounds during the process of ______________.
6. The shape of the land is its ______________.

Link Vocabulary and Key Concepts

Use vocabulary terms from the previous page to complete the concept map.

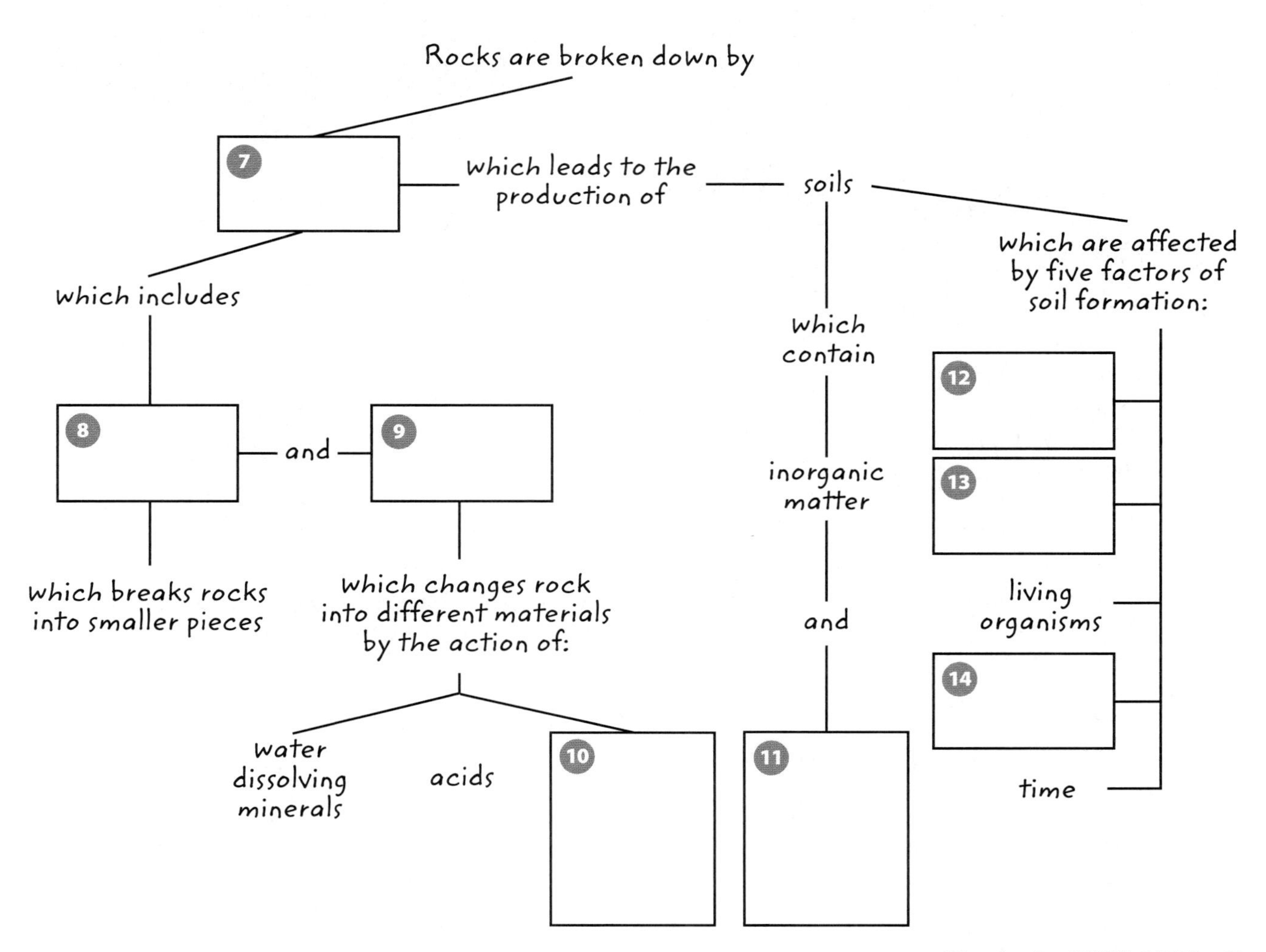

Chapter 2 Review

Fill in the correct answer choice.

Understand Key Concepts

1. Which is an example of chemical weathering? **SC.6.E.6.1**
   - Ⓐ abrasion
   - Ⓑ ice wedging
   - Ⓒ organisms
   - Ⓓ oxidation

2. A statue made of limestone is damaged by its environment. What most likely caused this damage? **SC.6.E.6.1**
   - Ⓐ acid
   - Ⓑ a root
   - Ⓒ topography
   - Ⓓ wind

3. The picture below shows how mechanical and chemical weathering can change rock.

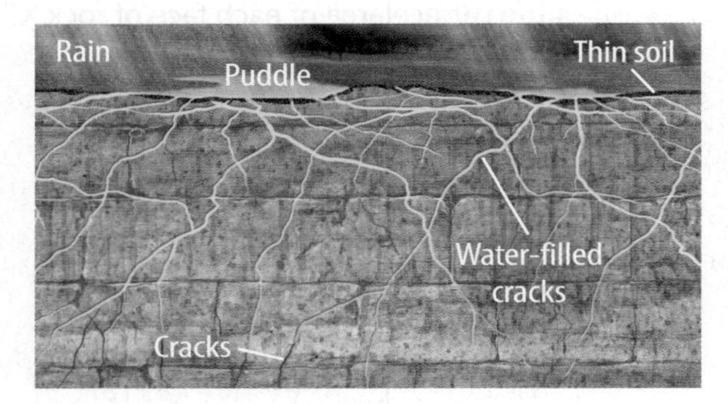

   What kind of chemical weathering is most likely illustrated above? **SC.6.E.6.1**
   - Ⓐ acid rain reactions
   - Ⓑ ice wedging
   - Ⓒ mineral absorption
   - Ⓓ root pressure

4. What kind of climate has the fastest weathering? **SC.6.E.6.1**
   - Ⓐ cold and dry
   - Ⓑ cold and wet
   - Ⓒ hot and dry
   - Ⓓ hot and wet

5. How does organic matter help soil? **SC.6.E.6.1**
   - Ⓐ It decomposes bacteria in the soil.
   - Ⓑ It holds water.
   - Ⓒ It weathers and forms clay.
   - Ⓓ It weathers nearby rocks.

Critical Thinking

6. **Infer** A student notices that when it rains, most of the water that falls on her yard runs off instead of soaking in. Is it more likely that the soil in her yard contains mostly clay or mostly sand? Explain. **SC.6.E.6.1**

7. **Explain** How do the biota shown in the image below help form soil? **SC.6.E.6.1**

8. **Explain** how climate helps to create soil. **SC.6.E.6.1**

9. **Describe** how soil horizons are produced and identified. **SC.6.E.6.1**

10 **Compare** Stone buildings near cities usually undergo more chemical weathering than buildings away from cities. Explain why this is true. SC.6.E.6.1

11 **Summarize** how soil is important to life. LA.6.2.2.3

12 **Identify** how chemical weathering and mechanical weathering make soil. SC.6.E.6.1

13 **Describe** how ice wedging and plant roots are similar in breaking down rocks. SC.6.E.6.1

Writing in Science

14 **Write** a short story on another piece of paper that explains how a large boulder becomes sand through weathering. In your story, include both mechanical and chemical weathering. Include main ideas and supporting details. SC.6.E.6.1

Math Skills

MA.6.A.3.6

Use the following data to answer the questions.

Rock Sample	Length	Width	Height
X	8 cm	8 cm	8 cm
Y	2 cm	16 cm	16 cm

15 How do the surface areas of rock sample X and rock sample Y compare?

16 What is the surface area of each face of rock X? Rock Y?

17 Rock sample X breaks into 8 equal cubes.

a. What is the surface area of each cube?

b. What is the total surface area of the broken rock?

c. How does this area compare with the original surface area?

Fill in the correct answer choice.

Multiple Choice

1 Which is true of oxidation? **SC.6.E.6.1**

Ⓐ It is a physical process.

Ⓑ No change occurs in the makeup of rock.

Ⓒ Rocks weather at different rates.

Ⓓ Water enters cracks in rock.

2 How can biota affect weathering? **SC.6.E.6.1**

Ⓕ Animals chemically weather bedrock.

Ⓖ Animal waste adds organic matter to soil.

Ⓗ Biota does not affect weathering.

Ⓘ Burrowing animals form passageways for water.

Use the table below to answer question 3.

Sample	Description
1	gravel
2	silt
3	sand
4	clay

3 Students collected four sediment samples and described them in the table above. Put the samples in order from most weathered to least weathered. **SC.6.E.6.1**

Ⓐ 1, 3, 2, 4

Ⓑ 3, 1, 2, 4

Ⓒ 4, 2, 3, 1

Ⓓ 1, 4, 2, 3

4 Which is LEAST likely to weather bedrock buried beneath layers of soil? **SC.6.E.6.1**

Ⓕ abrasion

Ⓖ acidic water

Ⓗ ice

Ⓘ plant roots

Use the diagram below to answer question 5.

5 At which spot in the landscape above would you most likely find a pile of broken, weathering rocks? **SC.6.E.6.1**

Ⓐ 1

Ⓑ 2

Ⓒ 3

Ⓓ 4

6 Which is NOT a process of chemical weathering? **SC.6.E.6.1**

Ⓕ abrasion

Ⓖ oxidation

Ⓗ dissolution by acid

Ⓘ dissolution by water

7 The grinding of rock by friction or impact is called **SC.6.E.6.1**

Ⓐ abrasion.

Ⓑ decomposition.

Ⓒ erosion.

Ⓓ infiltration.

8 Which does NOT cause abrasion? **SC.6.E.6.1**

Ⓕ animals

Ⓖ glaciers

Ⓗ water

Ⓘ wind

Use the diagram below to answer questions 9 and 10.

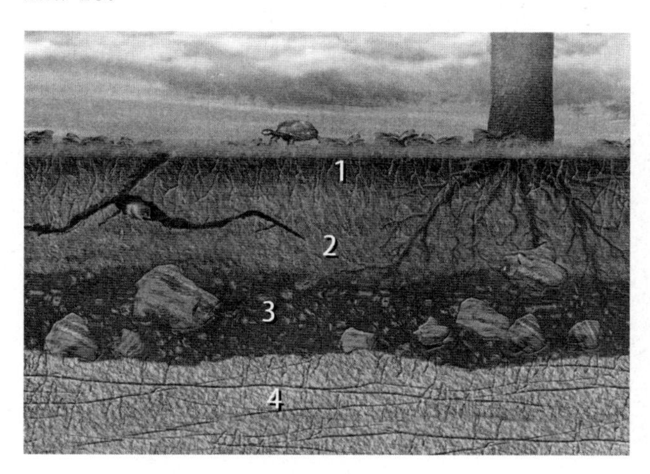

9 Which area pictured in the diagram above has been most affected by weathering? SC.6.E.6.1

Ⓐ 1

Ⓑ 2

Ⓒ 3

Ⓓ 4

10 Which area pictured in the diagram above has been least affected by weathering? SC.6.E.6.1

Ⓕ 1

Ⓖ 2

Ⓗ 3

Ⓘ 4

11 Physical weathering can reduce the surface area of rocks and sediment. Which type of sediment would have the greatest surface area? SC.6.E.6.1

Ⓐ clay

Ⓑ gravel

Ⓒ sand

Ⓓ silt

Use the table below to answer question 12.

Soil Horizon	Description
O	the top layer that contains mostly organic matter
A	the layer that contains a lot of organic matter and plant roots; usually the darkest of the soil horizons; excellent for growing plants
B	the clay-rich layer beneath the A-horizon
C	the layer of mixed sediment and parent material
R	unweathered bedrock that makes up the parent material for the soil

12 Using the information in the table above, which statement is true about the role of weathering in the formation of soil layers? SC.6.E.6.1

Ⓕ The O-horizon is the result of physical and chemical weathering.

Ⓖ Only chemical weathering altered the material in the C-horizon.

Ⓗ Only physical weathering altered the material in the B-horizon.

Ⓘ The sediment in the A-, B-, and C-horizons was likely weathered from the R-horizon.

NEED EXTRA HELP?

If You Missed Question...	1	2	3	4	5	6	7	8	9	10	11	12
Go to Lesson...	1	2	2	1	2	1	1	1	2	2	1	2

Name ______________________ Date ______________

Multiple Choice *Bubble the correct answer.*

1. To which material would the largest amount of water and minerals attach? **SC.6.E.6.1**

Ⓐ 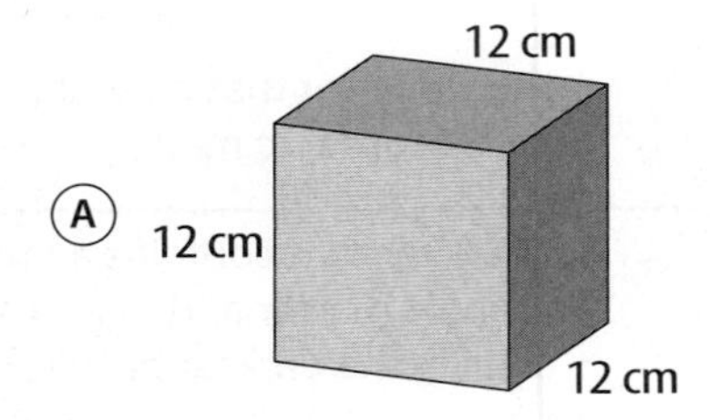

Ⓑ 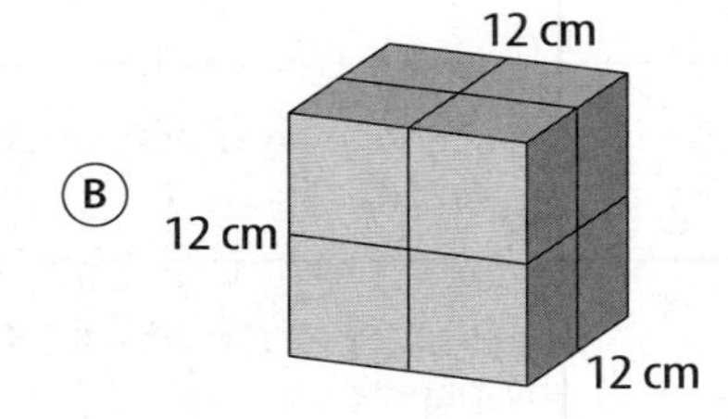

Ⓒ 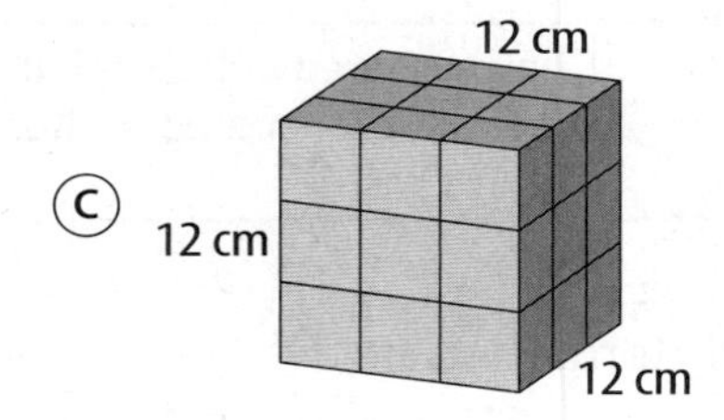

Ⓓ 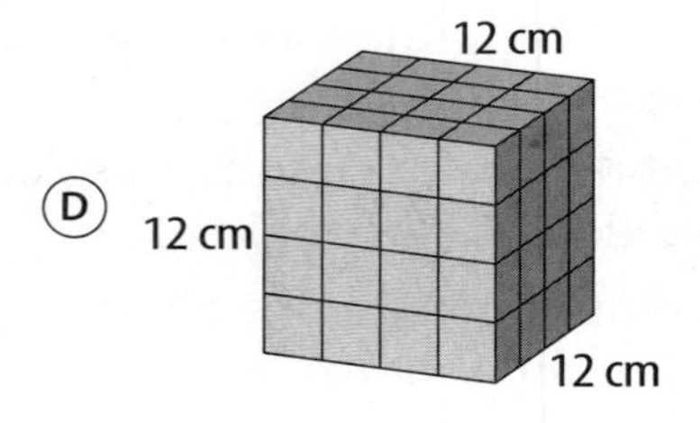

2. Which material might be involved in both mechanical and chemical weathering? **SC.6.E.6.1**

Ⓕ oxygen

Ⓖ water

Ⓗ carbon dioxide

Ⓘ carbonic acid

3. Which is NOT a cause of mechanical weathering? **SC.6.E.6.1**

Ⓐ A factory pollutes the atmosphere.

Ⓑ A mole burrows underground.

Ⓒ A plant grows into a crack in a rock.

Ⓓ Waves break along a shoreline.

4 One sample of rain has a pH of 4.5, and another sample of rain has a pH of 6.0. Which statement BEST describes these samples? **SC.6.E.6.1**

Ⓕ The pH levels of the two samples show that both are acid rain.

Ⓖ One sample is pure water and the other sample is acid rain.

Ⓗ One sample will cause more chemical weathering than the other sample.

Ⓘ One sample will cause more mechanical weathering than the other sample.

Name ______________________________ Date ______________

Multiple Choice *Bubble the correct answer.*

Sand	Silt	Clay

1. Which BEST describes the soil particles in the image above? **SC.6.E.6.1**

Ⓐ They affect drainage and water storage.

Ⓑ They help decompose leaves and twigs.

Ⓒ They hold minerals together in clusters.

Ⓓ They hold more water in the soil.

2. Which soil properties are BEST for farming? **SC.6.E.6.1**

Ⓕ few nutrients and organic material

Ⓖ large amounts of sand and clay

Ⓗ rocky with many horizons

Ⓘ young and deep with few horizons

3. Which process adds organic material to soils? **SC.6.E.6.1**

Ⓐ decomposition

Ⓑ infiltration

Ⓒ oxidation

Ⓓ weathering

4. How does the activity in the image above affect soil formation? **SC.6.E.6.1**

Ⓕ The earthworms break down parent material into smaller particles.

Ⓖ The earthworms remove organic matter from soil.

Ⓗ The tunnels allow water to pass through soil, which increases soil formation.

Ⓘ The tunnels break up pores, decreasing the flow of water and increasing soil formation.

Name ______________________ Date ____________

Notes

What is erosion?

Cheryl, Jane, and Marco were discussing processes that change the surface of Earth. They did not agree on the process of erosion. This is what they said:

Cheryl: I think erosion is when Earth materials are laid down or settle somewhere.

Jane: I think erosion is when Earth materials move from one area to another.

Marco: I think erosion is when Earth materials break down, such as when a big rock breaks into smaller rocks.

Circle the name of the student you agree with and describe why you agree. Describe your ideas about erosion.

Erosion and DEPOSITION

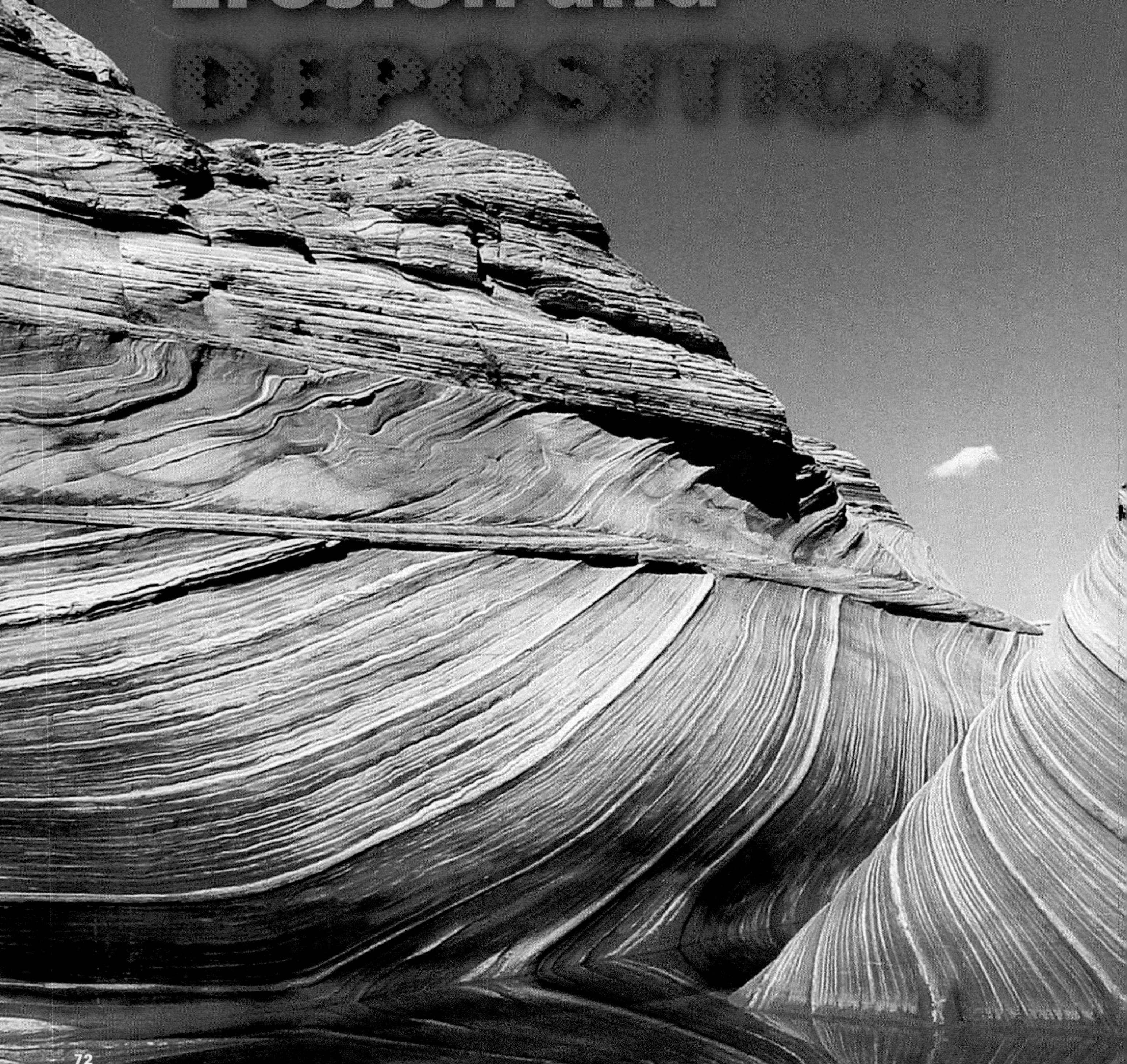

FLORIDA BIG IDEAS

1 The Practice of Science

6 Earth Structures

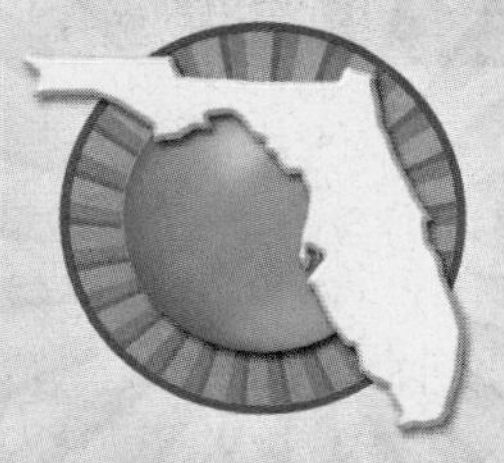

The Big Idea

Think About It!

How do erosion and deposition shape Earth's surface?

The swirling slopes of this ravine look as if heavy machines carved patterns in the rock. But nature formed these patterns.

1. What do you think caused the layers of colors in the rock?

2. Why do you think the rock has smooth curves instead of sharp edges?

3. How do you think erosion and deposition formed waves in the rock?

Get Ready to Read

What do you think about erosion and deposition?

Before you read, decide if you agree or disagree with each of these statements. As you read this chapter, see if you change your mind about any of the statements.

	AGREE	DISAGREE
1 Wind, water, ice, and gravity continually shape Earth's surface.	☐	☐
2 Different sizes of sediment tend to mix when being moved along by water.	☐	☐
3 A beach is a landform that does not change over time.	☐	☐
4 Windblown sediment can cut and polish exposed rock surfaces.	☐	☐
5 Landslides are a natural process that cannot be influenced by human activities.	☐	☐
6 A glacier leaves behind very smooth land as it moves through an area.	☐	☐

There's More Online!

Video • Audio • Review • ⓘLab Station • WebQuest • Assessment • Concepts in Motion • Multilingual eGlossary

Lesson 1

The Erosion-Deposition PROCESS

ESSENTIAL QUESTIONS

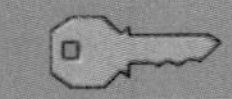
How can erosion shape and sort sediment?

How are erosion and deposition related?

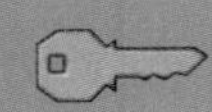
What features suggest whether erosion or deposition created a landform?

Vocabulary

erosion p. 77
deposition p. 79

Florida NGSSS

LA.6.2.2.3 The student will organize information to show understanding (e.g., representing main ideas within text through charting, mapping, paraphrasing, summarizing, or comparing/contrasting);

LA.6.4.2.2 The student will record information (e.g., observations, notes, lists, charts, legends) related to a topic, including visual aids to organize and record information and include a list of sources used;

SC.6.E.6.1 Describe and give examples of ways in which Earth's surface is built up and torn down by physical and chemical weathering, erosion, and deposition.

SC.6.E.6.2 Recognize that there are a variety of different landforms on Earth's surface such as coastlines, dunes, rivers, mountains, glaciers, deltas, and lakes and relate these landforms as they apply to Florida.

SC.6.N.1.1 Define a problem from the sixth grade curriculum, use appropriate reference materials to support scientific understanding, plan and carry out scientific investigation of various types, such as systematic observations or experiments, identify variables, collect and organize data, interpret data in charts, tables, and graphics, analyze information, make predictions, and defend conclusions.

Inquiry Launch Lab

SC.6.N.1.1

10 minutes

How do the shape and size of sediment differ?

Sediment forms when rocks break apart. Wind, water, and other factors move the sediment from place to place. As the sediment moves, its shape and size can change.

Procedure

1. Read and complete a lab safety form.
2. Obtain a **bag of sediment** from your teacher. Pour the sediment onto a sheet of **paper.**
3. Use a **magnifying lens** to observe the differences in shape and size of the sediment.
4. Divide the sediment into groups according to its size and whether it has rounded or sharp edges.

Data and Observations

Think About This

1. What were the different groups you used to sort the sediment?

2. Key Concept How do you think movement by wind and water might affect the shape and size of the sediment?

Inquiry How did the island form?

1. Florida's coast is peppered with barrier islands. The islands provide protection for Florida's beaches, habitats for wildlife, and recreation for visitors. What processes created landforms such as these barrier islands? How do you think these landforms will change in the future?

Reshaping Earth's Surface

Have you ever seen bulldozers, backhoes, and dump trucks at the construction site of a building project? You might have seen a bulldozer smoothing the land and making a flat surface or pushing soil around and forming hills. A backhoe might have been digging deep trenches for water or sewer lines. The dump trucks might have been dumping gravel or other building materials into small piles. The changes that people make to a landscape at a construction site are small examples of those that happen naturally to Earth's surface.

A combination of constructive **processes** and destructive processes produce landforms. Constructive processes build up features on Earth's surface. For example, lava erupting from a volcano hardens and forms new land on the area where the lava falls. Destructive processes tear down features on Earth's surface. A strong hurricane, for example, could wash part of Florida's shoreline into the sea. Constructive and destructive processes continually shape and reshape Earth's surface.

ACADEMIC VOCABULARY

process
(noun) an ongoing event or a series of related events

Active Reading **2. Differentiate** What is the difference between constructive and destructive processes?

Figure 1 The continual weathering, erosion, and deposition of sediment occurs from the top of a mountain and across Earth's surface to the distant ocean.

Active Reading **3. Analyze** Fill in the blanks with the name of each process described.

A Continual Process of Change

Imagine standing on a mountain, such as the one shown in **Figure 1.** In the distance you might see a river or an ocean. What was this area like thousands of years ago? Will the mountains still be here thousands of years from now? Landforms on Earth are constantly changing, but the changes often happen so slowly that you do not notice them. What causes these changes?

Weathering

Active Reading **4. Identify** Circle the agents of weathering.

One type of destructive process that changes Earth's surface is weathering, the breakdown of rock. Chemical weathering changes the chemical composition of rock. Physical weathering breaks rock into pieces, called sediment, but it does not change the chemical composition of rock. Gravel, sand, silt, and clay are different sizes of sediment.

Weathering Agents Water, wind, and ice are called agents, or causes, of weathering. Water, for example, can dissolve minerals in rock. Wind can grind and polish rocks by blowing particles against them. Also, a rock can break apart as ice expands or plant roots grow within cracks in the rock.

Different Rates of Weathering The mineral composition of some rocks makes them more resistant to weathering than other rocks. The differences in weathering rates can produce unusual landforms, as shown in **Figure 2.** Weathering can break away less resistant parts of the rock and leave behind the more resistant parts.

Figure 2 Different rates of weathering of rock can produce unusual rock formations.

Erosion

Weathered material is often transported away from its source rock. **Erosion** *is the removal of weathered material from one location to another.* Agents of erosion include water, wind, glaciers, and gravity.

The Rate of Erosion Like weathering, erosion occurs at different rates. For example, a rushing stream or strong wind can erode a large quantity of material quickly. However, a gentle stream or a light breeze might erode a small amount of material slowly. Factors that affect the rate of erosion include weather, climate, topography, and type of rock. Weathered rock moves faster down a steep hill than across a flat area. And erosion occurs faster on barren land than on land covered with vegetation.

Active Reading **5. Point Out** Circle the factors that can affect the rate of erosion.

Inquiry SC.6.N.1.1 SC.6.E.6.1

iLAB STATION **Try It!**

MiniLab *Can weathering be measured?* at connectED.mcgraw-hill.com

Apply It! After you complete the lab, answer these questions.

1. How did you model weathering?

2. Did you model chemical or physical weathering? Explain your answer.

Active Reading **Figure 3.**
6. Compare Label these rocks as either well rounded or poorly rounded.

Rate of Erosion and Rock Type The rate of erosion sometimes depends on the type of rock. Weathering can break some types of rock, such as sandstone, into large pieces. Other rock types, such as shale or siltstone, can easily break into smaller pieces. These smaller pieces can be removed and transported faster by agents of erosion. For example, large rocks in streams usually move only short distances every few decades, but silt particles might move a kilometer or more each day.

Rounding Rock fragments bump against each other during erosion. When this happens, the shapes of the fragments can change. Rock fragments can range from poorly rounded to well-rounded. The more spherical and well rounded a rock is, the more it has been polished during erosion. Rough edges break off as the rock fragments bump against each other. Differences in sediment rounding are shown in **Figure 3.**

7. NGSSS Check **Describe** How can erosion affect the shape of sediment? SC.6.E.6.1

Sorting Erosion also affects the level of sorting of sediment. Sorting is the separating of items into groups according to one or more properties. As sediment is transported, it can become sorted by grain size, as shown in **Figure 4.** Sediment is often well sorted when it has been moved a lot by wind or waves. Poorly sorted sediment often results from rapid transportation, perhaps by a storm, a flash flood, or a volcanic eruption. Sediment left at the edges of glaciers is also poorly sorted.

8. NGSSS Check **Recall** How can erosion sort sediment? SC.6.E.6.1

Figure 4 Erosion can sort sediment according to its size.

9. Visual Check **Differentiate** Write the correct letter on each photo that corresponds to the caption that best describes it.

A **Poorly sorted** sediment has a wide range of sizes.

B **Well-sorted** sediment is all about the same size.

C **Moderately sorted** sediment has a small range of sizes.

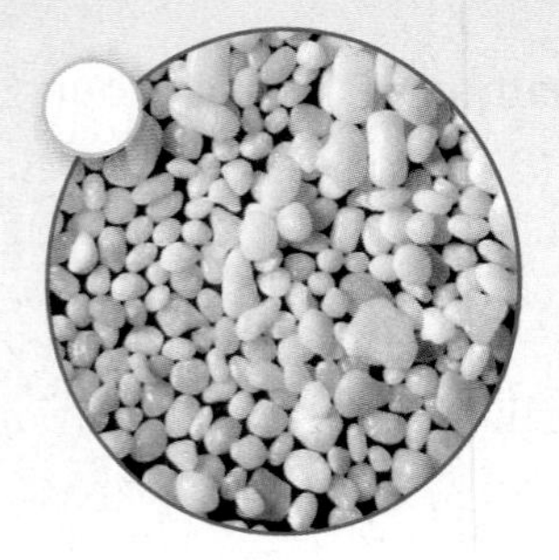

Deposition

You have read about two destructive processes that shape Earth's surface—weathering and erosion. After material has been eroded, a constructive process takes place. **Deposition** *is the laying down or settling of eroded material.* As water or wind slows down, it has less energy and can hold less sediment. Some of the sediment can then be laid down, or deposited.

Depositional Environments in Florida Sediment is deposited in locations called depositional environments. These locations are on land, along coasts, or in oceans. Within each depositional environment there may be certain landforms. For example, a coastal depositional environment may include swamps, deltas, beaches, and dunes.

The landforms that take shape in a depositional environment depend on the energy of the water or the wind in the environment. The amount of energy determines the size and the amount of sediment moved. High-energy environments transport large volumes of large sediment. The sediment is deposited when or where the energy decreases. Large, crashing waves, like the ones you might see at Daytona Beach, are high-energy and easily transport sand. If the waves lose their energy, some or all of their sediment is deposited. A swamp, such as the one shown in **Figure 5,** is a low-energy environment where fine-grained sediment is deposited.

Sediment Layers Sediment deposited in water typically forms layers called beds. Some examples of beds appear as "stripes" in the photo at the beginning of this lesson. Beds often form as layers of sediment at the bottom of rivers, lakes, and oceans. These layers can be preserved in sedimentary rocks.

10. NGSSS Check
Relate How are erosion and deposition related?
SC.6.E.6.1

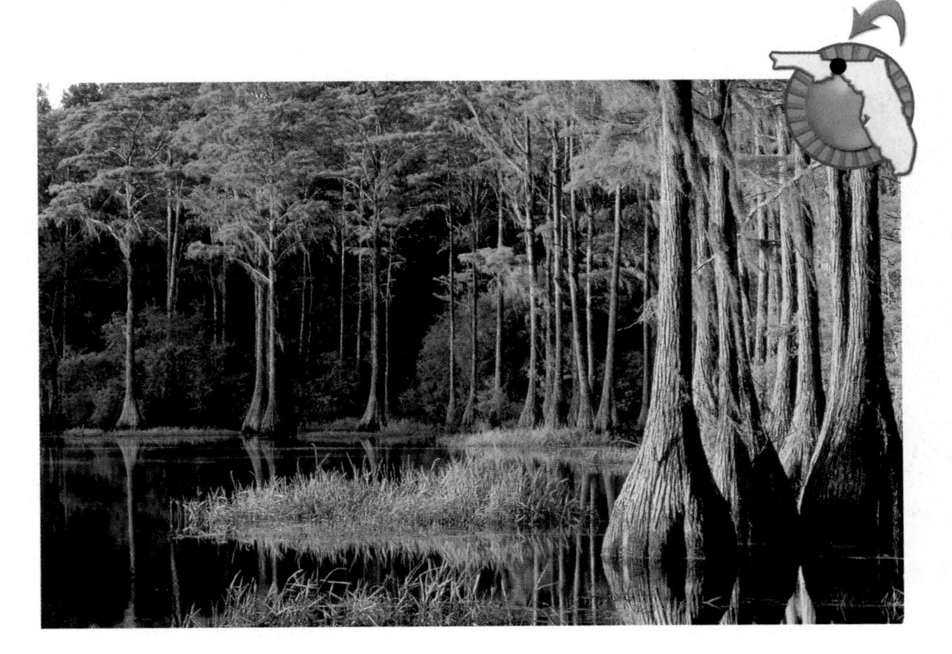

Figure 5 This swamp near Tallahassee, Florida, is a low-energy environment. Silt and clay are deposited here along with dark, organic matter from decaying plants.

Figure 6 The tall, steep, somewhat sharp features shown in these photographs are common in landforms carved by erosion.

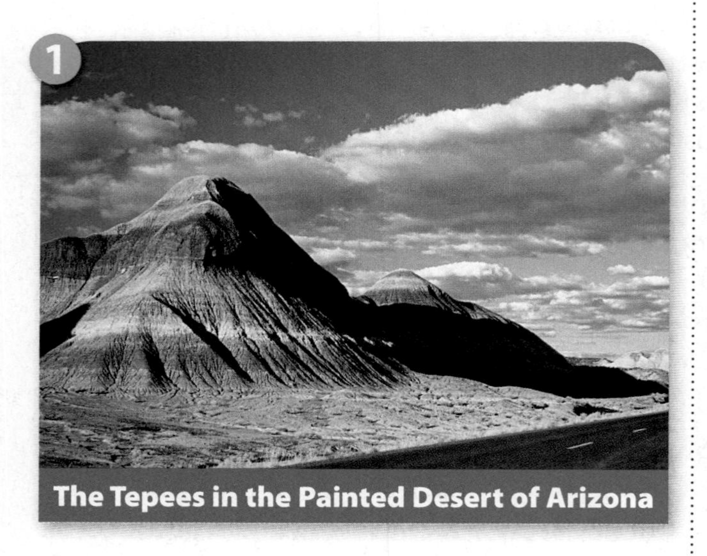

1 The Tepees in the Painted Desert of Arizona

2 Hoodoos at Bryce Canyon National Park

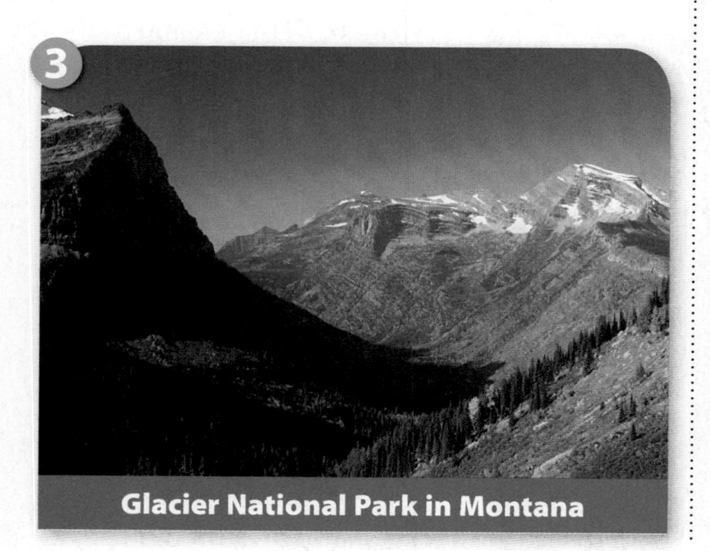

3 Glacier National Park in Montana

Interpreting Landforms

What do landform characteristics, such as structure, elevation, and rock exposure, suggest about the development of landforms? Examples of landforms include mountains, valleys, dunes, and lakes. These landforms are always changing, although you might not observe these changes in your lifetime. Landform characteristics can be observed to determine whether destructive forces, such as erosion, or constructive forces, such as deposition, produced the landforms.

Landforms Created by Erosion

Landforms can have features that are clearly produced by erosion. These landforms are often tall, jagged structures with cuts in layers of rock, as shown in the photographs in **Figure 6.**

1 Landforms formed by erosion can expose several layers of rock. The Tepees in the Painted Desert of Arizona contain several layers of different materials. Over time, erosion wore away parts of the land, leaving behind multicolored mounds.

2 Recall that different rates of erosion can result in unusual landforms when some rocks erode and leave more erosion-resistant rocks behind. For example, tall, protruding landforms called hoodoos are shown in the middle photograph of **Figure 6.** Over time, water and ice eroded the less-resistant sedimentary rock. The remaining rocks are more resistant. If you would like to examine hoodoos more closely, look back at **Figure 2.**

3 Glacial erosion and coastal erosion also form unique landforms. Glacial erosion produces jagged mountain peaks and U-shaped valleys, such as those in Glacier National Park, shown in the bottom photograph. Coastal erosion on the Oregon coast formed scenic sea cliffs and caves. In Florida, coastal erosion changes the size and the shape of beaches.

Landforms Created by Deposition

Landforms created by deposition are often flat and low-lying. Wind deposition, for example, can gradually form deserts of sand. Deposition also occurs where mountain streams reach the gentle slopes of wide, flat valleys. A gently sloping apron of sediment, called an alluvial fan, often forms where a stream flows from a steep, narrow canyon onto a flat plain at the foot of a mountain, as shown in **Figure 7.**

Water traveling in a river can slow due to friction with the edges and the bottom of the river channel. An increase in channel width or depth also can slow the current and promote deposition. Deposition along a riverbed or a coastline can occur where the speed of the water slows. This deposition can form a sandbar, as shown in **Figure 8.** The endpoint for most rivers is where they reach a lake or an ocean and deposit sediment under water. Wave action along shorelines also moves and deposits sediment.

Figure 7 An alluvial fan is a gently sloping mass of sediment.

Active Reading **11. Summarize** How does an alluvial fan form?

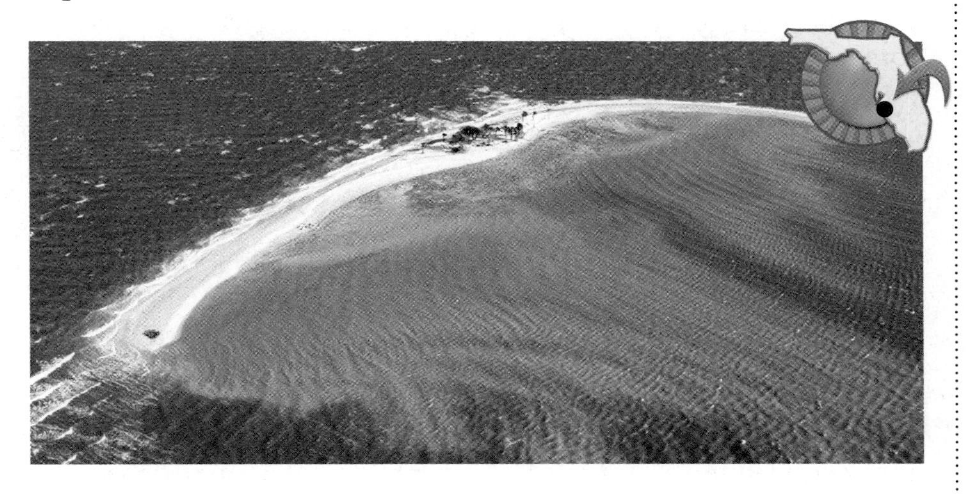

Figure 8 A sandbar is a depositional feature in rivers and near ocean shores, like this one off the coast of Tampa Bay.

As glaciers melt, they can leave behind piles of sediment and rock. For example, glaciers can create long, narrow deposits called eskers and moraines. In the United States, these features are best preserved in northern states such as Wisconsin and New York. You will read more about glacial deposition in Lesson 3.

Comparing Landforms

Look again at the landforms shown in **Figure 6, Figure 7,** and **Figure 8.** Notice how landforms produced by erosion and deposition are different. Erosion produces landforms that are often tall and jagged, but deposition usually produces landforms on flat, low land. By observing the features of a landform, you can infer whether erosion or deposition produced it.

12. NGSSS Check Compare What features suggest whether erosion or deposition produced a landform? SC.6.E.6.2

Active Reading FOLDABLES® LA.6.2.2.3

Make a two-tab book and label it as shown. Use your book to describe and identify some landforms created by the processes of erosion and deposition.

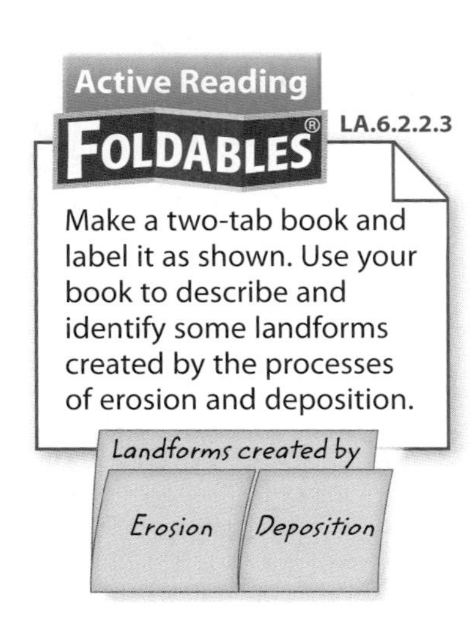

Lesson Review 1

Visual Summary

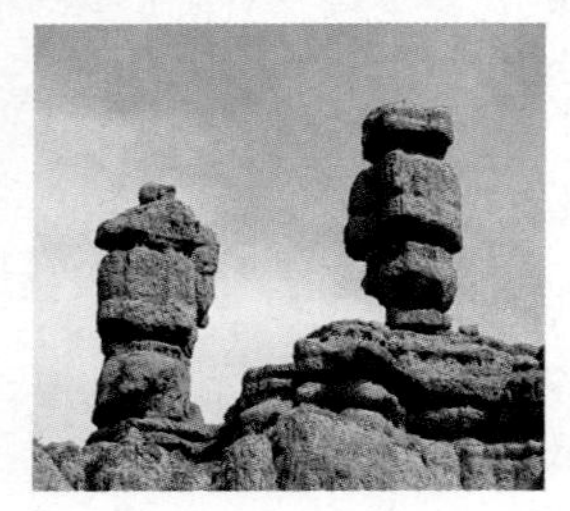

Erosion occurring at different rates can carve rock into interesting landforms.

Rock fragments with rough edges are rounded during transportation.

Landforms produced by deposition are often flat and low-lying.

Use Vocabulary

1. **Define** *deposition* in your own words.

2. **Use the term** *erosion* in a complete sentence. SC.6.E.6.1

Understand Key Concepts 

3. Which would most likely leave behind well-sorted sediment? SC.6.E.6.2
   - (A) flash flood
   - (B) melting glacier
   - (C) ocean waves
   - (D) volcanic eruption

4. **Describe** some features of an alluvial fan that suggest that it was formed by deposition.

5. **Explain** how erosion and deposition by a stream are related. SC.6.E.6.1

Interpret Graphics

6. **Sequence** Fill in the graphic organizer to describe the possible history of a grain of the mineral quartz that begins in a boulder at the top of a mountain and ends as a piece of sand on the Florida coast. SC.6.E.6.1

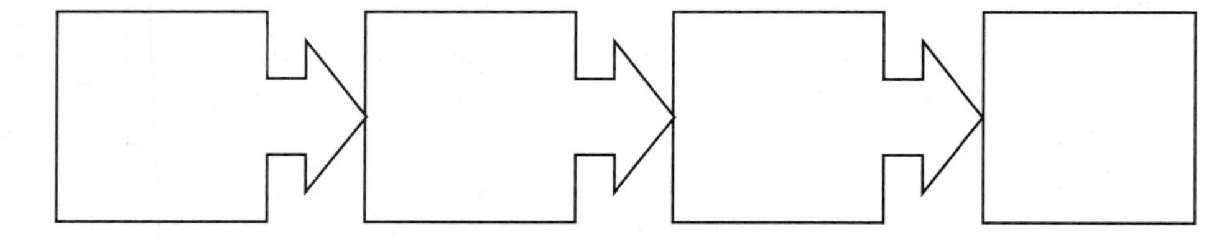

Critical Thinking

7. **Decide** Imagine a river that deposits only small particles where it flows into a sea. Is the river current most likely fast or slow? Why? SC.6.E.6.1

Coastal Erosion in Florida

SC.6.E.6.1
SC.6.E.6.2

FOCUS on FLORIDA

Can Florida's beaches be saved?

Florida has approximately 2,027 km of coastline, more than any other state in the continental United States. Florida's beaches are important resources for home owners and Florida's multibillion-dollar tourism industry. But coastal erosion is wearing away over half of Florida's beaches, threatening property, tourism, and natural habitats.

Coastal erosion can also happen quickly as a result of hurricanes or tropical storms, as shown along the coast of Jacksonville.

How are beaches being eroded?

Have you ever played in the waves on a Florida beach? Wind produces ocean waves that crash along the Florida coast. As waves break at an angle, they produce longshore currents. The waves and currents can erode sediment and deposit it farther down the beach.

The process of longshore currents transporting sediment along beaches is naturally balanced—the amount of sand eroded should be about the same as the amount of sand deposited. However, human-made structures, such as piers and groins, can affect the natural erosion-deposition process by trapping sand.

How can beach erosion be stopped?

Scientists and government agencies are working together to restore Florida's beaches and protect them from being washed away. Structures can be built along coastlines to help prevent erosion. For example, groins are structures built perpendicular to the beach that extend into the water to trap sand transported in longshore currents. While these structures can help widen beaches, they also remove sand from the longshore currents. As a result, locations farther down-current do not receive the sand deposits that would naturally replenish the beach.

An alternative to these structures is rebuilding the beach. Sand is dredged from offshore or trucked in from other locations and then deposited on the beach. Bulldozers spread the sand to build up the beach, helping to rebuild natural habitats and properties that are valuable to Florida's residents and visitors.

Miami Beach, Florida, is maintained through annual restoration.

It's Your Turn

RESEARCH How do erosion and deposition affect other Florida landforms? Using pictures or illustrations, describe how the landforms form and how they change through the erosion-deposition process. Share your findings with the class. LA.6.4.2.2

Lesson 2

Landforms Shaped by WATER AND WIND

ESSENTIAL QUESTIONS

What are the stages of stream development?

How do water erosion and deposition change Earth's surface?

How do wind erosion and deposition change Earth's surface?

Vocabulary

meander p. 86
longshore current p. 87
delta p. 88
abrasion p. 90
dune p. 90
loess p. 90

Florida NGSSS

LA.6.2.2.3 The student will organize information to show understanding (e.g., representing main ideas within text through charting, mapping, paraphrasing, summarizing, or comparing/contrasting);

SC.6.E.6.1 Describe and give examples of ways in which Earth's surface is built up and torn down by physical and chemical weathering, erosion, and deposition.

SC.6.E.6.2 Recognize that there are a variety of different landforms on Earth's surface such as coastlines, dunes, rivers, mountains, glaciers, deltas, and lakes and relate these landforms as they apply to Florida.

SC.6.N.1.1 Define a problem from the sixth grade curriculum, use appropriate reference materials to support scientific understanding, plan and carry out scientific investigation of various types, such as systematic observations or experiments, identify variables, collect and organize data, interpret data in charts, tables, and graphics, analyze information, make predictions, and defend conclusions.

SC.6.N.1.1

Inquiry Launch Lab

15 minutes

How do water and wind shape Earth?

Imagine a fast-moving river rushing over rocks or a strong wind blowing across a field. What changes on Earth do the water and the wind cause?

Procedure

1. Form groups and discuss the pictures below with others in your group.
2. Can you recognize evidence of ways water and wind have changed the land—through both erosion and deposition?

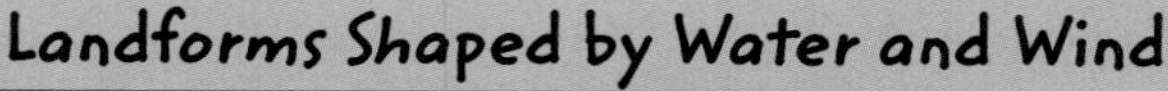
Landforms Shaped by Water and Wind

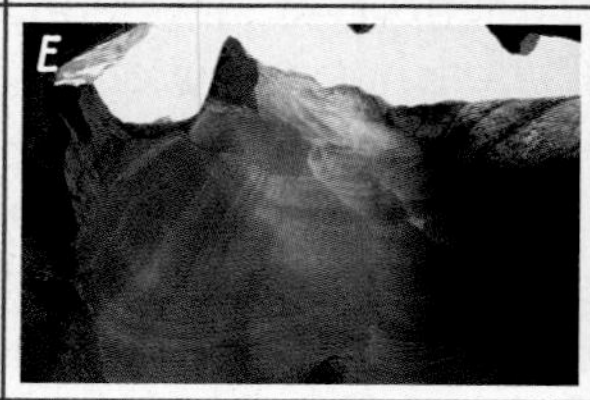

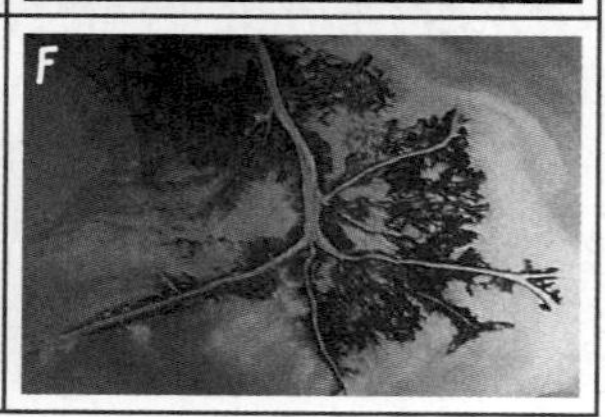

Think About This

1. What are some examples of erosion and deposition in the pictures?

2. **Key Concept** Describe ways you think water might have changed the land in the pictures. What are some ways wind might have changed the land?

Inquiry Twisted River?

1. As a river flows down a mountain, it usually flows in the same general direction. What do you think causes this river in the Everglades National Park to flow side-to-side? Why do you think it doesn't flow in a straight path?

Shaping the Land with Water and Wind

Recall that landforms on Earth's surface undergo continual change. Weathering and erosion are destructive processes that shape Earth's surface. These destructive processes often produce tall, jagged landforms. Deposition is a constructive process that also shapes Earth's surface. Constructive processes often produce flat, low-lying landforms.

What causes these processes that continually tear down and build up Earth's surface? In this lesson, you will read that water and wind are two important agents of weathering, erosion, and deposition. The rocks shown in **Figure 9** are an example of how erosion by water and wind have changed the shape of a coastline in Florida. In the next lesson, you will read about ways Earth's surface is changed by the downhill movement of rocks and soil and by the movement of glaciers.

Active Reading **Figure 9.** 2. **Explain** What agents of erosion shaped these rocks on Florida's Jupiter Island?

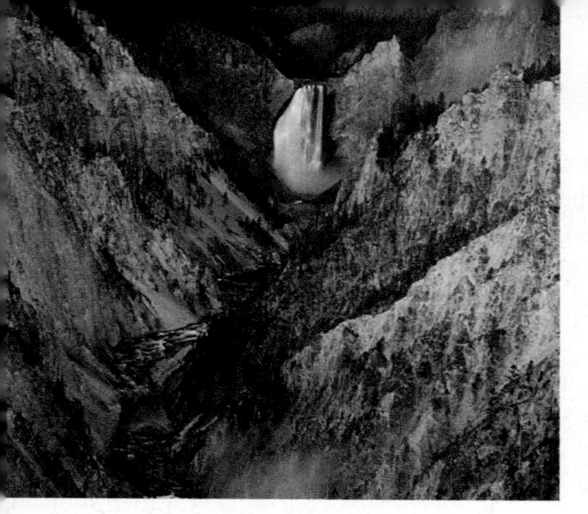

Figure 10 Water erosion carved this V-shaped valley at Lower Falls, Yellowstone National Park, in Wyoming.

Water Erosion and Deposition

Water can shape landforms on and below Earth's surface. The speed of water movement and the depositional environment often affect the shape of landforms.

Water Erosion

If you have ever waded into the Atlantic Ocean or the Gulf of Mexico from a Florida beach and felt the waves rushing toward shore, you know that moving water can be incredibly strong. Moving water causes erosion along streams, at beaches, and underground.

Stream Erosion Streams are active systems that erode land and transport sediment. The erosion produced by a stream depends on the stream's energy. This energy is usually greatest in steep, mountainous areas where young streams flow rapidly downhill. The rushing water often carves V-shaped valleys, such as the one shown in **Figure 10.** Waterfalls and river rapids are common in steep mountain streams.

Water in a young stream slows as it reaches gentler slopes. The stream is then called a mature stream, such as the one shown in **Figure 11.** Slower-moving water erodes the sides of a stream channel more than its bottom, and the stream develops curves. *A* **meander** *is a broad, C-shaped curve in a stream.*

When a stream reaches flat land, it moves even slower and is called an old stream. Over time, meanders change shape. More erosion occurs on the outside of bends where water flows faster. More deposition occurs on the inside of bends where water flows slower. Over time, the meander's size increases.

Active Reading FOLDABLES® LA.6.2.2.3

Make a two-tab book and label it as shown. Use your book to organize information about landforms and features created by erosion and deposition by water and wind.

Erosion and Deposition

Water | Wind

Figure 11 Streams change as they flow from steep slopes to gentle slopes and finally to flat plains.

3. **NGSSS Check** Compose Describe stream development using the photos below. SC.6.E.6.2

Figure 12 A longshore current erodes and deposits large amounts of sediment along a coastline.

Florida Coastal Erosion Like streams, coastlines continually change. Waves crashing onto shore erode loose sand, gravel, and rock along coastlines. One type of coastal erosion is shown in **Figure 12.** *A* **longshore current** *is a current that flows parallel to the shoreline.* This current moves sediment and continually changes the size and shape of beaches. Coastal erosion also occurs when the cutting action of waves along rocky shores forms sea cliffs, sea stacks (tall pillars just offshore), and sea arches. In Florida, government agencies developed a program to restore beaches eroded by longshore currents.

4. **NGSSS Check** Summarize How does water erosion change Earth's surface? SC.6.E.6.2

Groundwater Erosion Water that flows underground can also erode rock. Have you ever wondered how caverns form? When carbon dioxide in the air mixes with rainwater, a weak acid forms. Some of this rainwater becomes groundwater. As acidic groundwater seeps through rock and soil, it can pass through layers of limestone. The acidic water dissolves and washes away the limestone, forming a cavern, such as the one shown in **Figure 13.** Because Florida's limestone bedrock is easily dissolved by acidic rainwater, karst topography is common in Florida. Dissolving rock also can create sinkholes and aquifers.

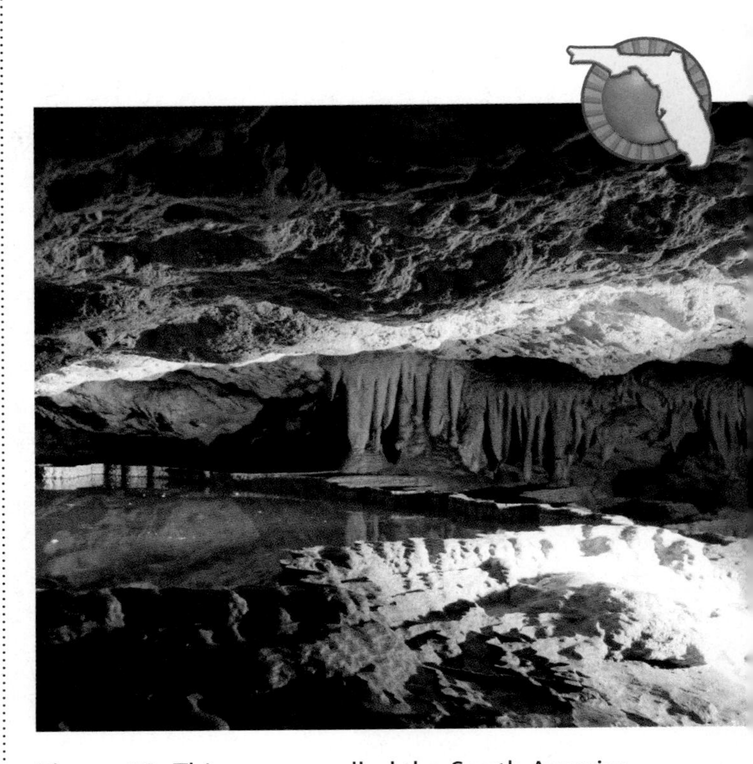

Figure 13 This cavern, called the South America Room, is part of the Florida Caverns State Park.

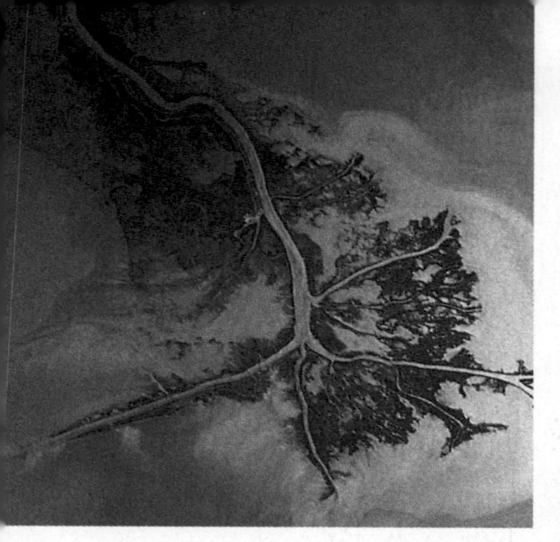

Figure 14 The Mississippi River forms a delta in the Gulf of Mexico.

Active Reading **5. Explain** How did this delta form?

Water Deposition

Flowing water deposits sediment as the water slows. A loss of speed reduces the amount of energy that the water has to carry sediment.

Deposition Along Streams Deposition by a stream can occur anywhere along its path where the water's speed decreases. As you read earlier, slower-moving water deposits sediment on the inside curves of meanders. A stream also slows and deposits sediment when it reaches flat land or a large body of water, such as a lake or an ocean. An example is the delta shown in **Figure 14.** *A* **delta** *is a large deposit of sediment that forms where a stream enters a large body of water.*

Deposition Along Coastlines Much of the sand on most ocean beaches was originally deposited by rivers. Longshore currents transport the sand along Florida coasts. Eventually, sand is deposited where currents are slower and have less energy. Sandy beaches often develop at those locations.

Groundwater Deposition Weathering and erosion produce caverns, but deposition forms many structures within caverns. Look again at the Florida cavern in **Figure 13.** The cavern contains landforms that dripping groundwater formed as it deposited minerals. Over time, the deposits developed into stalactites and stalagmites. Stalactites hang from the ceiling. Stalagmites build up on the cavern's floor.

6. NGSSS Check Recall How does water deposition change Earth's surface? SC.6.E.6.2

Inquiry LAB STATION **Try It!** SC.6.N.1.1 SC.6.E.6.1

MiniLab *How do stalactites form?* at connectED.mcgraw-hill.com

Apply It! After you complete the lab, answer this question.

1. Are stalactites erosional features or depositional features? Explain your answer.

Land Use Practices

Damage caused by water erosion can be affected by the ways people use land. Two areas of concern are erosion of beaches along Florida coasts and soil erosion.

Florida Beach Erosion Ocean waves can erode Florida beaches by removing sand and sediment. To reduce this erosion, people sometimes build structures such as retaining walls, or groins, like those shown in **Figure 15.** A row of groins is constructed at right angles to the shore. They are built to trap sediment and reduce the erosive effects of longshore currents.

Figure 15 These shoreline groins prevent beach erosion by trapping sediment.

Another way people can protect Florida coastlines from wave erosion is to protect dunes and mangrove trees. Storms that produce large, high-energy waves can wash away large areas of Florida's coastline. Dunes slow the landward erosion by these storms. Mangroves, because of their root system, also protect the coastline from wave erosion. Florida and other coastal states have laws protecting dunes and mangroves.

Some ways people affect beaches are unintended. For example, people build dams on rivers for various purposes. However, dams on rivers prevent river sand from reaching beaches. Then, beach sand that is washed out to sea by waves is not replaced.

Active Reading **7. Point out** (Circle) the natural and humanmade things that help prevent beach erosion.

Soil Erosion Reducing the amount of vegetation or removing it from the land increases surface erosion. Agricultural production, construction activities, and cutting trees for lumber and paper production are some reasons that people remove vegetation.

A floodplain is a wide, flat landform next to a river. It is usually dry land but can be flooded when the river overflows. Heavy rain or rapid melting of snow can cause a river to flood. Building within a floodplain is risky, as shown in **Figure 16.** However, floods supply mineral-rich soil that is ideal for farming.

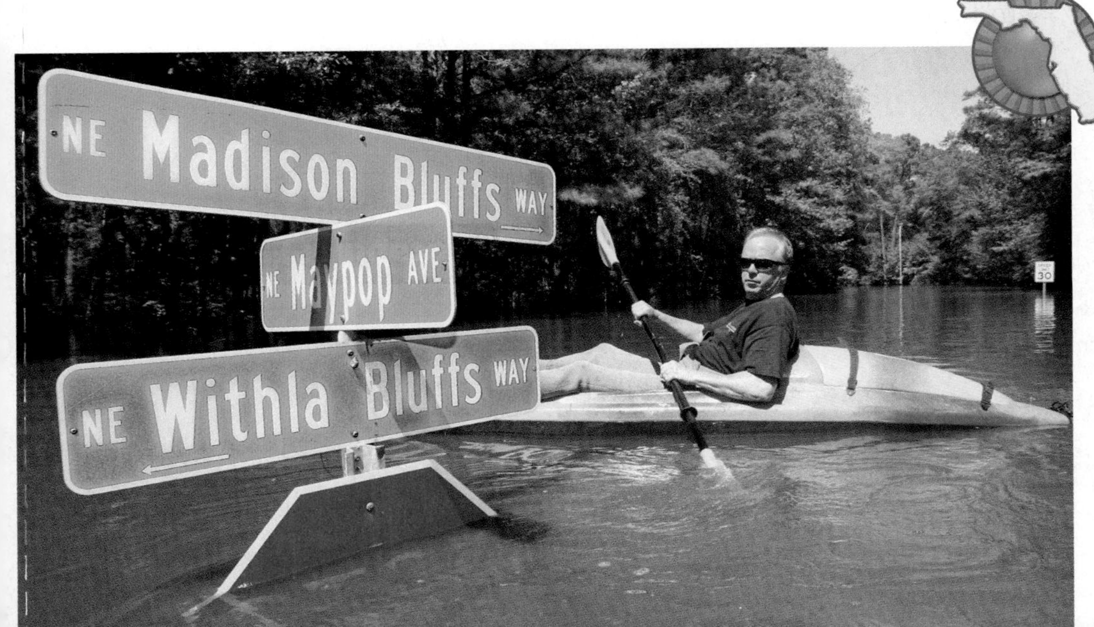

Figure 16 The Withlacoochee River in central Florida flooded in 2009, causing major damage to or destroying 200 homes and causing minor damage to more than 500 more.

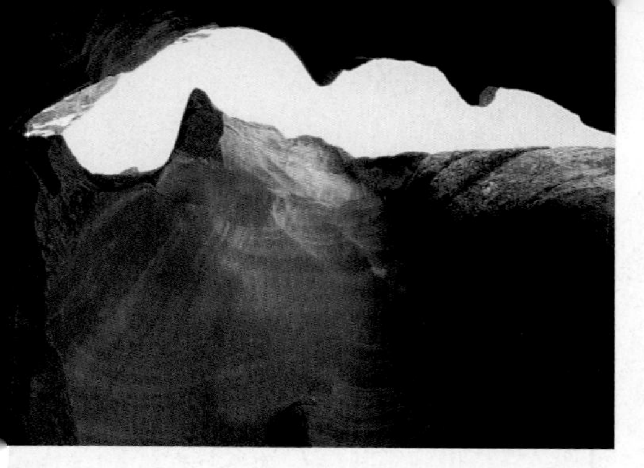

Figure 17 Wind abrasion carved this unusual landform in the red sandstone of Nevada's Valley of Fire region.

Wind Erosion and Deposition

If you think about a gentle wind that blows leaves in the autumn, it seems unlikely that the wind can cause land erosion and deposition. But strong or long-lasting winds can significantly change the land.

Wind Erosion

As wind carries sediment along, the sediment cuts and polishes exposed rock. **Abrasion** *is the grinding away of rock or other surfaces as particles carried by wind, water, or ice scrape against them.* Examples of rock surfaces carved by wind abrasion are shown in **Figure 17** and at the beginning of this chapter.

Wind Deposition

Two common types of wind-blown deposits are dunes and loess (LUHS). *A* **dune** *is a pile of windblown sand.* Over time, entire fields of dunes can travel across the land as wind continues to blow the sand. Some dunes are shown in **Figure 18.** **Loess** *is a crumbly, windblown deposit of silt and clay.* One type of loess forms from rock that was ground up and deposited by glaciers. Wind picks up this fine-grain sediment and redeposits it as thick layers of loess.

Word Origin
loess
from Swiss German *Lösch*, means "loose"

8. **NGSSS Check** **Analyze** How do wind erosion and deposition change Earth's surface? SC.6.E.6.1

Land Use Practices

People contribute to wind erosion. For example, plowed fields and dry, overgrazed pastures expose soil. Strong winds can remove topsoil that is not held in place by plants. One way to slow the effects of wind erosion is to leave fields unplowed after harvesting crops. Farmers can also plant rows of trees to slow wind and protect the soil.

Figure 18 Dunes, such as these on Grayton Beach, Florida, formed by the deposition of wind-blown sand.

Lesson Review 2

Visual Summary

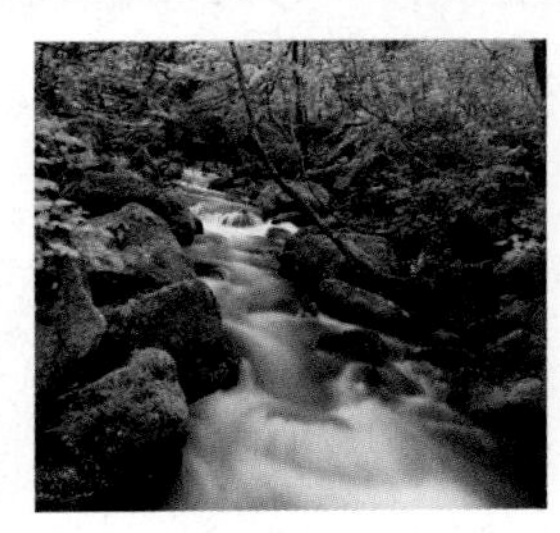

Water erosion is evident in the change in features of a stream over time.

Water transports sediment and deposits it in places where the speed of the water decreases.

Wind erosion can change Earth's surface by moving sediment.

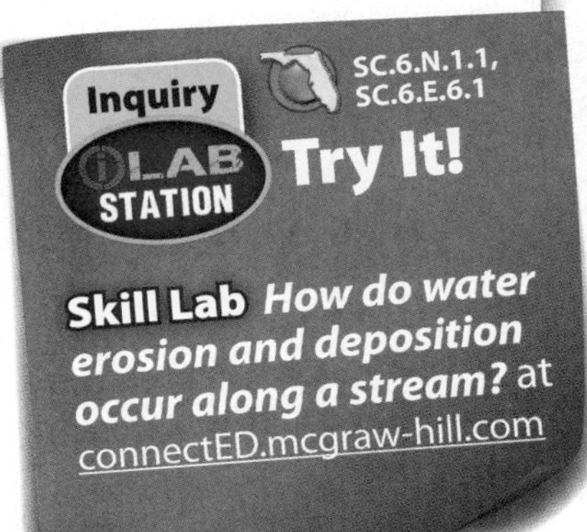

Use Vocabulary

1. **Distinguish** between loess and a dune. SC.6.E.6.2

2. **Use the term** *delta* in a complete sentence. SC.6.E.6.2

Understand Key Concepts

3. Which feature would a young river most likely have? SC.6.E.6.2
   - (A) meander
   - (B) slow movement
   - (C) waterfall
   - (D) wide channel

4. **Explain** how wind erosion might affect exposed rock. SC.6.E.6.1

5. **Compare and contrast** the advantages and disadvantages of farming on a floodplain. SC.6.E.6.1

Interpret Graphics

6. **Determine Cause and Effect** Fill in the graphic organizer below to identify two ways waves cause seashore erosion. SC.6.E.6.1

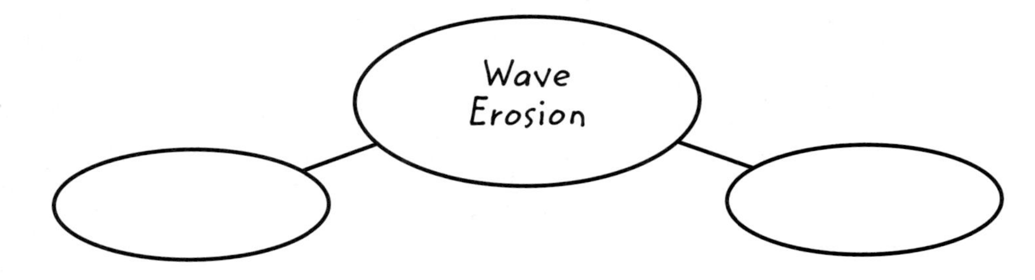

Critical Thinking

7. **Suppose** the amount of sand in front of a large, Florida beachfront hotel is slowly disappearing. Explain the process that is likely causing this problem. Suggest a way to avoid further loss of sand. SC.6.E.6.2

Landforms In Florida

FOCUS on FLORIDA

Gulf Islands National Seashore
The Gulf Islands National Seashore consists of 12 **barrier islands** off the coasts of Mississippi and Florida.

Britton Hill
Florida's highest elevation—106 m—is at Britton Hill.

Florida Caverns State Park
Limestone formations within the **caverns** include rimstones, flowstones, draperies, and soda straws—all depositional features.

Okefenokee Swamp
The Okefenokee Swamp is filled with peat, or decayed plant material. The abundance of peat makes the water in the **swamp** appear black.

Manatee Springs
Aquifers in Florida's limestone bedrock feed this **spring**. Manatees sometimes visit the spring.

Karst Topography
Florida's limestone bedrock is soluble, resulting in **karst topography**. Florida's **aquifer** system and **sinkholes** are associated with karst.

Lake Okeechobee
The **lake**, fed by Lake Kissimmee and the Kissimmee River, was once part of an ancient sea. It is one of the largest freshwater lakes in the United States.

Everglades National Park
The Everglades include **swamps**, **forests**, **rivers**, and **islands.** The 1.5 million acres of land remaining in the park are less than 20 percent of the Everglades' original size. The other 80 percent have been drained.

Florida Keys
The **islands** of the Florida Keys are ancient **coral reefs**. Farther south, Florida's living **coral reefs** provide habitats for about 650 species of fish, 40 species of coral, and numerous other invertebrates.

It's Your Turn

DESIGN Have you visited any Florida landforms? Design a travel brochure for the landforms in Florida that you have visited or that you would like to visit. Present your brochure to the class.

Lesson 3

Mass Wasting and GLACIERS

ESSENTIAL QUESTIONS

What are some ways gravity shapes Earth's surface?

How do glaciers erode Earth's surface?

Vocabulary

mass wasting p. 94
landslide p. 95
talus p. 95
glacier p. 97
till p. 98
moraine p. 98
outwash p. 98

Florida NGSSS

LA.6.2.2.3 The student will organize information to show understanding (e.g., representing main ideas within text through charting, mapping, paraphrasing, summarizing, or comparing/contrasting);

MA.6.A.3.6 Construct and analyze tables, graphs, and equations to describe linear functions and other simple relations using both common language and algebraic notation.

SC.6.E.6.1 Describe and give examples of ways in which Earth's surface is built up and torn down by physical and chemical weathering, erosion, and deposition.

SC.6.E.6.2 Recognize that there are a variety of different landforms on Earth's surface such as coastlines, dunes, rivers, mountains, glaciers, deltas, and lakes and relate these landforms as they apply to Florida.

SC.6.N.1.1 Define a problem from the sixth grade curriculum, use appropriate reference materials to support scientific understanding, plan and carry out scientific investigation of various types, such as systematic observations or experiments, identify variables, collect and organize data, interpret data in charts, tables, and graphics, analyze information, make predictions, and defend conclusions.

SC.6.N.1.1 Define a problem from the sixth grade curriculum using appropriate reference materials to support scientific understanding, plan and carry out scientific investigations of various types, such as systematic observations or experiments, identify variables, collect and organize data, interpret data in charts, tables, and graphics, analyze information, make predictions, and defend conclusions.

SC.6.N.1.1

Inquiry Launch Lab

15 minutes

How does a moving glacier shape Earth's surface?

A glacier is a huge mass of slow-moving ice. The weight of a glacier is so great that its movement causes significant erosion and deposition along its path. Use a model glacier to observe these effects.

Procedure

1. Read and complete a lab safety form.
2. Half-fill an **aluminum pan** with **dirt** and **gravel**. Mix enough **water** so that the dirt holds together easily. Use **two books** to raise one end of the pan.
3. Sprinkle **colored sand** at the top of the dirt hill.
4. Place a **model glacier** at the top of the hill. Slowly move the glacier downhill, pressing down gently.

Data and Observations

Think About This

1. What happened to the colored sand as the glacier moved downhill?

2. Key Concept What kinds of erosion and deposition did your model glacier cause?

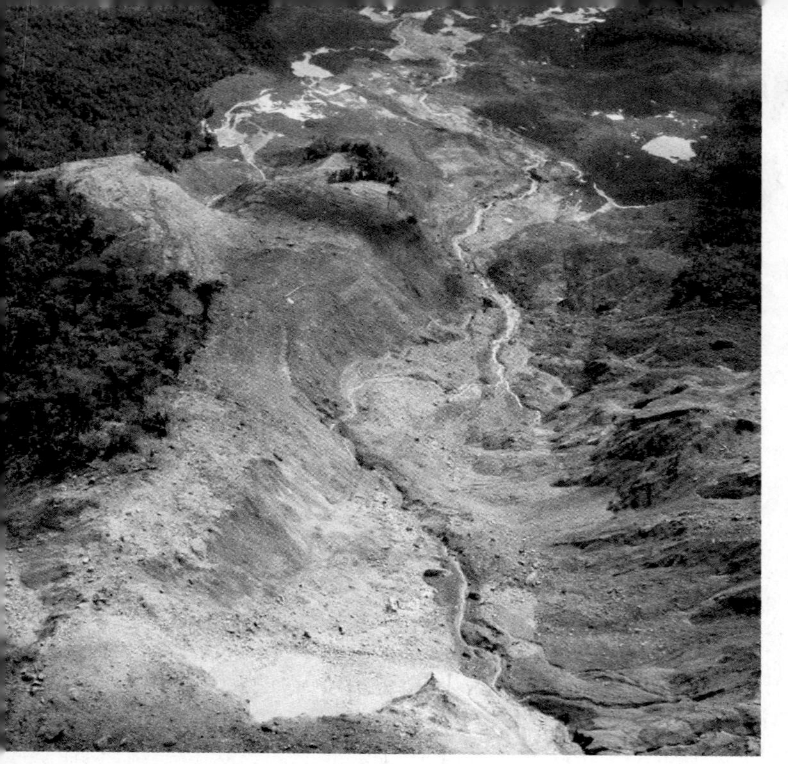

Inquiry **River of Mud?**

1. Heavy rains loosened the sediment on this mountain. Eventually the land collapsed and caused a river of mud to flow downhill. How do you think something like this can affect people?

Mass Wasting

Have you ever seen or heard a news report about a large pile of boulders that has fallen down a mountain onto a road? This is an example of a mass wasting event. **Mass wasting** *is the downhill movement of a large mass of rocks or soil because of the pull of gravity.* There are two important parts to this definition:

- material moves in bulk as a large mass
- gravity is the dominant cause of movement. For example, the mass moves all at once, rather than as separate pieces over a long period of time. Also, the mass is not moved by, in, on, or under a transporting agent such as water, ice, or air.

Active Reading **2. Identify** Underline the two main characteristics of a mass wasting event.

The photo above shows a mass wasting event called a mud flow. Even though water did not transport the mud, it did contribute to this mass wasting event. Mass wasting commonly occurs when soil on a hillside is soaked with rainwater. The water-soaked soil becomes so heavy that it breaks loose and slides down the hillside.

Recall that vegetation on a steep slope reduces the amount of water erosion during a heavy rainfall. The presence of thick vegetation on a slope also reduces the likelihood of a mass wasting event. Root systems of plants help hold sediment in place. Vegetation also reduces the force of falling rain. This minimizes erosion by allowing water to gently soak into the soil.

Active Reading **3. Explain** How can vegetation help prevent mass wasting?

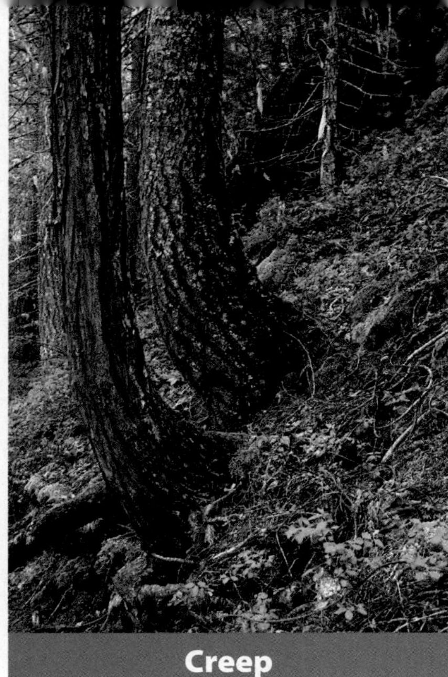

Figure 19 A rockfall, slump, and creep are examples of mass wasting.

Erosion by Mass Wasting

There are many types of mass wasting events. For example, *a* **landslide** *is the rapid downhill movement of soil, loose rocks, and boulders.* Two types of landslides are a rockfall, such as the one shown in **Figure 19,** and a mudslide, shown on the previous page. Slump is a type of mass wasting where the material moves slowly, in a large mass. If the material moves too slowly to be noticeable, causing trees and other objects to lean over, the event is called creep, also shown in **Figure 19.**

Active Reading **4. Examine** What evidence do you see in the figure that mass wasting has occurred?

The amount of erosion that occurs during a mass wasting event depends on factors such as the type of rock, the amount of water in the soil, and how strongly the rock and soil are held together. Erosion also tends to be more destructive when the mass wasting occurs on steep slopes. For example, landslides on a steep hillside can cause extensive damage because they transport large amounts of material quickly.

5. NGSSS Check Identify What are some ways gravity shapes Earth's surface? SC.6.E.6.1

Deposition by Mass Wasting

The erosion that occurs during mass wasting continues as long as gravity is greater than other forces holding the rock and soil in place. But when the material reaches a stable location, such as the base of a mountain, the material is deposited. **Talus** *is a pile of angular rocks and sediment from a rockfall,* like the pile of rock at the base of the hill in **Figure 19.**

Math Skills MA.6.A.3.6

Use Ratios

Slope is the ratio of the change in vertical height over the change in horizontal distance. The slope of the hill in the drawing is

$$\frac{(108\ \text{m} - 100\ \text{m})}{40\ \text{m}} = \frac{8\ \text{m}}{40\ \text{m}} = 0.2$$

Multiply the answer by 100 to calculate percent slope.

$$0.2 \times 100 = 20\%$$

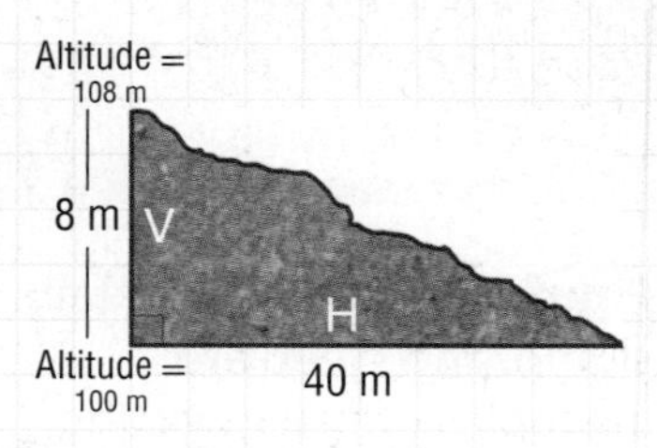

6. Practice

A mountain rises from 380 m to 590 m over a horizontal distance of 3,000 m. What is its percent slope?

Figure 20 Building on steep slopes can increase the risk of a landslide. Construction or removal of vegetation makes the hillside even less stable.

Land Use Practices

Human activities can affect both the severity of mass wasting and the tendency for it to occur. The homes in **Figure 20** were built on steep and unstable slopes and were damaged during a landslide. Removing vegetation increases soil erosion and can promote mass wasting. The use of heavy machines or blasting can shake the ground and trigger mass wasting. In addition, landscaping can make a slope steeper. A steep slope is more likely to undergo mass wasting. Because Florida is generally flat, landslides in Florida are rare. The only documented landslide in Florida occured in 1948 in Gasden County.

Active Reading **7. Summarize** What are some ways human activities can increase or decrease the risk of mass wasting?

Apply It! After you complete the lab, answer this question.

1. How could you change your model to make a slope more stable?

Glacial Erosion and Deposition

You have read about erosion and deposition caused by mass wasting events. Glaciers can also cause erosion and deposition. *A* **glacier** *is a large mass of ice that formed on land and moves slowly across Earth's surface.* Glaciers form on land in areas where the amount of snowfall is greater than the amount of snowmelt.

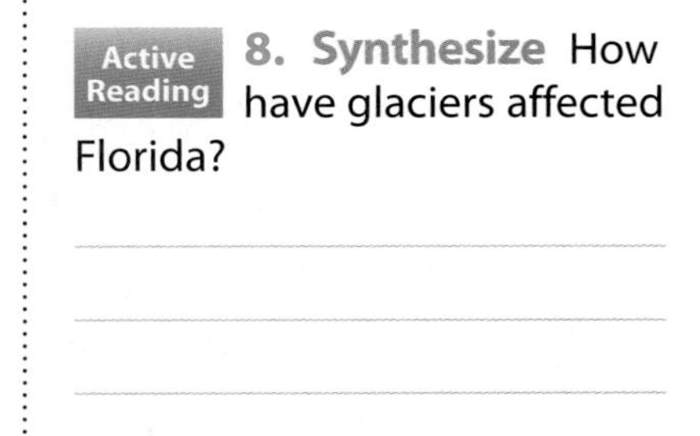

8. Synthesize How have glaciers affected Florida?

There are two main types of glaciers—alpine glaciers and ice sheets. Alpine glaciers form in mountains and flow downhill. Ice sheets cover large areas of land and move outward from central locations. Continental ice sheets exist today on Antarctica and Greenland.

Glaciers and Florida

You might not think glaciers have affected Florida because they form mainly in polar regions. But glaciers have affected Florida's coastline. During the last ice age, so much of Earth's water was frozen in glaciers that sea level fell. Florida's coastline was 92 m deeper than it is today. Rising and falling sea levels associated with periods of glaciation also influenced the cycles of erosion and deposition along Florida's coast. This, in turn, affected the rocks that formed in Florida. The Florida Keys formed 150,000 years ago, when sea level was higher than it is today. When sea level dropped during a period of glaciation, the islands were exposed.

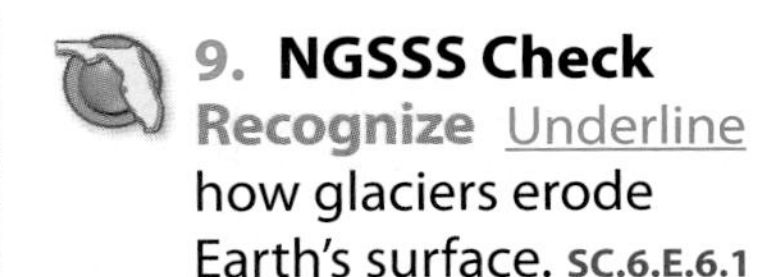

9. NGSSS Check
Recognize Underline how glaciers erode Earth's surface. **SC.6.E.6.1**

Glacial Erosion

Glaciers erode Earth's surface as they slide over it. They act as bulldozers, carving the land as they move. Rocks and grit frozen within the ice create grooves and scratches on underlying rocks. This is similar to the way sandpaper scratches wood. Alpine glaciers produce distinctive erosional features like the ones shown in **Figure 21.** Notice the U-shaped valleys that glaciers carved through the mountains.

Figure 21 Alpine glaciers produce distinctive erosion features.

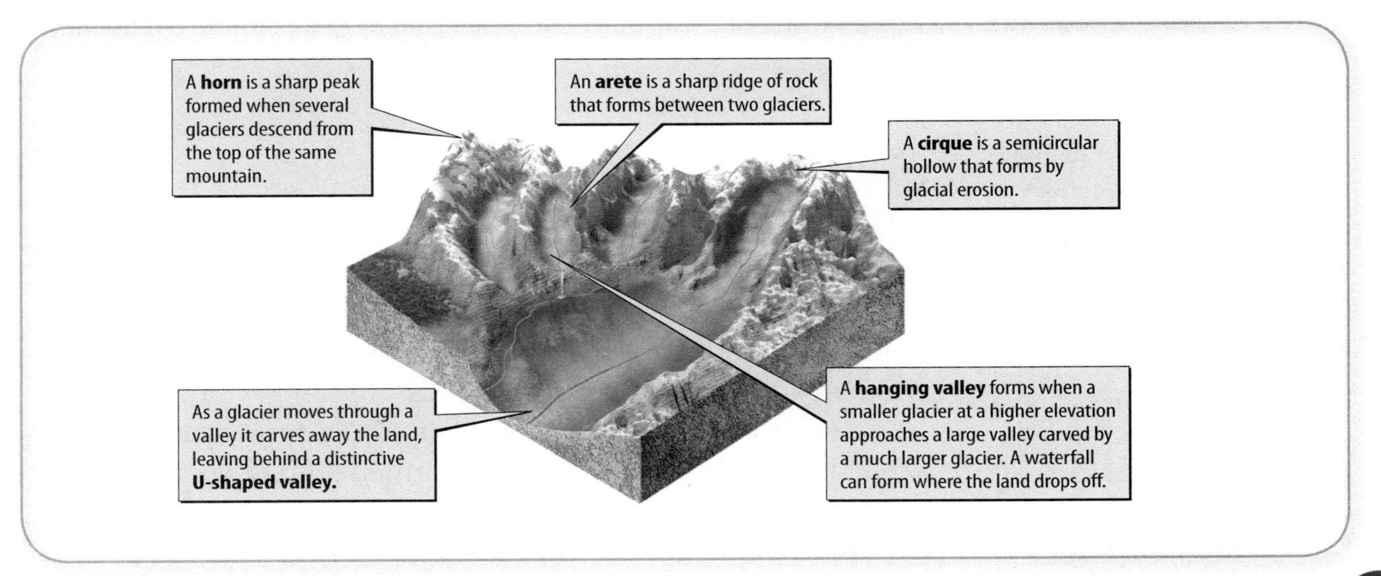

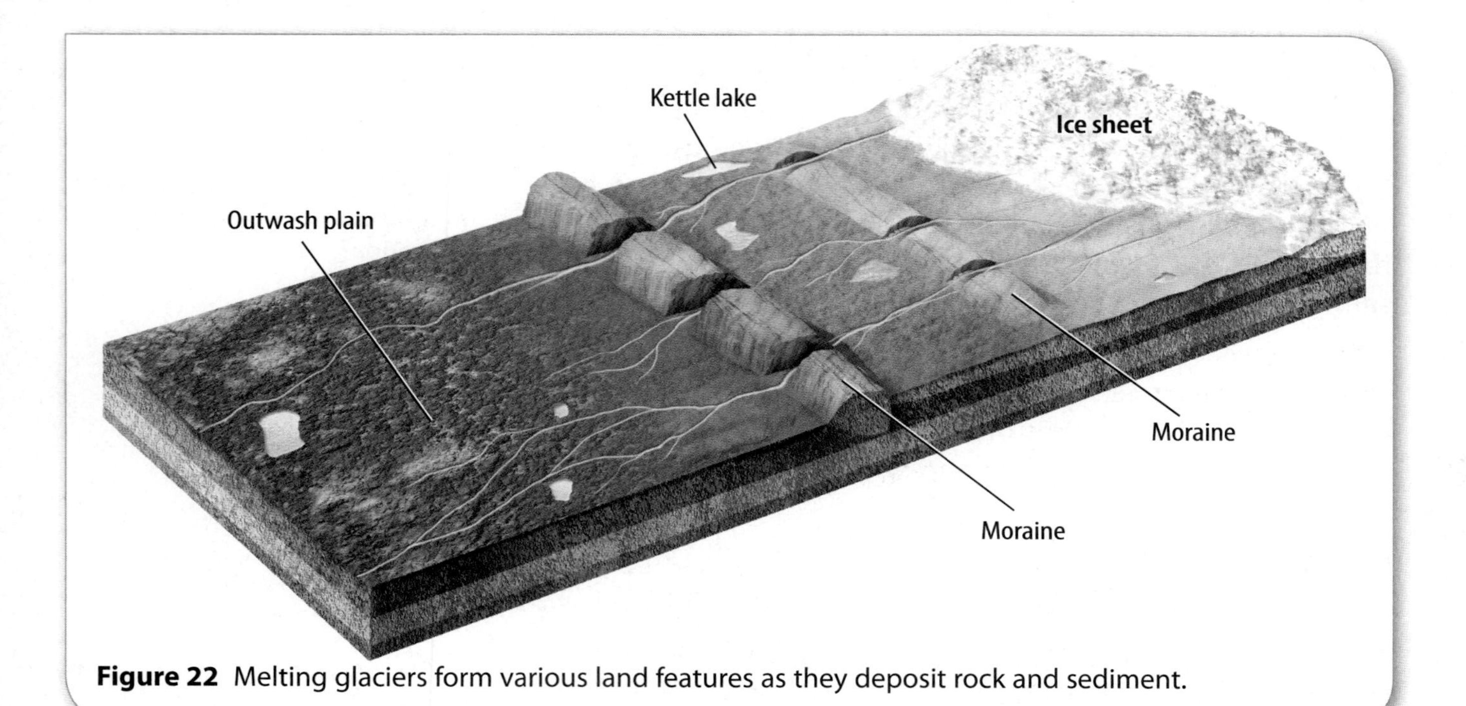

Figure 22 Melting glaciers form various land features as they deposit rock and sediment.

Glacial Deposition

Glaciers slowly melt as they move down from high altitudes or when the climate in the area warms. Sediment that was once frozen in the ice eventually is deposited in various forms, as illustrated in **Figure 22. Till** *is a mixture of various sizes of sediment deposited by a glacier.* Deposits of till are poorly sorted. They commonly contain particles that range in size from boulders to clay. Till often piles up along the sides and fronts of glaciers. It can be shaped and streamlined into many features by the moving ice. For example, *a* **moraine** *is a mound or ridge of unsorted sediment deposited by a glacier.* **Outwash** *is layered sediment deposited by streams of water that flow from a melting glacier.* Outwash consists mostly of well-sorted sand and gravel.

Science Use v. Common Use

till

Science Use rock and sediment deposited by a glacier

Common Use to work by plowing, sowing, and raising crops

Word Origin

moraine

from French *morena,* means "mound of earth"

Active Reading **10. Compare** How does outwash differ from a moraine?

Land Use Practices

At first, it might not seem that human activities affect glaciers. But in some ways, the effects are more significant than they are for other forms of erosion and deposition. For example, human activities contribute to global warming—the gradual increase in Earth's average temperature. This can cause considerable melting of glaciers. Glaciers contain about two-thirds of all the freshwater on Earth. As glaciers melt, sea level rises around the world and coastal flooding is possible.

Lesson Review 3

Visual Summary

Mass wasting can occur very fast, such as when a landslide occurs, or slowly over many years.

Material moved by a mass wasting event is deposited when it reaches a relatively stable location. An example is talus deposited at the base of this hill.

A glacier erodes Earth's surface as it moves and melts. Glaciers can form U-shaped valleys when they move past mountains.

Use Vocabulary

1. **Define** *mass wasting* in your own words. **SC.6.E.6.1**

2. **Use the term** *talus* in a sentence.

3. Erosion by the movement of a(n) __________ can produce a U-shaped valley. **SC.6.E.6.1**

Understand Key Concepts

4. Which is the slowest mass wasting event? **SC.6.E.6.1**
   - (A) creep
   - (B) landslide
   - (C) rockfall
   - (D) slump

5. **Classify** each of the following as features of either erosion or deposition: (a) arete, (b) outwash, (c) cirque, and (d) till. **SC.6.E.6.1**

Interpret Graphics

6. **Compare and Contrast** Fill in the table below to compare and contrast moraine and outwash. **LA.6.2.2.3**

Similarities	Differences

Critical Thinking

7. **Compose** a list of evidence for erosion and deposition that you might find in a mountain park that would indicate that glaciers once existed in the area. **SC.6.E.6.1**

Math Skills **MA.6.A.3.6**

8. A mountain's base is 2,500 m high. The peak is 3,500 m high. The horizontal distance covers 4,000 m. What is the percent slope?

Chapter 3 Study Guide

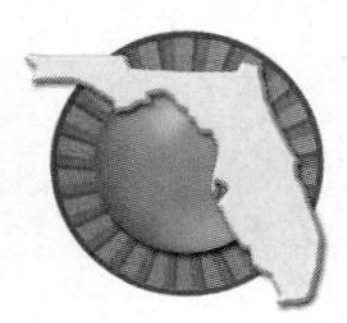

Think About It! Erosion and deposition are constructive and destructive forces that shape Earth's surface by building up and tearing down landforms such as coastlines, dunes, rivers, lakes, mountains, glaciers, and deltas.

Key Concepts Summary

LESSON 1 The Erosion-Deposition Process

- Erosion is the wearing away and transportation of weathered material. Deposition is the laying down of the eroded material.
- Erosion tends to make rocks more rounded. Erosion can sort sediment according to its grain size.
- Landforms produced by deposition are usually on flat, low land. Landforms produced by erosion are often tall and/or jagged.

Vocabulary

erosion p. 77
deposition p. 79

LESSON 2 Landform Shaped by Water and Wind

- Deforestation, desertification, habitat destruction, and increased rates of extinction are associated with using land as a resource.
- Landfills are constructed to prevent contamination of soil and water by pollutants from waste. Hazardous waste must be disposed of in a safe manner.
- Positive impacts on land include preservation, reforestation, and reclamation.

meander p. 86
longshore current p. 87
delta p. 88
abrasion p. 90
dune p. 90
loess p. 90

LESSON 3 Mass Wasting and Glaciers

- Gravity can shape Earth's surface through mass wasting. Creep is an example of mass wasting.
- A glacier erodes Earth's surface as it moves by carving grooves and scratches into rock.

mass wasting p. 94
landslide p. 95
talus p. 95
glacier p. 97
till p. 98
moraine p. 98
outwash p. 98

Active Reading FOLDABLES® Chapter Project

Assemble your lesson Foldables as shown to make a Chapter Project. Use the project to review what you have learned in this chapter.

Use Vocabulary

1. Water moving sediment down slopes and a glacier forming a U-shaped valley as it moves past mountains are examples of ______________.

2. Wind has less energy as it slows, and ______________ of sediment occurs.

3. The grinding of rock as water, wind, or glaciers move sediment is ______________.

4. An apron of sediment known as a(n) ______________ forms where a stream enters a lake or an ocean.

5. A landslide and creep are types of ______________.

6. A large pile of rocks formed from a rockfall is ______________.

Link Vocabulary and Key Concepts

Use vocabulary items from the previous page to complete the concept map.

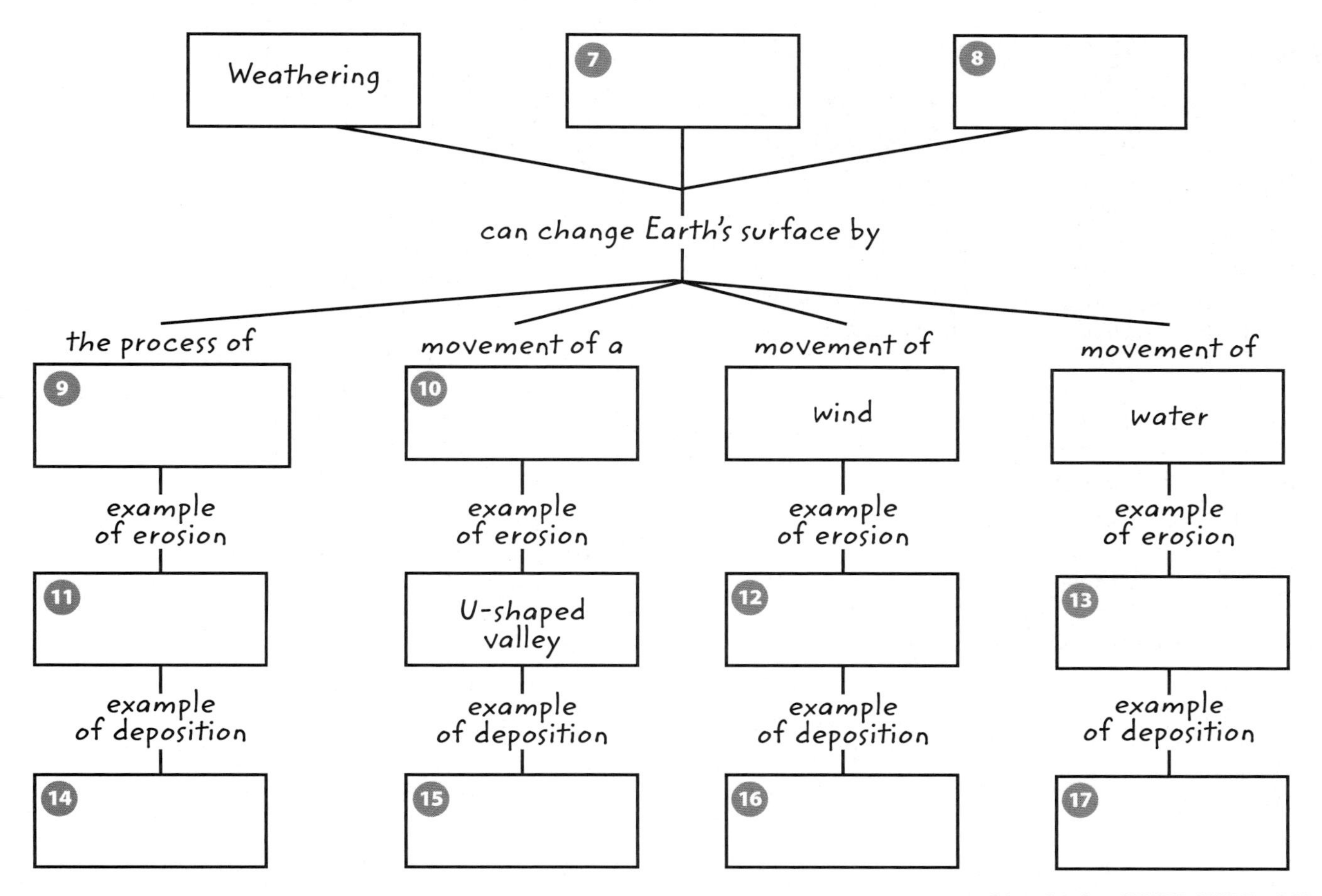

Chapter 3 Review

Fill in the correct answer choice.

Understand Key Concepts

1 Which is a structure created mostly by deposition? **SC.6.E.6.1**

Ⓐ cirque
Ⓑ hoodoo
Ⓒ sandbar
Ⓓ slump

2 Which shows an example of sediment that is both poorly rounded and well sorted? **SC.6.E.6.1**

Ⓐ
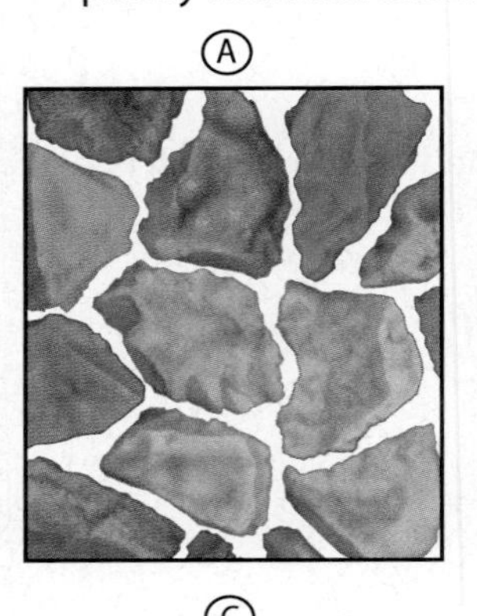

Ⓑ
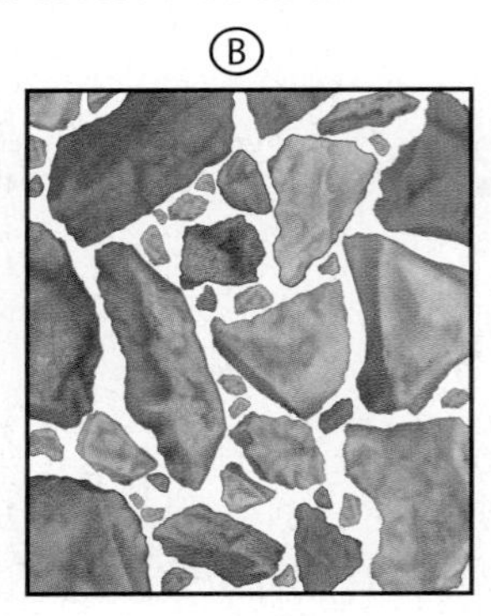

Ⓒ
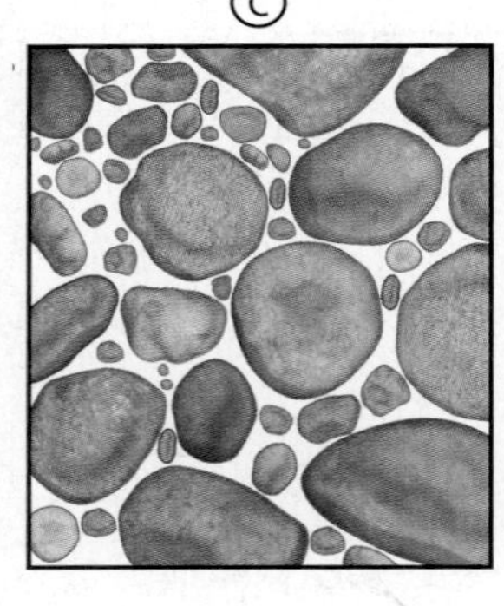

Ⓓ
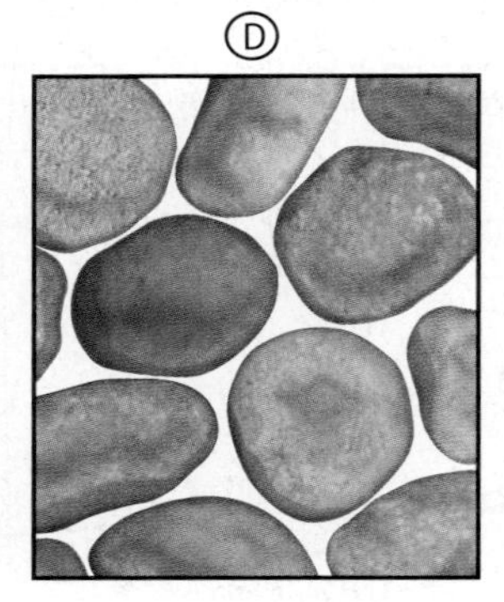

3 Which is typically a low-energy depositional environment? **SC.6.E.6.1**

Ⓐ a fast-moving river
Ⓑ an ocean shore with waves
Ⓒ a stream with meanders
Ⓓ a swamp with decaying trees

4 Which would most likely produce a moraine? **SC.6.E.6.1**

Ⓐ a glacier
Ⓑ an ocean
Ⓒ a river
Ⓓ the wind

Critical Thinking

5 **Describe** one erosion feature and one deposition feature you might expect to find (a) in a valley, (b) in a desert, and (c) high in the mountains. **SC.6.E.6.1**

6 **Classify** these landforms as formed mostly by erosion or deposition: (a) cirque, (b) sand dune, (c) alluvial fan, (d) hoodoo. **SC.6.E.6.1**

7 **Construct** On a separate piece of paper draw a chart that lists three careless land uses that result in mass wasting that could be dangerous to humans. Include in your chart details about how each land use could be changed to be safer. **LA.6.2.2.3**

8 **Produce** a list of at least three hazardous erosion or deposition conditions that would be worse during a particularly stormy, rainy season. **SC.6.E.6.1**

9 **Contrast** the rounding and sorting of sediment caused by a young stream to that caused by an old stream. **SC.6.E.6.1**

Writing in Science

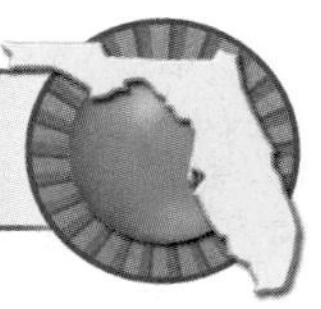

10 **Write** Imagine you are planning to build a home on a high cliff overlooking the sea. On a separate piece of paper, write a paragraph that assesses the potential for mass wasting along the cliff. Describe at least four features that would concern you. **SC.6.E.6.1**

Big Idea Review

11 How do erosion and deposition shape Earth's surface? **SC.6.E.6.1**

12 The photo at the beginning of the chapter shows a landform in Arizona known as The Wave. Explain how erosion and deposition might have produced this landform. **SC.6.E.6.1**

Math Skills MA.6.A.3.6

Use Ratios

13 **Calculate** the average percent slope of the mountains in parts **a** and **b.**

a. Mountain A rises from 3,200 m to 6,700 m over a horizontal distance of 10,000 m.

b. Mountain B rises from 1,400 m to 9,400 m over a horizontal distance of 2.5 km.

c. If mountains A and B are composed of the same materials, which mountain is more likely to experience mass wasting?

14 If the slope of a hill is 10 percent, how many meters does the hill rise for every 10 m of horizontal distance?

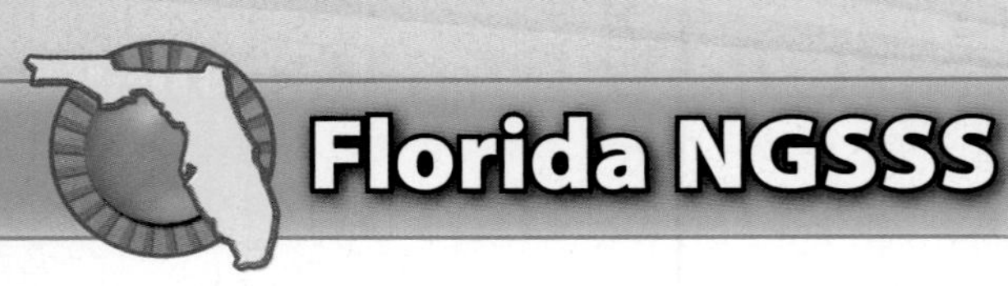

Fill in the correct answer choice.

Multiple Choice

1 Which landform is created by deposition? SC.6.E.6.1

Ⓐ alluvial fan

Ⓑ glacial valley

Ⓒ mountain range

Ⓓ river channel

Use the diagram below to answer question 2.

2 Which process formed the features shown in the diagram above? SC.7.E.6.1

Ⓕ A stream eroded and deposited sediment.

Ⓖ Groundwater deposited minerals in a cavern.

Ⓗ Groundwater dissolved several layers of rock.

Ⓘ Wind and ice wore away soft sedimentary rock.

3 Which causes movement in mass wasting? SC.6.E.6.1

Ⓐ gravity

Ⓑ ice

Ⓒ magnetism

Ⓓ wind

4 Which typically is NOT a depositional environment? SC.7.E.6.1

Ⓕ delta

Ⓖ mountain peak

Ⓗ ocean floor

Ⓘ swamp

Use the diagram below to answer questions 5 and 6.

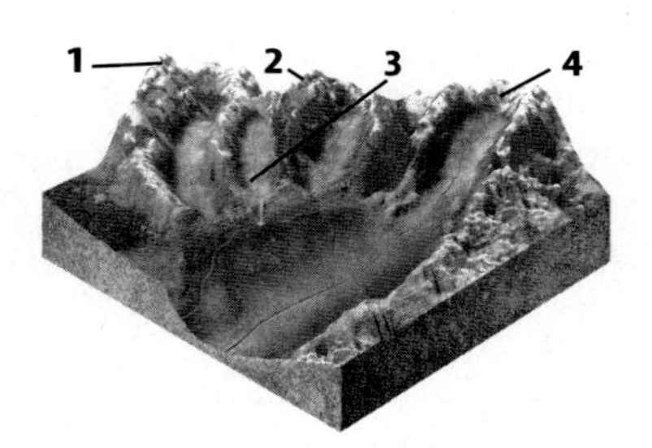

5 Which landform on the diagram above is a cirque? SC.6.E.6.1

Ⓐ 1

Ⓑ 2

Ⓒ 3

Ⓓ 4

6 How did structure 1 form in the diagram above? SC.6.E.6.1

Ⓕ A glacier deposited a large amount of land as it moved.

Ⓖ A small glacier approached a valley carved by a large glacier.

Ⓗ Several glaciers descended from the top of the same mountain.

Ⓘ Two glaciers formed on either side of a ridge.

7 Which agent of erosion can create a limestone cave? SC.6.E.6.1

Ⓐ acidic water

Ⓑ freezing and melting ice

Ⓒ growing plant roots

Ⓓ gusty wind

8 Which deposit does mass wasting create? SC.6.E.6.1

Ⓕ loess

Ⓖ outwash

Ⓗ talus

Ⓘ till

Multiple Choice *Bubble the correct answer.*

1. Look at the image above. Which of the following statements is true of this type of stream? **SC.6.E.6.1**

Ⓐ Water carves rock to form deep valleys.

Ⓑ Water flows rapidly downhill.

Ⓒ Water erodes banks faster than the stream bottom erodes.

Ⓓ Water flows slowly through wide, open areas.

2. A large deposit of sediment that forms where a stream enters a large body of water is called a **SC.6.E.6.2**

Ⓕ dam.

Ⓖ delta.

Ⓗ groin.

Ⓘ levee.

3. A landform with rocks that look like they have been polished was likely created by **SC.6.E.6.1**

Ⓐ groundwater deposition.

Ⓑ surface erosion.

Ⓒ water erosion.

Ⓓ wind erosion.

4. Which image represents the flow of a longshore current in relation to a sandy beach? **SC.6.E.6.2**

Ⓕ
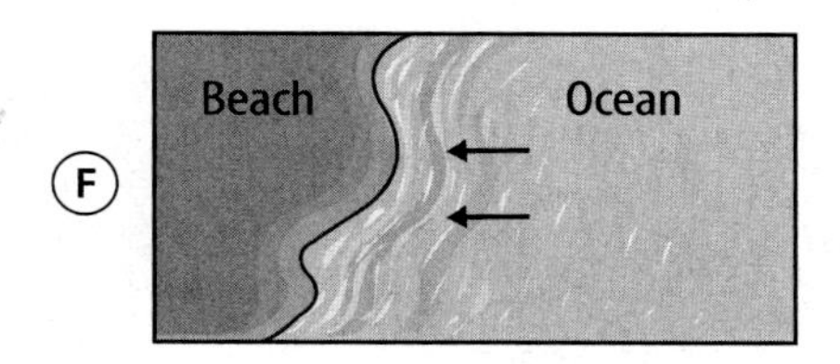

Ⓖ
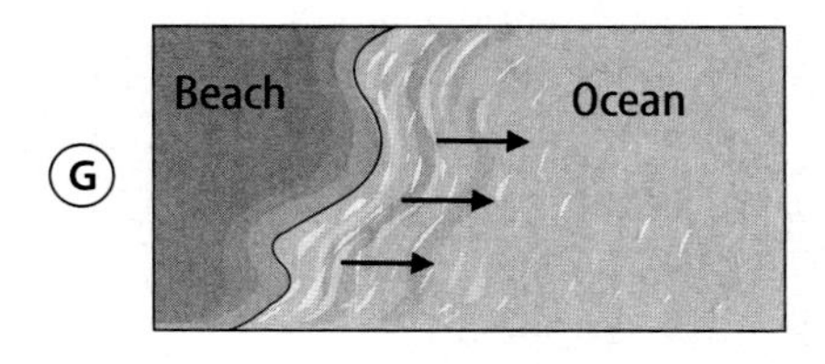

Ⓗ
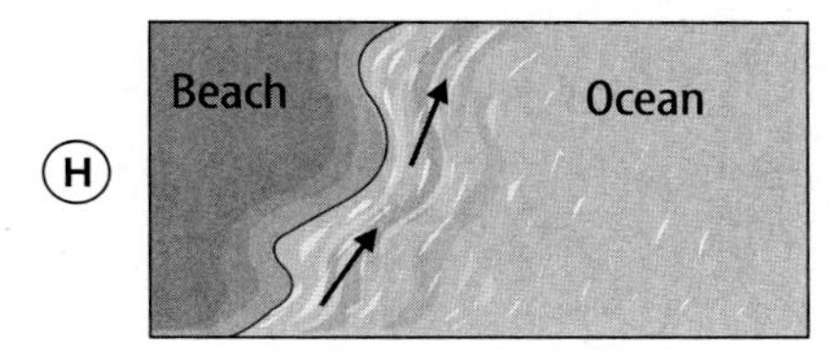

Ⓘ
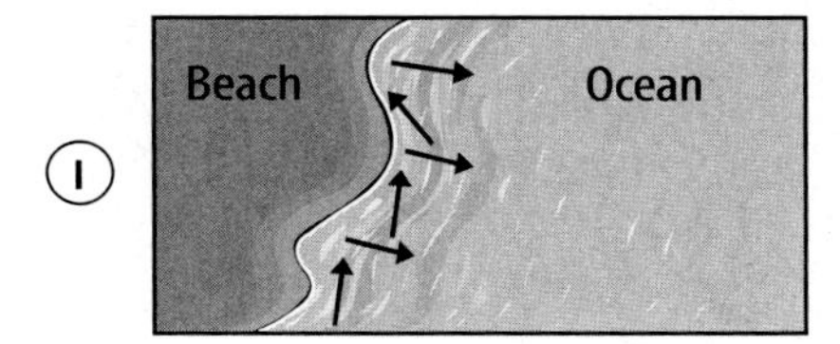

Multiple Choice *Bubble the correct answer.*

1. Which image shows mass wasting that is least likely to be triggered by an earthquake? **SC.6.E.6.1**

Ⓐ 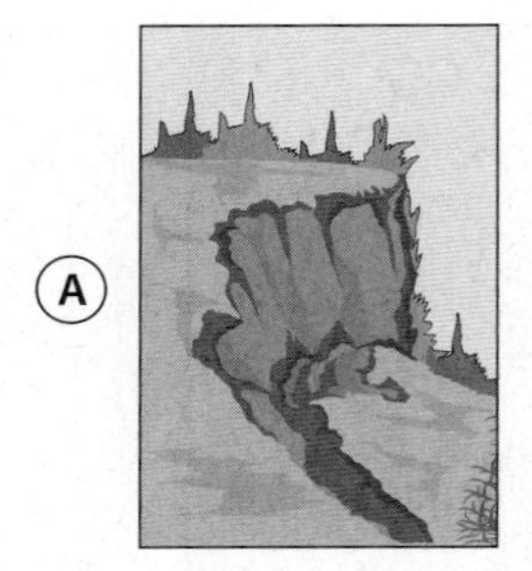

Ⓑ 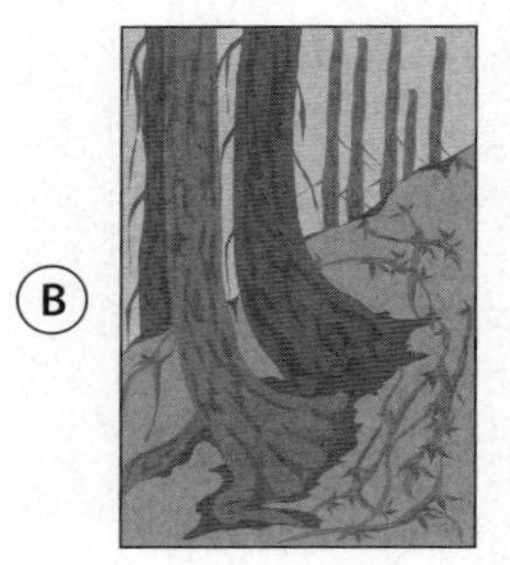

Ⓒ 

Ⓓ 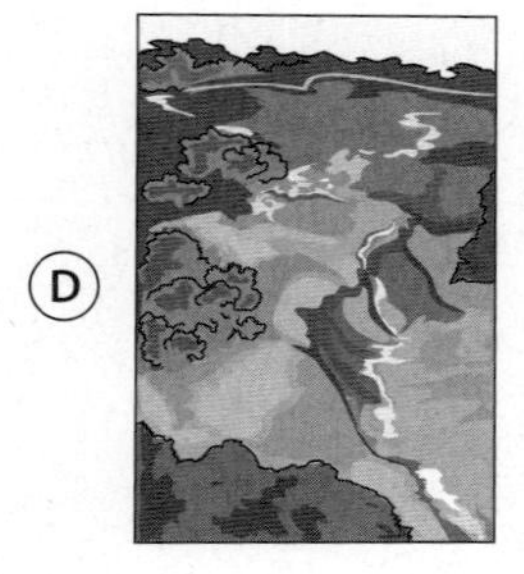

2. Which statement is true about glaciers? **SC.6.E.6.2**

Ⓕ Glaciers are fast-moving masses of ice.

Ⓖ Glaciers tend to be motionless.

Ⓗ Glaciers are slow-moving masses of ice.

Ⓘ Glaciers form on land where snow melts.

3. Darrin identifies a deposit of talus near the side of a landform. He determines that the deposit was formed by **SC.6.E.6.1**

Ⓐ creep.

Ⓑ mudslide.

Ⓒ rockfall.

Ⓓ slump.

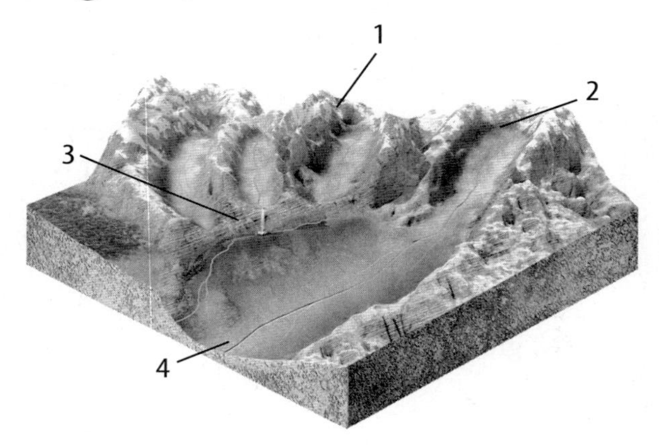

4. Look at the image above. Which type of glacial erosion is represented by 1? **SC.6.E.6.1**

Ⓕ arête

Ⓖ cirque

Ⓗ hanging valley

Ⓘ U-shaped valley

Name ______________________ Date ______________

Notes

1441
Prince Munjong of Korea invents the first rain gauge to gather and measure the amount of liquid precipitation over a period of time.

1450
The first anemometer, a tool to measure wind speed, is developed by Leone Battista Alberti.

1643
Italian physicist Evangelista Torricelli invents the barometer to measure pressure in the air. This tool improves meteorology, which relied on simple sky observations.

1714
German physicist Daniel Fahrenheit develops the mercury thermometer, making it possible to measure temperature.

1752
Swedish astronomer Andres Celsius proposes a centigrade temperature scale where 0° is the freezing point of water and 100° is the boiling point of water.

1800 — 1900 — 2000

1806
Francis Beaufort creates a system for naming wind speeds and aptly names it the Beaufort Wind Force Scale. This scale is used mainly to classify sea conditions.

1960
TIROS 1, the world's first weather satellite, is sent into space equipped with a TV camera.

1964
The U.S. National Severe Storms Laboratory begins experimenting with the use of Doppler radar for weather-monitoring purposes.

2006
Meteorologists hold 8,800 jobs in the United States alone. These scientists work in government and private agencies, in research services, on radio and television stations, and in education.

Inquiry
Visit ConnectED for this unit's **STEM** activity.

Unit 2

Nature of SCIENCE

Patterns

You might see the Moon as a large, glowing disk in the night sky. At other times, the Moon appears as a crescent. These shapes are the Moon's phases, or the changing portions of the Moon that are seen from Earth. The Moon's phases occur as a repeating pattern about every 28 days. A **pattern** is a consistent plan or model used as a guide for understanding and predicting things. You can predict the next phase of the Moon or you can determine the previous phase of the Moon if you know the pattern.

Active Reading **1. Differentiate** Underline the term *pattern* and its definition.

Patterns in Earth Science

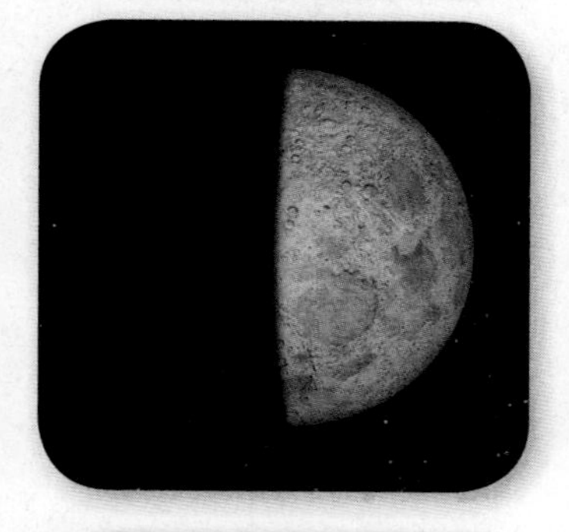

Patterns help scientists understand observations. This allows them to predict future events or understand past events. Geologists are Earth scientists who measure the chemical composition, age, and location of rocks. They look for patterns in these measurements to propose what processes formed rocks. Geologists also use patterns to draw conclusions about how Earth has changed over time and to estimate how it will change in the future.

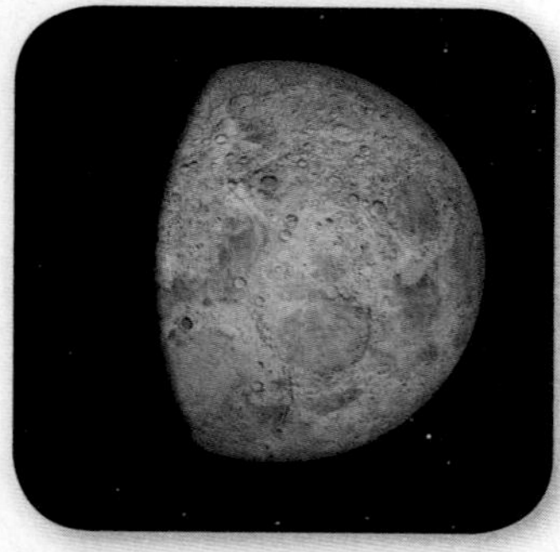

Meteorologists are scientists who study weather and climate. They study patterns of fronts, winds, cloud formation, precipitation, and ocean temperatures to make weather forecasts. They track patterns in hurricanes, such as wind speed, movements, and rotation velocity. Predicting the strength and the path of a storm can help save lives, buildings, and property. Meteorologists use weather patterns to predict the severity of a storm and when and where it will hit. Then they can send advance warnings for safe preparation.

Active Reading **2. Determine** Suggest ways patterns are useful to scientists.

Types of Patterns

Physical Patterns

A pattern you observe with your senses or scientific instruments is a physical pattern. Earthquake uplift and erosion reveal physical patterns in layers of rock. These patterns tell geologists many things, including the order in which the rocks formed, the different minerals and fossils the rocks contain, and the relative age and movement of landforms.

Patterns in Graphs

Scientists plot data on graphs and then analyze the graphs for patterns. This graph shows a pattern in sea level as it increased between 1994 and 2010. Graphic patterns help to predict events.

Active Reading **3. Predict** How might the pattern in the graph help to predict sea level in 2018?

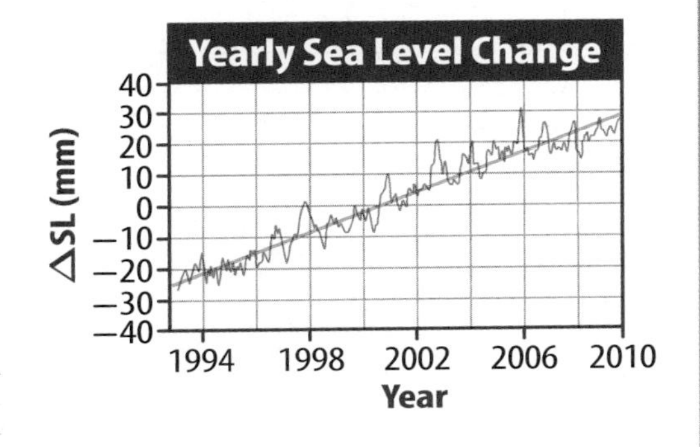

Cyclic Patterns

An event that repeats in a predictable order, such as the phases of the Moon, has a cyclic pattern. Water temperatures in both the North and South Atlantic Ocean rise and fall equally each year. The annual changes in the water temperature follow a cyclic pattern. How do the temperature patterns in the two oceans differ?

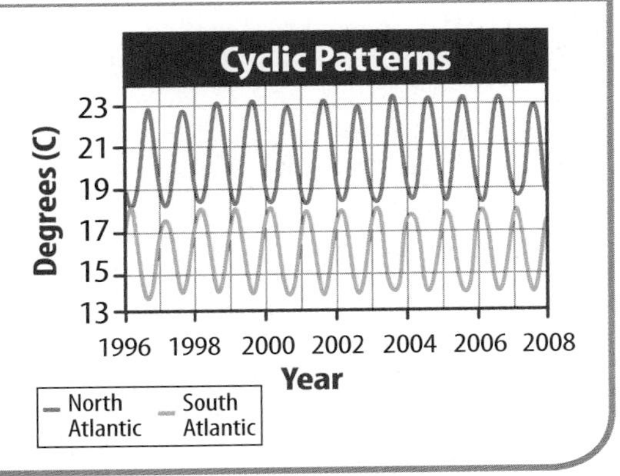

Inquiry LAB STATION SC.6.N.1.5 **Try It!**

MiniLab ***What patterns are in your year?*** at connectED.mcgraw-hill.com

Apply It! After you complete the lab, answer these questions.

1. **Analyze** Discuss how patterns are useful to scientists who study volcanoes, migration, chemical reactions, radio waves, or solar flares.

2. **Apply** Summarize how patterns are useful in a work environments for firefighters, surgeons, pilots, animal trainers, forest rangers, automobile engineers, or building construction workers.

Name ______________________ Date ____________

Notes

Temperature Changes in the Atmosphere

Six friends hiked to the top of a tall mountain. They noticed the air was cooler on top of the mountain than at the bottom of the mountain. They wondered what happens to the atmosphere's temperature the higher and higher it is above Earth's surface.

Frank: I think the farther the atmosphere is from Earth's surface, the colder it is.

Julia: I think the farther the atmosphere is from Earth's surface, the warmer it is.

Juanita: I think the atmosphere gradually cools to a certain altitude, then it gradually warms the farther it is from Earth's surface.

Tyson: I think the atmosphere gradually warms to a certain altitude, then it gradually cools the farther it is from Earth's surface.

Tilly: I think the atmosphere's temperature changes with each layer of the atmosphere. I think it can cool, warm, cool again, and warm again.

Lyndon: I think the atmosphere cools for a while, but once it is at a certain altitude, it is at a constant temperature.

Whom do you agree with the most? Explain why you agree.

Earth's ATMOSPHERE

The Big Idea

FLORIDA BIG IDEAS

1 **The Practice of Science**

2 **The Characteristics of Scientific Knowledge**

3 **The Role of Theories, Laws, Hypotheses, and Models**

6 **Earth Structures**

7 **Earth Systems and Patterns**

Think About It!

How does Earth's atmosphere affect life on Earth?

Earth's atmosphere is made up of gases and small amounts of liquid and solid particles. Earth's atmosphere surrounds and sustains life.

1. What type of particles do you think make up clouds in the atmosphere?

2. How might conditions in the atmosphere change as height above sea level increases?

3. How do you think Earth's atmosphere affects life on Earth?

Get Ready to Read

What do you think about Earth's atmosphere?

Before you read, decide if you agree or disagree with each of these statements. As you read this chapter, see if you change your mind about any of the statements.

	AGREE	DISAGREE
1 Air is empty space.		
2 Earth's atmosphere is important to living organisms.		
3 All the energy from the Sun reaches Earth's surface.		
4 Earth emits energy back into the atmosphere.		
5 Uneven heating in different parts of the atmosphere creates air circulation patterns.		
6 Warm air sinks and cold air rises.		
7 If no humans lived on Earth, there would be no air pollution.		
8 Pollution levels in the air are not measured or monitored.		

There's More Online!
Video • Audio • Review • ⓘLab Station • WebQuest • Assessment • Concepts in Motion • Multilingual eGlossary

Lesson 1

Describing Earth's ATMOSPHERE

ESSENTIAL QUESTIONS

 How did Earth's atmosphere form?

 What is Earth's atmosphere made of?

 What are the layers of the atmosphere?

 How do air pressure and temperature change as altitude increases?

Vocabulary

atmosphere p. 119
water vapor p. 120
troposphere p. 122
stratosphere p. 122
ozone layer p. 122
ionosphere p. 123

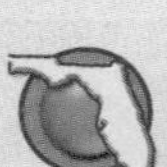

Florida NGSSS

LA.6.2.2.3 The student will organize information to show understanding (e.g., representing main ideas within text through charting, mapping, paraphrasing, summarizing, or comparing/contrasting); describe linear functions and other simple relations using both common language and algebraic notation;

LA.6.4.2.2 The student will record information (e.g., observations, notes, lists, charts, legends) related to a topic, including visual aids to organize and record information and include a list of sources used;

SC.6.E.7.9 Describe how the composition and structure of the atmosphere protects life and insulates the planet.

SC.6.N.1.4 Discuss, compare, and negotiate methods used, results obtained, and explanations among groups of students conducting the same investigation.

SC.6.N.2.1 Distinguish science from other activities involving thought.

Launch Lab

SC.6.N.2.1

20 minutes

Where does air apply pressure?

With the exception of Mercury, most planets in the solar system have some type of atmosphere. However, Earth's atmosphere provides what the atmospheres of other planets cannot: oxygen and water. Oxygen, water vapor, and other gases make up the gaseous mixture in the atmosphere called air. In this activity, you will explore air's effect on objects on Earth's surface.

Procedure

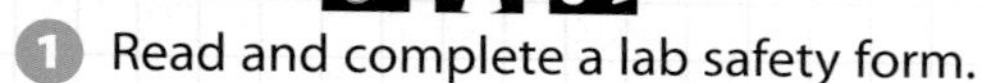

1. Read and complete a lab safety form.
2. Add **water** to a **cup** until it is two-thirds full.
3. Place a large **index card** over the opening of the cup so that it is completely covered.
4. Hold the cup over a tub or a large bowl.
5. Place one hand on the index card to hold it in place as you quickly turn the cup upside down. Remove your hand.

Think About This

1. What happened when you turned the cup over?

2. How did air play a part in your observation?

3. Key Concept How do you think these results might differ if you repeated the activity in a vacuum?

Why is the atmosphere important?

1. What do you think Earth would be like without its atmosphere? Earth's surface would be scarred with craters created from the impact of meteorites. Earth would experience extreme daytime-to-nighttime temperature changes. How would changes in the atmosphere affect life? What effect would atmospheric changes have on weather and climate?

Importance of Earth's Atmosphere

The photo above shows Earth's atmosphere as seen from space. How would you describe the atmosphere? *The* **atmosphere** (AT muh sfihr) *is a thin layer of gases surrounding Earth.* Earth's atmosphere is hundreds of kilometers high. However, when compared to Earth's size, it is about the same relative thickness as an apple's skin to an apple.

The atmosphere contains the oxygen, carbon dioxide, and water necessary for life on Earth. Earth's atmosphere also acts like insulation on a house. It helps keep temperatures on Earth within a range in which living organisms can survive. Without it, daytime temperatures would be extremely high and night-time temperatures would be extremely low.

The atmosphere helps protect living organisms from some of the Sun's harmful rays. It also helps protect Earth's surface from being struck by meteors. Most meteors that fall toward Earth burn up before reaching Earth's surface. Friction with the atmosphere causes them to burn. Only the very largest meteors strike Earth.

Active Reading **2. Recall** Underline why Earth's atmosphere is important to life on Earth.

WORD ORIGIN

atmosphere
from Greek *atmos,* means "vapor"; and Latin *sphaera,* means "sphere"

Review Vocabulary

liquid
matter with a definite volume but no definite shape that can flow from one place to another

Origins of Earth's Atmosphere

Most scientists agree that when Earth formed, it was a ball of molten rock. As Earth slowly cooled, its outer surface hardened. Erupting volcanoes emitted hot gases from Earth's interior. These gases surrounded Earth, forming an atmosphere.

Ancient Earth's atmosphere was thought to be water vapor with a little carbon dioxide (CO_2) and nitrogen. **Water vapor** *is water in its gaseous form.* This ancient atmosphere did not have enough oxygen to support life as we know it. As Earth and its atmosphere cooled, the water vapor condensed into **liquid**. Rain fell and then evaporated from Earth's surface repeatedly for thousands of years. Eventually, water accumulated on Earth's surface, forming oceans. Most of the original CO_2 that dissolved in rain is in rocks on the ocean floor. Today the atmosphere has more nitrogen than CO_2.

Earth's first organisms could undergo photosynthesis, which changed the atmosphere. Recall that photosynthesis uses light energy to produce sugar and oxygen from carbon dioxide and water. The organisms removed CO_2 from the atmosphere and released oxygen into it. Eventually the levels of CO_2 and oxygen supported the development of other organisms.

Apply It!

After you complete the lab, answer the questions below.

1. What kind of environments would cause dust to be harmful to life on Earth?

2. Why is regularly removing dust from your living area a healthy action?

Active Reading 3. **Summarize** How did Earth's present atmosphere form?

Active Reading 4. **Write** Sequence the number of each event on the time line to describe how Earth's atmosphere changed over time.

1. Photosynthetic organisms remove carbon dioxide from the air and release oxygen.
2. Water vapor cools and condenses. Rain falls, evaporates, and eventually accumulates in oceans.
3. Atmosphere contains present levels of carbon dioxide, oxygen, nitrogen, and other gases.
4. Atmosphere is mainly water vapor with a little carbon dioxide and nitrogen.

Early Atmosphere — Present Time

Figure 1 Oxygen and nitrogen make up most of the atmosphere, with the other gases making up only 1 percent.

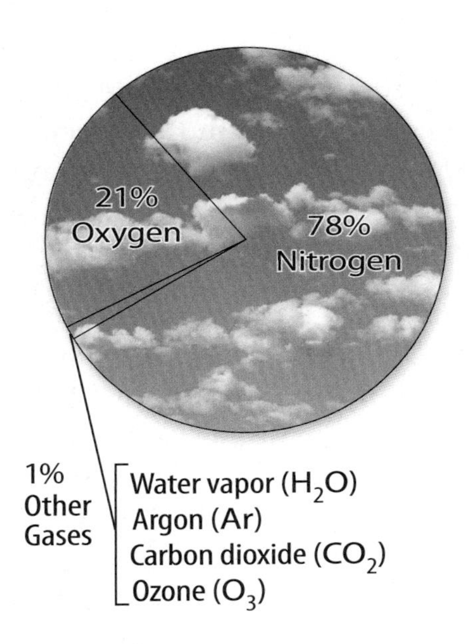

5. Visual Check **Interpret** What percent of the atmosphere is made up of oxygen and nitrogen?

Figure 2 One way solid particles enter the atmosphere is from volcanic eruptions.

Composition of the Atmosphere

Today's atmosphere is mostly made up of invisible gases, including nitrogen, oxygen, and carbon dioxide. Some solid and liquid particles, such as ash from volcanic eruptions and water droplets, are also present.

Gases in the Atmosphere

Study **Figure 1.** Which gas is the most abundant in Earth's atmosphere? Nitrogen makes up about 78 percent of Earth's atmosphere. About 21 percent of Earth's atmosphere is oxygen. Other gases, including argon, carbon dioxide, and water vapor, make up the remaining 1 percent of the atmosphere.

The amounts of water vapor, carbon dioxide, and ozone vary. The concentration of water vapor in the atmosphere ranges from 0 to 4 percent. Carbon dioxide is 0.038 percent of the atmosphere. A small amount of ozone is found at high altitudes. Ozone also occurs near Earth's surface in urban areas.

Solids and Liquids in the Atmosphere

Tiny solid particles are also in Earth's atmosphere. Many of these, such as pollen, dust, and salt, can enter the atmosphere through natural processes. **Figure 2** shows another natural source of particles in the atmosphere—ash from volcanic eruptions. Some solid particles enter the atmosphere because of human activities, such as driving vehicles that expel soot.

The most common liquid particles in the atmosphere are water droplets. Although microscopic in size, water droplets are visible when they form clouds. Other atmospheric liquids include acids that result when volcanoes erupt and fossil fuels are burned. Sulfur dioxide and nitrous oxide combine with water vapor in the air and form the acids.

6. Identify Underline what Earth's atmosphere is made of.

Layers of the Atmosphere

(km)
700
600
500
400
300
200
100
50
10
0

Exosphere
Satellite
Thermosphere
Meteor
Mesosphere
Stratosphere
Ozone layer
Weather balloon
Plane
Troposphere
Clouds

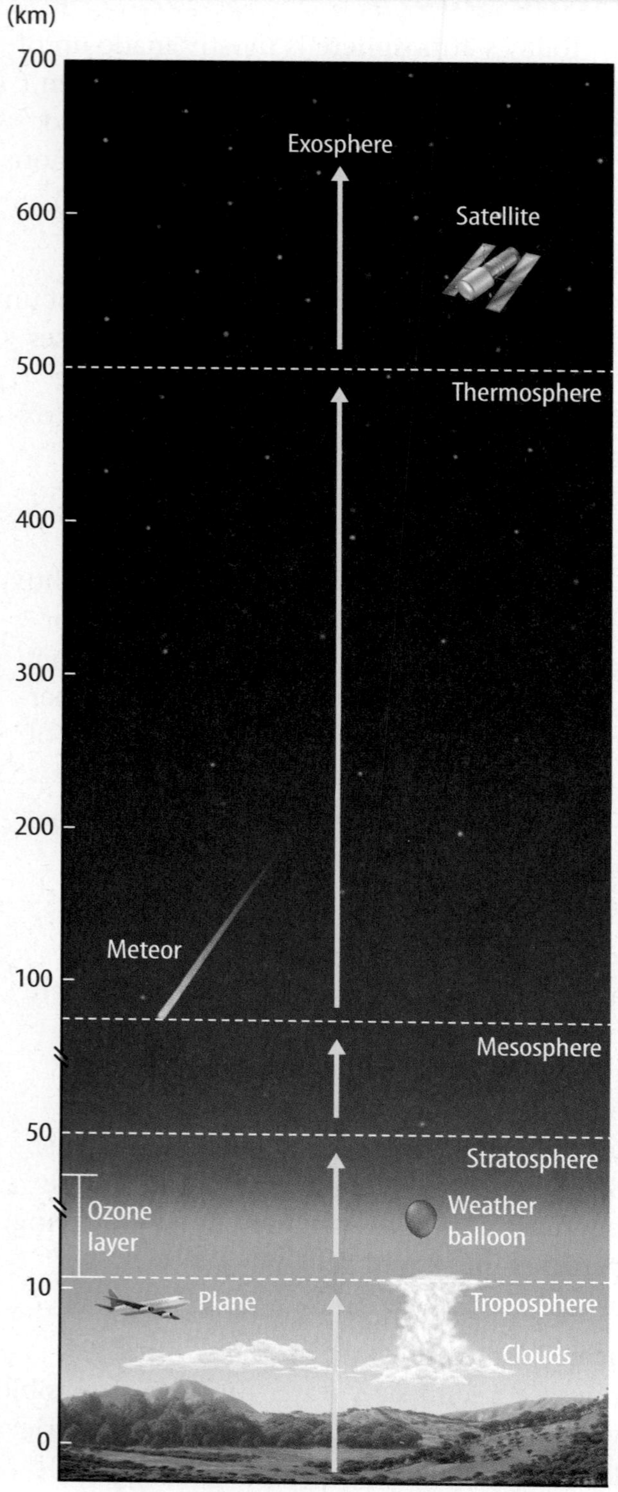

Figure 3 Scientists divide Earth's atmosphere into different layers.

7. Visual Check Identify Circle the layer of the atmosphere where planes fly.

Layers of the Atmosphere

The atmosphere has several different layers, as shown in **Figure 3.** Each layer has unique properties, including the composition of gases and how temperature changes with altitude. Notice that the scale between 0–100 km in **Figure 3** is not the same as the scale from 100–700 km. This is so all the layers can be shown in one image.

Troposphere

The atmospheric layer closest to Earth's surface is called the **troposphere** (TRO puh sfihr). Most people spend their entire lives within the troposphere. It extends from Earth's surface to altitudes between 8–15 km. Its name comes from the Greek word *tropos,* which means "change." The temperature in the troposphere decreases as you move away from Earth. The warmest part of the troposphere is near Earth's surface. This is because most sunlight passes through the atmosphere and warms Earth's surface. The warmth is radiated to the troposphere, causing weather.

Active Reading **8. Find** Highlight the height of Earth's troposphere.

Stratosphere

The atmospheric layer directly above the troposphere is the **stratosphere** (STRA tuh sfihr). The stratosphere extends from about 15 km to about 50 km above Earth's surface. The lower half of the stratosphere contains the greatest amount of ozone gas. *The area of the stratosphere with a high concentration of ozone is referred to as the* **ozone layer**. The presence of the ozone layer causes increasing stratospheric temperatures with increasing altitude.

An ozone (O_3) molecule differs from an oxygen (O_2) molecule. Ozone has three oxygen atoms instead of two. This difference is important because ozone absorbs the Sun's ultraviolet rays more effectively than oxygen does. Ozone protects Earth from ultraviolet rays that can kill plants, animals, and other organisms and cause skin cancer in humans.

Mesosphere and Thermosphere

As shown in **Figure 3,** the mesosphere extends from the stratosphere to about 85 km above Earth. The thermosphere can extend from the mesopshere to more than 500 km above Earth. Combined, these layers are much broader than the troposphere and the stratosphere, yet only 1 percent of the atmosphere's gas molecules are found in the mesosphere and the thermosphere. Most meteors burn up in these layers instead of striking Earth.

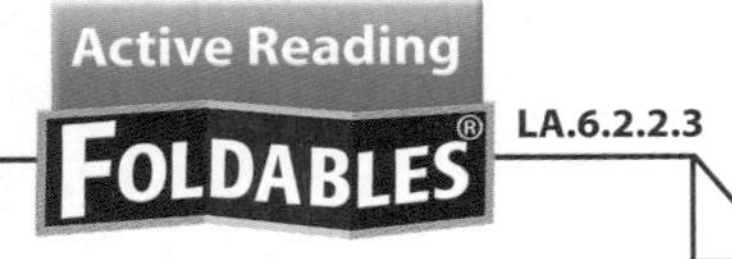

LA.6.2.2.3

Make a vertical four-tab book using the titles shown. Use it to record similarities and differences among these four layers of the atmosphere. Fold the top half over the bottom and label the outside *Layers of the Atmosphere.*

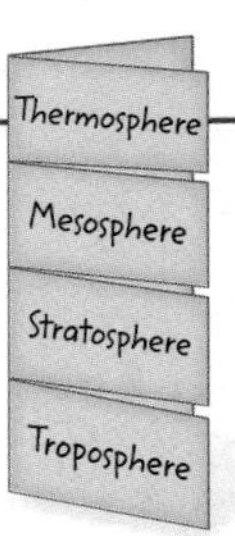

The Ionosphere *The* **ionosphere** *is a region within the mesosphere and thermosphere that contains ions.* Between 60 km and 500 km above Earth's surface, the ionosphere's ions reflect AM radio waves transmitted at ground level. After sunset when ions recombine, this reflection increases. **Figure 4** shows how AM radio waves can travel long distances, especially at night, by bouncing off Earth and the ionosphere.

Radio Waves and the Ionosphere

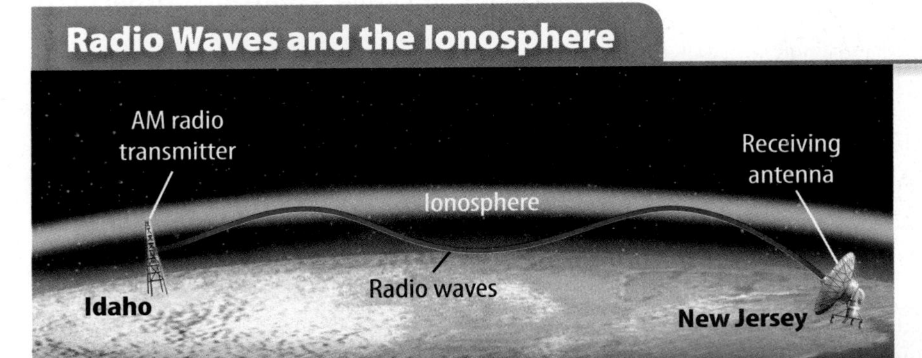

Figure 4 Radio waves can travel long distances in the atmosphere.

Auroras The ionosphere is where stunning displays of colored lights called auroras occur, as shown in **Figure 5.** Auroras are most frequent in the spring and fall but are best seen when the winter skies are dark. Auroras occur when ions from the Sun strike air molecules, causing them to emit vivid colors of light. People who live in the higher latitudes, nearer to the North Pole and the South Pole, are most likely to see auroras.

Figure 5 Auroras occur in the ionosphere.

Exosphere

The exosphere is the atmospheric layer farthest from Earth's surface. Here, pressure and density are so low that individual gas molecules rarely strike one another. The molecules move at incredibly fast speeds after absorbing the Sun's radiation. Because the atmosphere does not have a definite edge, molecules can escape the pull of gravity and travel into space.

Active Reading 9. **List** What are the layers of the atmosphere?

Figure 6 Molecules in the air are closer together near Earth's surface than they are at higher altitudes.

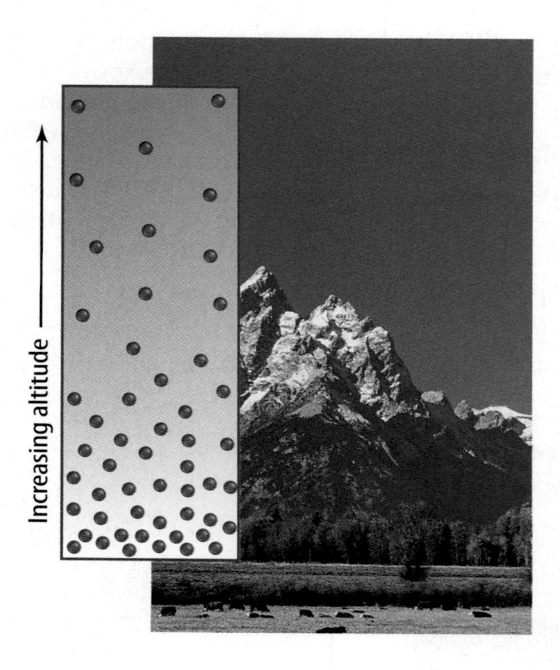

Figure 7 Temperature differences occur within the layers of the atmosphere.

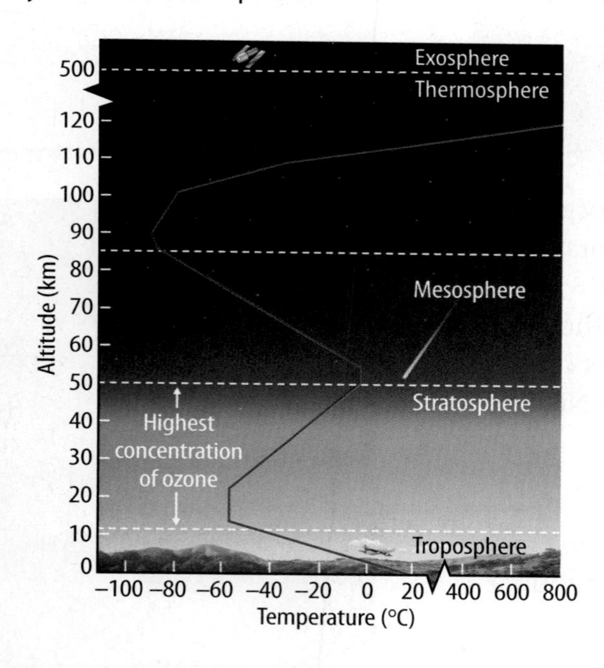

12. Visual Check Identify Name the part of the temperature pattern that is most like the troposphere's.

Air Pressure and Altitude

Gravity is the force that pulls all objects toward Earth. When you stand on a scale, you can read your weight. This is because gravity is pulling you toward Earth. Gravity also pulls the atmosphere toward Earth. The pressure that a column of air exerts on anything below it is called air pressure. Gravity's pull on air increases the density of the air. At higher altitudes, the air is less dense. **Figure 6** shows that air pressure is greatest near Earth's surface because the air molecules are closer together. This dense air exerts more force than the less dense air near the top of the atmosphere. Mountain climbers sometimes carry oxygen tanks at high altitudes because fewer oxygen molecules are in the air at high altitudes.

Active Reading **10. Generalize** Highlight how air pressure changes as altitude increases.

Temperature and Altitude

Figure 7 shows how temperature changes with altitude in the different layers of the atmosphere. If you have ever been hiking in the mountains, you have experienced the temperature cooling as you hike to higher elevations. In the troposphere, temperature decreases as altitude increases. Notice that the opposite effect occurs in the stratosphere. As altitude increases, temperature increases. This is because of the high concentration of ozone in the stratosphere. Ozone absorbs energy from sunlight, which increases the temperature in the stratosphere.

In the mesosphere, as altitude increases, temperature again decreases. In the thermosphere and exosphere, temperatures increase as altitude increases. These layers receive large amounts of energy from the Sun. This energy is spread across a small number of particles, creating high temperatures.

11. NGSSS Check Generalize Highlight how temperature changes as altitude increases. SC.6.E.7.5

Lesson Review 1

Visual Summary

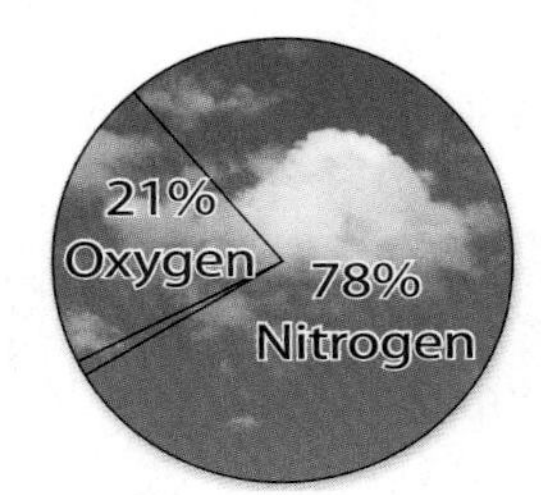

Earth's atmosphere consists of gases that make life possible.

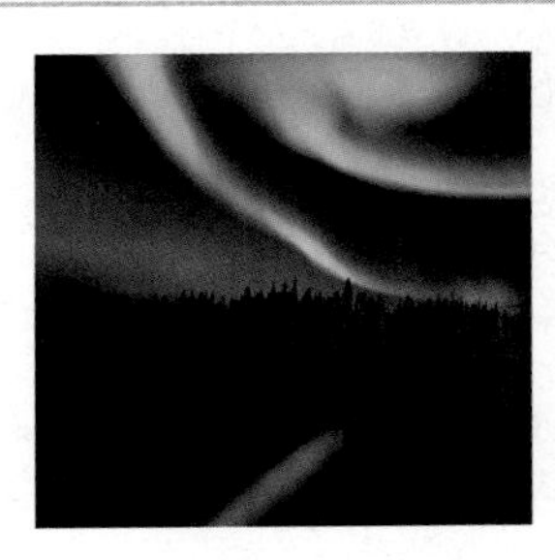

Layers of the atmosphere include the troposphere, the stratosphere, the mesosphere, the thermosphere, and the exosphere.

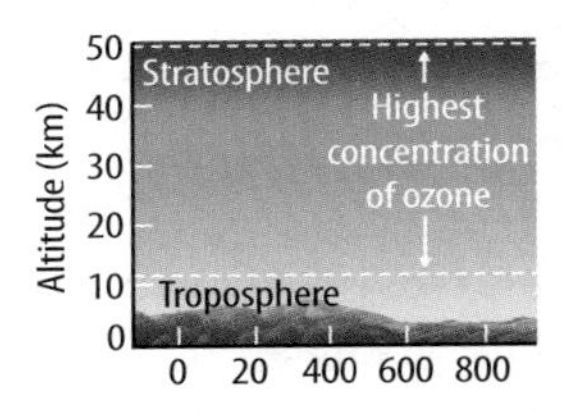

The ozone layer is the area in the stratosphere with a high concentration of ozone.

Use Vocabulary

1. The ______ is a thin layer of gases surrounding Earth.

2. The area of the stratosphere that helps protect Earth's surface from harmful ultraviolet rays is the ______.

3. **Define** Using your own words, define *water vapor.*

______

______

Understand Key Concepts

4. Which atmospheric layer is closest to Earth's surface?
   - (A) mesosphere
   - (B) stratosphere
   - (C) thermosphere
   - (D) troposphere

5. **Identify** the two atmospheric layers in which temperature decreases as altitude increases.

______

Interpret Graphics

6. **Contrast** Fill in the graphic organizer below to contrast the composition of gases in Earth's early atmosphere and its present-day atmosphere. **LA.6.2.2.3**

Atmosphere	Gases
Early	
Present-day	

Critical Thinking

7. **Explain** three ways the atmosphere is important to living things. **SC.6.E.7.9**

______

______

______

CAREERS in SCIENCE

AMERICAN MUSEUM of NATURAL HISTORY

A Crack in Earth's Shield

Scientists discover an enormous hole in the ozone layer that protects Earth.

The ozone layer is like sunscreen, protecting Earth from the Sun's ultraviolet rays. But not all of Earth is covered. Every spring since 1985, scientists have been monitoring a growing hole in the ozone layer above Antarctica.

This surprising discovery was the outcome of years of research from Earth and space. The first measurements of polar ozone levels began in the 1950s, when a team of British scientists began launching weather balloons in Antarctica. In the 1970s, NASA started using satellites to measure the ozone layer from space. Then, in 1985 a close examination of the British team's records indicated a large drop in ozone levels during the Antarctic spring. The levels were so low that the scientists checked and rechecked their instruments before they reported their findings. NASA scientists quickly confirmed the discovery—an enormous hole in the ozone layer over the entire continent of Antarctica. They reported that the hole might have originated as far back as 1976.

A hole in the ozone layer has developed over Antarctica.

Human-made compounds found mostly in chemicals called chlorofluorocarbons, or CFCs, are destroying the ozone layer. During cold winters, molecules released from these compounds are transformed into new compounds by chemical reactions on ice crystals that form in the ozone layer over Antarctica. In the spring, warming by the Sun breaks down the new compounds and releases chlorine and bromine. These chemicals break apart ozone molecules, slowly destroying the ozone layer.

In 1987, CFCs were banned in many countries around the world. Since then, the loss of ozone has slowed and possibly reversed, but a full recovery will take a long time. One reason is that CFCs stay in the atmosphere for more than 40 years. Still, scientists predict the hole in the ozone layer will eventually mend.

Global Warming and the Ozone

Drew Shindell is a NASA scientist investigating the connection between the ozone layer in the stratosphere and the buildup of greenhouse gases throughout the atmosphere. Surprisingly, while these gases warm the troposphere, they are actually causing temperatures in the stratosphere to become cooler. As the stratosphere cools above Antarctica, more clouds with ice crystals form—a key step in the process of ozone destruction. While the buildup of greenhouse gases in the atmosphere may slow the recovery, Shindell still thinks that eventually the ozone layer will heal itself.

It's Your Turn

NEWSCAST Work with a partner to develop three questions about the ozone layer. Conduct research to find the answers. Take the roles of reporter and scientist. Present your findings to the class in a newscast format. LA.6.4.2.2

Lesson 2

Energy Transfer in the ATMOSPHERE

ESSENTIAL QUESTIONS

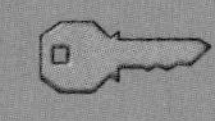
How does energy transfer from the Sun to Earth and the atmosphere?

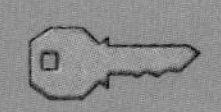
How are air circulation patterns within the atmosphere created?

Vocabulary

radiation p. 128
conduction p. 131
convection p. 131
stability p. 132
temperature inversion p. 133

Florida NGSSS

LA.6.2.2.3 The student will organize information to show understanding (e.g., representing main ideas within text through charting, mapping, paraphrasing, summarizing, or comparing/contrasting);

MA.6.A.3.6 Construct and analyze tables, graphs, and equations to describe linear functions and other simple relations using both common language and algebraic notation.

SC.6.E.7.1 Differentiate among radiation, conduction, and convection, the three mechanisms by which heat is transferred through Earth's system.

SC.6.E.7.5 Explain how energy provided by the sun influences global patterns of atmospheric movement and the temperature differences between air, water, and land.

SC.6.E.7.9 Describe how the composition and structure of the atmosphere protects life and insulates the planet.

SC.6.N.1.1 Define a problem from the sixth grade curriculum, use appropriate reference materials to support scientific understanding, plan and carry out scientific investigation of various types, such as systematic observations or experiments, identify variables, collect and organize data, interpret data in charts, tables, and graphics, analyze information, make predictions, and defend conclusions.

SC.6.N.2.1 Distinguish science from other activities involving thought.

SC.6.N.1.1, SC.6.N.2.1

Launch Lab

15 minutes

What happens to air as it warms?

Light energy from the Sun is converted to thermal energy on Earth. Thermal energy powers the weather systems that impact your everyday life.

Procedure

1. Read and complete a lab safety form.
2. Turn on a **lamp** with an incandescent lightbulb.
3. Place your hands under the light near the lightbulb. What do you feel?
4. Dust your hands with **powder.**
5. Place your hands below the lightbulb and clap them together once.
6. Observe and record what happens to the particles.

Data and Observations

Think About This

1. How might the energy in step 3 move from the lightbulb to your hand?

2. How did the particles move when you clapped your hands?

3. Key Concept How did particle motion show you how the air was moving?

What's really there?

1. Mirages are created as light passes through layers of air that have different temperatures. How do you think energy creates the reflections? What other effects does energy have on the atmosphere?

Active Reading **2. Relate** Contrast visible light and ultraviolet light.

Energy from the Sun

The Sun's energy travels 148 million km to Earth in only 8 minutes. How does the Sun's energy get to Earth? It reaches Earth through the process of radiation. **Radiation** *is the transfer of energy by electromagnetic waves.* Ninety-nine percent of the radiant energy from the Sun consists of visible light, ultraviolet light, and infrared radiation.

Visible Light

The majority of sunlight is visible light. Recall that visible light is light that you can see. The atmosphere is like a window to visible light, allowing it to pass through. At Earth's surface it is converted to thermal energy, commonly called heat.

Near-Visible Wavelengths

The wavelengths of ultraviolet (UV) light and infrared radiation (IR) are just beyond the range of visibility to human eyes. UV light has short wavelengths and can break chemical bonds. Excess exposure to UV light will burn human skin and can cause skin cancer. Infrared radiation (IR) has longer wavelengths than visible light. You can sense IR as thermal energy or warmth. As energy from the Sun is absorbed by Earth, it is also radiated from Earth into the atmosphere as IR.

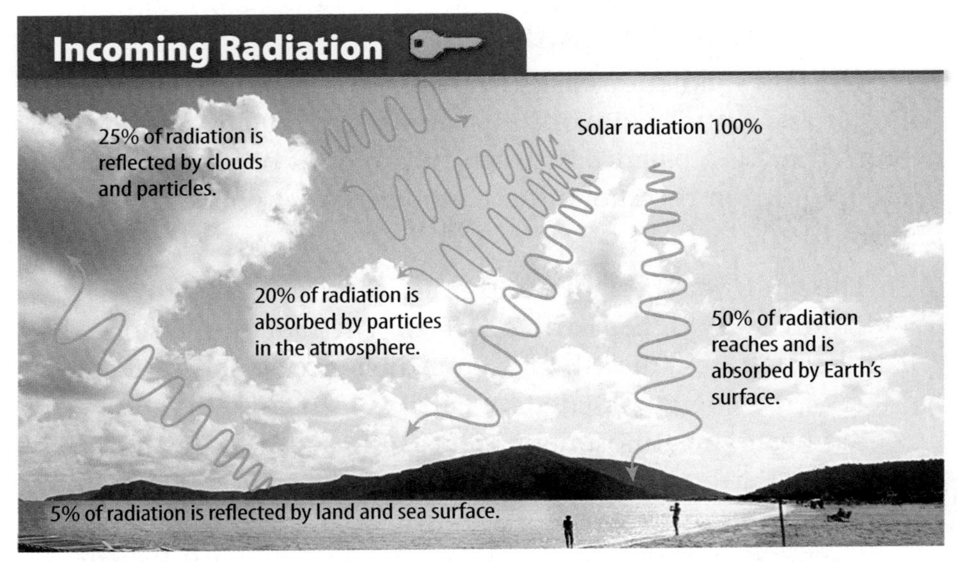

Figure 8 Some of the energy from the Sun is reflected or absorbed as it passes through the atmosphere.

3. Visual Check

Construct Use **Figure 8** to complete the circle graph below. Draw lines to divide the graph so that it indicates the percentage of solar radiation that is reflected by clouds and particles, absorbed by particles in the atmosphere, reflected by land and sea surface, and absorbed by Earth's surface. Create an identification key and color your graph.

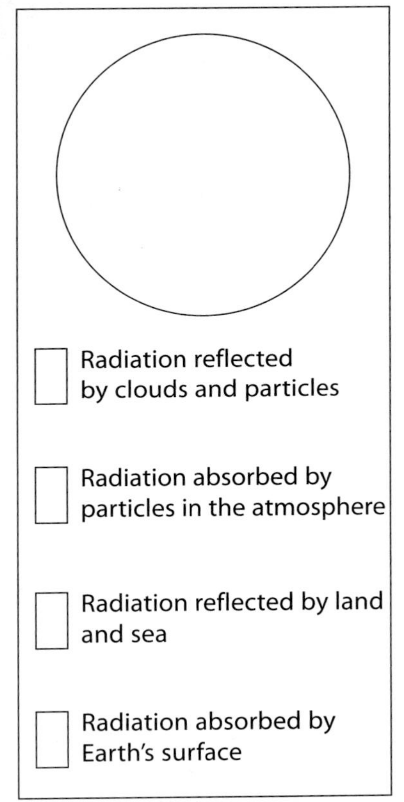

Energy on Earth

As the Sun's energy passes through the atmosphere, some of it is absorbed by gases and particles, and some of it is reflected back into space. As a result, not all the energy coming from the Sun reaches Earth's surface.

Absorption

Study **Figure 8.** Gases and particles in the atmosphere absorb about 20 percent of incoming solar radiation. Oxygen, ozone, and water vapor all absorb incoming ultraviolet light. Water and carbon dioxide in the troposphere absorb some infrared radiation from the Sun. Earth's atmosphere does not absorb visible light. Visible light must be converted to infrared radiation before it can be absorbed.

Reflection

Bright surfaces, especially clouds, **reflect** incoming radiation. Study **Figure 8** again. Clouds and other small particles in the air reflect about 25 percent of the Sun's radiation. Some radiation travels to Earth's surface and is then reflected by land and sea surfaces. Snow-covered, icy, or rocky surfaces are especially reflective. As shown in **Figure 8,** this accounts for about 5 percent of incoming radiation. In all, about 30 percent of incoming radiation is reflected into space. This means that, along with the 20 percent of incoming radiation that is absorbed in the atmosphere, Earth's surface only receives and absorbs about 50 percent of incoming solar radiation.

Science Use v. Common Use

reflect

Science Use to return light, heat, sound, etc., after it strikes a surface

Common Use to think quietly and calmly

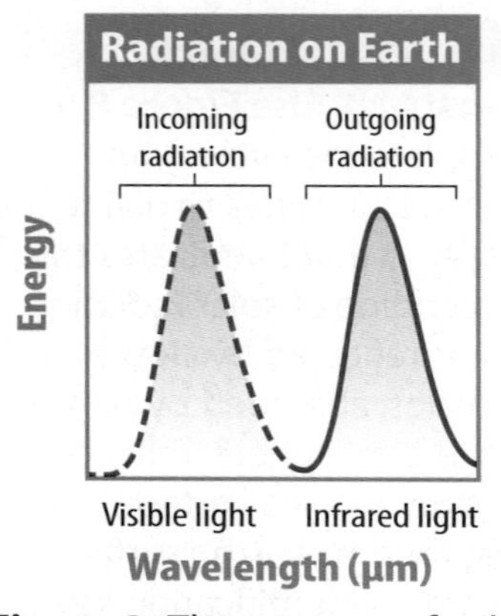

Figure 9 The amount of solar energy absorbed by Earth and its atmosphere is equal to the amount of energy Earth radiates back into space.

Radiation Balance

The Sun's radiation heats Earth. So, why doesn't Earth get hotter and hotter as it continues to receive radiation from the Sun? There is a balance between the amount of incoming radiation from the Sun and the amount of outgoing radiation from Earth.

The land, water, plants, and other organisms absorb solar radiation that reaches Earth's surface. The radiation absorbed by Earth is then re-radiated, or bounced back, into the atmosphere. Most of the energy radiated by Earth is infrared radiation, which heats the atmosphere. **Figure 9** shows that the amount of radiation Earth receives from the Sun is the same as the amount Earth radiates into the outer atmosphere. Earth absorbs the Sun's energy and then radiates that energy away until a balance is achieved.

The Greenhouse Effect

As shown in **Figure 10,** the glass of a greenhouse allows light to pass through, where it is converted to infrared energy. The glass prevents the IR from escaping and it warms the greenhouse. Some of the gases in the atmosphere, called greenhouse gases, act like the glass of a greenhouse. They allow sunlight to pass through, but they prevent some of Earth's IR energy from escaping. Greenhouse gases in Earth's atmosphere trap IR and direct it back toward Earth's surface. This causes additional heat to build up at Earth's surface. The gases that trap IR best are water vapor (H_2O), carbon dioxide (CO_2), and methane (CH_4).

Active Reading **4. Explain** What is the greenhouse effect?

____

____

____

The Greenhouse Effect

Figure 10 Some of the outgoing radiation is directed back toward Earth's surface by greenhouse gases.

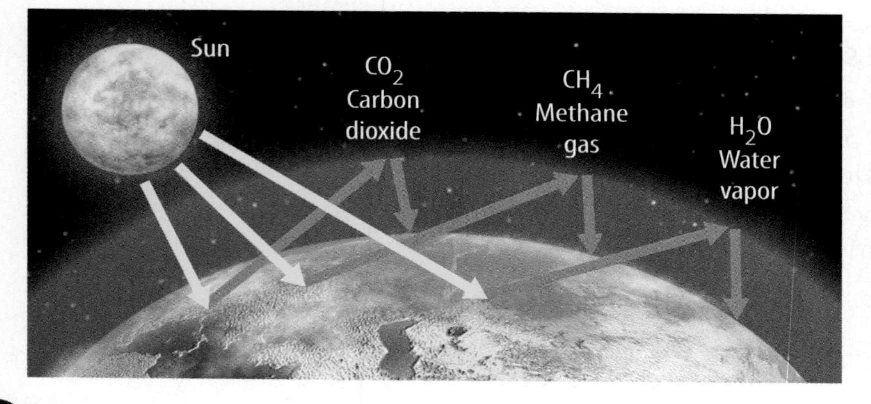

Thermal Energy Transfer

Recall that there are three types of thermal energy transfer—radiation, conduction, and convection. All three occur in the atmosphere. Recall that radiation is the process that transfers energy from the Sun to Earth.

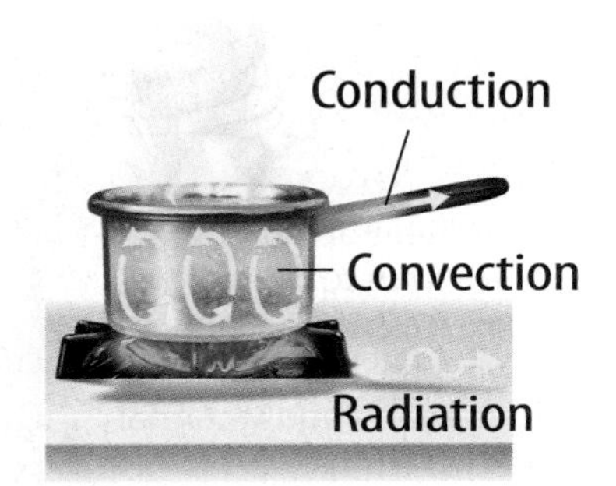

Figure 11 Energy is transferred through conduction, convection, and radiation.

Conduction

Thermal energy always moves from an object with a higher temperature to an object with a lower temperature. **Conduction** *is the transfer of thermal energy by collisions between particles of matter.* Particles must be close enough to touch to transfer energy by conduction. Touching the pot of water, shown in **Figure 11,** would transfer energy from the pot to your hand. Conduction occurs where the atmosphere touches Earth.

WORD ORIGIN

conduction
from Latin *conducere,* means "to bring together"

Convection

As molecules of air close to Earth's surface are heated by conduction, they spread apart, and air becomes less dense. Less dense air rises, transferring thermal energy to higher altitudes. *The transfer of thermal energy by the movement of particles within matter is called* **convection**. Convection can be seen in **Figure 11** as the boiling water circulates and steam rises.

Latent Heat

More than 70 percent of Earth's surface is covered by a highly unique substance—water! Water is the only substance that can exist as a solid, a liquid, and a gas at the temperature ranges on Earth. Recall that latent heat is exchanged when water changes from one phase to another, as shown in **Figure 12.** Latent heat energy is transferred from Earth's surface to the atmosphere.

5. NGSSS Check Explain How does energy transfer from the Sun to Earth and the atmosphere? SC.6.E.7.1

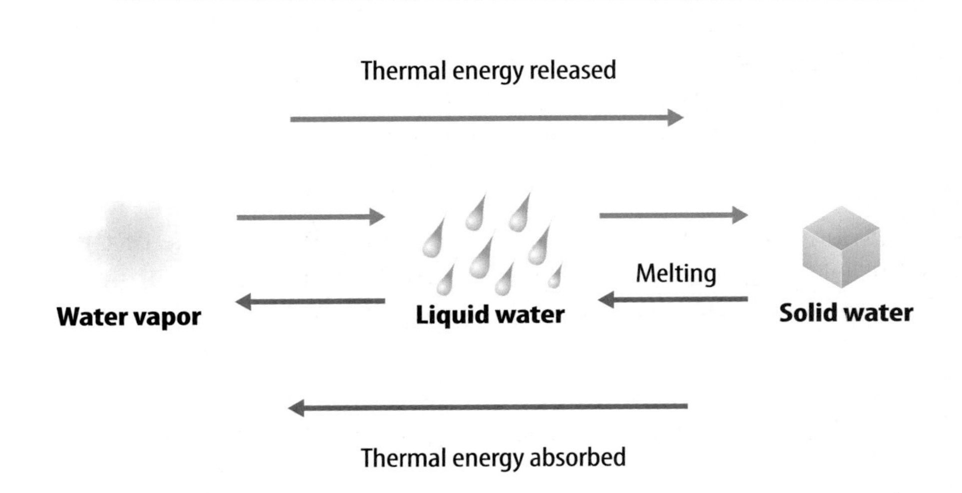

Figure 12 Water releases or absorbs thermal energy during phase changes.

6. Match Insert the words below on the arrows in **Figure 12** to indicate the correct action during the phase change. One word has already been placed.

- Freezing
- Evaporation
- Condensation

Active Reading

FOLDABLES® LA.6.2.2.3

Fold a sheet of paper to make a four-column, four-row table and label as shown. Use it to record information about thermal energy transfer.

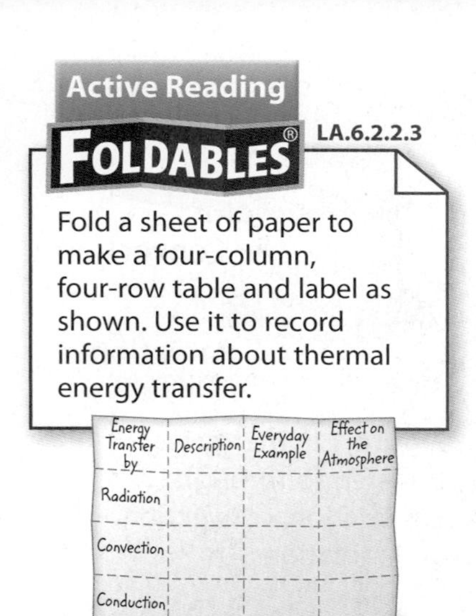

Circulating Air

Figure 13 Rising warm air is replaced by cooler, denser air that sinks beside it.

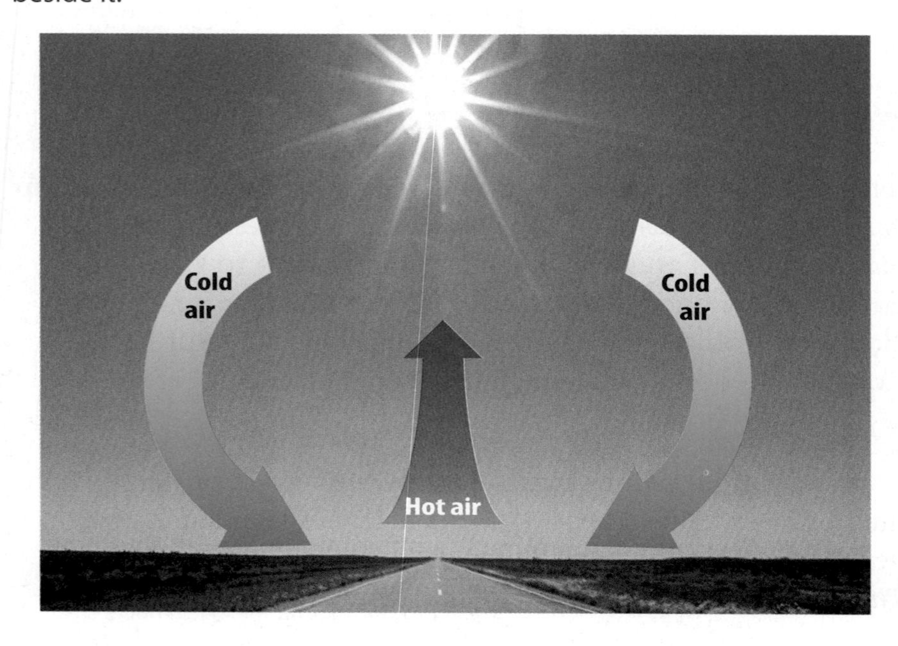

Figure 14 Lens-shaped lenticular clouds form when air rises with a mountain wave.

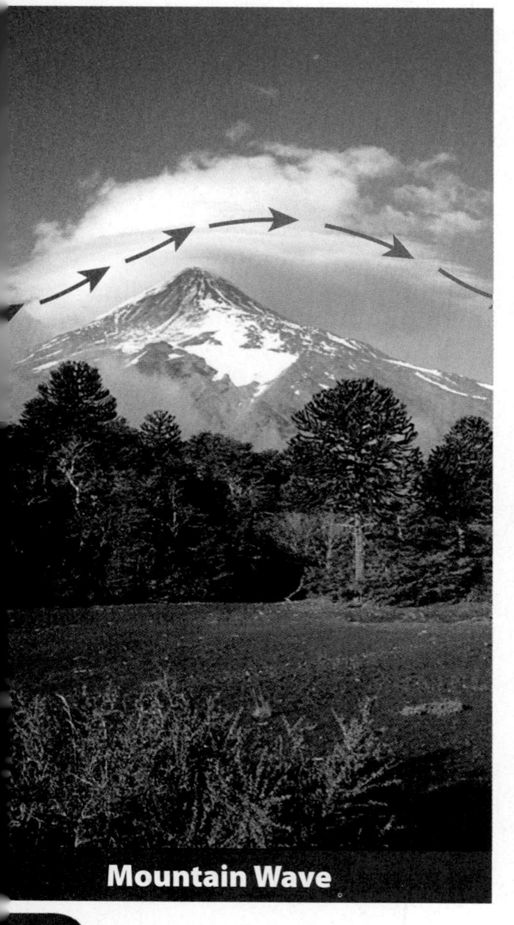

Circulating Air

You've read that energy is transferred through the atmosphere by convection. On a hot day, air that is heated becomes less dense. This creates a pressure difference. Cool, denser air pushes the warm air out of the way. The warm air is replaced by the more dense air, as shown in **Figure 13.** The warm air is often pushed upward. Warmer, rising air is always accompanied by cooler, sinking air.

Air is constantly moving. For example, wind flowing into a mountain range rises and flows over it. After reaching the top, the air sinks. This up-and-down motion sets up an atmospheric phenomenon called a mountain wave. The upward moving air within mountain waves creates lenticular (len TIH kyuh lur) clouds, shown in **Figure 14.** Circulating air affects weather and climate around the world.

7. NGSSS Check Recall Highlight how air circulation patterns within the atmosphere are created. SC.6.E.7.5

Stability

When you stand in the wind, your body forces some of the air to move above you. The same is true for hills and buildings. Conduction and convection also cause air to move upward. **Stability** *describes whether circulating air motions will be strong or weak.* When air is unstable, circulating motions are strong. During stable conditions, circulating motions are weak.

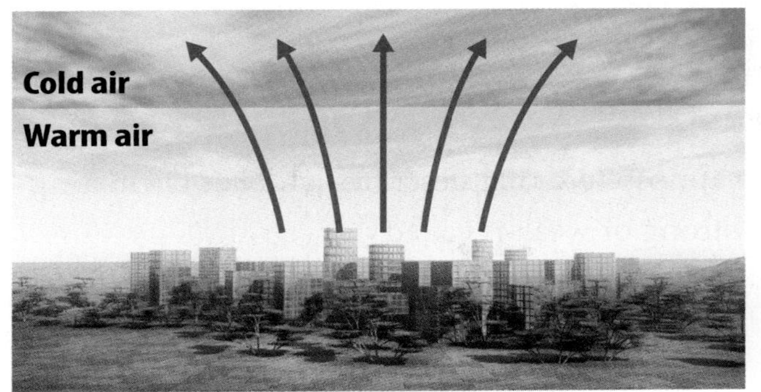

Figure 15 A temperature inversion occurs when cooler air is trapped beneath warmer air.

Unstable Air and Thunderstorms Unstable conditions often occur on warm, sunny afternoons. During unstable conditions, ground-level air is much warmer than higher-altitude air. As warm air rises rapidly in the atmosphere, it cools and forms large, tall clouds. Latent heat, released as water vapor changes from a gas to a liquid, adds to the instability, and produces a thunderstorm.

Stable Air and Temperature Inversions Sometimes ground-level air is nearly the same temperature as higher-altitude air. During these conditions, the air is stable, and circulating motions are weak. A temperature inversion can occur under these conditions. *A* **temperature inversion** *occurs in the troposphere when temperature increases as altitude increases.* During a temperature inversion, a layer of cooler air is trapped by a layer of warmer air above it, as shown in **Figure 15.** Temperature inversions prevent air from mixing and can trap pollution in the air close to Earth's surface.

Active Reading **8. Explain** Fill in the blanks to explain air movement during a thunderstorm.

During unstable conditions, ground-level air is much warmer than ____________. Air rises ____________, cools, and produces large, tall clouds. ____________, released as water vapor, changes from a ____________ to a ____________, adds to the instability, and produces a violent storm.

9. Visual Check
Describe How do conditions during a temperature inversion differ from normal conditions?

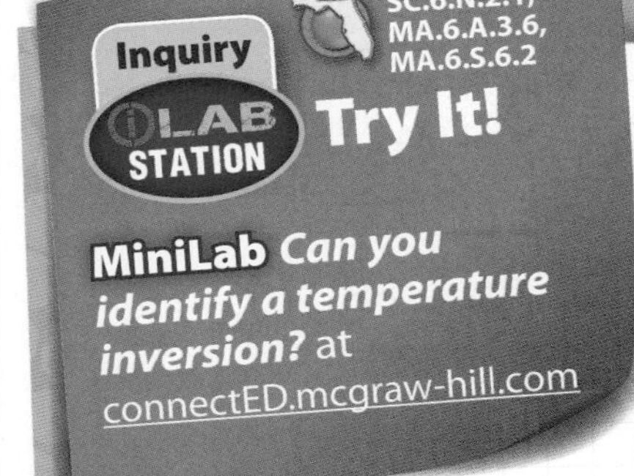

Apply It!

After you complete the lab, answer the questions below.

1. Describe what would happen to the temperature inversion line if the graph went above 1,000 m altitude.

2. Using the graph, determine how high the height of the warm air layer is during the temperature inversion.

Lesson Review 2

Visual Summary

Not all radiation from the Sun reaches Earth's surface.

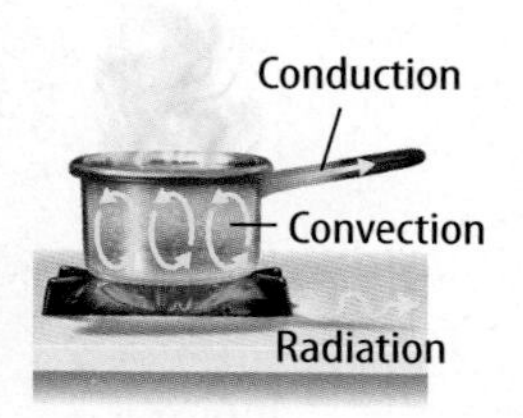

Thermal energy transfer in the atmosphere occurs through radiation, conduction, and convection.

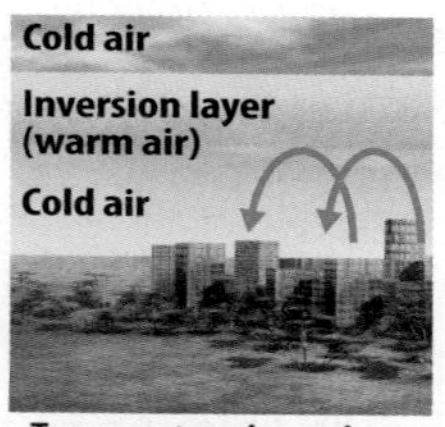

Temperature inversions prevent air from mixing and can trap pollution in the air close to Earth's surface.

Use Vocabulary

1. The property of the atmosphere that describes whether circulating air motions will be strong or weak is called ________.

2. **Define** *conduction* in your own words.

3. ________ is the transfer of thermal energy by the movement of particles within matter.

Understand Key Concepts

4. Which statement is true?
   - (A) The Sun's energy is completely blocked by Earth's atmosphere.
   - (B) The Sun's energy passes through the atmosphere without warming it significantly.
   - (C) The Sun's IR energy is absorbed by greenhouse gases.
   - (D) The Sun's energy is primarily in the UV range.

5. **Distinguish** between conduction and convection. SC.6.E.7.1

Interpret Graphics

6. **Sequence** Copy and fill in the graphic organizer below to describe how energy from the Sun is absorbed in Earth's atmosphere. LA.6.2.2.3

Energy Absorption

Critical Thinking

7. **Suggest** a way to keep a parked car cool on a sunny day.

8. **Relate** temperature inversions to air stability.

FOCUS on FLORIDA

Salads on Mars

Have you eaten a salad today?

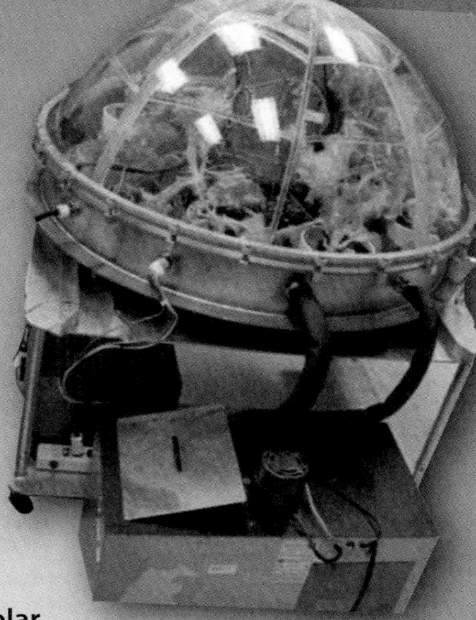

The plants that make up a salad have vitamins and nutrients that humans need to survive. Salads are hard to find on Mars, but scientists are working to fix that.

You may recall that as the Sun's energy passes through Earth's atmosphere, some of that energy is absorbed by gases and particles, and some is reflected back into space. Scientists call this interaction the greenhouse effect.

Unfortunately, the greenhouse effect barely occurs in the Martian atmosphere. The thin atmosphere is not able to absorb and reflect solar radiation like Earth's atmosphere. Although Mars is approximately 79 million kilometers farther from the Sun than Earth is, more solar radiation reaches the surface of Mars than the surface of Earth. This means more radiation also escapes the atmosphere instead of being trapped and warming the surface. The greenhouse effect that exists on Earth to help regulate its temperature and grow crops will need to be duplicated by astronauts on Mars for them to survive. Being able to live and work with solar radiation on Mars is an important problem facing scientists and astronauts at the National Aeronautics and Space Administration (NASA).

At NASA's Kennedy Space Center, near Orsino, Florida, scientists have already started experiments to address these problems. They are developing small, pressurized greenhouses for the Martian landscape.

These greenhouses will be able to maintain an atmospheric pressure like that of Earth. They also will regulate the temperature inside the greenhouse so the plants will feel like they are right at home. If the pressure and temperature aren't right, the plants will use too much water, wither, and die.

By mimicking Earth's atmosphere, Martian greenhouses will play a very important role in allowing astronauts to live successfully on Mars. The food and oxygen they provide will be essential to allowing astronauts to survive on Mars for long time periods. Hopefully, astronauts will be able to grow enough plants for a good salad.

It's Your Turn

RESEARCH What kinds of plants should astronauts take with them to Mars? Research the nutritional values of various edible plants. Create a menu of what the astronauts should take. Compare your menu with those of your classmates.

Lesson 3

AIR CURRENTS

ESSENTIAL QUESTIONS

How does uneven heating of Earth's surface result in air movement?

How are air currents on Earth affected by Earth's spin?

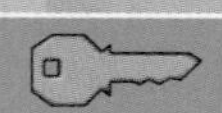
What are the main wind belts on Earth?

Vocabulary

wind p. 137
trade winds p. 139
westerlies p. 139
polar easterlies p. 139
jet stream p. 139
sea breeze p. 140
land breeze p. 140

Florida NGSSS

LA.6.2.2.3 The student will organize information to show understanding (e.g., representing main ideas within text through charting, mapping, paraphrasing, summarizing, or comparing/contrasting);

SC.6.E.7.3 Describe how global patterns such as the jet stream and ocean currents influence local weather in measurable terms such as temperature, air pressure, wind direction and speed, and humidity and precipitation.

SC.6.E.7.5 Explain how energy provided by the sun influences global patterns of atmospheric movement and the temperature differences between air, water, and land.

SC.6.N.1.1 Define a problem from the sixth grade curriculum, use appropriate reference materials to support scientific understanding, plan and carry out scientific investigation of various types, such as systematic observations or experiments, identify variables, collect and organize data, interpret data in charts, tables, and graphics, analyze information, make predictions, and defend conclusions.

SC.6.N.1.5 Recognize that science involves creativity, not just in designing experiments, but also in creating explanations that fit evidence.

SC.6.N.2.1 Distinguish science from other activities involving thought.

SC.6.N.3.4 Identify the role of models in the context of the sixth grade science benchmarks.

Inquiry Launch Lab

SC.6.N.1.5

15 minutes

Why does air move?

Early sailors relied on wind to move their ships around the world. Today, wind is used as a renewable source of energy. In the following activity, you will explore what causes air to move.

Procedure

1. Read and complete a lab safety form.
2. Inflate a **balloon.** Do not tie it. Hold the neck of the balloon closed.
3. Describe how the inflated balloon feels.
4. Open the neck of the balloon without letting go of the balloon. Record your observations of what happens below.

Data and Observations

Think About This

1. What caused the inflated balloon surface to feel the way it did when the neck was closed?

2. What caused the air to leave the balloon when the neck was opened?

3. Key Concept Why didn't outside air move into the balloon when the neck was opened?

How do you think air pushes these blades?

1. If you have ever ridden a bicycle into a strong wind, you know that the movement of air can be a powerful force. Some areas of the world have more wind than others. What do you think causes these differences? What do you think causes wind?

Global Winds

There are great wind belts that circle the globe. The energy that causes this massive movement of air originates at the Sun. However, wind patterns can be global or local.

Unequal Heating of Earth's Surface

The Sun's energy warms Earth. However, the same amount of energy does not reach all of Earth's surface. The amount of energy an area gets depends largely on the Sun's angle. For example, energy from the rising or setting Sun is not very intense. But Earth heats up quickly when the Sun is high in the sky.

In latitudes near the equator—an area referred to as the tropics—sunlight strikes Earth's surface at a high angle—nearly 90°—year round. As a result, in the tropics there is more sunlight per unit of surface area. This means that the land, the water, and the air at the equator are always warm.

At latitudes near the North Pole and the South Pole, sunlight strikes Earth's surface at a low angle. Sunlight is now spread over a larger surface area than in the tropics. As a result, the poles receive very little energy per unit of surface area and are cooler.

Recall that differences in density cause warm air to rise. Warm air puts less pressure on Earth than cooler air. Because it's so warm in the tropics, air pressure is usually low. Over colder areas, such as the North Pole and the South Pole, air pressure is usually high. This difference in pressure creates wind. **Wind** *is the movement of air from areas of high pressure to areas of low pressure.* Global wind belts influence both climate and weather on Earth.

2. **NGSSS Check** **Identify** Underline how the uneven heating of Earth's surface results in air movement. SC.6.E.7.5

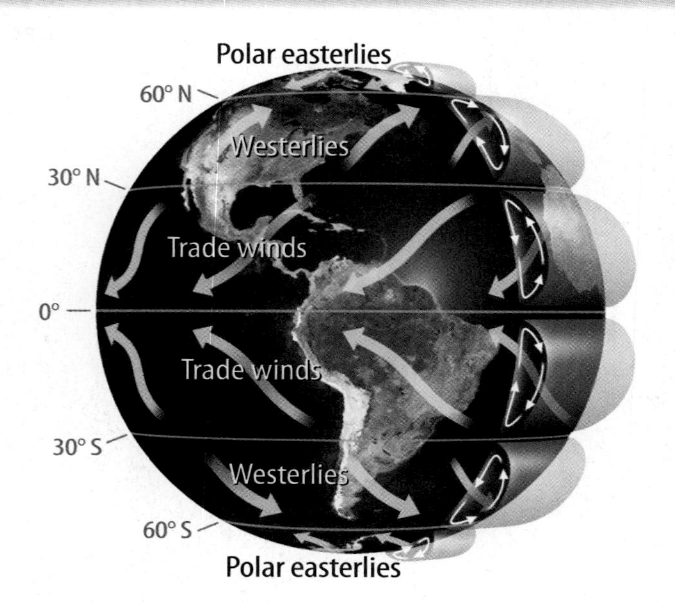

Figure 16 Three cells in each hemisphere move air through the atmosphere.

3. Visual Check
Identify Which wind belt is Florida located in?

Active Reading
FOLDABLES®
SC.6.E.7.5
LA.6.2.2.3

Make a shutterfold. As illustrated, draw Earth and the three cells found in each hemisphere on the inside of the shutterfold. Describe each cell and explain the circulation of Earth's atmosphere. On the outside, label the global wind belts.

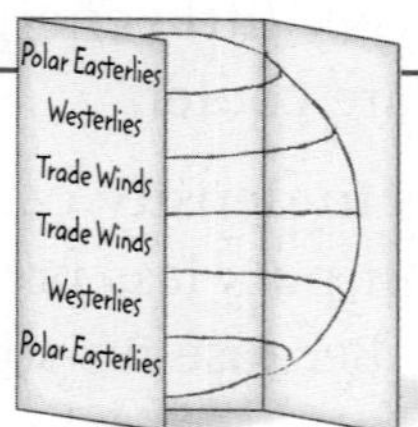

Global Wind Belts

Figure 16 shows the three-cell model of circulation in Earth's atmosphere. In the northern hemisphere, hot air in the cell nearest the equator moves to the top of the troposphere. There, the air moves northward until it cools and moves back to Earth's surface near 30° latitude. Most of the air in this convection cell then returns to Earth's surface near the equator.

The cell at the highest northern latitudes is also a convection cell. Air from the North Pole moves toward the equator along Earth's surface. The cooler air pushes up the warmer air near 60° latitude. The warmer air then moves northward and repeats the cycle. The cell between 30° and 60° latitude is not a convection cell. Its motion is driven by the other two cells, in a motion similar to a pencil that you roll between your hands. Three similar cells exist in the southern hemisphere. These cells help generate the global wind belts.

The Coriolis Effect

What happens when you throw a ball to someone across from you on a moving merry-go-round? The ball appears to curve because the person catching the ball has moved. Similarly, Earth's rotation causes moving air and water to appear to move to the right in the Northern Hemisphere and to the left in the Southern Hemisphere. This is called the Coriolis effect. The contrast between high and low pressure and the Coriolis effect creates distinct wind patterns, called prevailing winds.

4. NGSSS Check **Summarize** Highlight how air currents on Earth are affected by Earth's spin. SC.6.E.7.5

Prevailing Winds

The three global cells in each hemisphere create northerly and southerly winds. When the Coriolis effect acts on the winds, they blow to the east or the west, creating relatively steady, predictable winds. Locate the trade winds in **Figure 16.** *The* **trade winds** *are steady winds that flow from east to west between 30°N latitude and 30°S latitude.*

At about 30°N and 30°S air cools and sinks. This creates areas of high pressure and light, calm winds, called the doldrums. Sailboats without engines can be stranded in the doldrums.

The prevailing **westerlies** *are steady winds that flow from west to east between latitudes 30°N and 60°N, and 30°S and 60°S.* This region is also shown in **Figure 15.** *The* **polar easterlies** *are cold winds that blow from the east to the west near the North Pole and the South Pole.*

Active Reading **5. List** What are the main wind belts on Earth?

Jet Streams

Near the top of the troposphere is a narrow band of high winds called the **jet stream**. Shown in **Figure 17,** jet streams flow around Earth from west to east, often making large loops to the north or the south. Jet streams influence weather as they move cold air from the poles toward the tropics and warm air from the tropics toward the poles. Jet streams can move at speeds up to 300 km/h and are more unpredictable than prevailing winds.

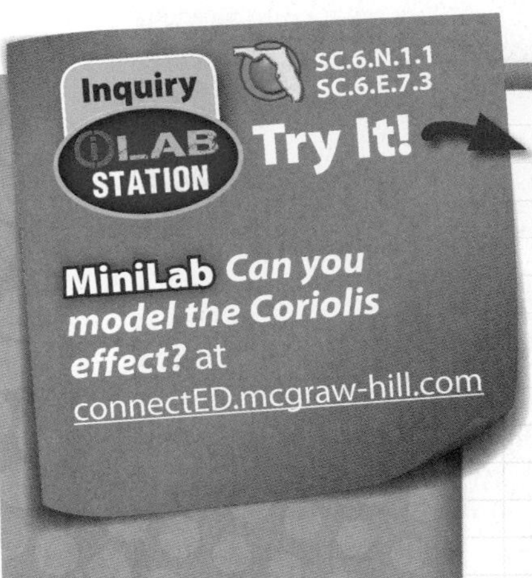

Apply It! After you complete the lab, answer the questions below.

1. How does the path of the ball on the spinning foamboard model the movement of wind on Earth?

2. Which way would the winds in the Northern Hemisphere and the Southern Hemisphere appear to move if Earth rotated towards the west, instead of towards the east?

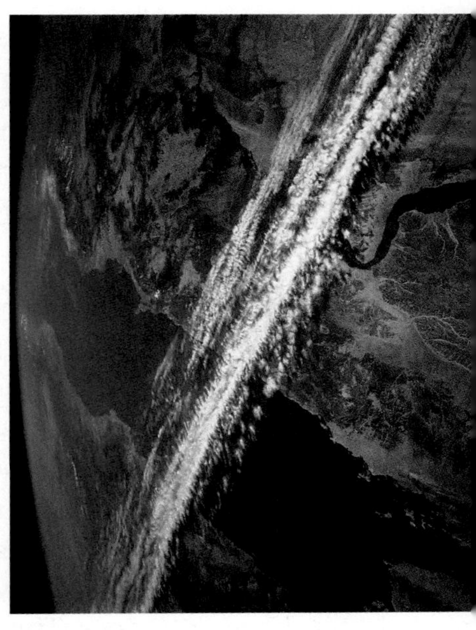

Figure 17 Jet streams are thin bands of high wind speed. The clouds seen here have condensed within a cooler jet stream.

Active Reading **6. Name** Use the descriptions below to help name the type of wind. Place the name in the box to the left of the description.

Winds

A.

B.

C.

D.

E.

Description

A. Steady winds that flow toward the equator from east to west between 30°N latitude and 30°S latitude
B. Areas of high pressure and light, calm winds at about 30°N latitude and 30°S latitude
C. Steady winds that flow from west to east between latitudes 30°N and 60°N and between latitudes 30°S and 60°S
D. Cold winds that blow from east to west near the North Pole and the South Pole
E. A narrow band of high winds, commonly located near the top of the troposphere; influences weather

Local Winds

You have just read that global winds occur because of pressure differences around the globe. In the same way, local winds occur whenever air pressure is different from one location to another.

Sea and Land Breezes

Anyone who has spent time near a lake or an ocean shore has probably experienced the connection between temperature, air pressure, and wind. *A* **sea breeze** *is wind that blows from the sea to the land due to local temperature and pressure differences.* **Figure 18** shows how sea breezes form. On sunny days, land warms up faster than water does. The air over the land warms by conduction and rises, creating an area of low pressure. The air over the water sinks, creating an area of high pressure because it is cooler. The differences in pressure over the warm land and the cooler water result in a cool wind that blows from the sea onto land.

A **land breeze** *is a wind that blows from the land to the sea due to local temperature and pressure differences.* **Figure 18** shows how land breezes form. At night, the land cools more quickly than the water. Therefore, the air above the land cools more quickly than the air over the water. As a result, an area of lower pressure forms over the warmer water. A land breeze then blows from the land toward the water.

Active Reading **7. Explain** Compare and contrast sea breezes and land breezes using the Venn diagram below.

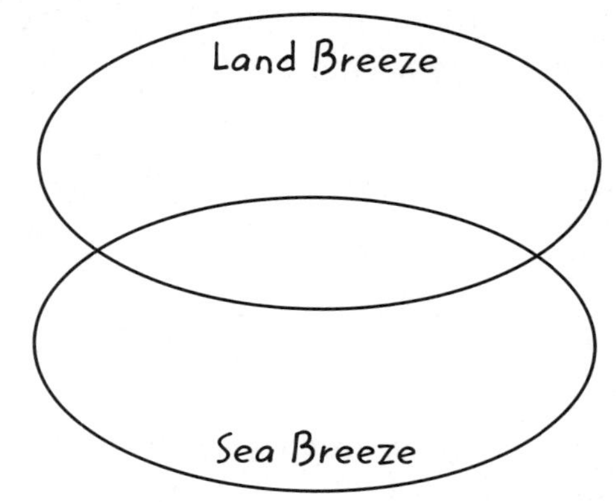

Figure 18 Sea breezes and land breezes are created as part of a large reversible convection current.

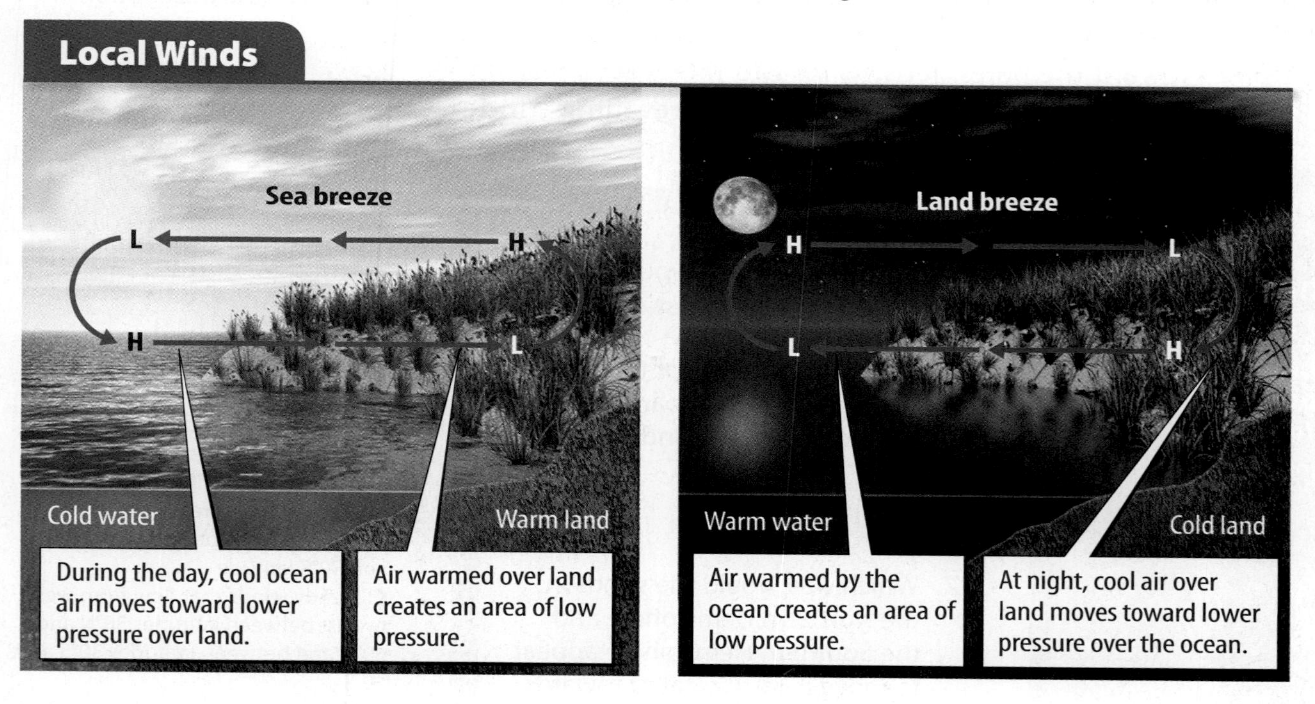

8. Visual Check Arrange Write the steps involved in the formation of a land breeze.

Lesson Review 3

Visual Summary

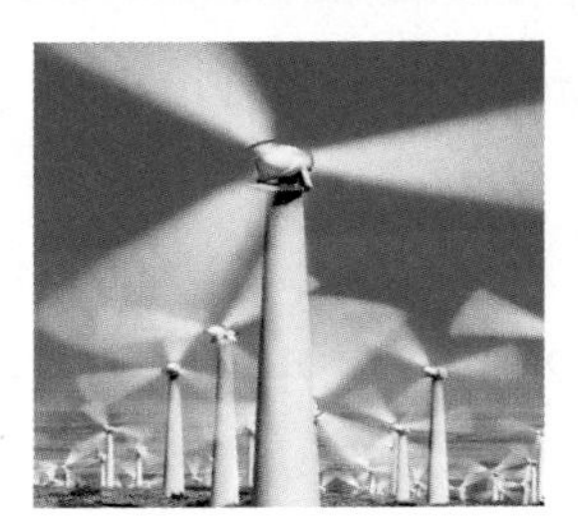

Wind is created by pressure differences between one location and another.

Prevailing winds in the global wind belts are the trade winds, the westerlies, and the polar easterlies.

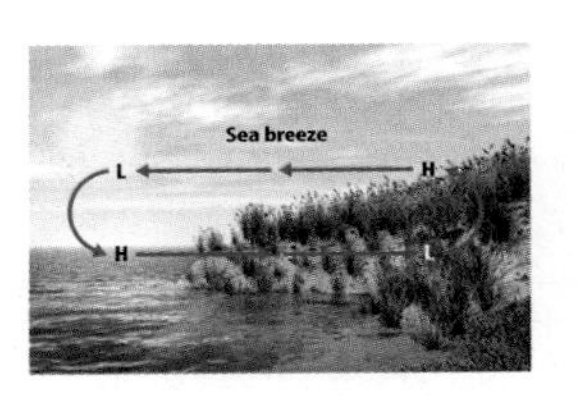

Sea breezes and land breezes are examples of local winds.

Use Vocabulary

1. The movement of air from areas of high pressure to areas of low pressure is ______________.

2. A(n) ______________ is wind that blows from the sea to the land due to local temperature and pressure differences.

3. **Distinguish** between westerlies and trade winds.

Understand Key Concepts 

4. Which does NOT affect global wind belts? SC.6.E.7.5
   (A) air pressure
   (B) land breezes
   (C) the Coriolis effect
   (D) the Sun

5. **Relate** Earth's spinning motion to the Coriolis effect.

Interpret Graphics

6. **Organize** Fill in the graphic organizer below to summarize Earth's global wind belts. LA.6.2.2.3

Wind Belt	Description
Trade winds	
Westerlies	
Polar easterlies	

Critical Thinking

7. **Infer** what would happen without the Coriolis effect.

Assess Read the sentences about air currents and pressure in Earth's atmosphere. If the statement is true, mark the box in the true column. If the statement is false, mark the box in the false column and rewrite the underlined portion so that it is true.

	True	False	
1	☐	☐	Two of the three cells that scientists use to describe circulation of Earth's atmosphere are <u>conduction</u> cells.
2	☐	☐	The first belt begins with warm air rising at the equator and dropping back to Earth near <u>30° latitude</u>.
3	☐	☐	The third cell, at the <u>lowest</u> latitude, is also a convection cell.
4	☐	☐	The amount of energy an area receives depends largely on the Sun's <u>brightness</u>.
5	☐	☐	Over colder areas, such as the North Pole and the South Pole, air pressure is usually <u>high</u>.
6	☐	☐	<u>The jet stream</u> causes moving air and water to appear to move to the right in the Northern Hemisphere and to the left in the Southern Hemisphere.
7	☐	☐	<u>The trade winds</u> are steady winds that flow from east to west between 30° N latitude and 30° S latitude.
8	☐	☐	The polar easterlies are cold winds that blow from east to west near the <u>equator</u>.
9	☐	☐	Near the top of the troposphere is a narrow band of high winds called the <u>Coriolis Effect</u>.
10	☐	☐	A <u>land breeze</u> is wind that blows from the sea to the land due to local temperature and pressure differences.
11	☐	☐	At night, the land cools <u>more</u> quickly than the water.
12	☐	☐	Warm air puts <u>more</u> pressure on Earth than cooler air.

Lesson 4

AIR QUALITY

ESSENTIAL QUESTIONS

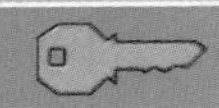

- How do humans impact air quality?
- Why do humans monitor air quality standards?

Vocabulary

air pollution p. 144
acid precipitation p. 145
photochemical smog p. 145
particulate matter p. 146

Florida NGSSS

HE.6.C.1.3 Identify environmental factors that affect personal health.

LA.6.2.2.3 The student will organize information to show understanding (e.g., representing main ideas within text through charting, mapping, paraphrasing, summarizing, or comparing/contrasting);

MA.6.A.3.6 Construct and analyze tables, graphs, and equations to describe linear functions and other simple relations using both common language and algebraic notation.

SC.6.E.7.5 Explain how energy provided by the sun influences global patterns of atmospheric movement and the temperature differences between air, water, and land.

SC.6.E.7.9 Describe how the composition and structure of the atmosphere protects life and insulates the planet.

SC.6.N.1.1 Define a problem from the sixth grade curriculum, use appropriate reference materials to support scientific understanding, plan and carry out scientific investigation of various types, such as systematic observations or experiments, identify variables, collect and organize data, interpret data in charts, tables, and graphics, analyze information, make predictions, and defend conclusions.

SC.6.N.1.4 Discuss, compare, and negotiate methods used, results obtained, and explanations among groups of students conducting the same investigation.

SC.6.N.1.5 Recognize that science involves creativity, not just in designing experiments, but also in creating explanations that fit evidence.

Inquiry Launch Lab

SC.6.N.1.5

20 minutes

How does acid rain form?

Vehicles, factories, and power plants release chemicals into the atmosphere. When these chemicals combine with water vapor, they can form acid rain.

Substances	pH
Hydrochloric acid	0.0
Lemon juice	2.[illegible]
Vinegar	2.9
Tomato juice	4.1
Coffee (black)	5.0
Acid rain	5.6
Rainwater	6.5
Milk	6.6
Distilled water	7.0
Blood	7.4
Baking soda solution	8.4
Toothpaste	9.9
Household ammonia	11.9
Sodium hydroxide	14.0

Procedure

1. Read and complete a lab safety form.
2. Half-fill a **plastic cup** with **distilled water.**
3. Dip a strip of **pH paper** into the water. Use a **pH color chart** to determine the pH of the distilled water. Record the pH.
4. Use a **dropper** to add **lemon juice** to the water until the pH equals that of acid rain. Swirl and test the pH each time you add 5 drops of the lemon juice to the mixture.

Data and Observations

Think About This

1. A strong acid has a pH between 0 and 2. How does the pH of lemon juice compare to the pH of other substances? Is acid rain a strong acid?
2. **Key Concept** Why might scientists monitor the pH of rain?

Inquiry **How do you think the dark cloud layer formed?**

1. Air pollution can be trapped near Earth's surface during a temperature inversion. This is especially common in cities located in valleys and surrounded by mountains. What do you think the quality of the air is like on a day like this one? Where do you think pollution comes from?

Active Reading 2. **Relate** Compare point-source and nonpoint-source pollution.

Sources of Air Pollution

The contamination of air by harmful substances including gases and smoke is called **air pollution**. Air pollution is harmful to humans and other living things. Years of exposure to polluted air can weaken a human's immune system. Respiratory diseases such as asthma can be caused by air pollution.

Air pollution comes from many sources. Point-source pollution is pollution that comes from an identifiable source. Examples of point sources include smokestacks of large factories, such as the one shown in **Figure 19,** and electric power plants that burn fossil fuels. They release tons of polluting gases and particles into the air each day. An example of natural point-source pollution is an erupting volcano.

Nonpoint-source pollution is pollution that comes from a widespread area. One example of pollution from a nonpoint-source is air pollution in a large city. This is considered nonpoint-source pollution because it cannot be traced back to one source. Some bacteria found in swamps and marshes are examples of natural sources of nonpoint-source pollution.

Figure 19 One example of point-source pollution is a factory smoke stack.

Causes and Effects of Air Pollution

The harmful effects of air pollution are not limited to human health. Some pollutants, including ground-level ozone, can damage plants. Air pollution can also cause serious damage to human-made structures. Sulfur dioxide pollution can discolor stone, corrode metal, and damage paint on cars.

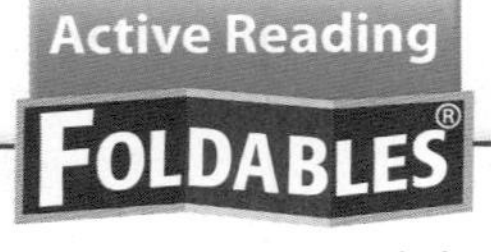

LA.6.2.2.3
HE.6.C.1.3

Make a horizontal three-tab Foldable and label it as shown. Use it to organize your notes about the formation of air pollution and its effects. Fold the right and left thirds over the center and label the outside *Types of Air Pollution.*

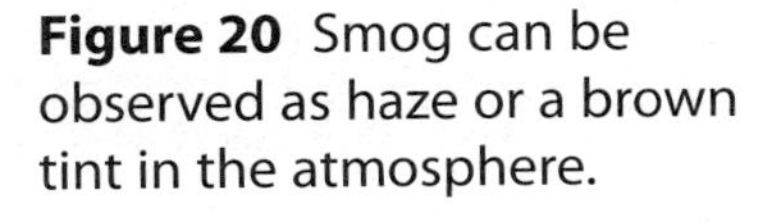

Acid Precipitation

When sulfur dioxide and nitrogen oxides combine with moisture in the atmosphere and form precipitation that has a pH lower than that of normal rainwater, it is called **acid precipitation**. Acid precipitation includes acid rain, snow, and fog. It affects the chemistry of water in lakes and rivers. This can harm the organisms living in the water. Acid precipitation damages buildings and other structures made of stone. Natural sources of sulfur dioxide include volcanoes and marshes. However, the most common sources of sulfur dioxide and nitrogen oxides are automobile exhausts and factory and power plant smoke.

Smog

Photochemical smog *is air pollution that forms from the interaction between chemicals in the air and sunlight.* Smog forms when nitrogen dioxide, released in gasoline engine exhaust, reacts with sunlight. A series of chemical reactions produces ozone and other compounds that form smog. Recall that ozone in the stratosphere helps protect organisms from the Sun's harmful rays. However, ground-level ozone can damage the tissues of plants and animals. Ground-level ozone is the main component of smog. Smog in urban areas reduces visibility and makes air difficult to breathe. **Figure 20** shows New York City on a clear day and on a smoggy day.

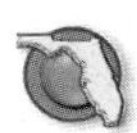

3. **NGSSS Check** **Describe** How do humans impact air quality?
HE.6.C.1.3

Figure 20 Smog can be observed as haze or a brown tint in the atmosphere.

Particulate Pollution

Although you can't see them, over 10,000 solid or liquid particles are in every cubic centimeter of air. A cubic centimeter is about the size of a sugar cube. This type of pollution is called particulate matter. **Particulate** (par TIH kyuh lut) **matter** *is a mixture of dust, acids, and other chemicals that can be hazardous to human health.* The smallest particles are the most harmful. These particles can be inhaled and can enter your lungs. They can cause asthma, bronchitis, and lead to heart attacks. Children and older adults are most likely to experience health problems due to particulate matter.

WORD ORIGIN

particulate
from Latin *particula,* means "small part"

Particulate matter in the atmosphere absorbs and scatters sunlight. This can create haze. Haze particles scatter light, make things blurry, and reduce visibility.

Movement of Air Pollution

Wind can influence the effects of air pollution. Because air carries pollution with it, some wind patterns cause more pollution problems than others. Weak winds or no wind prevents pollution from mixing with the surrounding air. During weak wind conditions, pollution levels can become dangerous.

For example, the conditions in which temperature inversions form are weak winds, clear skies, and longer winter nights. As land cools at night, the air above it also cools. Calm winds, however, prevent cool air from mixing with warm air above it. **Figure 21** shows how cities located in valleys experience a temperature inversion. Cool air, along with the pollution it contains, is trapped in valleys. More cool air sinks down the sides of the mountain, further preventing layers from mixing. The pollution in the photo at the beginning of the lesson was trapped due to a temperature inversion.

Figure 21 At night, cool air sinks down the mountain sides, trapping pollution in the valley below.

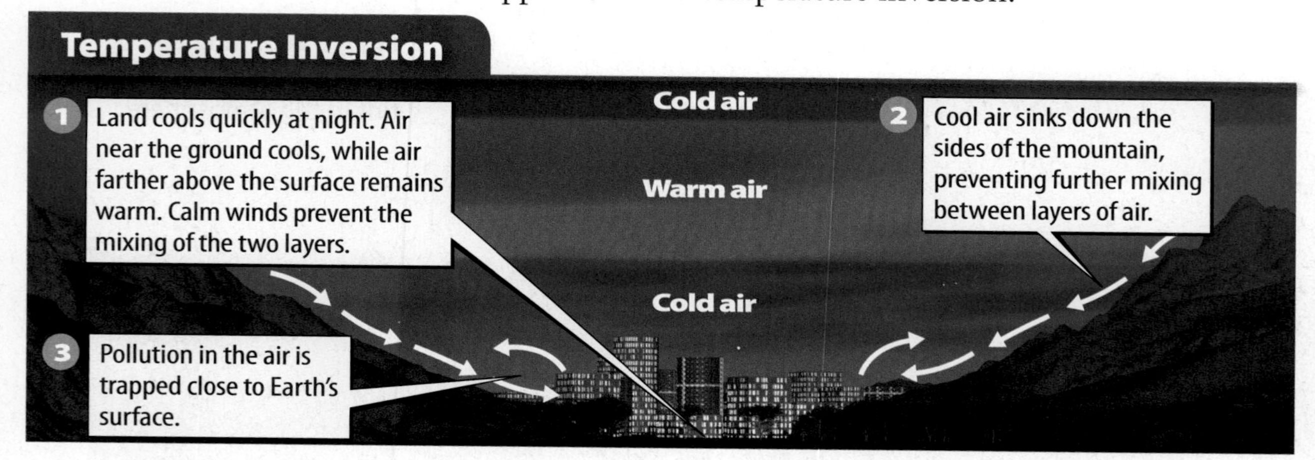

4. **Visual Check** Describe How is pollution trapped by a temperature inversion?

Maintaining Healthful Air Quality

Preserving the quality of Earth's atmosphere requires the cooperation of government officials, scientists, and the public. The Clean Air Act is an example of how government can help fight pollution. Since the Clean Air Act became law in 1970, steps have been taken to reduce automobile emissions. Pollutant levels have decreased significantly in the United States. Despite these advances, serious problems still remain. The amount of ground-level ozone is still too high in many large cities. Also, acid precipitation produced by air pollutants continues to harm organisms in lakes, streams, and forests.

Air Quality Standards

The Clean Air Act gives the U.S. government the power to set air quality standards. The standards protect humans, animals, plants, and buildings from the harmful effects of air pollution. All states are required to make sure that pollutants, such as carbon monoxide, nitrogen oxides, particulate matter, ozone, and sulfur dioxide, do not exceed harmful levels.

Active Reading **5. Define** Underline the purpose of the Clean Air Act.

Monitoring Air Pollution

Pollution levels are continuously monitored by hundreds of instruments in all major U.S. cities. If the levels are too high, authorities may advise people to limit outdoor activities.

Active Reading **6. Describe** Complete the sentences to describe two aspects of the Clean Air Act.

The Clean Air Act gives the U.S. government ______________________.

______________________________________.

The standards require states to ______________________.

Active Reading **7. Summarize** Complete the statement to explain how monitoring air quality helps people.

If air pollution levels are too high . . .

. . . then the public is notified of the danger and

______________________________.

______________________________.

Apply It!

After you complete the lab, answer the questions below.

1. At what value is the air quality unhealthful for a normal, healthy individual?

2. Why does the Air Quality Index (AQI) give a range of values?

Air Quality Trends

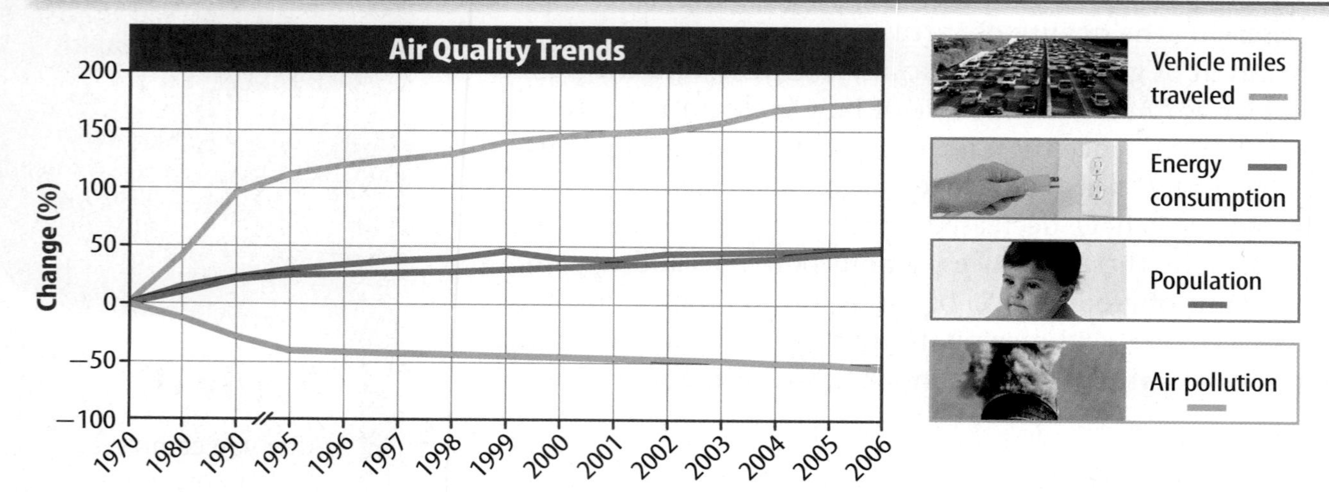

Figure 22 Pollution emissions have declined, even though the population is increasing.

Math Skills

MA.6.A.3.6

Use Graphs

The graph above shows the percent change in four different pollution factors from 1970 through 2006. All values are based on the 0 percent amount in 1970. For example, from 1970 to 1990, the number of vehicle miles driven increased by 100 percent, or the vehicle miles doubled. Use the graph to infer which factors might be related.

Practice

8. What was the percent change in population between 1970 and 2006?

9. What other factor changed by about the same amount during that period?

Air Quality Trends

Over the last several decades, air quality in U.S. cities has improved, as shown in **Figure 22.** Even though some pollution-producing processes have increased, such as burning fossil fuels and traveling in automobiles, levels of certain air pollutants have decreased. Airborne levels of lead and carbon monoxide have decreased the most. Levels of sulfur dioxide, nitrogen oxide, and particulate matter have also decreased.

However, ground-level ozone has not decreased much. Why do ground-level ozone trends lag behind those of other pollutants? Recall that ozone can be created from chemical reactions involving automobile exhaust. Because the number of vehicle miles traveled has increased, more automobile exhaust has created ground-level ozone.

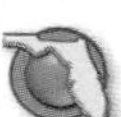

10. **NGSSS Check** **Summarize** Why do humans monitor air quality standards? HE.6.C.1.3

Indoor Air Pollution

Not all air pollution is outdoors. The air inside homes and other buildings can be as much as 50 times more polluted than outdoor air! The quality of indoor air can impact human health much more than outdoor air quality.

Indoor air pollution comes from many sources. Tobacco smoke, cleaning products, pesticides, and fireplaces are some common sources. Furniture upholstery, carpets, and foam insulation also add pollutants to the air. Another indoor air pollutant is radon, an odorless gas given off by some soil and rocks. Radon leaks through cracks in a building's foundation and sometimes builds up to harmful levels inside homes. Harmful effects of radon come from breathing its particles.

Lesson Review 4

Visual Summary

Air pollution comes from point sources, such as factories, and nonpoint sources, such as automobiles.

Photochemical smog contains ozone, which can damage tissues in plants and animals.

Use Vocabulary

1. **Define** *acid precipitation* in your own words.

2. ______________ forms when chemical reactions combine pollution with sunlight.

3. The contamination of air by harmful substances, including gases and smoke, is ______________.

Understand Key Concepts

4. Which is NOT true about smog?
   (A) It contains nitrogen oxide.
   (B) It contains ozone.
   (C) It reduces visibility.
   (D) It is produced only by cars.

5. **Describe** two ways humans add pollution to the atmosphere. HE.6.C.1.3

Interpret Graphics

6. **Compare and Contrast** Fill in the graphic organizer below to compare and contrast details of smog and acid precipitation. LA.6.2.2.3

	Similarities	Differences
Smog		
Acid Precipitation		

Critical Thinking

7. **Describe** how conduction and convection are affected by paving over a grass field. SC.6.E.7.1

Math Skills MA.6.A.3.6

8. Based on the graph on the opposite page, what was the total percent change in air pollution between 1970 and 2006?

Chapter 4 Study Guide

Think About It! **The gases in Earth's atmosphere help sustain life on Earth and regulate the cycling of thermal energy among Earth's systems.**

Key Concepts Summary

LESSON 1: Describing Earth's Atmosphere

- Earth's **atmosphere** formed as Earth cooled and chemical and biological processes took place.
- Earth's atmosphere consists of nitrogen, oxygen, and a small amount of other gases, such as CO_2 and **water vapor**.
- The atmospheric layers are the **troposphere**, the **stratosphere**, the mesosphere, the thermosphere, and the exosphere.
- Air pressure decreases as altitude increases. Temperature either increases or decreases as altitude increases, depending on the layer of the atmosphere.

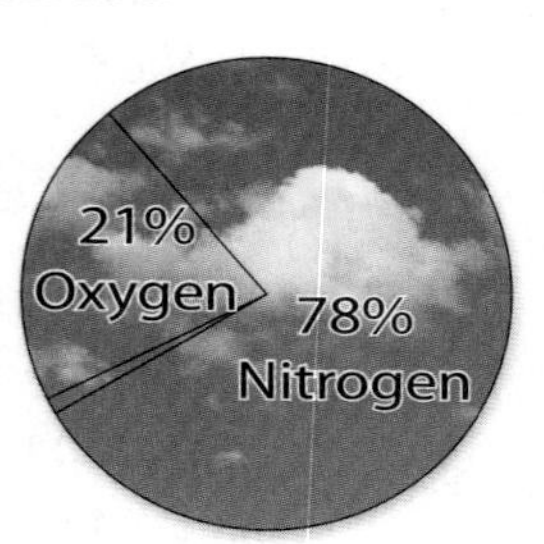

Vocabulary

atmosphere p. 119
water vapor p. 120
troposphere p. 122
stratosphere p. 122
ozone layer p. 122
ionosphere p. 123

LESSON 2: Energy Transfer in the Atmosphere

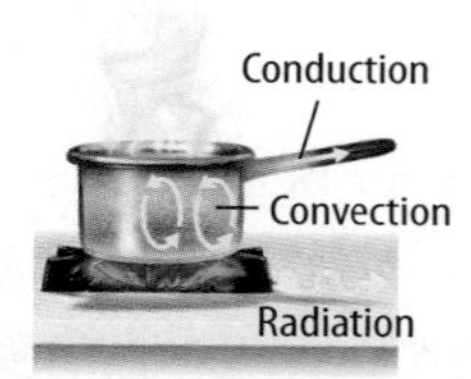

- The Sun's energy is transferred to Earth's surface and the atmosphere through **radiation**, **conduction**, **convection**, and latent heat.
- Air circulation patterns are created by convection currents.

radiation p. 128
conduction p. 131
convection p. 131
stability p. 132
temperature inversion p. 133

LESSON 3: Air Currents

- Uneven heating of Earth's surface creates pressure differences. **Wind** is the movement of air from areas of high pressure to areas of low pressure.
- Air currents curve to the right or to the left due to the Coriolis effect.
- The main wind belts on Earth are the **trade winds**, the **westerlies**, and the **polar easterlies**.

wind p. 137
trade winds p. 139
westerlies p. 139
polar easterlies p. 139
jet stream p. 139
sea breeze p. 140
land breeze p. 140

LESSON 4: Air Quality

- Some human activities release pollution into the air.
- Air quality standards are monitored for the health of organisms and to determine if anti-pollution efforts are successful.

air pollution p. 144
acid precipitation p. 145
photochemical smog p. 145
particulate matter p. 146

Active Reading

FOLDABLES® Chapter Project

Assemble your lesson Foldables as shown to make a Chapter Project. Use the project to review what you have learned in this chapter.

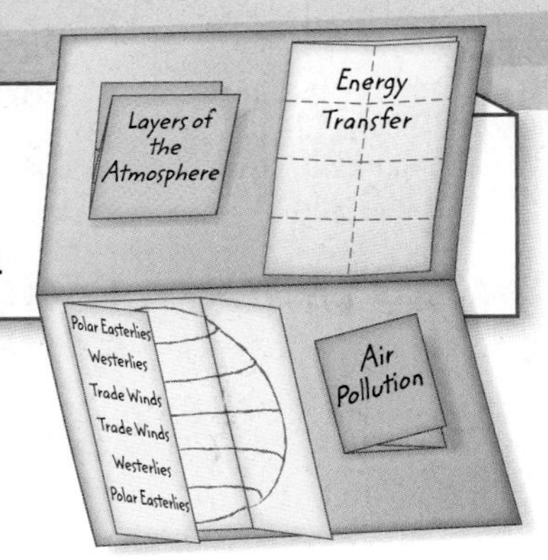

Use Vocabulary

1. Radio waves travel long distances by bouncing off electrically charged particles in the ____________.

2. The Sun's thermal energy is transferred to Earth through space by ____________.

3. Rising currents of warm air transfer energy from Earth to the atmosphere through ____________.

4. A narrow band of winds located near the top of the troposphere is a(n) ____________.

5. ____________ are steady winds that flow from east to west between 30°N latitude and 30°S latitude.

6. In large urban areas, ____________ forms when pollutants in the air interact with sunlight.

7. A mixture of dust, acids, and other chemicals that can be hazardous to human health is called ____________.

Link Vocabulary and Key Concepts

Use vocabulary terms from the previous page to complete the concept map.

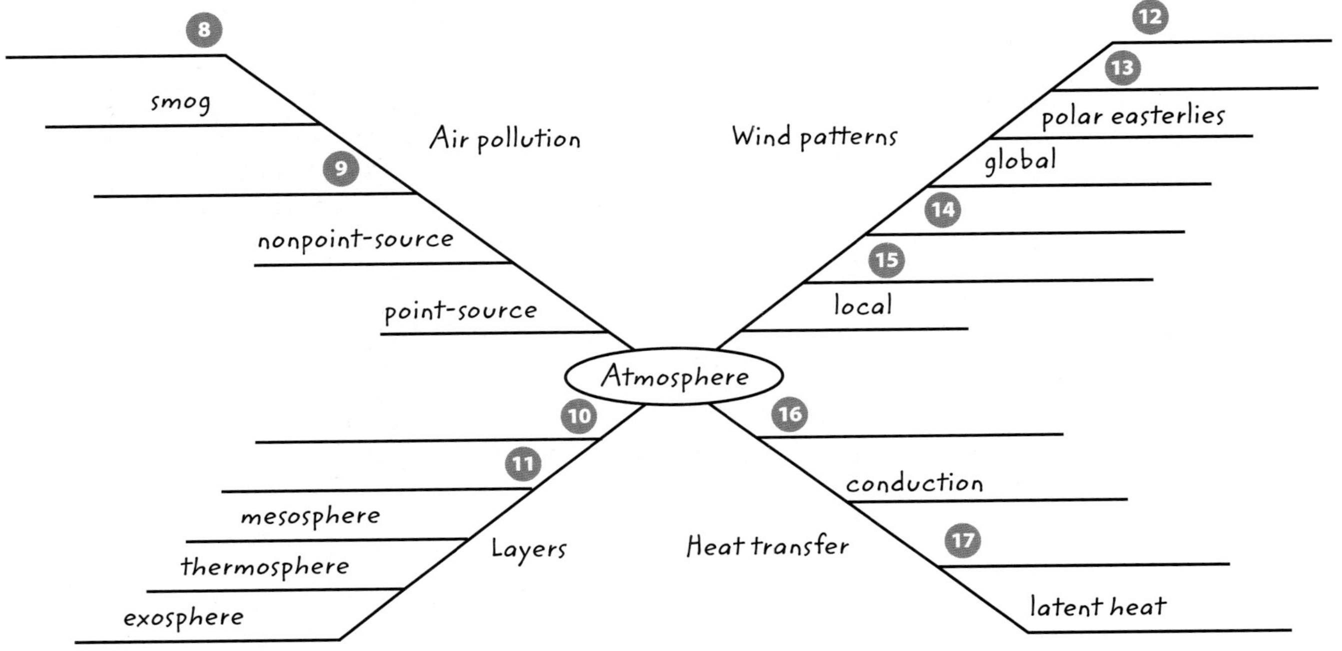

Chapter 4 Review

Understand Key Concepts

1 Air pressure is greatest SC.6.E.7.9

Ⓐ at a mountain base.
Ⓑ on a mountain top.
Ⓒ in the stratosphere.
Ⓓ in the ionosphere.

2 In which layer of the atmosphere is the ozone layer found? SC.6.E.7.9

Ⓐ troposphere
Ⓑ stratosphere
Ⓒ mesosphere
Ⓓ thermosphere

Use the image below to answer question 3.

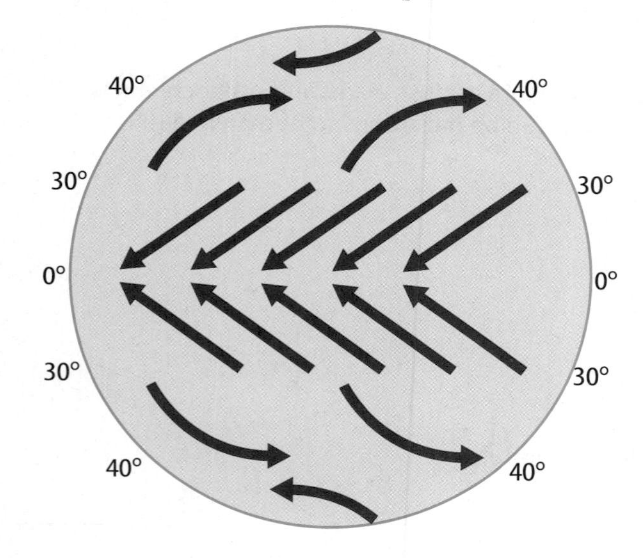

3 This diagram represents the atmosphere's SC.6.E.7.5

Ⓐ air masses.
Ⓑ global wind belts.
Ⓒ inversions.
Ⓓ particulate motion.

4 The Sun's energy SC.6.E.7.5

Ⓐ is completely absorbed by the atmosphere.
Ⓑ is completely reflected by the atmosphere.
Ⓒ is in the form of latent heat.
Ⓓ is transferred to the atmosphere after warming Earth.

Critical Thinking

5 **Predict** how atmospheric carbon dioxide levels might change if more trees were planted on Earth. Explain your prediction. SC.6.E.7.9

6 **Compare** visible and infrared radiation. SC.6.E.7.9

7 **Assess** whether your home is heated by conduction or convection. SC.6.E.7.1

8 **Interpret Graphics** What are the top three sources of particulate matter in the atmosphere? What could you do to reduce particulate matter from any of the sources shown here? LA.6.2.2.3

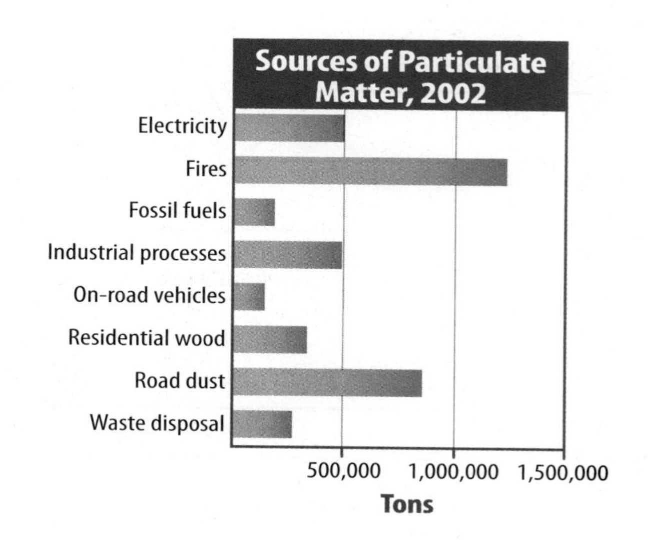

9 **Sequence** how the unequal heating of Earth's surface leads to the formation of wind. SC.6.E.7.5

10 **Evaluate** whether a sea breeze could occur at night. SC.6.E.7.5

Writing in Science

11 **Write** a paragraph on another sheet of paper explaining whether you think it would be possible to permanently pollute the atmosphere with particulate matter. LA.6.2.2.3

Big Idea Review

12 Review the title of each lesson in the chapter. List all of the characteristics and components of the troposphere and the stratosphere that affect life on Earth. Describe how life is impacted by each one. SC.6.E.7.9

13 Discuss how energy is transferred from the Sun throughout Earth's atmosphere. SC.6.E.7.5

Math Skills MA.6.A.3.6

Use Graphs

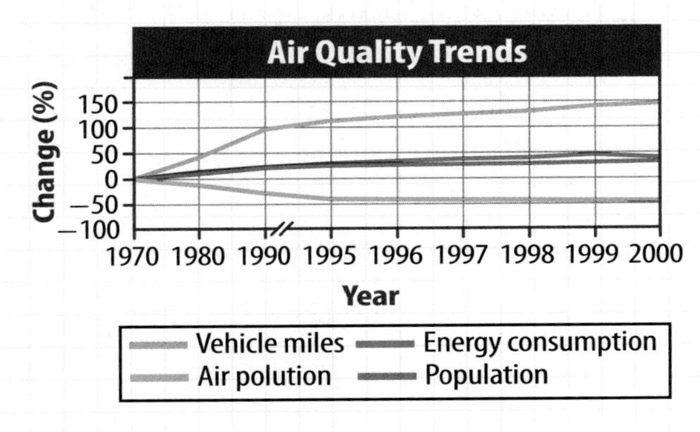

14 What was the percent change in energy usage between 1996 and 1999?

15 What happened to energy usage between 1999 and 2000?

16 What was the total percentage change between vehicle miles traveled and air pollution from 1970 to 2000?

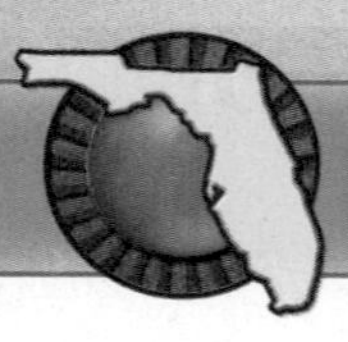

Florida NGSSS Benchmark Practice

Fill in the correct answer choice.

Multiple Choice

1 What causes the phenomenon known as a mountain wave? **SC.6.E.7.5**

Ⓐ radiation imbalance

Ⓑ rising and sinking air

Ⓒ temperature inversion

Ⓓ the greenhouse effect

Use the diagram below to answer question 2.

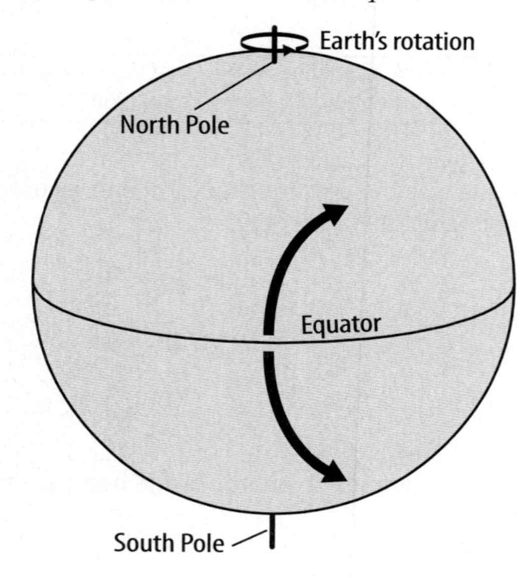

2 What phenomenon does the diagram above illustrate? **SC.6.E.7.5**

Ⓕ radiation balance

Ⓖ temperature inversion

Ⓗ the Coriolis effect

Ⓘ the greenhouse effect

3 What type of energy transfer occurs when the heat in the sand transfers to a person's feet? **SC.6.E.7.1**

Ⓐ convection

Ⓑ conduction

Ⓒ radiation

Ⓓ absorption

4 In which direction does moving air appear to turn in the Northern Hemisphere? **SC.6.E.7.5**

Ⓕ down

Ⓖ up

Ⓗ right

Ⓘ left

Use the diagram below to answer question 5.

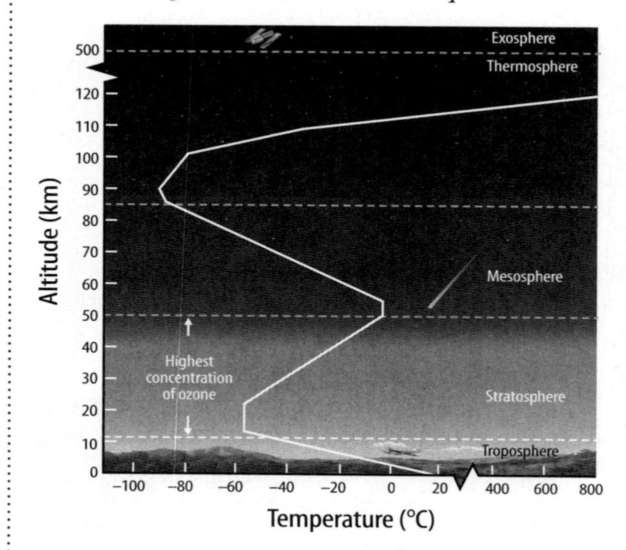

5 Which layer of the atmosphere has the widest range of temperatures? **SC.6.E.7.5**

Ⓐ mesosphere

Ⓑ stratosphere

Ⓒ thermosphere

Ⓓ troposphere

6 Plants and animals thrived on Earth after the ozone layer was formed. What does the ozone protect the Earth from? **SC.6.E.7.9**

Ⓕ solar radiation

Ⓖ meteor showers

Ⓗ acid rain

Ⓘ volcanic gas

7 Which is the primary cause of the global wind patterns on Earth? **SC.6.E.7.5**

Ⓐ ice cap melting
Ⓑ uneven heating
Ⓒ weather changing
Ⓓ waves breaking

Use the diagram below to answer question 8.

Energy Transfer Methods

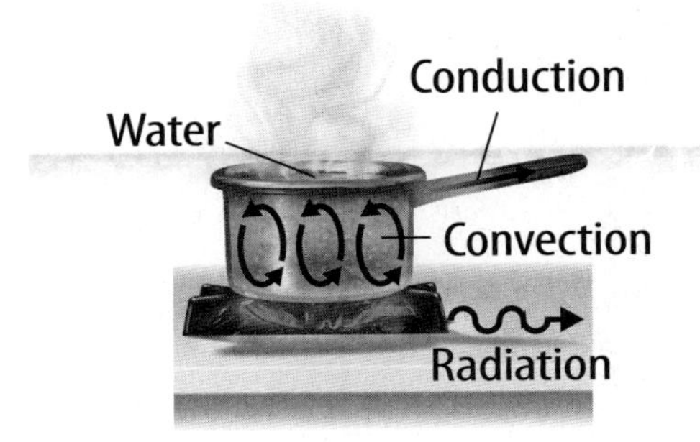

8 In the diagram above, which transfers thermal energy in the same way the Sun's energy is transferred to Earth? **SC.6.E.7.1**

Ⓕ the boiling water
Ⓖ the burner flame
Ⓗ the hot handle
Ⓘ the rising steam

9 Which substance in the air of U.S. cities has decreased least since the Clean Air Act began? **HE.6.C.1.3**

Ⓐ carbon monoxide
Ⓑ ground-level ozone
Ⓒ particulate matter
Ⓓ sulfur dioxide

10 Which heat-transfer process can be described as currents of warm air rising from Earth's surface as currents of cool air descend? **SC.6.E.7.1**

Ⓕ conduction
Ⓖ convection
Ⓗ latent heat
Ⓘ radiation

11 Which atmospheric layer absorbs the Sun's harmful ultraviolet rays? **SC.7.E.7.9**

Ⓐ mesosphere
Ⓑ stratosphere
Ⓒ thermosphere
Ⓓ troposphere

Use the table below to answer question 12.

Heat Transfer	
Type of Transfer	**How It Transfers**
Radiation	with rays or waves
Conduction	contact of material
Convection	flow of material

12 Heat can be transferred in several ways. The table describes types of heat transfer. Which is an example of conduction? **SC.6.E.7.1**

Ⓕ sunlight shining on a metal chair
Ⓖ fire heating a room
Ⓗ a metal pan burning a hand
Ⓘ hair dryer blowing hair

NEED EXTRA HELP?

If You Missed Question . . .	1	2	3	4	5	6	7	8	9	10	11	12
Go to Lesson . . .	2	3	2	3	1	1	3	2	4	2	1	2

Name ______________________ Date ______________

Multiple Choice *Bubble the correct answer.*

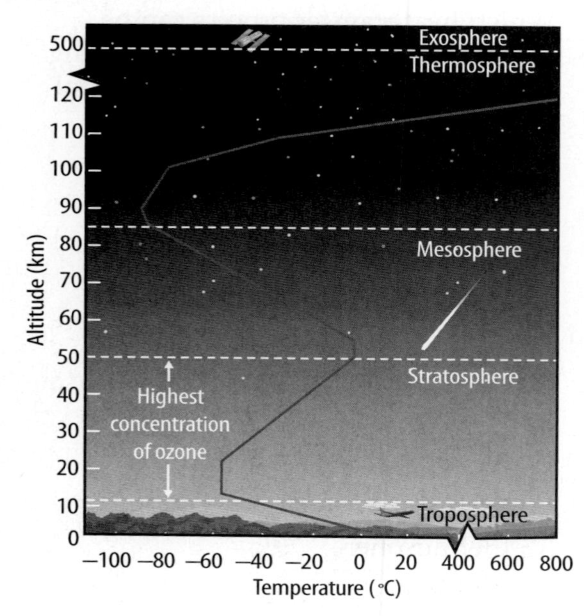

1. Look at the image above. In which layer of the atmosphere are the warmest air temperatures found? **SC.6.E.7.9**

Ⓐ exosphere

Ⓑ stratosphere

Ⓒ thermosphere

Ⓓ troposphere

2. In which layer of the atmosphere do most people spend their entire lives? **SC.6.E.7.9**

Ⓕ exosphere

Ⓖ mesosphere

Ⓗ stratosphere

Ⓘ troposphere

3. Which image shows solid particles that have been put into the atmosphere by human processes? **SC.6.E.7.9**

Ⓐ

Ⓑ

Ⓒ

Ⓓ

Name ______________________ Date ____________

Multiple Choice *Bubble the correct answer.*

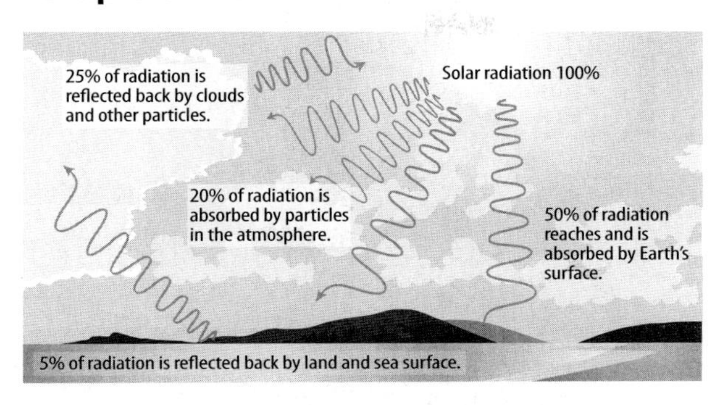

1. In the image above, what percentage of radiation is reflected back into the atmosphere by clouds and other particles? **SC.6.E.7.1**

(A) 20 percent

(B) 25 percent

(C) 50 percent

(D) 55 percent

2. Which gas is NOT an efficient greenhouse gas? **SC.6.E.7.9**

(F) helium

(G) methane

(H) carbon dioxide

(I) water vapor

3. In the image above, which type of heat transfer is represented by the number 1? **SC.6.E.7.1**

(A) conduction

(B) convection

(C) latent heat

(D) radiation

4. In what form does most solar radiation reach Earth? **SC.6.E.7.5**

(F) infrared light

(G) microwaves

(H) ultraviolet light

(I) visible light

Name ______________________ Date ______________

Multiple Choice *Bubble the correct answer.*

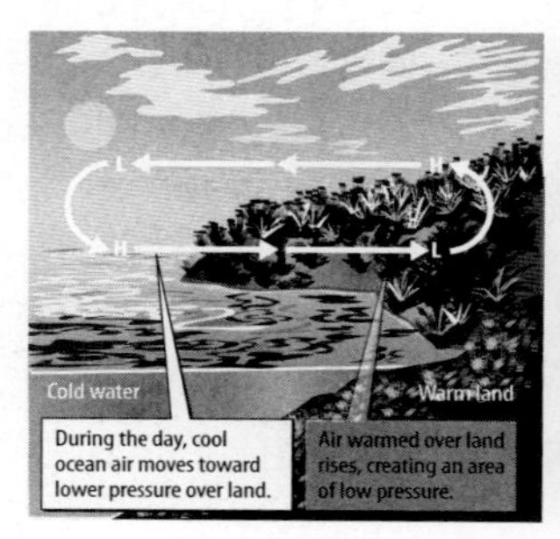

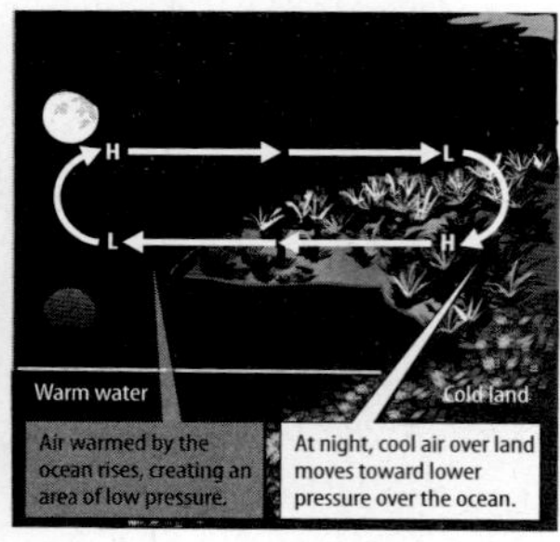

1. Look at the images above. What is the name of the nighttime circulation shown? **SC.6.E.7.3**

Ⓐ cold land

Ⓑ land breeze

Ⓒ sea breeze

Ⓓ warm water

2. What is the difference between the trade winds and the prevailing westerlies? **SC.6.E.7.5**

Ⓕ Only the prevailing westerlies are driven by the Coriolis effect.

Ⓖ Trade winds flow away from the equator; westerlies are steady and flow toward the equator.

Ⓗ Trade winds flow from west to east, and westerlies flow from east to west.

Ⓘ Unlike the trade winds, westerlies are not a result of a convection cell.

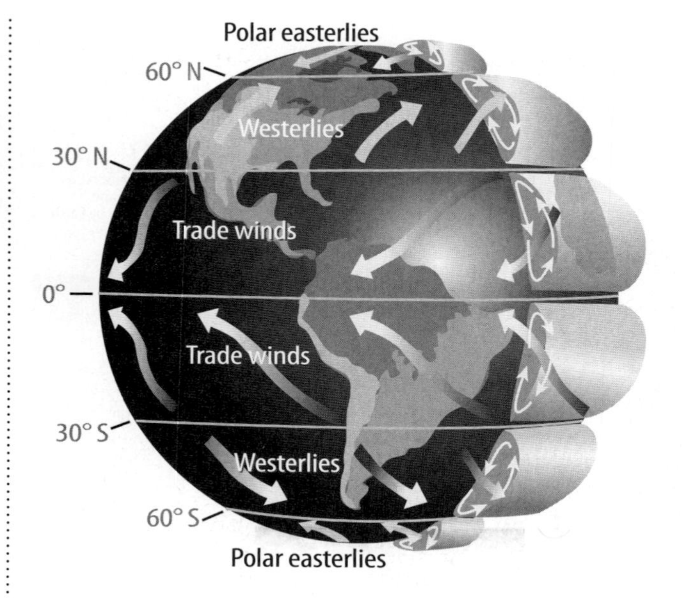

3. In the illustration above, between which latitudes is the first convection cell located? **SC.6.E.7.3**

Ⓐ at 0°

Ⓑ between 0° and 30°

Ⓒ between 30° and 60°

Ⓓ between 60° and the pole

Name ______________________________ Date ______________

Multiple Choice *Bubble the correct answer.*

1. Which image shows an example of nonpoint-source pollution? **SC.6.E.7.9**

Ⓐ

Ⓑ

Ⓒ

Ⓓ

2. Which type of pollution is a mixture of dust, acids, and other chemicals? **HE.6.C.1.3**

Ⓕ acid precipitation

Ⓖ ground-level ozone

Ⓗ particulate matter

Ⓘ photochemical smog

3. Which list is an example of indoor air pollutants? **HE.6.C.1.3**

Ⓐ carpets, furniture upholstery, radon

Ⓑ cleaning products, insects, rugs

Ⓒ fireplaces, foam insulation, trees

Ⓓ pesticides, grass, tobacco smoke

Name ______________________ Date ______________

Notes

Air Pressure Ideas

Six students looked at a barometer, a weather device that measures air pressure. They had different ideas about air pressure. This is what they thought:

Kimberly: I think air has to be moving to create air pressure.

Glenn: Air pressure is a downward force.

Rae: The higher in the atmosphere, the greater the air pressure.

Jeff: Air pressure is the same in all directions.

Oliver: Air pressure increases as the number of air particles increases.

Cameron: Warm air masses have higher pressure than cool air masses.

Jay: When air pressure decreases, it might rain or snow.

Circle the names of the students you agree with and describe why you agree. Describe your ideas about air pressure.

FLORIDA
Chapter 5

FLORIDA BIG IDEAS

1 **The Practice of Science**
2 **The Characteristics of Scientific Knowledge**
3 **The Role of Theories, Laws, Hypotheses, and Models**
7 **Earth Systems and Patterns**

Think About It!

How do scientists describe and predict weather?

Florida orange growers spray water onto their oranges to protect them from the cold weather. This icy protection prevents damage to the oranges.

1. Why do you think some parts of the country get colder than other areas?

2. How do you think scientists describe and predict weather?

Get Ready to Read

What do you think about weather?

Before you read, decide if you agree or disagree with each of these statements. As you read this chapter, see if you change your mind about any of the statements.

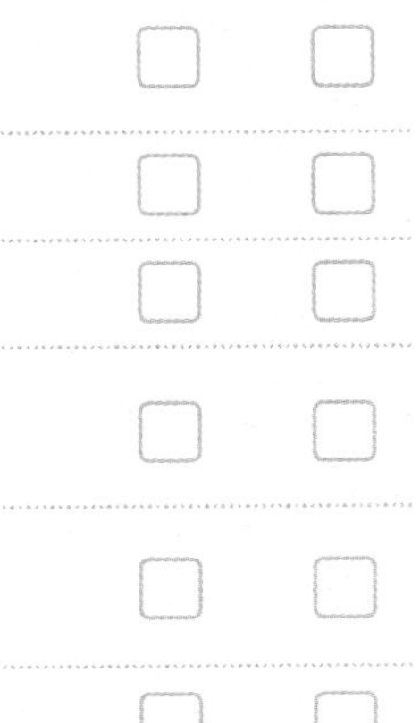

	AGREE	DISAGREE
1. Weather is the long-term average of atmospheric patterns of an area.		
2. All clouds are at the same altitude within the atmosphere.		
3. Precipitation often occurs at the boundaries of large air masses.		
4. There are no safety precautions for severe weather, such as tornadoes and hurricanes.		
5. Weather variables are measured every day at locations around the world.		
6. Modern weather forecasts are done using computers.		

There's More Online!
Video • Audio • Review • ⓘLab Station • WebQuest • Assessment • Concepts in Motion • Multilingual eGlossary

Lesson 1

Describing WEATHER

ESSENTIAL QUESTIONS

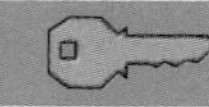
What is weather?

What variables are used to describe weather?

How is weather related to the water cycle?

Vocabulary

weather p. 165
air pressure p. 166
humidity p. 166
relative humidity p. 167
dew point p. 167
precipitation p. 169
water cycle p. 169

Florida NGSSS

LA.6.2.2.3 The student will organize information to show understanding (e.g., representing main ideas within text through charting, mapping, paraphrasing, summarizing, or comparing/contrasting);

SC.6.E.7.2 Investigate and apply how the cycling of water between the atmosphere and hydrosphere has an effect on weather patterns and climate.

SC.6.E.7.6 Differentiate between weather and climate.

SC.6.E.7.7 Investigate how natural disasters have affected human life in Florida.

SC.6.N.1.1 Define a problem from the sixth grade curriculum, use appropriate reference materials to support scientific understanding, plan and carry out scientific investigation of various types, such as systematic observations or experiments, identify variables, collect and organize data, interpret data in charts, tables, and graphics, analyze information, make predictions, and defend conclusions.

SC.6.N.1.3 Explain the difference between an experiment and other types of scientific investigation, and explain the relative benefits and limitations of each.

Launch Lab

SC.6.N.1.1, SC.6.N.1.3

15 minutes

How can you model the cycling of water?

Earth's water constantly cycles between the hydrosphere and atmosphere. How does this happen?

Procedure

1. Read and complete a lab safety form.
2. Half-fill a **500-mL beaker** with **ice** and **cold water.**
3. Pour 125 mL of **warm water** into a **resealable plastic bag** and seal the bag.
4. Carefully lower the bag into the ice water. Record your observations.

Data and Observations

Think About This

1. What do you observe when the warm water in the bag is put into the beaker?

2. What might happen next if the bag were placed into a beaker of warm water?

3. Key Concept What examples in the natural world result from the same process?

Why are clouds different?

1. If you look closely at the photo, you'll see that there are different types of clouds in the sky. How do clouds form? If all clouds consist of water droplets and ice crystals, why do they look different? Are clouds weather?

What is weather?

Everybody talks about the weather. "Nice day, isn't it?" "How was the weather during your vacation?" Talking about weather is so common that we even use weather terms to describe unrelated topics. "That homework assignment was a breeze." Or "I'll take a rain check."

Weather *is the atmospheric conditions, along with short-term changes, of a certain place at a certain time.* If you have ever been caught in a rainstorm on what began as a sunny day, you know the weather can change quickly. Sometimes it changes in just a few hours. But other times your area might have the same sunny weather for several days in a row.

2. **NGSSS Check**
Define What is weather? SC.6.E.7.6

Weather Variables

Perhaps some of the first things that come to mind when you think about weather are temperature and rainfall. As you dress in the morning, you need to know what the temperature will be throughout the day to help you decide what to wear. If it is raining, you might cancel your picnic.

Temperature and rainfall are just two of the **variables** used to describe weather. Meteorologists, scientists who study and predict weather, use several specific variables that describe a variety of atmospheric conditions. These include air temperature, air pressure, wind speed and direction, humidity, cloud coverage, and precipitation.

Review Vocabulary
variable
a quantity that can change

REVIEW VOCABULARY

kinetic energy
energy an object has due to its motion

Air Temperature

The measure of the average **kinetic energy** of molecules in the air is air temperature. When the temperature is high, molecules have a high kinetic energy. Therefore, molecules in warm air move faster than molecules in cold air. Air temperatures vary with time of day, season, location, and altitude.

Air Pressure

The pressure that a column of air exerts on the air or a surface below it is called **air pressure**. Study **Figure 1.** Is air pressure at Earth's surface more or less than air pressure at the top of the atmosphere? Air pressure decreases as altitude increases. Therefore, air pressure is greater at low altitudes than at high altitudes.

You might have heard the term *barometric pressure* during a weather forecast. Barometric pressure refers to air pressure. Air pressure is measured with an instrument called a barometer, shown in **Figure 2.** Air pressure is typically measured in millibars (mb). Knowing the barometric pressure of different areas helps meteorologists predict the weather.

Active Reading **3. Identify** Within the text, (circle) the instrument that measures air pressure.

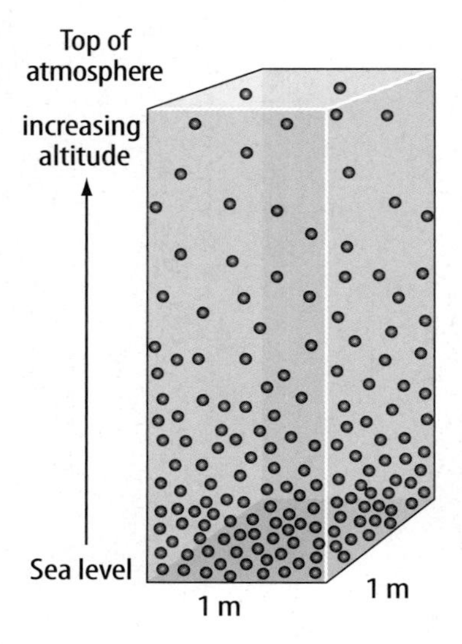

Figure 1 Increasing air pressure comes from having more molecules overhead.

Active Reading **4. Explain** What happens to air pressure as altitude decreases?

____

____

____

Wind

As air moves from areas of high pressure to areas of low pressure, it creates wind. Wind direction is the direction from which the wind is blowing. For example, the westerlies blow from west to east. Meteorologists measure wind speed using an instrument called an anemometer (a nuh MAH muh tur). An anemometer is also shown in **Figure 2.**

Humidity

The amount of water vapor in the air is called **humidity** (hyew MIH duh tee). Humidity can be measured in grams of water per cubic meter of air (g/m^3). When the humidity is high, there is more water vapor in the air. On a day with high humidity, your skin might feel sticky, and sweat might not evaporate from your skin as quickly.

Figure 2 Barometers, left, and anemometers, right, are used to measure weather variables.

Active Reading **5. Identify** (Circle) the instrument that measures wind speed.

Relative Humidity

Think about how a sponge can absorb water. At some point, it becomes full and cannot absorb any more water. In the same way, air can contain only a certain amount of water vapor. When air is saturated, it contains as much water vapor as possible. Temperature determines the maximum amount of water vapor air can contain. Warm air can contain more water vapor than cold air. *The amount of water vapor present in the air compared to the maximum amount of water vapor the air could contain at that temperature is called* **relative humidity**.

Relative humidity is measured using an instrument called a psychrometer and is given as a percent. For example, air with a relative humidity of 100 percent cannot hold any more moisture and dew or rain will form. Air that contains only half the water vapor it could hold has a relative humidity of 50 percent.

6. Explain Compare and contrast humidity and relative humidity.

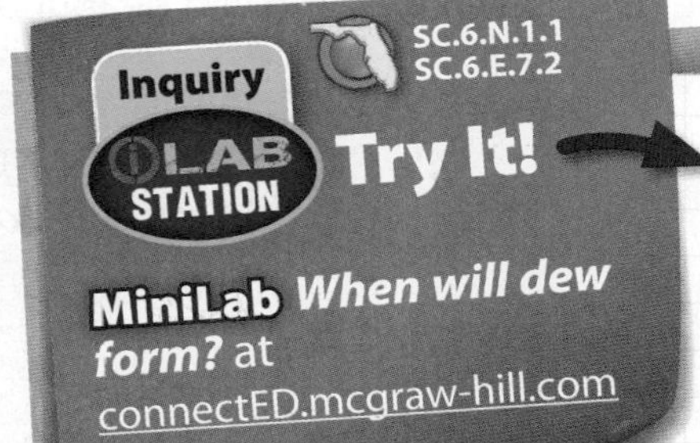

Apply It!

After you complete the lab, answer the questions below.

1. Use the greater than (>) symbol or less than (<) symbol to indicate which air temperature can hold the greater amount of water vapor.

   warm air ◯ cool air

2. Explain the meaning of the phrase, *relative humidity of 75 percent.*

Dew Point

When a sponge becomes saturated with water, the water starts to drip from the sponge. Similarly, when air becomes saturated with water vapor, the water vapor will condense and form water droplets. When air near the ground becomes saturated, the water vapor in air will condense to a liquid. If the temperature is above 0°C, dew forms. If the temperature is below 0°C, ice crystals, or frost, form. Higher in the atmosphere clouds form. The graph in **Figure 3** shows the total amount of water vapor that air can contain at different temperatures.

When the temperature decreases, the air can hold less moisture. As you just read, the air becomes saturated and dew forms. *The* **dew point** *is the temperature at which air is fully saturated because of decreasing temperatures while holding the amount of moisture constant.*

Figure 3 As air temperature increases, the air can hold more water vapor.

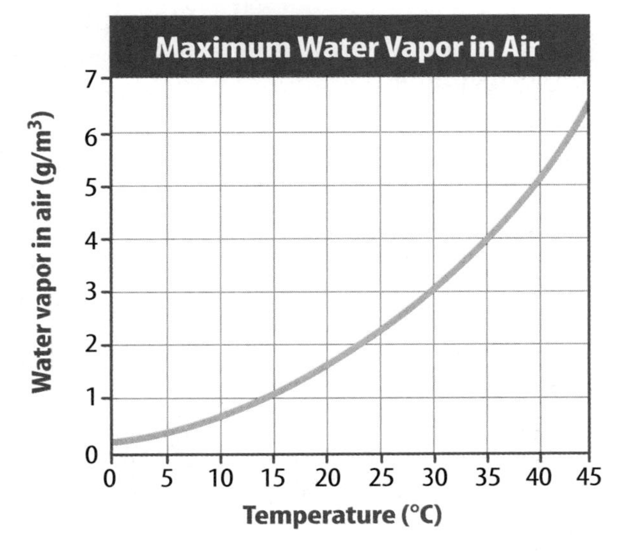

Figure 4 Clouds have different shapes and can be found at different altitudes.

Stratus clouds
- flat, white, and layered
- altitude up to 2,000 m

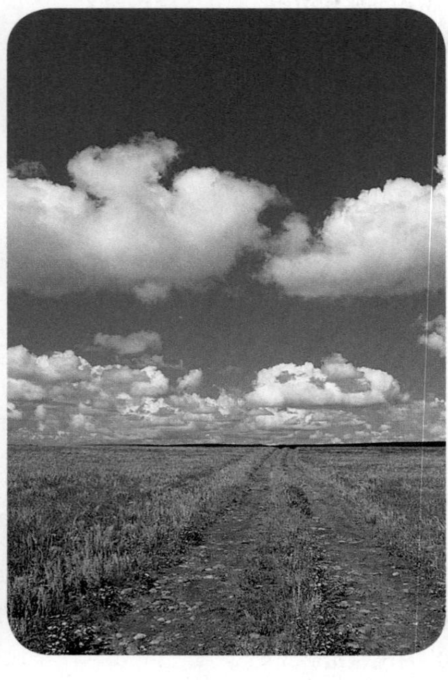

Cumulus clouds
- fluffy, heaped, or piled up
- 2,000 to 6,000 m altitude

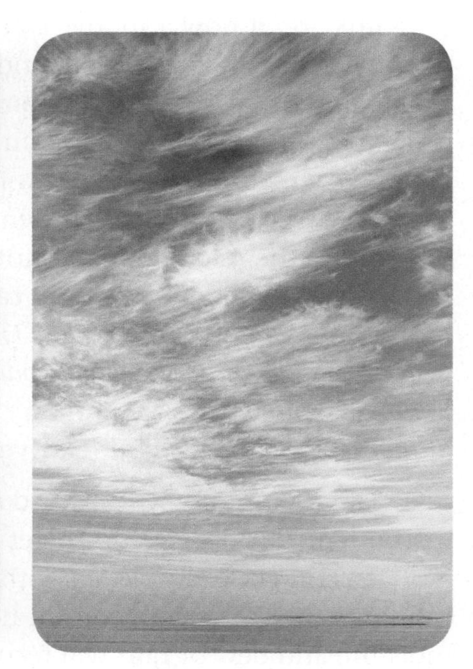

Cirrus clouds
- wispy
- above 6,000 m

Clouds and Fog

When you exhale outside on a cold winter day, you can see the water vapor in your breath condense into a foggy cloud in front of your face. This also happens when warm air containing water vapor cools as it rises in the atmosphere. When the cooling air reaches its dew point, water vapor condenses on small particles in the air and forms droplets. Surrounded by thousands of other droplets, these small droplets block and reflect light. This makes them visible as clouds.

Clouds are water droplets or ice crystals suspended in the atmosphere. Clouds can have different shapes and be present at different altitudes within the atmosphere. Different types of clouds are shown in **Figure 4.** Because we observe that clouds move, we recognize that water and thermal energy are transported from one location to another. Recall that clouds are also important in reflecting some of the Sun's incoming radiation.

A cloud that forms near Earth's surface is called fog. Fog is a suspension of water droplets or ice crystals close to or at Earth's surface. Fog reduces visibility, the distance a person can see into the atmosphere.

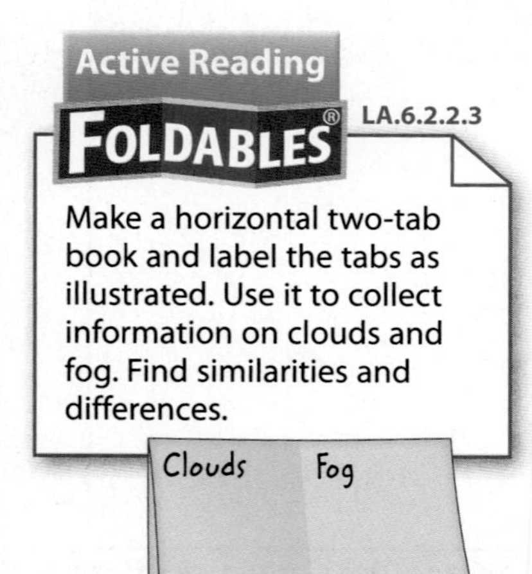

Active Reading 7. **Define** What is fog?

Precipitation

Recall that droplets in clouds form around small solid particles in the atmosphere. These particles might be dust, salt, or smoke. Precipitation occurs when cloud droplets combine and become large enough to fall back to Earth's surface. **Precipitation** *is water, in liquid or solid form, that falls from the atmosphere.* Examples of precipitation—rain, snow, sleet, and hail—are shown in **Figure 5.**

Rain is precipitation that reaches Earth's surface as droplets of water. Snow is precipitation that reaches Earth's surface as solid, frozen crystals of water. Sleet may originate as snow. The snow melts as it falls through a layer of warm air and refreezes when it passes through a layer of below-freezing air. Other times it is just freezing rain. Hail reaches Earth's surface as large pellets of ice. Hail starts as a small piece of ice that is repeatedly lifted and dropped by an updraft within a cloud. A layer of ice is added with each lifting. When it finally becomes too heavy for the updraft to lift, it falls to Earth.

8. NGSSS Check List What variables are used to describe weather? SC.6.E.7.6

The Water Cycle

Precipitation is an important process in the water cycle. Evaporation and condensation are phase changes that are also important to the water cycle. *The* **water cycle** *is the series of natural processes by which water continually moves among oceans, land, and the atmosphere.* As illustrated in **Figure 6,** most water vapor enters the atmosphere when water at the ocean's surface is heated and evaporates. Water vapor cools as it rises in the atmosphere and condenses back into a liquid. Eventually, droplets of liquid and solid water form clouds. Clouds produce precipitation, which falls to Earth's surface and later evaporates, continuing the cycle.

Types of Precipitation

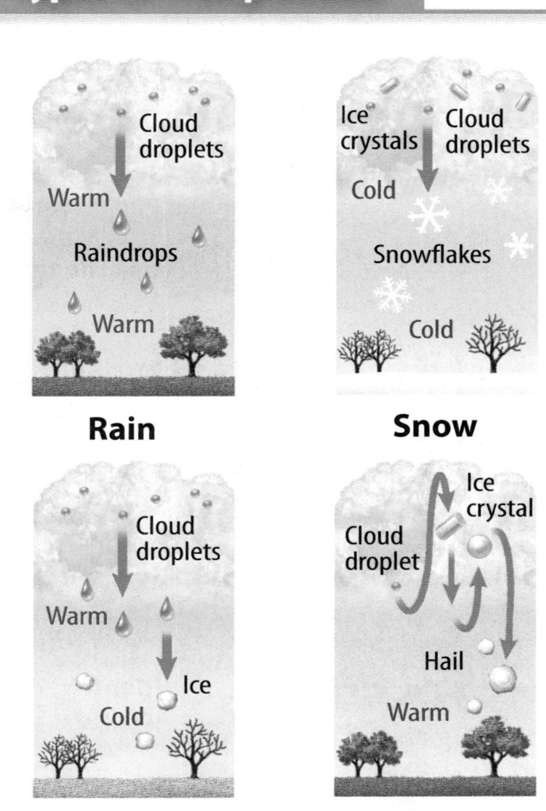

Figure 5 Rain, snow, sleet, and hail are forms of precipitation.

9. NGSSS Check Explain How is weather related to the water cycle? SC.6.E.7.2

The Water Cycle

Figure 6 The Sun's energy powers the water cycle, which is the continual movement of water between the ocean, the land, and the atmosphere.

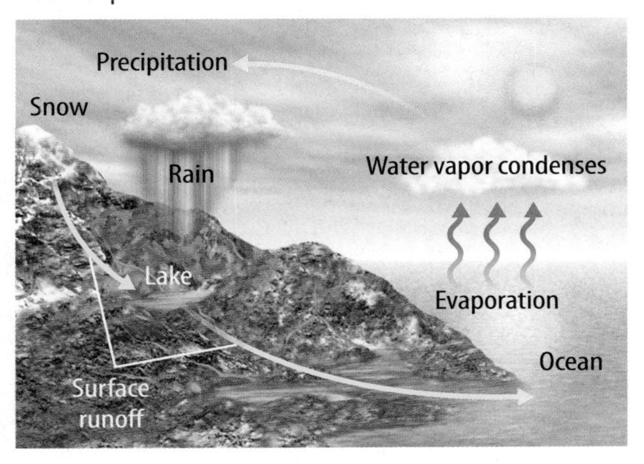

Lesson Review 1

Visual Summary

Weather is the atmospheric conditions, along with short-term changes, of a certain place at a certain time.

Weather variables include air temperature, air pressure, wind, humidity, and relative humidity.

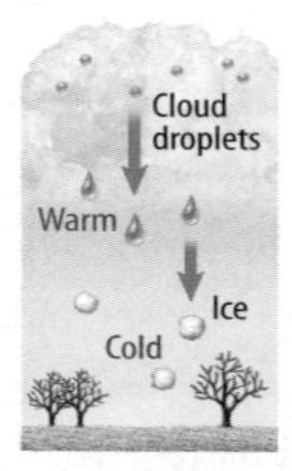

Forms of precipitation include rain, sleet, snow, and hail.

Use Vocabulary

1. **Define** *humidity* in your own words.

2. **Use the term** *precipitation* in a sentence.

3. The pressure that a column of air exerts on the surface below it is called ____________.

Understand Key Concepts

4. Which is NOT a standard weather variable?
   - (A) air pressure
   - (B) moon phase
   - (C) temperature
   - (D) wind speed

5. **Identify** and describe the different variables used to describe weather. SC.6.E.7.6

6. **Relate** humidity to cloud formation. SC.6.E.7.2

7. **Describe** how processes in the water cycle are related to weather. SC.6.E.7.2

Interpret Graphics

8. **Identify** Which type of precipitation is shown in the diagram below? How does this precipitation form?

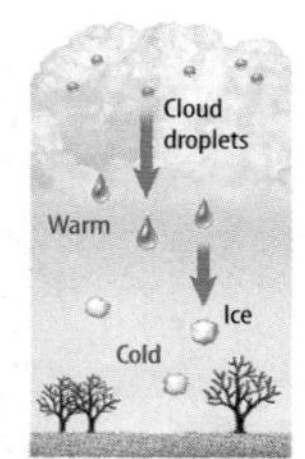

Critical Thinking

9. **Differentiate** among cloud formation, fog formation, and dew point.

FOCUS on FLORIDA

Is there a link between Hurricanes and Global Warming?

Scientists worry that hurricanes might be getting bigger and happening more often.

In the wake of the numerous hurricanes that have hit Florida and the Gulf Coast, many wonder whether global warming is responsible. If warm oceans are the fuel for hurricanes, could rising temperatures cause stronger or more frequent hurricanes?

Climatologists, scientists that study the climate, have several ways to investigate this question. They examine past hurricane activity, sea surface temperature, and other climate data. They compare these different types of data and look for patterns. Then the scientists put climate and hurricane data into equations. A computer solves these equations and makes computer models. Scientists analyze the models to see whether there is a connection between hurricane activity and different climate variables.

What have scientists learned? So far they have not found a link between warming oceans and the frequency of hurricanes. However, they have found a connection between warming oceans and hurricane strength. Models suggest that rising ocean temperatures might create more destructive hurricanes with stronger winds and more rainfall.

But global warming is not the only cause of warming oceans. As the ocean circulates, it goes through cycles of warming and cooling. Data show that the Atlantic Ocean has been in a warming phase for the past few decades.

Whether due to global warming or natural cycles, ocean temperatures are expected to rise even more in coming years. While rising ocean temperatures might not produce more hurricanes, climate research shows they could produce more powerful hurricanes.

It's Your Turn

RESEARCH Florida's geographical location leaves them in the path of many storms. Investigate how natural disasters have affected Florida residents. SC.6.E.7.7

Lesson 2

Weather PATTERNS

ESSENTIAL QUESTIONS

- What are two types of pressure systems?
- What drives weather patterns?
- Why is it useful to understand weather patterns?
- What are some examples of severe weather?

Vocabulary

high-pressure system p. 173
low-pressure system p. 173
air mass p. 174
front p. 176
tornado p. 179
hurricane p. 180
blizzard p. 181

Florida NGSSS

LA.6.2.2.3 The student will organize information to show understanding (e.g., representing main ideas within text through charting, mapping, paraphrasing, summarizing, or comparing/contrasting);

MA.6.A.3.6 Construct and analyze tables, graphs, and equations to describe linear functions and other simple relations using both common language and algebraic notation.

SC.6.E.7.2 Investigate and apply how the cycling of water between the atmosphere and hydrosphere has an effect on weather patterns and climate.

SC.6.E.7.8 Describe ways human beings protect themselves from hazardous weather and sun exposure.

SC.6.N.1.1 Define a problem from the sixth grade curriculum, use appropriate reference materials to support scientific understanding, plan and carry out scientific investigation of various types, such as systematic observations or experiments, identify variables, collect and organize data, interpret data in charts, tables, and graphics, analyze information, make predictions, and defend conclusions.

SC.6.N.2.1 Distinguish science from other activities involving thought.

SC.6.N.2.2 Explain that scientific knowledge is durable because it is open to change as new evidence or interpretations are encountered.

SC.6.N.2.3 Recognize that scientists who make contributions to scientific knowledge come from all kinds of backgrounds and possess varied talents, interests, and goals.

Inquiry Launch Lab

LA.6.2.2.3

10 minutes

How can temperature affect pressure?

Air molecules that have low energy can be packed closely together. As energy is added to the molecules, they begin to move and bump into one another.

Procedure

1. Read and complete a lab safety form.
2. Close a **resealable plastic bag** except for a small opening. Insert a **straw** through the opening and blow air into the bag until it is as firm as possible. Remove the straw and quickly seal the bag.
3. Submerge the bag in a **container** of **ice water** and hold it there for 2 minutes. Record your observations.
4. Remove the bag from the ice water and submerge it in **warm water** for 2 minutes. Record your observations.

Data and Observations

Think About This

1. What do the results tell you about the movement of air molecules in cold air and in warm air?

2. Key Concept What property of the air is demonstrated in this activity?

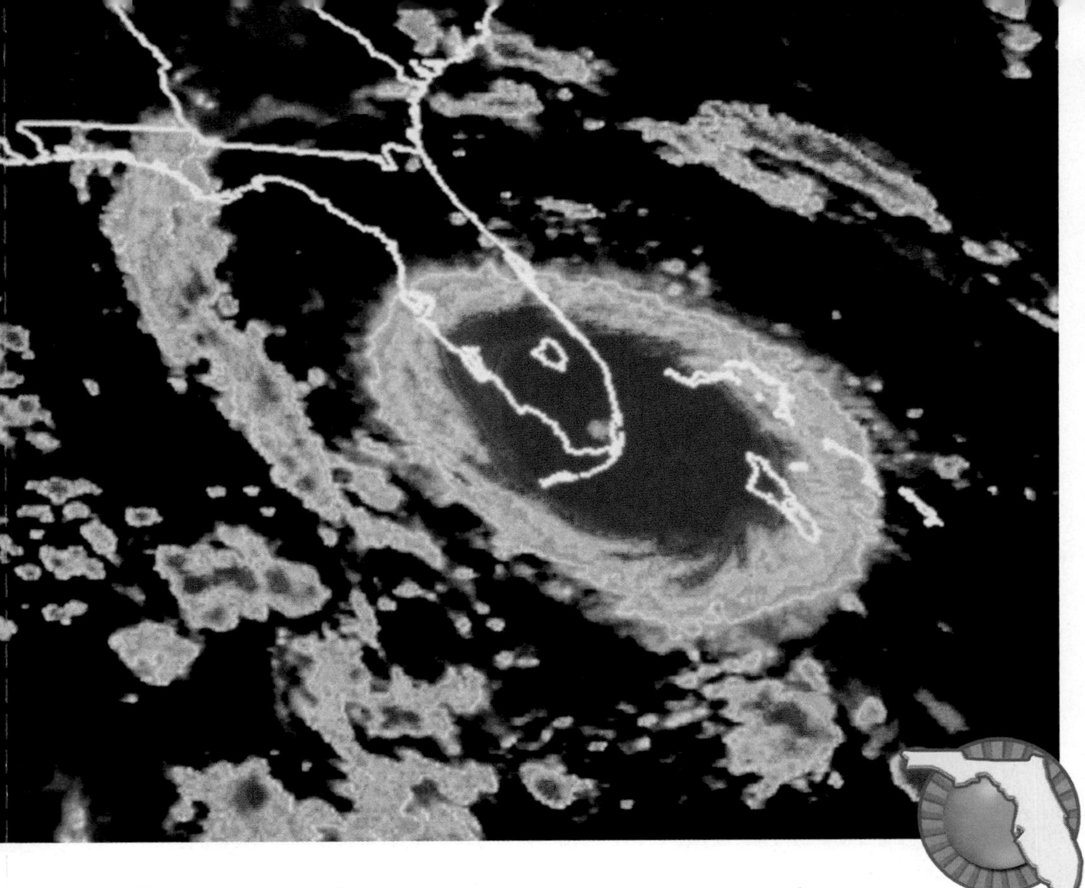

Inquiry **Is this a storm?**

1. Surging waves, strong winds, and heavy rain from Hurricane Andrew caused flooding in Florida. Why are hurricanes and other types of severe weather dangerous? How does severe weather form?

Pressure Systems

Weather is often associated with pressure systems. Recall that air pressure is the weight of the molecules in air. Cool air molecules are closer together than warm air molecules. Cool air has higher pressure than warm air.

A **high-pressure system**, shown in **Figure 7,** *is a large body of circulating air with high pressure at its center and lower pressure outside the system.* Because dense air sinks, it moves away from the center to areas of low pressure. High-pressure systems bring sunny skies and fair weather.

A **low-pressure system**, also shown in **Figure 7,** *is a large body of circulating air with low pressure at its center and higher pressure outside the system.* This causes air inside the low pressure system to rise. The rising air cools and the water vapor condenses, forming clouds and sometimes precipitation—rain or snow.

Active Reading **2. Describe** Compare and contrast two types of pressure systems.

Figure 7 Air moving from areas of high pressure to areas of low pressure is called wind.

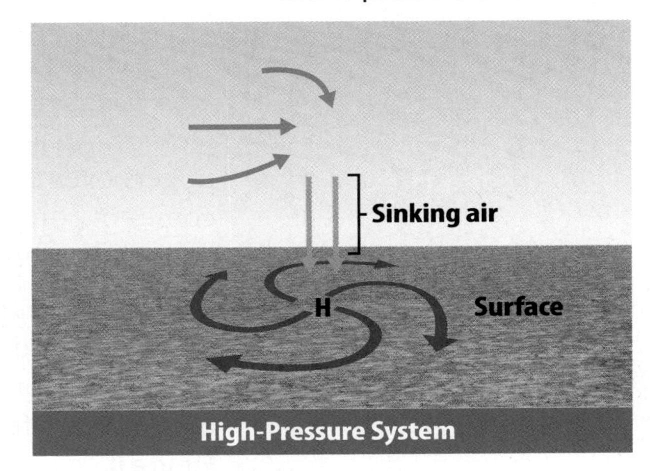

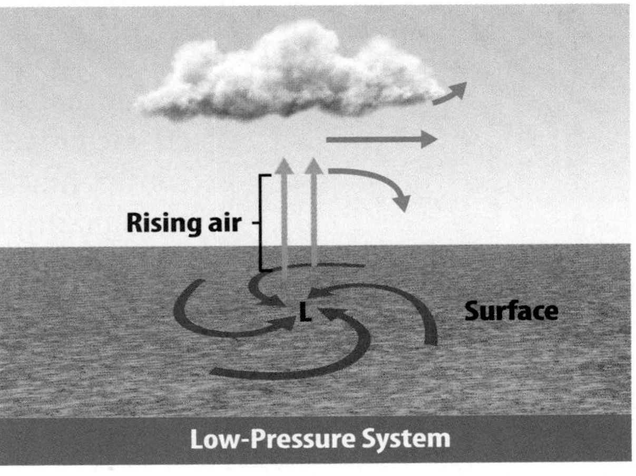

Figure 8 Five main air masses impact climate across North America.

Active Reading **3. Identify** Where does continental polar air come from?

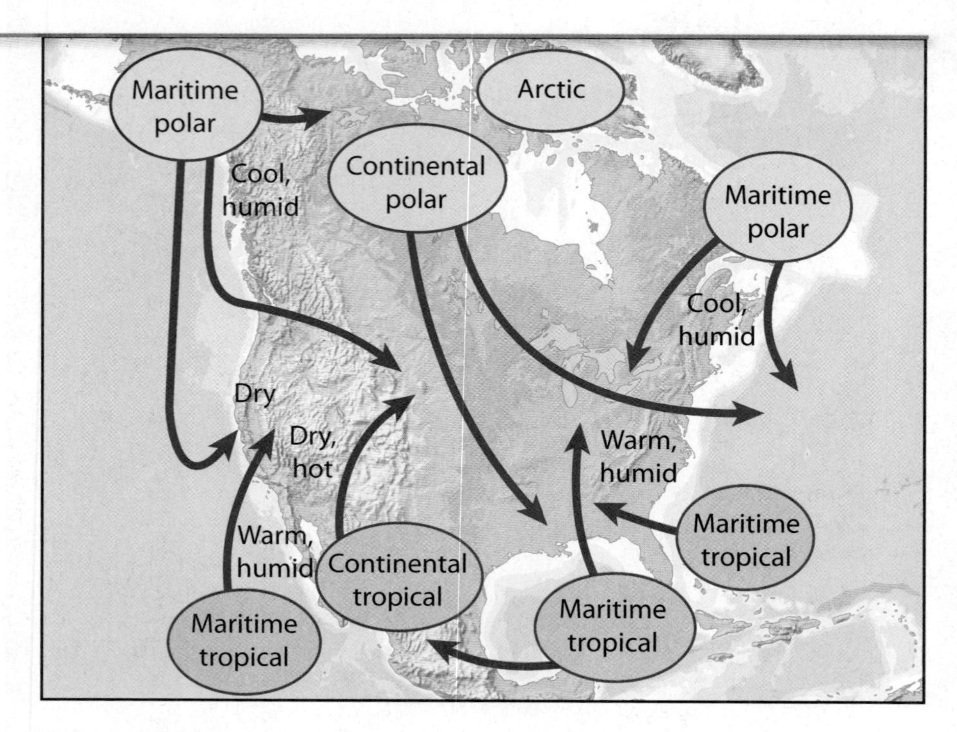

LA.6.2.2.3

Fold a sheet of paper into thirds along the long axis. Label the outside *Air Masses.* Make another fold about 2 inches from the long edge of the paper to make a three-column chart. Label as shown.

Active Reading **4. Identify** Underline the term that names air masses that form over water.

Air Masses

Have you ever noticed that the weather sometimes stays the same for several days in a row? For example, during winter in the northern United States, extremely cold temperatures often last for three or four days in a row. Afterward, several days might follow with warmer temperatures and snow showers.

Air masses are responsible for this pattern. **Air masses** *are large bodies of air that have uniform temperature, humidity, and pressure.* An air mass forms when a large high pressure system lingers over an area for several days. As a high pressure system comes in contact with Earth, the air in the system takes on the temperature and moisture characteristics of the surface below it.

Like high- and low-pressure systems, air masses can extend for a thousand kilometers or more. Sometimes one air mass covers most of the United States. Examples of the main air masses that affect weather in the United States are shown in **Figure 8.**

Air Mass Classification

Air masses are classified by their temperature and moisture characteristics. Air masses that form over land are referred to as continental air masses. Those that form over water are referred to as maritime masses. Warm air masses that form in the equatorial regions are called tropical. Those that form in cold regions are called polar. Air masses near the poles, over the coldest regions of the globe, are called arctic and antarctic air masses.

Arctic Air Masses Forming over Siberia and the Arctic are arctic air masses. They contain bitterly cold, dry air. During winter, an arctic air mass can bring temperatures down to −40°C.

Continental Polar Air Masses Because land cannot transfer as much moisture to the air as oceans can, air masses that form over land are drier than air masses that form over oceans. Continental polar air masses are fast-moving and bring cold temperatures in winter and cool weather in summer. Find the continental polar air masses over Canada in **Figure 8.**

Maritime Polar Air Masses Forming over the northern Atlantic and Pacific Oceans, maritime polar air masses are cold and humid. They often bring cloudy, rainy weather.

Continental Tropical Air Masses Because they form in the tropics over dry, desert land, continental tropical air masses are hot and dry. They bring clear skies and high temperatures. Continental tropical air masses usually form during summer.

Maritime Tropical Air Masses As shown in **Figure 8,** maritime tropical air masses form over the western Atlantic Ocean, the Gulf of Mexico, and the eastern Pacific Ocean. These moist air masses bring hot, humid air to Florida and the southeastern United States during summer. In winter, they can bring heavy snowfall.

Air masses can change as they move over the land and ocean. Warm, moist air can move over land and become cool and dry. Cold, dry air can move over water and become moist and warm.

5. **NGSSS Check** **Explain** What drives weather patterns?
SC.6.E.7.2

Math Skills

MA.6.A.3.6

Conversions

To convert Fahrenheit (°F) units to Celsius (°C) units, use this equation:

$$°C = \frac{(°F - 32)}{1.8}$$

Convert **76°F** to °C

1. Always perform the operation in parentheses first.
   (**76°F** − 32 = **44°F**)
2. Divide the answer from Step 1 by 1.8.
   $\frac{44°F}{1.8} = 24°C$

To convert °C to °F, follow the same steps using the following equation:

$$°F = (°C \times 1.8) + 32$$

Practice

6. Convert 86°F to °C.

7. Convert 37°C to °F.

Inquiry LAB STATION SC.6.N.2.1, LA.6.2.2.3

Try It!

MiniLab *How can you observe air pressure?* at connectED.mcgraw-hill.com

Apply It! After you complete the lab, answer this question.

1. Write the five types of air masses and the conditions they create.

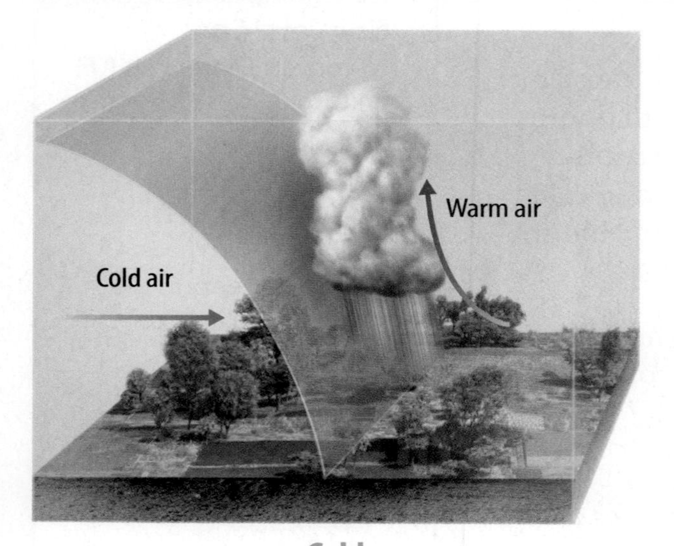

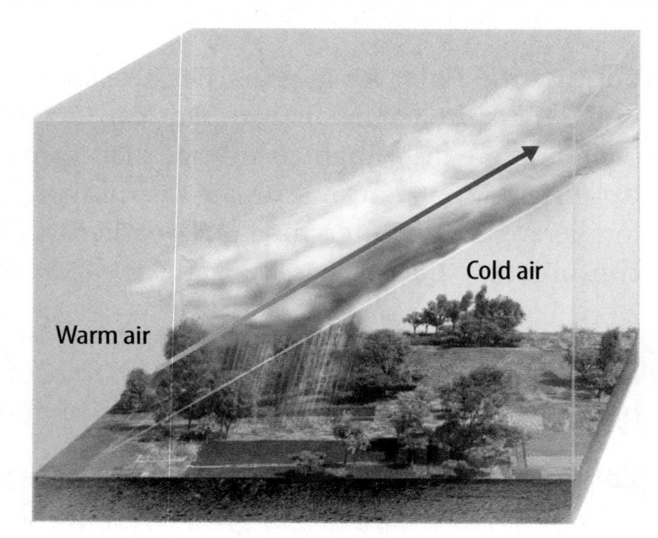

Figure 9 Certain types of fronts are associated with specific weather.

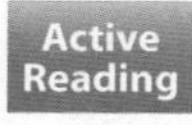
8. Describe What is the difference between a cold front and a warm front?

Science Use v. Common Use
front
Science Use a boundary between two air masses
Common Use the foremost part or surface of something

Fronts

In 1918, Norwegian meteorologist Jacob Bjerknes (BYURK nehs) and his coworkers were busy developing a new method for forecasting the weather. Bjerknes noticed that specific types of weather occur at the boundaries between different air masses. Because he was trained in the army, Bjerknes used a military term to describe this boundary—front.

A military front is the boundary between opposing armies in a battle. *A weather* **front**, *however, is a boundary between two air masses.* Drastic weather changes often occur at fronts. As wind carries an air mass away from the area where it formed, the air mass will eventually interact with another air mass. Changes in temperature, humidity, clouds, wind, and precipitation are common at fronts.

Cold Fronts

When a colder air mass moves toward a warmer air mass, a cold front forms, as shown in **Figure 9.** The cold air, which is denser than the warm air, pushes underneath the warm air mass. The warm air rises and cools. Water vapor in the air condenses and clouds form. Showers and thunderstorms often form along cold fronts. It is common for temperatures to decrease as much as 10°C when a cold front passes through. The wind becomes gusty and changes direction. In many cases, cold fronts give rise to severe storms.

Active Reading **9. State** What types of weather are associated with cold fronts?

Stationary

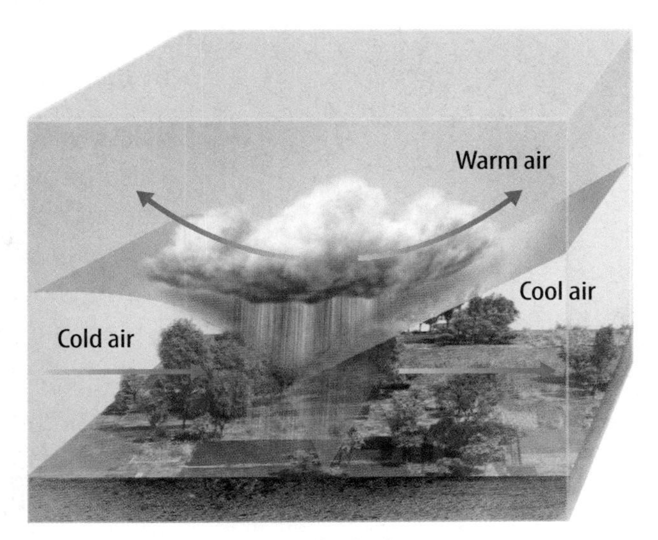

Occluded

Warm Fronts

As shown in **Figure 9,** a warm front forms when less dense, warmer air moves toward colder, denser air. The warm air rises as it glides above the cold air mass. When water vapor in the warm air condenses, it creates a wide blanket of clouds. These clouds often bring steady rain or snow for several hours or even days. A warm front not only brings warmer temperatures, but it also causes the wind to shift directions.

Both a cold front and a warm front form at the edge of an approaching air mass. Because air masses are large, the movement of fronts is used to make weather forecasts. When a cold front passes through your area, temperatures will remain low for the next few days. When a warm front arrives, the weather will become warmer and more humid.

Stationary and Occluded Fronts

Sometimes an approaching front will stall for several days with warm air on one side of it and cold air on the other side. When the boundary between two air masses stalls, the front is called a stationary front. Study the stationary front shown in **Figure 9.** Cloudy skies and light rain are found along stationary fronts.

Cold fronts move faster than warm fronts. When a fast-moving cold front catches up with a slow-moving warm front, an occluded or blocked front forms. Occluded fronts, shown in **Figure 9,** usually bring precipitation.

10. NGSSS Check Explain Why is it useful to understand weather patterns associated with fronts? **SC.6.E.7.8**

Severe Weather

Some weather events can cause major damage, injuries, and death. These events, such as thunderstorms, tornadoes, hurricanes, and blizzards, are called severe weather.

Thunderstorms

Also known as electrical storms because of their lightning, thunderstorms have warm temperatures, moisture, and rising air, which may be supplied by a low-pressure system. When these conditions occur, a cumulus cloud can grow into a 10-km-tall thundercloud, or cumulonimbus cloud, in as little as 30 minutes.

Academic Vocabulary

dominate
(verb) to exert the guiding influence on

A typical thunderstorm has a three-stage life cycle, shown in **Figure 10.** The cumulus stage is **dominated** by cloud formation and updrafts. Updrafts are air currents moving vertically away from the ground. After the cumulus cloud has been created, downdrafts begin to appear. Downdrafts are air currents moving vertically toward the ground. In the mature stage, heavy winds, rain, and lightning dominate the area. Within 30 minutes of reaching the mature stage, the thunderstorm begins to fade, or dissipate. In the dissipation stage, updrafts stop, winds die down, lightning ceases, and precipitation weakens.

Active Reading **11. Describe** What happens during each stage of a thunderstorm?

Strong updrafts and downdrafts within a thunderstorm cause millions of tiny ice crystals to rise and sink, crashing into each other. This creates positively and negatively charged particles in the cloud. The difference in the charges of particles between the cloud and the charges of particles on the ground eventually creates electricity. This is seen as a bolt of lightning. Lightning can move from cloud to cloud, cloud to ground, or ground to cloud. Lightening can heat the nearby air to more than 27,000°C. Air molecules near the bolt rapidly expand and then contract, creating the sound identified as thunder. Florida has the highest number of lightening strikes in the United States. There are more injuries from lightening in Florida that all other states combined.

Figure 10 Thunderstorms have distinct stages characterized by the direction in which air is moving.

Thunderstorms

Cumulus Stage

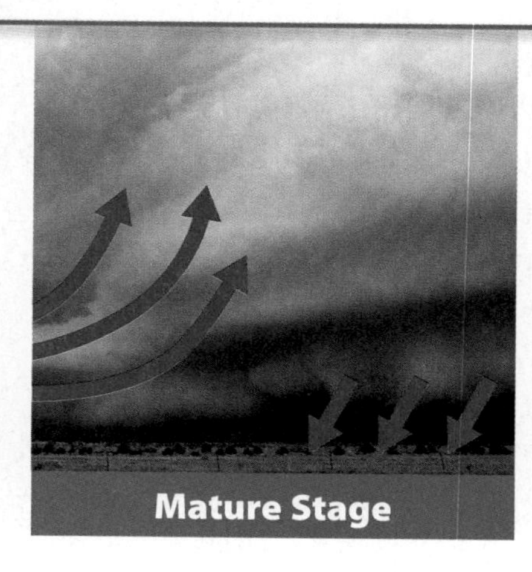
Mature Stage

Dissipation Stage

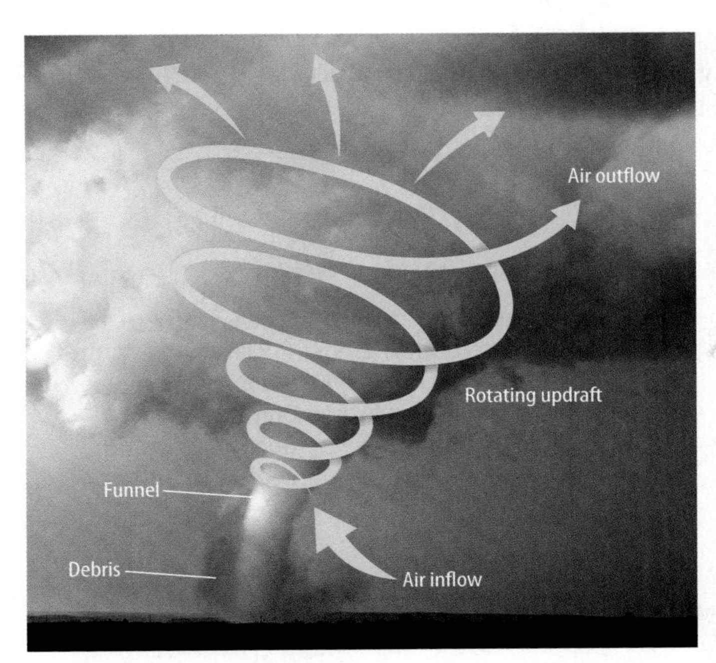

Figure 11 A funnel cloud forms when updrafts within a thunderstorm begin rotating.

Tornadoes

Perhaps you have seen photos of the damage from a tornado. *A* **tornado** *is a violent, whirling column of air in contact with the ground.* Most tornadoes have a diameter of several hundred meters. The largest tornadoes exceed 1,500 m in diameter. The intense, swirling winds within tornadoes can reach speeds of more than 400 km/h. These winds are strong enough to send cars, trees, and even entire houses flying through the air. Tornadoes usually last only a few minutes. More destructive tornadoes, however, can last for several hours.

Formation of Tornadoes When thunderstorm updrafts begin to rotate, as shown in **Figure 11,** tornadoes can form. Swirling winds spiral downward from the thunderstorm's base, creating a funnel cloud. When the funnel reaches the ground, it becomes a tornado. Although the swirling air is invisible, you can easily see the debris lifted by the tornado.

Active Reading **12. Describe** How do tornadoes form?

Tornado Alley More tornadoes occur in the United States than anywhere else on Earth. The central United States, from Nebraska to Texas, experiences the most tornadoes. This area has been nicknamed Tornado Alley. In this area, cold air blowing southward from Canada frequently collides with warm, moist air moving northward from the Gulf of Mexico. These conditions are ideal for severe thunderstorms and tornadoes.

Florida Tornadoes Florida tornadoes generally occur between the months of June through September. About 60 percent of all Florida tornadoes occur during this time. However, the most severe tornadoes tend to occur in Florida during the months of January through April. These months tend to have more powerful tornadoes due to the presence of the jet stream. When a jet stream moves south into Florida and is accompanied by a strong cold front and thunderstorms, the jet stream can strengthen the thunderstorm. This type of powerful storm can produce dangerous downbursts of winds, hail, and deadly tornadoes. Florida tornadoes that are strong and violent can take place any time of day or night. A tornado occurring in the evening could be very dangerous because most people are asleep and cannot receive weather warnings.

Active Reading **13. Identify** Underline why tornadoes are more dangerous at night?

Classifying Tornadoes Dr. Ted Fujita developed a method to classify tornadoes based on the damage they cause. On the modified Fujita intensity scale, F0 tornadoes cause light damage, breaking tree branches and damaging billboards. F1 though F4 tornadoes cause moderate to devastating damage, including tearing roofs from homes, derailing trains, and throwing vehicles in the air. F5 tornadoes cause incredible damage, such as demolishing concrete and steel buildings and pulling the bark from trees.

Active Reading **14. Identify** Circle the name in the text given to the area in the central United States that experiences the most tornadoes.

Figure 12 Hurricanes consist of alternating bands of heavy precipitation and sinking air.

Hurricane Formation

1. As warm, moist air rises, it cools, water vapor condenses, and clouds form. As more air rises, it creates an area of low pressure over the ocean.
2. As air continues to rise, a tropical depression forms. Tropical depressions bring thunderstorms with winds between 37–62 km/h.
3. Air continues to rise, rotating counterclockwise. The storm builds to a tropical storm with winds in excess of 63 km/h. It produces strong thunderstorms.
4. When winds exceed 119 km/h, the storm becomes a hurricane. Only one percent of tropical storms become hurricanes.

Inside a Hurricane

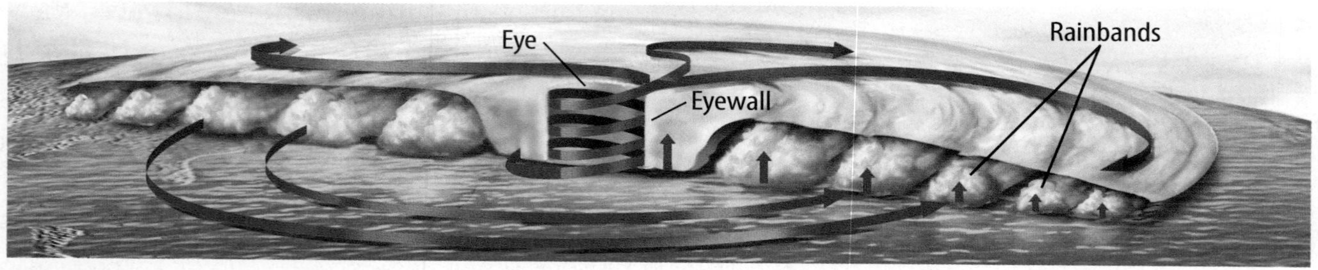

Active Reading **15. Describe** How do hurricanes that strike Florida form?

____

____

Hurricanes

An intense tropical storm with winds exceeding 119 km/h is a **hurricane**. Hurricanes are the most destructive storms on Earth. Like tornadoes, hurricanes have a circular shape with intense, swirling winds. However, hurricanes do not form over land. Hurricanes typically form in late summer over warm, tropical ocean water. Florida's peak hurricane season is between the months of August and October. **Figure 12** sequences the steps in hurricane formation. A typical hurricane is 480 km across, more than 150 thousand times larger than a tornado. At the center of a hurricane is the eye, an area of clear skies and light winds.

Word Origin

hurricane
from Spanish *huracan*, means "tempest"

Damage from hurricanes occurs as a result of strong winds and flooding. While still out at sea, hurricanes create high waves that can flood coastal areas. As a hurricane crosses the coastline, strong rains intensify and can flood entire areas. But once a hurricane moves over land or colder water, it loses its energy and dissipates.

Active Reading **16. Recognize** Highlight the terms in the text that cause damage during a hurricane.

Florida's unique geographical location leaves them in a vulnerable position. Developing storms often steer toward the state. Almost every year tropical storms threaten Florida, and most years Florida experiences a direct strike from a tropical storm. Hurricanes, on the other hand, are not as frequent, but they're still rather common in this area.

Winter Storms

Winter weather can be severe and hazardous. Ice storms, as shown in **Figure 13,** can down power lines and tree branches and make driving dangerous. *A* **blizzard** *is a violent winter storm characterized by freezing temperatures, strong winds, and blowing snow.* During a blizzard, blowing snow reduces visibility, and freezing temperatures can cause frostbite and hypothermia (hi poh THER mee uh).

Active Reading **17. Find** Underline in the text the results of freezing temperatures on human beings.

Freezing Rain

Figure 13 The weight of ice from freezing rain can cause trees, power lines, and other structures to break.

Active Reading **18. State** What are some examples of severe weather?

Severe Weather Safety

To help keep people safe, the U.S. National Weather Service issues watches and warnings during severe weather events. A watch means that severe weather is possible. A warning means that severe weather is already occurring. Heeding severe weather watches and warnings is important and could save your life.

It is also important to know how to protect yourself during dangerous weather. During thunderstorms, you should stay inside if possible, and stay away from metal objects and electrical cords. If you are outside, stay away from water, high places, and isolated trees. Dressing properly is important in all kinds of weather. When windchill temperatures are below −20°C, you should dress in layers, keep your head and fingers covered, and limit your time outdoors.

Not all weather safety pertains to bad weather. The Sun's ultraviolet (UV) radiation can cause health risks, including skin cancer. The U.S. National Weather Service issues a daily UV Index Forecast. Precautions on sunny days include covering up, using sunscreen, and wearing a hat and sunglasses. Surfaces such as snow, water, and beach sand can double the effects of the Sun's UV radiation.

Active Reading **19. Locate** Circle in the text a health risk created by the Sun.

Lesson Review 2

Visual Summary

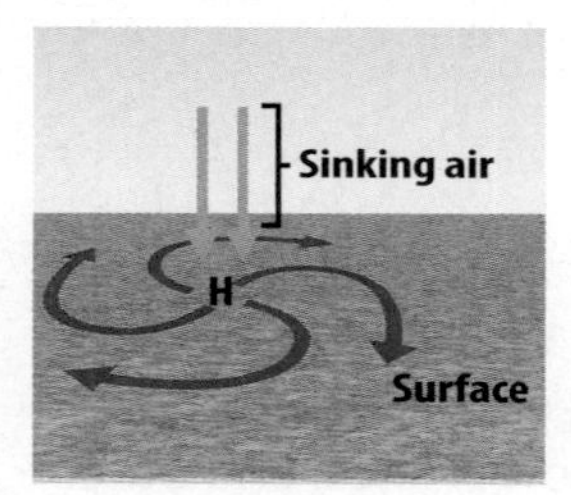

Low-pressure systems, high-pressure systems, and air masses all influence weather.

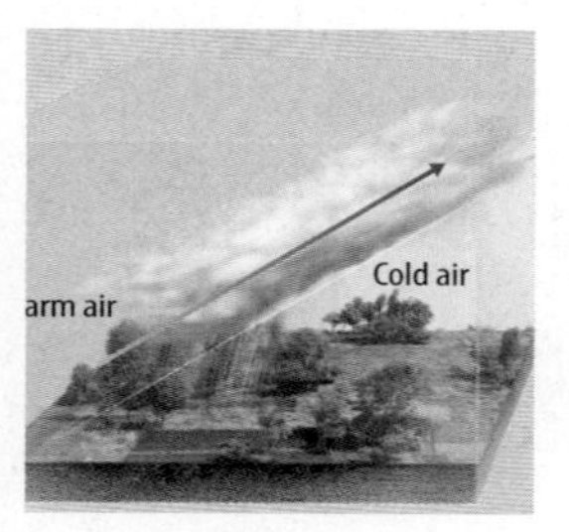

Weather often changes as a front passes through an area.

The National Weather Service issues warnings about severe weather such as thunderstorms, tornadoes, hurricanes, and blizzards.

Use Vocabulary

1. **Distinguish** between an air mass and a front.

2. **Define** *low-pressure system* using your own words.

Understand Key Concepts

3. Which air mass is humid and warm? SC.6.E.7.2
   - (A) continental polar
   - (B) continental tropical
   - (C) maritime polar
   - (D) maritime tropical

4. **Compare and contrast** hurricanes and tornadoes. SC.6.E.7.8

Interpret Graphics

5. **Compare and Contrast** Fill in the graphic organizer below to compare and contrast high-pressure and low-pressure systems.

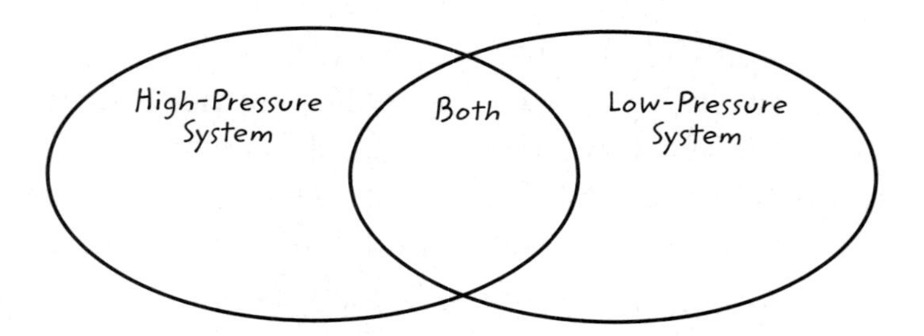

Critical Thinking

6. **Design** a pamphlet that contains tips on how to stay safe during different types of severe weather. SC.6.E.7.8

Math Skills MA.6.A.3.6

7. Convert 212°F to °C.

8. Convert 20°C to °F.

Hurricane Andrew Hits the Florida Coasts

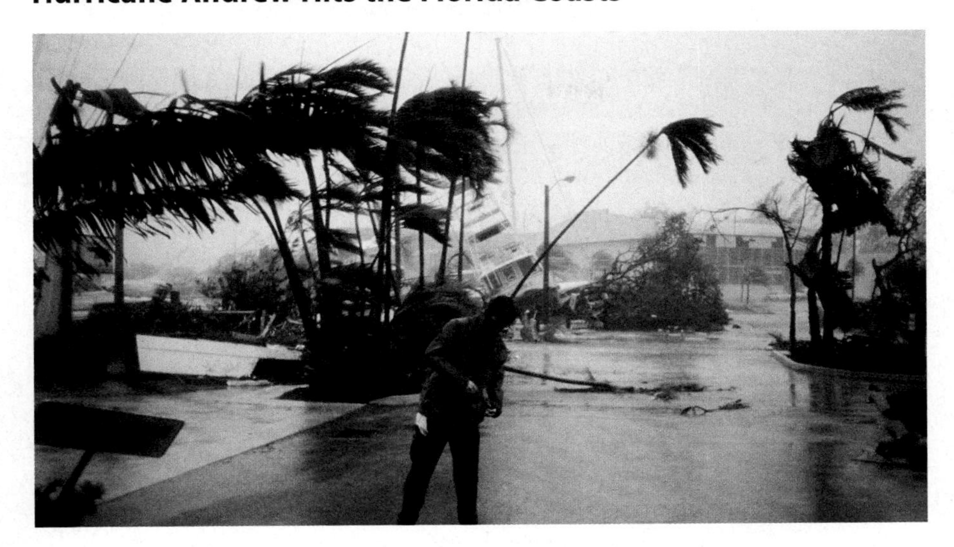

Sequence the steps in the formation of a hurricane.

1. Warm, moist air ____________ and ____________. Water vapor ____________, and clouds form. As more air rises, an area of ____________ forms over the ocean.

⇩

2. As air ____________, a ____________ forms. Air begins to turn ____________ because of the ____________. Winds are between ____________.

⇩

3. As air continues to rise and ____________, the storm builds to a ____________. Winds are greater than ____________ but less than ____________.

⇩

4. When winds reach ____________, the storm becomes a ____________.

Lesson 3

Weather FORECASTS

ESSENTIAL QUESTIONS

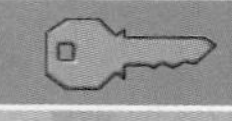 What instruments are used to measure weather variables?

 How are computer models used to predict the weather?

Vocabulary

surface report p. 185
upper-air report p. 185
Doppler radar p. 186
isobar p. 187
computer model p. 188

Florida NGSSS

LA.6.2.2.3 The student will organize information to show understanding (e.g., representing main ideas within text through charting, mapping, paraphrasing, summarizing, or comparing/contrasting);

MA.6.S.6.2 Select and analyze the measures of central tendency or variability to represent, describe, analyze, and/or summarize a data set for the purposes of answering questions appropriately.

SC.6.E.7.7 Investigate how natural disasters have affected human life in Florida.

SC.6.N.1.1 Define a problem from the sixth grade curriculum, use appropriate reference materials to support scientific understanding, plan and carry out scientific investigation of various types, such as systematic observations or experiments, identify variables, collect and organize data, interpret data in charts, tables, and graphics, analyze information, make predictions, and defend conclusions.

SC.6.N.2.1 Distinguish science from other activities involving thought.

SC.6.N.3.4 Identify the role of models in the context of the sixth grade science benchmarks.

Inquiry Launch Lab

LA.6.2.2.3

10 minutes

Can you understand the weather report?

Weather reports use numbers and certain vocabulary terms to help you understand the weather conditions in a given area for a given time period. Listen to a weather report for your area. Can you record all the information reported?

Procedure

1. Make a list of data you would expect to hear in a weather report.
2. Listen carefully to a **recording of a weather report** and jot down numbers and measurements you hear next to those on your list.
3. Listen a second time and make adjustments to your original notes, such as adding more data, if necessary.
4. Listen a third time, then share the weather forecast as you heard it.

Think About This

1. What measurements were difficult for you to apply to understanding the weather report?

2. Why are so many different types of data needed to give a complete weather report?

3. List the instruments that might be used to collect each kind of data.

4. Key Concept Where do meteorologists obtain the data they use to make a weather forecast?

What's inside?

1. Information about weather variables is collected by the weather radar station shown below. Data, such as the amount of rain falling in a weather system, help meteorologists make accurate predictions about severe weather. What other instruments do meteorologists use to forecast weather? How do they collect and use data?

Measuring the Weather

Being a meteorologist is like being a doctor. Using specialized instruments and visual observations, the doctor first measures the condition of your body. The doctor later combines these measurements with his or her knowledge of medical science. The result is a forecast of your future health, such as "You'll feel better in a few days if you rest and drink plenty of fluids."

Similarly, meteorologists, scientists who study weather, use specialized instruments to measure the conditions of the atmosphere, as you read in Lesson 1. These instruments include thermometers to measure temperature, barometers to measure air pressure, psychrometers to measure relative humidity, and anemometers to measure wind speed.

Surface and Upper-Air Reports

A **surface report** *describes a set of weather measurements made on Earth's surface.* Weather variables are measured by a weather station—a collection of instruments that report temperature, air pressure, humidity, precipitation, and wind speed and direction. Cloud amounts and visibility are often measured by human observers.

An **upper-air report** *describes wind, temperature, and humidity conditions above Earth's surface.* These atmospheric conditions are measured by a radiosonde (RAY dee oh sahnd), a package of weather instruments carried many kilometers above the ground by a weather balloon. Radiosonde reports are made twice a day simultaneously at hundreds of locations around the world.

Satellite and Radar Images

Images taken from satellites orbiting about 35,000 km above Earth provide information about weather conditions on Earth. A visible light image, such as the one shown in **Figure 14,** shows white clouds over Earth. The infrared image, also shown in **Figure 14,** shows infrared energy in false color. The infrared energy comes from Earth and is stored in the atmosphere as latent heat. Monitoring infrared energy provides information about cloud height and atmospheric temperature.

Figure 14 Meteorologists use visible light and infrared satellite images to identify fronts and air masses.

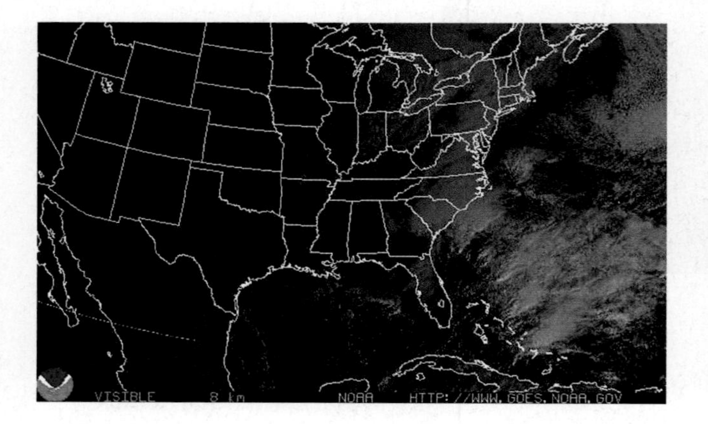

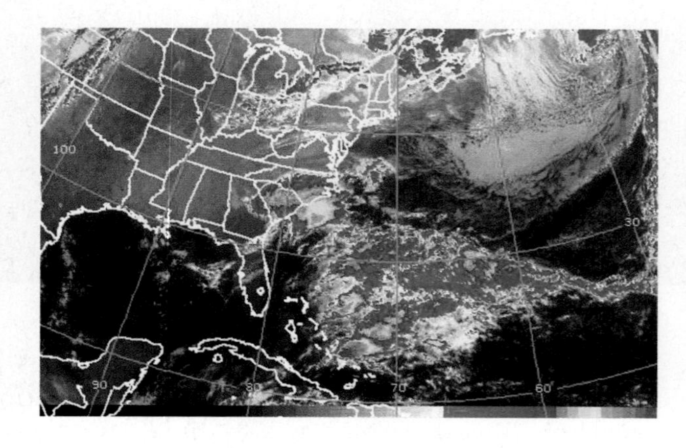

Active Reading **2. Identify** Circle the infrared satellite image and place an *X* on the light satellite image.

Radar measures precipitation when radio waves bounce off raindrops and snowflakes. **Doppler radar** *is a specialized type of radar that can detect precipitation as well as the movement of small particles, which can be used to approximate wind speed.* Because the movement of precipitation is caused by wind, Doppler radar can be used to estimate wind speed. This can be especially important during severe weather, such as tornadoes or thunderstorms.

Active Reading **3. Identify** What are the weather variables that radiosondes, infrared satellites, and Doppler radar measure?

Active Reading FOLDABLES® LA.6.2.2.3

Make a horizontal two-tab book and label the tabs as illustrated. Use it to collect information on satellite and radar images. Compare and contrast these information tools.

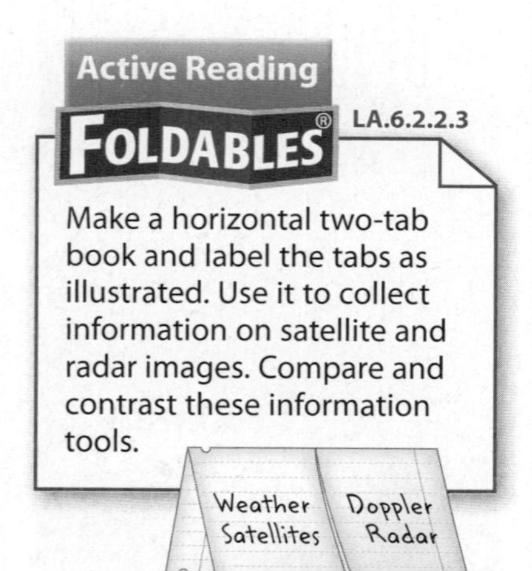

Weather Maps

Every day, thousands of surface reports, upper-air reports, and satellite and radar observations are made around the world. Meteorologists have developed tools that help them simplify and understand this enormous amount of weather data.

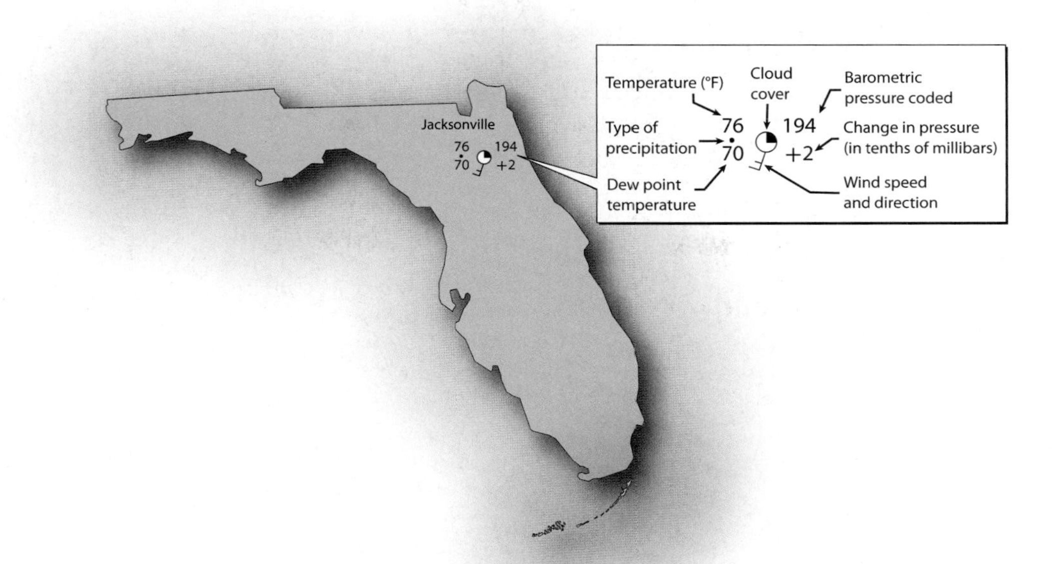

The Station Model

As shown in **Figure 15,** the station model diagram displays data from many different weather measurements for a particular location. It uses numbers and symbols to display data and observations from surface reports and upper-air reports.

Figure 15 Station models contain information about weather variables.

Mapping Temperature and Pressure

In addition to station models, weather maps also have other symbols. For example, **isobars** *are lines that connect all places on a map where pressure has the same value.* Locate an isobar on the map in **Figure 16.** Isobars show the location of high- and low-pressure systems. Isobars also provide information about wind speed. Winds are strong when isobars are close together. Winds are weaker when isobars are farther apart.

In a similar way, isotherms (not shown) are lines that connect places with the same temperature. Isotherms show which areas are warm and which are cold. Fronts are represented as lines with symbols on them, as indicated in **Figure 16.**

WORD ORIGIN

isobar
from Greek *isos,* means "equal"; and *baros,* means "heavy"

Active Reading **4. Analyze** Compare isobars and isotherms.

Weather Map

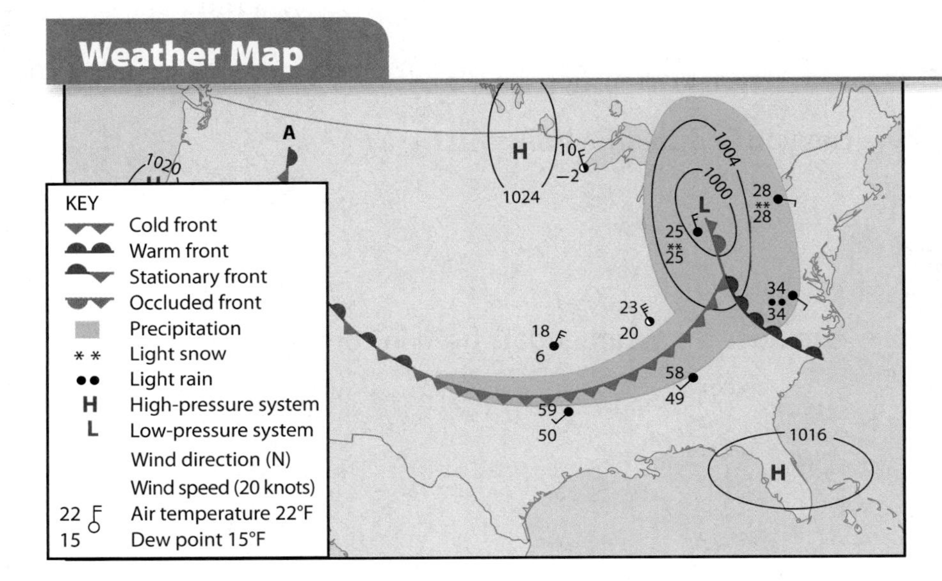

Figure 16 Weather maps contain symbols that provide information about the weather.

Active Reading **5. Locate** (Circle) the symbols that represent high-pressure and low-pressure systems.

Figure 17 Meteorologists analyze data from various sources—such as radar and computer models—in order to prepare weather forecasts.

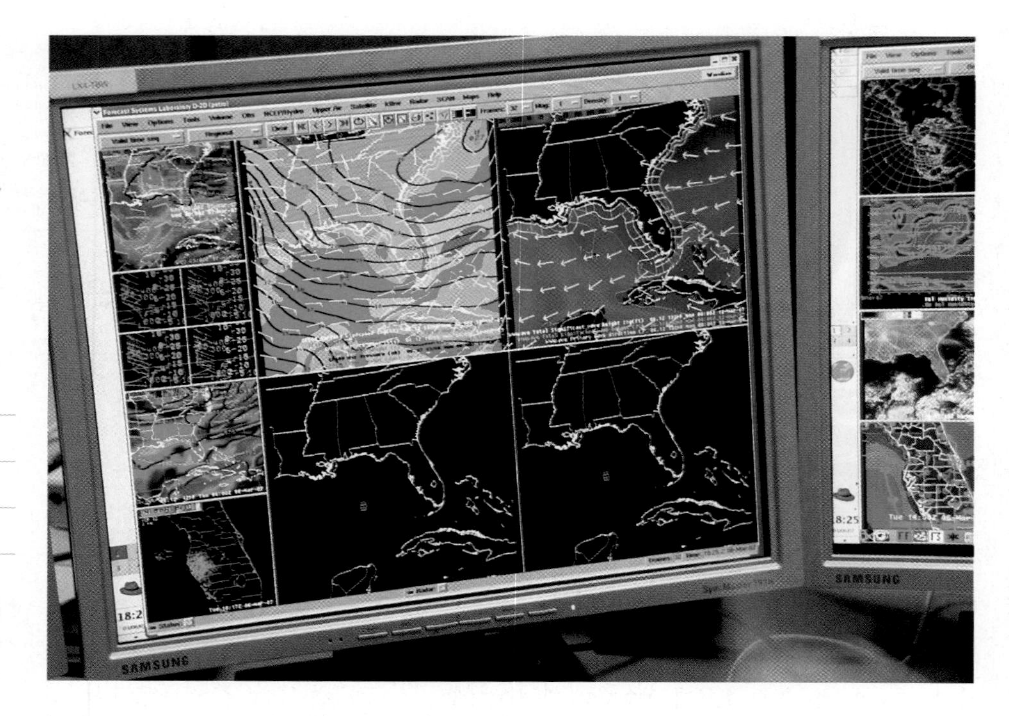

6. NGSSS Check
Describe How are computers used to predict the weather? SC.6.N.3.4

Predicting the Weather

Modern weather forecasts are made with the help of computer models, such as the ones shown in **Figure 17.** **Computer models** *are detailed computer programs that solve a set of complex mathematical formulas.* The formulas predict what temperatures and winds might occur, when and where it will rain and snow, and what types of clouds will form.

Government meteorological offices also use computers and the Internet to exchange weather measurements continuously throughout the day. Weather maps are drawn and forecasts are made using computer models. Then, through television, radio, newspapers, and the Internet, the maps and forecasts are made available to the public.

Inquiry SC.6.N.1.1, LA.6.2.2.3

iLAB STATION Try It!

MiniLab *How is weather represented on a map?* at connectED.mcgraw-hill.com

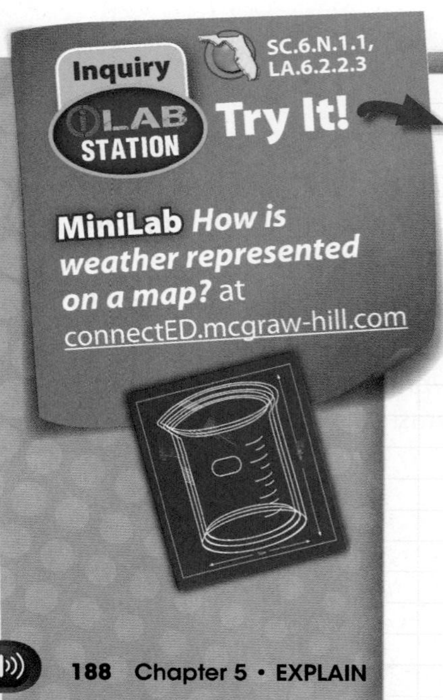

Apply It! After you complete the lab, answer these questions.

1. Where can you find weather forecasts?

2. From where does the weather data used in computer models come ?

Lesson Review 3

Visual Summary

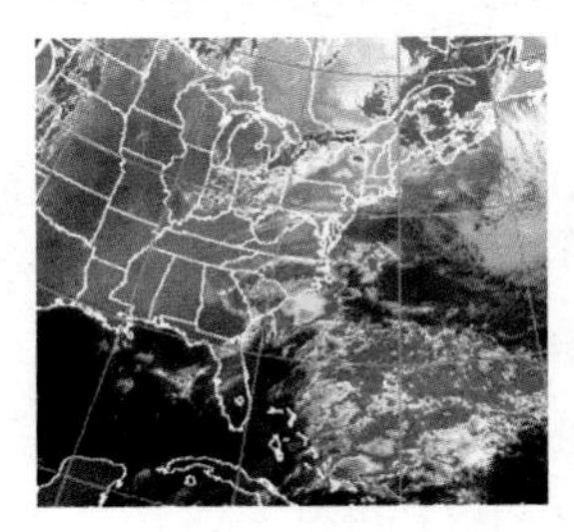

Weather variables are measured by weather stations, radiosondes, satellites, and Doppler radar.

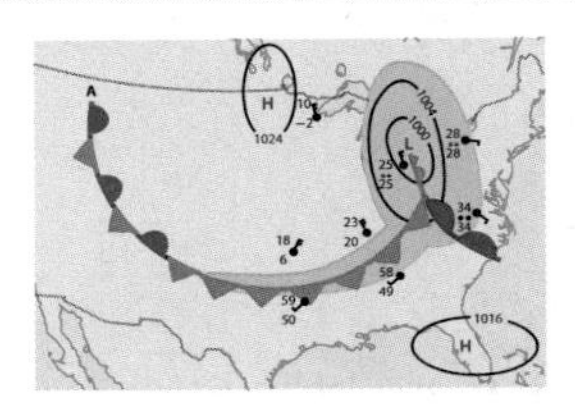

Weather maps contain information in the form of a station model, isobars and isotherms, and symbols for fronts and pressure systems.

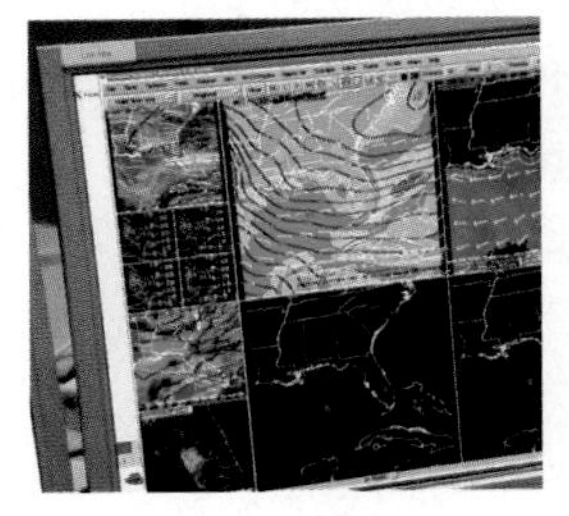

Meteorologists use computer models to help forecast the weather.

Use Vocabulary

1. **Define** *computer model* in your own words. SC.6.N.3.4

____

2. A line connecting places with the same pressure is called a(n) _______________.

3. **Use the term** *surface report* in a sentence.

____

Understand Key Concepts

4. Which diagram shows surface weather measurements? SC.6.N.3.4
   - (A) an infrared satellite image
   - (B) an upper-air chart
   - (C) a station model
   - (D) a visible light satellite image

5. **List** two ways that upper-air weather conditions are measured.

____

____

6. **Describe** how computers are used in weather forecasting.

____

____

7. **Distinguish** between isobars and isotherms.

____

____

Interpret Graphics

8. **Identify** Fill in the graphic organizer below to identify the components of a surface map.

Symbol	Meaning
(stationary front symbol)	
H	

Critical Thinking

9. **Suggest** ways to forecast the weather without using computers.

____

____

Chapter 5 Study Guide

Think About It! Scientists use weather variables such as temperature, air pressure, and wind direction and speed to describe weather and study weather systems. Scientists use computers to predict the weather and model interactions between Earth's systems.

Key Concepts Summary

LESSON 1 Describing Weather

- **Weather** is the atmospheric conditions, along with short-term changes, of a certain place at a certain time.
- Variables used to describe weather are air temperature, **air pressure**, wind, **humidity**, and **relative humidity**.
- The processes in the water cycle—evaporation, condensation, and **precipitation**—are all involved in the formation of different types of weather.

Vocabulary

weather p. 165
air pressure p. 166
humidity p. 166
relative humidity p. 167
dew point p. 167
precipitation p. 169
water cycle p. 169

LESSON 2 Weather Patterns

- **Low-pressure systems** and **high-pressure systems** are two systems that influence weather.
- Weather patterns are driven by the movement of **air masses**.
- Understanding weather patterns helps make weather forecasts more accurate.
- Severe weather includes thunderstorms, **tornadoes**, **hurricanes**, and **blizzards**.

high-pressure system p. 173
low-pressure system p. 173
air mass p. 174
front p. 176
tornado p. 179
hurricane p. 180
blizzard p. 181

LESSON 3 Weather Forecasts

- Thermometers, barometers, anemometers, radiosondes, satellites, and **Doppler radar** are used to measure weather variables.
- **Computer models** use complex mathematical formulas to predict temperature, wind, cloud formation, and precipitation.

surface report p. 185
upper-air report p. 185
Doppler radar p. 186
isobar p. 187
computer model p. 188

Active Reading

FOLDABLES® Chapter Project

Assemble your lesson Foldables as shown to make a Chapter Project. Use the project to review what you have learned in this chapter.

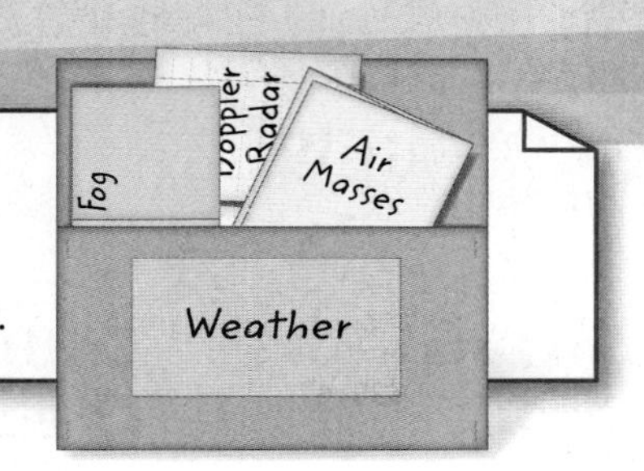

Use Vocabulary

1. The pressure that a column of air exerts on the area below it is called ________.
2. The amount of water vapor in the air is called ________.
3. The natural process in which water constantly moves among oceans, land, and the atmosphere is called the ________.
4. A(n) ________ is a boundary between two air masses.
5. At the center of a(n) ________, air rises and forms clouds and precipitation.
6. A continental polar ________ brings cold temperatures during winter.
7. When the same ________ passes through two locations on a weather map, both locations have the same pressure.
8. The humidity in the air compared to the amount air can hold is the ________.

Link Vocabulary and Key Concepts

Use vocabulary terms from the previous page to complete the concept map.

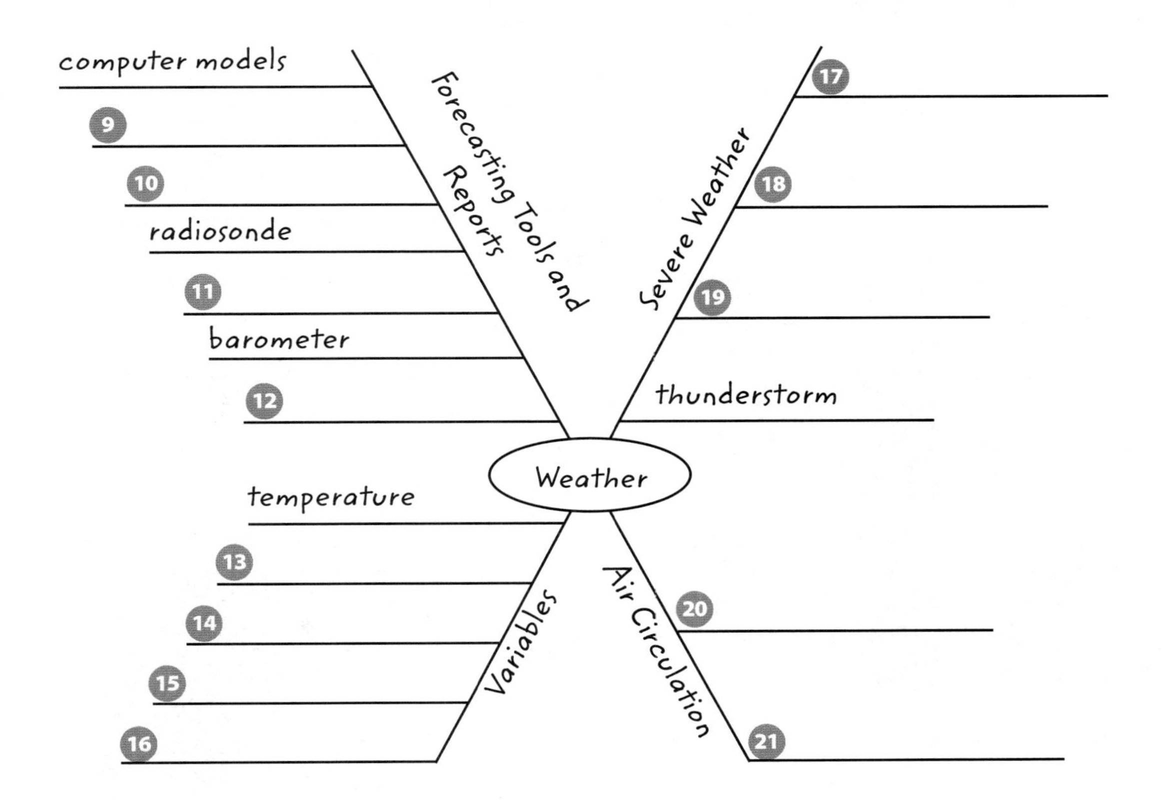

Chapter 5 Review

Fill in the correct answer choice.

Understand Key Concepts

1. Clouds form when water changes from SC.6.E.7.2
   - Ⓐ gas to liquid.
   - Ⓑ liquid to gas.
   - Ⓒ solid to gas.
   - Ⓓ solid to liquid.

2. Which type of precipitation reaches Earth's surface as large pellets of ice? SC.6.E.7.2
   - Ⓐ hail
   - Ⓑ rain
   - Ⓒ sleet
   - Ⓓ snow

3. Which of these sinking-air situations usually brings fair weather? SC.6.E.7.2
   - Ⓐ air mass
   - Ⓑ cold front
   - Ⓒ high-pressure system
   - Ⓓ low-pressure system

4. Which air mass contains cold, dry air? SC.6.E.7.2
   - Ⓐ continental polar
   - Ⓑ continental tropical
   - Ⓒ maritime tropical
   - Ⓓ maritime polar

5. Study the front below.

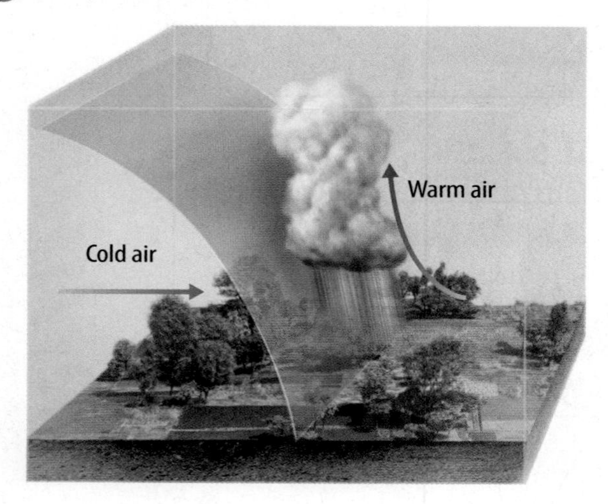

How does this type of front form? SC.6.E.7.2
   - Ⓐ A cold front overtakes a warm front.
   - Ⓑ Cold air moves toward warmer air.
   - Ⓒ The boundary between two fronts stalls.
   - Ⓓ Warm air moves toward colder air.

Critical Thinking

6. **Predict** Suppose you are on a ship near the equator in the Atlantic Ocean. You notice that the barometric pressure is dropping. Predict what type of weather you might experience. SC.6.E.7.2

7. **Compare** a continental polar air mass with a maritime tropical air mass. SC.6.E.7.2

8. **Assess** why clouds usually form in the center of a low-pressure system. SC.6.E.7.2

9. **Predict** how maritime air masses would change if the oceans froze.

10. **Compare** two types of severe weather. SC.6.E.7.8

11. **Interpret Graphics** Identify the front crossing the state on the weather map below. Predict the weather for areas along that front. SC.6.E.7.8

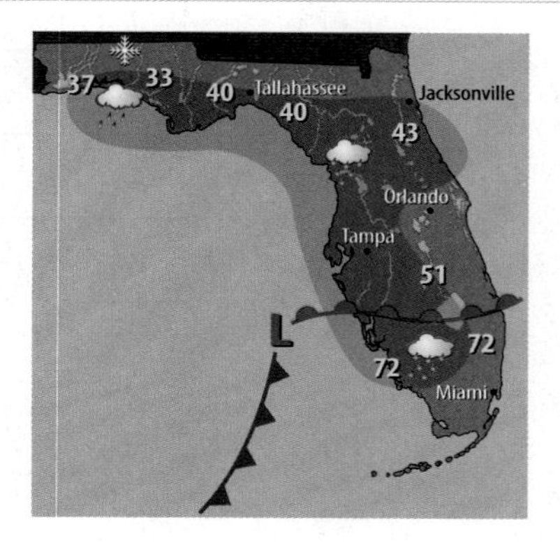

12 **Assess** the validity of the weather forecast: "Tomorrow's weather will be similar to today's weather." LA.6.2.2.3

13 **Compare and contrast** surface weather reports and upper-air reports. Why is it important for meteorologists to monitor weather variables high above Earth's surface? LA.6.2.2.3

Writing in Science

14 **Write** a paragraph on a separate sheet of paper about the ways computers have improved weather forecasts. Be sure to include a topic sentence and a concluding sentence. SC.6.N.3.4

Big Idea Review

15 Identify the instruments used to measure weather variables. SC.6.E.7.6

16 How do scientists use weather variables to describe and predict weather? SC.6.E.7.6

17 Describe the factors that influence weather. SC.6.E.7.2

18 Use the factors listed in question 17 to describe how a continental polar air mass can change to a maritime polar air mass. SC.6.E.7.2

Math Skills MA.6.A.3.6

Use Conversions

19 Convert from Fahrenheit to Celsius.

a. Convert 0°F to °C. **b.** Convert 104°F to °C.

20 Convert from Celsius to Fahrenheit.

a. Convert 0°C to °F. **b.** Convert −40°C to °F.

21 The Kelvin scale of temperature measurement starts at zero and has the same unit size as Celsius degrees. Zero degrees Celsius is equal to 273 kelvin.

Convert 295 K to Fahrenheit.

Florida NGSSS Benchmark Practice

Fill in the correct answer choice.

Multiple Choice

1 How does most of the water in the atmosphere get there? **SC.6.E.7.2**

Ⓐ evaporation

Ⓑ condensation

Ⓒ precipitation

Ⓓ surface runoff

Use the diagram below to answer question 2.

2 If you are walking home and observe a cloud forming like the one in the above diagram, which type of hazardous weather should you prepare for? **SC.6.E.7.8**

Ⓕ hurricane

Ⓖ winter storm

Ⓗ thunderstorm

Ⓘ ultraviolet radiation

3 What is likely to happen between high-pressure and low-pressure systems? **SC.6.E.7.2**

Ⓐ a line of thunderstorms

Ⓑ formation of an air mass

Ⓒ a warm front extending from the high-pressure system to the low-pressure system

Ⓓ wind blowing from the high-pressure system to the low-pressure system

4 Which hazard do you most likely face on a sunny day at the beach? **SC.6.E.7.7**

Ⓕ tornado

Ⓖ hurricane

Ⓗ thunderstorm

Ⓘ ultraviolet radiation

5 Which is NOT a measurement used in reporting weather? **SC.6.E.7.6**

Ⓐ air pressure

Ⓑ annual rainfall

Ⓒ relative humidity

Ⓓ wind speed and direction

6 Which natural disaster is most likely to occur in Florida? **SC.6.E.7.7**

Ⓕ tornado

Ⓖ tsunami

Ⓗ warm front

Ⓘ winter storm

Use the diagram below to answer question 7.

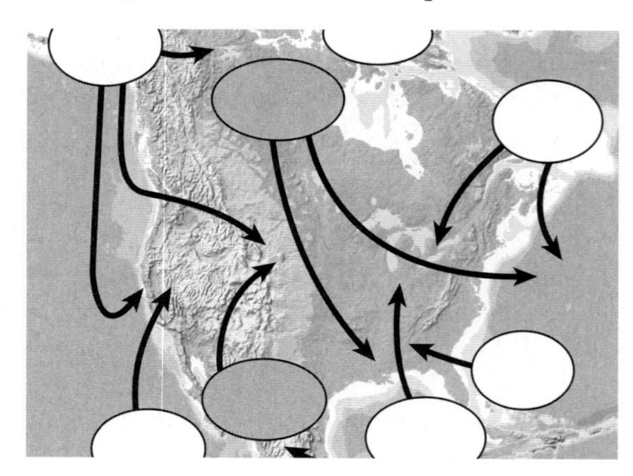

7 Which type of air masses do the shaded ovals in the diagram depict? **SC.6.E.7.2**

Ⓐ antarctic

Ⓑ arctic

Ⓒ continental

Ⓓ maritime

8 What is the movement of water between land, ocean, and air called? SC.6.E.7.2

Ⓕ front

Ⓖ isobar

Ⓗ air mass

Ⓘ water cycle

Use the diagram below to answer question 9.

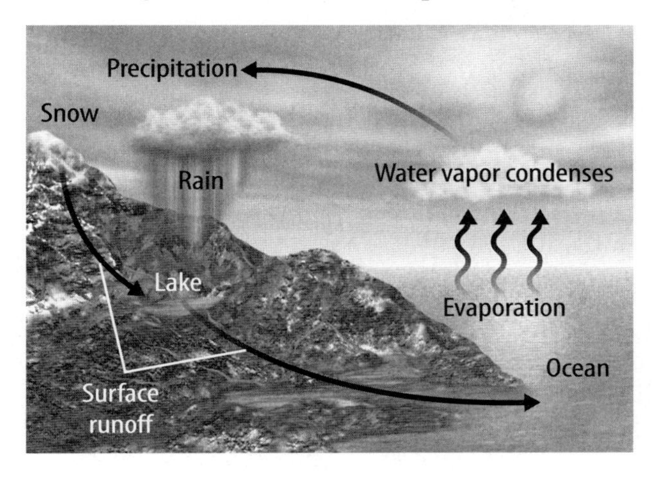

9 What does the above image show? SC.6.E.7.2

Ⓐ formation of an air mass

Ⓑ warm front precipitation

Ⓒ formation of a thunderstorm

Ⓓ cycling of water in and out of the atmosphere

10 Which term best describes current atmospheric conditions? SC.6.E.7.6

Ⓕ wind

Ⓖ weather

Ⓗ humidity

Ⓘ temperature

11 Which provides energy for the water cycle? SC.6.E.7.2

Ⓐ air currents

Ⓑ Earth's core

Ⓒ ocean currents

Ⓓ the Sun

Use the table below to answer questions 12 and 13.

Severe Weather	Characteristic
1.	lightning, strong winds, and heavy rain
2.	swirling winds that can reach speeds of 400 km/h
3.	strong winds and flooding from a storm over 480 km across
4.	cold temperatures, wind, and swirling snow

12 In the table above, what severe weather is number 2 describing? SC.6.E.7.8

Ⓕ tornado

Ⓖ blizzard

Ⓗ hurricane

Ⓘ thunderstorm

13 What safety precaution would you take for number 1 in the table above? SC.6.E.7.8

Ⓐ evacuate coastal areas

Ⓑ dress warmly if you must go outside

Ⓒ go to an indoor area without windows

Ⓓ stay indoors and away from metal objects

NEED EXTRA HELP?

If You Missed Question...	1	2	3	4	5	6	7	8	9	10	11	12	13
Go to Lesson...	1	2	1, 2	2	1	2	2	1	1	1	1	2	2

Name ______________________ Date ______________

Multiple Choice *Bubble the correct answer.*

1. Which type of cloud is shown in the image above? **SC.6.E.7.2**

   (A) altostratus

   (B) cumulonimbus

   (C) cumulus

   (D) stratus

2. During which season does the air at sea level in Florida have the most kinetic energy? **SC.6.E.7.6**

   (F) summer

   (G) fall

   (H) winter

   (I) spring

3. Which image shows the formation of snow? **SC.6.E.7.2**

(A) 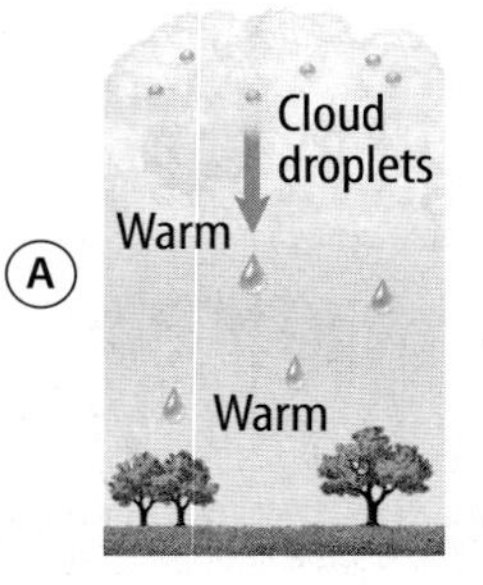

(B) 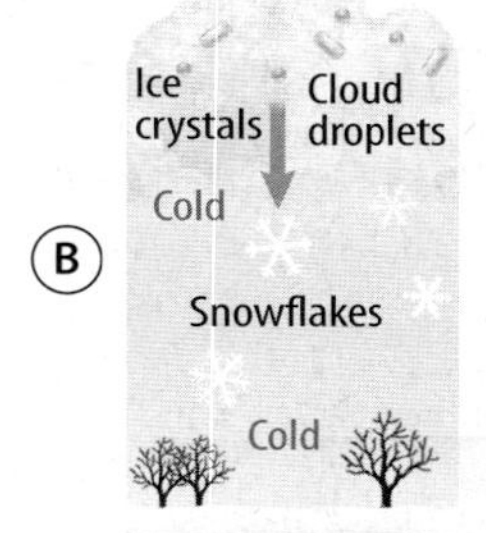

(C) 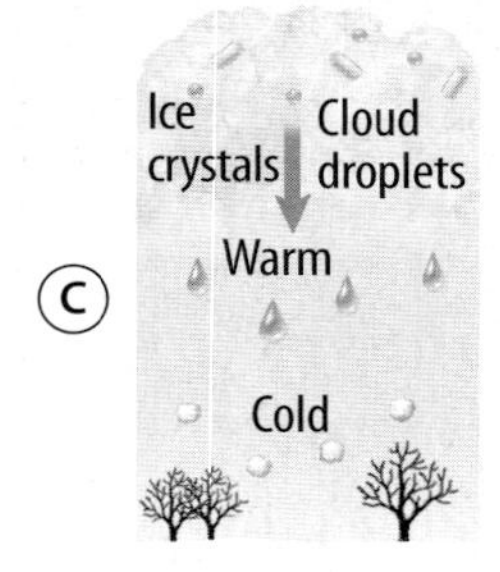

(D) 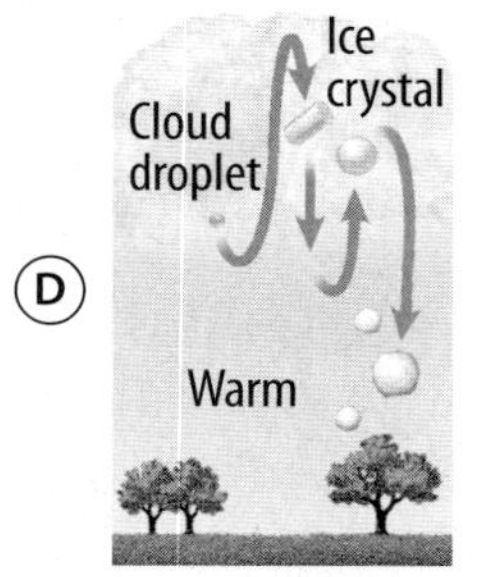

Name ______________________________ Date ______________

Multiple Choice *Bubble the correct answer.*

1. Which image shows an approaching cold front? **SC.6.E.7.2**

Ⓐ

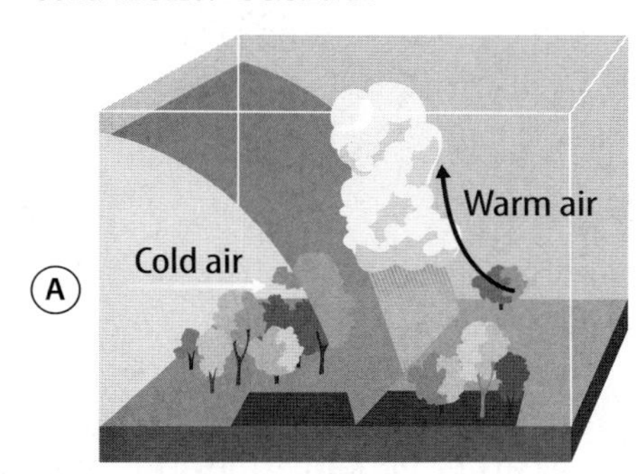

Ⓑ

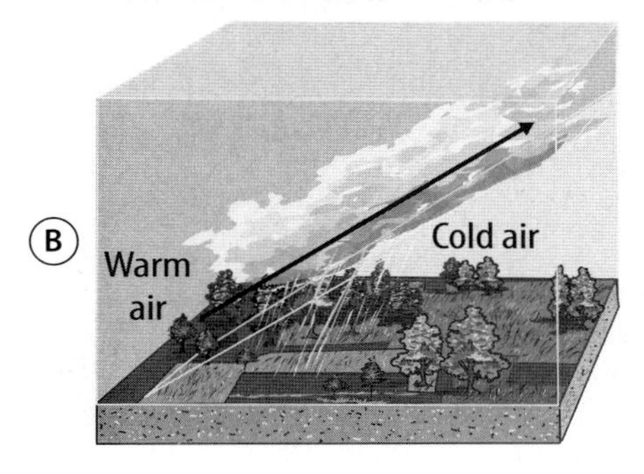

Ⓒ

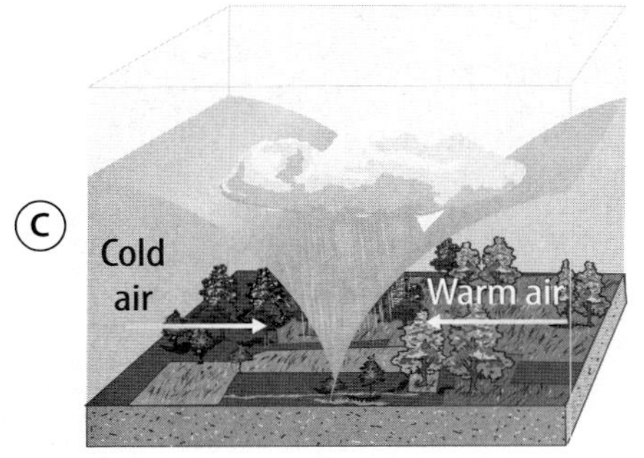

Ⓓ

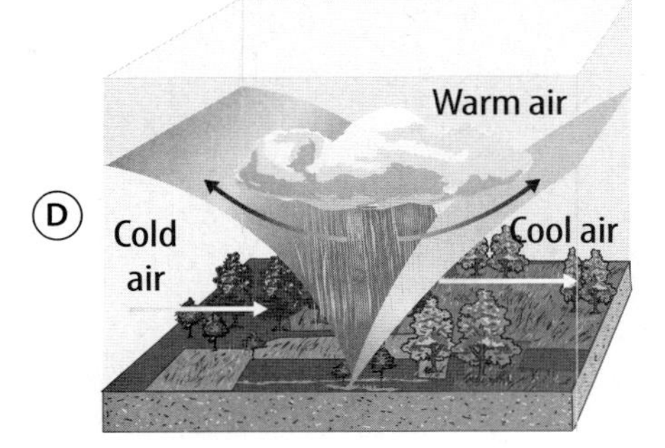

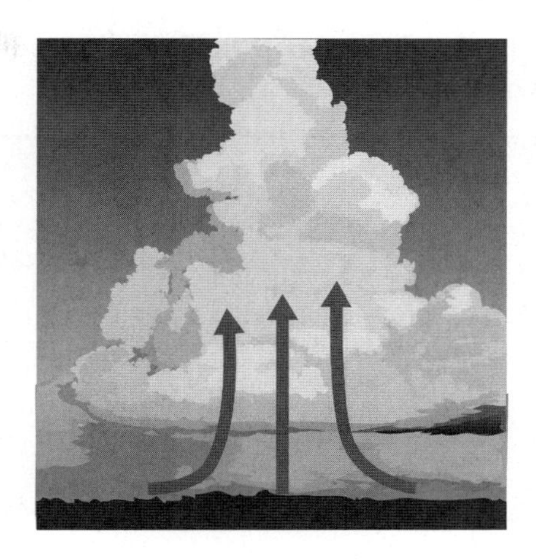

2. Which stage of a thunderstorm's development is shown above? **SC.6.E.7.2**

Ⓕ arrival stage

Ⓖ cumulus stage

Ⓗ dissipation stage

Ⓘ mature stage

3. Which air mass contains cold, humid air? **SC.6.E.7.2**

Ⓐ continental polar

Ⓑ continental tropical

Ⓒ maritime polar

Ⓓ maritime tropical

Multiple Choice *Bubble the correct answer.*

Use the weather map below to answer questions 1 and 2.

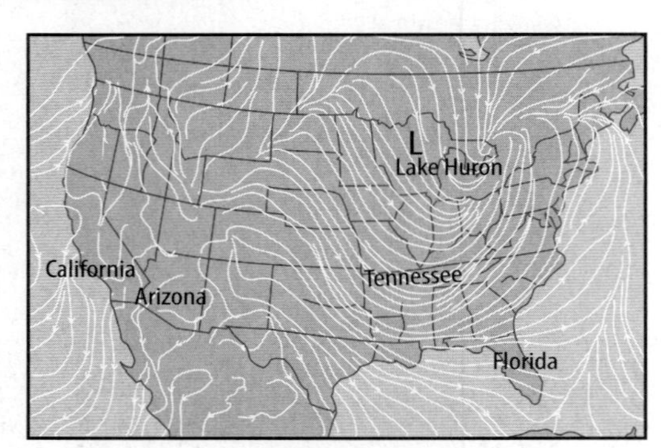

1. Look at the isobars in the weather map. In which of these states are the winds strongest? **SC.6.E.7.7**

Ⓐ Arizona

Ⓑ California

Ⓒ Florida

Ⓓ Tennessee

2. What does the symbol over Lake Huron indicate? **SC.6.N.3.4**

Ⓕ an arctic air mass

Ⓖ a high-pressure system

Ⓗ a low-pressure system

Ⓘ a maritime polar air mass

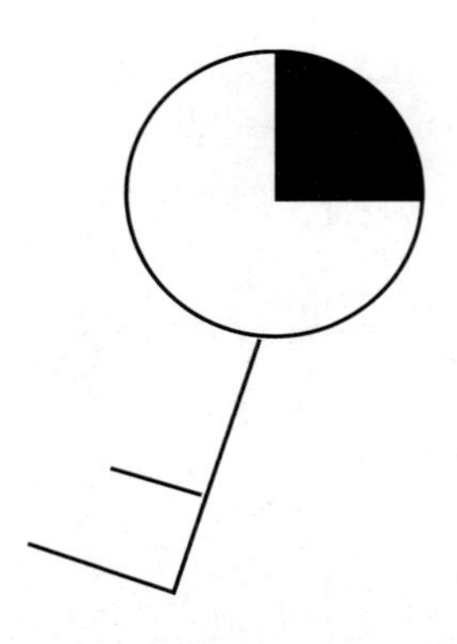

3. Which weather variable does the symbol above represent? **SC.6.N.3.4**

Ⓐ dew point

Ⓑ temperature

Ⓒ barometric pressure

Ⓓ wind speed and direction

4. Which tool is used to determine wind speed in a potential tornado? **SC.6.E.7.7**

Ⓕ Doppler radar

Ⓖ ground radar

Ⓗ infrared satellite imagery

Ⓘ visual satellite imagery

Name ______________________________ Date ______________

Notes

Name ______________________ Date ______________

Notes

Is Earth getting warmer?

Two students were talking about how cold it was during the past winter. They argued about climate change and whether they thought Earth was getting warmer.

Kirby: I think Earth is getting warmer. There is evidence that Earth's climate is changing, resulting in global warming.

Jan: I don't think Earth is getting warmer. It was freezing in Florida this year and there were big snowstorms in some places that normally get very little snow.

Circle the student you most agree with and explain why you agree. Describe your ideas about Earth's climate.

FLORIDA
Chapter 6
CLIMATE

FLORIDA BIG IDEAS

1 **The Practice of Science**

2 **The Characteristics of Scientific Knowledge**

7 **Earth Systems and Patterns**

The Big Idea

Think About It!

What is climate and how does it impact life on Earth?

Climate differs from one area of Earth to another. Some areas have little rain and high temperatures. Other areas have low temperatures and lots of snow. Where this tree grows—on Humphrey Head Point in England—there is constant wind.

1. What do you think are the characteristics of different climates?

2. What factors might affect the climate of a region?

3. What do you think climate is and how does it impact life on Earth?

Get Ready to Read

What do you think about climate?

Before you read, decide if you agree or disagree with each of these statements. As you read this chapter, see if you change your mind about any of the statements.

	AGREE	DISAGREE
1 Locations at the center of large continents usually have the same climate as locations along the coast.		
2 Latitude does not affect climate.		
3 Climate on Earth today is the same as it has been in the past.		
4 Climate change occurs in short-term cycles.		
5 Human activities can impact climate.		
6 You can help reduce the amount of greenhouse gases released into the atmosphere.		

There's More Online!
Video • Audio • Review • ⓘLab Station • WebQuest • Assessment • Concepts in Motion • Multilingual eGlossary

Lesson 1

Climates of EARTH

ESSENTIAL QUESTIONS

 What is climate?

 Why is one climate different from another?

 How are climates classified?

Vocabulary

climate p. 205
rain shadow p. 207
specific heat p. 207
microclimate p. 209

Florida NGSSS

LA.6.2.2.3 The student will organize information to show understanding (e.g., representing main ideas within text through charting, mapping, paraphrasing, summarizing, or comparing/contrasting);

MA.6.S.6.2 Select and analyze the measures of central tendency or variability to represent, describe, analyze, and/or summarize a data set for the purposes of answering questions appropriately.

SC.6.E.7.2 Investigate and apply how the cycling of water between the atmosphere and hydrosphere has an effect on weather patterns and climate.

SC.6.E.7.4 Differentiate and show interactions among the geosphere, hydrosphere, cryosphere, atmosphere, and biosphere.

SC.6.E.7.5 Explain how energy provided by the sun influences global patterns of atmospheric movement and the temperature differences between air, water, and land.

SC.6.E.7.6 Differentiate between weather and climate.

SC.6.N.1.1 Define a problem from the sixth grade curriculum, use appropriate reference materials to support scientific understanding, plan and carry out scientific investigation of various types, such as systematic observations or experiments, identify variables, collect and organize data, interpret data in charts, tables, and graphics, analyze information, make predictions, and defend conclusions.

SC.6.N.2.1 Distinguish science from other activities involving thought.

 LA.6.2.2.3

Inquiry Launch Lab

20 minutes

How do climates compare?

Procedure

1. Read and complete a lab safety form.
2. Select a location on a **globe.**
3. Research the average monthly temperatures and levels of precipitation for this location.
4. Record your data in a chart like the one shown here.

Omsk, Russia 73.5° E, 55° N		
Month	Average Monthly Temperature	Average Monthly Level of Precipitation
January	−14°C	13 mm
February	−12°C	9 mm
March	−5°C	9 mm
April	8°C	18 mm
May	18°C	31 mm
June	24°C	52 mm
July	25°C	61 mm
August	22°C	50 mm
September	17°C	32 mm
October	7°C	26 mm
November	−4°C	19 mm
December	−12°C	15 mm

Data and Observations

Think About This

1. Describe the climate of your selected location in terms of temperature and precipitation.

2. Compare your data to Omsk, Russia. How do the climates differ?

3. Key Concept Mountains, oceans, and latitude can affect climates. Do any of these factors account for the differences you observed? Explain.

What makes a desert a desert?

1. How much precipitation do you think a desert gets? Are deserts always hot? What types of plants might grow in the desert? Scientists look at the answers to all of these questions to determine if an area is a desert.

What is climate?

You probably already know that the term *weather* describes the atmospheric conditions and short-term changes of a certain place at a certain time. The weather changes from day to day in many places on Earth. Other places on Earth have more constant weather. For example, temperatures in Antarctica rarely are above 0°C, even in the summer. Areas in Africa's Sahara, shown above, have temperatures above 20°C year-round.

Climate *is the long-term average weather conditions that occur in a particular region.* A region's climate depends on average temperature and precipitation, as well as how these variables change throughout the year.

What affects climate?

Several factors determine a region's climate. The latitude of a location affects climate. For example, areas close to the equator have the warmest climates. Large bodies of water, including lakes and oceans, also influence the climate of a region. Along coastlines, weather is more constant throughout the year. Hot summers and cold winters typically happen in the center of continents. The altitude and location of mountains affect climate. The prevailing wind direction is also a factor. For example, the prevailing wind direction across most of North America and Europe is west to east. That is why temperatures are more moderate on the west coast of the United States.

2. **NGSSS Check** Describe What is climate? SC.6.E.7.6

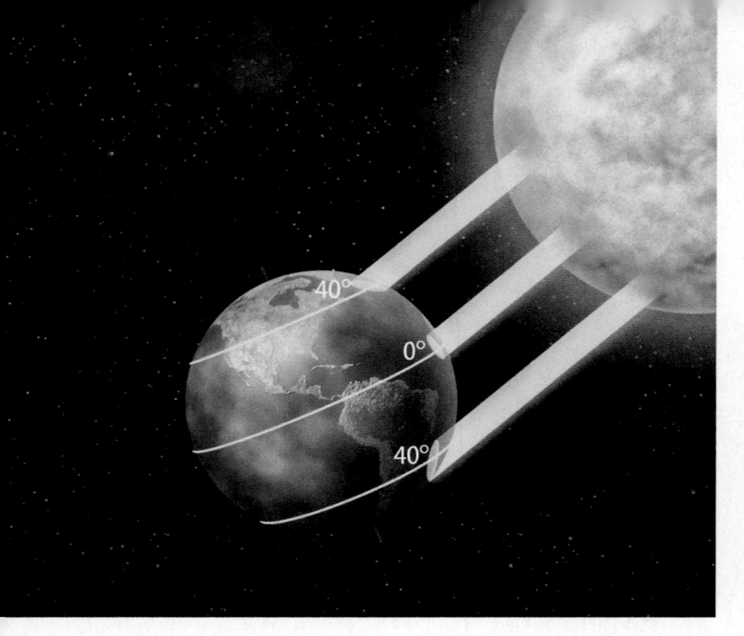

Figure 1 Latitudes near the poles receive less solar energy and have lower average temperatures.

3. Visual Check The Florida state capital of Tallahassee is located at latitude 30.5°N.
Describe How is Florida's climate affected by latitude?

Latitude

Recall that, starting at the equator, latitude increases from 0° to 90° as you move toward the North Pole or the South Pole. The amount of solar energy per unit of Earth's surface area depends on latitude. **Figure 1** shows that locations close to the equator receive more solar energy per unit of surface area than areas located farther north or south. This difference is due mainly to the fact that Earth's curved surface causes the angle of the Sun's rays to spread out over a larger area. Polar regions are colder because annually they receive less solar energy per unit of surface area. In the middle latitudes, between 30° and 60°, summers are generally hot and winters are usually cold. Warmer regions near the equator have more evaporation and rainfall than polar regions.

Altitude

Climate is also influenced by altitude. Recall that temperature decreases as altitude increases in the troposphere. So, as you climb a tall mountain you might experience the same cold, snowy climate that is near the poles. **Figure 2** shows the difference in average temperatures between two cities in Colorado at different altitudes.

Altitude and Climate

Figure 2 As altitude increases, temperature decreases.

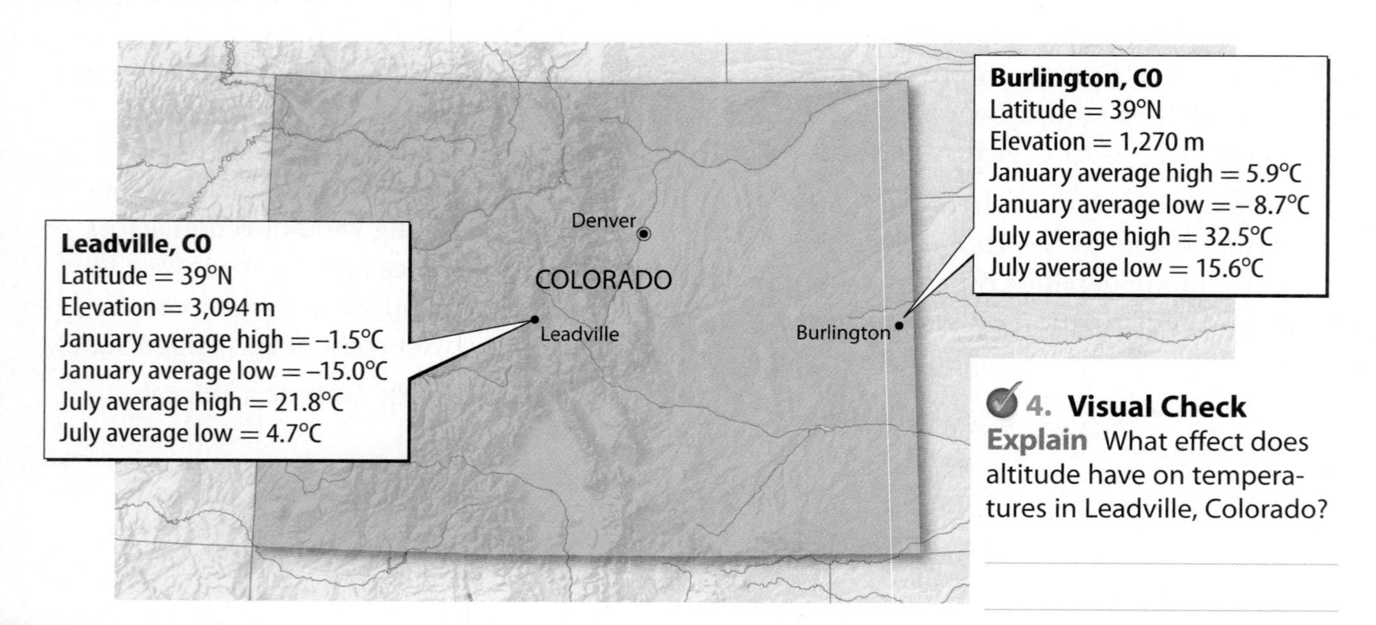

4. Visual Check
Explain What effect does altitude have on temperatures in Leadville, Colorado?

Figure 3 Key Concept Rain shadows form on the downwind slope of a mountain.

Rain Shadows

Mountains influence climate because they are barriers to prevailing winds. This leads to unique **precipitation** patterns called rain shadows. *An area of low rainfall on the downwind slope of a mountain is called a* **rain shadow**, as shown in **Figure 3.** Different amounts of precipitation on either side of a mountain range influence the types of vegetation that grow. Abundant amounts of vegetation grow on the side of the mountain exposed to the precipitation. The amount of vegetation on the downwind slope is sparse due to the dry weather.

Active Reading **5. Identify** Circle the area on the illustration where large amounts of vegetation will grow.

REVIEW VOCABULARY

precipitation
water, in liquid or solid form, that falls from the atmosphere

Active Reading **6. Explain** Why do rain shadows only form on the downwind slope of mountains?

Large Bodies of Water

On a sunny day at the beach, why does the sand feel warmer than the water? It is because water has a high specific heat. **Specific heat** *is the amount (joules) of thermal energy needed to raise the temperature of 1 kg of a material by 1°C.* The specific heat of water is about six times higher than the specific heat of sand. This means the ocean water would have to absorb six times as much thermal energy to be the same temperature as the sand.

The high specific heat of water causes the climates along coastlines to remain more constant than those in the middle of a continent. Prevailing winds extend the moderate temperatures farther inland. For example, much of the western United States and western Europe have moderate temperatures year-round.

Ocean currents can also modify climate. The Gulf Stream is a warm current flowing northward along the coast of eastern North America. It brings warmer temperatures to Europe.

Active Reading **7. Restate** How do large bodies of water influence climate?

World Climates 

Figure 4 The map shows a modified version of Köppen's climate classification system.

Active Reading 8. **Utilize** Based on the key of climates, identify how the majority of Florida's climate is classified.

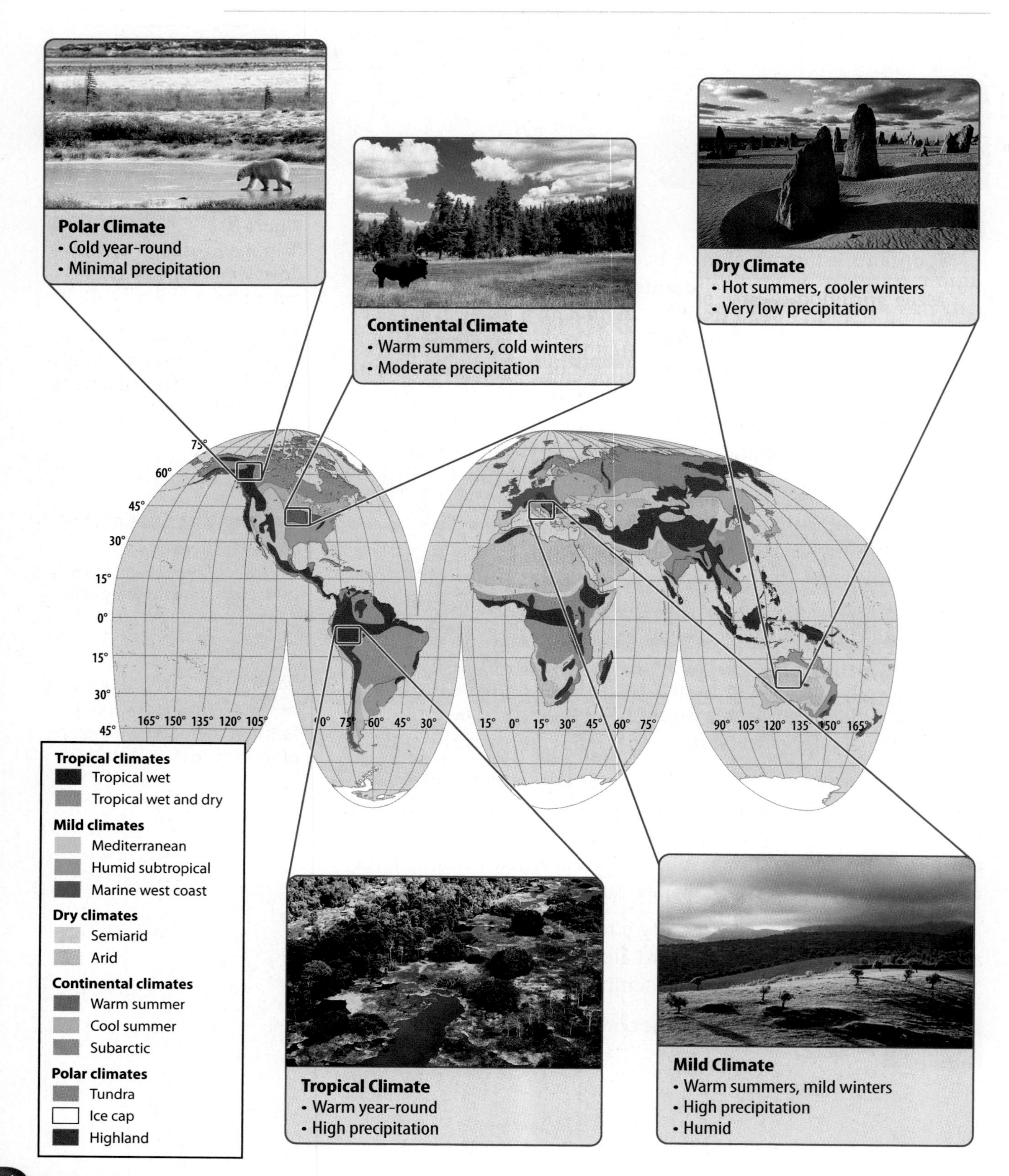

Classifying Climates

What is the climate of any particular region on Earth? This can be a difficult question to answer because many factors affect climate. In 1918 German scientist Wladimir Köppen (vlah DEE mihr • KAWP pehn) developed a system for classifying the world's many climates. Köppen classified a region's climate by studying its temperature, precipitation, and native vegetation. Native vegetation is often limited to particular climate conditions. For example, you would not expect to find a warm-desert cactus growing in the cold, snowy arctic. Köppen identified five climate types. A modified version of Köppen's classification system is shown in **Figure 4.**

Active Reading **9. Find** Underline how climates are classified.

Microclimates

Roads and buildings in cities have more concrete than surrounding rural areas. The concrete absorbs solar radiation, causing warmer temperatures than those of the surrounding countryside. The result is a common microclimate called the urban heat island, as shown in **Figure 5.** *A* **microclimate** *is a localized climate that is different from the climate of the larger area surrounding it.* Other examples of microclimates include forests, which are often cooler and less windy than the surrounding countryside, and hilltops, which are windier than nearby lower land.

Active Reading **10. Identify** What are two examples of microclimates?

Active Reading

FOLDABLES® LA.6.2.2.3

Use three sheets of notebook paper to make a layered book. Label it as shown. Use it to organize your notes on the factors that determine a region's climate.

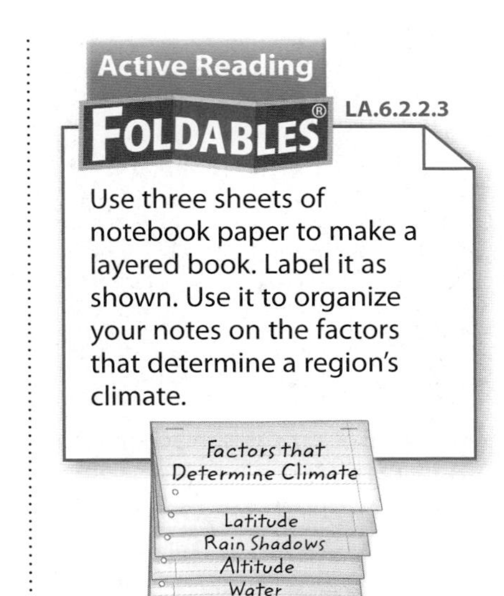

WORD ORIGIN

microclimate
from Greek *mikros,* means "small"; and *klima,* means "region, zone"

Microclimate

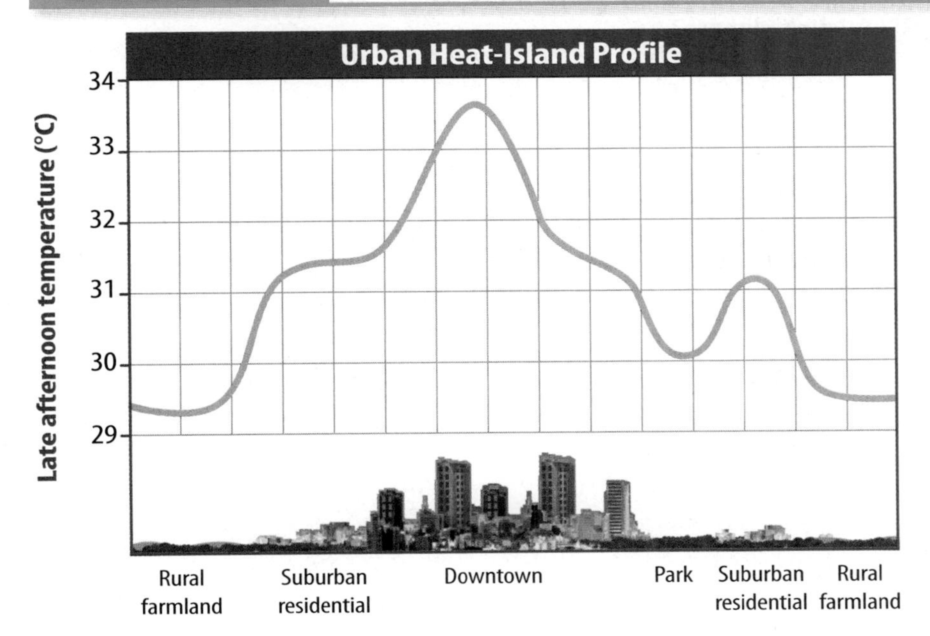

Figure 5 The temperature is often warmer in urban areas when compared to temperatures in the surrounding countryside.

11. Visual Check
Identify What is the temperature difference between downtown and rural farmland?

Figure 6 Camels are adapted to dry climates and can survive up to three weeks without drinking water.

How Climate Affects Living Organisms

Organisms have adaptations for the climates where they live. For example, polar bears have thick fur and a layer of fat that helps keep them warm in the Arctic. Many animals that live in deserts, such as the camels in **Figure 6,** have adaptations for surviving in hot, dry conditions. Some desert plants have extensive shallow root systems that collect rainwater. Deciduous trees, found in continental climates, lose their leaves during the winter, which reduces water loss when soils are frozen.

Climate also influences humans in many ways. Average temperature and rainfall in a location help determine the type of crops humans grow there. Thousands of orange trees grow in Florida, where the climate is mild. Wisconsin's continental climate is ideal for growing cranberries.

Climate also influences the way humans design buildings. In polar climates, the soil is frozen year-round—a condition called permafrost. Humans build houses and other buildings in these climates on stilts. This is done so that thermal energy from the building does not melt the permafrost.

Active Reading **12. Paraphrase** How are organisms adapted to different climates?

Inquiry SC.6.N.1.1, SC.6.E.7.2, LA.7.2.2.3

iLAB STATION **Try It!**

MiniLab ***What factors affect climate?*** at connectED.mcgraw-hill.com

Apply It! After you complete the lab, answer these questions.

1. How does the climate in your town in Florida compare to Austin's climate?

2. Which prediction was the furthest from the actual average temperature and precipitation? Why?

Lesson Review 1

Visual Summary

Climate is influenced by several factors including latitude, altitude, and an area's location relative to a large body of water or mountains.

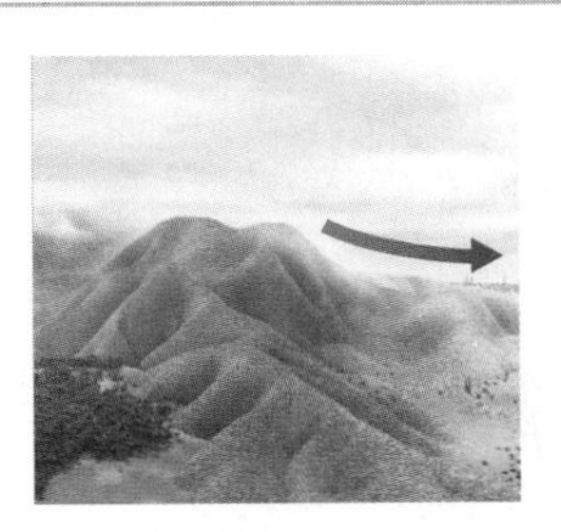

Rain shadows occur on the downwind slope of mountains.

Microclimates can occur in urban areas, forests, and hilltops.

Use Vocabulary

1. The amount of thermal energy needed to raise the temperature of 1 kg of a material by 1°C is called ______________.

2. **Distinguish** between *climate* and *microclimate.* SC.6.E.7.6

3. **Use the term** *rain shadow* in a sentence. SC.6.E.7.4

Understand Key Concepts 

4. How are climates classified? SC.6.E.7.6
   - (A) by cold- and warm-water ocean currents
   - (B) by latitude and longitude
   - (C) by measurements of temperature and humidity
   - (D) by temperature, precipitation, and vegetation

5. **Describe** the climate of an island in the tropical Atlantic Ocean.

6. **Distinguish** between weather and climate. SC.6.E.7.6

Interpret Graphics

7. **Summarize** Fill in the graphic organizer to summarize information about the different types of climate worldwide. LA.6.2.2.3

Climate Type	Description
Tropical	
Dry	
Mild	
Continental	
Polar	

Critical Thinking

8. **Distinguish** between the climates of a coastal location and a location in the center of a large continent. SC.6.E.7.3

Sequence the events that result in a rain shadow.

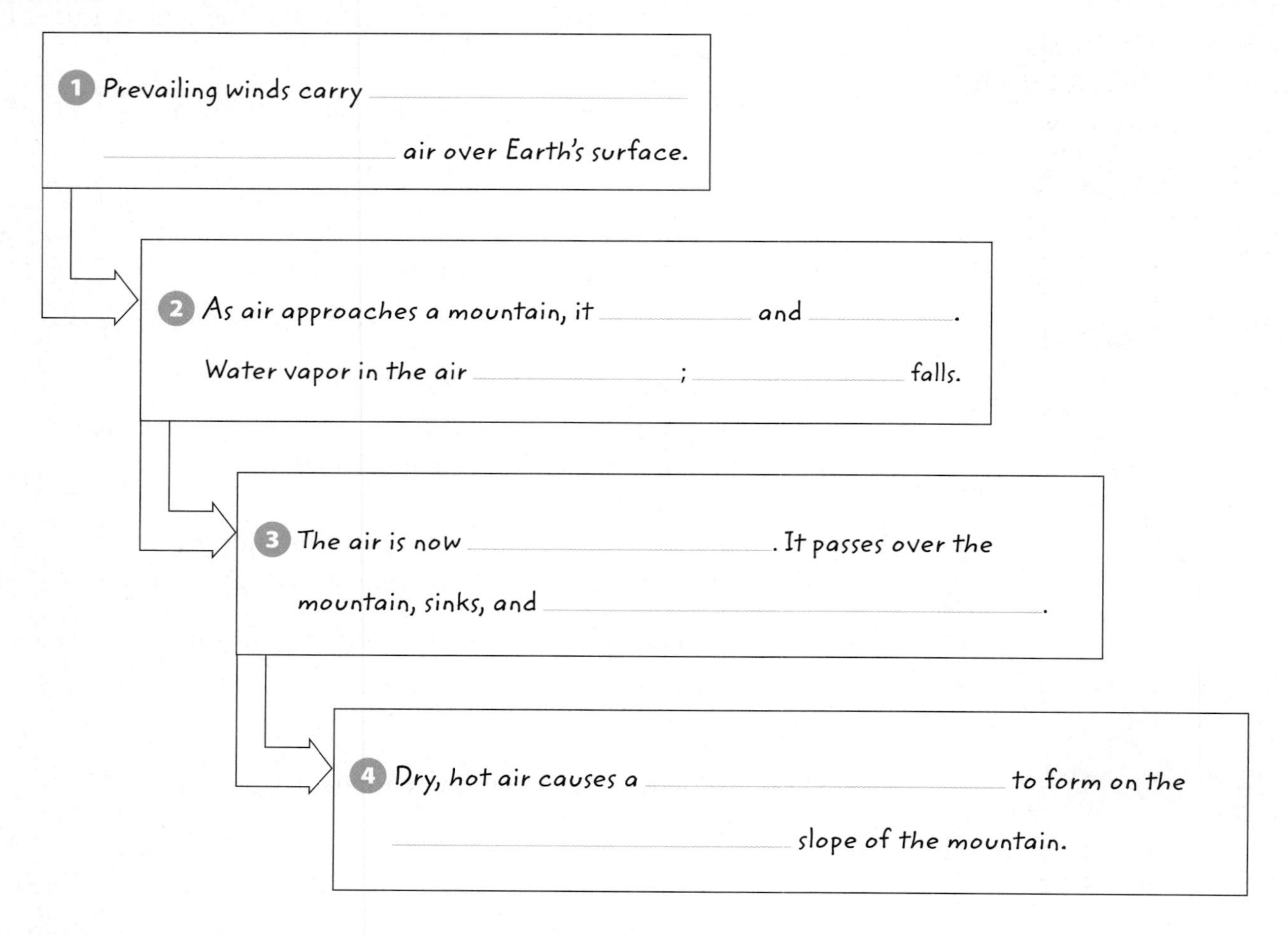

Identify and describe Köppen's five climate types.

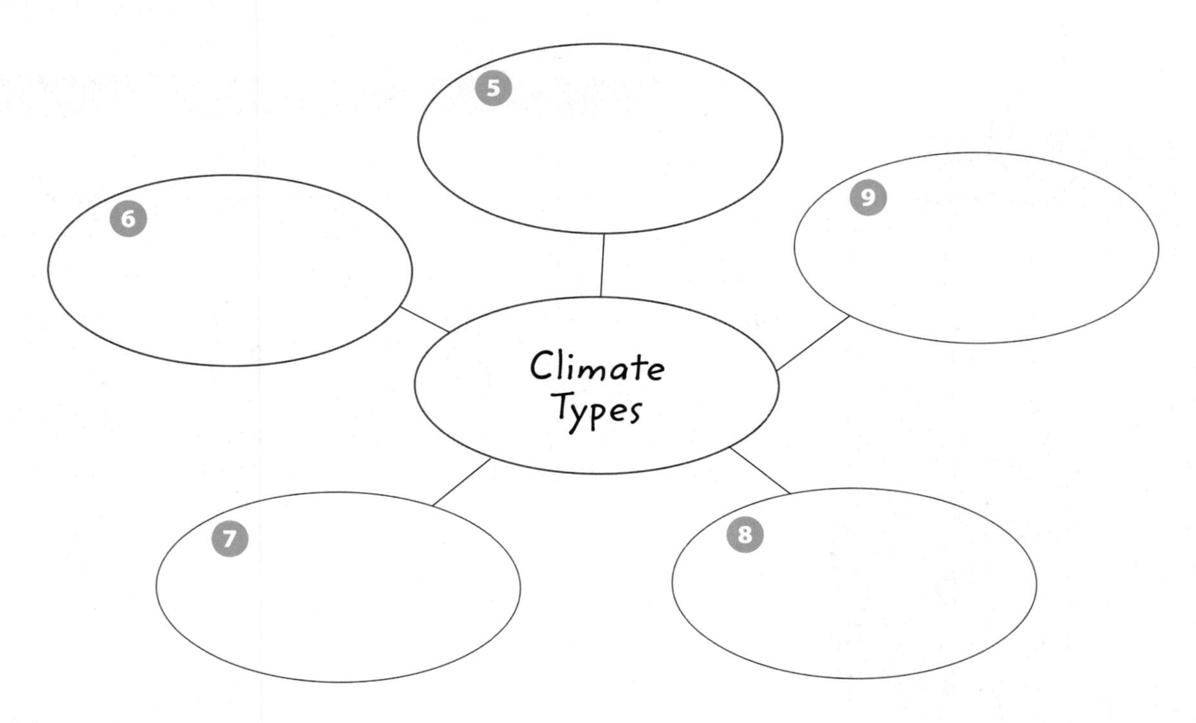

Lesson 2

CLIMATE CYCLES

ESSENTIAL QUESTIONS

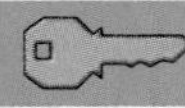
How has climate varied over time?

What causes seasons?

How does the ocean affect climate?

Vocabulary

ice age p. 214
interglacial p. 214
El Niño/Southern Oscillation p. 218
monsoon p. 219
drought p. 219

Florida NGSSS

LA.6.2.2.3 The student will organize information to show understanding (e.g., representing main ideas within text through charting, mapping, paraphrasing, summarizing, or comparing/contrasting);

SC.6.E.7.3 Describe how global patterns such as the jet stream and ocean currents influence local weather in measurable terms such as temperature, air pressure, wind direction and speed, and humidity and precipitation.

SC.6.E.7.4 Differentiate and show interactions among the geosphere, hydrosphere, cryosphere, atmosphere, and biosphere.

SC.6.N.1.1 Define a problem from the sixth grade curriculum, use appropriate reference materials to support scientific understanding, plan and carry out scientific investigation of various types, such as systematic observations or experiments, identify variables, collect and organize data, interpret data in charts, tables, and graphics, analyze information, make predictions, and defend conclusions.

SC.6.N.1.5 Recognize that science involves creativity, not just in designing experiments, but also in creating explanations that fit evidence.

Inquiry Launch Lab

SC.6.N.1.5

20 minutes

How does Earth's tilted axis affect climate?

Earth's axis is tilted at an angle of 23.5°. This tilt influences climate by affecting the amount of sunlight that reaches Earth's surface.

Procedure

1. Read and complete a lab safety form.
2. Hold a **penlight** about 25 cm above a sheet of paper at a 90° angle. Use a **protractor** to check the angle.
3. Turn off the overhead lights and turn on the penlight. Your partner should trace the circle of light cast by the penlight onto the paper.
4. Repeat steps 2 and 3, but this time hold the penlight at an angle of 23.5° from perpendicular.

Think About This

1. How did the circles of light change during each trial?

2. Which trial represented the tilt of Earth's axis?

3. Key Concept How might changes in the tilt of Earth's axis affect climate? Explain.

How did this lake form?

1. A melting glacier formed this lake. How long ago do you think this happened? What type of climate change occurred to cause a glacier to melt? Will it happen again?

Long-Term Cycles

Weather and climate have many cycles. In most areas on Earth, temperatures increase during the day and decrease at night. Each year, the air is warmer during summer and colder during winter. You will experience many of these cycles in your lifetime. But climate also changes in cycles that take much longer than a lifetime to complete.

Much of our knowledge about past climates comes from natural records of climate. Scientists study ice cores, shown in **Figure 7,** drilled from ice layers in glaciers and ice sheets. Fossilized pollen, ocean sediments, and the growth rings of trees also are used to gain information about climate changes in the past. Scientists use the information to compare present-day climates to those that occurred many thousands of years ago.

Active Reading **2. Cite** Highlight the ways scientists find information about past climates on Earth.

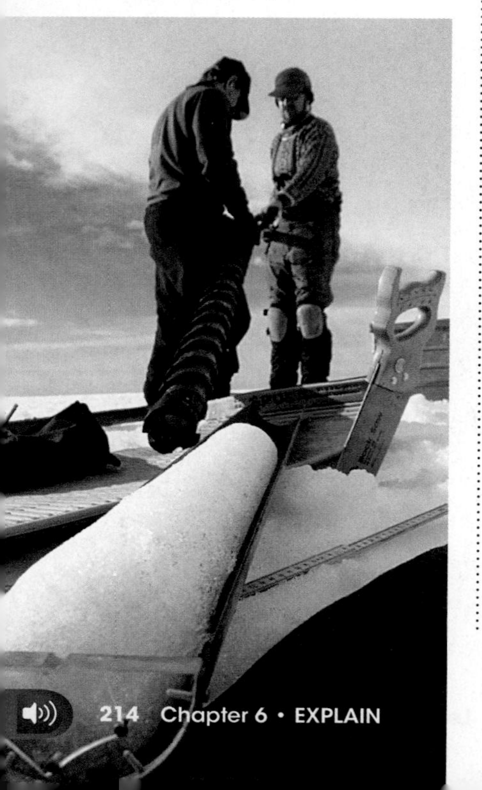

Figure 7 Scientists study the different layers in an ice core to learn more about climate changes in the past.

Ice Ages and Interglacials

Earth has experienced many major atmospheric and climate changes in its history. **Ice ages** *are cold periods lasting from hundreds to millions of years when glaciers cover much of Earth.* Glaciers and ice sheets advance during cold periods and retreat during **interglacials**—*the warm periods that occur during ice ages.*

Major Ice Ages and Warm Periods

The most recent ice age began about 2 million years ago. The ice sheets reached maximum size about 20,000 years ago. At that time, about half the northern hemisphere was covered by ice. About 10,000 years ago, Earth entered its current interglacial period, called the Holocene Epoch.

Temperatures on Earth have fluctuated during the Holocene. For example, the period between A.D. 950 and 1100 was one of the warmest in Europe. The Little Ice Age, which lasted from 1250 to about 1850, was a period of bitterly cold temperatures.

WORD ORIGIN

interglacial
from Latin *inter-*, means "among, between"; and *glacialis,* means "icy, frozen"

Causes of Long-Term Climate Cycles

As the amount of solar energy reaching Earth changes, Earth's climate also changes. One factor that affects how much energy Earth receives is the shape of its orbit. The shape of Earth's orbit appears to vary between elliptical and circular over the course of about 100,000 years. As shown in **Figure 8,** when Earth's orbit is more circular, Earth averages a greater distance from the Sun. This results in below-average temperatures on Earth.

Another factor that scientists suspect influences climate change on Earth is changes in the tilt of Earth's axis. The tilt of Earth's axis changes in 41,000-year cycles. Changes in the angle of Earth's tilt affect the range of temperatures throughout the year. For example, a decrease in the angle of Earth's tilt, as shown in **Figure 8,** could result in a decrease in temperature differences between summer and winter. Long-term climate cycles are also influenced by the slow movement of Earth's continents, as well as changes in ocean circulation.

Active Reading **3. Summarize** How has climate varied over time?

Figure 8 This image shows how the shape of Earth's orbit varies between elliptical and circular. The angle of the tilt varies from 22° to 24.5° about every 41,000 years. Earth's current tilt is 23.5°.

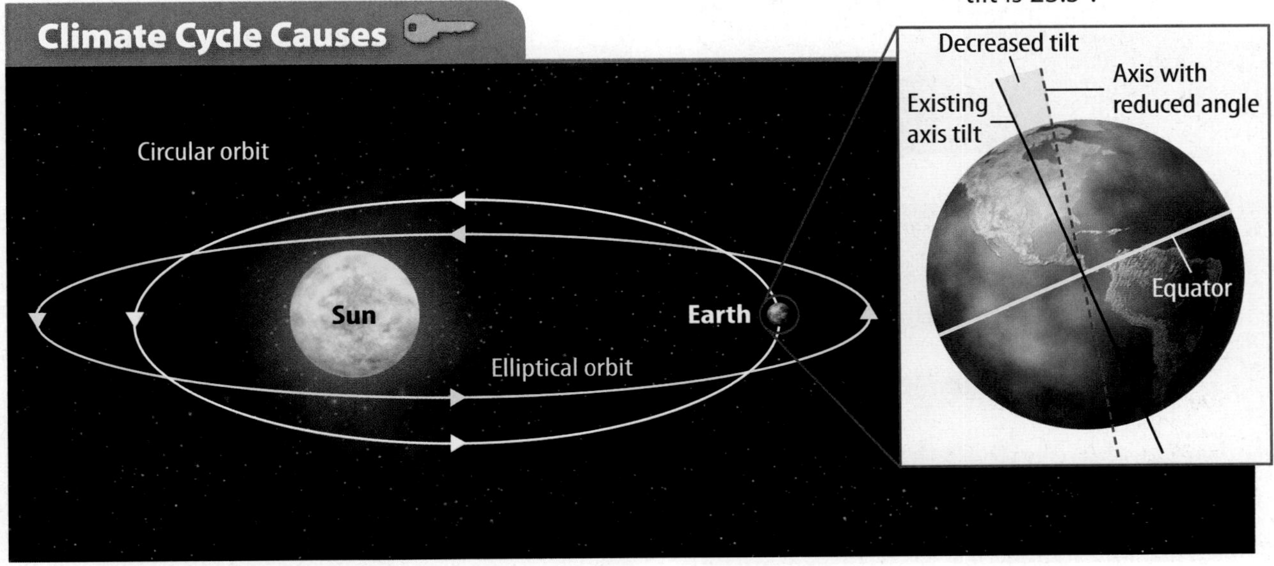

Short-Term Cycles

In addition to its long-term cycles, climate also changes in short-term cycles. Seasonal changes and changes that result from the interaction between the ocean and the atmosphere are some examples of short-term climate change.

Seasons

Changes in the amount of solar energy received at different latitudes during different times of the year give rise to the seasons. Seasonal changes include regular changes in temperature and the number of hours of day and night.

Recall from Lesson 1 that the amount of solar energy per unit of Earth's surface is related to latitude. Another factor that affects the amount of solar energy received by an area is the tilt of Earth's axis. **Figure 9** shows that when the northern hemisphere is tilted toward the Sun, there are more daylight hours than dark hours, and temperatures are warmer. The northern hemisphere receives more direct solar energy and it is summer. At the same time, the southern hemisphere receives less overall solar energy and it is winter there.

Figure 9 shows that the opposite occurs when six months later the northern hemisphere is tilted away from the Sun. Daylight hours are fewer than nighttime hours, and temperatures are colder. Indirect solar energy reaches the northern hemisphere, resulting in winter. The southern hemisphere receives more direct solar energy and it is summer.

Active Reading **5. Explain** What causes seasons?

Figure 9 The solar energy rays reaching a given area of Earth's surface is more intense when tilted toward the Sun.

Active Reading **4. Label** Insert the correct season for each hemisphere in this figure.

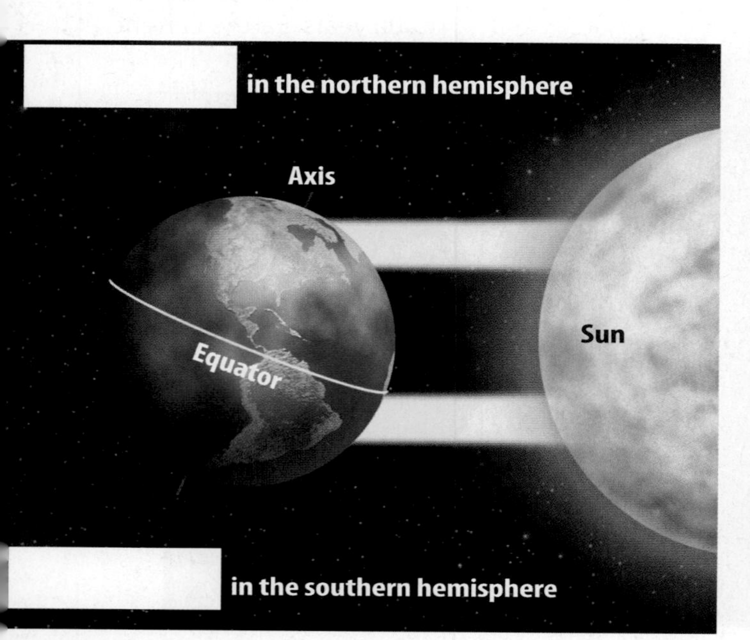

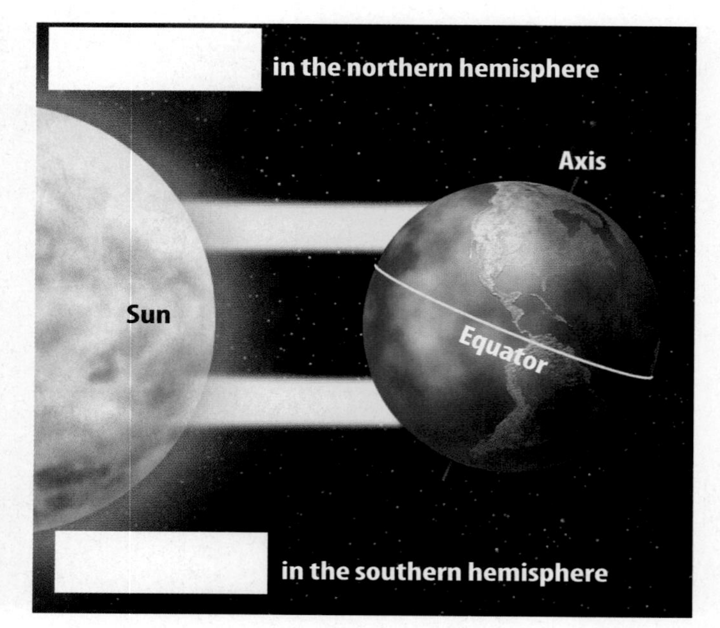

Northern Hemisphere Seasons

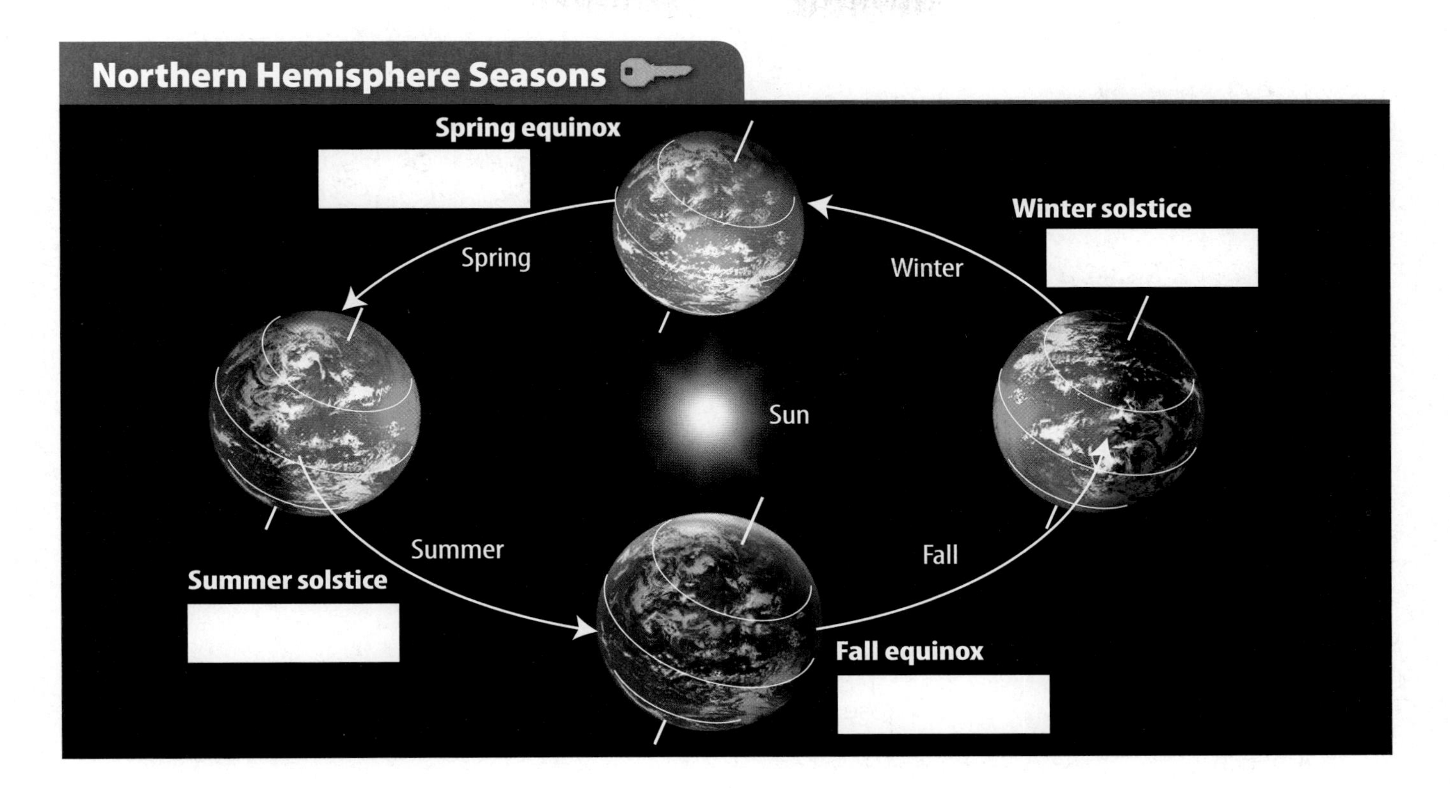

Figure 10 Seasons change as Earth completes its yearly revolution around the Sun.

Active Reading **6. Identify** Label the dates of the solstices and equinoxes in the diagram.

Solstices and Equinoxes

Earth revolves around the Sun once about every 365 days. During Earth's **revolution**, there are four days that mark the beginning of each of the seasons. These days are a summer solstice, a fall equinox, a winter solstice, and a spring equinox.

As shown in **Figure 10,** the solstices mark the beginnings of summer and winter. In the northern hemisphere, the summer solstice occurs on June 21 or 22. On this day, the northern hemisphere is tilted toward the Sun. In the southern hemisphere, this day marks the beginning of winter. The winter solstice begins on December 21 or 22 in the northern hemisphere. On this day, the northern hemisphere is tilted away from the Sun. In the southern hemisphere, this day marks the beginning of summer.

Equinoxes, also shown in **Figure 10,** are days when Earth is positioned so that neither the northern hemisphere nor the southern hemisphere is tilted toward or away from the Sun. The equinoxes are the beginning of spring and fall. On equinox days, the number of daylight hours almost equals the number of nighttime hours everywhere on Earth. In the northern hemisphere, the spring equinox occurs on March 21 or 22. This is the beginning of fall in the southern hemisphere. On September 22 or 23, fall begins in the northern hemisphere and spring begins in the southern hemisphere.

Active Reading **7. Distinguish** Contrast solstices and equinoxes.

SCIENCE USE V. COMMON USE

revolution

Science Use the action by a celestial body of going around in an orbit or an elliptical course

Common Use a sudden, radical, or complete change

El Niño

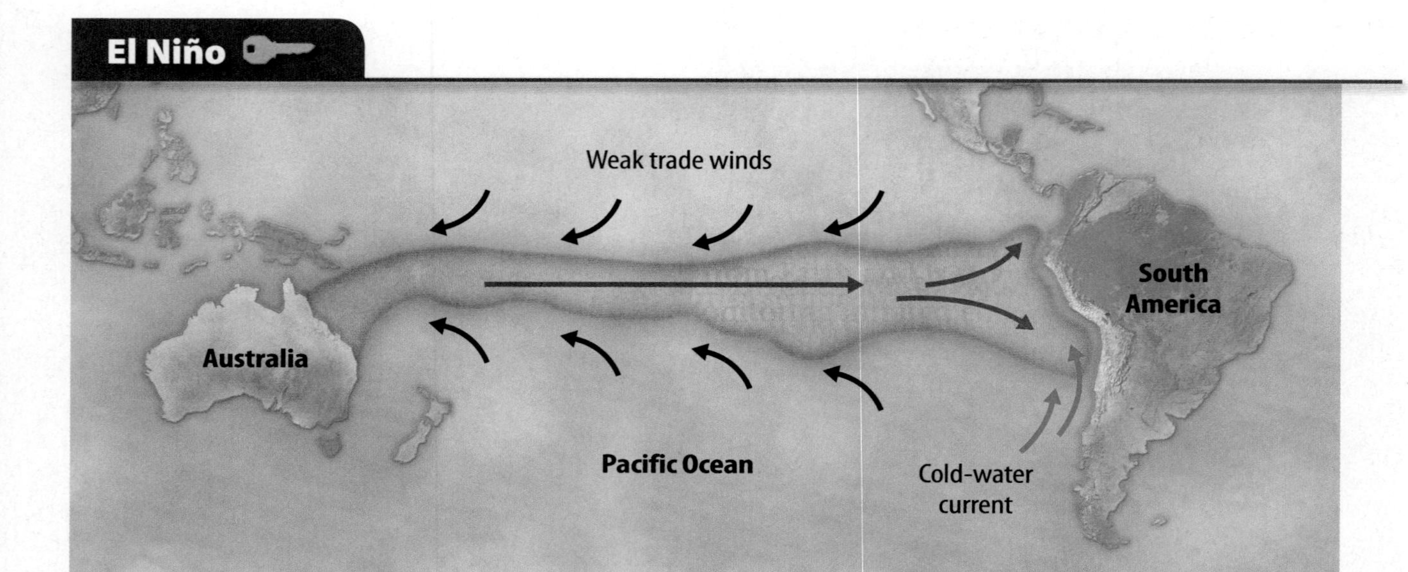

Figure 11 During El Niño, the trade winds weaken and warm water surges toward South America.

8. Identify Circle the source cause of El Niño.

Academic Vocabulary

phenomenon
(noun) an observable fact or event

9. Describe How do conditions in the Pacific Ocean differ from normal during El Niño?

El Niño and the Southern Oscillation

Close to the equator, the trade winds blow from east to west. These steady winds push warm surface water in the Pacific Ocean away from the western coast of South America. This allows cold water to rush upward from below—a process called upwelling. The air above the cold, upwelling water cools and sinks, creating a high-pressure area. On the other side of the Pacific Ocean, air rises over warm, equatorial waters, creating a low-pressure area. This difference in air pressures across the Pacific Ocean helps keep the trade winds blowing.

As **Figure 11** shows, sometimes the trade winds weaken, reversing the normal pattern of high and low pressures across the Pacific Ocean. Warm water surges back toward South America, preventing cold water from upwelling. This **phenomenon**, called El Niño, shows the connection between the atmosphere and the ocean. During El Niño, the normally dry, cool western coast of South America warms and receives lots of precipitation. Climate changes can be seen around the world. Droughts occur in areas that are normally wet. The number of violent storms in California and the southern United States increases.

The combined ocean and atmospheric cycle that results in weakened trade winds across the Pacific Ocean is called **El Niño/Southern Oscillation**, or ENSO. A complete ENSO cycle occurs every three to eight years. The North Atlantic Oscillation (NAO) is another cycle that can change the climate for decades at a time. The NAO affects the strength of storms throughout North America and Europe by changing the position of the jet stream.

Monsoons

Another climate cycle involving both the atmosphere and the ocean is a monsoon. A **monsoon** *is a wind circulation pattern that changes direction with the seasons.* Temperature differences between the ocean and the land cause winds, as shown in **Figure 12.** During summer, warm air over land rises and creates low pressure. Cooler, heavier air sinks over the water, creating high pressure. The winds blow from the water toward the land, bringing heavy rainfall. During winter, the pattern reverses and winds blow from the land toward the water.

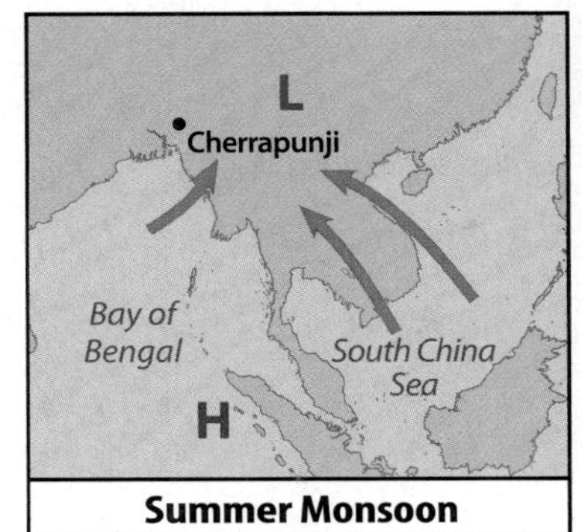

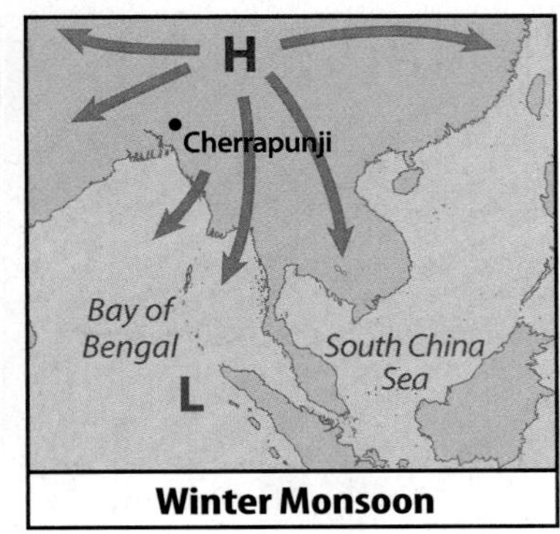

Figure 12 Monsoon winds reverse with the change of seasons.

The world's largest monsoon is found in Asia. Cherrapunji, India, is one of the world's wettest locations—receiving an average of 10 m of monsoon rainfall each year. Precipitation is even greater during El Niño events. A smaller monsoon occurs in southern Arizona. As a result, weather is dry during spring and early summer with thunderstorms occurring more often from July to September.

10. NGSSS Check State How does the ocean affect climate? SC.6.E.7.3

Droughts, Heat Waves, and Cold Waves

A **drought** *is a period with below-average precipitation.* A drought can cause crop damage and water shortages.

Droughts are often accompanied by heat waves—periods of unusually high temperatures. Droughts and heat waves occur when large hot-air masses remain in one place for weeks or months. Cold waves are long periods of unusually cold temperatures. These events occur when a large continental polar air mass stays over a region for days or weeks. Severe weather of these kinds can be the result of climatic changes on Earth or just extremes in the average weather of a climate.

Apply It!

After you complete the lab, answer the questions below.

1. What was the average temperature difference in Florida because of La Niña?

2. How else might El Niño and La Niña impact a city or town?

Lesson Review 2

Visual Summary

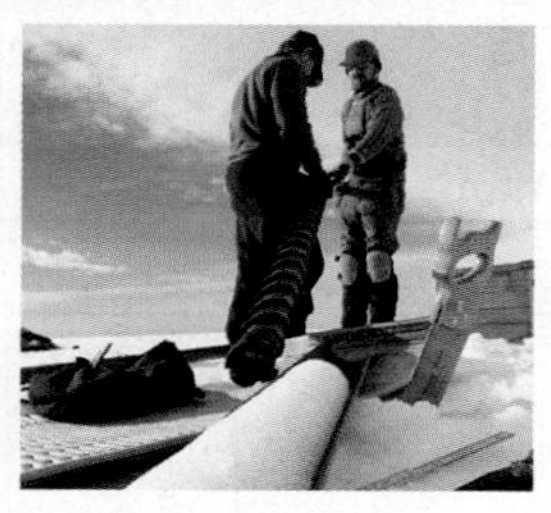

Scientists learn about past climates by studying natural records of climate, such as ice cores, fossilized pollen, and growth rings of trees.

Long-term climate changes, such as ice ages and interglacials, can be caused by changes in the shape of Earth's orbit and the tilt of its axis.

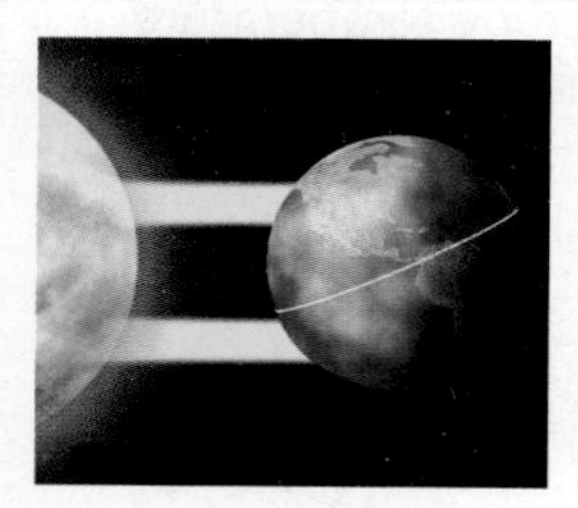

Short-term climate changes include seasons, El Niño/Southern Oscillation, and monsoons.

Use Vocabulary

1. **Distinguish** an ice age from an interglacial.

2. A(n) ______________ is a period of unusually high temperatures.

3. **Define** *drought* in your own words.

Understand Key Concepts

4. What happens during El Niño/Southern Oscillation? SC.6.E.7.3
   - (A) An interglacial climate shift occurs.
   - (B) The Pacific pressure pattern reverses.
   - (C) The tilt of Earth's axis changes.
   - (D) The trade winds stop blowing.

5. **Identify** causes of long-term climate change.

6. **Describe** how upwelling can affect climate. SC.6.E.7.3

Interpret Graphics

7. **Sequence** Fill in the graphic organizer to describe the sequence of events during El Niño/Southern Oscillation. SC.6.E.7.3

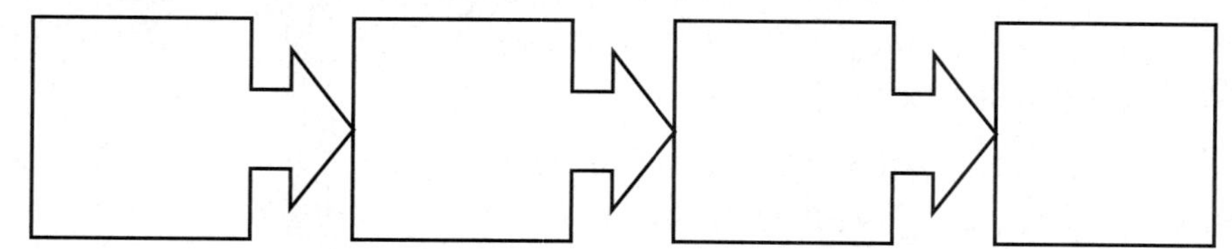

Critical Thinking

8. **Assess** the possibility that Earth will soon enter another ice age.

9. **Evaluate** the relationship between heat waves and drought.

CAREERS in SCIENCE

Frozen in Time

Looking for clues to past climates, Lonnie Thompson races against the clock to collect ancient ice from melting glaciers.

Earth's climate is changing. To understand why, scientists investigate how climates have changed throughout Earth's history by looking at ancient ice that contains clues from past climates. Scientists collected these ice samples only from glaciers at the North Pole and the South Pole. Then, in the 1970s, geologist Lonnie Thompson began collecting ice from a new location—the tropics.

Thompson has led expeditions to 15 countries and Antarctica.

Thompson, a geologist from the Ohio State University, and his team scale glaciers atop mountains in tropical regions. On the Quelccaya ice cap in Peru, they collect ice cores—columns of ice layers that built up over hundreds to thousands of years. Each layer is a capsule of a past climate, holding dust, chemicals, and gas that were trapped in the ice and snow during that period.

To collect ice cores, they drill hundreds of feet into the ice. The deeper they drill, the further back in time they go. One core is nearly 12,000 years old!

Collecting ice cores is not easy. The team hauls heavy equipment up rocky slopes in dangerous conditions—icy windstorms, thin air, and avalanche threats. Thompson's greatest challenge is the warming climate. The Quelccaya ice cap is melting. It has shrunk by 30 percent since Thompson's first visit in 1974. It's a race against time to collect ice cores before the ice disappears. When the ice is gone, so are the secrets it holds about climate change.

Thousands of ice-core samples are stored in deep freeze at Thompson's lab. One core from Antarctica is over 700,000 years old, which is well before the existence of humans.

Secrets in the Ice

In the lab, Thompson and his team analyze the ice cores to determine

- Age of ice: Every year, snow accumulations form a new layer. Layers help scientists date the ice and specific climate events.
- Precipitation: Each layer's thickness and composition help scientists determine the amount of snowfall that year.
- Atmosphere: As snow turns to ice, it traps air bubbles, providing samples of Earth's atmosphere. Scientists can measure the trace gases from past climates.
- Climate events: The concentration of dust particles helps scientists determine periods of increased wind, volcanic activity, dust storms, and fires.

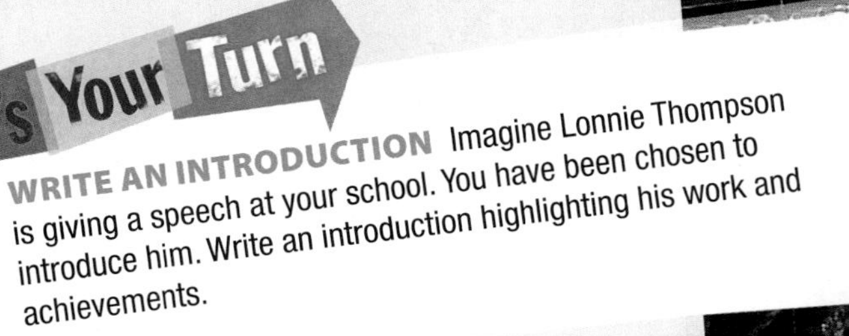

WRITE AN INTRODUCTION Imagine Lonnie Thompson is giving a speech at your school. You have been chosen to introduce him. Write an introduction highlighting his work and achievements.

Lesson 3

Recent Climate CHANGE

ESSENTIAL QUESTIONS

How can human activities affect climate?

How are predictions for future climate change made?

Vocabulary

global warming p. 224
greenhouse gas p. 224
deforestation p. 225
global climate model p. 227

Florida NGSSS

LA.6.2.2.3 The student will organize information to show understanding (e.g., representing main ideas within text through charting, mapping, paraphrasing, summarizing, or comparing/contrasting);

MA.6.A.3.6 Construct and analyze tables, graphs, and equations to describe linear functions and other simple relations using both common language and algebraic notation.

SC.6.E.7.9 Describe how the composition and structure of the atmosphere protects life and insulates the planet.

SC.6.N.1.1 Define a problem from the sixth grade curriculum, use appropriate reference materials to support scientific understanding, plan and carry out scientific investigation of various types, such as systematic observations or experiments, identify variables, collect and organize data, interpret data in charts, tables, and graphics, analyze information, make predictions, and defend conclusions.

SC.6.N.1.4 Discuss, compare, and negotiate methods used, results obtained, and explanations among groups of students conducting the same investigation.

SC.6.N.1.5 Recognize that science involves creativity, not just in designing experiments, but also in creating explanations that fit evidence.

SC.6.N.2.1 Distinguish science from other activities involving thought.

SC.6.N.2.1

Inquiry Launch Lab

20 minutes

What changes climates?

Natural events such as volcanic eruptions spew dust and gas into the atmosphere. These events can cause climate change.

Procedure

1. Read and complete a lab safety form.
2. Place a **thermometer** on a sheet of **paper.**
3. Hold a **flashlight** 10 cm above the paper. Shine the light on the thermometer bulb for 5 minutes. Observe the light intensity. Record the temperature below.
4. Use a **rubber band** to secure three to four layers of **cheesecloth or gauze** over the bulb end of the flashlight. Repeat step 3.

Data and Observations

Think About This

1. Describe the effect of the cheesecloth on the flashlight in terms of brightness and temperature.

2. Key Concept Would a volcanic eruption cause temperatures to increase or decrease? Explain.

Will Key West sink or swim?

1. The Florida Keys are a collection of close to 1,700 islands. These islands are just several feet above sea level. The Keys have been affected by a change in sea level. What might cause a change in sea level? How would these islands and the residents be impacted?

Regional and Global Climate Change

Average temperatures on Earth have been increasing for the past 100 years. As the graph in **Figure 13** shows, the warming has not been steady. Globally, average temperatures were fairly steady from 1880 to 1900. From 1900 to 1945, they increased by about 0.5°C. A cooling period followed, ending in 1975. Since then, average temperatures have steadily increased. The greatest warming has been in the northern hemisphere. However, temperatures have been steady in some areas of the southern hemisphere. Parts of Antarctica have cooled.

Active Reading **2. Summarize** Underline how temperatures have changed over the last 100 years.

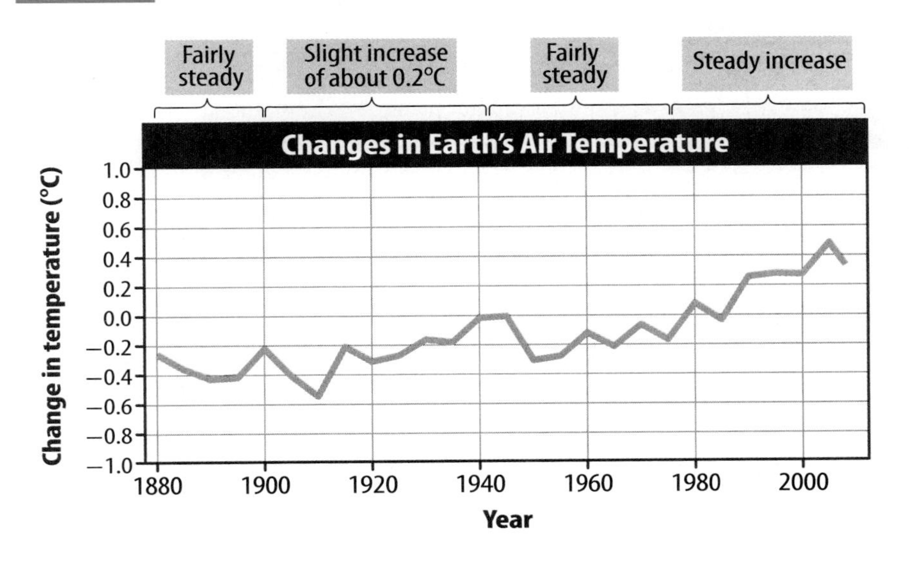

Active Reading **FOLDABLES®** LA.6.2.2.3

Make a tri-fold book from a sheet of paper. Label it as shown. Use it to organize your notes about climate change and the possible causes.

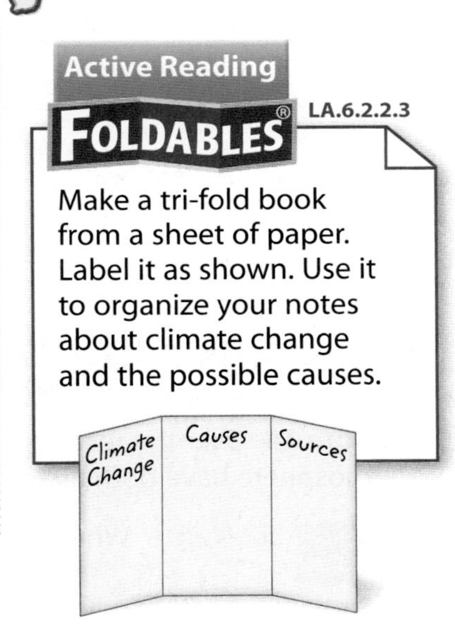

Figure 13 Temperature change has not been constant throughout the past 100 years.

3. Visual Check
Find Which 20-year period has seen the most change?

Human Impact on Climate Change

The rise in Earth's average surface temperature during the past 100 years is often referred to as **global warming**. Scientists have been studying this change and the possible causes of it. In 2007, the Intergovernmental Panel on Climate Change (IPCC), an international organization created to study global warming, concluded that most of this temperature increase is due to human activities. These activities include the release of increasing amounts of greenhouse gases into the atmosphere due to burning fossil fuels and large-scale cutting and burning of forests. Although many scientists agree with the IPCC, some scientists propose that global warming is due to natural climate cycles.

WORD ORIGIN

deforestation
from Latin *de-*, means "down from, concerning"; and *forestum silvam,* means "the outside woods"

Greenhouse Gases

Gases in the atmosphere that absorb Earth's outgoing infrared radiation are **greenhouse gases**. Greenhouse gases help keep temperatures on Earth warm enough for living things to survive. Recall that this phenomenon is referred to as the greenhouse effect. Without greenhouse gases, the average temperature on Earth would be much colder, about –18°C. Carbon dioxide (CO_2), methane, and water vapor are all greenhouse gases.

Active Reading **4. Describe** How do greenhouse gases affect temperatures on Earth?

Study the graph in **Figure 14.** What has happened to the levels of CO_2 in the atmosphere over the last 120 years? Levels of CO_2 have been increasing. Higher levels of greenhouse gases create a greater greenhouse effect. Most scientists suggest that global warming is due to the greater greenhouse effect. What are some sources of the excess CO_2?

Climate Change

Figure 14 Over the recent past, globally averaged temperatures and carbon dioxide concentration in the atmosphere have both increased.

Active Reading **5. Name** What are two likely causes of the rise in global temperatures?

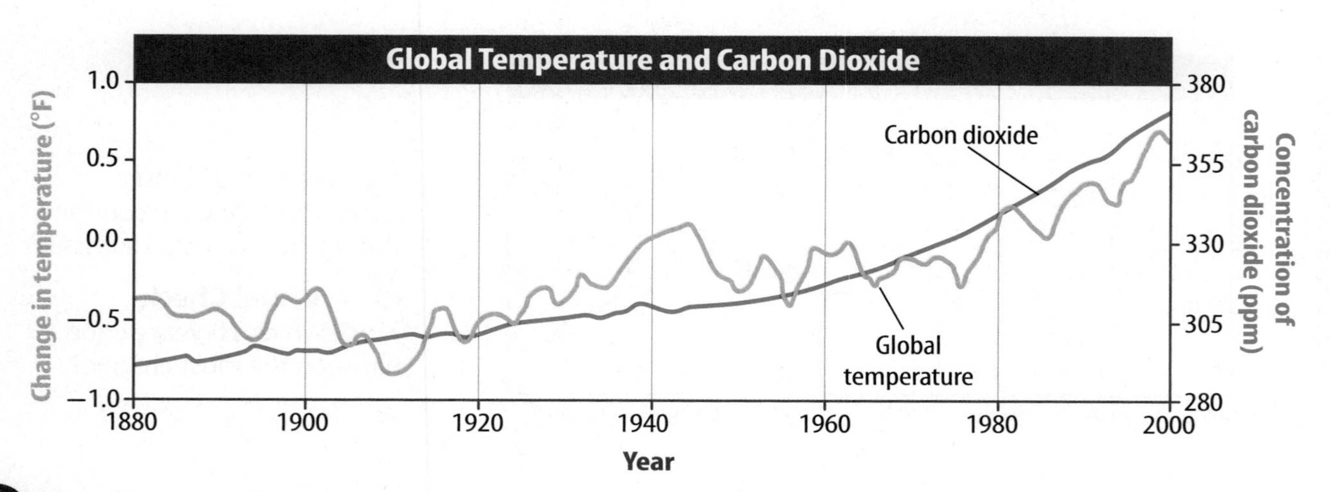

Human-Caused Sources Carbon dioxide enters the atmosphere when fossil fuels, such as coal, oil, and natural gas, burn. Burning fossil fuels releases energy that provides electricity, heats homes and buildings, and powers automobiles.

Deforestation *is the large-scale cutting and/or burning of forests.* Forest land is often cleared for agricultural and development purposes. Deforestation, shown in **Figure 15,** affects global climate by increasing carbon dioxide in the atmosphere in two ways. Living trees remove carbon dioxide from the air during photosynthesis. Cut trees, however, do not. Sometimes cut trees are burned to clear a field, adding carbon dioxide to the atmosphere as the trees burn. According to the Food and Agriculture Organization of the United Nations, deforestation makes up about 25 percent of the carbon dioxide released from human activities.

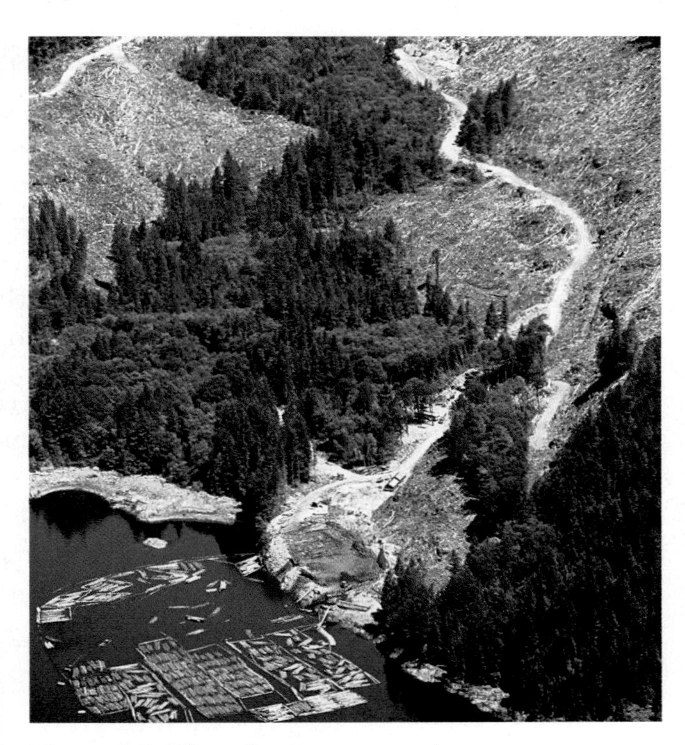

Figure 15 When forests are cut down, trees can no longer use carbon dioxide from the atmosphere. In addition, any wood that is left rots and releases more carbon dioxide into the atmosphere.

Natural Sources Carbon dioxide occurs naturally in the atmosphere. Its sources include volcanic eruptions and forest fires. Cellular respiration in organisms contributes additional CO_2.

Aerosols

The burning of fossil fuels releases more than just greenhouse gases into the atmosphere. Aerosols, tiny liquid or solid particles, are also released. Most aerosols reflect sunlight back into space. This prevents some of the Sun's energy from reaching Earth, potentially cooling the climate over time.

Aerosols also cool the climate in another way. When clouds form in areas with large amounts of aerosols, the cloud droplets are smaller. Clouds with small droplets, as shown in **Figure 16,** reflect more sunlight than clouds with larger droplets. By preventing sunlight from reaching Earth's surface, small-droplet clouds help cool the climate.

6. NGSSS Check **Identify** Highlight the human activities that affect climate. SC.6.E.7.4

Figure 16 Clouds made up of small droplets reflect more sunlight than clouds made up of larger droplets.

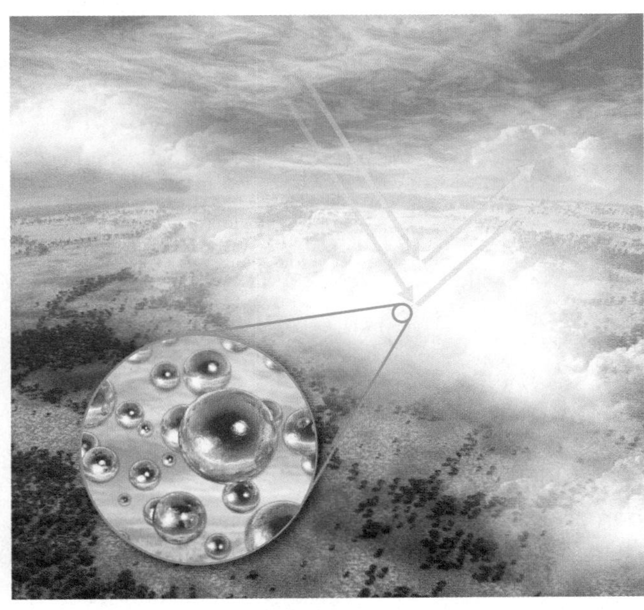

Math Skills MA.6.A.3.6

Use Percents

If Earth's population increases from 6 billion to 9 billion, what percent is this increase?

1. Subtract the initial value from the final value:
   9 billion − 6 billion = 3 billion
2. Divide the difference by the starting value:
   $\frac{3 \text{ billion}}{6 \text{ billion}} = 0.50$
3. Multiply by 100 and add a % sign:
   $0.50 \times 100 = 50\%$

7. Practice

If a climate's mean temperature changes from 18.2°C to 18.6°C, what is the percentage of increase?

Climate and Society

A changing climate can present challenges and benefits to society. Heat waves and droughts can cause food and water shortages. Excessive rainfall can cause flooding and mudslides. Warmer temperatures can mean longer growing seasons. Farmers can grow crops in areas that were previously too cold. Governments throughout the world are responding to the problems and opportunities created by climate change.

Environmental Impacts of Climate Change

Warmer temperatures can cause more water to evaporate from the ocean surface. The increased water vapor in the atmosphere has resulted in heavy rainfall and frequent storms in parts of North America. Precipitation has decreased over parts of southern Africa, the Mediterranean, and southern Asia.

Increasing temperatures can impact the environment in many ways. Melting glaciers and polar ice sheets can cause the sea level to rise. Ecosystems in Florida can be disrupted as coastal areas flood. Coastal flooding is a serious concern for the residents of Florida and the other one billion people living in low-lying areas on Earth.

Extreme weather events are also becoming more common. What effect will heat waves, droughts, and heavy rainfall have on infectious disease, existing plants and animals, and other systems of nature? Will increased CO_2 levels work similarly?

The annual thawing of frozen ground has caused the building shown in **Figure 17** to slowly sink as the ground becomes soft and muddy. Permanently higher temperatures would create similar events worldwide. This and other ecosystem changes can affect migration patterns of insects, birds, fish, and mammals.

Figure 17 Buildings in the Arctic that were built on frozen soil are now being damaged by the constant freezing and thawing of the soil.

Predicting Climate Change

Weather forecasts help people make daily choices about their clothing and activities. In a similar way, climate forecasts help governments decide how to respond to future climate changes.

A **global climate model**, *or GCM, is a set of complex equations used to predict future climates.* GCMs are similar to models used to forecast the weather. GCMs and weather forecast models are different. GCMs make long-term, global predictions, but weather forecasts are short-term and can be only regional predictions. GCMs combine mathematics and physics to predict temperature, amount of precipitation, wind speeds, and other characteristics of climate. Powerful supercomputers solve mathematical equations, and the results are displayed as maps. GCMs include the effects of greenhouse gases and oceans in their calculations. In order to test climate models, past records of climate change can and have been used.

Active Reading **8. Define** What is a GCM?

One drawback of GCMs is that the forecasts and predictions cannot be immediately compared to real data. A weather forecast model can be analyzed by comparing its predictions with meteorological measurements made the next day. GCMs predict climate conditions for several decades in the future. For this reason, it is difficult to evaluate the accuracy of climate models.

Most GCMs predict further global warming as a result of greenhouse gas emissions. By the year 2100, temperatures are expected to rise by between 1°C and 4°C. The polar regions are expected to warm more than the tropics. Summer arctic sea ice is expected to completely disappear by the end of the twenty-first century. Global warming and increases in sea-level are predicted to continue for several centuries.

9. NGSSS Check Explain How are predictions for future climate change made?

Apply It! After you complete the lab, answer these questions.

1. What is the difference in CO_2 emissions between the SUV and the hybrid?

2. What are some alternative fuel options for new vehicles?

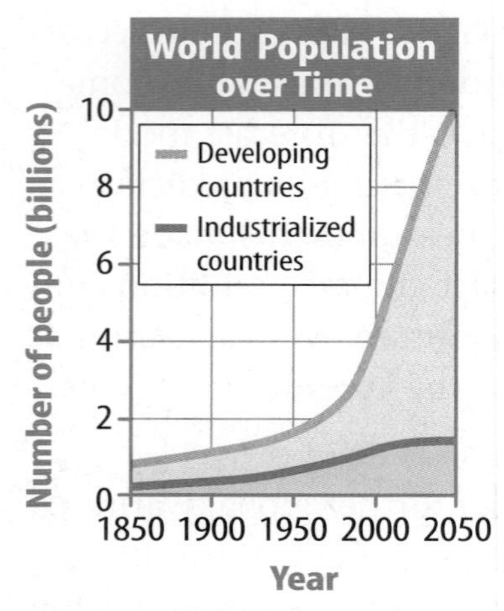

Figure 18 Earth's population is predicted to increase to more than 9 billion people by 2050.

Human Population

In 2000, more than 6 billion people inhabited Earth. As shown in **Figure 18,** Earth's population is expected to increase to more than 9 billion by the year 2050. What effects will a 50-percent increase in population have on Earth's atmosphere?

It is predicted that by the year 2030, two of every three people on Earth will live in urban areas. Many of these areas will be in developing countries in Africa and Asia. Large areas of forests are already being cleared to make room for expanding cities. As a result, significant amounts of greenhouse gases and other air pollutants will be added to the atmosphere.

Active Reading **10. Examine** How might an increase in human population affect climate change?

Ways to Reduce Greenhouse Gases

People have many options for reducing levels of pollution and greenhouse gases. One way is to develop alternative sources of energy that do not release carbon dioxide into the atmosphere, such as solar energy or wind energy. Automobile emissions can be reduced by as much as 35 percent by using hybrid vehicles. Hybrid vehicles use an electric motor part of the time, which reduces fuel use.

Emissions can be further reduced by green building. Green building is the practice of creating energy-efficient buildings, such as the one shown in **Figure 19.** People can also help remove carbon dioxide from the atmosphere by planting trees in deforested areas.

You can also help control greenhouse gases and pollution by conserving fuel and recycling. Turning off lights and electronic equipment when you are not using them reduces the amount of electricity you use. Recycling metal, paper, plastic, and glass reduces the amount of fuel required to manufacture these materials.

Figure 19 Solar heating, natural lighting, and water recycling are some of the technologies used in green buildings.

Lesson Review 3

Visual Summary

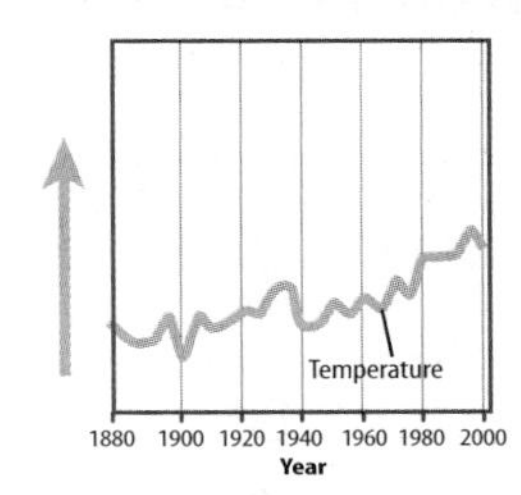

Many scientists suggest that global warming is due to increased levels of greenhouse gases.

Human activities, such as deforestation can contribute to global warming.

Ways to reduce greenhouse gas emissions include creating energy-efficient buildings.

Use Vocabulary

1. **Define** *global warming* in your own words.

2. A set of complex equations used to predict future climates is called ________.

3. **Use the term** *deforestation* in a sentence.

Understand Key Concepts 

4. Which human activity can have a cooling effect on climate? **SC.6.E.7.4**
   - (A) release of aerosols
   - (B) global climate models
   - (C) greenhouse gas emission
   - (D) large-area deforestation

5. **Describe** how human activities can impact climate. **SC.6.E.7.4**

6. **Describe** two ways deforestation contributes to the greenhouse effect. **SC.6.E.7.4**

Interpret Graphics

7. **Cause and Effect** Use the graphic organizer to identify two ways burning fossil fuels impacts climate. **SC.6.E.7.4**

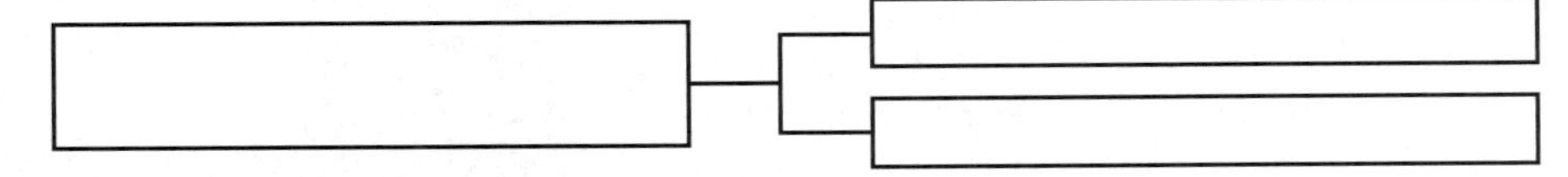

Critical Thinking

8. **Assess** the effects of global warming in the area where you live.

Math Skills MA.6.A.3.6

9. A 32-inch LCD flat-panel TV uses about 125 watts of electricity. If the screen size is increased to 40 inches, the TV uses 200 watts of electricity. What is the percent reduction of electricity if you use a 32-inch TV instead of a 40-inch TV?

Chapter 6 Study Guide

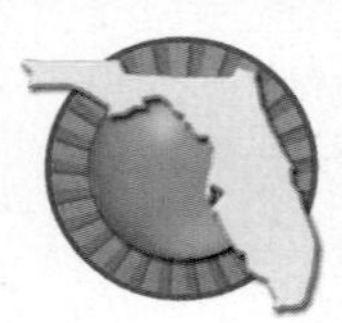

Think About It! **Climate is the long-term average weather conditions that occur in an area. It is influenced by the interactions between Earth's systems. Living things have adaptations to the climate in which they live.**

Key Concepts Summary

LESSON 1 Climates of Earth

- **Climate** is the long-term average weather conditions that occur in a particular region.
- Climate is affected by factors such as latitude, altitude, **rain shadows** on the downwind slope of mountains, vegetation, and the **specific heat** of water.
- Climate is classified based on precipitation, temperature, and native vegetation.

Vocabulary

climate p. 205
rain shadow p. 207
specific heat p. 207
microclimate p. 209

LESSON 2 Climate Cycles

- Over the past 4.6 billion years, climate on Earth has varied between **ice ages** and warm periods. **Interglacials** marked warm periods on Earth during ice ages.
- Earth's axis is tilted. This causes seasons as Earth revolves around the Sun.
- The **El Niño/Southern Oscillation** and **monsoons** are two climate patterns that result from interactions between oceans and the atmosphere.

ice age p. 214
interglacial p. 214
El Niño/Southern Oscillation p. 218
monsoon p. 219
drought p. 219

LESSON 3 Recent Climate Change

- Releasing carbon dioxide and aerosols into the atmosphere through burning fossil fuels and **deforestation** are two ways humans can affect climate change.
- Predictions about future climate change are made using computers and **global climate models**.

global warming p. 224
greenhouse gas p. 224
deforestationp. 225
global climate model p. 227

Active Reading

FOLDABLES® Chapter Project

Assemble your lesson Foldables as shown to make a Chapter Project. Use the project to review what you have learned in this chapter.

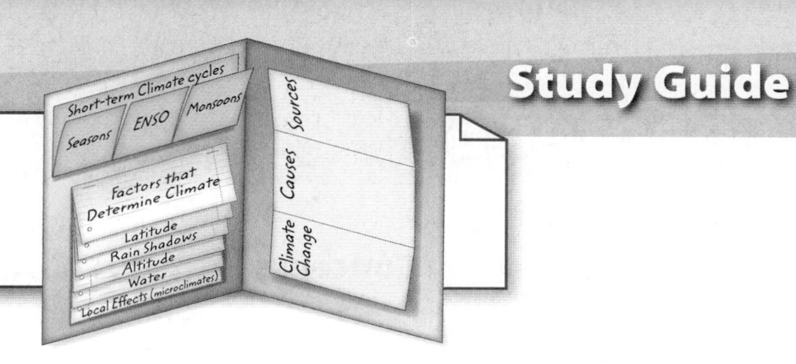

Use Vocabulary

1. A(n) ______________ is an area of low rainfall on the downwind slope of a mountain.

2. Forests often have their own ______________, with cooler temperatures than the surrounding countryside.

3. The lower ______________ of land causes it to warm up faster than water.

4. A wind circulation pattern that changes direction with the seasons is a(n) ______________.

5. Upwelling, trade winds, and air-pressure patterns across the Pacific Ocean change during a(n) ______________.

6. Earth's current ______________ is called the Holocene Epoch.

7. A(n) ______________ such as carbon dioxide absorbs Earth's infrared radiation and warms the atmosphere.

8. Additional CO_2 is added to the atmosphere when ______________ of large land areas occurs.

Link Vocabulary and Key Concepts

Use vocabulary terms from the previous page and other terms in this chapter to complete the concept map.

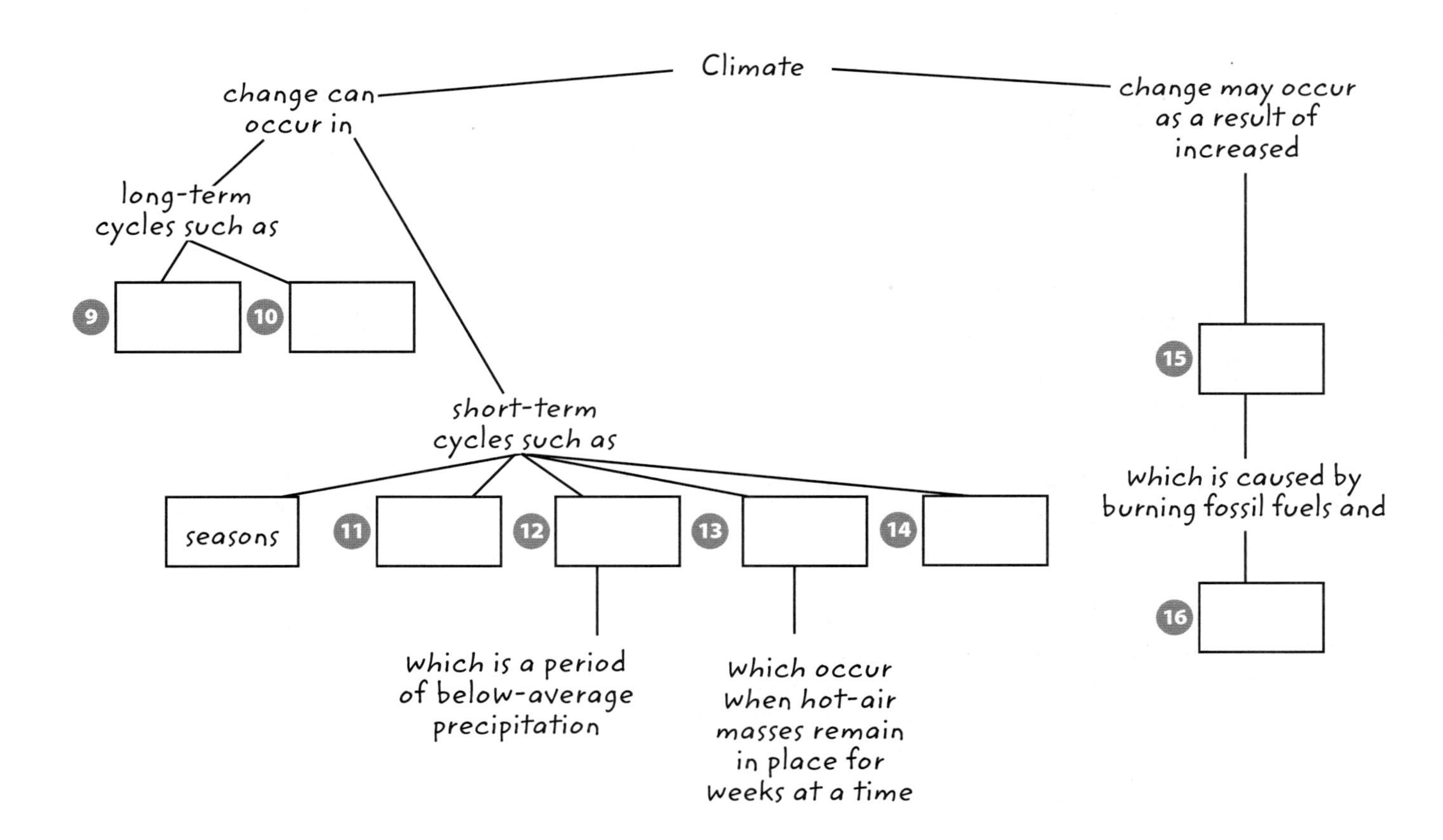

Chapter 6 Review

Fill in the correct answer choice.

Understand Key Concepts

1 The specific heat of water is ________ than the specific heat of land. SC.6.E.7.5

Ⓐ higher
Ⓑ lower
Ⓒ less efficient
Ⓓ more efficient

2 The graph below shows average monthly temperature and precipitation of an area over the course of a year.

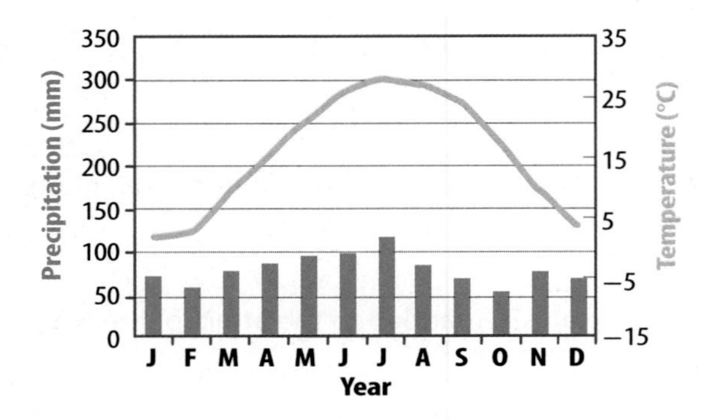

Which is the most likely location of the area? SC.6.E.7.3

Ⓐ in the middle of a large continent
Ⓑ in the middle of the ocean
Ⓒ near the North Pole
Ⓓ on the coast of a large continent

3 Which are warm periods during ice ages? SC.6.E.7.2

Ⓐ ENSO
Ⓑ interglacials
Ⓒ monsoons
Ⓓ Pacific oscillations

4 Long-term climate cycles are caused by all of the following EXCEPT SC.6.E.7.4

Ⓐ changes in ocean circulation.
Ⓑ Earth's revolution of the Sun.
Ⓒ the slow movement of the continents.
Ⓓ variations in the shape of Earth's orbit.

5 A rain shadow is created by which factor that affects climate? SC.6.E.7.4

Ⓐ a large body of water
Ⓑ buildings and concrete
Ⓒ latitude
Ⓓ mountains

Critical Thinking

6 **Hypothesize** how the climate of your town would change if North America and Asia moved together and became one enormous continent. SC.6.E.7.3

7 **Interpret Graphics** Identify the factor that affects climate, as shown in this graph. How does this factor affect climate? SC.6.E.7.5

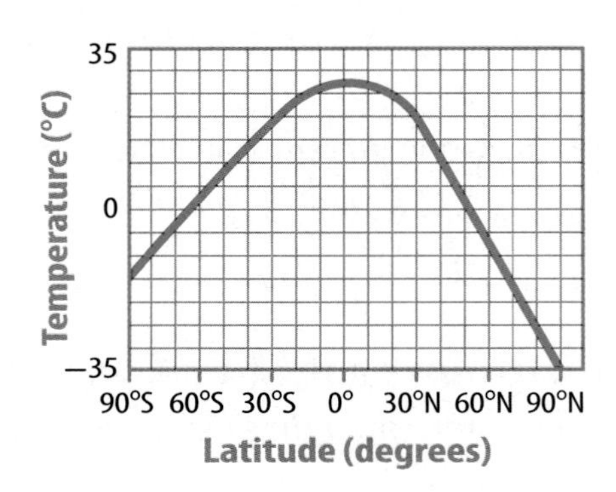

8 **Diagram** Draw a diagram that explains the changes that occur during an El Niño/Southern Oscillation event. SC.6.E.7.3

9 **Evaluate** which would cause more problems for your city or town: a drought, a heat wave, or a cold wave. Explain. SC.6.E.7.7

10 **Recommend** a life change you could make if the climate in your city were to change. LA.6.2.2.3

11 **Formulate** your opinion about the cause of global warming. Use facts to support your opinion. LA.6.2.2.3

12 **Predict** the effects of population increase on the climate where you live. SC.6.E.7.4

13 **Compare** how moisture affects the climate on either side of a mountain range. SC.6.E.7.3

Writing in Science

14 **Write** on a separate sheet of paper a short paragraph that describes a microclimate near your school or your home. What is the cause of the microclimate? SC.6.E.7.2

Big Idea Review

15 Explain what factors affect climate and give three examples of different types of climate. SC.6.E.7.6

16 Explain how life on Earth is affected by climate. SC.6.E.7.4

Math Skills MA.6.A.3.6

Use Percentages

17 Fred switches from a sport-utility vehicle that uses 800 gal of gasoline a year to a compact car that uses 450 gal.

a. By what percent did Fred reduce the amount of gasoline used?

b. If each gallon of gasoline released 20 pounds of CO_2, by what percent did Fred reduce the released CO_2?

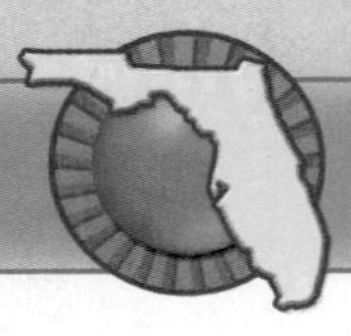

Florida NGSSS Benchmark Practice

Fill in the correct answer choice.

Multiple Choice

1 Why is the climate at the equator hot and humid? **SC.6.E.7.5**

Ⓐ the low altitude

Ⓑ The Sun strikes at a nearly 90° angle.

Ⓒ The Sun shines more than 16 hours a day.

Ⓓ Warm ocean currents come from high latitudes.

Use the diagram below to answer question 2.

2 What kind of climate would you expect to find at position 4? **SC.6.E.7.4**

Ⓕ mild

Ⓖ continental

Ⓗ tropical

Ⓘ dry

3 Which factor moderates coastal climates? **SC.6.E.7.3**

Ⓐ closeness to the ocean

Ⓑ lower altitude

Ⓒ lower latitude

Ⓓ global patterns

4 Which does NOT help explain climate differences? **SC.6.E.7.6**

Ⓕ altitude

Ⓖ latitude

Ⓗ oceans

Ⓘ organisms

5 What is the primary cause of seasonal changes on Earth? **SC.6.E.7.5**

Ⓐ Earth's distance from the Sun

Ⓑ Earth's ocean currents

Ⓒ Earth's prevailing winds

Ⓓ Earth's tilt on its axis

Use the diagram below to answer question 6.

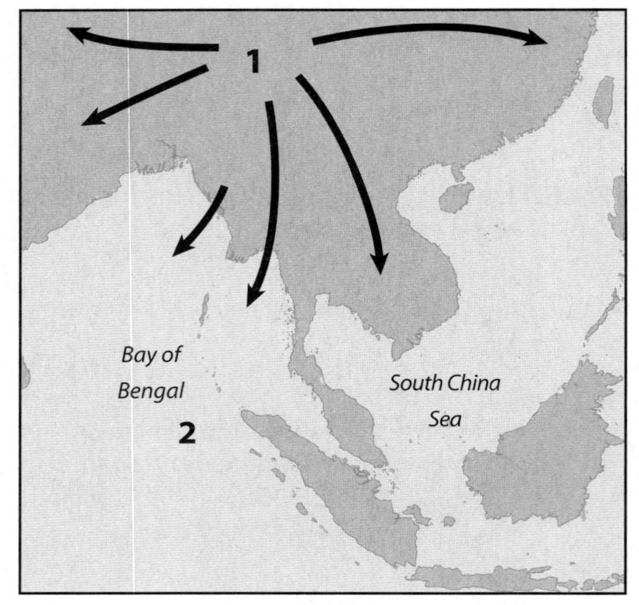

6 In the above diagram of the Asian winter monsoon, what does 1 represent? **SC.6.E.7.3**

Ⓕ high pressure

Ⓖ increased precipitation

Ⓗ low temperatures

Ⓘ wind speed

7 Climate is the ________ average weather conditions that occur in a particular region. Which completes the definition of *climate?* **SC.6.E.7.6**

Ⓐ global

Ⓑ long-term

Ⓒ mid-latitude

Ⓓ seasonal

Use the diagram below to answer question 8.

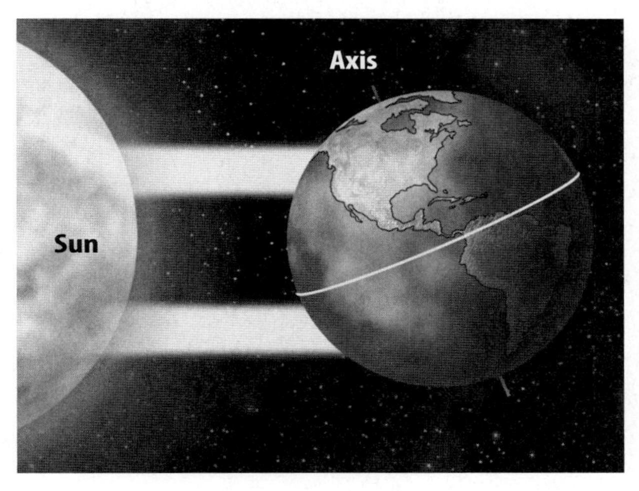

8 In the diagram above, what season is North America experiencing? **SC.6.E.7.5**

Ⓕ fall

Ⓖ spring

Ⓗ summer

Ⓘ winter

9 A rare blizzard in the Florida Panhandle would be an example of what? **SC.6.E.7.6**

Ⓐ climate

Ⓑ weather

Ⓒ El Niño

Ⓓ monsoon

10 Which is a short-term climate cycle? **SC.6.E.7.6**

Ⓕ droughts

Ⓖ monsoons

Ⓗ heat waves

Ⓘ cold waves

Use the graph below to answer question 11.

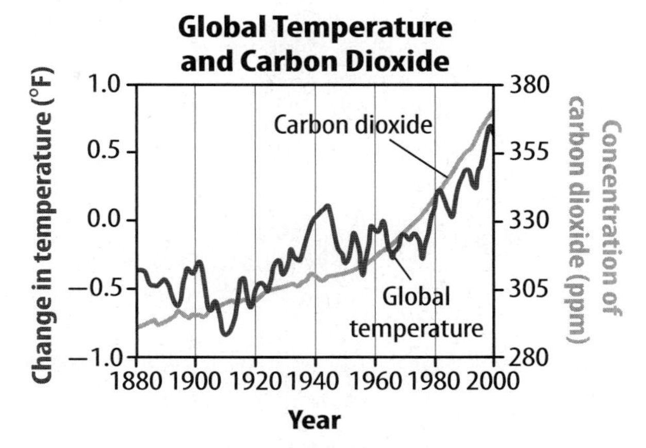

11 What does the graph above suggest about the relationship between global temperature and carbon dioxide in the atmosphere? **SC.6.E.7.4**

Ⓐ The two have only been related since 1980.

Ⓑ A global temperature increase causes an increase in carbon dioxide.

Ⓒ Global temperature and carbon dioxide did not exist before 1800.

Ⓓ As carbon dioxide increased in the atmosphere, global temperatures increased.

12 Which climate factor moves Atlantic hurricanes that form off the west coast of Africa toward Florida? **SC.6.E.7.3**

Ⓕ ocean water

Ⓖ prevailing wind

Ⓗ latitude

Ⓘ altitude

NEED EXTRA HELP?

If You Missed Question...	1	2	3	4	5	6	7	8	9	10	11	12
Go to Lesson...	1	1	1	1	2	2	1	2	1	2	3	1

Multiple Choice *Bubble the correct answer.*

1. In the image above, for what climate is the plant adapted? **SC.6.E.7.6**

Ⓐ dry

Ⓑ mild

Ⓒ polar

Ⓓ tropical

2. Which is NOT a factor that affects climate? **SC.6.E.7.6**

Ⓕ altitude

Ⓖ latitude

Ⓗ lakes and oceans

Ⓘ living organisms

3. Which image shows a microclimate? **SC.6.E.7.6**

Ⓐ 

Ⓑ 

Ⓒ 

Ⓓ 

Name ______________________ Date ____________

Benchmark Mini-Assessment

Multiple Choice *Bubble the correct answer.*

Use the image below to answer questions 1 and 2.

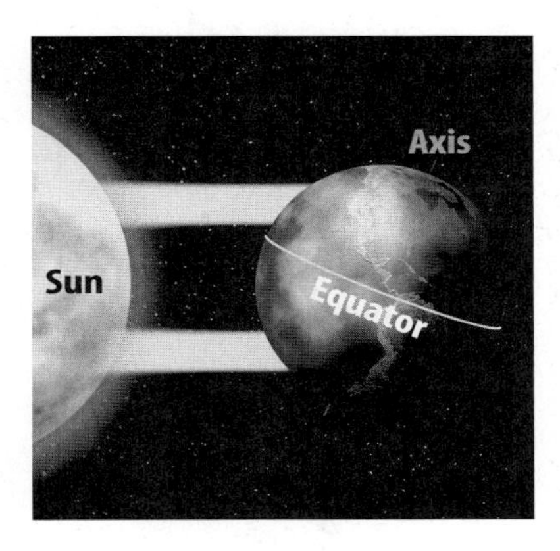

1. Which season is shown in the northern hemisphere? **SC.6.E.7.3**

Ⓐ spring

Ⓑ summer

Ⓒ autumn

Ⓓ winter

2. Which season is shown in the southern hemisphere? **SC.6.E.7.3**

Ⓕ spring

Ⓖ summer

Ⓗ autumn

Ⓘ winter

3. Which is an influence on long-term climate cycles? **SC.6.E.7.3**

Ⓐ ocean currents

Ⓑ trade winds

Ⓒ shape of Earth's orbit

Ⓓ shape of winter jet stream

4. What is a period with below-average precipitation called? **SC.6.E.7.4**

Ⓕ drought

Ⓖ equinox

Ⓗ monsoon

Ⓘ upwelling

Name ______________________ Date ____________

FLORIDA NGSSS

mini BAT

Multiple Choice *Bubble the correct answer.*

Use the image below to answer questions 1 and 2.

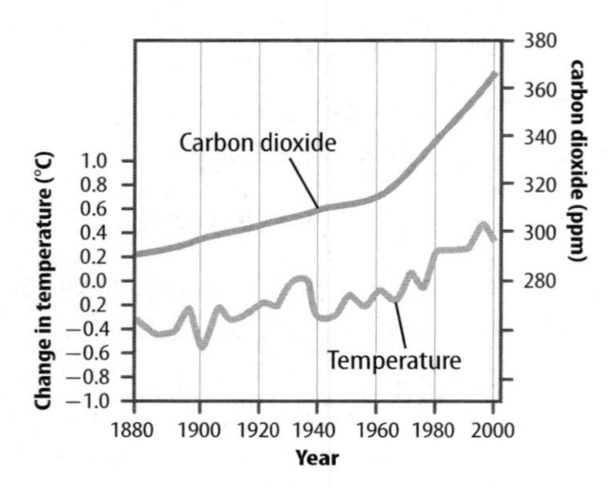

1. By what amount did carbon dioxide in Earth's atmosphere increase between 1960 and 2000? **MA.6.A.3.6**

Ⓐ 40–50 ppm

Ⓑ 60–70 ppm

Ⓒ 110–120 ppm

Ⓓ 360–370 ppm

2. In what year was the average global temperature lowest between 1880 and 2000? **MA.6.A.3.6**

Ⓕ 1880

Ⓖ 1900

Ⓗ 1940

Ⓘ 2000

3. Which is a natural source of greenhouse gases? **SC.6.E.7.9**

Ⓐ aerosols

Ⓑ deforestation

Ⓒ burning fossil fuels

Ⓓ volcanic eruptions

4. Which is a method for reducing global carbon emissions? **SC.6.E.7.9**

Ⓕ

Ⓖ

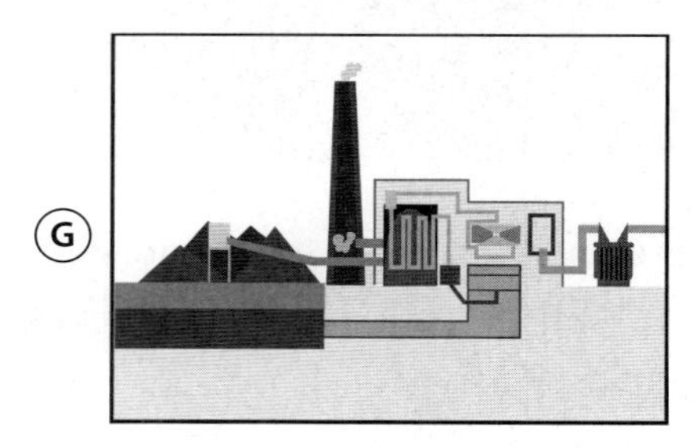

Ⓗ

Ⓘ

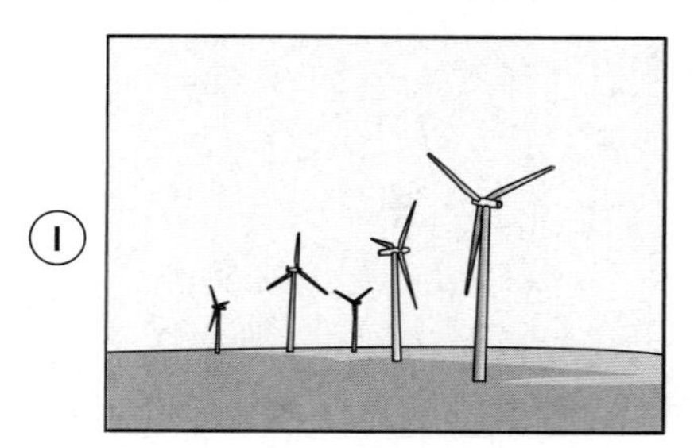

Name ____________________ Date ____________

Notes

3500 B.C.
The oldest wheeled vehicle is depicted in Mesopotamia, near the Black Sea.

400 B.C.
The Greeks invent the stone-hurling catapult.

1698
English military engineer Thomas Savery invents the first crude steam engine while trying to solve the problem of pumping water out of coal mines.

1760–1850
The Industrial Revolution results in massive advances in technology and social structure in England.

1769
The first vehicle to move under its own power is designed by Nicholas Joseph Cugnot and constructed by M. Breszin. A second replica is built that weighs 3,629 kg and has a top speed of 3.2 km per hour.

1794
Eli Whitney receives a patent for the mechanical cotton gin.

1817
Baron von Drais invents a machine to help him quickly wander the grounds of his estate. The machine is made of two wheels on a frame with a seat and a pair of pedals. This machine is the beginning design of the modern bicycle.

1903
Wilbur and Orville Wright build their airplane, called the Flyer, and take the first successful, powered, piloted flight.

1976
The first computer for home use is invented by college dropouts Steve Wozniak and Steve Jobs, who go on to found Apple Computer, Inc.

? Inquiry
Visit ConnectED for this unit's STEM activity.

Models

Have you ridden on an amusement park roller coaster such as the one in **Figure 1?** Did you think to yourself, "I hope I don't fly off this thing"? Before construction begins on a roller coaster, engineers build models of the thrill ride to ensure proper construction and safety. A **model** is a representation of an object, an idea, or a system that is similar to the physical object or idea being studied.

Figure 1 Engineers use various models to design roller coasters.

Using Models in Physical Science

Models are used to study things that are too big or too small, happen too quickly or too slowly, or are too dangerous or too expensive to study directly. Different types of models serve different purposes. Roller-coaster engineers build physical models for new, daring coasters. Mathematical and computer models calculate measurements of hills, angles, and loops to ensure a safe ride. Finally, engineers create a blueprint drawing that details the construction of the ride. Studying the various models allows engineers to predict how the actual coaster will behave when it travels through a loop or down a giant hill.

Active Reading **1. Point out** Highlight the purpose of using a model.

Types of Models

Physical Model

A physical model is a model that you can see and touch. It shows how parts relate to one another, how something is built, or how complex objects work. Physical models are built to scale. A limitation of a physical model is that it might not reflect the physical behavior of the full-sized object.

Mathematical Model

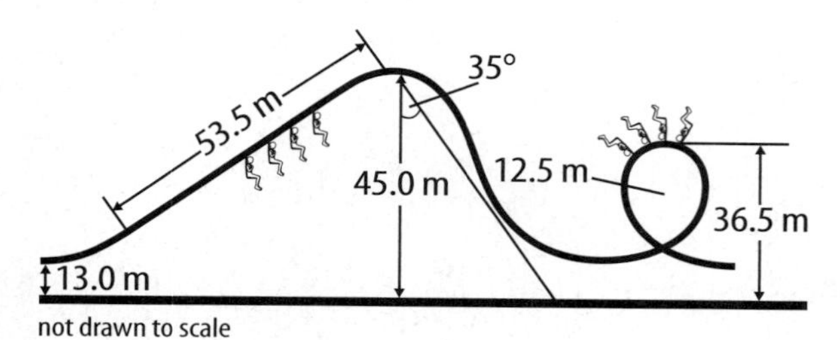

A mathematical model uses numerical data and equations to model an event or idea. When designing a thrill ride, engineers use mathematical models to calculate the heights, the angles of loops and turns, and the forces that affect the ride. One limitation of a mathematical model is that it cannot be used to model how different parts are assembled.

Making Models

An important factor in making a model is determining its purpose. You might need a model that physically represents an object, or a model that includes only important elements of an object or a process. When you build a model, first determine the function of the model. What materials should you use? What do you need to communicate to others? **Figure 2** shows two models of a glucose molecule, each with a different purpose.

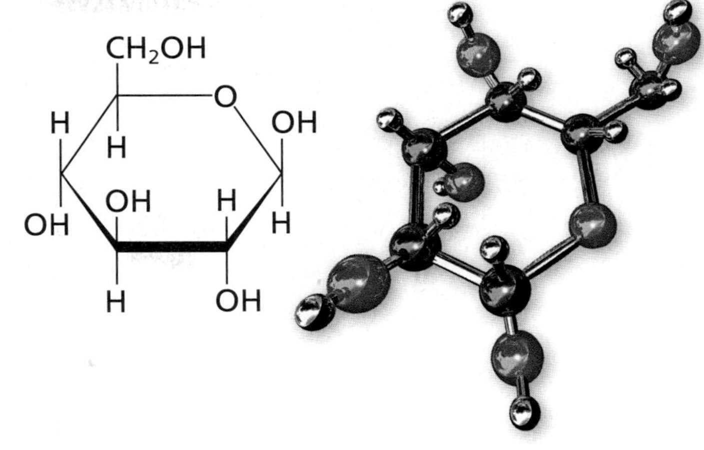

Figure 2 The model on the left is used to represent how the atoms in a glucose molecule bond together. The model on the right is a 3-D representation of the molecule, which shows how atoms might interact.

Limitations of Models

It is impossible to include all the details about an object or an idea into one model. All models have limitations. An engineer must be aware of the information each model does and does not provide. A blueprint of a roller coaster does not show the maximum weight that a car can support. However, a mathematical model would include this information. Scientists and engineers consider the purpose and the limitations of the model they use to ensure they draw accurate conclusions from models.

Active Reading **2. Determine** Highlight what scientists and engineers must consider when using models in their work.

Computer Simulation

A computer simulation is a model that combines mathematical models with computer graphic and animation programs. Simulations can contain thousands of complex mathematical models. When engineers change variables in mathematical models, they use computer simulation to view the effects of the change.

Apply It!

After you complete the lab, answer these questions.

1. **Analyze** How can each type of model be helpful in designing your safe, yet thrilling, coaster ride?

   Physical model:

   Mathematical model:

   Computer model:

2. **Recall** Think of a time you constructed a model.

   What was the purpose of the model?

   What materials did you use?

   How did the model help communicate information to you and your peers?

Name ______________________ Date ______________

Notes

What is energy?

Four friends argued about energy. They each had different ideas about energy. This is what they said:

Polly: I think energy gets used up like fuel.

Sophie: I think energy causes things to happen around us.

Trey: I think energy is a type of force that makes things move.

Chad: I think energy is a waste product given off when something is active.

With whom do you most agree? __________. Explain why you agree with that person.

Energy and Energy TRANSFORMATIONS

FLORIDA BIG IDEAS

1 **The Practice of Science**

2 **The Characteristics of Scientific Knowledge**

11 **Energy Transfer and Transformations**

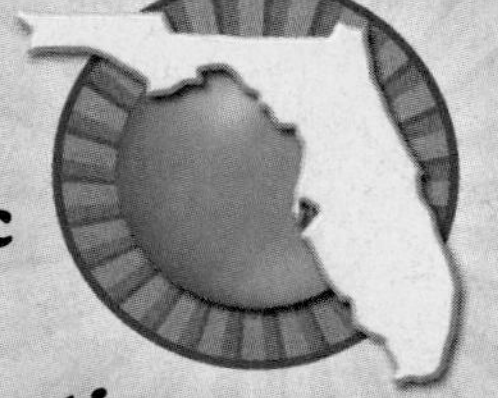

The Big Idea

Think About It!

What is energy, and what are energy transformations?

All objects in the photo contain energy. Some objects contain more energy than other objects. The Sun contains so much energy that it is considered an energy resource.

1. From where do you think the energy that powers the cars comes?

2. Do you think the energy in the Sun and the energy in the green plants are related?

3. What do the terms *energy* and *energy transformations* mean to you?

Get Ready to Read

What do you think about energy?

Before you read, decide if you agree or disagree with each of these statements. As you read this chapter, see if you change your mind about any of the statements.

	AGREE	DISAGREE
1 A fast-moving baseball has more kinetic energy than a slow-moving baseball.	☐	☐
2 A large truck and a small car moving at the same speed have the same kinetic energy.	☐	☐
3 Energy can change from one form to another.	☐	☐
4 Energy is destroyed when you apply the brakes on a moving bicycle or a moving car.	☐	☐
5 Warm water sinks below cool water when they are in the same container.	☐	☐
6 Wearing a coat slows the loss of thermal energy from your body.	☐	☐

There's More Online!

Video • Audio • Review • ⓘLab Station • WebQuest • Assessment • Concepts in Motion • Multilingual eGlossary

Lesson 1

Forms of ENERGY

ESSENTIAL QUESTIONS

 What is energy?

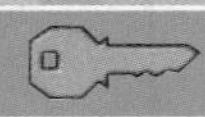 What are potential and kinetic energy?

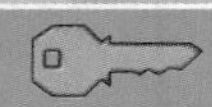 How is energy related to work?

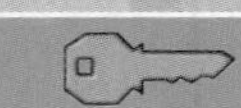 What are different forms of energy?

Vocabulary

energy p. 249
kinetic energy p. 250
potential energy p. 250
work p. 252
mechanical energy p. 253
sound energy p. 253
thermal energy p. 253
electric energy p. 253
radiant energy p. 253
nuclear energy p. 253

Florida NGSSS

LA.6.2.2.3 The student will organize information to show understanding (e.g., representing main ideas within text through charting, mapping, paraphrasing, summarizing, or comparing/contrasting);

SC.6.P.11.1 Explore the Law of Conservation of Energy by differentiating between potential and kinetic energy. Identify situations where kinetic energy is transformed into potential energy and vice versa.

SC.6.P.13.2 Explore the Law of Gravity by recognizing that every object exerts gravitational force on every other object and that the force depends on how much mass the objects have and how far apart they are.

SC.6.N.1.5 Recognize that science involves creativity, not just in designing experiments, but also in creating explanations that fit evidence.

SC.6.N.2.1 Distinguish science from other activities involving thought.

Launch Lab

SC.6.N.1.5, SC.6.N.2.1

20 minutes

Can you change matter?

You observe many things changing. Birds change their positions when they fly. Bubbles form in boiling water. The filament in a lightbulb glows when you turn on a light. How can you cause a change in matter?

Procedure

1. Read and complete the lab safety form.
2. Half-fill a **foam cup** with **sand.** Place the bulb of a **thermometer** about halfway into the sand. *Do not stir.* Record the temperature in the space below.
3. Remove the thermometer and place a lid on the cup. Hold down the lid and shake the cup vigorously for 10 min.
4. Remove the lid. Measure and record the temperature of the sand.

Data and Observations

Think About This

1. What change did you observe in the sand?

2. How could you change your results?

3. Key Concept What do you think caused the change you observed in the sand?

Inquiry **Why is the cat glowing?**

1. A camera that detects temperature made this image. Dark colors represent cooler temperatures and light colors represent warmer temperatures. Temperatures are cooler where the cat's body emits less radiant energy and warmer where the cat's body emits more radiant energy. What are some of the cat's warmest areas?

What is energy?

It might be exciting to watch a fireworks display, such as the one shown in **Figure 1.** Over and over, you hear the crack of explosions and see bursts of colors in the night sky. Fireworks release energy when they explode. **Energy** *is the ability to cause change.* The energy in the fireworks causes the changes you see as bursting flashes of light and hear as loud booms.

Energy also causes other changes. The plant in **Figure 1** uses the energy from the Sun and makes food that it uses for growth and other processes. Energy can cause changes in the motions and positions of objects, such as the nail in **Figure 1.** Can you think of other ways energy might cause changes?

Word Origin

energy
from Greek *energeia,* means "activity"

Active Reading **2. Define** What is energy?

Active Reading **3. Write** Fill in the missing information about energy in the graphic organizer below.

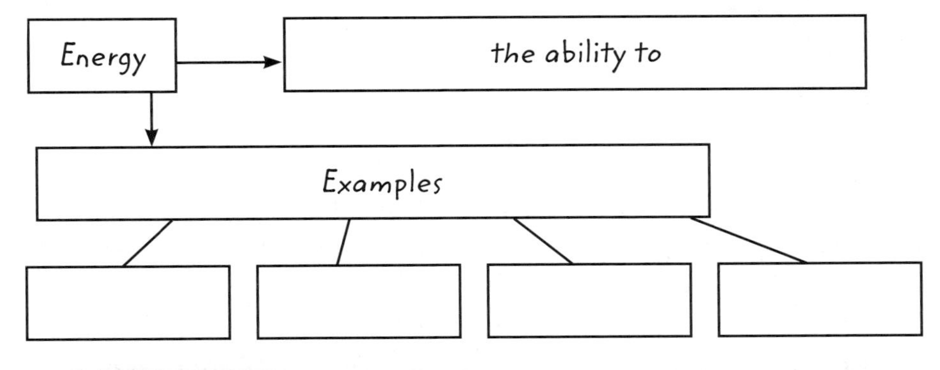

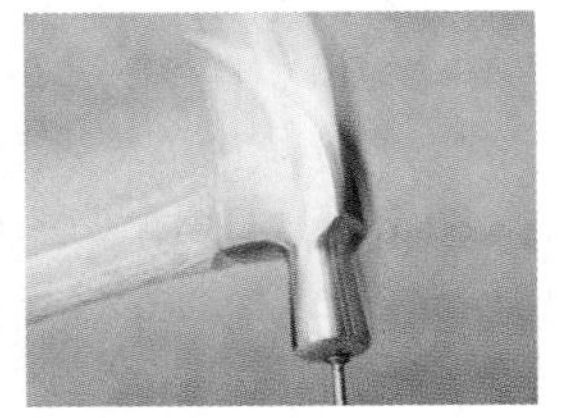

Figure 1 The explosion of fireworks, the growth of a plant, and the motion of a hammer all involve energy.

4. Visual Check **Deduce** What might be the energy source of the hammer in **Figure 1**?

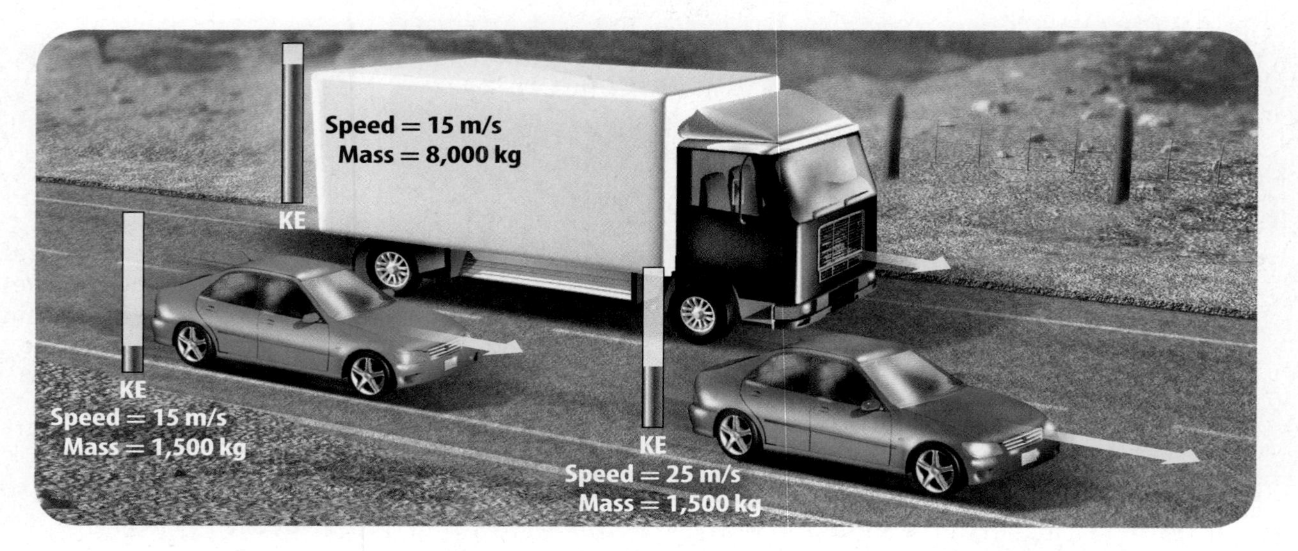

Figure 2 The kinetic energy (KE) of an object depends on its speed and its mass. The vertical bars show the kinetic energy of each vehicle.

5. Visual Check
Interpret Circle the car in the figure that has the most kinetic energy.

Active Reading FOLDABLES® SC.6.P.11.1 LA.6.2.2.3

Make a 18-cm fold along the long edge of a sheet of paper to make a two-pocket book. Label it as shown. Organize information about the forms of energy on quarter sheets of paper, and put them in the pockets.

Kinetic Energy—Energy of Motion

Have you ever been to a bowling alley? When you rolled the ball and it hit the pins, a change occurred—the pins fell over. This change occurred because the ball had a form of energy called kinetic (kuh NEH tik) energy. **Kinetic energy** *is energy due to motion.* All moving objects have kinetic energy.

Kinetic Energy and Speed

An object's kinetic energy depends on its speed. The faster an object moves, the more kinetic energy it has. For example, the blue car has more kinetic energy than the green car in **Figure 2** because the blue car is moving faster.

Kinetic Energy and Mass

A moving object's kinetic energy also depends on its mass. If two objects move at the same speed, the object with more mass has more kinetic energy. For example, the truck and the green car in **Figure 2** are moving at the same speed, but the truck has more kinetic energy because it has more mass.

6. NGSSS Check Define What is kinetic energy? SC.6.P.11.1

Potential Energy—Stored Energy

Energy can be present even if objects are not moving. If you hold a ball in your hand and then let it go, the gravitational interaction between the ball and Earth causes a change to occur. Before you dropped the ball, it had a form of energy called potential (puh TEN chul) energy. **Potential energy** *is stored energy due to the interactions between objects or particles.* Gravitational potential energy, elastic potential energy, and chemical potential energy are all forms of potential energy.

Gravitational Potential Energy

Even when you are just holding a book, gravitational potential energy is stored between the book and Earth. The girl shown in **Figure 3** increases the gravitational potential energy between her backpack and Earth by lifting the backpack higher from the ground.

The gravitational potential energy stored between an object and Earth depends on the object's weight and height. Dropping a bowling ball from a height of 1 m causes a greater change than dropping a tennis ball from 1 m. Similarly, dropping a bowling ball from 3 m causes a greater change than dropping the same bowling ball from only 1 m.

7. Describe What factors determine the gravitational potential energy stored between an object and Earth?

Elastic Potential Energy

When you stretch a rubber band, as in **Figure 3,** another form of potential energy, called elastic (ih LAS tik) potential energy, is being stored in the rubber band. Elastic potential energy is energy stored in objects that are compressed or stretched, such as springs and rubber bands. When you release the end of a stretched rubber band, the stored elastic potential energy is transformed into kinetic energy. This transformation is obvious when it flies across the room.

Chemical Potential Energy

Food, gasoline, and other substances are made of atoms joined together by chemical bonds. Chemical potential energy is energy stored in the chemical bonds between atoms, as shown in **Figure 3.** Chemical potential energy is released when chemical reactions occur. Your body uses the chemical potential energy in foods for all its activities. People also use the chemical potential energy in gasoline to power cars and buses.

Potential Energy

Figure 3 There are different forms of potential energy.

Gravitational Potential Energy

Gravitational potential energy increases when the girl lifts her backpack.

Elastic Potential Energy

The rubber band's elastic potential energy increases when it is stretched.

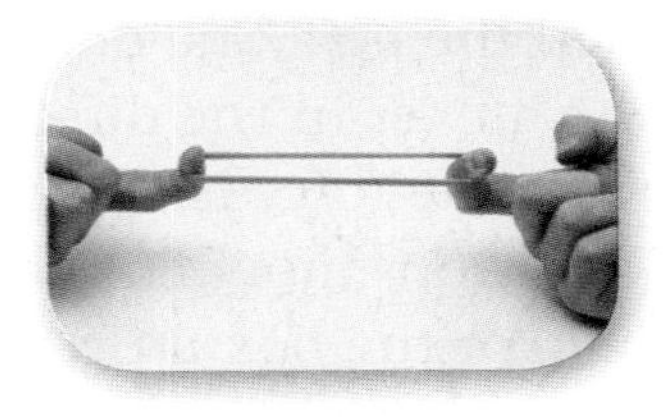

Chemical Potential Energy

Foods and other substances, including glucose, have chemical potential energy stored in the bonds between atoms.

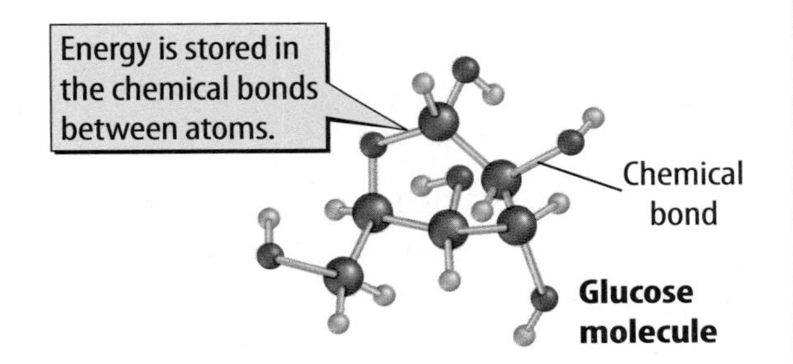

8. NGSSS Check Review In what ways are all forms of potential energy the same? SC.6.P.11.1

Figure 4 The girl does work on the box as she lifts it and increases its gravitational potential energy. The colored bars show the work that the girl does (W) and the box's potential energy (PE).

9. Visual Check Determine When did the transfer of energy take place between the girl and the box?

Energy and Work

You can transfer energy by doing work. **Work** *is the transfer of energy that occurs when a force makes an object move in the direction of the force while the force is acting on the object.* For example, as the girl lifts the box onto the shelf in **Figure 4,** she transfers energy from herself to the box. She does work only while the box moves in the direction of the force and while the force is applied to the box. If the box stops moving, the force is no longer applied, or the box movement and the applied force are in different directions, work is not done on the box.

10. Summarize How is energy related to work?

An object that has energy also can do work. For example, when a bowling ball collides with a bowling pin, the bowling ball does work on the pin. Some of the ball's kinetic energy is transferred to the bowling pin. Because of this connection between energy and work, energy is sometimes described as the ability to do work.

Other Forms of Energy

Some other forms of energy are shown in **Table 1.** All energy can be measured in joules (J). A softball dropped from a height of about 0.5 m has about 1 J of kinetic energy just before it hits the floor.

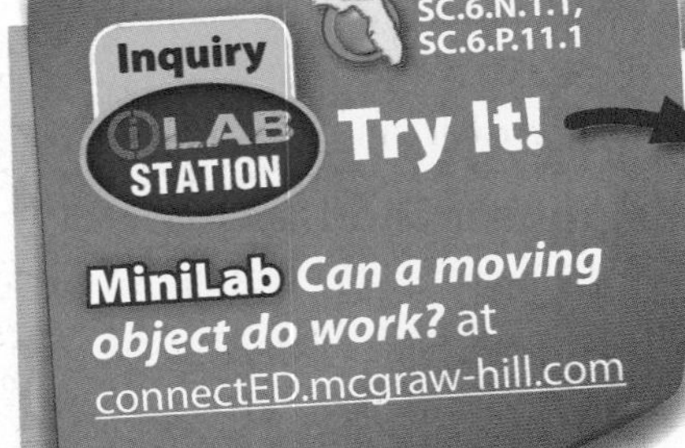

Apply It!

After you complete the lab, answer these questions.

1. What is the relationship between work and energy?

2. Does a person do work when he or she lifts a box? Does a person do work when he or she holds a box in the air?

3. Highlight the examples in which work is occurring.

pushing on a wall	sending a text message
grocery shopping	ice skating
thinking	sitting in a car

Table 1 Forms of Energy

Forms of Energy	
Mechanical Energy *The sum of potential energy and kinetic energy in a system of objects is* **mechanical energy**. For example, the mechanical energy of a basketball increases when a player shoots the basketball. Both the kinetic energy and gravitational potential energy of the ball increases in the player-ball-ground system.	
Sound Energy When you pluck a guitar string, the string vibrates and produces sound. *The energy that sound carries is* **sound energy**. Vibrating objects emit sound energy. However, sound energy cannot travel through a vacuum, such as the space between Earth and the Sun.	
Thermal Energy All objects and materials are made of particles that have energy. **Thermal energy** *is the sum of kinetic energy and potential energy of the particles that make up an object.* Mechanical energy is due to large-scale motions and interactions in a system and thermal energy is due to atomic-scale motions and interactions of particles. Thermal energy moves from warmer objects, such as burning logs, to cooler objects, such as air.	
Electric Energy An electrical fan uses another form of energy—electric energy. When you turn on a fan, there is an electric current through the fan's motor. **Electric energy** *is the energy an electric current carries.* Electrical appliances, such as fans and dishwashers, change electric energy into other forms of energy.	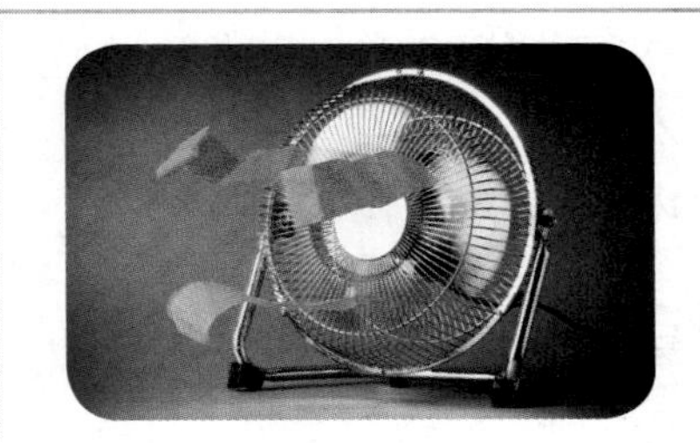
Radiant Energy—Light Energy The Sun gives off energy that travels to Earth as electromagnetic waves. Unlike sound waves, electromagnetic waves can travel through a vacuum. Light waves, microwaves, and radio waves are all electromagnetic waves. *The energy that electromagnetic waves carry is* **radiant energy**. Radiant energy sometimes is called light energy.	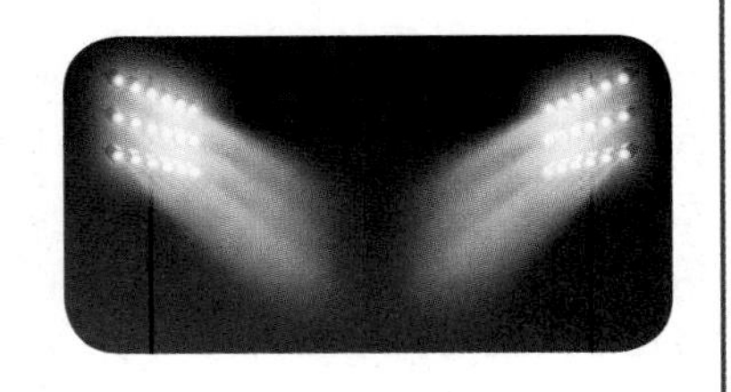
Nuclear Energy At the center of every atom is a nucleus. **Nuclear energy** *is energy that is stored and released in the nucleus of an atom.* In the Sun, nuclear energy is released when nuclei join together. In a nuclear power plant, nuclear energy is released when the nuclei of uranium atoms are split apart.	

Active Reading **11. Locate** (Circle) the form of energy that an MP3 player generates.

Lesson Review 1

Visual Summary

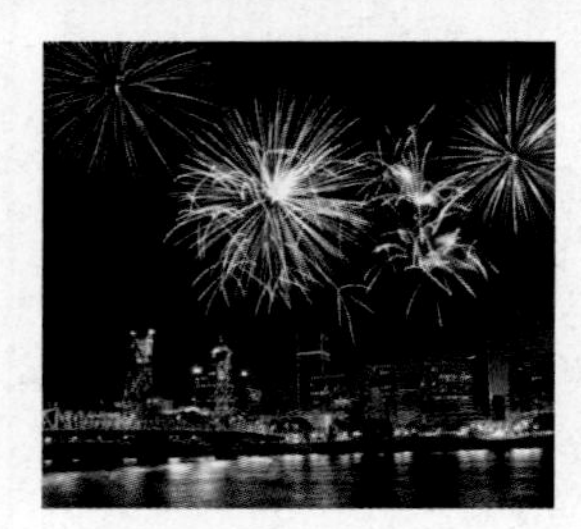

Energy is the ability to cause change.

The gravitational potential energy between an object and Earth increases when you lift the object.

You do work on an object when you apply a force to that object over a distance.

Skill Lab ***Can you identify potential and kinetic energy?*** at connectED.mcgraw-hill.com

Use Vocabulary

1. **Distinguish** between kinetic energy and potential energy. SC.6.P.11.1

2. **Write** a definition of *work*.

Understand Key Concepts

3. Which type of energy increases when you compress a spring?
   - (A) elastic potential energy
   - (B) kinetic energy
   - (C) radiant energy
   - (D) sound energy

4. **Infer** How could you increase the gravitational potential energy between yourself and Earth? SC.6.P.13.2

5. **Infer** how a bicycle's kinetic energy changes when that bicycle slows down. SC.6.P.11.1

6. **Compare and contrast** radiant energy and sound energy.

Interpret Graphics

7. **Identify** Use the graphic organizer below to identify three types of potential energy. LA.6.2.2.3

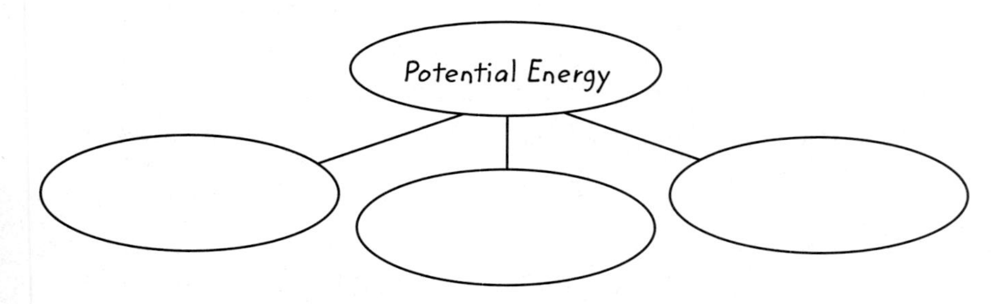

Critical Thinking

8. **Analyze** Will pushing on a car always change the car's mechanical energy? What must happen for the car's kinetic energy to increase?

Review Summarize what you have learned in this lesson on energy by completing the graphic organizer below.

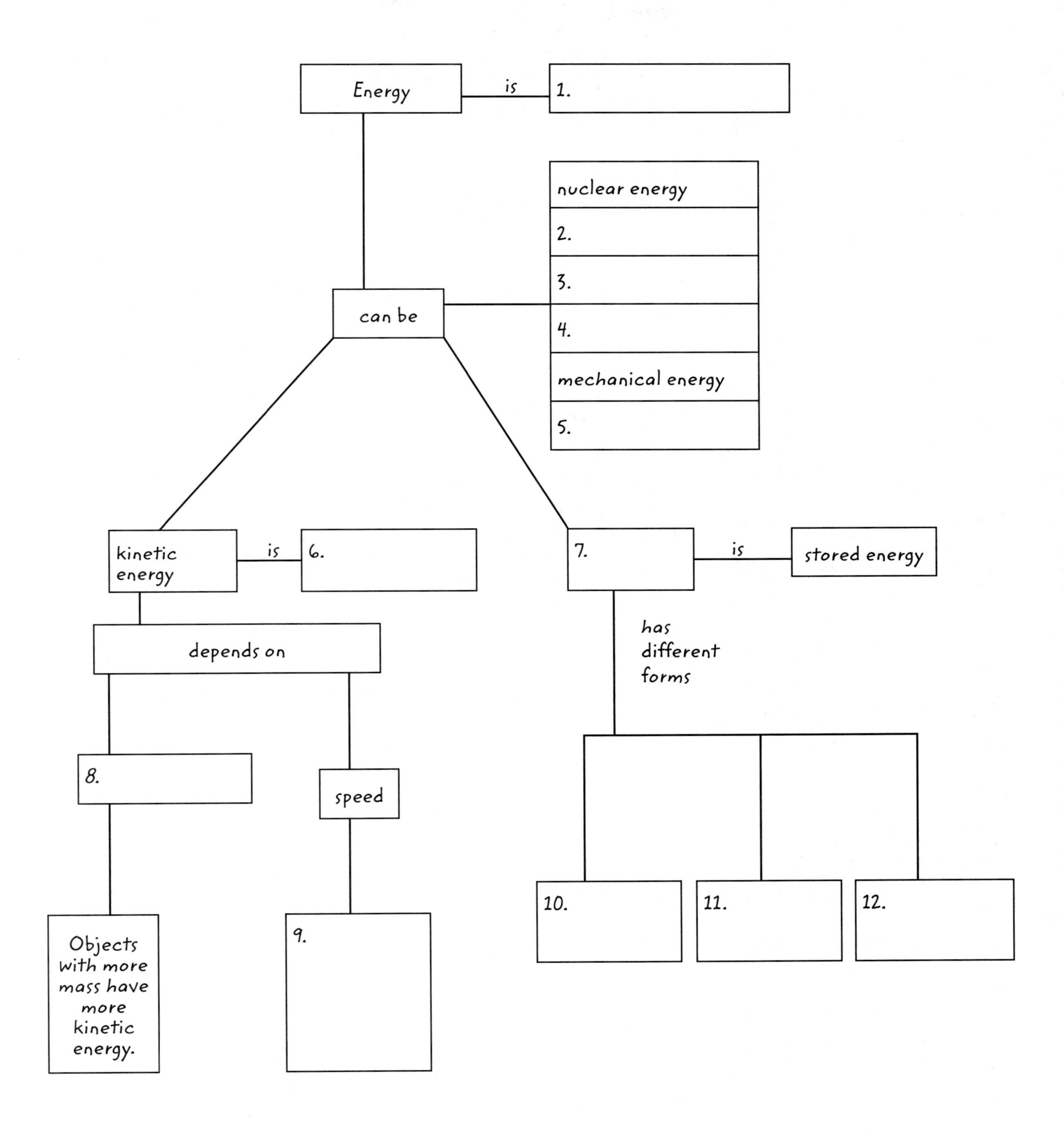

Lesson 2

Energy TRANSFORMATIONS

ESSENTIAL QUESTIONS

 What is the law of conservation of energy?

 How does friction affect energy transformations?

 How are different types of energy used?

Vocabulary

law of conservation of energy p. 258

friction p. 259

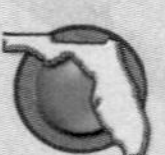

Florida NGSSS

LA.6.2.2.3 The student will organize information to show understanding (e.g., representing main ideas within text through charting, mapping, paraphrasing, summarizing, or comparing/contrasting);

MA.6.A.3.6 Construct and analyze tables, graphs, and equations to describe linear functions and other simple relations using both common language and algebraic notation.

SC.6.P.11.1 Explore the Law of Conservation of Energy by differentiating between potential and kinetic energy. Identify situations where kinetic energy is transformed into potential energy and vice versa.

SC.6.N.1.1 Define a problem from the sixth grade curriculum, use appropriate reference materials to support scientific understanding, plan and carry out scientific investigation of various types, such as systematic observations or experiments, identify variables, collect and organize data, interpret data in charts, tables, and graphics, analyze information, make predictions, and defend conclusions.

SC.6.N.2.1 Distinguish science from other activities involving thought.

SC.6.N.3.1 Recognize and explain that a scientific theory is a well-supported and widely accepted explanation of nature and is not simply a claim posed by an individual. Thus, the use of the term theory in science is very different than how it is used in everyday life.

SC.6.N.3.2 Recognize and explain that a scientific law is a description of a specific relationship under given conditions in the natural world. Thus, scientific laws are different from societal laws.

Inquiry Launch Lab

SC.6.N.1.1, SC.6.N.2.1

15 minutes

Is energy lost when it changes form?

Energy can have different forms. What happens when energy changes from one form to another?

Procedure

1. Read and complete a lab safety form.
2. Three students should sit in a circle. One student has 30 **buttons,** one has 30 **pennies,** and one has 30 **paper clips.**
3. Each student should exchange 10 items with the student to the right and 10 items with the student to the left.
4. Repeat step 3.

Think About This

1. If the buttons, the pennies, and the paper clips represent different forms of energy, what represents changes from one form of energy to another?

2. Key Concept If each button, penny, and paper clip represents one unit of energy, does the total amount of energy increase, decrease, or stay the same? Explain your answer.

Inquiry What's that sound?

1. Blocks of ice breaking off the front of this glacier can be bigger than a car. Imagine the loud rumble they make as they crash into the sea. But after the ice falls into the sea, it will melt gradually. All these processes involve energy transformations—energy changing from one form to another. Describe another series of energy transformations.

Changes Between Forms of Energy

It is the weekend and you are ready to make some popcorn in the microwave and watch a movie. Energy changes form when you make popcorn and watch TV. As shown in **Figure 5,** a microwave changes electric energy into **radiant** energy. Radiant energy changes into thermal energy in the popcorn kernels.

The changes from electric energy to radiant energy to thermal energy are called energy transformations. As you watch the movie, energy transformations also occur in the television. A television transforms electric energy into sound energy and radiant energy.

Science Use v. Common Use

radiant

Science Use energy transmitted by electromagnetic waves

Common Use bright and shining; glowing

1. Electric energy is transferred from the electric outlet to the microwave.
2. The microwave oven transforms electric energy into radiant energy.
3. Radiant energy is transformed into thermal energy as the popcorn kernels absorb the microwaves. This causes the kernels to become hot and pop.

Figure 5 Energy changes from one form to another when you use a microwave oven to make popcorn.

2. Visual Check
Identify Which energy transformation pops the corn kernels?

Conservation of Energy

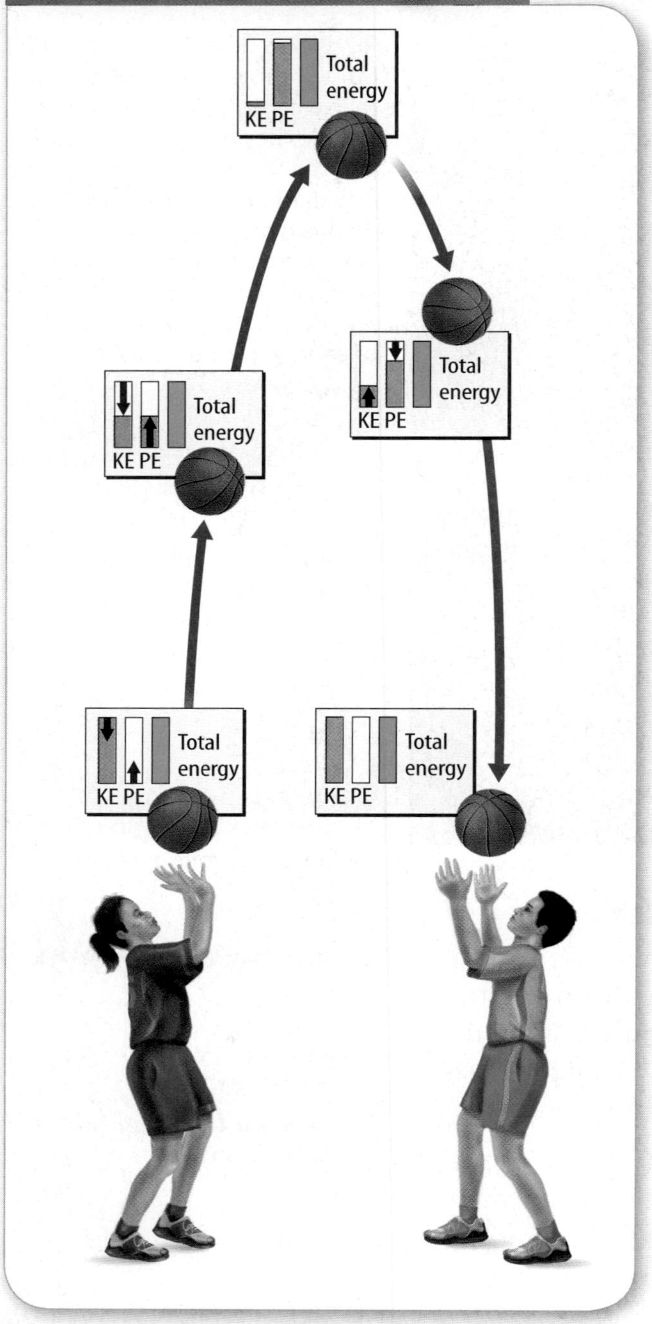

Figure 6 The ball's kinetic energy (KE) and potential energy (PE) change as it moves.

Active Reading **4. Interpret** Circle the ball that has the greatest gravitational potential energy.

5. NGSSS Check Define What is the law of conservation of energy? SC.6.P.11.1

Changes Between Kinetic and Potential Energy

Energy transformations also occur when you toss a ball upward, as shown in **Figure 6.** The ball slows down as it moves upward and then speeds up as it moves downward. The ball's speed and height change as energy changes from one form to another.

Kinetic Energy to Potential Energy

The ball is moving fastest and has the most kinetic energy as it leaves your hand, as shown in **Figure 6.** As the ball moves upward, its speed and kinetic energy decrease. However, the potential energy is increasing because the ball's height is increasing. Kinetic energy is changing into potential energy. At the ball's highest point, the gravitational potential energy is at its greatest, and the ball's kinetic energy is at its lowest.

Potential Energy to Kinetic Energy

As the ball moves downward, its potential energy decreases. At the same time, the ball's speed increases. Therefore, the ball's kinetic energy increases. Potential energy is transformed into kinetic energy. When the ball reaches the other player's hand, its kinetic energy is at the maximum value again.

3. Explain Why does the potential energy decrease as the ball falls?

The Law of Conservation of Energy

The total energy in the universe is the sum of all the different forms of energy everywhere. According to the **law of conservation of energy**, *energy can be transformed from one form into another or transferred from one region to another, but energy cannot be created or destroyed.* The total amount of energy in the universe does not change.

Friction and the Law of Conservation of Energy

Sometimes it may seem as if the law of conservation of energy is not accurate. Imagine riding a bicycle, as in **Figure 7.** The moving bicycle has mechanical energy. What happens to this mechanical energy when you apply the brakes and the bicycle stops?

When you apply the brakes, the bicycle's mechanical energy is not destroyed. Instead, the bicycle's mechanical energy is transformed into thermal energy, as shown in **Figure 7.** The total amount of energy never changes. The additional thermal energy causes the brakes, the wheels, and the air around the bicycle to become slightly warmer.

Friction between the bicycle's brake pads and the moving wheels transforms mechanical energy into thermal energy. **Friction** *is a force that resists the sliding of two surfaces that are touching.*

Active Reading **6. Explain** How does friction affect energy?

____

____

There is always some friction between any two surfaces that are rubbing against each other. As a result, some mechanical energy is always transformed into thermal energy when two surfaces rub against each other.

It is easier to pedal a bicycle if there is less friction between the bicycle's parts. With less friction, less of the bicycle's mechanical energy is transformed into thermal energy. One way to reduce friction is to apply a lubricant, such as oil, grease, or graphite, to surfaces that rub against each other.

Active Reading FOLDABLES® SC.6.P.11.1 LA.6.2.2.3

Cut three sheets of paper in half. Use the six half sheets to make a side-tab book with five tabs and a cover. Use your book to organize your notes on energy transformations.

Energy Transformations | Potential Energy | Kinetic Energy | Chemical Energy | Radiant Energy | Electric Energy

WORD ORIGIN

friction
from Latin *fricare,* means "to rub"

Friction and Thermal Energy

Figure 7 As the brakes transform mechanical energy to thermal energy, the total amount of energy does not change.

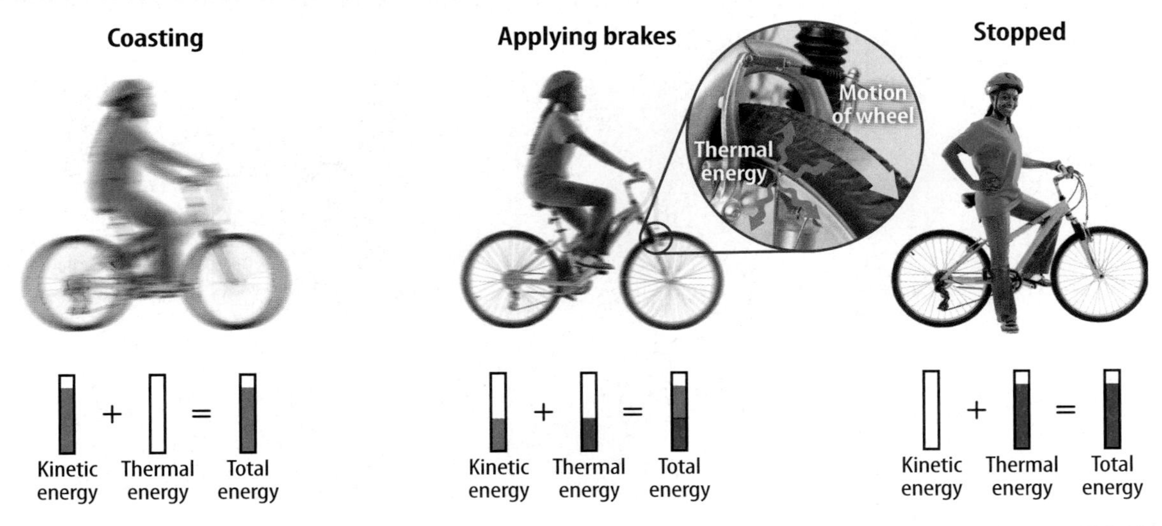

Math Skills MA.6.A.3.6

Solve a One-Step Equation

Electric energy often is measured in units called kilowatt-hours (kWh). To calculate the electric energy used by an appliance in kWh, use this equation:

$$\text{kWh} = \left(\frac{\text{watts}}{1{,}000}\right) \times \text{hours}$$

Appliances typically have a power rating measured in watts (W)

Practice

7. A hair dryer is rated at 1,200 W. If you use the dryer for 0.25 h, how much electric energy do you use? Use the space below for your calculations.

8. **Name** What are two uses of thermal energy?

Using Energy

Every day you use different forms of energy to do different things. You might use the radiant energy from a lamp to light a room, or you might use the chemical energy stored in your body to run a race. When you use energy, you usually change it from one form into another. For example, the lamp changes electric energy into radiant energy and thermal energy.

Using Thermal Energy

All forms of energy can be transformed into thermal energy. People often use thermal energy to cook food or provide warmth. A gas stove transforms the chemical energy stored in natural gas into the thermal energy that cooks food. An electric space heater transforms the electric energy from a power plant into the thermal energy that warms a room. In a jet engine, burning fuel releases thermal energy that the engine transforms into mechanical energy.

Using Chemical Energy

During photosynthesis, a plant transforms the Sun's radiant energy into chemical energy that it stores in chemical compounds. Some of these compounds become food for other living things. Your body transforms the chemical energy from your food into the kinetic energy necessary for movement. Your body also transforms chemical energy into the thermal energy necessary to keep you warm.

Using Radiant Energy

The cell phone in **Figure 8** sends and receives radiant energy using microwaves. When you are listening to someone on a cell phone, that cell phone is transforming radiant energy into electric energy and then into sound energy. When you are speaking into a cell phone, it is transforming sound energy into electric energy and then into radiant energy.

Figure 8 A cell phone changes sound energy into radiant energy when you speak.

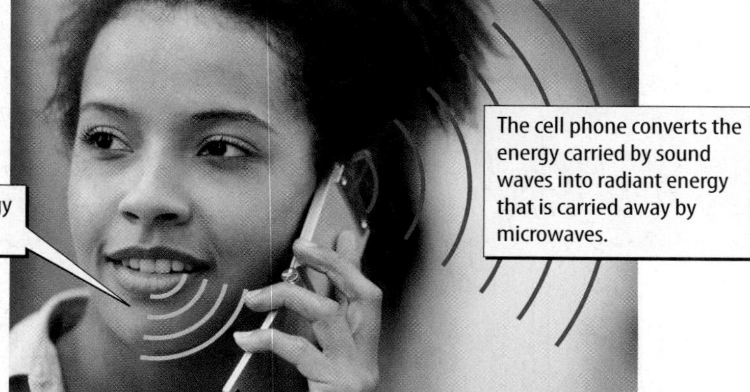

Using Electric Energy

Many devices such as handheld video games, MP3 players, and hair dryers use electric energy. Some devices, such as hair dryers, use electric energy from electric power plants. Other appliances, such as handheld video games, transform the chemical energy stored in batteries into electric energy.

Active Reading **9. Describe** How are different types of energy used?

Waste Energy

When energy changes form, some thermal energy is always released. For example, a lightbulb converts some electric energy into radiant energy. However, the lightbulb also transforms some electric energy into thermal energy. This is what makes the lightbulb hot. Some of this thermal energy moves into the air and cannot be used.

Scientists often refer to thermal energy that cannot be used as waste energy. Whenever energy is used, some energy is transformed into useful energy and some is transformed into waste energy. For example, we use the chemical energy in gasoline to make cars, such as those in **Figure 9,** move. However, most of that chemical energy ends up as waste energy—thermal energy that moves into the air.

Active Reading **10. Define** What is waste energy?

Apply It!

After you complete the lab, answer the questions below.

1. How could the lighter marble create the same depression as the one made by the heavier marble?

2. When meteors strike Earth, craters are formed. Why can two meteors create craters of different depth?

Figure 9 Cars transform most of the chemical energy in gasoline into waste energy.

Lesson Review 2

Visual Summary

Energy can change form, but according to the law of conservation of energy, energy can never be created or destroyed.

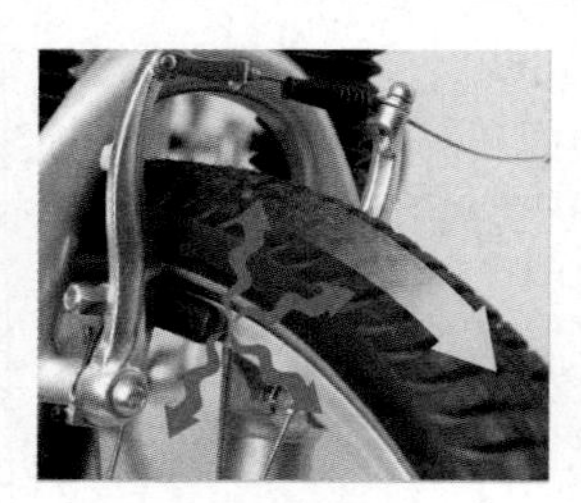

Friction transforms mechanical energy into thermal energy.

Different forms of energy, such as sound and radiant energy, are used when someone talks on a cell phone.

Use Vocabulary

1. **Use the term** *friction* in a complete sentence.

Understand Key Concepts

2. **Explain** the law of conservation of energy in your own words. SC.6.P.11.1

3. **Describe** the energy transformations that occur when a piece of wood burns.

4. **Identify** the energy transformation that takes place when you apply the brakes on a bicycle.

5. Which energy transformation occurs in a toaster?
   - (A) chemical to electric
   - (B) kinetic to chemical
   - (C) electric to thermal
   - (D) thermal to potential

Interpret Graphics

6. **Organize Information** To show how kinetic and potential energy change when a ball is thrown straight up and then falls down. SC.6.P.1.1

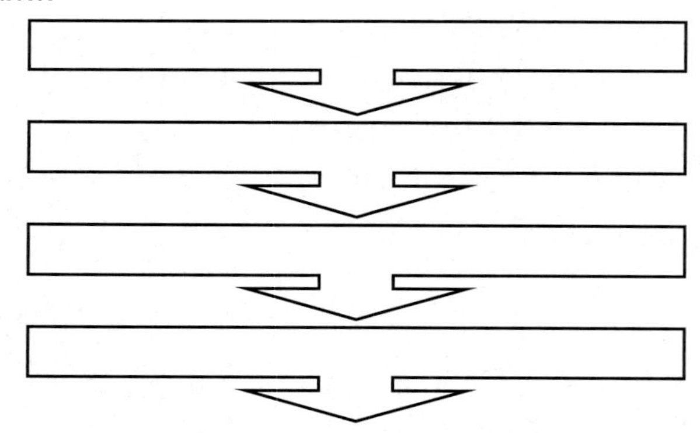

Critical Thinking

7. **Judge** An advertisement states that a machine with moving parts will continue moving forever without having to add any energy. Can this be correct? Explain. SC.6.N.2.1

Fossil Fuels and Rising CO_2

AMERICAN MUSEUM OF NATURAL HISTORY

Investigate the link between energy use and carbon dioxide in the atmosphere.

You use energy every day—when you ride in a car or on a bus, turn on a television or a radio, and even when you send an e-mail.

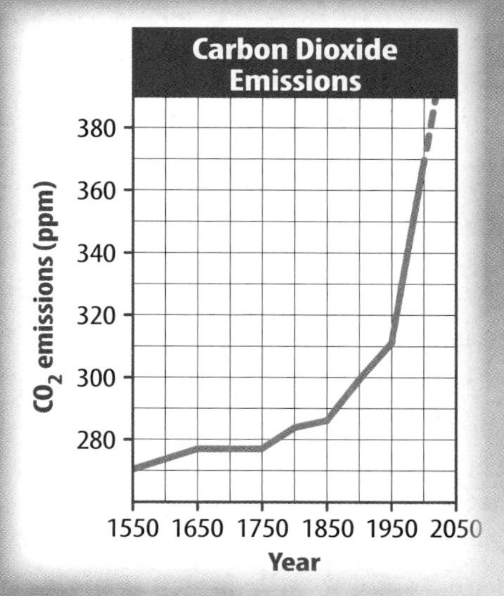

Much of the energy that produces electricity, heats and cools buildings, and powers engines comes from burning fossil fuels—coal, oil, and natural gas. When fossil fuels burn, the carbon in them combines with oxygen in the atmosphere and forms carbon dioxide gas (CO_2). Carbon dioxide is a greenhouse gas. Greenhouse gases absorb energy. This causes the atmosphere and Earth's surface to become warmer. Greenhouse gases make Earth warm enough to support life. Without greenhouse gases, Earth's surface would be frozen.

However, over the past 150 years the amount of CO_2 in the atmosphere has increased faster than at any time in the past 800,000 years. Most of this increase is the result of burning fossil fuels. More carbon dioxide in the atmosphere might cause average global temperatures to increase. As temperatures increase, weather patterns worldwide could change. More storms and heavier rainfall could occur in some areas, while other regions could become drier. Increased temperatures could also cause more of the polar ice sheets to melt and raise sea levels. Higher sea levels would cause more flooding in coastal areas.

Developing other energy sources such as geothermal, solar, nuclear, wind, and hydroelectric power would reduce the use of fossil fuels and slow the increase in atmospheric CO_2.

It's Your Turn

MAKE A LIST How can CO_2 emissions be reduced? Work with a partner. List five ways people in your home, school, or community could reduce their energy consumption. Combine your list with your classmates' lists to make a master list. LA.6.2.2.3

GREEN SCIENCE

300 Years OF CARBON DIOXIDE

1712

A new invention, the steam engine, is powered by burning coal that heats water to produce steam.

Early 1800s

Coal-fired steam engines, able to pull heavy trains and power steamboats, transform transportation.

1882

Companies make and sell electricity from coal for everyday use. Electricity is used to power the first lightbulbs, which give off 20 times the light of a candle.

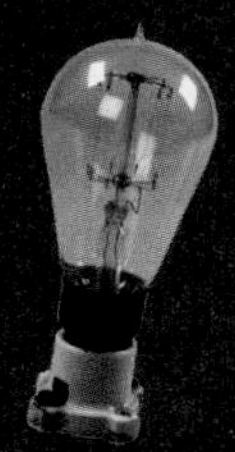

1908

The first mass-produced automobiles are made available. By 1915, Ford is selling 500,000 cars a year. Oil becomes the fuel of choice for car engines.

Late 1900s

Electrical appliances transform the way we live, work, and communicate. Most electricity is generated by coal-burning power plants.

2007

There are more than 800 million cars and light trucks on the world's roads.

Lesson 3

Thermal Energy on THE MOVE

ESSENTIAL QUESTIONS

 What is heat?

 How is thermal energy transferred?

Vocabulary

heat p. 265
conduction p. 267
thermal conductor p. 267
thermal insulator p. 267
convection p. 268
radiation p. 269

Florida NGSSS

LA.6.2.2.3 The student will organize information to show understanding (e.g., representing main ideas within text through charting, mapping, paraphrasing, summarizing, or comparing/contrasting);

MA.6.A.3.6 Construct and analyze tables, graphs, and equations to describe linear functions and other simple relations using both common language and algebraic notation.

MA.6.S.6.2 Select and analyze the measures of central tendency or variability to represent, describe, analyze, and/or summarize a data set for the purposes of answering questions appropriately.

SC.6.N.1.1 Define a problem from the sixth grade curriculum, use appropriate reference materials to support scientific understanding, plan and carry out scientific investigation of various types, such as systematic observations or experiments, identify variables, collect and organize data, interpret data in charts, tables, and graphics, analyze information, make predictions, and defend conclusions.

SC.6.N.1.5 Recognize that science involves creativity, not just in designing experiments, but also in creating explanations that fit evidence.

SC.6.N.2.1 Distinguish science from other activities involving thought.

Launch Lab

SC.6.N.1.1

20 minutes

How does thermal energy move?

This activity investigates one way thermal energy moves.

Procedure

1. Read and complete a lab safety form.
2. Insert a **stoppered test tube** filled with **rubbing alcohol** through holes in each **foam cup** as shown in the photo. Note that the holes are at different heights. A portion of the test tube must extend into each cup.
3. Add **ice water** to the cup that holds the test tube at the lower height. The ice water should just cover the test tube. Press a **lid** firmly onto each cup.
4. Place a **thermometer** through the lid of the cup that does not contain ice water. Read the thermometer at 1-min intervals for 15 min.

Data and Observations

Think About This

1. Describe what you think caused the temperature inside the cup to change.

2. Key Concept Diagram how thermal energy moved between the two cups.

Thermal Energy?

1. This meerkat is enjoying the transfer of thermal energy from a heat lamp. Warm air does not blow on the meerkat, nor is the meerkat in contact with the heat lamp. How do you think thermal energy move from the heat lamp to the meerkat?

Active Reading 2. **Find** As you read, underline the main ideas under each heading. After you finish reading, review the main ideas that you have underlined.

Heat

Have you ever seen a glassblower, like the one in **Figure 10,** at work? A glowing, molten blob of glass at nearly 1,000°C is on the end of a hollow pipe. The glassblower inflates the glass blob by blowing air into the other end of the pipe. The glass is then worked into the desired shape. Gradually, the glass loses thermal energy to the cool, surrounding air. This cooling process involves the loss of **heat**—*thermal energy moving from a region of higher temperature to a region of lower temperature.* The movement of thermal energy occurs in several ways.

Active Reading 3. **Review** How is heat related to thermal energy?

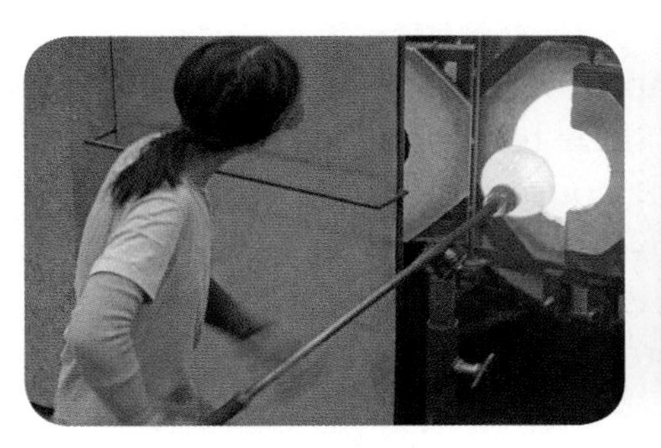

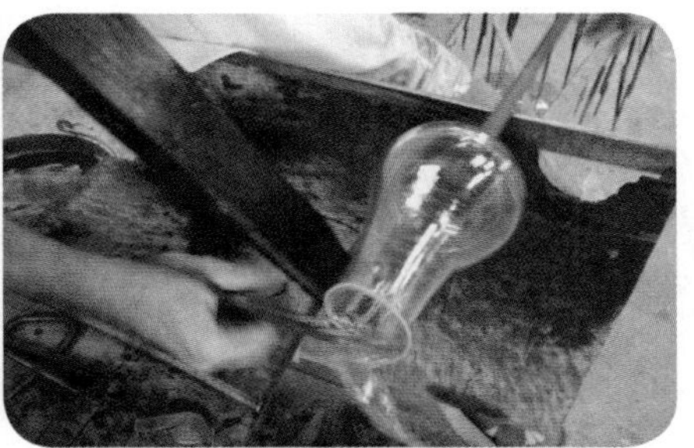

Figure 10 Thermal energy flows from the hot glass to the cooler, surrounding air.

4. **Visual Check**
Apply Why do you think heating an object increases its thermal energy?

Heat

Figure 11 Thermal energy flows from areas of higher temperature to areas of lower temperature.

5. Identify Circle the area of warmer temperature in the illustration.

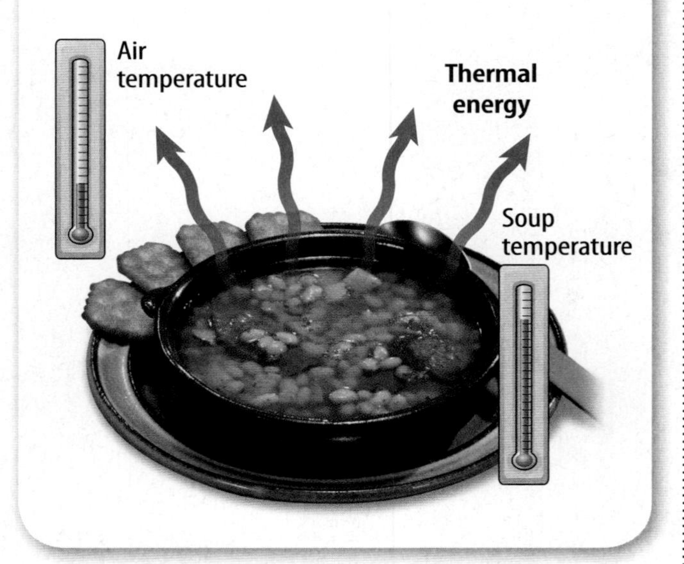

Math Skills MA.6.A.3.6

Solve an Equation

A *rate* tells how quickly something changes. The rate of temperature change is calculated using this equation:

$$\textbf{rate} = \frac{T_{final} - T_{initial}}{t}$$

where,

T_{final} = final temperature (in °C)

$T_{initial}$ = initial temperature (in °C)

t = time (in s, min, or hr)

The rate will have units of °C/s, °C/min, or °C/hr.

Practice

7. Water at 22°C is heated to 76°C in 10 min. What is the rate of temperature change?

Heat and Thermal Energy

The everyday meaning of heat often differs from its scientific meaning. It is important that you do not confuse heat and thermal energy. Objects contain thermal energy, not heat. Heat is thermal energy moving from a warmer object to a cooler object. Once the thermal energy moves, it is no longer considered heat.

Temperature Change

Thermal energy always moves from a higher temperature to a lower temperature. Thus, thermal energy flows from the hot soup shown in **Figure 11** to the cooler air surrounding it. The particles in the soup, on average, are moving faster than the particles in the air. Faster-moving particles on the surface of the soup collide with slower-moving particles in the air. The collisions transfer kinetic energy to particles in the air.

The movement of thermal energy that occurs in **Figure 11** causes changes in temperature. Particles on the surface of the soup collide with particles in the air. These collisions transfer kinetic energy from the soup to particles in the air. The result? The soup cools and the air warms.

6. Label Underline how the transfer of kinetic energy cools the soup.

What happens to the temperature and the thermal energy of the soup if half of it is poured into another bowl? The temperature does not change because the average kinetic energy of the remaining soup is unchanged. The soup's thermal energy, however, decreases by half. Why? Because the amount of the soup decreased by half.

Thermal Equilibrium

As the soup cools and the air warms, the difference in their temperatures decreases. Eventually the soup and the air around it reach the same temperature. Two objects in contact with each other at the same temperature are said to be in thermal equilibrium.

Conduction

Have you ever burned your fingers on a hot pan? This painful experience was caused by **conduction** (kuhn DUK shun)—*the transfer of thermal energy due to collisions between particles in matter.* Your fingers burned as thermal energy moved from the hot pan to your cooler skin. Conduction occurs in solids, liquids, and gases.

Active Reading **8. Infer** Why does conduction occur more slowly in gases than in solids?

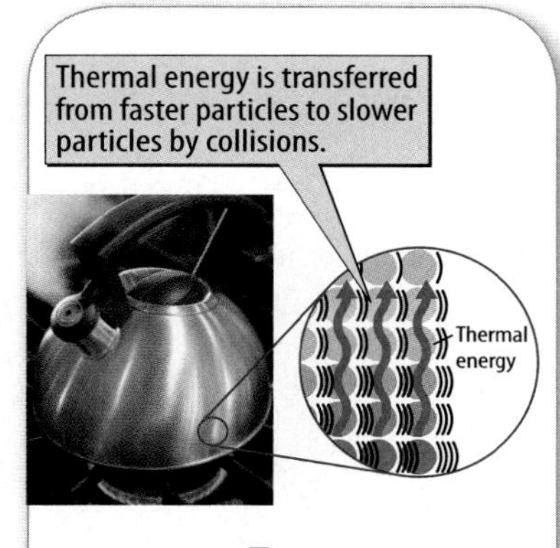

Figure 12 Collisions between particles transfer thermal energy from the warmer areas to cooler areas within the pot.

The Process of Conduction

Conduction through a metal pot is illustrated in **Figure 12.** The flame heats the bottom surface of the pot. Faster-moving particles in this warmer area of the pot collide with nearby particles with slower speeds. Thermal energy is transferred during the collisions. The slower-moving particles gain energy and speed up. As the collisions continue, thermal energy is conducted through the pot.

Thermal Conductors and Insulators

Thermal energy moves at different rates through different materials. *A material in which thermal energy moves quickly is called a* **thermal conductor**. In general, solids are better thermal **conductors** than liquids and gases. Most metals are excellent thermal conductors. *A material in which thermal energy moves slowly is a* **thermal insulator**. In general, gases are better thermal insulators than liquids and solids.

SCIENCE USE V. COMMON USE

conductor

Science Use a substance that transmits heat, light, sound, or electric charge

Common Use a person who leads, guides, directs, or manages, such as an orchestra conductor or a train conductor

Using Conductors and Insulators

Thermal conductors and thermal insulators have many uses. Metals, for example, are excellent conductors and are often used in cookware. Thermal insulators are used in winter coats like the one in **Figure 13.** Air, which is a mixture of gases, is a good thermal insulator. The trapped air slows the movement of thermal energy away from your warm body to the colder outside air.

Inquiry **LAB STATION** SC.6.N.1.1 MA.6.S.6.2 **Try It!**

Skill Lab ***Does the temperature of a liquid affect how quickly it warms or cools?*** at connectED.mcgraw-hill.com

Figure 13 A winter coat filled with an air-trapping material is a good thermal insulator.

Convection

WORD ORIGIN
convection
from Latin *convehere,* means "to carry together"

Earth scientists know that air currents and water currents move matter across Earth's surface. These currents also transfer thermal energy. **Convection** *is the transfer of thermal energy by the movement of particles from one part of a material to another.* Convection occurs in liquids and gases, but not in solids.

The Process of Convection

When part of a liquid or gas becomes warmer than the rest of it, convection begins. Inside a teapot, for example, the water closest to the burner is heated by conduction. This heated water expands. Recall that the density of a substance is equal to its mass divided by its volume. Thus, the density of the heated water decreases as it expands. A current forms as the cooler denser water sinks, pushing the warmer, less-dense water upward. The current carries particles of matter and thermal energy through the material. Convection does not occur in solids because the particles in solids cannot flow.

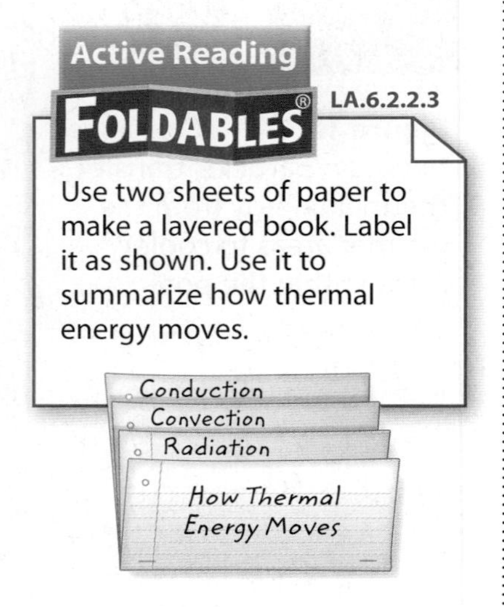

Convection Currents

Examine the heated beaker of water in **Figure 14.** Note how the areas of rising and sinking water form a loop. This loop is a convection current. The convection current carries thermal energy throughout the beaker, greatly increasing the rate at which the water is heated. Similarly, metal radiators in classrooms use convection currents in the air to warm the room.

Active Reading **9. Summarize** Describe a process that transfers thermal energy.

____

____

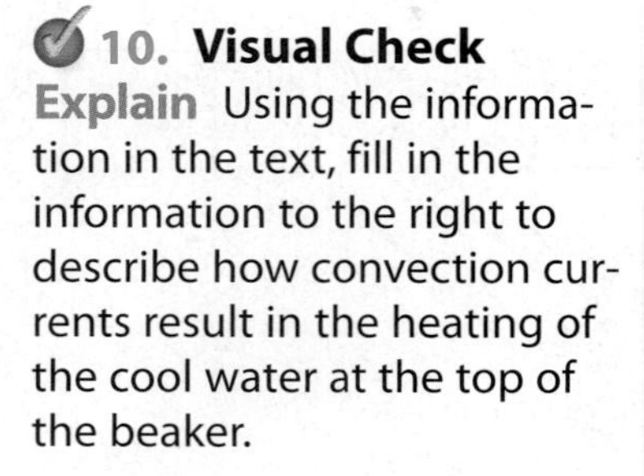

10. Visual Check
Explain Using the information in the text, fill in the information to the right to describe how convection currents result in the heating of the cool water at the top of the beaker.

Figure 14 Warm water rises from the bottom of the beaker and forms a convection current. The current moves thermal energy through the liquid.

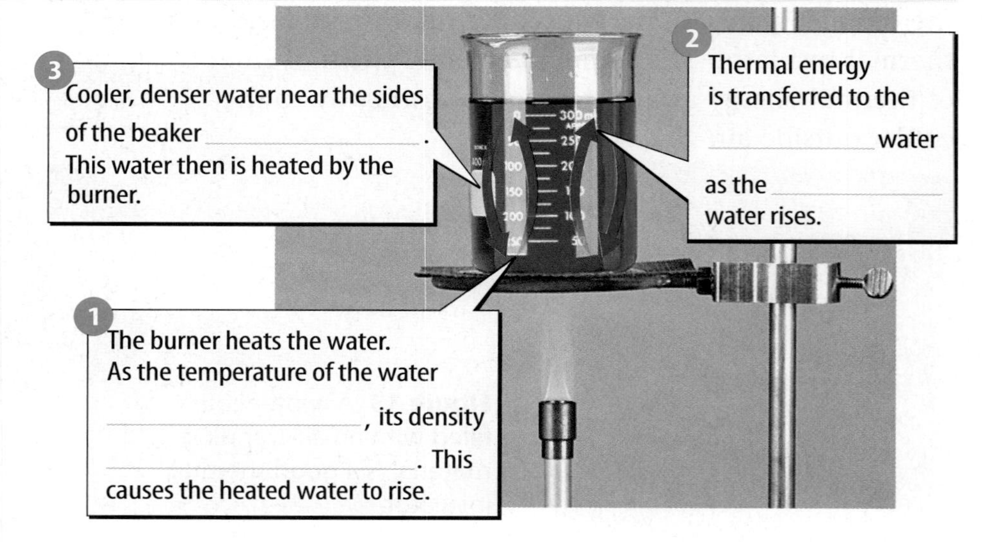

Radiation

Sunlight feels warm on the skin. The warmth is due to **radiation**, *the transfer of thermal energy from one object to another by electromagnetic waves.* Radiation is unique because it transfers thermal energy through matter or through space, where no matter exists. Thus, radiation can occur between objects that are not in contact. Electromagnetic waves carry thermal energy between objects. This is how thermal energy from the Sun reaches Earth and how the meerkat in the photo at the beginning of this lesson feels warmth.

Electromagnetic Waves

All objects give off electromagnetic waves, but most are not visible. Objects at extremely high temperatures, however, emit visible light, a type of electromagnetic wave. The Sun and glowing molten glass are examples. Except for light waves, all other types of electromagnetic waves are invisible.

The Process of Radiation

Electromagnetic waves carry energy. Radiation transfers this thermal energy from objects at higher temperatures to objects at lower temperatures. As shown in **Figure 15,** electromagnetic waves travel from a heat lamp to food. Particles in the food begin to move faster. As a result, the kinetic energy and the temperature of the particles in the food increase.

Active Reading **11. Differentiate** Contrast convection and radiation.

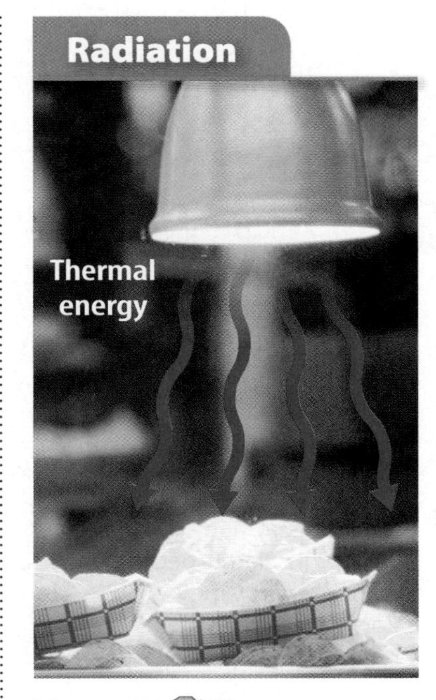

Figure 15 Electromagnetic waves carry thermal energy from the lamp to the food. The temperature of the food increases as it absorbs the waves.

Apply It! After you complete the lab, answer these questions.

1. Model the flow of heat from one material to another. Draw an example, use arrows to indicate the direction of heat flow, and label all the elements. Use red to indicate warmer objects and blue to indicate cooler objects.

2. What is the gain and loss of kinetic energy as shown in your drawing above?

Lesson Review 3

Visual Summary

This hat, coat, and scarf are thermal insulators because thermal energy moves slowly through them.

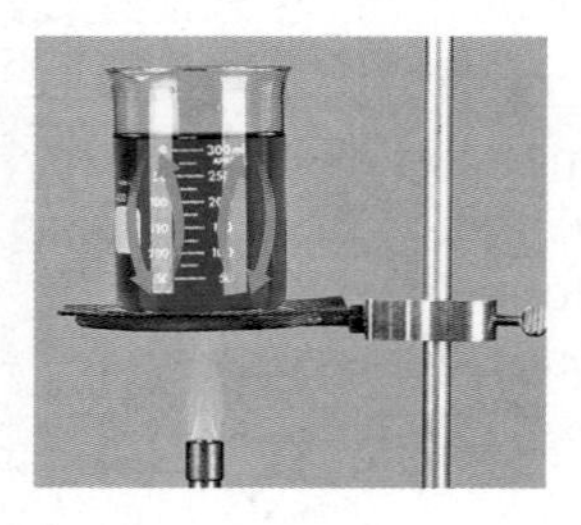

The transfer of thermal energy by the movement of particles from one part of a material to another is convection.

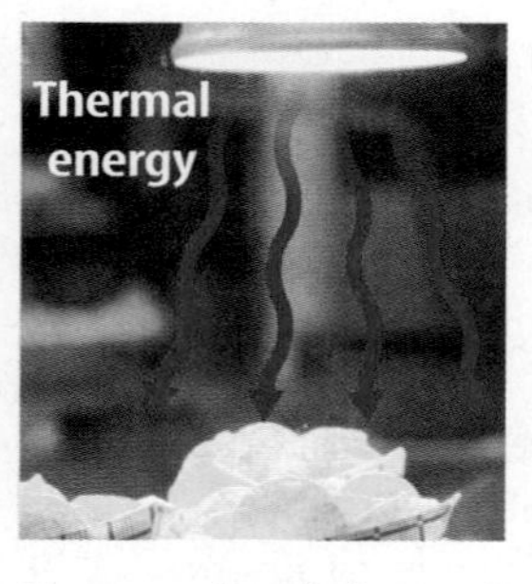

The transfer of thermal energy by electromagnetic waves is radiation.

Use Vocabulary

1 **Distinguish** between thermal conductor and thermal insulator.

Understand Key Concepts

2 **Explain** the relationship between heat and thermal energy.

3 Which is an example of conduction?

- (A) currents in a pot of heated water
- (B) warm air rising above hot pavement
- (C) warmth you feel standing near a fire
- (D) warmth you feel when touching a metal spoon in a hot cup of tea

4 **Describe** examples of conduction, convection, and radiation that you have experienced.

Interpret Graphics

5 **Sequence** the events that produces a convection current. LA.6.2.2.3

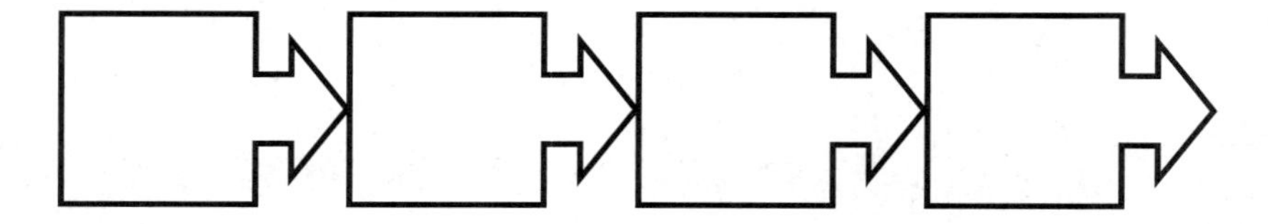

Critical Thinking

6 **Suggest** a way to keep a pizza warm for a picnic an hour later.

7 **Hypothesize** why windows have two panes of glass separated by a layer of air.

Math Skills MA.6.A.3.6

8 Water cools from 68°C to 23°C in 30 min. What is the rate of temperature change?

Solar Energy in Florida

FOCUS on FLORIDA

Using Sunshine to Power Our Lives

How many different ways have you used energy today? You might have ridden in a car or a bus on the way to school, used a hair dryer to get ready, or used a toaster to make breakfast. If you did, you used energy. Furnaces and stoves use thermal energy to heat buildings and to cook food. Cars and other vehicles use mechanical energy to carry people and materials from one part of the country to another.

Energy cannot be made; it must come from the natural world. The surface of Earth receives energy from two sources—the Sun and Earth's interior. Nearly all the energy you used today can be traced to the Sun.

Solar energy is energy from the Sun. Many devices use solar energy for power, including solar-powered calculators, home water-heating tanks, and solar-powered cookers for cooking outdoors. If you look around your city, you might see large, rectangular panels attached to the roofs of buildings or houses. These panels convert radiant energy from the Sun directly into electric energy. These panels are called solar panels.

Solar panels are built from several individual solar cells. The more solar cells in a panel, the more total electric output the panel can produce. Factors that can affect electric output are barriers to direct sunlight, such as buildings, and weather conditions, such as cloudy days.

Solar energy is the reason why the Florida Solar Energy Center (FSEC) exists. In 1975, the Florida legislature created the FSEC after the oil embargo of the 1970s. FSEC is located in the city of Cocoa, near the University of Central Florida. A few of their responsibilities are to conduct research, test and certify solar systems, and develop educational programs. FSEC has worked with the Department of Energy and Habitat for Humanity to construct energy-efficient homes.

This house, located in Tallahasee Florida, is completely solar powered.

It's Your Turn

RESEARCH AND REPORT Find out what solar cells are made of and how they work. Create a model of a solar cell to share with your classmates. **LA.6.4.2.2**

Chapter 7 Study Guide

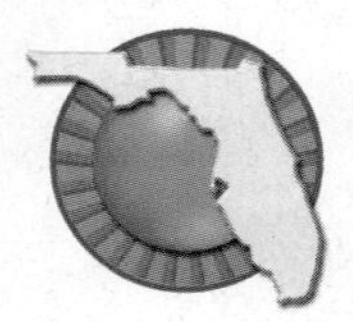

Think About It! Energy is the ability to cause change. Energy transformations occur when one form of energy changes into another form of energy. Energy is conserved during energy transformations.

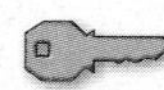

Key Concepts Summary

LESSON 1 Forms of Energy

- **Energy** is the ability to cause change.
- **Kinetic energy** is the energy an object has because of its motion. **Potential energy** is stored energy.
- **Work** is the transfer of energy that occurs when a force makes an object move in the direction of the force while the force is acting on the object.
- Different forms of energy include **thermal energy** and **radiant energy**.

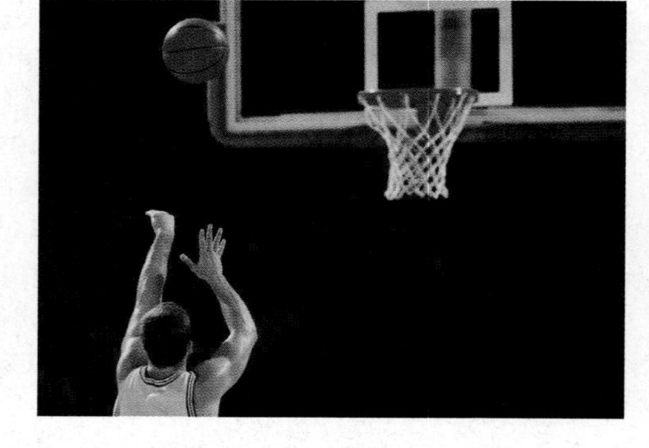

Vocabulary

energy p. 249
kinetic energy p. 250
potential energy p. 250
work p. 252
mechanical energy p. 253
sound energy p. 253
thermal energy p. 253
electric energy p. 253
radiant energy p. 253
nuclear energy p. 253

LESSON 2 Energy Transformations

- According to the **law of conservation of energy**, energy can be transformed from one form into another or transferred from one region to another, but energy cannot be created or destroyed.
- **Friction** transforms mechanical energy into thermal energy.
- Different types of energy are used in many ways including providing energy to move your body, to light a room, and to make and to receive cell phone calls.

law of conservation of energy p. 258
friction p. 259

LESSON 3 Thermal Energy on the Move

- **Heat** is thermal energy that moves from matter at a higher temperature to matter at a lower temperature.
- There are three ways in which thermal energy is transferred–**conduction**, **convection**, and **radiation**.
- A material through which thermal energy moves quickly is a **thermal conductor**.

heat p. 265
conduction p. 267
thermal conductor p. 267
thermal insulator p. 267
convection p. 268
radiation p. 269

FOLDABLES® Chapter Project

Assemble your Lesson Foldables as shown to make a Chapter Project. Use the project to review what you have learned in this chapter.

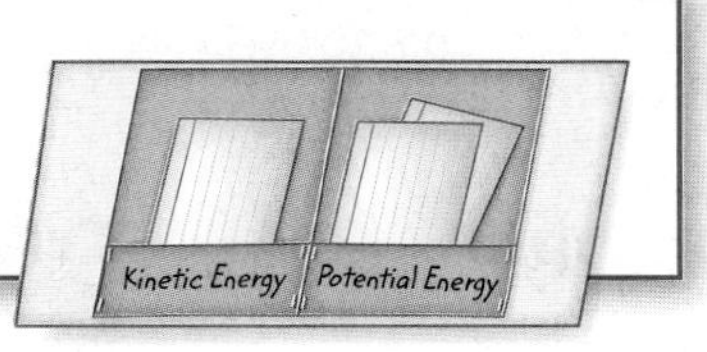

Use Vocabulary

Each of the following sentences is false. Make the sentence true by replacing the italicized word with a vocabulary term.

1. *Thermal energy* is the form of energy carried by an electric current.

2. The *chemical potential energy* of an object depends on its mass and its speed.

3. *Friction* is the transfer of energy that occurs when a force is applied over a distance.

4. *Radiation* is the transfer of thermal energy due to collisions between particles in matter.

5. *Convection* is the transfer of thermal energy from one object to another by electromagnetic waves.

Link Vocabulary and Key Concepts

Use vocabulary terms from the previous page to complete the concept map.

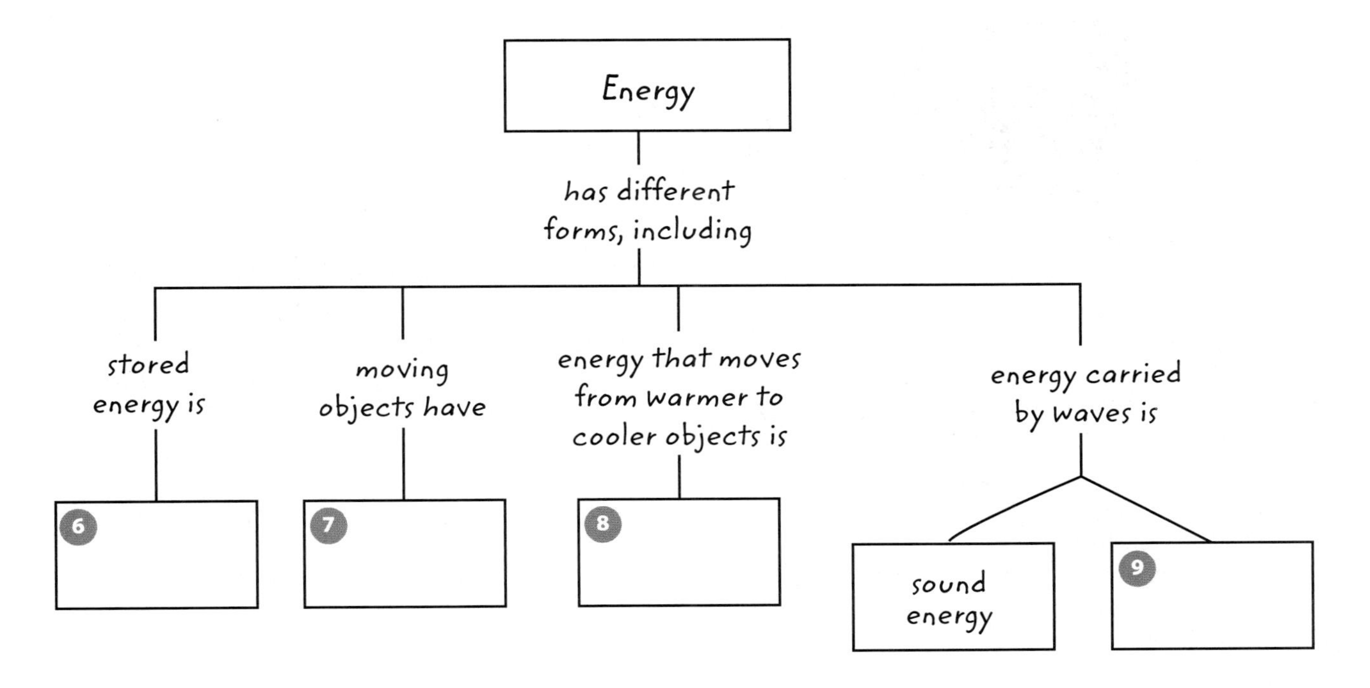

Chapter 7 Review

Fill in the correct answer choice.

Understand Key Concepts

1. What factors determine an object's kinetic energy? SC.6.P.11.1
   - Ⓐ its height and its mass
   - Ⓑ its mass and its speed
   - Ⓒ its size and its weight
   - Ⓓ its speed and its height

2. The gravitational potential energy stored between an object and Earth depends on SC.6.P.13.2
   - Ⓐ the object's height and weight.
   - Ⓑ the object's mass and speed.
   - Ⓒ the object's size and weight.
   - Ⓓ the object's speed and height.

3. When a ball is thrown upward, where does it have the least kinetic energy? SC.6.P.11.1
   - Ⓐ at its highest point
   - Ⓑ at its lowest point when it is moving downward
   - Ⓒ at its lowest point when it is moving upward
   - Ⓓ midway between its highest point and its lowest point

4. Which type of energy is released when the string in the photo below is plucked? SC.6.P.11.1

   - Ⓐ electric energy
   - Ⓑ nuclear energy
   - Ⓒ radiant energy
   - Ⓓ sound energy

5. According to the law of conservation of energy, which is always true? SC.6.P.11.1
   - Ⓐ Energy can never be created or destroyed.
   - Ⓑ Energy is always converted to friction in moving objects.
   - Ⓒ The universe is always gaining energy in many different forms.
   - Ⓓ Work is done when a force is exerted on an object.

Critical Thinking

6. **Determine** if work is done on the nail shown below if a person pulls the handle to the left and the handle moves. Explain. SC.6.P.11.1

7. **Contrast** the energy transformations that occur in a electrical toaster oven and in an electrical fan. SC.6.P.11.1

8. **Explain** why conduction occurs in all materials but convection occurs only in liquids and gases. SC.6.P.11.1

9. **Infer** Juanita moves a round box and a square box from a lower shelf to a higher shelf. The gravitational potential energy for the round box increases by 50 J. The gravitational potential energy for the square box increases by 100 J. On which box did Juanita do more work? Explain your reasoning. SC.6.P.11.1

10 **Explain** why a skateboard coasting on a flat surface slows down and comes to a stop. SC.6.P.11.1

11 **Describe** how energy is conserved when a basketball is thrown straight up into the air and falls back into your hands. SC.6.P.11.1

12 **Decide** Harold stretches a rubber band and lets it go. The rubber band flies across the room. Harold says this demonstrates the transformation of kinetic energy to elastic potential energy. Is Harold correct? Explain. SC.6.P.11.1

Writing in Science

13 **Write** a short essay on a separate sheet of paper explaining the energy transformations that occur in an incandescent lightbulb. LA.6.2.2.3

Big Idea Review

14 Write an explanation of energy and energy transformations for a fourth grader who has never heard of these terms. LA.6.2.2.3

15 Identify five energy transformations occurring in the photo below. LA.6.2.2.3

Math Skills MA.6.A.3.6

Solve One-Step Equations

16 An electrical water heater is rated at 5,500 W and operates for 106 h per month. How much electric energy in kWh does the water heater use each month? MA.6.A.3.6

17 A family uses 1,303 kWh of electric energy in a month. If the power company charges $0.08 cents per kilowatt hour, what is the total electric energy bill for the month? MA.6.A.3.6

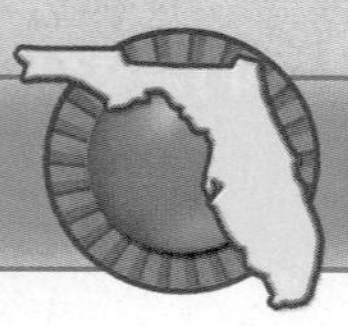

Florida NGSSS Benchmark Practice

Fill in the correct answer choice.

Multiple Choice

1 According to the law of conservation of energy, what happens to the total amount of energy in the universe? **SC.6.P.11.1**

Ⓐ It remains constant.

Ⓑ It changes constantly.

Ⓒ It increases.

Ⓓ It decreases.

Use the diagram below to answer questions 2 and 3.

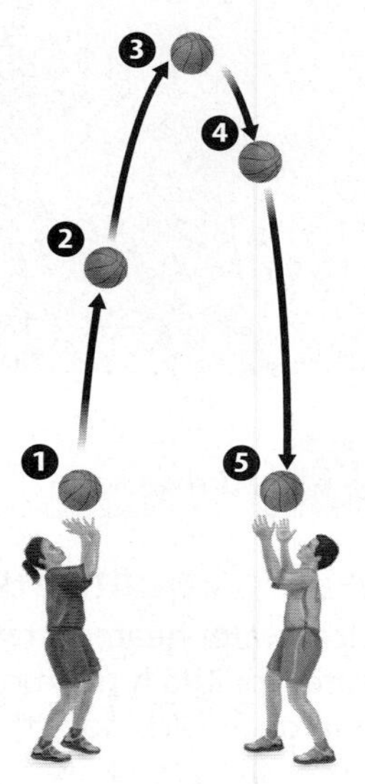

2 At which points is the kinetic energy of the basketball greatest? **SC.6.P.11.1**

Ⓕ 1 and 5

Ⓖ 2 and 3

Ⓗ 2 and 4

Ⓘ 3 and 4

3 At which point is the gravitational potential energy at its maximum? **SC.6.P.11.1**

Ⓐ 1

Ⓑ 2

Ⓒ 3

Ⓓ 4

Use the diagram below to answer question 4.

Vehicle	Mass	Speed
Car 1	1,200 kg	20 m/s
Car 2	1,500 kg	20 m/s
Truck 1	4,800 kg	20 m/s
Truck 2	6,000 kg	20 m/s

4 Which vehicle has the most kinetic energy? **SC.6.P.11.1**

Ⓕ car 1

Ⓖ car 2

Ⓗ truck 1

Ⓘ truck 2

5 A girl drops the ball she is holding. Which best explains the energy transformation taking place in the system? **SC.6.P.11.1**

Ⓐ The ball in her hands has potential energy. When it falls, the potential energy changes to kinetic energy.

Ⓑ The ball in her hands has kinetic energy. When it falls, the kinetic energy changes to potential energy.

Ⓒ The ball in her hands has potential energy. When it falls, the potential energy remains constant.

Ⓓ The ball in her hands has kinetic energy. When it falls, the kinetic energy remains constant.

6 A rock falls from a ledge to the ground. What energy conversion occurs during this process? **SC.6.P.11.1**

Ⓕ Potential energy is converted into kinetic energy.

Ⓖ Kinetic energy is converted into potential energy.

Ⓗ Thermal energy is converted into mechanical energy.

Ⓘ Chemical energy is converted into mechanical energy.

7 A person jumps from an airplane with a parachute. Which form of energy is decreasing as the person falls to the ground? SC.6.P.11.1

Ⓐ chemical

Ⓑ kinetic

Ⓒ mechanical

Ⓓ potential

Use the graph below to answer question 8.

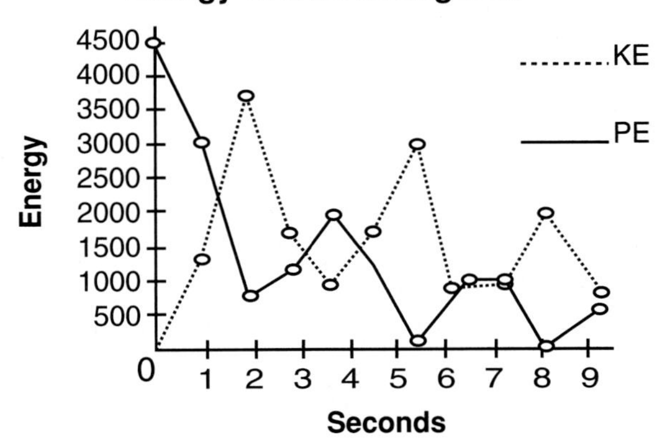

8 The graph shows the kinetic energy (KE) and potential energy (PE) of a bouncing ball over a period of 9 seconds. What is reasonable hypothesis based on these data? SC.6.P.11.1

Ⓕ As the kinetic energy decreases, the ball will stop bouncing.

Ⓖ As the kinetic energy decreases, the potential energy will remain unchanged.

Ⓗ As the kinetic energy decreases, the potential energy will also decrease.

Ⓘ As the kinetic energy decreases, the potential energy will increase.

9 In the diagram, a roller-coaster car starts from rest at point A and moves along the track. At which point does the roller coaster have the greatest kinetic energy? SC.6.P.11.1

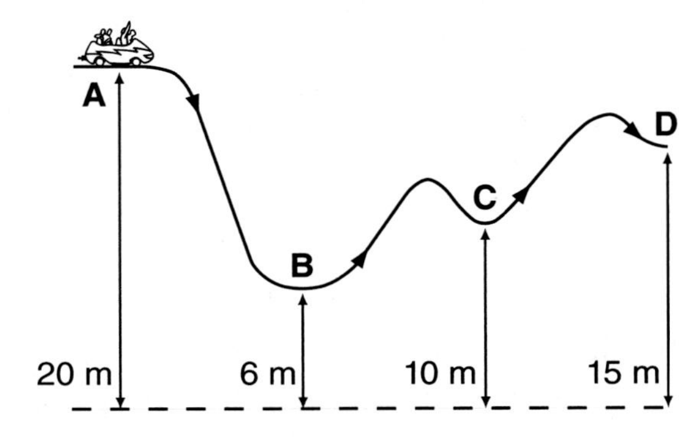

Ⓐ A

Ⓑ B

Ⓒ C

Ⓓ D

10 A pendulum bob swings back and forth in an arc. Which describes the energy conversion process as the pendulum moves from left to right? SC.6.P.11.1

Ⓕ kinetic energy to chemical energy to potential energy

Ⓖ kinetic energy to potential energy back to kinetic energy

Ⓗ potential energy to kinetic energy back to potential energy

Ⓘ potential energy to kinetic energy to electromagnetic energy

NEED EXTRA HELP?

If You Missed Question...	1	2	3	4	5	6	7	8	9	10
Go to Lesson...	3	2	2	1	1	3	2	3	3	3

Name ______________________ Date ______________

Multiple Choice *Bubble the correct answer.*

	Mass	Diameter	Height of Drop
Ball A	180 g	9 cm	3 m
Ball B	180 g	8 cm	4 m

1. The table above gives information about two balls dropped from different windows. Which statement correctly compares the balls when they reach the ground? **SC.6.P.11.1**

Ⓐ Ball A has greater potential energy.
Ⓑ Ball B has greater kinetic energy.
Ⓒ Both balls have the same amount of potential energy.
Ⓓ Both balls have the same amount of kinetic energy.

2. Radiant energy is also known as **SC.6.P.11.1**

Ⓕ electric energy.
Ⓖ kinetic energy.
Ⓗ light energy.
Ⓘ thermal energy.

3. When would you have the most potential energy? **SC.6.P.11.1**

Ⓐ walking up the hill
Ⓑ sitting at the top of the hill
Ⓒ running up the hill
Ⓓ sitting at the bottom of the hill

4. An oven changes electric energy into which other kind of energy in order to cook food? **SC.6.P.11.1**

Ⓕ chemical
Ⓖ mechanical
Ⓗ nuclear
Ⓘ thermal

Name ______________________________ Date ______________

Benchmark Mini-Assessment Chapter 7 • Lesson 2

Multiple Choice *Bubble the correct answer.*

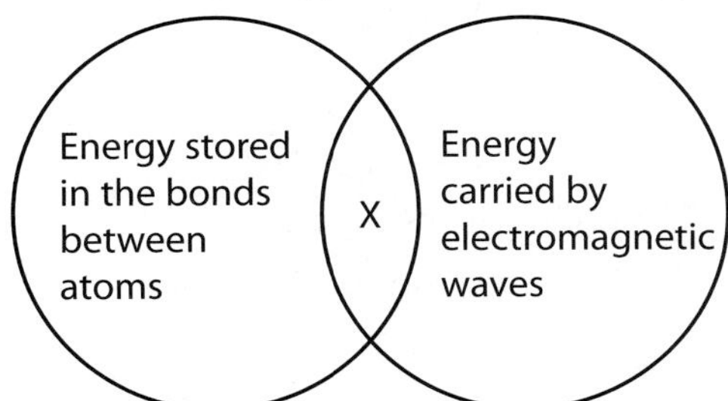

1. Which is X in the Venn diagram above? **SC.6.P.11.1**

(A) can be a form of potential energy

(B) can be a form of waste energy

(C) can be transformed into thermal energy

(D) can be used to create energy

2. Which energy transformation occurs during photosynthesis? **SC.6.P.11.1**

(F) chemical to thermal

(G) radiant to chemical

(H) radiant to kinetic

(I) thermal to chemical

3. Which energy transformation allows food to be cooked in the device shown above? **SC.6.P.11.1**

(A) electric to radiant

(B) potential to radiant

(C) radiant to nuclear

(D) radiant to thermal

4. Which statement about energy transformations is NOT true? **SC.6.P.11.1**

(F) All forms of energy can be transformed into thermal energy.

(G) Energy is never destroyed or created during transformations.

(H) Potential chemical energy always transforms into kinetic energy.

(I) Some thermal energy is always released during a transformation.

Name _______________ Date _______________

Multiple Choice *Bubble the correct answer.*

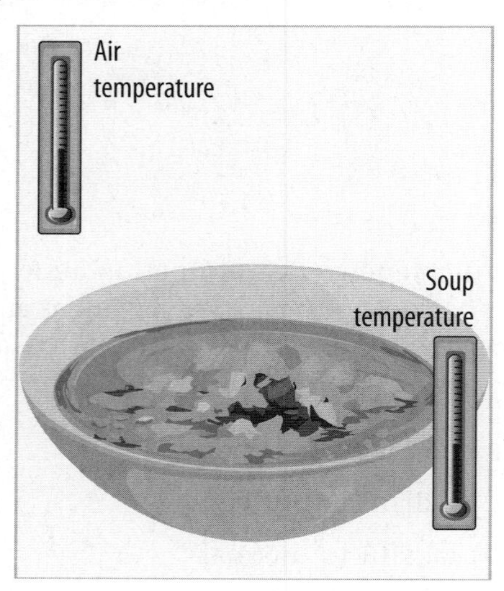

1. What is the description of heat flow in the situation above? **SC.6.P.11.1**

Ⓐ Heat will flow from the air to the soup.

Ⓑ Heat will flow from the soup to the air.

Ⓒ Heat will not flow between the air and the soup.

Ⓓ Heat will flow back and forth, between the air and the soup.

2. Which of these substances is the BEST thermal conductor? **SC.6.E.7.1**

Ⓕ air

Ⓖ fiberglass

Ⓗ steel

Ⓘ wood

3. Which type of heat transfer causes your face to feel warm when you sit in the sun? **SC.6.E.7.1**

Ⓐ conduction

Ⓑ radiation

Ⓒ kinetic energy

Ⓓ potential energy

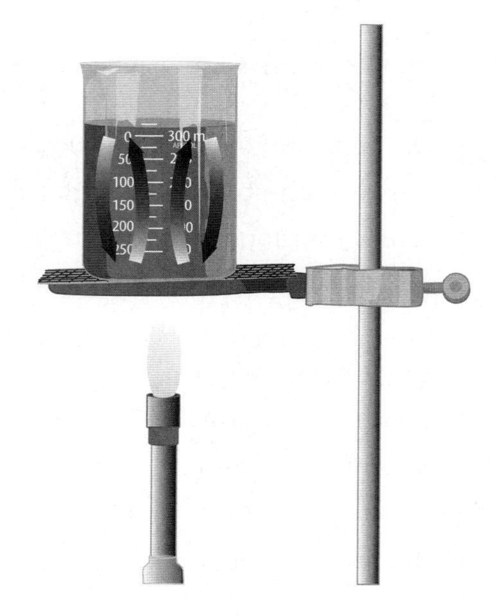

4. The image above shows **SC.6.E.7.1**

Ⓕ cold water rising through conduction.

Ⓖ warm water rising through conduction.

Ⓗ cold water rising through convection currents.

Ⓘ warm water rising through convection currents.

Name ______________________ Date ______________________

Notes

Name ______________________ Date ______________

Notes

Beach Ball

When does gravitational force act on a beach ball? Check off all the descriptions that are examples of gravity acting on a beach ball.

_____ A. Beach ball tossed up into the air, moving upward

_____ B. Beach ball falling downward after it is tossed into the air

_____ C. Beach ball floating in a swimming pool

_____ D. Person holding a beach ball

_____ E. Beach ball resting on the ground, not moving

Explain your thinking. What rule or reasoning did you use to decide when gravity acts on a beach ball?

Motion and FORCES

FLORIDA BIG IDEAS

1 **The Practice of Science**

12 **Motion of Objects**

13 **Forces and Changes in Motion**

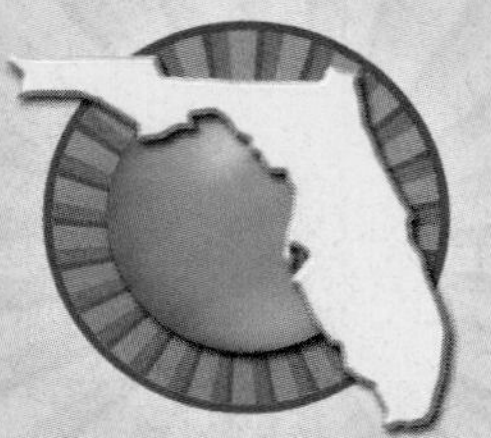

Think About It!

What is the relationship between motion and forces?

At first, you might think this image shows several riders, but it is just one. Several photos of the rider at different points during his jump help you see the ways in which his motion changes.

1. What do you think might happen if the rider isn't going fast enough when he takes off?

2. What are some ways you could describe the motion of the rider?

3. How do you think forces are related to the motion of the rider?

Get Ready to Read

What do you think about motion and forces?

Before you read, decide if you agree or disagree with each of these statements. As you read this chapter, see if you change your mind about any of the statements.

	AGREE	DISAGREE
1 If an object's distance from a starting point changes, the object is in motion.	☐	☐
2 Speed describes how fast something is going and the direction in which it is moving.	☐	☐
3 You can show the path an object takes using a graph of distance and time.	☐	☐
4 You can tell how fast objects are moving if you look at a graph of speed and time.	☐	☐
5 To apply a force, one object must be touching another object.	☐	☐
6 If an object is at rest, there are no forces acting on it.	☐	☐

ConnectED **There's More Online!**
Video • Audio • Review • ⓘLab Station • WebQuest • Assessment • Concepts in Motion • Multilingual eGlossary

Lesson 1

Describing MOTION

ESSENTIAL QUESTIONS

- How do you describe an object's position?
- How do you describe an object's motion?
- How do speed and velocity differ?
- What is acceleration?

Vocabulary

reference point p. 287
position p. 287
displacement p. 288
motion p. 288
speed p. 289
velocity p. 290
acceleration p. 292

Florida NGSSS

LA.6.2.2.3 The student will organize information to show understanding (e.g., representing main ideas within text through charting, mapping, paraphrasing, summarizing, or comparing/contrasting);

MA.6.A.3.6 Construct and analyze tables, graphs, and equations to describe linear functions and other simple relations using both common language and algebraic notation.

SC.6.P.12.1 Measure and graph distance versus time for an object moving at a constant speed. Interpret this relationship.

SC.6.N.1.1 Define a problem from the sixth grade curriculum, use appropriate reference materials to support scientific understanding, plan and carry out scientific investigation of various types, such as systematic observations or experiments, identify variables, collect and organize data, interpret data in charts, tables, and graphics, analyze information, make predictions, and defend conclusions.

SC.6.N.1.4 Discuss, compare, and negotiate methods used, results obtained, and explanations among groups of students conducting the same investigation.

SC.6.N.1.4

Launch Lab

10 minutes

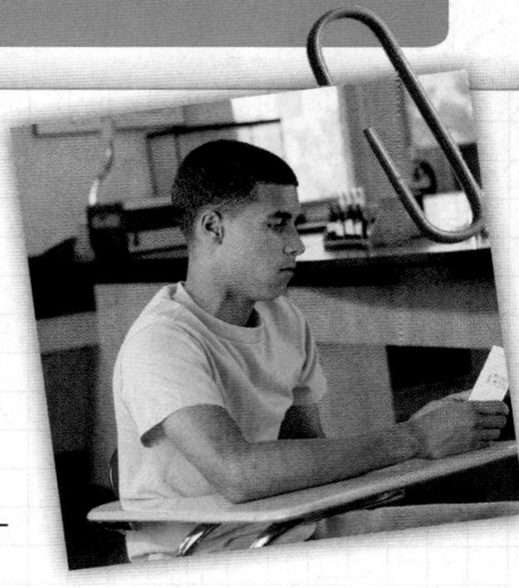

Where are you right now?

Suppose you place a pencil next to your textbook. How could you describe where the pencil is? You could say it is next to your textbook. A clearer description might be 5 cm to the right of your textbook. How can you write a clear description of where an object is located?

Procedure

1. Choose an **object** in your classroom, such as a chair. Work with others in your group to describe the location of the object.
2. Discuss the description with your group. Does it clearly describe the object's location? If not, work together to change the description.
3. Get a **card** from your teacher that names a place, such as the school or the city. Write a description of where you are within the place written on your card. Could someone find you from your description? Share and discuss descriptions with others in your group.

Think About This

1. Was there only one possible description in each case, or more than one? Why?

2. Key Concept What three things do you think make up a good description of where an object or a person is located?

Inquiry How are they moving?

1. As they move about the mall, the motion of each person changes. They speed up, slow down, and change direction. How would you describe the position of a person at any moment in time? What words could you use to describe the person's motion?

Describing Position

Think about calling a friend on his or her cell phone. One of the first questions you might ask is, "Where are you?" Your friend might answer, "I'm at the mall" or "I'm two blocks north of school" or "I'm 3 m away from you. Look down the hall."

What do all these answers have in common? Each one has a reference point that helps describe your friend's position. A **reference point** is *the starting point you choose to describe the location, or position, of an object.* Your friend describes his or her location by starting with a reference point, such as the mall, the school, or even you. Then your friend compares his or her position to that reference point.

Your friend also gives you other information to describe his or her location. Your friend might tell you a distance, such as 2 blocks or 3 m. Your friend might also mention a direction, such as north or down the hall. Your friend just described his or her position. **Position** *describes an object's distance and direction from a reference point.* Position always includes a distance, a direction, and a reference point.

Active Reading **3. Analyze** the position in the sentence below. Label the distance, direction, and reference point.

The park is 3 km west of the school.

WORD ORIGIN

reference
from Latin *referre,* means "to carry or direct back"

2. NGSSS Check
Describe How do you describe an object's position? SC.6.P.12.1

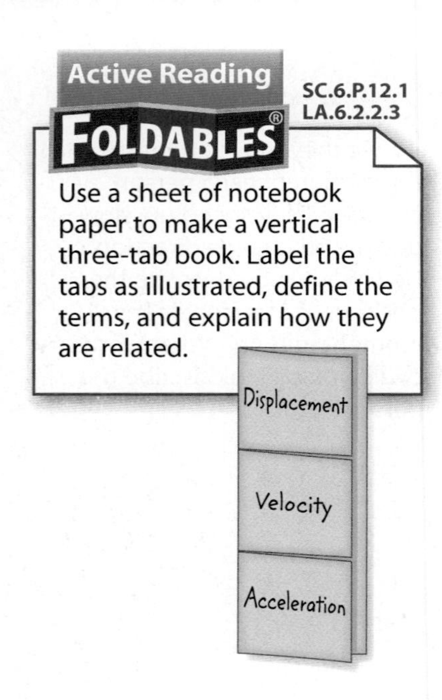

SC.6.P.12.1
LA.6.2.2.3

Use a sheet of notebook paper to make a vertical three-tab book. Label the tabs as illustrated, define the terms, and explain how they are related.

Using Reference Points

Suppose you are watching a soccer game. The position of a player depends on a reference point, as shown in **Figure 1.** If the reference point is the goal, or point A, the player's position is 10 m in front of the goal. If the reference point is center field, point B, the position of the player is 40 m toward the goal. Notice that the actual location of the player has not changed. Only the description of the position changed because the reference point changed.

Active Reading **4. Explain** Why does the description of an object's motion depend on a reference point?

Distance and Displacement

During one play in the soccer game shown, the player runs 41.2 m from position D to position C. Then she runs 10 m to position B. Her path is shown by the blue dotted lines. The total distance the player travels is 41.2 m + 10 m = 51.2 m.

The solid blue arrow in **Figure 1** shows the player's displacement. **Displacement** *is the difference between the initial, or starting, position and the final position.* The player starts at point D and finishes at point B. Her displacement is 40 m in front of her initial position. An object's displacement and the distance it travels are not always equal.

Motion

With 5 s left on the clock, the soccer ball is 50 m from the goal. When the game ends, the ball is in the goal. What happened to the ball during the last 5 s of the game? The ball was in motion. **Motion** *is the process of changing position.*

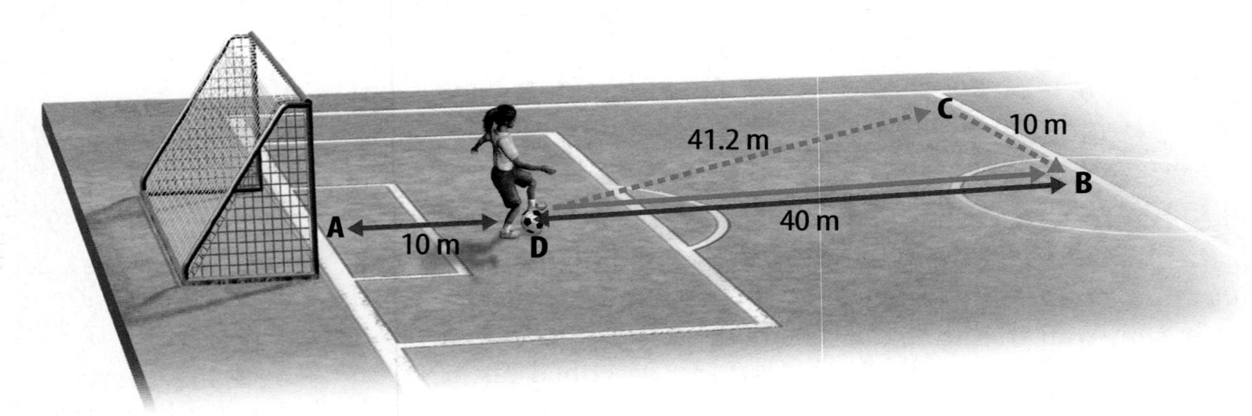

Figure 1 The position of the ball depends on the reference point. The distance traveled by the player and her displacement are not the same.

5. Visual Check Infer How would you describe the position of point B without using the words *center line*?

Describing Speed

Figure 2 The bus moves at a constant speed from position 1 to position 4. After that, the bus's speed increases.

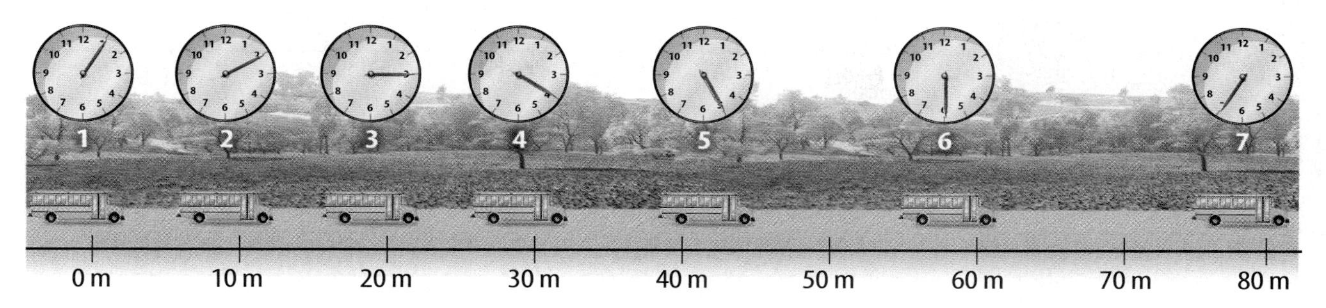

Speed

How fast does the bus that takes you to a soccer game move? On the highway, it might move fast. In traffic, it probably moves more slowly. **Speed** *is the distance an object moves in a unit of time.* Any unit of time, such as 1 s, 1 min, 1 h, or 1 y, may be used to calculate average speed. For example, you can say that a bus travels 15 km/h or 0.25 km/min.

Constant Speed

The bus in **Figure 2** moves from positions 1 to 2 to 3 to 4 at the same speed of 10 m/s. When an object moves the same distance over a given unit of time, it is said to have a constant speed. The bus has a constant speed from positions 1 to 4.

Changing Speed

How is the motion of the bus between positions 4 and 7 different from its earlier motion in **Figure 2?** The bus moves a greater distance each second. When the distance an object covers increases or decreases over a given unit of time, the object is said to be changing speed.

Average Speed

The speed of most moving objects is not constant, which is why the speedometer in a car is always slightly changing. Therefore, when you describe your speed over an entire trip to someone, you are describing average speed. Average speed is equal to the total distance traveled divided by the total time.

$$\text{average speed} = \frac{\text{total distance}}{\text{total time}}$$

The bus in **Figure 2** travels 80 m from second 1 to second 7 on the stopwatch. Therefore, the average speed of the bus is 80 m/6 s or 13.3 m/s.

Math Skills

MA.6.A.3.6

Use a Formula

The formula for determining average speed is

$$\text{average speed} = \frac{\text{distance}}{\text{time}}.$$

For example, a bus carrying students to a soccer game traveled **10 km** in **30 min.** What was the average speed of the bus in $\frac{\text{km}}{\text{h}}$?

1. Change minutes to hours.

   30 min = 0.5 h

2. Replace the terms in the formula with the given terms.

   $$\text{average speed} = \frac{10\ \text{km}}{0.5\text{h}}$$

3. Divide to get the answer.

   10 km/0.5 h = 20 km/h

Practice

6. On a hike, you travel 2,800 m in 2 h. What is your average speed?

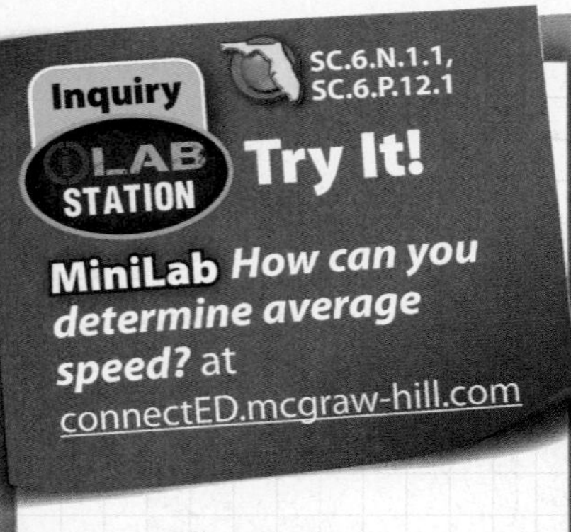

Apply It!

After you complete the lab, answer the questions below.

1. Did the ball speed up or slow down as it rolled along the track? Explain why the ball behaved as it did.

2. What are two variables in the MiniLab that could lower the accuracy of your calculations?

Science Use v. Common Use

speed

Science Use the distance an object covers in a given unit of time

Common Use the sense of going fast or to go fast

Active Reading 7. **Distinguish** Relate the concepts below to speed.

Term	What It Means
Speed	
Constant Speed	
Changing Speed	
Average Speed	

Velocity

If you talk to your friends about how fast the bus was traveling as it took you to the game, you probably describe your **speed**. What else do you need to know to understand the motion of the bus? Which way was it going? **Velocity** *is the speed and the direction of a moving object.* For example, the speed of an object might be 5 m/s, while its velocity is 5 m/s to the east. The words *speed* and *velocity* have different meanings, just as *distance* and *displacement* have different meanings.

You can use arrows to show the velocity of the object. The length of an arrow represents the speed of an object. The longer the arrow is, the faster the object is moving. The head of an arrow points in the direction the object moves. In **Figure 3,** the boy skates east at a greater speed than the girl skateboards west.

Figure 3 The lengths of the velocity arrows show speed. The arrows point in the direction of motion.

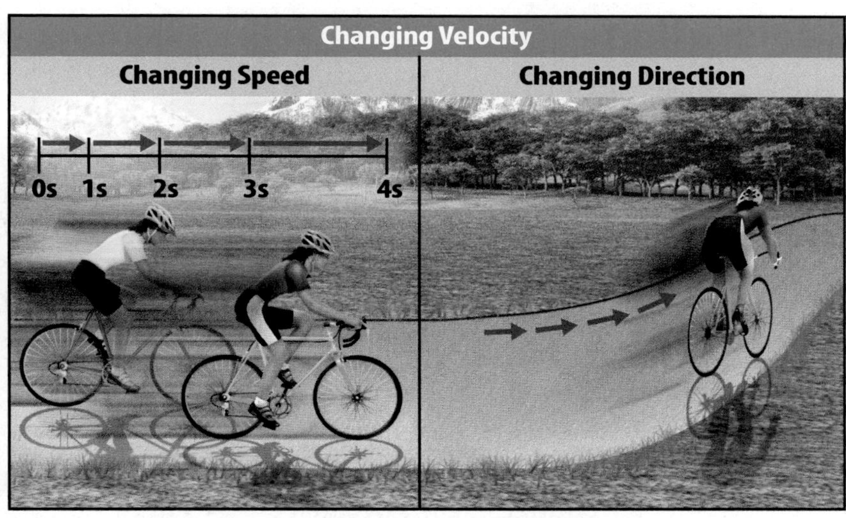

Figure 4 Velocity is constant if both speed and direction are constant.

Constant Velocity

Suppose a road biker rides along a flat straight road, as shown in the first panel of **Figure 4.** He or she moves with **constant** velocity. Constant velocity means that an object moves with constant speed and its direction does not change. If the biker's velocity is constant, then the arrows that represent the biker's velocity are all the same length and point in the same direction.

Changing Velocity

How is the motion of the bikers in the second and third panels of **Figure 4** different from the motion of the bikers in the first panel? In the second and third panels, the arrows that represent velocity are different from each other. In the second panel, the arrows have different lengths. In the third panel, the arrows point in different directions. Because the arrows are not identical, you know that the velocity is changing. Velocity changes when either an object's speed or direction of motion changes.

Change in Speed Imagine that the biker in the blue jersey in the second panel wants to pass the other biker. She goes faster. Each second, the rider moves a greater and greater distance. Her speed changes. Therefore, her velocity also changes.

Change in Direction The biker in the third panel of **Figure 4** rides along a turn in the road. Even though the biker's speed is constant, her direction changes, as shown by the different angles of the arrows. Because the rider's direction changes, her velocity changes.

8. Visual Check
Explain How do you know that the bikers in the first box are moving at a constant velocity?

Active Reading **9. Identify** What are two ways to change an object's velocity?

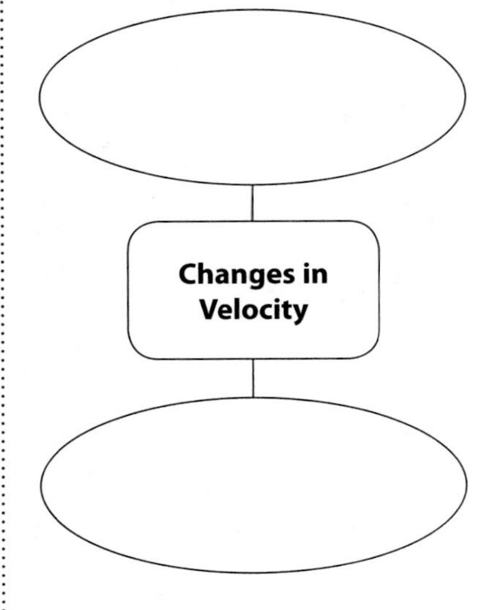

Acceleration

Imagine that you are on a roller coaster like the one in **Figure 5.** As your roller-coaster car goes down a hill, you move faster and faster. You feel as if you are being pushed back against the seat. Because the speed of the car increases, the velocity of the car increases. Next, your car climbs a hill. The car moves slower and slower as it climbs. Because the speed of the car decreases, the velocity of the car decreases. Suddenly the roller-coaster cars enter a curve. The velocity changes again because the direction of the car's motion changes. When the velocity of an object changes, it accelerates. **Acceleration** *is the measure of the change in velocity during a period of time.*

Speeding Up

As the roller-coaster car at the left of **Figure 5** travels downhill, it covers a greater distance each second. The velocity increases. The roller-coaster car's acceleration is in the same direction as its motion. This is called positive acceleration. The green arrow shows the acceleration.

Slowing Down

As the roller-coaster car climbs the next hill, it moves a shorter and shorter distance each second. The velocity decreases. This means that the roller-coaster car's acceleration is in the opposite direction of its motion. This is shown by the green arrow on the uphill part of **Figure 5.** The action of slowing down is called negative acceleration, also known as deceleration.

Changing Direction of Motion

As the roller-coaster car travels through the loop, the direction of its motion constantly changes. If the car is changing direction, then its velocity is changing and the car is accelerating. Because the car is accelerating, there must be unbalanced forces acting on it. The track applies an unbalanced force on the roller-coaster car by pushing it toward the center of the loop. This creates the roller-coaster car's circular motion. The green arrows show that the roller coaster cars accelerate toward the center of the loop.

Figure 5 A roller coaster can accelerate many times throughout the ride.

Active Reading **10. Label** Using arrows, indicate on **Figure 5** the three ways the roller-coaster can accelerate.

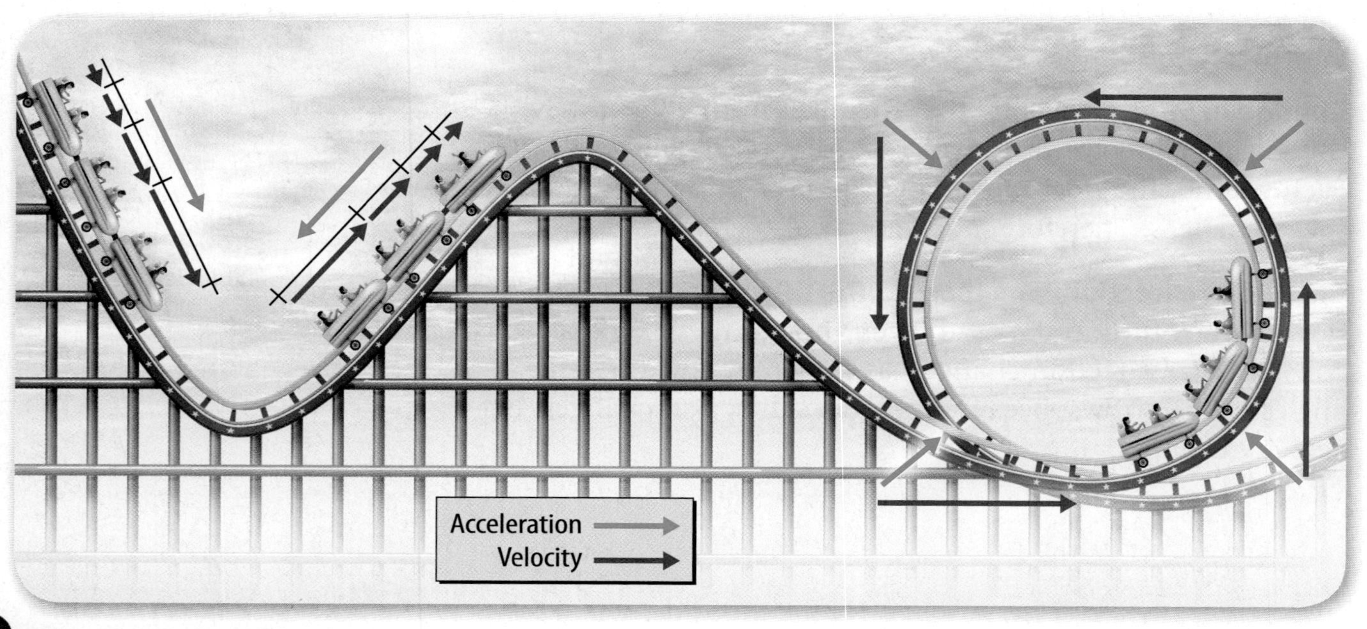

Lesson Review 1

Visual Summary

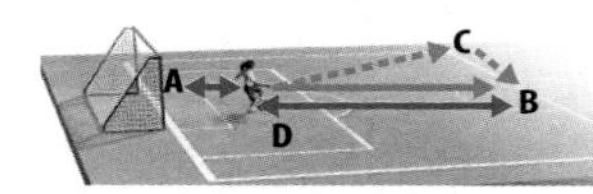

The position of an object is described by a reference point, a reference direction, and a distance. Motion is a change of position.

Speed is the distance traveled by an object during a unit of time. Velocity includes both speed and direction of motion.

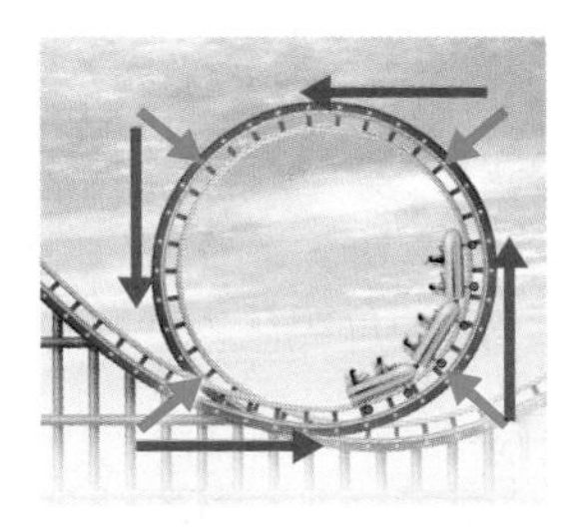

Acceleration is a change in velocity. Velocity changes when either the speed or the direction changes.

Use Vocabulary

1. A ______________ is the starting point from which you describe an object's position.

2. If a bird's velocity is 3 m/s south, its ______________ is 3 m/s.

Understand Key Concepts 

3. **Compare** distance and displacement.

4. **Apply** A jet airliner flew 4,100 km from New York City to San Francisco in 4.25 h. Its average speed was about
   (A) 965 km/h.
   (B) 1,000 km/h.
   (C) 965 km/h to the west.
   (D) 1,000 km/h to the west.

5. **Evaluate** A race car moves at a constant speed around an oval track. Is the car accelerating? Why or why not?

Interpret Graphics

6. **Interpret Data** The table below summarizes the motion of a train along a straight track. Circle the part of the trip where the train is moving with constant velocity? **LA.6.2.2.3**

Time	1 P.M.	2 P.M.	3 P.M.	4 P.M.
Distance traveled	50 km	40 km	40 km	35 km

Critical Thinking

7. **Imagine** Describe a situation in which the displacement of a roller skater is 0 km but the distance traveled is 2 km.

Math Skills MA.6.A.3.6

8. What is the average speed of a soccer ball that travels 34 m in 2 s?

9. How long would it take a bus traveling at 50 km/h to travel 125 km?

Describing Motion

Find three vocabulary words on the map. Connect the words and phrases with arrows to complete the definitions. Use a different color for each definition. *Displacement* has been started for you.

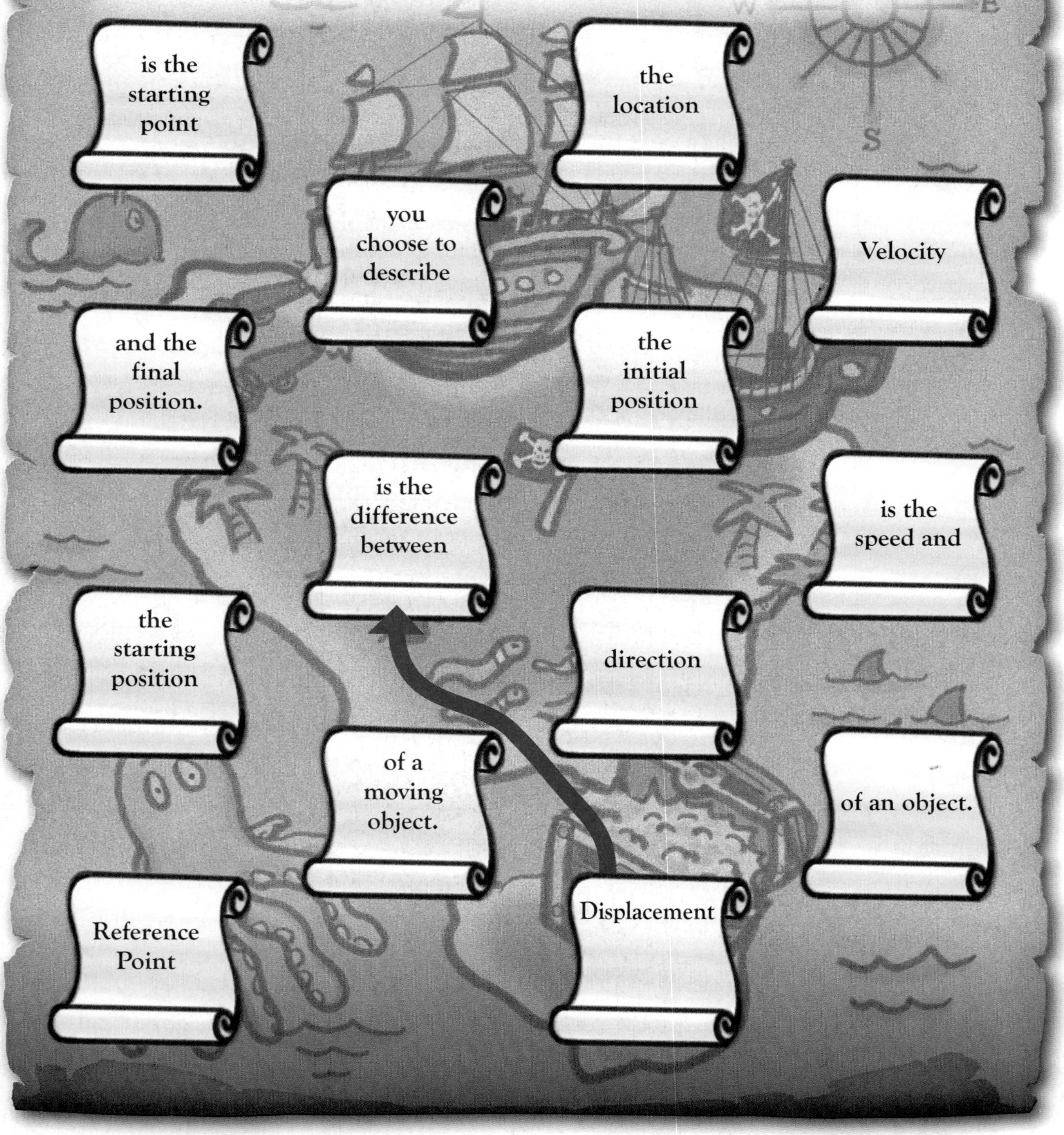

is the starting point

the location

you choose to describe

Velocity

and the final position.

the initial position

is the difference between

is the speed and

the starting position

direction

of a moving object.

of an object.

Reference Point

Displacement

Lesson 2

Graphing MOTION

ESSENTIAL QUESTIONS

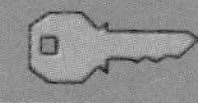

How can you graph an object's motion?

How can a graph help you understand an object's motion?

Vocabulary

distance-time graph p. 296

speed-time graph p. 299

Florida NGSSS

LA.6.2.2.3 The student will organize information to show understanding (e.g., representing main ideas within text through charting, mapping, paraphrasing, summarizing, or comparing/contrasting);

MA.6.A.3.6 Construct and analyze tables, graphs, and equations to describe linear functions and other simple relations using both common language and algebraic notation.

MA.6.S.6.2 Select and analyze the measures of central tendency or variability to represent, describe, analyze, and/or summarize a data set for the purposes of answering questions appropriately.

SC.6.P.12.1 Measure and graph distance versus time for an object moving at a constant speed. Interpret this relationship.

SC.6.N.1.1 Define a problem from the sixth grade curriculum, use appropriate reference materials to support scientific understanding, plan and carry out scientific investigation of various types, such as systematic observations or experiments, identify variables, collect and organize data, interpret data in charts, tables, and graphics, analyze information, make predictions, and defend conclusions.

SC.6.N.2.1 Distinguish science from other activities involving thought.

SC.6.N.1.1, SC.6.N.2.1

Launch Lab

10 minutes

What does a graph show?

Measurements you make during an investigation can sometimes be difficult to understand. Graphing can help make the meaning of the data clearer. The data table here shows the distance an ant traveled each second. What does this look like on a graph?

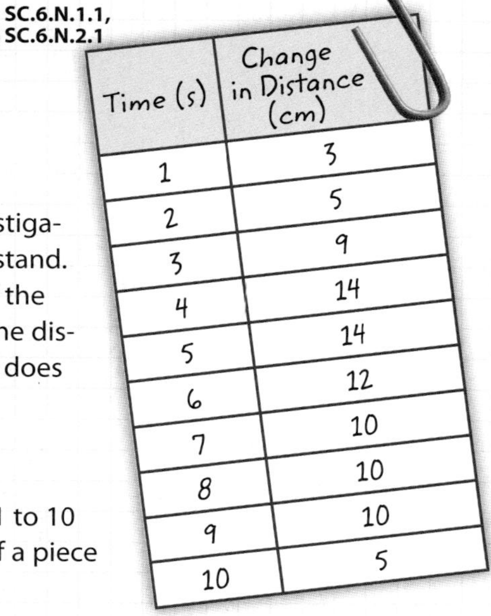

Time (s)	Change in Distance (cm)
1	3
2	5
3	9
4	14
5	14
6	12
7	10
8	10
9	10
10	5

Procedure

1. Use a **marker** to write the numbers 1 to 10 across the bottom of the long side of a piece of **construction paper.**
2. For each change in distance on the table, use **scissors** to cut a strip of **masking tape** that length. Place each strip vertically above the corresponding number of seconds.

Think About This

1. What does the area represented by the strips of masking tape represent?

2. Key Concept What do you think happened to the speed of the ant during each second represented on the graph?

Inquiry What is the turtle wearing?

1. Some animals, such as this sea turtle, travel long distances out of the sight of humans. Biologists use tracking devices to help them better understand the lives of turtles. How might graphing the motion of turtles help save their lives?

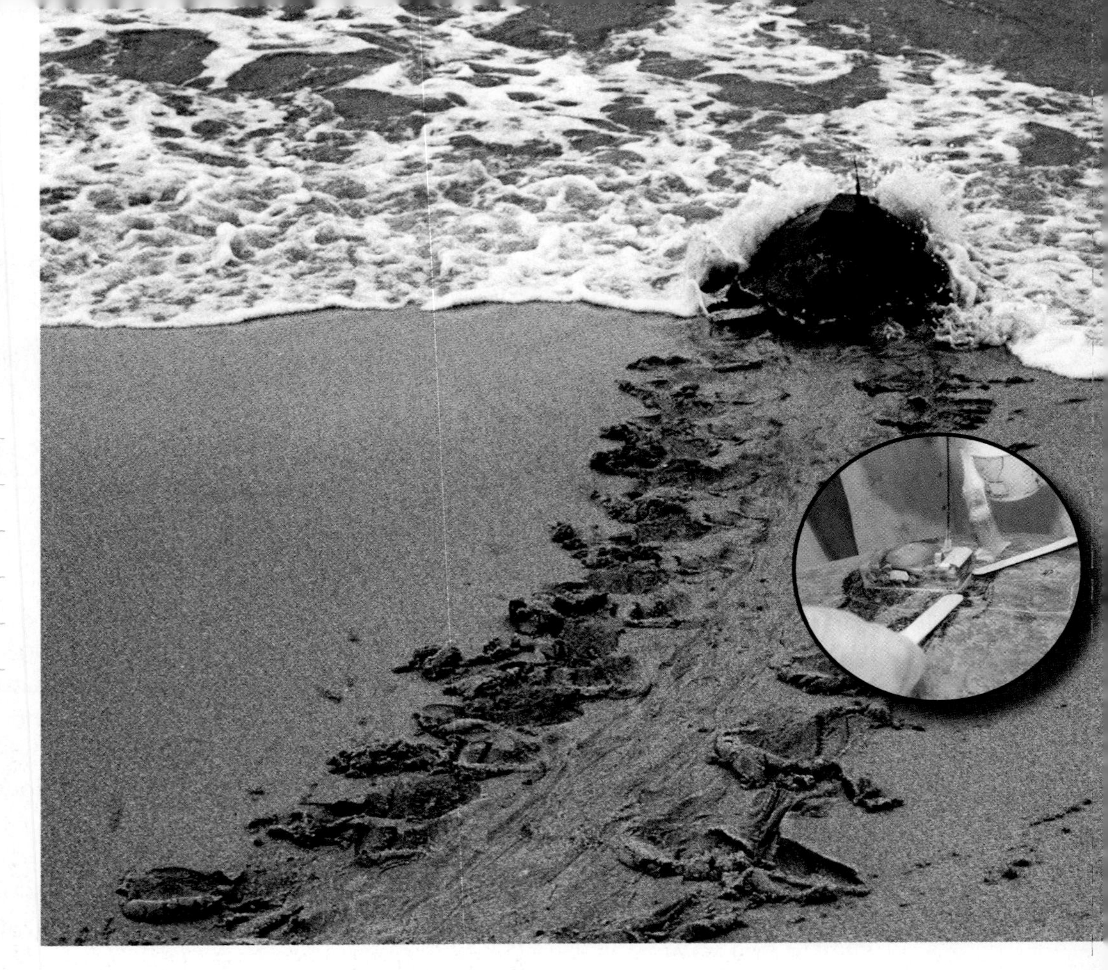

Word Origin

biologist
from Greek *bios* and *logia*, means "to study life"

Describing Motion with Graphs

How could you describe the motion of the sea turtle above? You could measure the total distance the turtle traveled or the time it took the turtle to reach the ocean, but that wouldn't give you a complete picture of its motion. When you study motion, you need to know how position changes as time passes. In this lesson, you will read how to draw and interpret graphs of motion.

Distance-Time Graphs

A graph that shows how distance and time are related is a **distance-time graph.** The y-axis shows the distance an object travels from a reference point. Time is on the x-axis. The line on a distance-time graph, such as the one in **Figure 6,** shows how an object's position changes during each time interval. From this, you can figure out an object's speed. For example, between 2 s and 3 s, the object traveled from 20 m to 30 m, a distance of 10 m. The speed of the object during that time interval was 10 m/s. If the slope of the line changes, you know that the speed changes. A distance-time graph does not show you the actual path the object took.

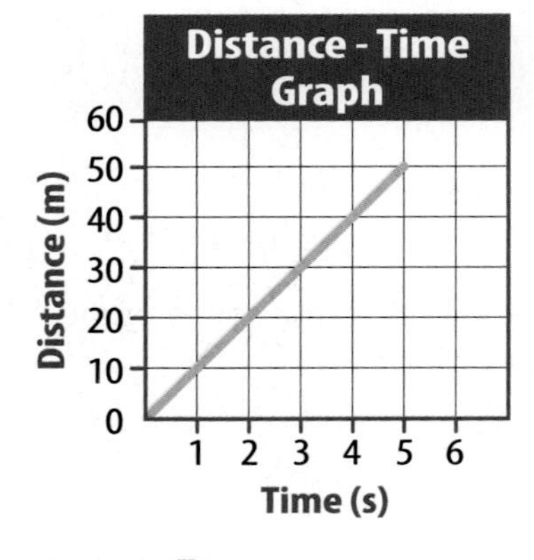

Figure 6 The line on this distance-time graph represents an object traveling at a constant speed.

Active Reading **2. Organize** Underline what the x- and y- axis shows on a distance time graph.

Making a Distance-Time Graph

In order to better understand how sea turtles migrate through the oceans, marine **biologists** attach satellite-tracking devices to turtles' shells. When turtles come to the ocean's surface for air, the devices send information to satellites orbiting Earth. The tracking devices record the turtles' positions and the times that they surface. Marine biologists can download and examine the data from the satellites. This helps them understand turtle behavior and factors that can affect the health of the turtles.

Table 1 shows satellite-tracking data that was gathered for a green sea turtle off the coast of Florida. The first column is the time since tracking began. The second column shows the distance the turtle swam from a reference point. **Figure 7** shows you how to use the data in **Table 1** to make a distance-time graph of the turtle's motion. As you read each step, study the part of the graph to which it refers. By following the same steps, you can make your own graph from similar data.

Active Reading 3. **Describe** Why might marine biologists track the motion of a sea turtle?

Table 1 Green Sea Turtle's Distance and Time Data

Time (days)	Distance (km)
0	0
1	16
2	32
3	48
4	64
5	80
6	96

4. **Visual Check**
Identify How far did the turtle travel between days three and four?

Distance-Time Graph

Figure 7 The graph shows the steps in making a distance-time graph.

Use the following steps to make a distance-time graph.

1. Draw x- and y-axes.
2. Label the x-axis for time measured in days. Label the y-axis for distance measured in kilometers.
3. Make tick marks on the axes and number them. Be sure the values you choose allow you to plot all the data.
4. Plot the data from each row of your data table. Move across the x-axis to the correct time and up the y-axis to the correct distance. Draw a small circle.
5. Connect data points with a line.

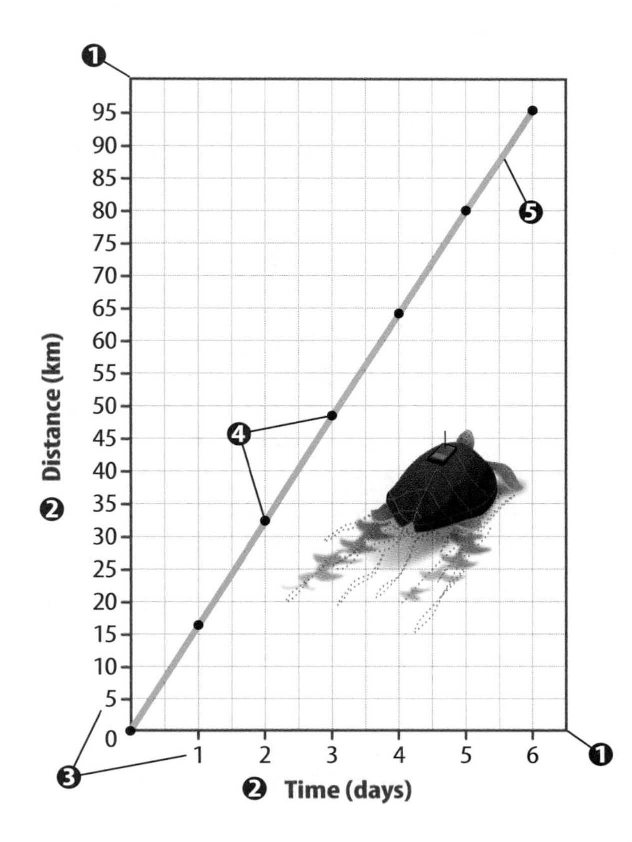

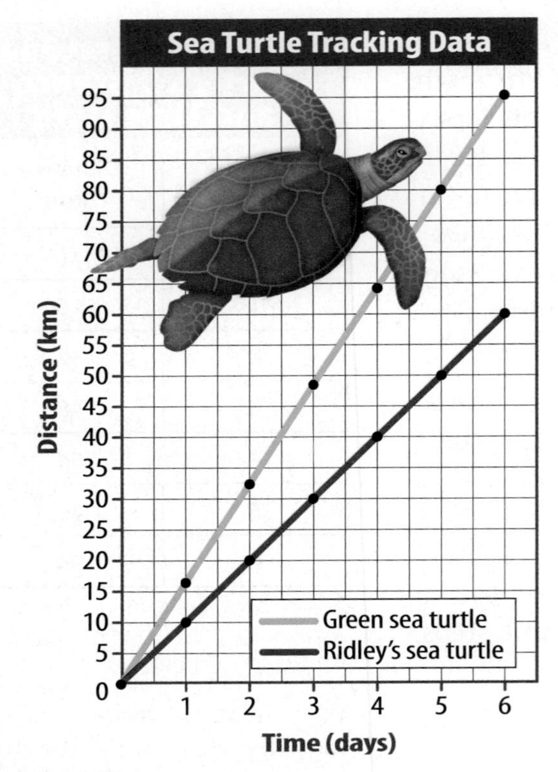

Figure 8 The slope of the green line is greater than the slope of the red line. The green sea turtle swam faster than the Ridley's sea turtle.

Comparing Speed

You can also use distance-time graphs to compare the motion of two different objects. **Figure 8** is a distance-time graph that shows satellite-tracking data for two sea turtles. The line showing the motion of the green sea turtle from **Table 1** on the previous page is green. The line showing the motion of a Ridley's sea turtle is red. After 6 days, the green sea turtle has traveled 96 km. The Ridley's sea turtle, on the other hand, has traveled 60 km.

Recall that average speed is total distance traveled divided by total time. The green sea turtle traveled a greater distance than the Ridley's sea turtle in the same amount of time. Therefore, the green sea turtle's average speed was greater. Notice that the green line is steeper than the red line. Steeper lines, or a greater slope, on distance-time graphs mean that the average speed is greater.

Constant Speed Look again at **Table 1** on the previous page. Each day, the green sea turtle traveled 16 km. You read in Lesson 1 that an object moving the same distance in the same amount of time moves at constant speed. The green sea turtle moved with constant speed.

The two lines in **Figure 8** are straight. You can tell that an object moves with constant speed if the line representing its motion on a position-time graph is straight.

Changing Speed The distance-time graph of a hawksbill sea turtle is shown in **Figure 9.** How is the line on this graph different from the ones in **Figure 8**? The line is not straight. Its slope changes. Each change in slope means that the average speed of the object changed during that time interval.

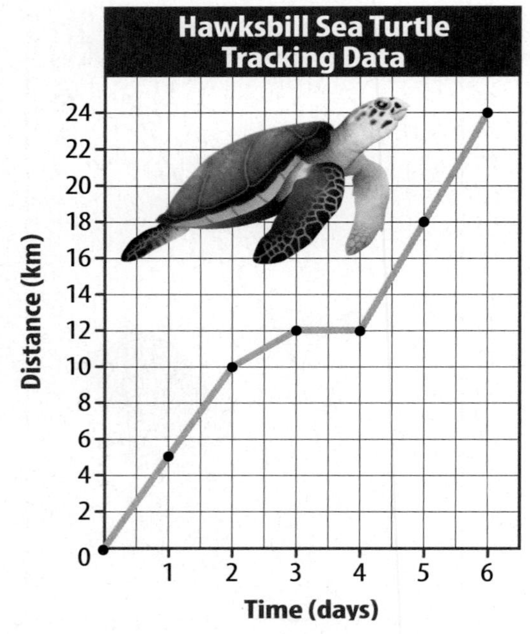

Figure 9 The hawksbill sea turtle changed speed. The line is not straight.

5. **NGSSS Check Illustrate** How can a graph show you if the motion of an object is constant? SC.6.P.12.1

Speed-Time Graphs

You have read how distance-time graphs can help you describe an object's motion. Distance-time graphs show how the distance that the object travels changes during each unit of time. Another type of graph, called a speed-time graph, shows motion in a different way. *A* **speed-time graph** *shows the speed of an object on the y-axis and time on the x-axis.* A speed-time graph shows how the speed of the object changes during each interval of time.

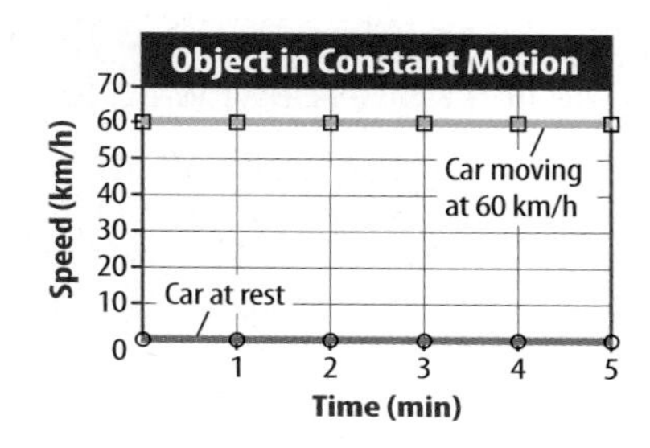

Figure 10 The line representing an object at rest on a speed-time graph is a horizontal line at $y = 0$. The line representing an object moving at constant speed is a horizontal line at that speed.

Resting

Suppose you are in a parked car. What is your speed? You read in Lesson 1 that speed describes how much an object changes position in a unit of time. Because a parked car does not change position, your speed is zero. Your speed remains zero as long as the car remains at rest. The speed-time graph for an object at rest is a horizontal line at $y = 0$, shown as the orange line in **Figure 10.**

Constant Speed

Have you ever been in a car that had the cruise control turned on? The cruise control keeps the car moving at a constant speed— for instance, 60 km/h. On a speed-time graph, an object moving with constant average speed is a horizontal line similar to the green line in **Figure 10.** A horizontal line farther from the *x*-axis represents an object moving at a faster speed than an object represented by a horizontal line closer to the *x*-axis.

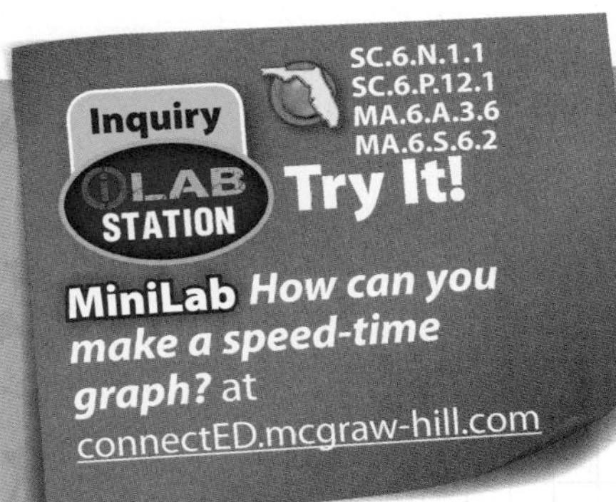

Apply It!

After you complete the lab, answer the questions below.

1. Explain what your graph does NOT tell you about the path of the ant?

2. Predict what the horizontal portions of the graph tell you about the ant's motion.

Active Reading **6. Characterize** What do the *x*-and *y*-axis in a speed-time graph describe?

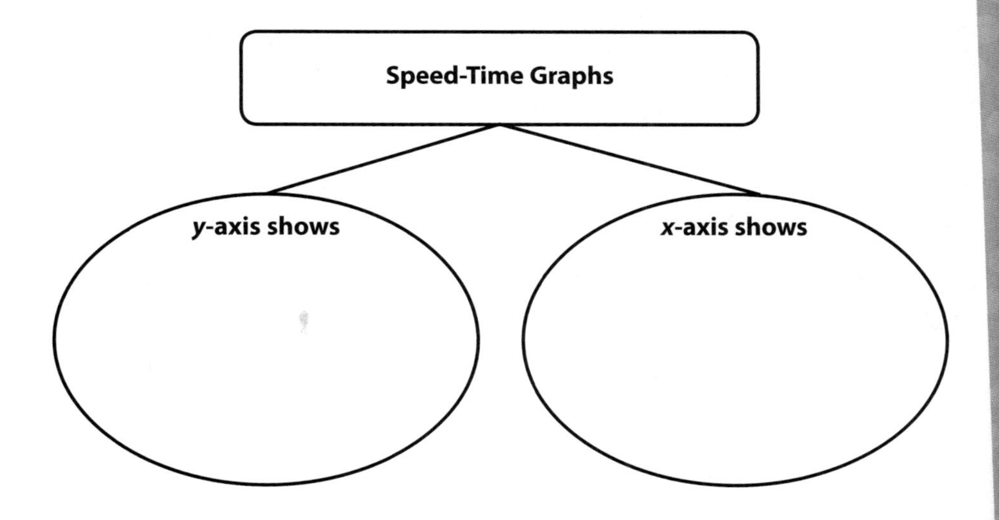

Figure 11 The speed-time graph of an object that is speeding up slopes upward. The line slopes downward for an object that is slowing down.

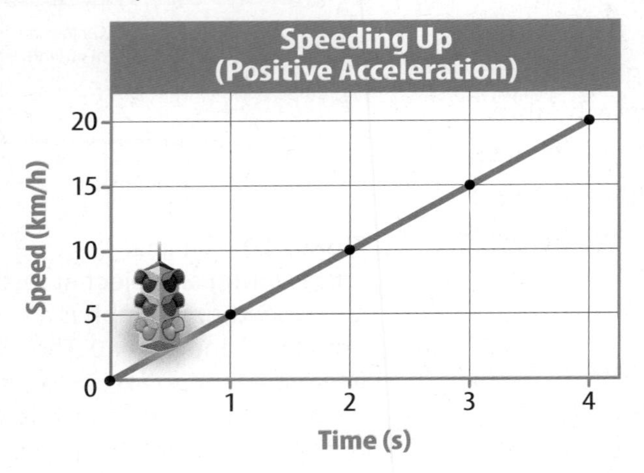

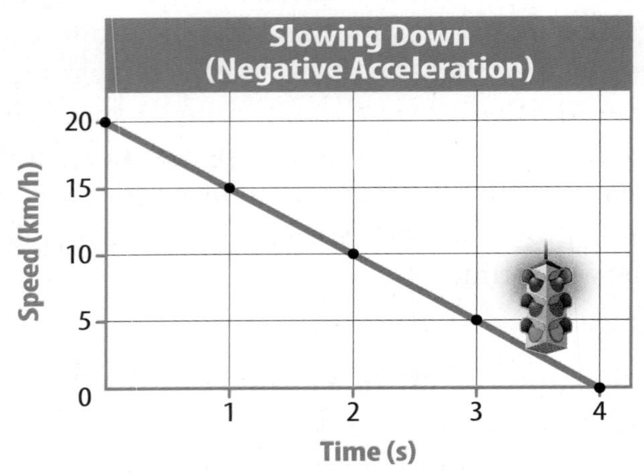

Changing Speed

Suppose you are in a car that is pulling away from a green traffic light. When the driver steps on the accelerator, the car's speed increases. Now the driver sees a red traffic light. The driver takes his or her foot off the accelerator, applies the brakes, and the car's speed decreases. What does the speed-time graph look like when the speed of an object changes?

SC.6.P.12.1
LA.6.2.2.3

Use a sheet of grid paper to make a vertical three-tab book. Draw a Venn diagram on the front and label the tabs. Record what you have learned about distance-time graphs and speed-time graphs under the tabs. Draw examples of each on the front tabs. Under the middle tab, explain graphs and describe what these two graphs have in common.

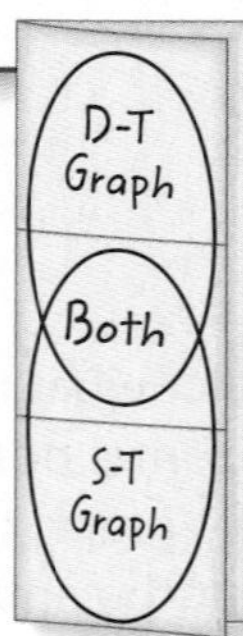

Speeding Up As a car pulls away from the traffic light, its speed increases. The car covers a greater distance each second. As you can see on the left graph in **Figure 11,** a line on the graph that shows the motion of an object with increasing speed slopes upward from left to right. A line that slopes upward on a speed-time graph shows positive acceleration.

Slowing Down When a car slows down, its speed decreases and the car covers a smaller distance each second. The car's motion is a line that slopes downward from left to right, as shown on the graph on the right in **Figure 11.** This represents negative acceleration.

In Lesson 1, you read that when the speed of an object changes, it accelerates. Therefore, if the line on a speed-time graph is not horizontal, you know that the object is accelerating. Objects also accelerate when their direction changes, but that is not shown on the speed-time graph. The line on the graph only shows the change in speed, not the direction in which the object is traveling.

Active Reading **7. Explain** What is the difference between an upward-sloping line on a speed-time graph and one on a distance-time graph?

Lesson Review 2

Visual Summary

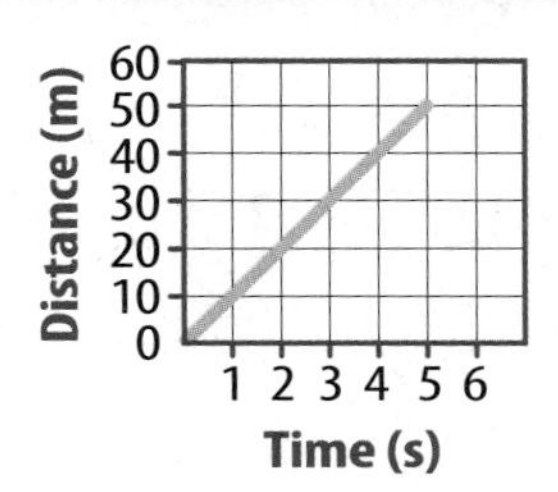

The slope on a distance-time graph shows the motion of an object traveling at a constant speed.

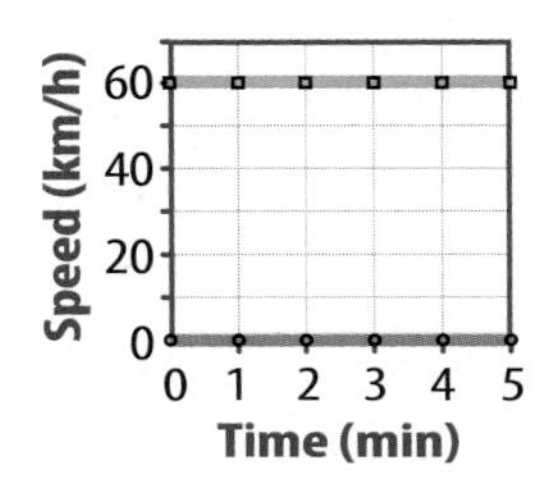

A horizontal line on a speed-time graph shows the motion of an object moving at a constant speed.

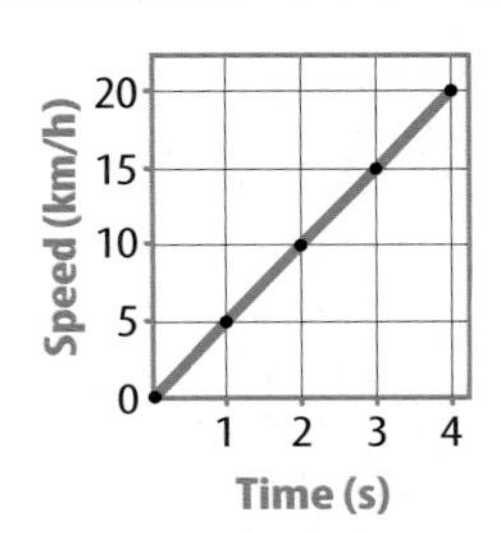

An upward-sloping line on a speed-time graph shows the motion of an object that is speeding up. A downward-sloping line shows the motion of an object that is slowing down.

Use Vocabulary

1. **Describe** a distance-time graph in your own words.

2. **Distinguish** between a distance-time graph and a speed-time graph.

Understand Key Concepts

3. **Deduce** A distance-time graph shows a horizontal line at a distance of 3 km. The object shown on the graph is SC.6.P.12.1
   - (A) at rest.
   - (B) moving at a constant speed.
   - (C) slowing down.
   - (D) speeding up.

4. **Apply** Draw a speed-time graph of a cat that moved at a constant speed of 1 m/s for 10 s, slowed down, and then laid down for a nap. SC.6.P.12.1

Interpret Graphics

5. **Interpret Data** The graph below shows the motion of an elevator. Explain its motion. LA.6.2.2.3

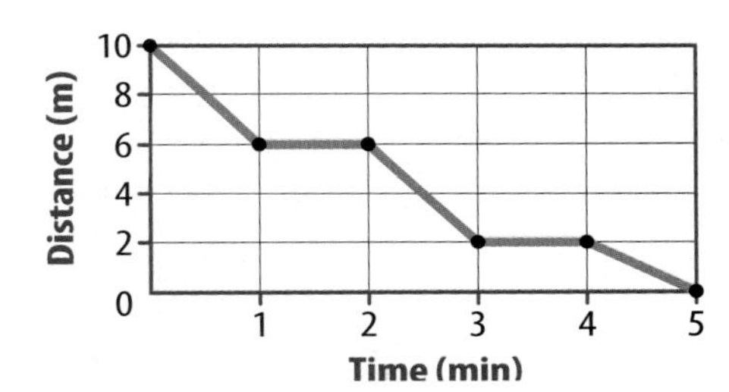

Critical Thinking

6. **Construct** Draw a distance-time graph and a speed-time graph on a separate sheet of paper using the data listed in the table. SC.6.P.12.1

Time (s)	Distance (m)	Time (s)	Distance (m)
0	0	3	30
1	10	4	40
2	20	5	50

Using Satellites to Track the Florida Panther

Using Technology to Help Wild Animals

Just as a GPS can provide information about the location of a car, satellite tracking technology can track wild animals. Data from the tracking provide scientists with information about animals' locations, migrations, and movement patterns. This information is useful in determining ways to help protect animals in the wild.

The Florida panther needs large, rich, forested lands in which to raise its young and hunt for food. Over the past 50 years, the vast wooded areas of southern Florida have been cleared for agricultural and residential development. Besides the loss of forestation, panthers also have experienced population loss due to inbreeding and diseases. By knowing a panther's home ranges through the use of tracking individual animals, scientists can gather demographic, biomedical, and genetic information. This information helps to guide management actions, access responses to natural events and human-caused impacts, and enhance panther recovery.

1. **Collecting Data** Researchers capture and attach satellite tags or collars to animals and then release them back into their habitats. Animals are tagged with one of several devices that transmits signals to satellites. The satellites transmit location data back to a receiver.

2. **Studying Data** Researchers combine location data with maps. Pinpointing exact latitude and longitude helps researchers plot an animal's home range.

3. **Using Data** Collected and analyzed data are useful tools in helping to protect some species. The data can assist in mapping an appropriately sized protected area. Data about animals' travels and habitats are useful in making informed decisions about how humans can better manage and protect the same areas.

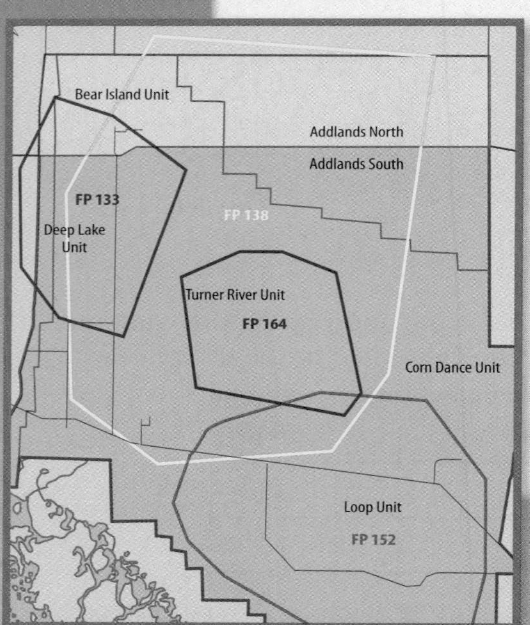

Home ranges of adult male Florida panthers monitored in SBICY from July 2007–June 2008.

It's Your Turn

RESEARCH What are some other factors contributing to the Florida Panthers' decreasing numbers? What can you do? Develop an action plan for how to help stabilize the Florida Panther habitat and share it with your class.

Lesson 3

FORCES

ESSENTIAL QUESTIONS

- What are different types of forces?
- What factors affect the force of gravity?
- What happens when forces combine?
- How are balanced and unbalanced forces related to motion?

Vocabulary

force p. 304
contact force p. 304
noncontact force p. 304
gravity p. 305
friction p. 306
air resistance p. 306
Newton's first law of motion p. 309
Newton's second law of motion p. 310
Newton's third law of motion p. 310

Florida NGSSS

SC.6.P.13.1 Investigate and describe types of forces including contact forces and forces acting at a distance, such as electrical, magnetic, and gravitational.

SC.6.P.13.2 Explore the Law of Gravity by recognizing that every object exerts gravitational force on every other object and that the force depends on how much mass the objects have and how far apart they are.

SC.6.P.13.3 Investigate and describe that an unbalanced force acting on an object changes its speed, or direction of motion, or both.

SC.6.N.1.1 Define a problem from the sixth grade curriculum, use appropriate reference materials to support scientific understanding, plan and carry out scientific investigation of various types, such as systematic observations or experiments, identify variables, collect and organize data, interpret data in charts, tables, and graphics, analyze information, make predictions, and defend conclusions.

SC.6.N.1.5 Recognize that science involves creativity, not just in designing experiments, but also in creating explanations that fit evidence.

SC.6.N.1.1

Inquiry Launch Lab

10 minutes

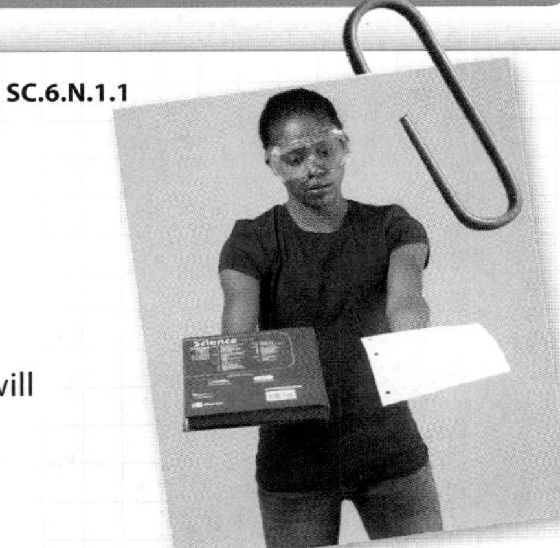

What affects the way objects fall?

If you drop a piece of paper and a book, will they fall in the same way? Let's find out!

Procedure

1. Read and complete a lab safety form.
2. Rest a **sheet of paper** on one hand and a **book** on the other hand with your palms up. Drop both hands at the same time. Observe how the objects fall.
3. Wad an identical sheet of paper into a ball. Repeat step 2.
4. Place the flat sheet of paper on top of the book so that the edges are even. Drop them, and observe how they fall.

Think About This

1. Compare and contrast the speeds of the objects as they fell.

2. Key Concept Why do you think the objects fell at the same or different speeds?

Doesn't that hurt?

1. When a karate expert kicks a concrete block, the energy in his or her foot is transferred to the block. It only hurts if he or she doesn't break the block! What are other ways in which people change the motion of objects?

Figure 12 The karate expert exerts a contact force on the boards.

2. NGSSS Check
List What are some examples of contact and noncontact forces you have experienced today? SC.6.P.13.1

What is force?

What have you pushed or pulled today? You might have pushed open the classroom door or pulled the zipper on your backpack. *A* **force** *is a push or a pull on an object.*

Force has both size and direction. Just as you used arrows to show the size and the direction of velocity and acceleration, arrows can show the size and direction of a force, as shown in **Figure 12.** The unit for force is the newton (N). You use about 1 N of force to lift a medium-sized apple.

Contact Forces

In **Figure 12,** the hand touches the wood as the karate expert applies a force. *A* **contact force** *is a push or a pull one object applies to another object that is touching it.* Contact forces can be small, such as a finger pushing a button on a phone, or they can be very large, such as a wrecking ball crashing into a building.

Noncontact Forces

The balloons in **Figure 13** are pulling the girl's hair toward them. *A force that one object applies to another object without touching it is a* **noncontact force**. The force that attracts the girl's hair to the balloon is an electric force. The force acting between iron and a magnet is also a noncontact force.

Gravity—A Noncontact Force

When you jump off a step, the force of Earth's gravity pulls you toward Earth. **Gravity** *is an attractive force that exists between all objects that have mass.* Mass is the amount of matter in an object. Both you and Earth have mass, so both you and Earth pull on each other.

Figure 13 Noncontact forces attract the girl's hair.

The Law of Gravity

In the late 1600s, Sir Isaac Newton developed the law of universal gravitation, also known as the law of gravity. This law states that all objects are attracted to each other by a gravitational force. The strength of the force depends on the mass of each object and the distance between them.

3. NGSSS Check Recall Underline the law of gravity. SC.6.P.13.2

Gravitational force depends on mass.

Figure 14 shows that, if the mass of an object increases, the gravitational force increases between it and another object. The gravitational force between you and Earth is large because of Earth's large mass. The force holds you to Earth's surface. The gravitational force between you and your pencil is small because you both have relatively small masses compared to Earth. You don't feel the attraction, even though it is present.

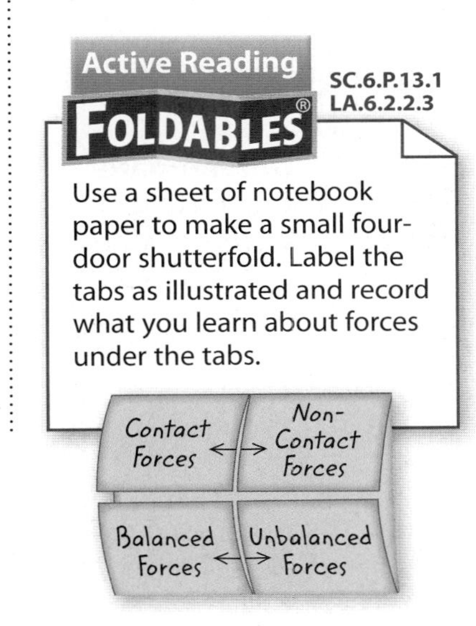

Gravitational force depends on distance.

As two objects move apart, the gravitational force between them decreases. **Figure 14** shows that the gravitational force between two objects 1 m apart is four times greater than the gravitational force between the same objects that are 2 m apart.

Figure 14 This figure shows the effects of distance and mass in the force of gravity.

Active Reading **4. Identify** Fill in the blanks on the figure below to show the effects of distance and mass have on the force of gravity.

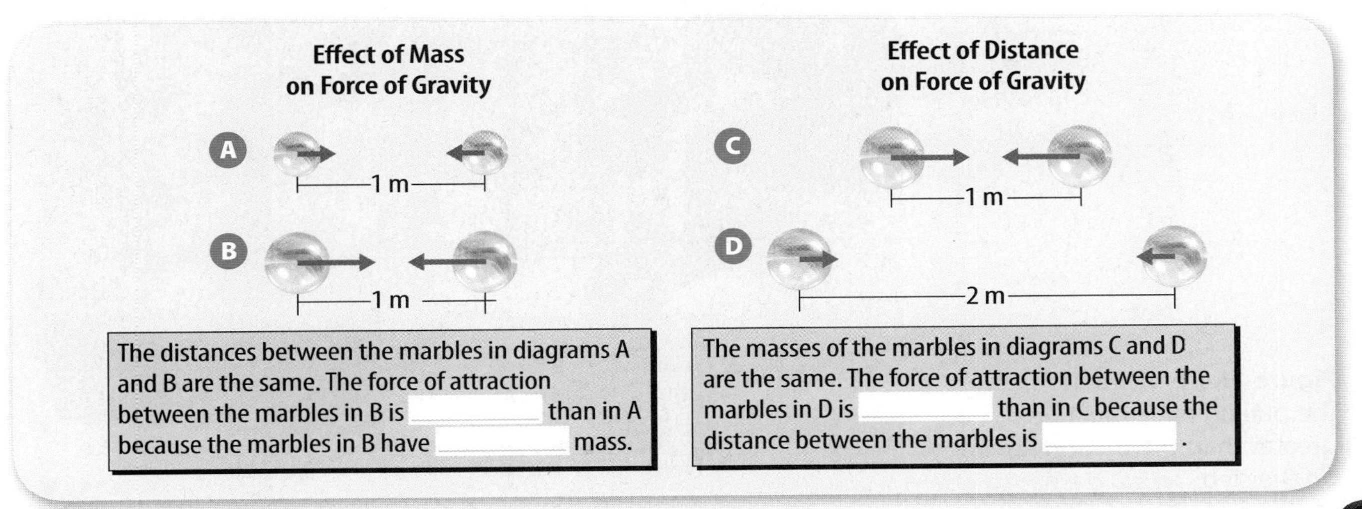

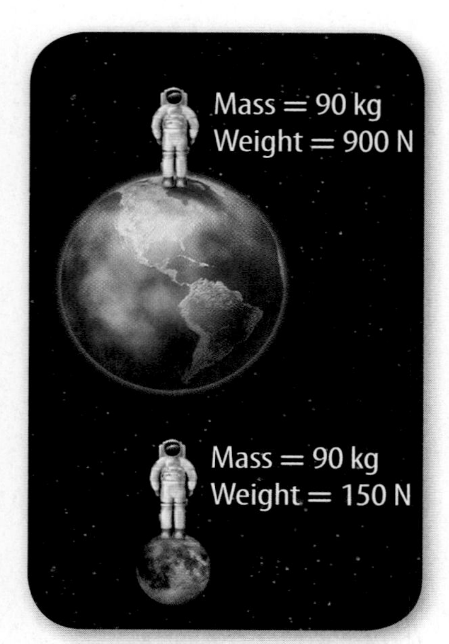

Figure 15 Because of its smaller mass, the Moon's gravity is only 1/6 that of Earth's. An astronaut's weight on the Moon is 1/6 his or her weight on Earth.

Active Reading **5. Differentiate** What is the difference between mass and weight?

Word Origin
friction
from Latin *fricare,* means "to rub"

Mass and weight are different.

Weight is a measure of the gravitational force acting on an object's mass. Therefore, weight depends on the masses of the objects and the distance between them. When comparing the weight of two objects at the same location on Earth, the object with more mass has a greater weight. The weights of the same objects on the Moon are less because the mass of the Moon is less, as shown in **Figure 15.**

Friction—A Contact Force

Rub your finger across your desk. Then rub it across a piece of your clothing. What did you feel? It's easy to run your finger over your desk because it is smooth. On your clothing, you felt a force called friction. **Friction** *is a contact force that resists the sliding motion of two surfaces that are touching.* Rough surfaces or materials tend to produce more friction.

Effects of Friction

Slide your book across the floor. The book slows down when you stop pushing it. The force of friction acts in the opposite direction of the book's motion. A heavier book is more affected by friction than a lighter one. In **Figure 16,** the force of friction between the sled and the ground acts in the opposite direction to the force of the man pulling the sled.

Air Resistance

When you drop a piece of paper, it slowly drifts downward. Friction between the air and the paper's surface slows its motion. **Air resistance** *is the frictional force between air and objects moving through it.* When you crumple the paper into a ball, less surface area is in contact with the air. As air resistance decreases, the ball falls more quickly.

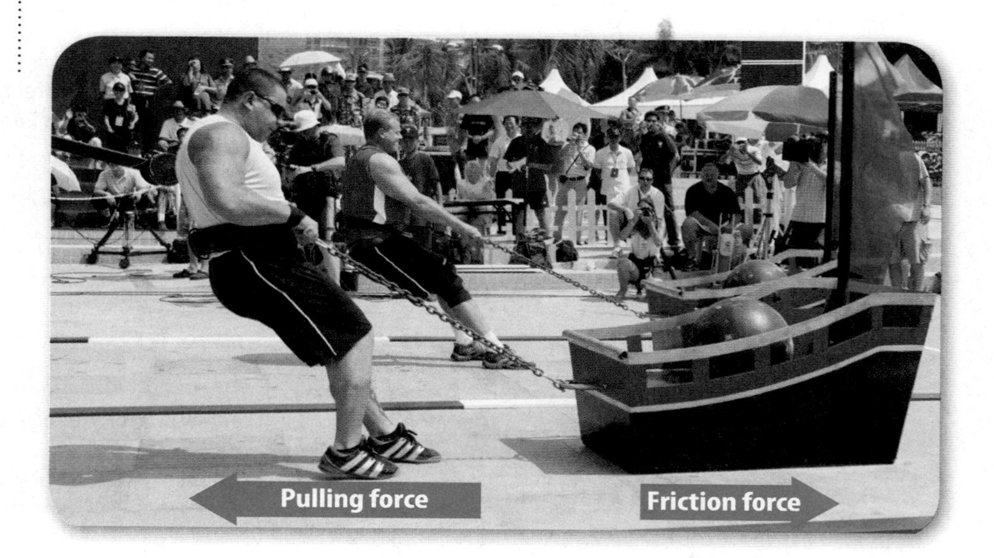

Figure 16 To move the sled, the pulling force must be greater than the friction force on the sled.

Combining Forces

Suppose you need to pull your desk away from the wall to get something that fell behind it. When you pull, the desk will not move, so you ask a friend to help you. With both of you pulling, you have enough force to overcome the force of friction, and the desk moves. When more than one force acts on an object, the forces combine and act as one force. The sum of all the forces acting on an object is called the net force. **Figure 17** shows how forces acting in the same direction form one net force.

When two forces act on the same object in opposite directions, as shown in **Figure 18,** you must include the direction of the forces when you add them. The positive direction is usually to the right. In the left photo of **Figure 18,** the girl's force on the dog's leash is +50 N. The dog's force on the leash is the same size as the girl's, but in the negative direction. The dog's force is −50 N. The net force on the leash is 50 N + (−50 N) = 0 N. The dog doesn't move.

Balanced Forces

If the net force on an object is 0 N, the forces acting on the object are called balanced forces. The net force on the leash in **Figure 18** is 0 N. The forces acting on the leash are balanced.

Unbalanced Forces

When the net force on an object is not 0, the forces acting on the object are unbalanced forces. The net force on the sled in the right photo of **Figure 18** is 100 N to the right. The forces acting on the sled are unbalanced, so it accelerates.

Figure 17 The net force acting on an object is the sum of the two forces and acts in the same direction.

6. NGSSS Check
Explain What can happen when forces combine? SC.6.P.13.2

Figure 18 To calculate the net force, add the forces acting on the object. Acceleration is in the direction of the larger force.

Balanced Forces

The forces are balanced. They are equal in size and opposite in direction.
50 N + (−50N) = 0

Unbalanced Forces

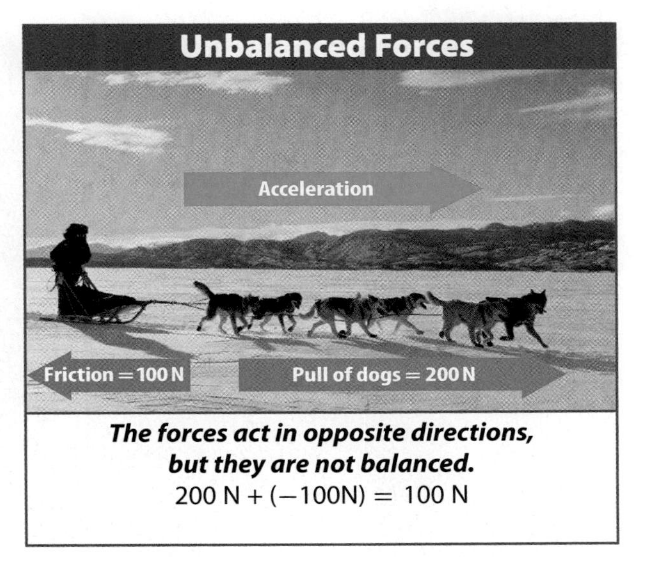

The forces act in opposite directions, but they are not balanced.
200 N + (−100N) = 100 N

Unbalanced Forces and Acceleration

When you kick a soccer ball, its motion changes. The forces on the ball are unbalanced. When unbalanced forces act on an object, the object's velocity changes. Unbalanced forces can change either the speed or the direction of motion.

Change in Speed

The train in the top left image of **Figure 19** is pulling away from the station. The force of the engine is greater than the force of friction. The forces on the train are unbalanced, so it accelerates. The train speeds up.

Change in Direction

When the train goes around a curve, as shown in the bottom left image of **Figure 19,** the track exerts a sideways force on the train's wheels. These unbalanced forces change the train's motion by changing its direction and, therefore, its velocity. The train accelerates.

Balanced Forces and Constant Motion

How do balanced forces affect an object's motion? The forces acting on the train sitting still on a track in **Figure 19** are balanced. The force of gravity pulls the train down toward Earth. The track pushes upward with an equal force.

The train in the bottom right image moves along a straight track. The force from the engine moves the train forward. The force of friction between the wheels and the track is equal in size to the engine's force, but in the opposite direction. The forces acting on the train are balanced. It does not accelerate, but it moves at a constant velocity. When balanced forces act on an object, the motion is constant. The object is either at rest or moving at a constant velocity.

7. NGSSS Check Describe How do balanced and unbalanced forces affect motion? SC.6.P.13.3

Figure 19 An object accelerates if forces acting on it are unbalanced. An object's velocity is constant if the forces acting on it are balanced.

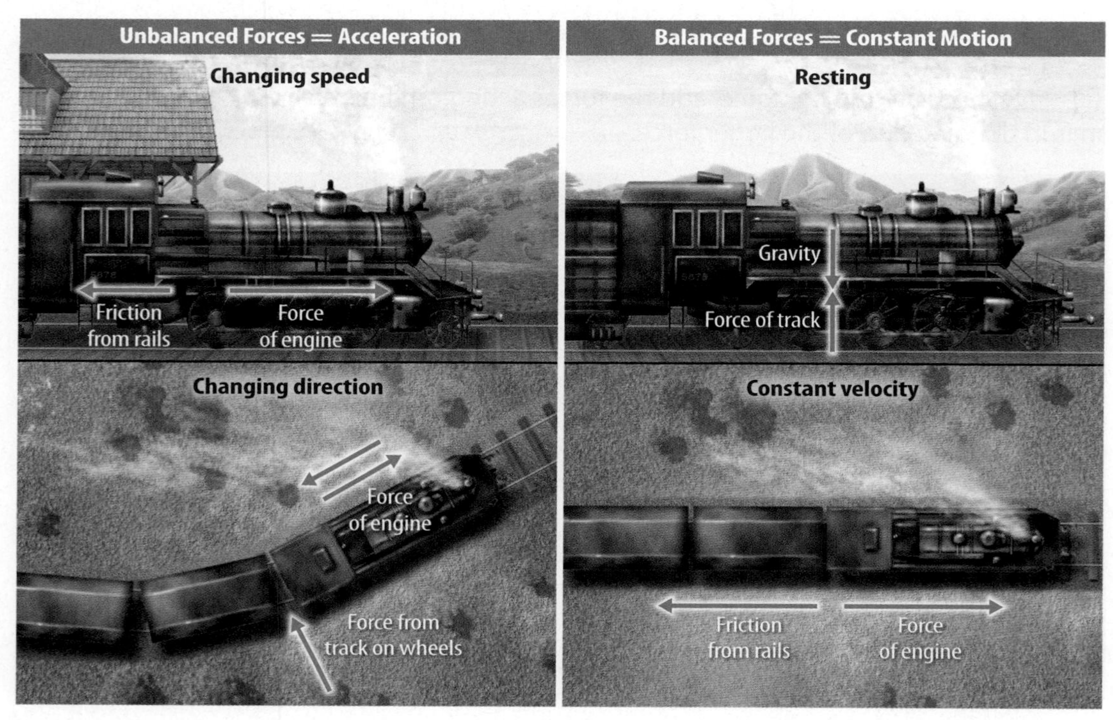

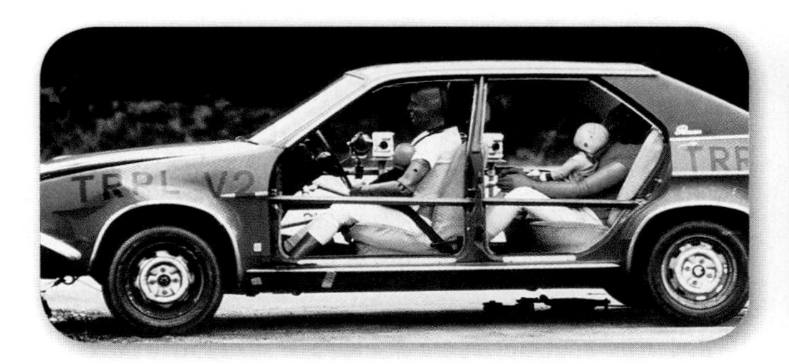

Figure 20 Because of inertia, the crash-test dummies without safety belts keep moving forward after the car stops at a barrier.

Forces and Newton's Laws of Motion

Isaac Newton was an English scientist and mathematician who lived in the late 1600s. He developed three important rules about motion called Newton's laws of motion.

Newton's First Law of Motion

As you just read, when balanced forces act on an object, the object's motion is constant. According to **Newton's first law of motion**, *if the net force acting on an object is zero, the motion of the object does not change.* Newton's first law of motion sometimes is called the law of **inertia**. Inertia is the tendency of an object to resist a change in its motion.

In the first photo in **Figure 20,** the car and the test dummies move with a constant velocity. When the car crashes into the wall, unbalanced forces act on the car, and it stops. However, the dummies, which are not attached to the car, continue to move with a constant velocity because of their inertia.

Word Origin

inertia
from Latin *inertia,* means "idle" or "inactive"

Apply It! After you complete the lab, answer these questions.

1. Explain one way you could have changed the motion of the skateboard.

2. Describe a situation in your life that is similar to what you observed in the MiniLab.

Figure 21 The greater the force the bowler exerts on the ball, the faster it will accelerate as long as the mass of the ball remains constant.

Active Reading **8. Calculate** How much force would you need to accelerate the ball at the rate of 10 m/s^2?

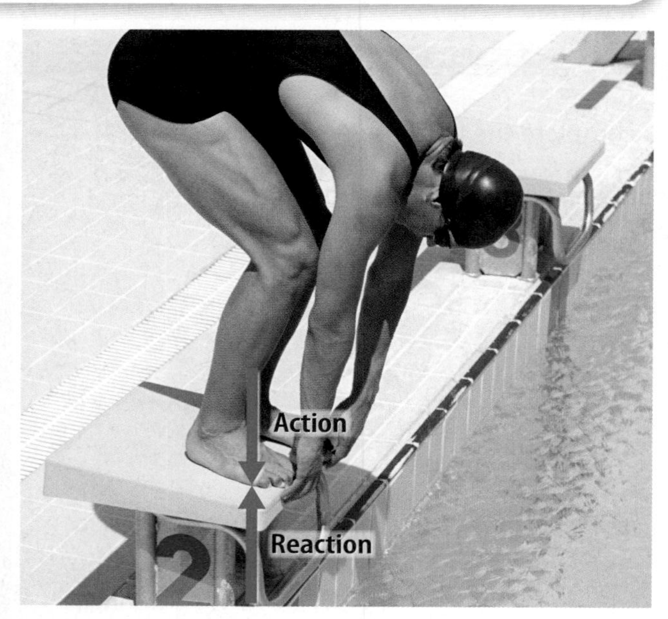

Figure 22 The force of the swimmer's feet pushing on the starting block and the force of the starting block pushing on the swimmer's feet are the force pair.

Newton's Second Law of Motion

Recall when an unbalanced force acts on an object, the object accelerates. **Newton's second law of motion** *states that acceleration of an object is equal to the net force exerted on the object divided by the object's mass.* This means that the greater the mass, the greater the force needed to accelerate the object at the same rate. Acceleration is in the same direction as net force. You can write Newton's second law of motion as an equation:

$$\text{acceleration } (a) = \frac{\text{force } (F)}{\text{mass } (m)}$$

For example, **Figure 21** shows that the bowler exerted a force of 80 N on the ball before releasing it. Eighty N of force causes a mass of 4 kg to accelerate 20 m/s every second. To accelerate a mass of 8 kg at the same rate, you would need a force of 160 N.

Newton's Third Law of Motion

The swimmer's feet in **Figure 22** exert a downward force on the starting block. The starting block also exerts an upward force on his feet. **Newton's third law of motion** *says that for every action there is an equal and opposite reaction.* When one object exerts a force on a second object, the second object exerts a force of the same size but in the opposite direction on the first object. These equal and opposite forces are called force pairs.

In **Figure 22,** the swimmer's feet exert a downward force on the starting block. The starting block exerts an upward force on the swimmer's feet. This is the force pair.

Force pairs are not the same as balanced forces. Balanced forces act on the same object. Recall **Figure 19.** The force from gravity and the force from the train track are balanced and they act on the same object—the train.

Newton's laws help you understand why objects move as they do.

Lesson Review 3

Visual Summary

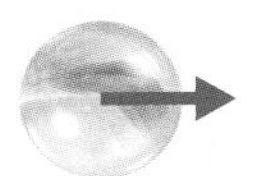
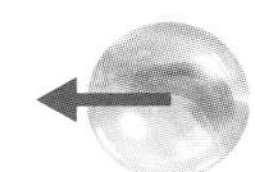

Gravity is a noncontact force between two objects. Gravity depends on the mass of the objects and the distance between them.

Balanced forces produce constant motion. Unbalanced forces produce acceleration.

Newton's laws of motion describe relationships among forces and their effect on motion.

Use Vocabulary

1. The force of ________________ depends on the mass of objects and the distance between them.

2. **Describe** Newton's first law of motion in your own words.

3. **Distinguish** between contact forces and noncontact forces.

Understand Key Concepts 

4. **Recall** Name two noncontact forces. SC.6.P.13.1

5. The force of gravity on a skydiver is 500 N. If the net force acting on him or her is 0 N, what other force is acting on him or her? SC.6.P.13.3
   (A) a 500-N force toward the ground
   (B) a 250-N force toward the ground
   (C) a 500-N force away from the ground
   (D) a 0-N force away from the ground

6. **Describe** what might happen to the mass and/or distance between two objects to increase the gravitational force of attraction. SC.6.P.13.3

Interpret Graphics

7. **Summarize** Use the graphic organizer below to summarize the two factors that affect the force of gravity. LA.6.2.2.3

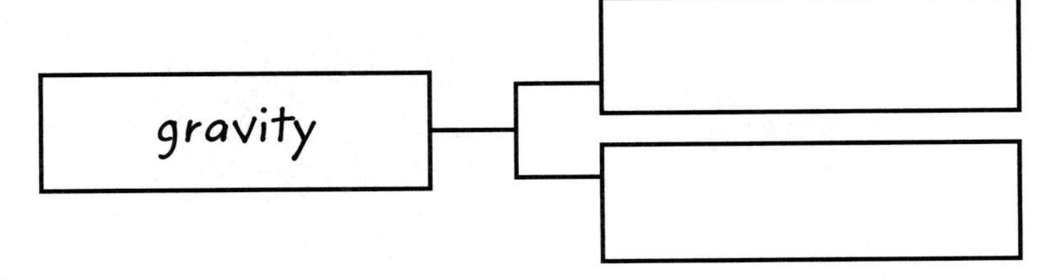

Critical Thinking

8. **Elaborate** Explain why the gravitational force between you and Earth is larger than the gravitational force between you and the Moon. SC.6.P.13.2

Chapter 8 Study Guide

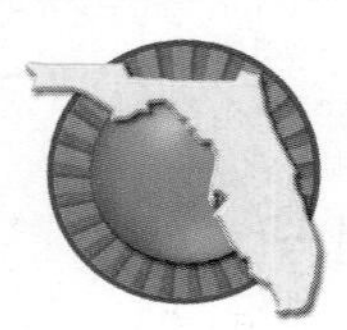

Think About It! The motion of an object is changed by a pushing or a pulling force. The force that moves an object can be either in physical contact with the object or at a distance.

Key Concepts Summary

LESSON 1 Describing Motion

- An object's **position** depends on a **reference point**, a distance, and a direction.
- An object's **motion** can be described using **speed, velocity**, or **acceleration**.
- Speed is how fast an object moves. Velocity describes an object's speed and the direction it moves.
- Acceleration is how fast an object's velocity changes.

Vocabulary

reference point p. 287
position p. 287
displacement p. 288
motion p. 288
speed p. 289
velocity p. 290
acceleration p. 292

LESSON 2 Graphing Motion

- Motion data can be plotted as points on a graph. The line connecting the points shows changes in the motion of the object.
- The line on a **distance-time graph** allows you to calculate an object's speed at any moment in time. A **speed-time graph** helps you understand both how fast an object moves and how fast the object's speed changes.

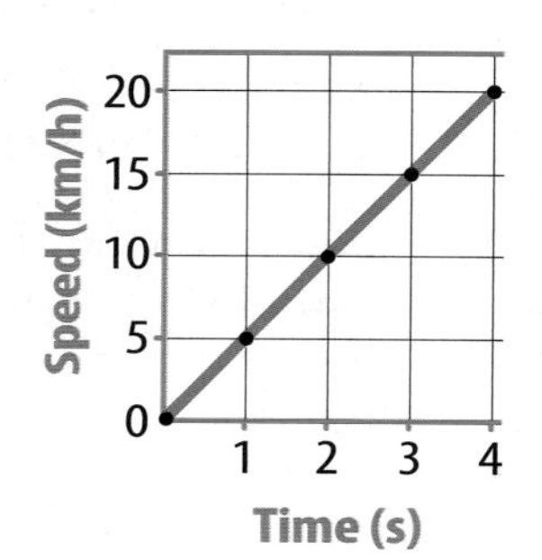

distance-time graph p. 296
speed-time graph p. 299

LESSON 3 Forces

- **Contact forces** include **friction** and **air resistance. Noncontact forces** include **gravity**, electricity, and magnetism.
- Mass and distance affect gravitational force.
- Both the size and the direction of forces must be used when combining forces to determine the net force acting on an object.
- **Newton's laws of motion** describe the relationships among forces, mass, and motion.

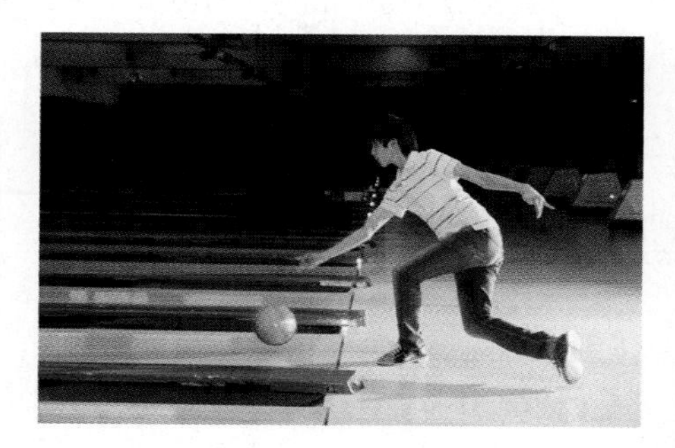

force p. 304
contact force p. 304
noncontact force p. 304
gravity p. 305
friction p. 306
air resistance p. 306
Newton's first law of motion p. 309
Newton's second law of motion p. 310
Newton's third law of motion p. 310

Assemble your lesson Foldables as shown to make a Chapter Project. Use the project to review what you have learned in this chapter.

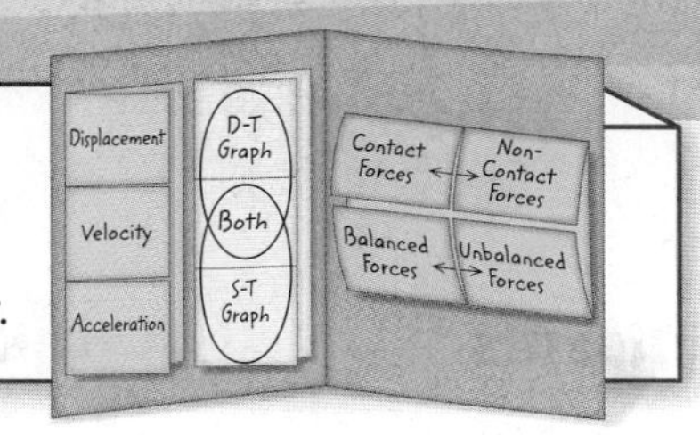

Use Vocabulary

1. An object is in ____________ if its position is changing in some way.

2. Distinguish between the terms *displacement* and *position.*

3. Describe a distance-time graph in your own words.

4. Distinguish between the *y*-axis of a speed-time graph and the *y*-axis of a distance-time graph.

5. If an object is accelerating, the line representing the motion on a ____________ will not be horizontal.

6. Explain Newton's third law of motion in your own words.

7. A ____________ is a push or a pull on an object exerted by another object.

8. Newton's ____________ describes the relationship among force, mass, and acceleration.

Link Vocabulary and Key Concepts

Use vocabulary terms from the previous page to complete the concept map.

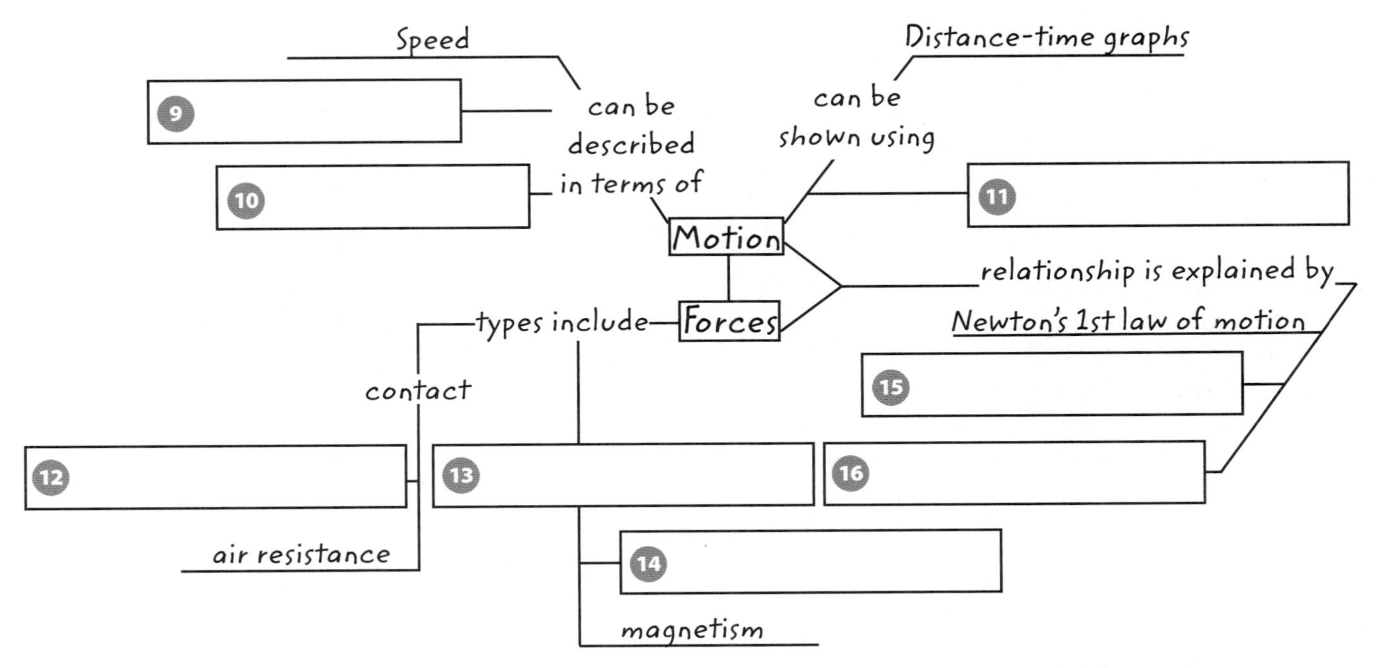

Chapter 8 Study Guide

Fill in the correct answer choice.

Understand Key Concepts

1. Which is NOT necessary when describing the position of an object? **SC.6.P.12.1**
   Ⓐ a reference point for the measurement
   Ⓑ its distance from the reference point
   Ⓒ its speed compared to the reference point
   Ⓓ the direction from the reference point

2. A horse walks along a straight path. It travels 3 m in one second, 4 m in the next second, and 5 m in the third second. Which statement is correct? **MA.6.A.3.6**
   Ⓐ The horse's acceleration is 0.
   Ⓑ The horse is accelerating at 3 m/s^2.
   Ⓒ The horse's velocity is changing.
   Ⓓ The horse's speed is constant.

3. The following is a speed-time graph of a skateboarder. Which statement is true? **MA.6.A.3.6**

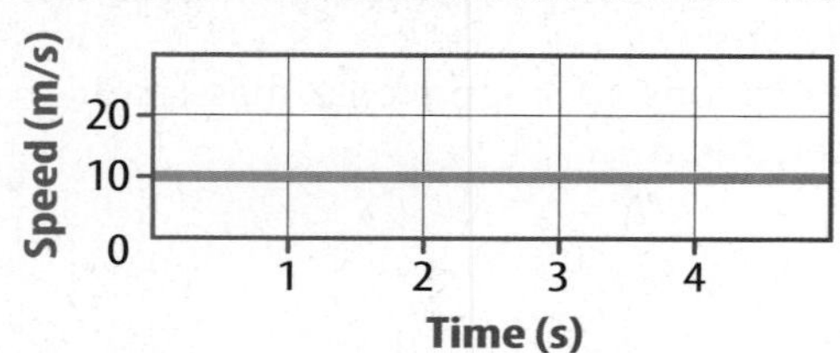

   Ⓐ The speed of the skateboarder is changing.
   Ⓑ The speed of the skateboarder is 10 m/s.
   Ⓒ The skateboarder is at rest.
   Ⓓ The skateboarder's acceleration is 10 m/s^2.

4. An airplane flies 1,000 km in 4 h. What is its average speed? **MA.6.A.3.6**
   Ⓐ 250 km/h
   Ⓑ 500 km/h
   Ⓒ 1,000 km/h
   Ⓓ 4,009 km/h

5. How would you best describe the motion of a sea turtle if the distance-time graph of its motion shows a horizontal line at 12 km? **SC.6.P.12.1**
   Ⓐ The turtle is positively accelerating.
   Ⓑ The turtle is at rest.
   Ⓒ The turtle is moving at a constant speed.
   Ⓓ The turtle is negatively accelerating.

Critical Thinking

6. **Construct** a diagram in which a person's displacement is the same as the distance he or she travels. **MA.6.A.3.6**

7. **Summarize** Imagine you are riding a bike. What are three ways that you can change your velocity? **SC.6.P.13.1**

8. **Calculate** The table below shows the position of a train at different times. What was the train's average speed? **MA.6.A.3.6**

Distance (km)	Time (h)	Distance (km)	Time (h)
0	0	60	2
20	0.5	80	2.5
40	1	100	3

9. **Sketch** a diagram of the forces acting on a skydiver who is falling at a constant speed. **SC.6.P.12.1**

10 **Describe** the force pair involved when you sit on a chair. SC.6.P.13.1

11 **Conclude** A ship is sailing due west. It sails 3 knots the first hour of its journey and 5 knots the second hour. Are the forces on the ship balanced or unbalanced? Explain how you came to your conclusion. SC.6.P.13.3

12 **Assess** A student pushes a heavy box across the floor at a constant velocity of 4 m/s. The sliding friction of the box is 50 N. How hard is the student pushing on the box? SC.6.P.13.3

Writing in Science

13 **Write** Consider the balanced and unbalanced forces that act on you from the time you get up until you arrive at school. On a separate sheet of paper, describe the forces and their effects on your motion as you lie in bed, stand at the sink, walk at a constant speed, and speed up to catch the bus. SC.6.P.13.3

Big Idea Review

14 How does Newton's first law explain the effect of balanced and unbalanced forces on motion? SC.6.P.13.3

15 How would you describe the motion of the bicycle rider in the picture in terms of forces? SC.6.P.13.2

Math Skills MA.6.A.3.6

Calculate Speed

16 On a hike, you walk 3.5 km the first hour, 3 km the second hour, and 2.5 km the third hour. What was your average speed?

17 If a school bus travels at an average speed of 55 km/h for 3 h, what is the total distance covered by the bus?

18 How long would it take you to bike 20 km at an average speed of 5 km/h?

Florida NGSSS Benchmark Practice

Fill in the correct answer choice.

Multiple Choice

1 Which would cause the gravitational force between object A and object B to increase? SC.6.P.13.1

Ⓐ The mass of object A decreases.

Ⓑ The mass of object B decreases.

Ⓒ The objects move closer together.

Ⓓ The objects move farther apart.

2 What changes when unbalanced forces act on an object? SC.6.P.13.3

Ⓕ mass

Ⓖ motion

Ⓗ inertia

Ⓘ weight

Use the graph below to answer question 3.

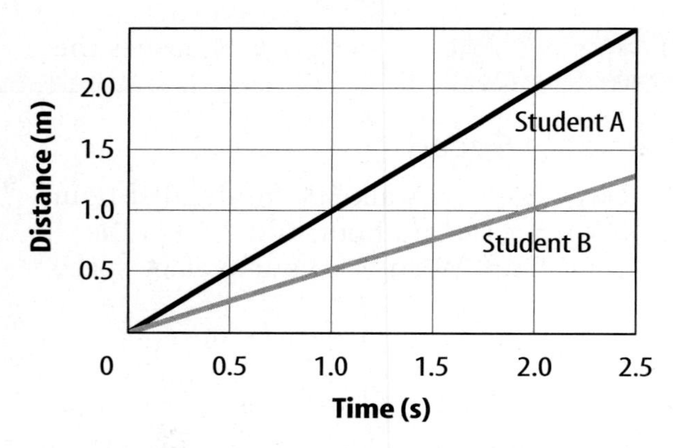

3 How does the speed of student A compare to the speed of student B? SC.6.P.12.1

Ⓐ It is half as large.

Ⓑ It is the same.

Ⓒ It is twice as large

Ⓓ It is three times as large.

4 A distance-time graph of a bicyclist has zero slope. What can you infer about the bicyclist? SC.6.P.12.1

Ⓕ The rider is speeding up.

Ⓖ The rider is slowing down.

Ⓗ The rider is going at a steady pace.

Ⓘ The rider is at rest.

Use the graph to answer questions 5–7.

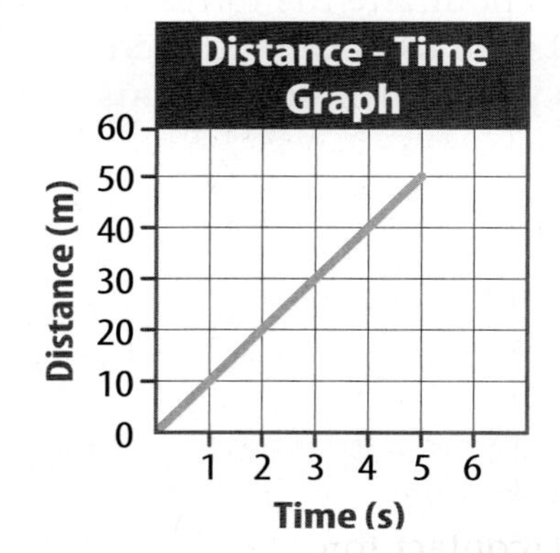

5 What is shown by the line in the graph? SC.6.P.12.1

Ⓐ acceleration

Ⓑ average speed

Ⓒ displacement

Ⓓ instantaneous speed

6 What can you conclude about the motion of the object from the graph? SC.6.P.12.1

Ⓕ Its speed was constant.

Ⓖ Its distance was greater than its displacement.

Ⓗ Its position was constant.

Ⓘ It moved toward the reference point.

7 If the object's speed does not change, what distance will the object travel after 6 s? SC.6.P.12.1

Ⓐ 10 m

Ⓑ 50 m

Ⓒ 60 m

Ⓓ 70 m

Use the figure below to answer question 8.

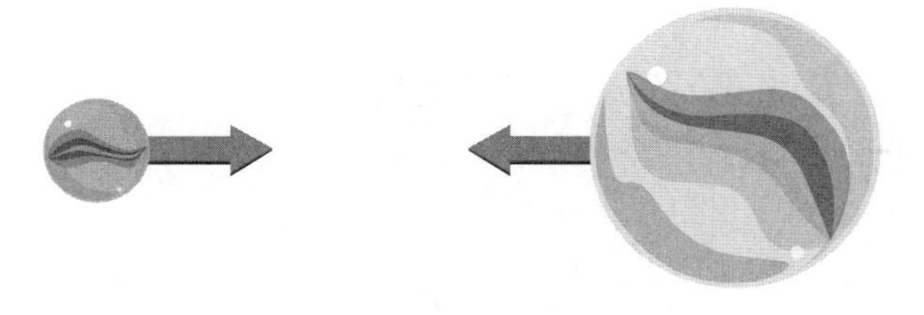

8 What would be the effect of decreasing the distance between the objects? SC.6.P.13.1

Ⓕ The force would remain the same.

Ⓖ The force would increase because the objects are closer together.

Ⓗ The force would decrease because the objects are closer together.

Ⓘ The force only changes if the masses of the objects change.

9 Which pairing matches a force with its correct type? SC.6.P.13.2

Ⓐ air resistance (contact force)

Ⓑ friction (noncontact force)

Ⓒ karate chop (noncontact force)

Ⓓ magnetic force (contact force)

10 Why do you notice the pull of Earth's gravity but not the pull of the Sun's gravity? SC.6.P.13.2

Ⓕ Gravity only pulls on objects touching each other.

Ⓖ Earth is much heavier than the Sun.

Ⓗ The Sun is very far away.

Ⓘ The Sun's gravity only pulls on you during the day.

Use the figure below to answer question 11.

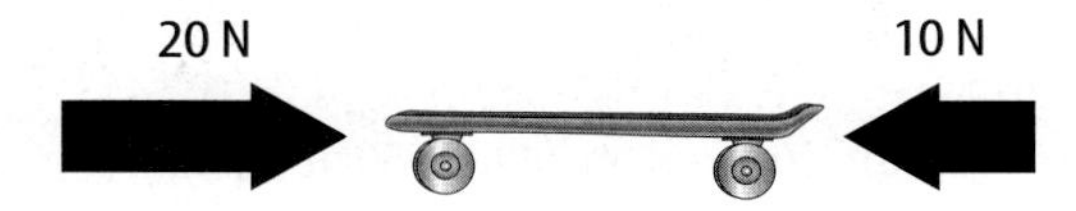

11 The skateboard above moves at a constant velocity before the two forces shown suddenly act on it. What happens to the skateboard? SC.6.P.13.3

Ⓐ The forces are balanced, so the skateboard accelerates.

Ⓑ The forces are unbalanced, so the skateboard accelerates.

Ⓒ The forces are balanced, so the skateboard continues moving at the same velocity.

Ⓓ The forces are unbalanced, so the skateboard continues moving at the same velocity.

12 A person pushes a cart forward with a force of 20 N. A friction force of 5 N pushes the cart backward. What is the net force on the cart? SC.6.P.13.3

Ⓕ 15 N backward

Ⓖ 15 N forward

Ⓗ 25 N backward

Ⓘ 25 N forward

13 Which is NOT an example of unbalanced forces acting on an object? SC.6.P.13.3

Ⓐ an acorn falling from a tree

Ⓑ a car moving at a constant speed of 40 km/h

Ⓒ a motorcycle changing speed from 20 km/h to 35 km/h

Ⓓ a truck slowing down as it approaches a red light

NEED EXTRA HELP?

If You Missed Question...	1	2	3	4	5	6	7	8	9	10	11	12	13
Go to Lesson...	3	3	2	2	2	2	2	2	3	3	3	3	3

Name ______________________________ Date ______________

Multiple Choice *Bubble the correct answer.*

1. In the image above, between which points is the bus moving at constant speed? **SC.6.P.12.1**

Ⓐ 1 and 4

Ⓑ 1 and 7

Ⓒ 3 and 6

Ⓓ 4 and 7

2. A friend describes his position by including distance, direction, and **SC.6.P.12.1**

Ⓕ acceleration.

Ⓖ displacement.

Ⓗ average speed.

Ⓘ reference point.

Use the image below to answer questions 3 and 4.

3. What happens to cyclist 1 between the first and second panels in the figure above? **SC.6.P.12.1**

Ⓐ The speed of the cyclist decreases.

Ⓑ The speed of the cyclist remains constant.

Ⓒ The velocity of the cyclist changes.

Ⓓ The velocity of the cyclist remains constant.

4. Which statement describes the velocity of cyclist 1 in the third panel above? **SC.6.P.12.1**

Ⓕ The cyclist's velocity is changing because her direction is changing.

Ⓖ The cyclist's velocity is changing because her speed is changing.

Ⓗ The cyclist's velocity is constant because her speed is constant.

Ⓘ The cyclist's velocity is constant because speed and direction are constant.

Name ______________________ Date ______________

Multiple Choice *Bubble the correct answer.*

Table 1 Green Sea Turtle's Distance and Time Data

Time (days)	Distance (km)
0	0
1	16
2	32
3	48
4	64
5	80
6	96

1. The data in the table above show how far a sea turtle travels over several days. What would the line on a graph of this data look like? **SC.6.P.12.1**

(A) The line would curve upward and to the right.

(B) The line would go up and down.

(C) The line would point straight upward to the right.

(D) The line would point upward then downward.

2. Which statement is true of a distance-time graph? **SC.6.P.12.1**

(F) A line with a curve indicates that the object has constant speed.

(G) The line in the graph points downward if the object is speeding up.

(H) The line shows a change in speed if its angle changes.

(I) The line shows the path of the object that is moving.

Use the image below to answer questions 3 and 4.

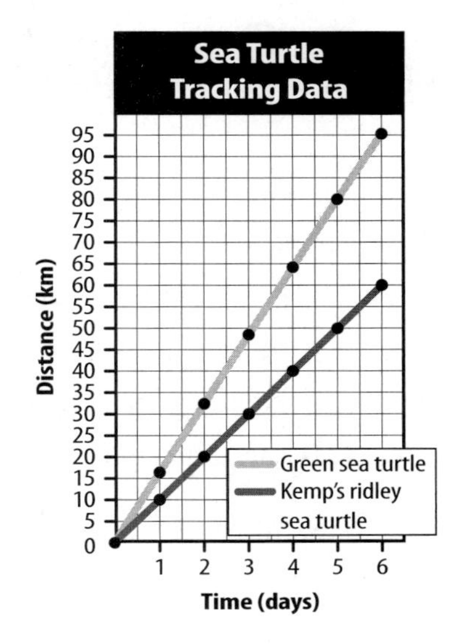

3. Which statement best describes the data in the graph? **MA.6.A.3.6**

(A) Both types of sea turtles moved at the same speed.

(B) The green sea turtle moved faster than the Kemp's ridley sea turtle did.

(C) The Kemp's ridley sea turtle changed direction while the green sea turtle did not change direction.

(D) The Kemp's ridley sea turtle moved at constant speed while the speed of the green sea turtle changed.

4. Scientists measured the speed of a third turtle. Starting at zero, it had constant velocity and was 75 km out on day 6. Where on the graph would the line for this turtle fall? **SC.6.P.12.1**

(F) above the green sea turtle line

(G) below the Kemp's ridley sea turtle line

(H) between the other two lines

(I) on top of the green sea turtle line

Name ______________________ Date ______________

Benchmark Mini-Assessment

Multiple Choice *Bubble the correct answer.*

A: 1m
C: 1m
B: 1m
D: 2m

1. Which correctly relates the relative size of the force of gravity for the figures in the image above? **SC.6.P.13.2**

- (A) Figure A is greater than figure B.
- (B) Figure A is the same as figure B.
- (C) Figure C is greater than figure D.
- (D) Figure C is the same as figure D.

2. Which is an example of a noncontact force? **SC.6.P.13.1**

- (F) A boy lifts a chair.
- (G) A girl pulls a wagon.
- (H) A boy's weight pushes a seesaw down.
- (I) A girl's hair is pulled toward a balloon.

Use the table below to answer questions 3 and 4.

	Size of Force (in newtons)	Direction of Force
Q	200 N	to the left
R	200 N	to the left
S	250 N	to the left
T	250 N	to the right
U	275 N	to the right

3. In the table, which two forces are balanced? **SC.6.P.13.1**

- (A) Q and R
- (B) R and S
- (C) S and T
- (D) T and U

4. Based on the data, which pair of forces produces the greatest net force in any direction? **SC.6.P.13.3**

- (F) Q and R
- (G) R and S
- (H) S and T
- (I) T and U

Name ______________________ Date ______________

Notes

Unit 4

Organization and Development of Organisms

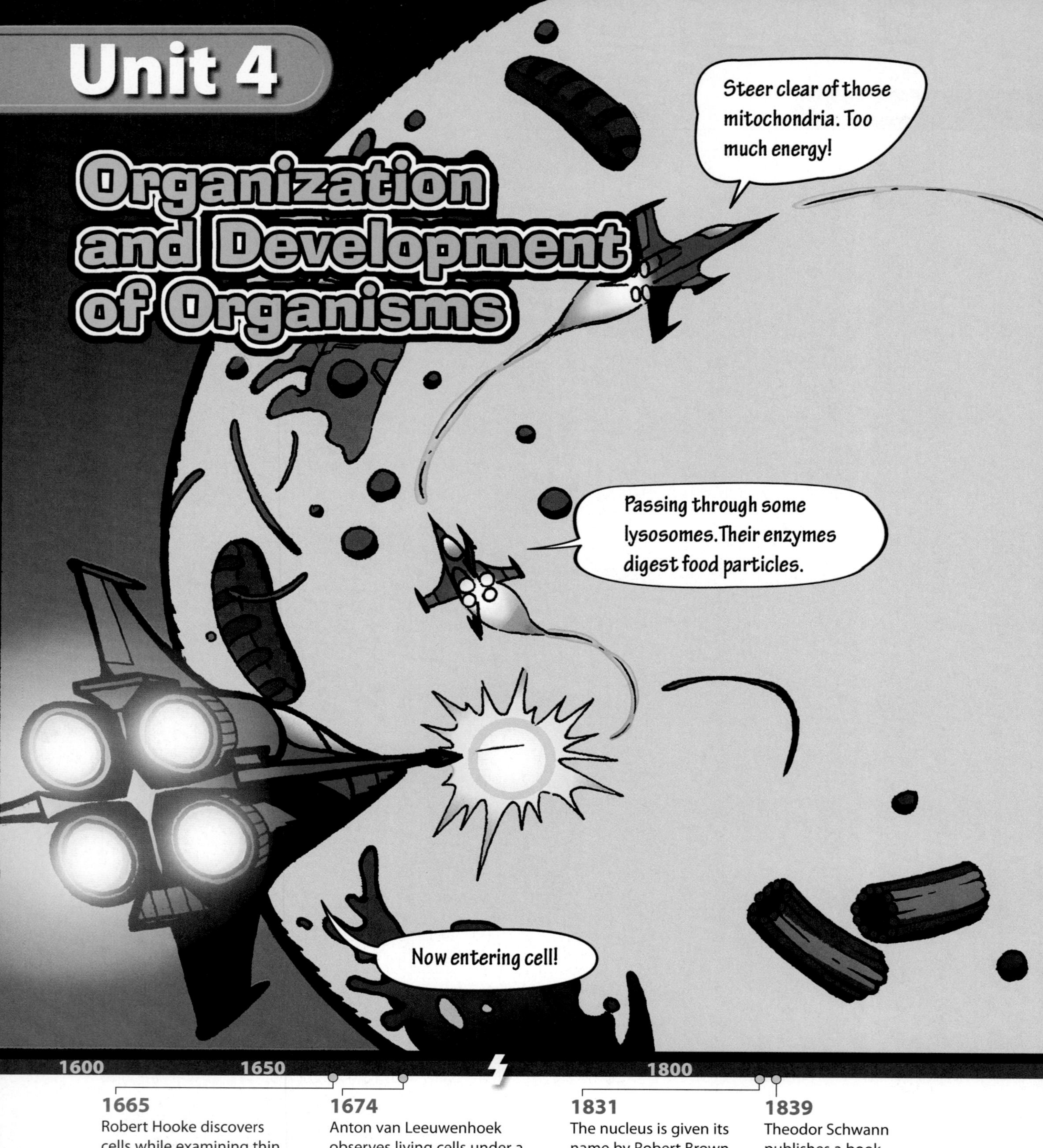

1665
Robert Hooke discovers cells while examining thin slices of cork under a microscope.

1674
Anton van Leeuwenhoek observes living cells under a microscope and names the moving organisms *animalcules.*

1831
The nucleus is given its name by Robert Brown.

1839
Theodor Schwann publishes a book suggesting that the cell is the basic unit of life.

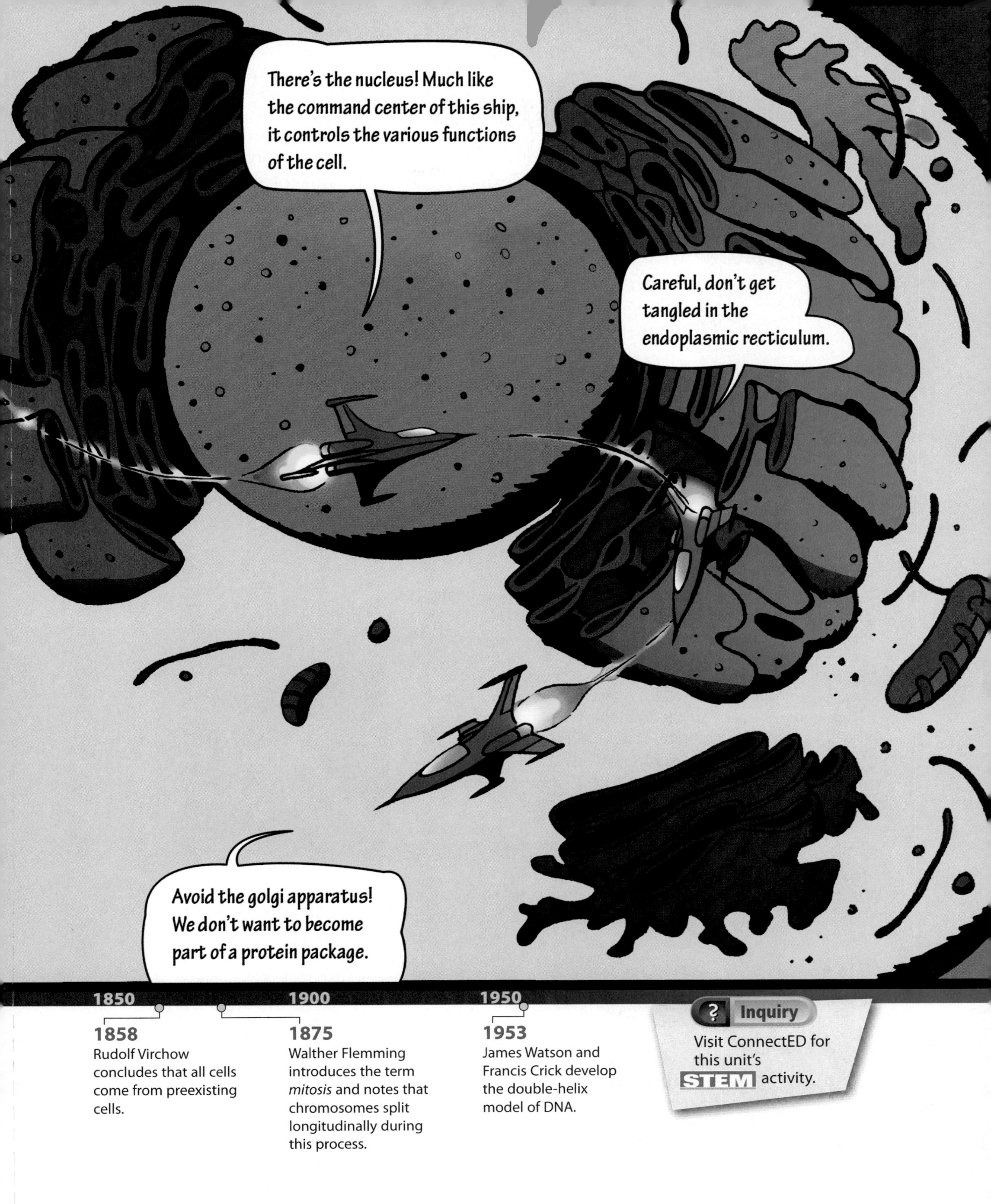

1850 · 1900 · 1950

1858
Rudolf Virchow concludes that all cells come from preexisting cells.

1875
Walther Flemming introduces the term *mitosis* and notes that chromosomes split longitudinally during this process.

1953
James Watson and Francis Crick develop the double-helix model of DNA.

Inquiry
Visit ConnectED for this unit's **STEM** activity.

Unit 4

Nature of SCIENCE

Technology

Some people wonder why governments have invested so much money to explore space. Why not solve problems here on Earth using that money? What did all of that money buy?

Technology is the practical application of science to commerce or industry. Once scientists understand a scientific concept, they can apply the science to new technologies. Many technologies originally developed for space are solving problems for people worldwide. For example, sensors designed to measure the temperature of distant stars led to the development of the thermometer shown in **Figure 1.** When pointed toward the ear canal, the thermometer provides an accurate temperature reading. **Figure 2** shows other technologies developed for space that are now used on Earth.

Figure 1 Space technology developed to determine the temperature of stars now enables a parent to easily and quickly determine whether a child has a fever.

Active Reading **1. Summarize** Restate in your own words the definition of technology.

Figure 2 Medical professionals use technologies originally developed for space to help improve the health of patients on Earth.

To monitor the health of astronauts during space walks, space suits contain tiny sensors that measure astronauts' temperature, respiration, and cardiac activity. The technology that led to the development of these sensors is now used on Earth to monitor people's health. ▶

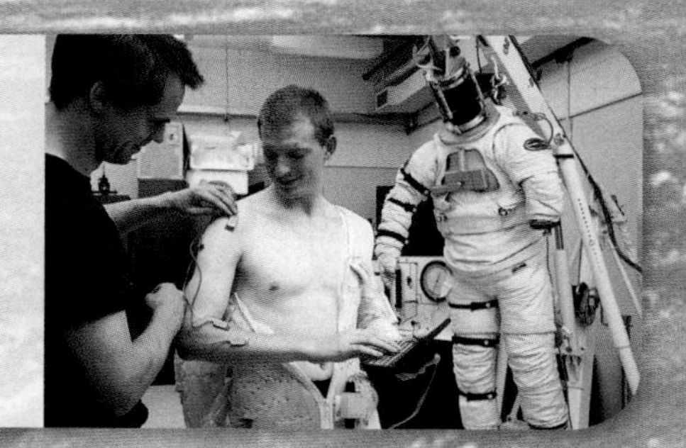

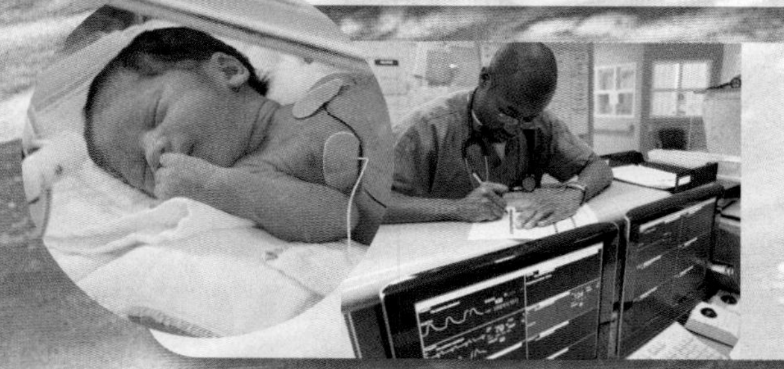

◀ Hospitals can monitor patients from a central nurses' station using electronics similar to those used in space suits. Sensors in infants' clothes can monitor a baby's breathing while the baby sleeps.

Doctors use sensors developed to study the effect of weightlessness on the muscles of astronauts to monitor repeated muscle movements that can lead to carpal tunnel syndrome of patients on Earth. ▶

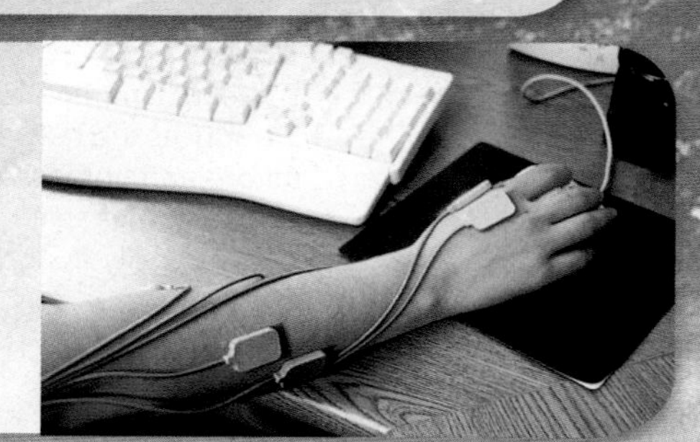

Space technology improves health worldwide.

People contribute to and benefit from space technologies. For example, a Canadian company produced robotic arms for the space shuttle. They then developed instruments that enable surgeons to perform microscopic surgery. Chinese scientists developed a high-resolution X-ray imaging system for spacecraft that is safer, faster, and more accurate than previous X-rays. Now doctors can use the system to more accurately diagnose diseases on Earth. A Spanish company developed a navigation system for the blind based on space navigation technology.

Science and technology working together, help medical professionals keep people healthy and improve the quality of life on Earth.

Active Reading **2. Analyze** Infer how space technology has advanced communications between businesses around the world.

MiniLab ***What can you invent from space technology?*** at connectED.mcgraw-hill.com

Apply It!

After you complete the lab, investigate and answer these questions.

1. **Explore** Choose and research a space technology spinoff used on Earth today.

2. **Construct** Design an advertisement for the space technology product you researched.

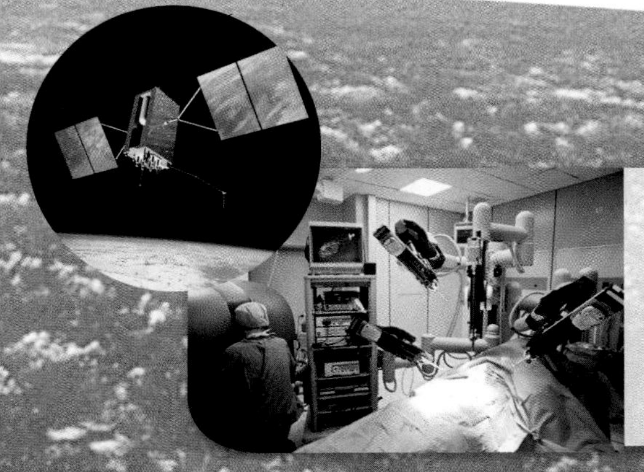

Surgeons use joysticks, similar to those used to control the Lunar Rover, to perform surgery on a patient who is thousands of miles away.

The temperature inside an astronaut's space suit can increase significantly. So NASA developed a technology that circulates a cool fluid through tubes built into the suit.

Scientists applied this technology and developed a therapy for people with multiple sclerosis (MS). MS is a disease that slows the transfer of nerve signals from the brain. MS can affect the ability to think, speak, and control movement. Studies show that a slight decrease in body temperature can restore the transfer of some nerves signals. Therefore, scientists developed cooling suits for patients with MS based on NASA's cooling space suits.

Name ______________________ Date ______________

Notes

Classification Systems

Mrs. Kenner's life science class discussed their ideas about classification systems. The class has different ideas about how classification systems are developed. Here are their ideas:

Group A thinks classification systems are developed naturally by the way organisms are grouped in nature.

Group B thinks classification systems are developed according to scientists' purposes for grouping organisms.

Which group best matches your thinking about classification? __________.
Explain why you agree with that group.

Classifying and Exploring LIFE

The Big Idea

FLORIDA BIG IDEAS

1 The Practice of Science

2 The Characteristics of Scientific Knowledge

14 Organization and Development of Living Organisms

15 Diversity and Evolution of Living Organisms

Think About It!

What are living things, and how can they be classified?

At first glance, you might think someone dropped dinner rolls on a pile of rocks. These objects might look like dinner rolls, but they're not.

1 What do you think the objects are? Do you think they are alive?

2 Why do you think they look like this?

3 What are living things, and how can they be classified?

Get Ready to Read

What do you think about exploring life?

Before you read, decide if you agree or disagree with each of these statements? As you read this chapter, see if you change your mind about any of the statements.

	AGREE	DISAGREE
1 All living things move.	☐	☐
2 The Sun provides energy for almost all organisms on Earth.	☐	☐
3 A dichotomous key can be used to identify an unknown organism.	☐	☐
4 Physical similarities are the only traits used to classify organisms.	☐	☐
5 Most cells are too small to be seen with the unaided eye.	☐	☐
6 Only scientists use microscopes.	☐	☐

There's More Online!
Video • Audio • Review • ⓘLab Station • WebQuest • Assessment • Concepts in Motion • Multilingual eGlossary

Lesson 1

Characteristics OF LIFE

ESSENTIAL QUESTION

What characteristics do all living things share?

Vocabulary

organism p. 331

cell p. 332

unicellular p. 332

multicellular p. 332

homeostasis p. 335

Florida NGSSS

LA.6.2.2.3 The student will organize information to show understanding (e.g., representing main ideas within text through charting, mapping, paraphrasing, summarizing, or comparing/contrasting);

SC.6.L.14.3 Recognize and explore how cells of all organisms undergo similar processes to maintain homeostasis, including extracting energy from food, getting rid of waste, and reproducing.

SC.6.N.1.1 Define a problem from the sixth grade curriculum, use appropriate reference materials to support scientific understanding, plan and carry out scientific investigation of various types, such as systematic observations or experiments, identify variables, collect and organize data, interpret data in charts, tables, and graphics, analyze information, make predictions, and defend conclusions.

SC.6.N.2.1 Distinguish science from other activities involving thought.

SC.6.N.2.1

Inquiry Launch Lab

15 minutes

Is it alive?

Living organisms have specific characteristics. Is a rock a living organism? Is a dog? What characteristics describe something that is living?

Procedure

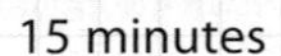

1. Read and complete a lab safety form.
2. Place three pieces of **pasta** in the bottom of a **clear plastic cup.**
3. Add **carbonated water** to the cup until it is 2/3 full.
4. Observe the contents of the cup for 5 minutes. Record your observations below.

Data and Observations

Think About This

1. Think about living things. How do you know they are alive?

2. Which characteristics of life do you think you are observing in the cup?

3. Key Concept Is the pasta alive? How do you know?

What do you think is missing?

1. This toy looks like a dog and can move, but it is a robot. What characteristics are missing that would make it alive?

Characteristics of Life

Look around your classroom and then at **Figure 1.** You might see many nonliving things, such as lights and books. Look again, and you might see many living things, such as your teacher, your classmates, and plants. What makes people and plants different from lights and books?

People and plants, like all living things, have all the characteristics of life. All living things are organized, grow and develop, reproduce, respond, maintain certain internal conditions, and use energy. Nonliving things might have some of these characteristics, but they do not have all of them. Books might be organized into chapters, and lights use energy. However, only those things that have all the characteristics of life are living. *Things that have all the characteristics of life are called* **organisms**.

Active Reading **2. Recall** Underline how living things differ from nonliving things.

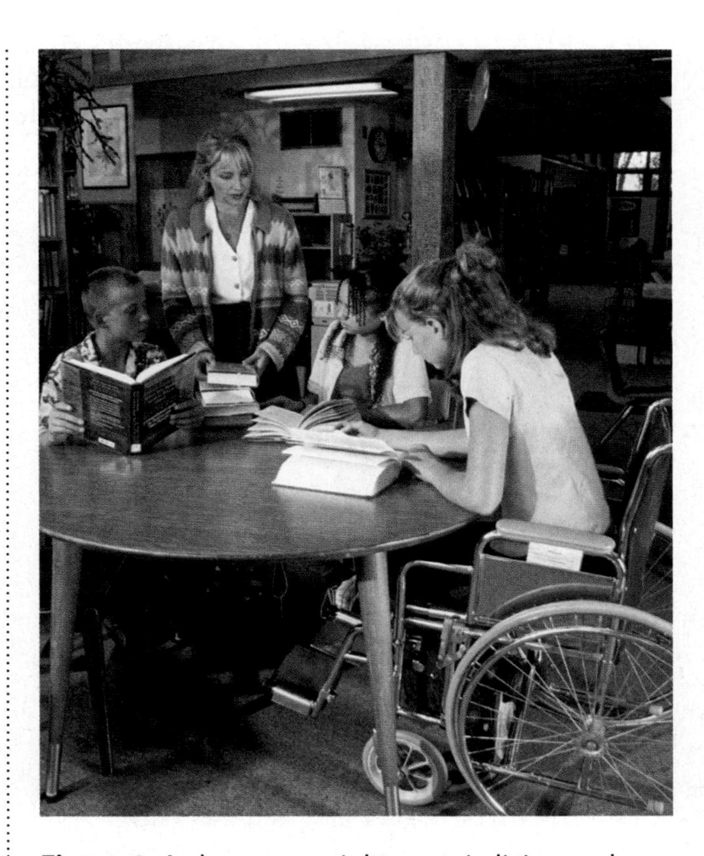

Figure 1 A classroom might contain living and nonliving things.

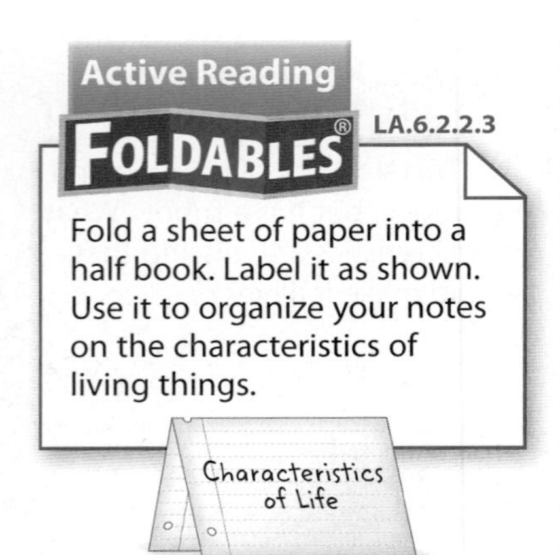

Fold a sheet of paper into a half book. Label it as shown. Use it to organize your notes on the characteristics of living things.

Organization

Your home is probably organized in some way. For example, the kitchen is for cooking, and the bedrooms are for sleeping. Living things are also organized. Whether an organism is made of one **cell**—*the smallest unit of life*—or many cells, all living things have structures that have specific functions.

Living things that are made of only one cell are called **unicellular** *organisms.* Within a unicellular organism are structures with specialized functions just like a house has rooms for different activities. Some structures take in nutrients or control cell activities. Other structures enable the organism to move.

Living things that are made of two or more cells are called **multicellular** *organisms.* Some multicellular organisms only have a few cells, but others have trillions of cells. The different cells of a multicellular organism usually do not perform the same function. Instead, the cells are organized into groups that have specialized functions, such as digestion or movement.

Growth and Development

The tadpole in **Figure 2** is not a frog, but it will soon lose its tail, grow legs, and become an adult frog. This happens because the tadpole, like all organisms, will grow and develop. When organisms grow, they increase in size. A unicellular organism grows as the cell increases in size. Multicellular organisms grow as the number of their cells increases.

Figure 2 A tadpole in Florida grows in size while developing into an adult frog.

Growth and Development

3. Visual Check State What characteristics of life can you identify in this figure?

1 **A frog egg develops into a tadpole.**

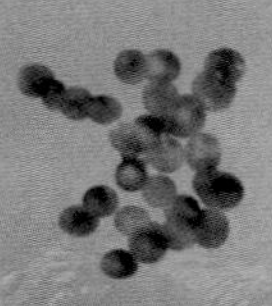

2 **As the tadpole grows, it develops legs.**

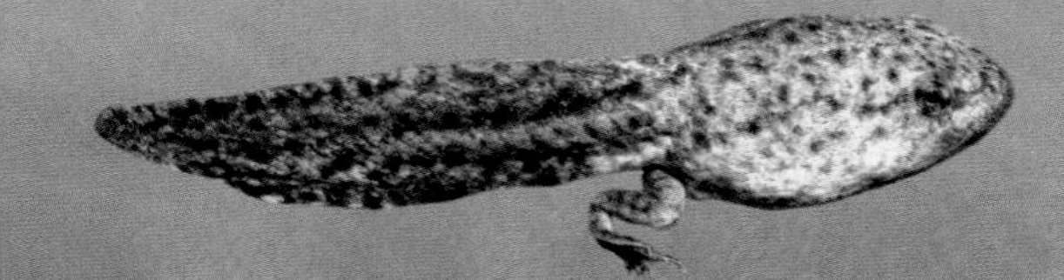

Changes that occur in an organism during its lifetime are called development. In multicellular organisms, development happens as cells become specialized into different cell types, such as skin cells or muscle cells. Some organisms undergo dramatic developmental changes over their lifetime, such as a tadpole developing into a frog.

Active Reading **4. Find** (Circle) what happens in development.

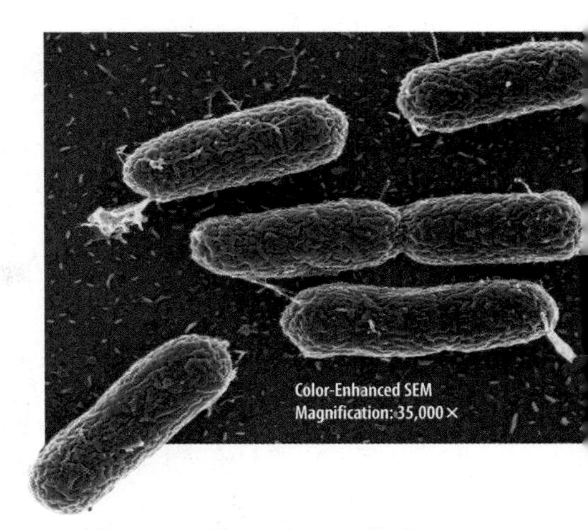

Figure 3 Some unicellular organisms, like the bacteria shown here, reproduce by dividing. The two new organisms are identical to the original organism.

Reproduction

As organisms grow and develop, they usually are able to reproduce. Reproduction is the process by which one organism makes one or more new organisms. In order for living things to continue to exist, organisms must reproduce. Some organisms within a population might not reproduce, but others must reproduce if the species is to survive.

Organisms do not all reproduce in the same way. Some organisms, like the ones in **Figure 3,** can reproduce by dividing and become two new organisms. Other organisms have specialized cells for reproduction. Some organisms must have a mate to reproduce, but others can reproduce without a mate. The number of offspring produced varies. Humans usually produce only one or two offspring at a time. Other organisms, such as the frog in **Figure 2,** can produce hundreds of offspring at one time.

Responses to Stimuli

If someone throws a ball toward you, you might react by trying to catch it. This is because you, like all living things, respond to changes in the environment. These changes can be internal or external and are called stimuli (STIHM yuh li).

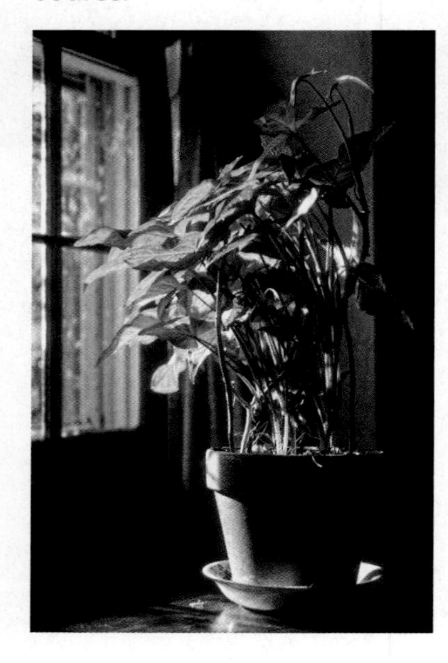

Figure 4 The leaves and stems of plants like this one will grow toward a light source.

Internal Stimuli

You respond to internal stimuli (singular, stimulus) every day. If you feel hungry and then look for food, you are responding to an internal stimulus—the feeling of hunger. The feeling of thirst that causes you to find and drink water is another example of an internal stimulus.

External Stimuli

Changes in an organism's environment that affect the organism are external stimuli. Some examples of external stimuli are light and temperature.

Many plants, like the one in **Figure 4,** will grow toward light. You respond to light, too. Your skin's response to sunlight might be to darken, turn red, or freckle.

Some animals respond to changes in temperature. The response can be more or less blood flowing to the skin. For example, if the temperature increases, the diameter of an animal's blood vessels increases. This allows more blood to flow to the skin, cooling an animal.

Apply It! After you complete the lab, answer these questions.

1. Think of another time you had a reflex. Write it below.

2. How was your reaction similar to the ball toss?

3. How was it different?

Homeostasis

Have you ever noticed that if you drink more water than usual, you have to go to the bathroom more often? That is because your body is working to keep your internal environment under normal conditions. *An organism's ability to maintain steady internal conditions when outside conditions change is called* **homeostasis** (hoh mee oh STAY sus).

WORD ORIGIN

homeostasis
from Greek *homoios,* means "like, similar"; and *stasis,* means "standing still"

The Importance of Homeostasis

Are there certain conditions you need to do your homework? Maybe you need a quiet room with a lot of light. Cells also need certain conditions to function properly. Maintaining certain conditions—homeostasis—ensures that cells can function. If cells cannot function normally, then an organism might become sick or even die.

Methods of Regulation

A person might not survive if his or her body temperature changes more than a few degrees from 37°C. When your outside environment becomes too hot or too cold, your body responds. It sweats, shivers, or changes the flow of blood to maintain a body temperature of 37°C.

Unicellular organisms, such as the paramecium in **Figure 5,** also have ways of regulating homeostasis. A structure called a contractile vacuole (kun TRAK tul • VA kyuh wohl) collects and pumps excess water out of the cell.

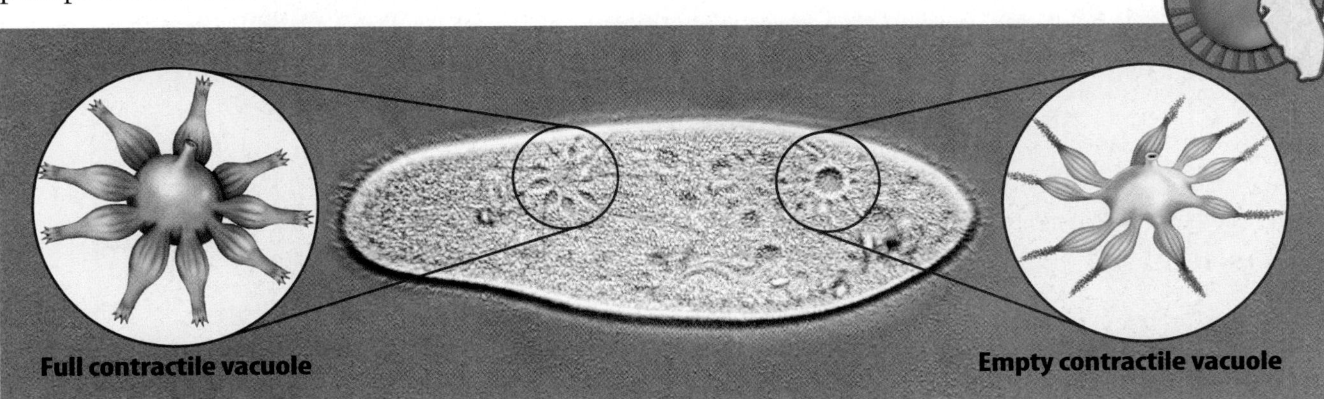

Figure 5 This paramecium lives in freshwater in Florida and many other states. Water continuously enters its cell and collects in contractile vacuoles. The vacuoles contract and expel excess water from the cell. This maintains normal water levels in the cell.

There is a limit to the amount of change that can occur within an organism. For example, you are able to survive only a few hours in water that is below 10°C. No matter what your body does, it cannot maintain steady internal conditions, or homeostasis, under these circumstances. As a result, your cells lose their ability to function.

5. NGSSS Check Infer Underline why maintaining homeostasis is important to organisms. SC.6.L.14.2

Energy

Everything you do requires energy. Digesting your food, sleeping, thinking, reading and all of the characteristics of life shown in **Table 1** on the next page require energy. Cells continuously use energy to transport substances, make new cells, and perform chemical reactions. Where does this energy come from?

For most organisms, this energy originally came to Earth from the Sun, as shown in **Figure 6.** For example, energy in the cactus came from the Sun. The squirrel gets energy by eating the cactus, and the coyote gets energy by eating the squirrel.

Active Reading **6. List** What characteristics do all living things share?

Energy Use

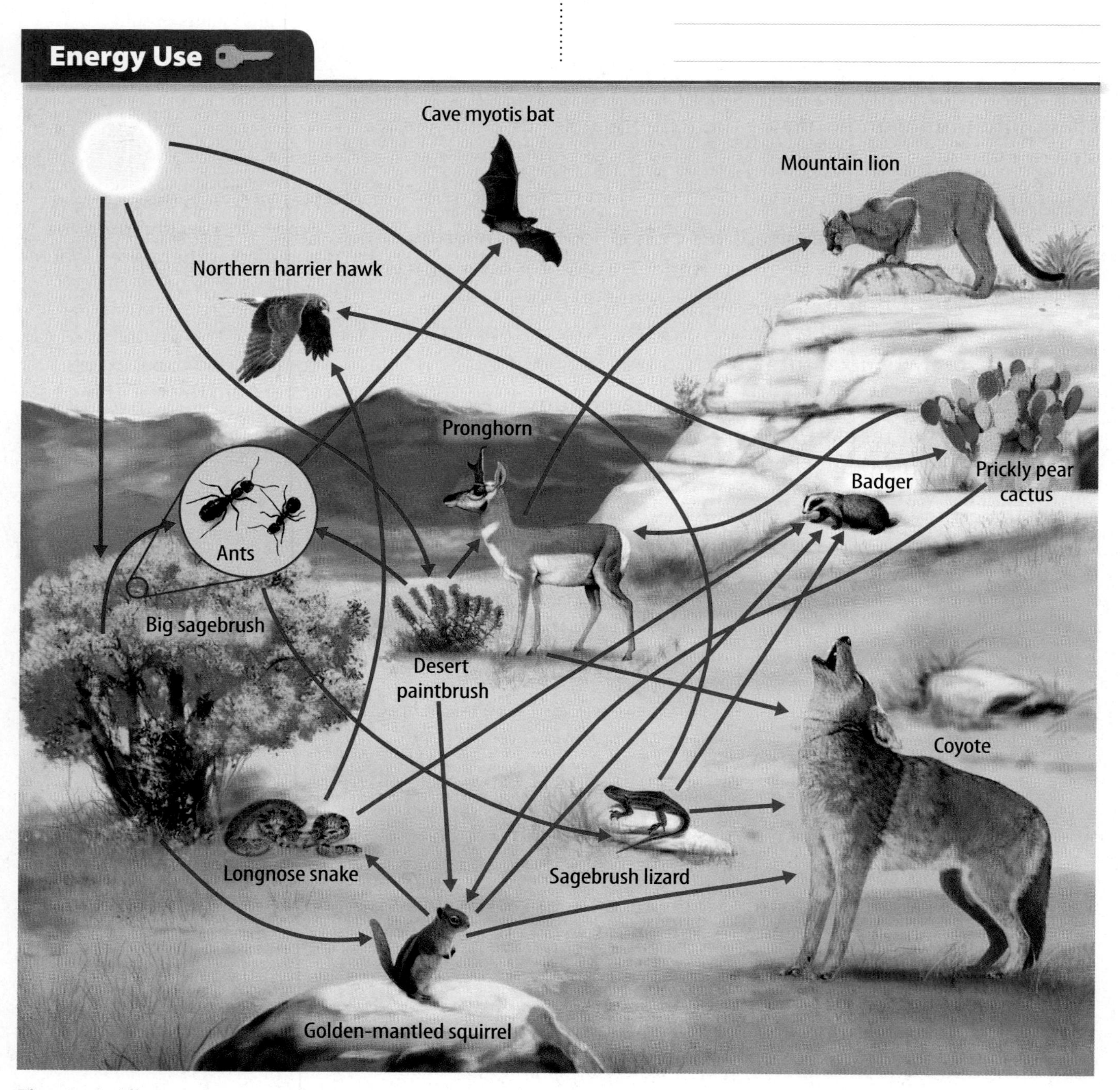

Figure 6 All organisms require energy to survive. In this food web, energy passes from one organism to another and to the environment.

7. Visual Check Identify (Circle) From which food sources does the badger get energy.

Table 1 Characteristics of Life

Characteristic	Definition	Example
Organization	Living things have specialized structures with specialized functions. Living things with more than one cell have a greater level of organization because groups of cells function together.	
Growth and development	Living things grow by increasing cell size and/or increasing cell number. Multicellular organisms develop as cells develop specialized functions.	
Reproduction	Living things make more living things through the process of reproduction.	
Response to stimuli	Living things adjust and respond to changes in their internal and external environments.	
Homeostasis	Living things maintain a stable internal environment.	
Use of energy	Living things use energy for all the processes they perform. Living things get energy by making their own food, eating food, or absorbing food.	

Lesson Review 1

Visual Summary

An organism has all the characteristics of life.

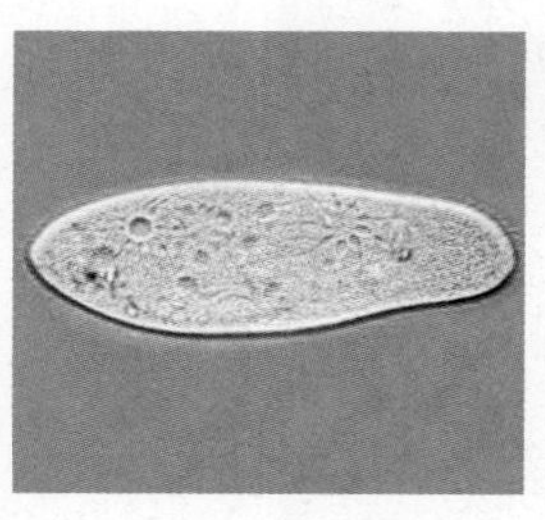

Unicellular organisms have specialized structures, much like a house has rooms for different activities.

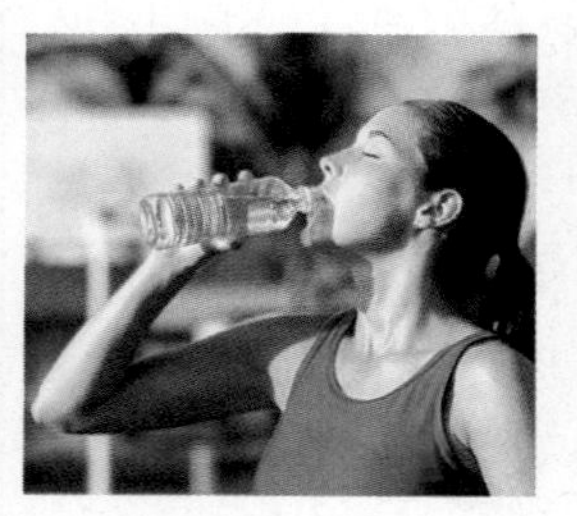

Homeostasis enables living things to maintain a steady internal environment.

Use Vocabulary

1. A(n) ____________ is the smallest unit of life. SC.6.L.14.2

2. **Distinguish** between unicellular and multicellular. SC.6.L.14.2

3. **Define** the term *homeostasis* in your own words. SC.6.L.14.3

Understand Key Concepts

4. Which is NOT a characteristic of all living things?
   (A) breathing
   (B) growing
   (C) reproducing
   (D) using energy

5. **Compare** the processes of reproduction and growth.

6. **Choose** the characteristic of living things that you think is most important. Explain why you chose that characteristic.

7. **Critique** the following statement: A candle flame is a living thing.

Interpret Graphics

8. **Summarize** Fill in the graphic organizer below to summarize the characteristics of living things. LA.6.2.2.3

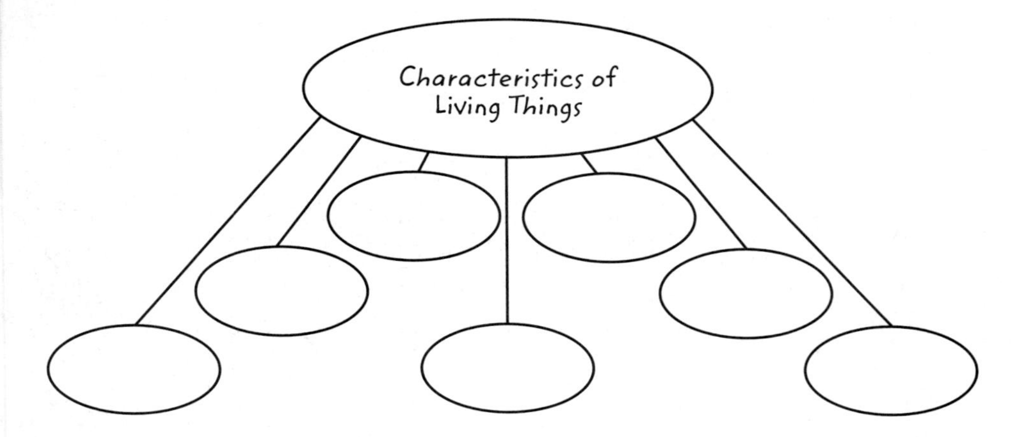

Critical Thinking

9. **Hypothesize** what would happen if living things could not reproduce.

AMERICAN MUSEUM OF NATURAL HISTORY

CAREERS in SCIENCE

The Amazing Adaptation of an Air-Breathing Catfish

Discover how some species of armored catfish breathe air.

Have you ever thought about why animals need oxygen? All animals, including you, get their energy from food. When you breathe, the oxygen you take in is used in your cells. Chemical reactions in your cells use oxygen and change the energy in food molecules into energy that your cells can use. Mammals and many other animals get oxygen from air. Most fish get oxygen from water. Either way, after an animal takes in oxygen, red blood cells carry oxygen to cells throughout its body.

Adriana Aquino is an ichthyologist (IHK thee AH luh jihst) at the American Museum of Natural History in New York City. She discovers and classifies species of fish, such as the armored catfish in family Loricariidae from South America. It lives in freshwater rivers and pools in the Amazon. Its name comes from the bony plates covering its body. Some armored catfish can take in oxygen from water and from air!

Some armored catfish live in fast-flowing rivers. The constant movement of the water evenly distributes oxygen throughout it. The catfish can easily remove oxygen from this oxygen-rich water.

But other armored catfish live in pools of still water, where most oxygen is only at the water's surface. This makes the pools low in oxygen. To maintain a steady level of oxygen in their cells, these fish have adaptations that enable them to take in oxygen directly from air. These catfish can switch from removing oxygen from water through their gills to removing oxygen from air through the walls of their stomachs. They can only do this when they do not have much food in their stomachs. Some species can survive up to 30 hours out of water!

Meet an Ichthyologist

Aquino examines hundreds of catfish specimens. Some she collects in the field, and others come from museum collections. She compares the color, the size, and the shape of the various species. She also examines their internal and external features, such as muscles, gills, and bony plates.

Some armored catfish remove oxygen from air.

It's Your Turn

BRAINSTORM Work with a group. Choose an animal and list five physical characteristics. Brainstorm how these adaptations help the animal be successful in its habitat. Present your findings to the class.

Lesson 2

Classifying ORGANISMS

ESSENTIAL QUESTIONS

What methods are used to classify living things into groups?

Why does every species have a scientific name?

Vocabulary

binomial nomenclature p. 343

species p. 343

genus p. 343

dichotomous key p. 344

cladogram p. 345

Florida NGSSS

LA.6.2.2.3 The student will organize information to show understanding (e.g., representing main ideas within text through charting, mapping, paraphrasing, summarizing, or comparing/contrasting);

SC.6.L.15.1 Analyze and describe how and why organisms are classified according to shared characteristics with emphasis on the Linnaean system combined with the concept of Domains.

SC.6.N.1.1 Define a problem from the sixth grade curriculum, use appropriate reference materials to support scientific understanding, plan and carry out scientific investigation of various types, such as systematic observations or experiments, identify variables, collect and organize data, interpret data in charts, tables, and graphics, analyze information, make predictions, and defend conclusions.

SC.6.N.1.4 Discuss, compare, and negotiate methods used, results obtained, and explanations among groups of students conducting the same investigation.

SC.6.N.2.1 Distinguish science from other activities involving thought.

SC.6.N.2.2 Explain that scientific knowledge is durable because it is open to change as new evidence or interpretations are encountered.

SC.6.N.2.3 Recognize that scientists who make contributions to scientific knowledge come from all kinds of backgrounds and possess varied talents, interests, and goals.

Launch Lab

SC.6.N.1.1

15 minutes

How do you identify similar items?

Do you separate your candies by color before you eat them? When your family does laundry, do you sort the clothes by color first? Identifying characteristics of items can enable you to place them into groups.

Procedure

1. Read and complete a lab safety form
2. Examine twelve **leaves.** Choose a characteristic that you could use to separate the leaves into two groups. Record the characteristic below.
3. Place the leaves into two groups, A and B, using the characteristic you chose in step 2.
4. Choose another characteristic that you could use to further divide group A. Record the characteristic, and divide the leaves.
5. Repeat step 4 with group B.

Data and Observations

Think About This

1. What types of characteristics did other groups in class choose to separate the leaves?

2. Key Concept Why would scientists need rules for separating and identifying items?

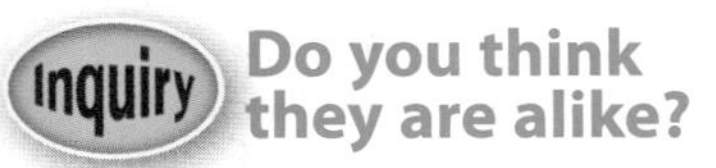

Do you think they are alike?

1. In a band, instruments are organized into groups, such as brass and woodwinds. The instruments in a group are alike in many ways. In a similar way, living things are classified into groups. Why are living things classified?

Classifying Living Things

How would you find your favorite fresh fruit or vegetable in the grocery store? You might look in the produce section, such as the one shown in **Figure 7.** Different kinds of peppers are displayed in one area. Citrus fruits such as oranges, lemons, and grapefruits are stocked in another area. There are many different ways to organize produce in a grocery store. In a similar way, there have been many different ideas about how to organize, or classify, living things.

A Greek philosopher named Aristotle (384 B.C.–322 B.C.) was one of the first people to classify organisms. Aristotle placed all organisms into two large groups, plants and animals. He classified animals based on the presence of "red blood," the animal's environment, and the shape and size of the animal. He classified plants according to the structure and size of the plant and whether the plant was a tree, a shrub, or an herb.

Figure 7 The produce in this store is classified into groups.

2. **Visual Check** **State** What other ways can you think of to classify and organize produce?

Determining Kingdoms

In the 1700s, Carolus Linnaeus, a Swedish physician and botanist, classified organisms based on similar structures. Linnaeus placed all organisms into two main groups, called **kingdoms**. Over the next 200 years, people learned more about organisms and discovered new organisms. In 1969 American biologist Robert H. Whittaker proposed a five-kingdom system for classifying organisms. His system included kingdoms Monera, Protista, Plantae, Fungi, and Animalia.

SCIENCE USE V. COMMON USE

kingdom

Science Use a classification category that ranks above phylum and below domain

Common Use a territory ruled by a king or a queen

Determining Domains

The classification system of living things is still changing. The current classification method is called systematics. Systematics uses all the evidence that is known about organisms to classify them. This evidence includes an organism's cell type, its habitat, the way an organism obtains food and energy, structure and function of its features, and the common ancestry of organisms. Systematics also includes molecular analysis—the study of molecules such as DNA within organisms.

Using systematics, scientists identified two distinct groups in Kingdom Monera—Bacteria and Archaea (ar KEE uh). This led to the development of another level of classification called domains. All organisms are now classified into one of three domains—Bacteria, Archaea, or Eukarya (yew KER ee uh)—and then into one of six kingdoms, as shown in **Table 2.**

3. NGSSS Check Select Underline evidence used to classify living things into groups. SC.6.L.15.1

Table 2 Domains and Kingdoms

Domain	Bacteria	Archaea	Eukarya			
Kingdom	**Bacteria**	**Archaea**	**Protista**	**Fungi**	**Plantae**	**Animalia**
Example						
Characteristics	Bacteria are simple unicellular organisms.	Archaea are simple unicellular organisms that often live in extreme environments.	Protists are unicellular and are more complex than bacteria or archaea.	Fungi are unicellular or multicellular and absorb food.	Plants are multicellular and make their own food.	Animals are multicellular and take in their food.

Scientific Names

Suppose you did not have a name. What would people call you? All organisms, just like people, have names. When Linnaeus grouped organisms into kingdoms, he also developed a system for naming organisms. This naming system, called binomial nomenclature (bi NOH mee ul · NOH mun klay chur), is the system we still use today.

Binomial Nomenclature

Linneaus's naming system, **binomial nomenclature**, *gives each organism a two-word scientific name,* such as *Ursus arctos* for a brown bear. This two-word scientific name is the name of an organism's species (SPEE sheez). *A* **species** *is a group of organisms that have similar traits and are able to produce fertile offspring.* In binomial nomenclature, the first word is the organism's genus (JEE nus) name, such as *Ursus. A* **genus** *is a group of similar species.* The second word might describe the organism's appearance or its behavior.

How do species and genus relate to kingdoms and domains? Similar species are grouped into one genus (plural, genera). Similar genera are grouped into families, then orders, classes, phyla, kingdoms, and finally domains, as shown for the grizzly bear in **Table 3.**

WORD ORIGIN
genus
from Greek *genos,* means "race, kind"

Table 3 The classification of the brown bear or grizzly bear shows that it belongs to the order Carnivora.

Table 3 Classification of the Brown Bear

Taxonomic Group	Number of Species	Examples
Domain Eukarya	About 4–10 million	
Kingdom Animalia	About 2 million	
Phylum Chordata	About 50,000	
Class Mammalia	About 5,000	
Order Carnivora	About 270	
Family Ursidae	8	
Genus *Ursus*	4	
Species *Ursus arctos*	1	

4. Visual Check Identify What domain does the brown bear belong to?

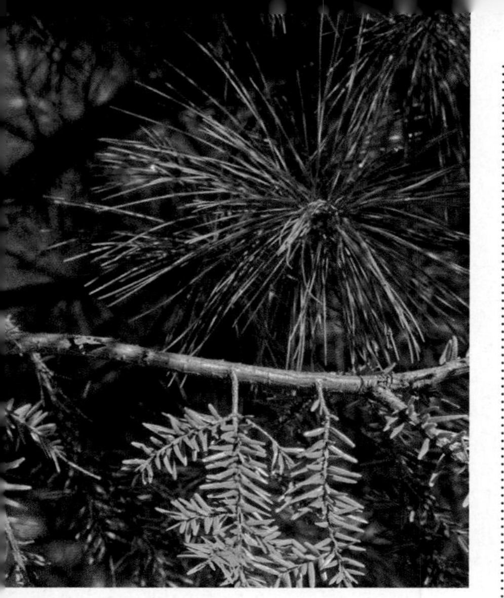

Figure 8 These trees are two different species. *Pinus alba* has long needles, and *Tsuga canadensis* has short needles.

Uses of Scientific Names

When you talk about organisms, you might use names such as bird, tree, or mushroom. However, these are common names for a number of different species. Sometimes there are several common names for one organism. The animal in **Table 3** on the previous page might be called a brown bear or a grizzly bear, but it has only one scientific name, *Ursus arctos*.

Other times, a common name might refer to several different types of organisms. For example, you might call both of the trees in **Figure 8** pine trees. But these trees are two different species. How can you tell? Scientific names are important for many reasons. Each species has its own scientific name. Scientific names are the same worldwide. This makes communication about organisms more effective because everyone uses the same name for the same species.

5. NGSSS Check Explain Why does every species have a scientific name? SC.6.L.15.1

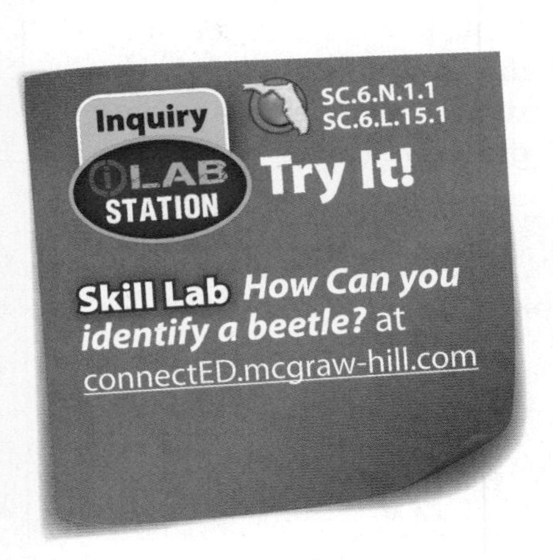

Classification Tools

Suppose you go fishing and catch a fish you don't recognize. How could you figure out what type of fish you have caught? There are several tools you can use to identify organisms.

Dichotomous Keys

A **dichotomous key** *is a series of descriptions arranged in pairs that lead the user to the identification of an unknown organism.* The chosen description leads to either another pair of statements or the identification of the organism. Choices continue until the organism is identified. The dichotomous key shown in **Figure 9** identifies several species of fish.

Dichotomous Key

1. a. This fish has a mouth that extends past its eye. It is an arrow goby.
1. b. This fish does not have a mouth that extends past its eye. Go to step 2.
2. a. This fish has a dark body with stripes. It is a chameleon goby.
2. b. This fish has a light body with no stripes. Go to step 3.
3. a. This fish has a black-tipped dorsal fin. It is a bay goby.
3. b. This fish has a speckled dorsal fin. It is a yellowfin goby.

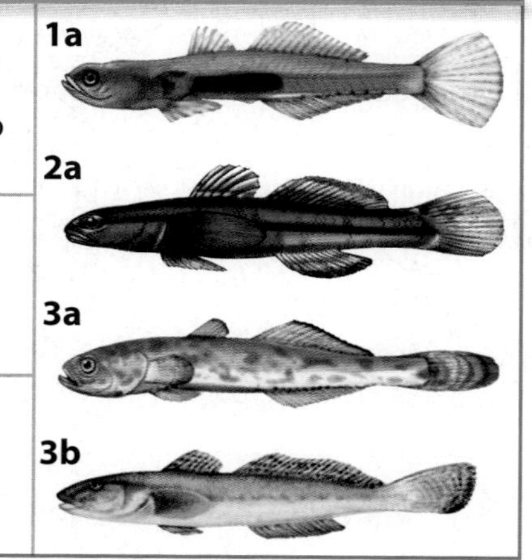

Figure 9 Dichotomous keys include a series of questions to identify organisms.

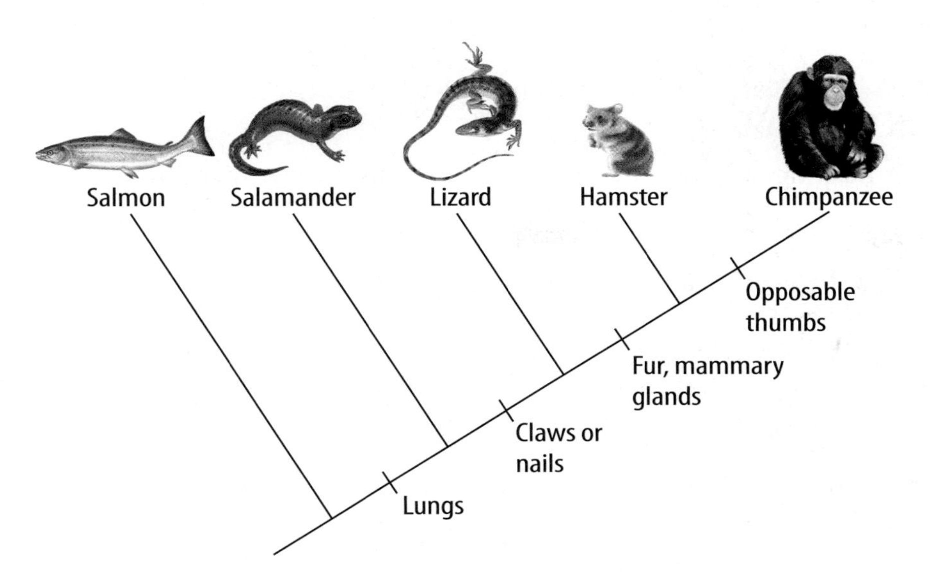

Figure 10 A cladogram shows relationships among species. In this cladogram, salamanders and lizards are more closely related to each other than they are to salmon.

Cladograms

A family tree shows the relationships among family members, including common ancestors. Biologists use a similar diagram, called a cladogram. *A* **cladogram** *is a branched diagram that shows the relationships among organisms, including common ancestors.* A cladogram, as shown in **Figure 10,** has a series of branches. Notice that each branch follows a new characteristic. Each characteristic is observed in all the species to its right. For example, the salamander, lizard, hamster, and chimpanzee have lungs, but the salmon does not. Therefore, they are more closely related to each other than they are to the salmon.

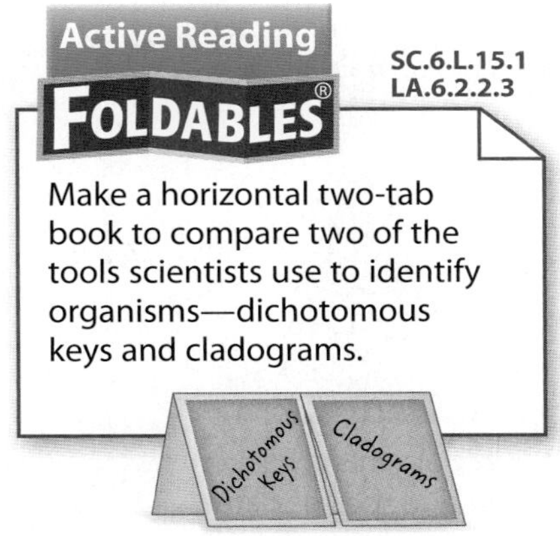

Apply It! After you complete the lab, answer thls question.

1. Use the organisms from the lab and the example cladogram above to explain how to place the organisms from the lab into a cladogram.

Lesson Review 2

Visual Summary

All organisms are classified into one of three domains: Bacteria, Archaea, or Eukarya.

Every organism has a unique species name.

A dichotomous key helps to identify an unknown organism through a series of paired descriptions.

Use Vocabulary

1. A naming system that gives every organism a two-word name is ____________________. SC.6.L.15.1

2. **Use the term** *dichotomous key* in a sentence. SC.6.L.15.1

3. **Organisms** of the same __________ are able to produce fertile offspring.

Understand Key Concepts 

4. **Describe** how you write a scientific name.

5. Which is NOT used to classify organisms? SC.6.L.15.1
   - (A) ancestry
   - (B) habitat
   - (C) age of the organism
   - (D) molecular evidence

Interpret Graphics

6. **Organize Information** Fill in the graphic organizer below to show how organisms are classified. SC.6.L.15.1

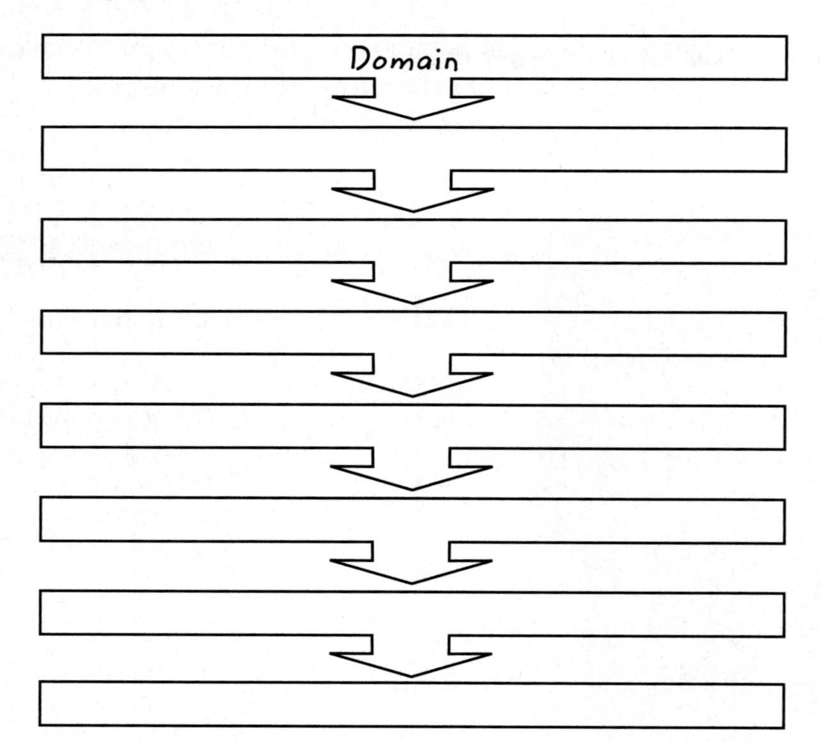

Critical Thinking

7. **Evaluate** the importance of scientific names.

The Name Game

Taxonomy, the science of organism identification and classification, is a discipline in which conclusions change as advances in technology result in new information.

Dr. Karen Steidinger is a taxonomist at the Fish and Wildlife Research Institute in St. Petersburg, Florida. Using improved technology, she revises species descriptions and classifications as new organism characteristics are discovered. Dr. Steidinger works with the toxic *Karenia brevis* shown below. This organism can produce algal blooms with harmful effects on public health and marine resources. Due to these effects, assigning this organism to the accurate taxonomic category is especially important. The genus *Karenia* was established in November 2000 to honor Dr. Steidinger. The species in this genus is the Florida red tide organism, *Karenia brevis*.

K. brevis "Tide Line"

1948: *Gymnodinium brevis* was named by Charles C. Davis from cells collected during a red tide event in Florida in 1946–1947.

1979: Dr. Steidinger transferred the species to the genus *Ptychodiscus* because she discovered it has a resistant cell covering.

1989: Scientists agreed that the original name, *Gymnodinium breve*, would be used until the type species of *Gymnodinium*, *Gyrodinium*, *Balechina*, and *Ptychodiscus* were investigated further to adequately characterize them.

2000: Scientists came to a conclusion about the need for a new genus based on algal cell pigments. The genus *Karenia* was established the following November.

2010: *K. brevis* continues to be classified under the genus *Karenia*. Dr. Steidinger currently investigates Florida Red Tide levels all around the coast each week.

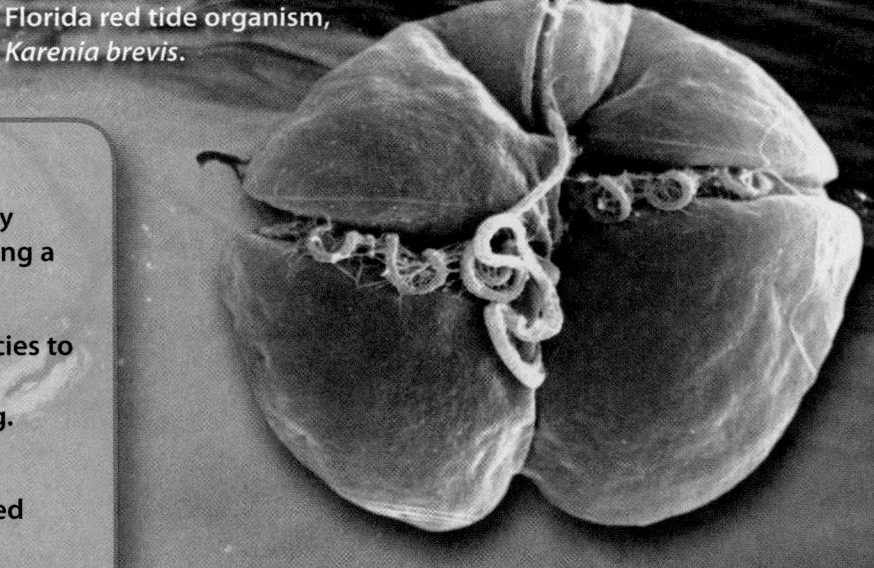

Karenia brevis

It's Your Turn

RESEARCH an organism that has been reclassified based on new scientific evidence. What was the reason for the reclassification? Report your results to the class.

Lesson 3

EXPLORING LIFE

ESSENTIAL QUESTIONS

 How did microscopes change our ideas about living things?

 What are the types of microscopes, and how do they compare?

Vocabulary

light microscope p. 350

compound microscope p. 350

electron microscope p. 351

Florida NGSSS

LA.6.2.2.3 The student will organize information to show understanding (e.g., representing main ideas within text through charting, mapping, paraphrasing, summarizing, or comparing/contrasting);

MA.6.A.3.6 Construct and analyze tables, graphs, and equations to describe linear functions and other simple relations using both common language and algebraic notation.

SC.6.L.15.1 Analyze and describe how and why organisms are classified according to shared characteristics with emphasis on the Linnaean system combined with the concept of Domains.

SC.6.N.1.1 Define a problem from the sixth grade curriculum, use appropriate reference materials to support scientific understanding, plan and carry out scientific investigation of various types, such as systematic observations or experiments, identify variables, collect and organize data, interpret data in charts, tables, and graphics, analyze information, make predictions, and defend conclusions.

SC.6.N.1.5 Recognize that science involves creativity, not just in designing experiments, but also in creating explanations that fit evidence.

SC.6.N.1.5

Launch Lab

15 minutes

Can a water drop make objects appear bigger or smaller?

For centuries, people have been looking for ways to see objects in greater detail. How can something as simple as a drop of water make this possible?

Procedure

1. Read and complete a lab safety form.
2. Lay a sheet of **newspaper** on your desk. Examine a line of text, noting the size and shape of each letter. Record your observations.
3. Add a large drop of **water** to the center of a piece of **clear plastic.** Hold the plastic about 2 cm above the same line of text.
4. Look through the water at the line of text you viewed in step 2. Record your observations.

Data and Observations

Think About This

1. Describe how the newsprint appeared through the drop of water.

2.  Key Concept How might microscopes change your ideas about living things?

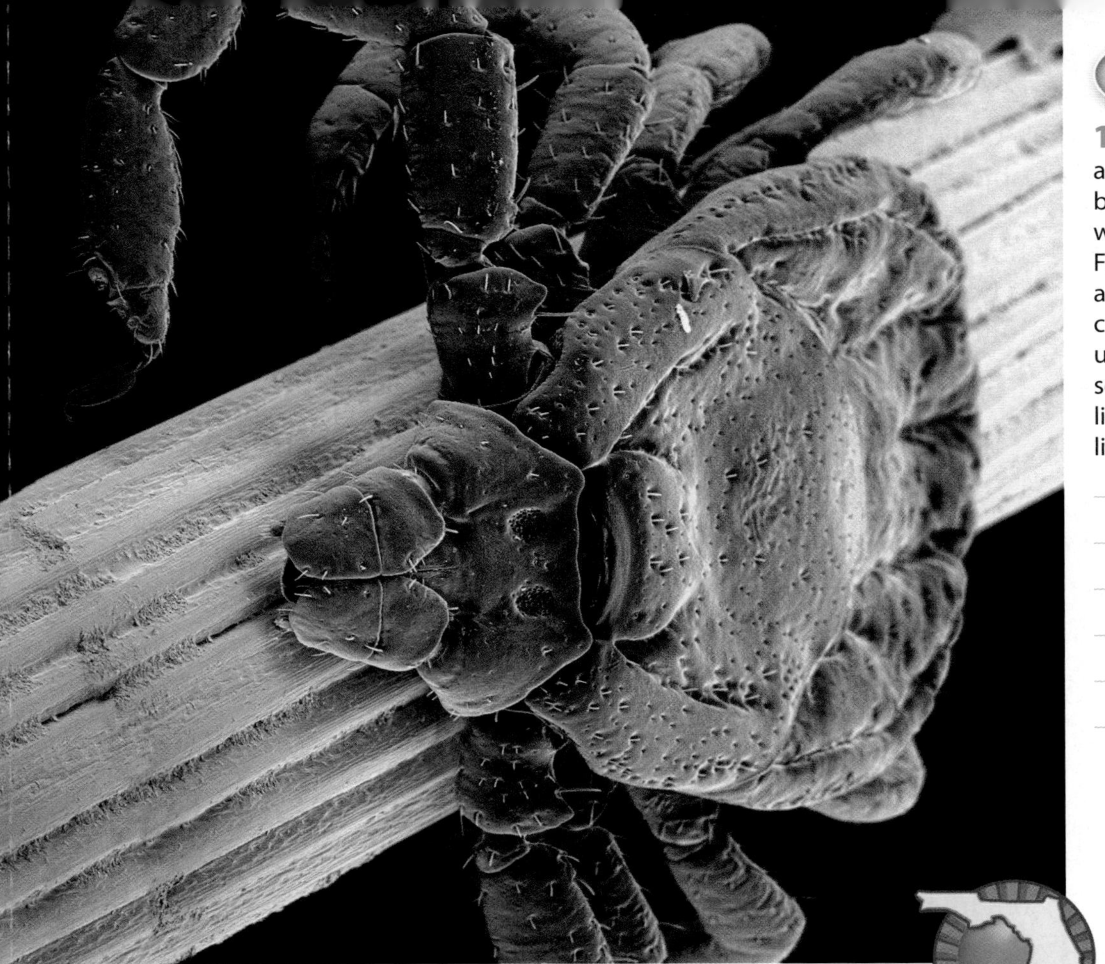

Inquiry Giant Insect?

1. Although this might look like a giant insect, it is a photo of the brown dog tick, common in warmer climate states such as Florida. This photo was taken with a high-powered microscope that can enlarge an image of an object up to 200,000 times. How can seeing an enlarged image of a living thing help you understand life?

The Development of Microscopes

Have you ever used a magnifying lens to see details of an object? If so, then you have used a tool similar to the first microscope. The invention of microscopes enabled people to see details of living things that they could not see with the unaided eye. The microscope also enabled people to make many discoveries about living things.

In the late 1600s the Dutch merchant Anton van Leeuwenhoek (LAY vun hook) made one of the first microscopes. His microscope, similar to the one shown in **Figure 11,** had one lens and could magnify an image about 270 times its original size. Another inventor of microscopes was Robert Hooke. In the early 1700s Hooke made one of the most significant discoveries using a microscope. He observed and named cells. Before microscopes, people did not know that living things are made of cells.

Active Reading **2. Suggest** Underline how microscopes changed our ideas about living things.

Figure 11 Anton van Leeuwenhoek observed pond water and insects using a microscope like the one shown above.

Math Skills MA.6.A.3.6

Use Multiplication

The magnifying power of a lens is expressed by a number and a multiplication symbol (×). For example, a lens that makes an object look ten times larger has a power of 10×. To determine a microscope's magnification, multiply the power of the ocular lens by the power of the objective lens. A microscope with a 10× ocular lens and a 10× objective lens magnifies an object 10 × 10, or 100 times.

Practice

3. What is the magnification of a compound microscope with a 10× ocular lens and a 4× objective lens?

Types of Microscopes

One characteristic of all microscopes is that they magnify objects. Magnification makes an object appear larger than it really is. Another characteristic of microscopes is resolution—how clearly the magnified object can be seen. The two main types of microscopes—light microscopes and electron microscopes—differ in magnification and resolution.

Light Microscopes

If you have used a microscope in school, then you have probably used a light microscope. **Light microscopes** *use light and lenses to enlarge an image of an object.* A simple light microscope has only one lens. *A light microscope that uses more than one lens to magnify an object is called a* **compound microscope.** A compound microscope magnifies an image first by one lens, called the ocular lens. The image is then further magnified by another lens, called the objective lens. The total magnification of the image is equal to the magnifications of the ocular lens and the objective lens multiplied together.

Light microscopes can enlarge images up to 1,500 times their original size. The resolution of a light microscope is about 0.2 micrometers (μm), or two-millionths of a meter. A resolution of 0.2 μm means you can clearly see points on an object that are at least 0.2 μm apart.

Light microscopes can be used to view living or nonliving objects. In some light microscopes, an object is placed directly under the microscope. For other light microscopes, an object must be mounted on a slide. In some cases, the object, such as the white blood cells in **Figure 12,** must be stained with a dye in order to see any details.

Active Reading 4. **Find** Highlight some ways an object can be examined under a light microscope.

Compound Light Microscope

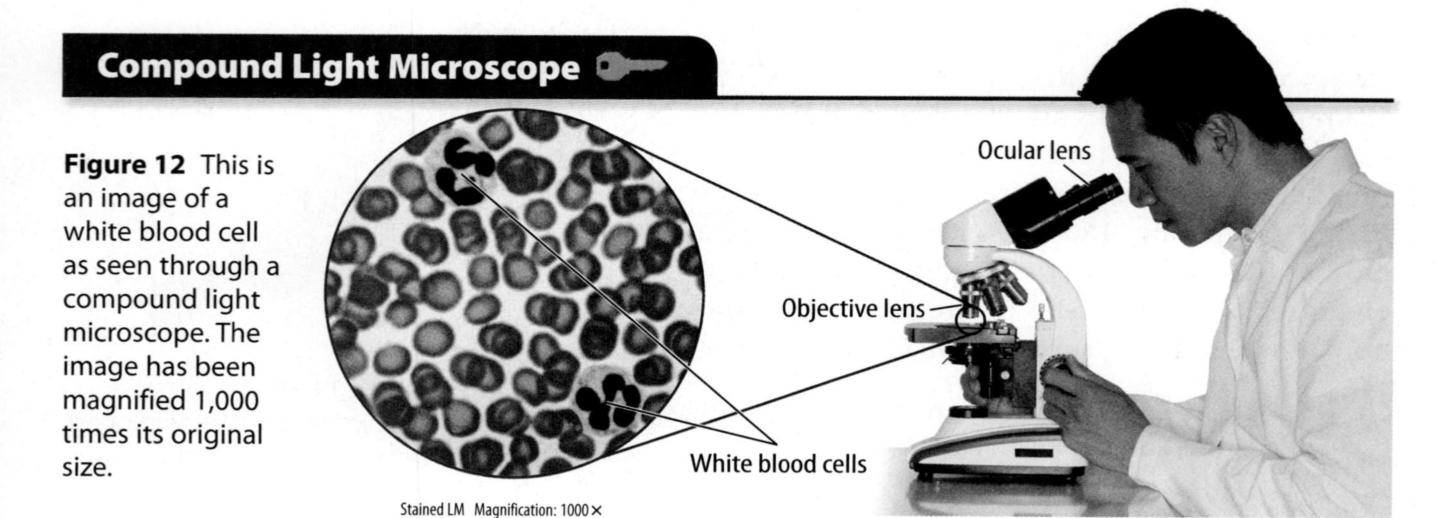

Figure 12 This is an image of a white blood cell as seen through a compound light microscope. The image has been magnified 1,000 times its original size.

Electron Microscopes

You might know that electrons are tiny particles inside atoms. **Electron microscopes** *use a magnetic field to focus a beam of electrons through an object or onto an object's surface.* An electron microscope can magnify an image up to 100,000 times or more. The resolution of an electron microscope can be as small as 0.2 nanometers (nm), or two-billionths of a meter. This resolution is up to 1,000 times greater than a light microscope. The two main types of electron microscopes are transmission electron microscopes (TEMs) and scanning electron microscopes (SEMs).

TEMs are usually used to study extremely small things such as cell structures. Because objects must be mounted in plastic and then very thinly sliced, only dead organisms can be viewed with a TEM. In a TEM, electrons pass through the object and a computer produces an image of the object. A TEM image of a white blood cell is shown in **Figure 13.**

SEMs are usually used to study an object's surface. In an SEM, electrons bounce off the object and a computer produces a three-dimensional image of the object. An image of a white blood cell from an SEM is shown in **Figure 13.** Note the difference in detail in this image compared to the image in **Figure 12** of a white blood cell from a light microscope.

Active Reading **5. Group** Underline the types of microscopes and how they compare.

REVIEW VOCABULARY

atom
the building block of matter that is composed of protons, neutrons, and electrons

Active Reading FOLDABLES® LA.6.2.2.3

Make a two-column folded chart. Label the front *Types of Microscopes,* and label the inside as shown. Use it to organize your notes about microscopes.

Figure 13 A TEM greatly magnifies thin slices of an object. An SEM is used to view a three-dimensional image of an object.

Electron Microscopes

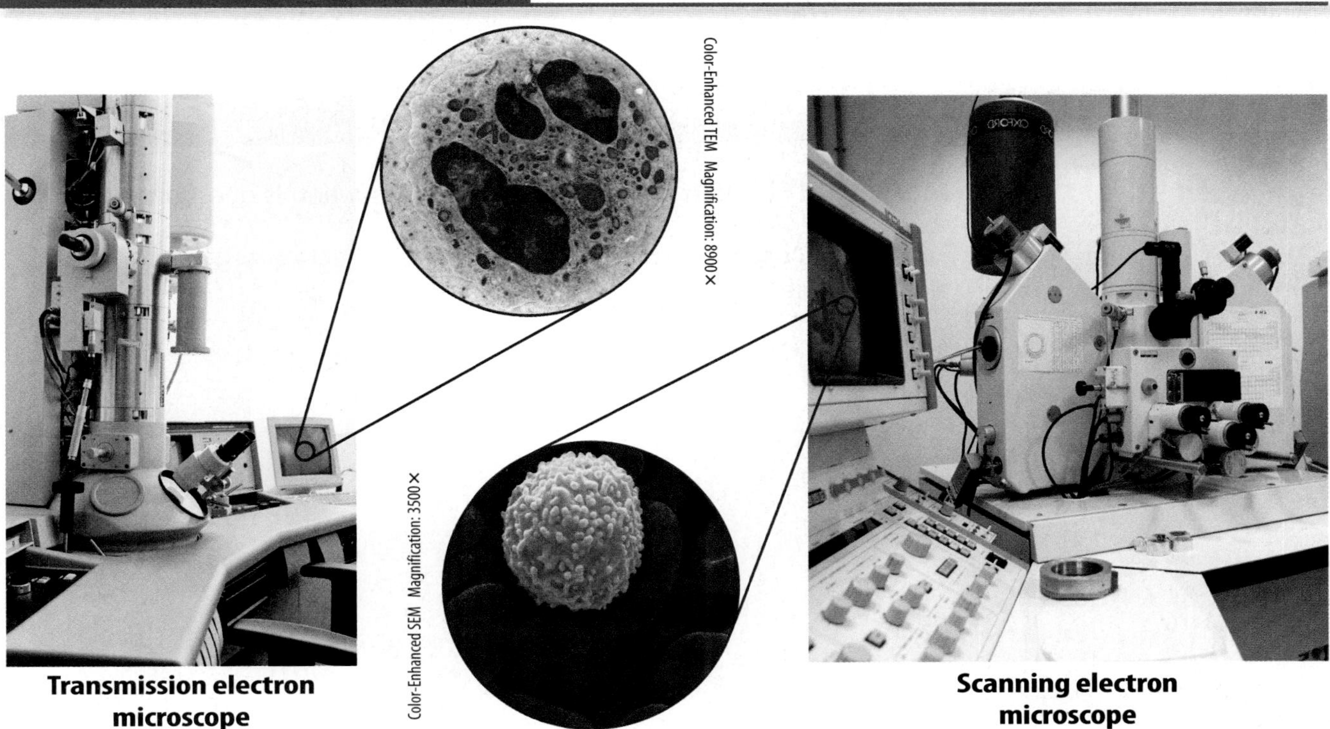

Transmission electron microscope

Scanning electron microscope

Using Microscopes

The **microscopes** used today are more advanced than the microscopes used by Leeuwenhoek and Hooke. The quality of today's light microscopes and the invention of electron microscopes have made the microscope a useful tool in many fields.

Health Care

People in health-care fields, such as doctors and laboratory technicians, often use microscopes. Microscopes are used in surgeries, such as cataract surgery and brain surgery. They enable doctors to view the surgical area in greater detail. The area being viewed under the microscope can also be displayed on a TV monitor so that other people can watch the procedure. Laboratory technicians use microscopes to analyze body fluids, such as blood and urine. They also use microscopes to determine whether tissue samples are healthy or diseased.

ACADEMIC VOCABULARY

identify
(verb) to determine the characteristics of a person or a thing

WORD ORIGIN

microscope
from Latin *microscopium,* means "an instrument for viewing what is small"

Other Uses

Health care is not the only field that uses microscopes. Have you ever wondered how police determine how and where a crime happened? Forensic scientists use microscopes to study evidence from crime scenes. The presence of different insects can help identify when and where a homicide happened. Microscopes might be used to **identify** the type and age of the insects.

People who study fossils might use microscopes. They might examine a fossil and other materials from where the fossil was found.

Some industries also use microscopes. The steel industry uses microscopes to examine steel for impurities. Microscopes are used to study jewels and identify stones. Stones have some markings and impurities that can be seen only by using a microscope.

6. List What are some uses of microscopes?

Apply It! After you complete the lab, answer this question.

1. Have you used a microscope to view other objects or organisms? Record them below.

Lesson Review 3

Visual Summary

Living organisms can be viewed with light microscopes.

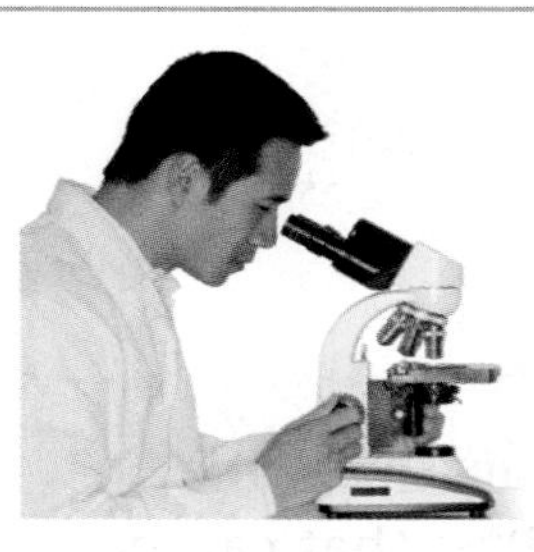

A compound microscope is a type of light microscope that has more than one lens.

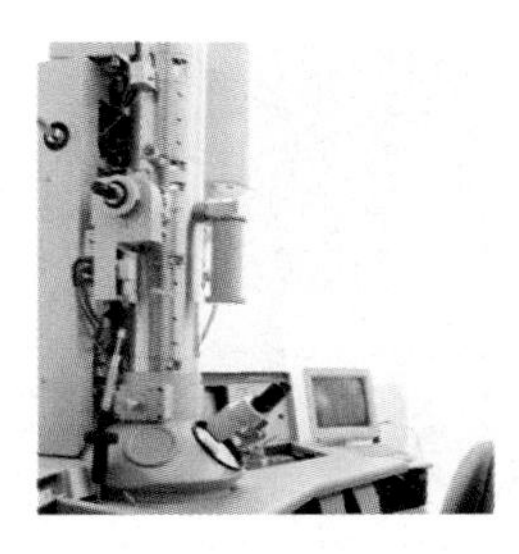

Living organisms cannot be viewed with a transmission electron microscope.

Use Vocabulary

1. **Define** the term *light microscope* in your own words.

2. A(n) ________ focuses a beam of electrons through an object or onto an object's surface.

Understand Key Concepts 

3. **Explain** how the discovery of microscopes has changed what we know about living things.

4. Which microscope would you use if you wanted to study the surface of an object?
   - (A) compound microscope
   - (B) light microscope
   - (C) scanning electron microscope
   - (D) transmission electron microscope

Interpret Graphics

5. **Identify** Fill in the graphic organizer below to identify four uses of microscopes. LA.6.2.2.

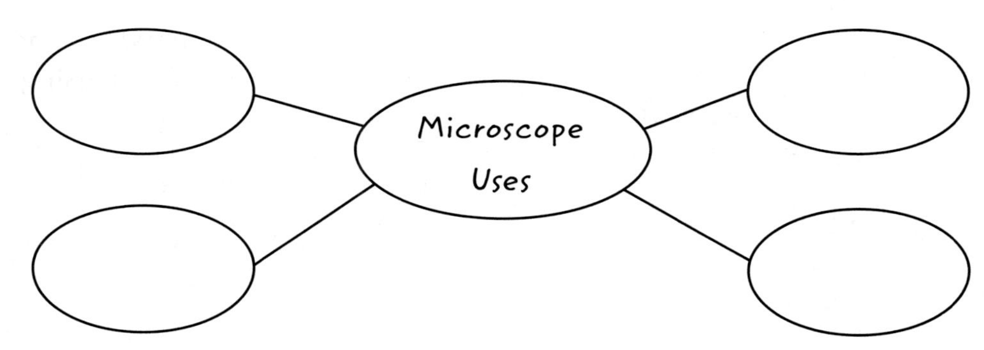

Critical Thinking

6. **Develop** a list of guidelines for choosing a microscope to use.

Math Skills

7. A student observes a blood sample with a compound microscope that has a 10× ocular lens and a 40× objective lens. How much larger do the blood cells appear under the microscope?

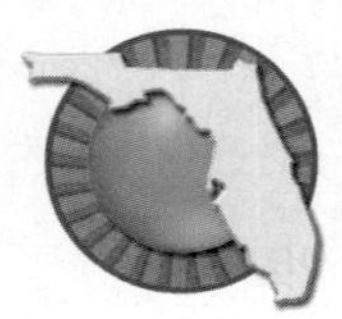

Chapter 9 Study Guide

Think About It! All living things share certain characteristics and can be organized in a functional and structural hierarchy. The invention of the microscope has enabled us to explore life further, which has led to changes in classification.

Key Concepts Summary

LESSON 1 Characteristics of Life

- An **organism** is classified as a living thing because it has all the characteristics of life.
- All living things are organized, grow and develop, reproduce, respond to stimuli, maintain **homeostasis**, and use energy.

Vocabulary

organism p. 331
cell p. 332
unicellular p. 332
multicellular p. 332
homeostasis p. 335

LESSON 2 Classifying Organisms

- Living things are classified into different groups based on physical or molecular similarities.
- Some **species** are known by many different common names. To avoid confusion, every species has a scientific name based on a system called **binomial nomenclature**.

Vocabulary

binomial nomenclature p. 343
species p. 343
genus p. 343
dichotomous key p. 344
cladogram p. 345

LESSON 3 Exploring Life

- The invention of microscopes allowed scientists to view cells, which enabled them to further explore and classify life.
- A **light microscope** uses light and has one or more lenses to enlarge an image up to about 1,500 times its original size. An **electron microscope** uses a magnetic field to direct beams of electrons, and it enlarges an image 100,000 times or more.

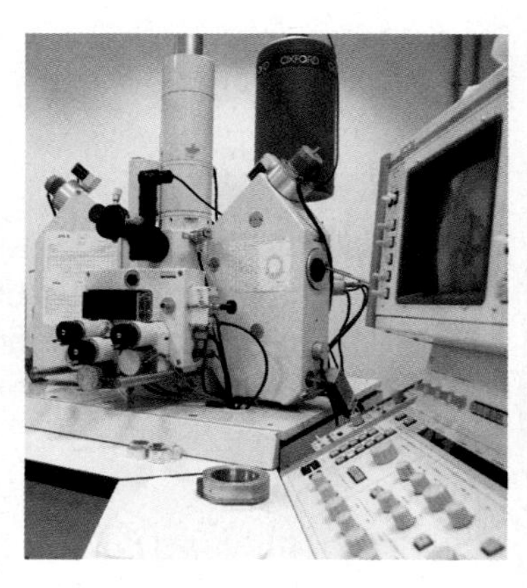

Vocabulary

light microscope p. 350
compound microscope p. 350
electron microscope p. 351

Active Reading

FOLDABLES® Chapter Project

Assemble your lesson Foldables as shown to make a Chapter Project. Use the project to review what you have learned in this chapter.

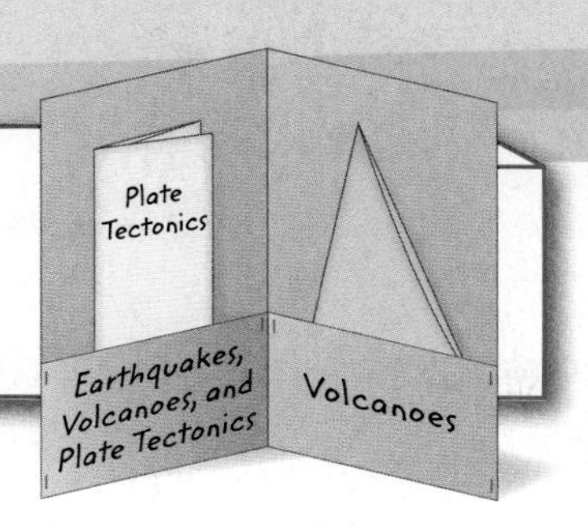

Use Vocabulary

1. A(n) ______________ organism is made of only one cell.

2. Something with all the characteristics of life is a(n) ______________.

3. A(n) ______________ shows the relationships among species.

4. A group of similar species is a(n) ______________.

5. A(n) ______________ has a resolution up to 1,000 times greater than a light microscope.

6. A(n) ______________ is a light microscope that uses more than one lens to magnify an image.

Link Vocabulary and Key Concepts

Use vocabulary terms from the previous page to complete the concept map.

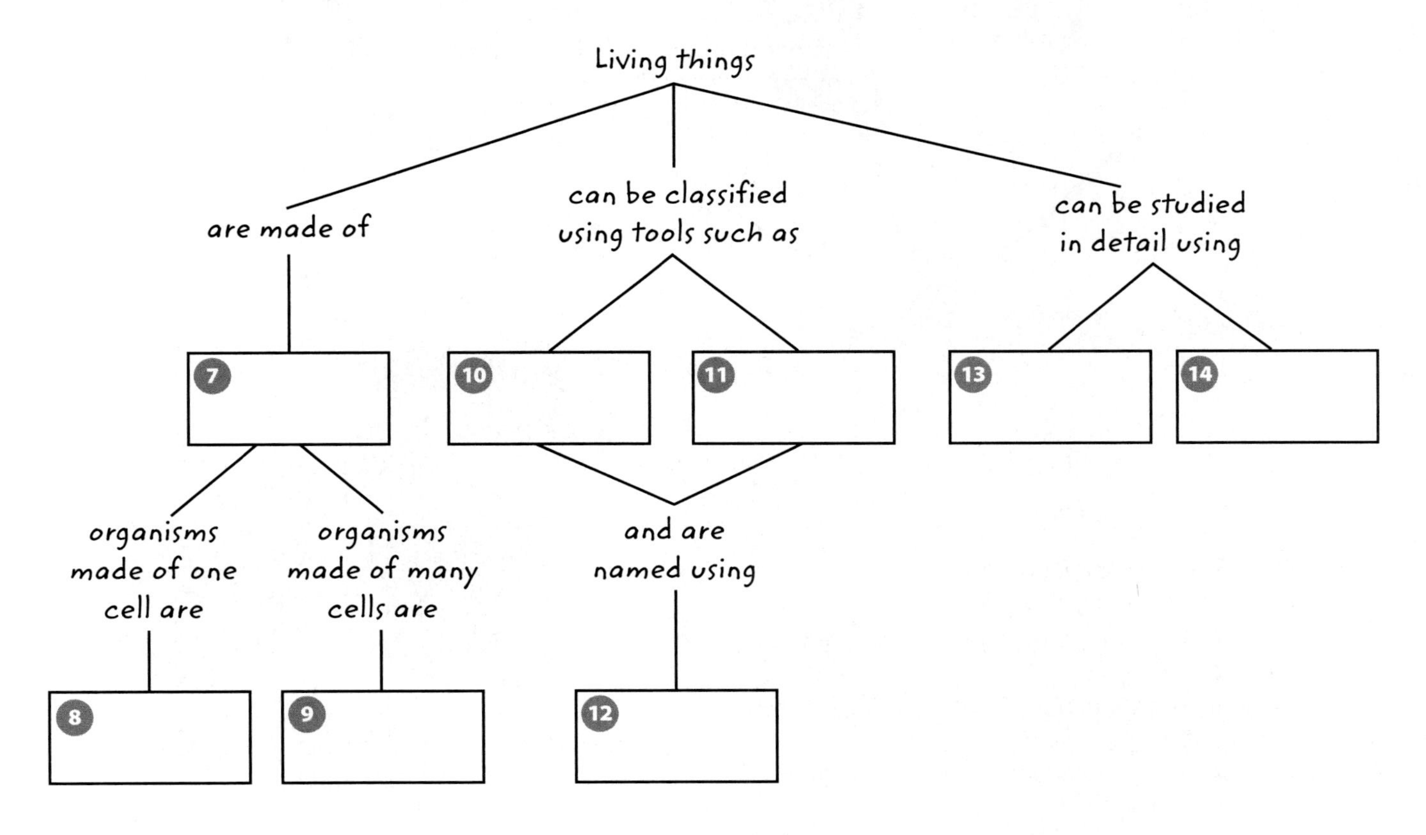

Chapter 9 Review

Fill in the correct answer.

Understand Key Concepts

1. Which is an internal stimulus? **SC.6.L.14.3**
   - Ⓐ an increase in moisture
   - Ⓑ feelings of hunger
   - Ⓒ number of hours of daylight
   - Ⓓ the temperature at night

2. Which is an example of growth and development? **SC.6.L.14.3**
   - Ⓐ a caterpillar becoming a butterfly
   - Ⓑ a chicken laying eggs
   - Ⓒ a dog panting
   - Ⓓ a rabbit eating carrots

3. Based on the food web below, what is an energy source for the mouse? **SC.6.L.14.3**

   - Ⓐ fox
   - Ⓑ grass
   - Ⓒ owl
   - Ⓓ snake

4. Which shows the correct order for the classification of species? **SC.6.L.15.1**
   - Ⓐ domain, kingdom, class, order, phylum, family, genus, species
   - Ⓑ domain, kingdom, phylum, class, order, family, genus, species
   - Ⓒ domain, kingdom, phylum, class, order, family, species, genus
   - Ⓓ domain, kingdom, phylum, order, class, family, genus, species

Critical Thinking

5. **Distinguish** between a unicellular organism and a multicellular organism. **SC.6.L.14.2**

6. **Critique** the following statement: An organism that is made of only one cell does not need organization. **SC.6.L.14.3**

7. **Infer** In the figure below, which plant is responding to a lack of water in its environment? Explain your answer. **SC.6.L.14.3**

8. **Explain** how using a dichotomous key can help you identify an organism. **SC.6.L.15.1**

9. **Describe** how the branches on a cladogram show the relationships among organisms. **SC.6.L.15.1**

10 **Assess** the effect of molecular evidence on the classification of organisms. SC.6.L.15.1

11 **Compare** light microscopes and electron microscopes. LA.6.2.2.3

12 **State** how microscopes have changed the way living things are classified. SC.6.L.15.1

13 **Compare** magnification and resolution. LA.6.2.2.3

14 **Evaluate** the impact microscopes have on our daily lives. SC.6.N.2.1

Writing in Science

15 **Write** a five-sentence paragraph on a separate sheet of paper explaining the importance of scientific names. Be sure to include a topic sentence and a concluding sentence in your paragraph. LA.6.2.2.3

Big Idea Review

16 **Define** the characteristics that all living things share. SC.6.L.14.3

Math Skills MA.6.A.3.6

Use Multiplication

17 A microscope has an ocular lens with a power of 5× and an objective lens with a power of 50×. What is the total magnification of the microscope?

18 A student observes a unicellular organism with a microscope that has a 10× ocular lens and a 100× objective lens. How much larger does the organism look through this microscope?

Florida NGSSS Benchmark Practice

Fill in the correct answer choice.

Multiple Choice

1 What feature of living things do the terms *unicellular* and *multicellular* describe? **SC.6.L.14.2**

Ⓐ how they are organized

Ⓑ how they reproduce

Ⓒ how they maintain temperature

Ⓓ how they produce macromolecules

Use the diagram below to answer question 2.

2 Which characteristic of life does the diagram show? **SC.6.L.14.3**

Ⓕ homeostasis

Ⓖ organization

Ⓗ growth and development

Ⓘ response to stimuli

3 A newly discovered organism is 1 m tall, multicellular, green, and it grows on land and performs photosynthesis. To which kingdom does it most likely belong? **SC.6.L.15.1**

Ⓐ Animalia

Ⓑ Fungi

Ⓒ Plantae

Ⓓ Protista

4 Unicellular organisms are members of which kingdoms? **SC.6.L.15.1**

Ⓕ Animalia, Archaea, Plantae

Ⓖ Archaea, Bacteria, Protista

Ⓗ Bacteria, Fungi, Plantae

Ⓘ Fungi, Plantae, Protista

5 Which are living things made of only one cell? **SC.6.L.14.2**

Ⓐ multicellular organisms

Ⓑ unicellular organisms

Ⓒ tricellular organisms

Ⓓ offspring

Use the diagram below to answer questions 6 and 7.

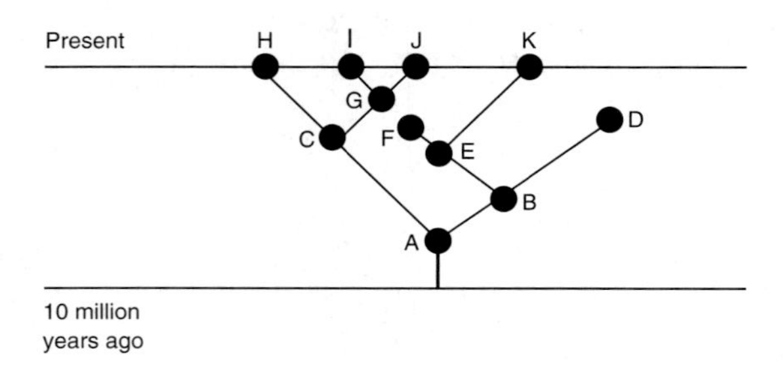

6 What type of diagram is shown in the illustration above? **SC.6.L.15.1**

Ⓕ dichotomous key

Ⓖ cladogram

Ⓗ DNA profile

Ⓘ pedigree

7 In the illustration, I and J are close together. What does this indicate? **SC.6.L.15.1**

Ⓐ They share few characteristics.

Ⓑ They are closest to the common ancestor.

Ⓒ They share many characteristics.

Ⓓ They share the same characteristics as H and K.

Use the diagram below to answer question 8.

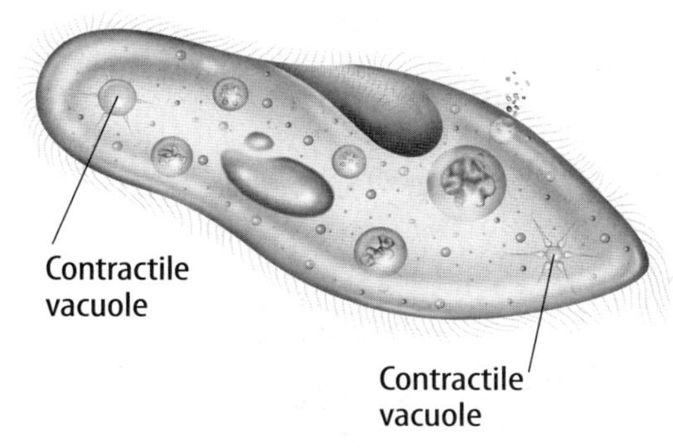

8 Which is the function of the structures in this paramecium? **SC.6.L.14.3**

Ⓕ growth

Ⓖ homeostasis

Ⓗ locomotion

Ⓘ reproduction

9 Which sequence is from the smallest group of organisms to the largest group of organisms? **SC.6.L.15.1**

Ⓐ genus → family → species

Ⓑ genus → species → family

Ⓒ species → family → genus

Ⓓ species → genus → family

10 Which information about organisms is excluded in the study of systematics? **SC.6.L.15.1**

Ⓕ calendar age

Ⓖ molecular analysis

Ⓗ energy source

Ⓘ normal habitat

11 Which of the following characteristics of life is shared by a tree growing a new branch and a duck having babies? **SC.6.L.14.3**

Ⓐ response to stimuli

Ⓑ homeostasis

Ⓒ organization

Ⓓ reproduction

12 Which statement is NOT true? **SC.6.L.15.1**

Ⓕ Binomial names are given to all unknown organisms.

Ⓖ Binomial names are less precise than common names.

Ⓗ Binomial names are different from common names.

Ⓘ Binomial names enable scientists to communicate accurately.

Use the diagram below to answer question 13.

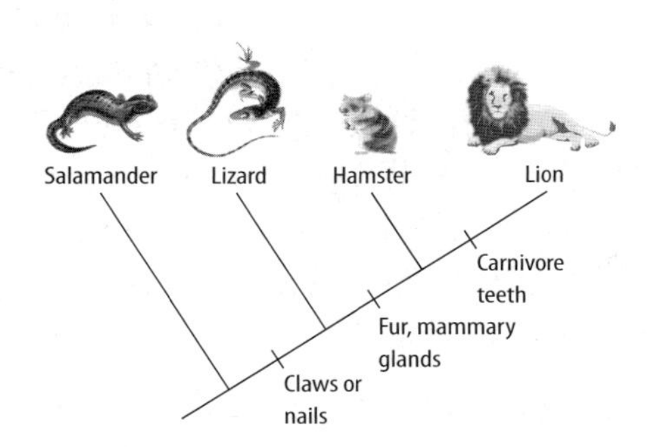

13 According to the diagram above, which of the following all have claws? **SC.6.L.15.1**

Ⓐ lizard, lion, salamander

Ⓑ hamster, salamander, lion

Ⓒ lizard, hamster, salamander

Ⓓ lizard, hamster, lion

Need Extra Help?

If You Missed Question...	1	2	3	4	5	6	7	8	9	10	11	12	13
Go to Lesson...	1	1	2	2	1	2	2	1	2	2	1	2	2

Name ______________________ Date ______________

FLORIDA NGSSS

mini BAT

Multiple Choice *Bubble the correct answer.*

Use the image to answer questions 1–3.

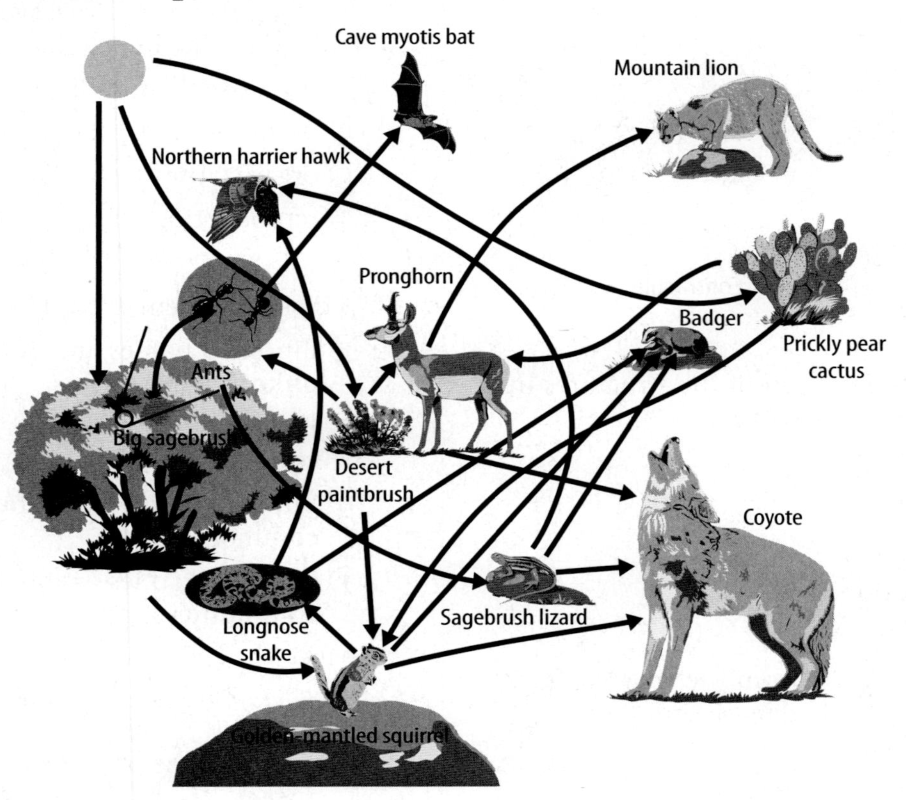

1. Which part of the food web is NOT a living thing? **SC.6.L.14.3**

(A) an ant

(B) a bat

(C) soil bacteria

(D) the Sun

2. Which characteristic of all living things is most obvious in the food web? **SC.6.L.14.3**

(F) All living things reproduce.

(G) All living things use energy.

(H) All living things grow and develop.

(I) All living things have homeostasis.

3. Which is a source of energy for a coyote? **SC.6.L.14.3**

(A) an ant

(B) a cactus

(C) a pronghorn

(D) the Sun

Name ______________________ Date ______________

Multiple Choice *Bubble the correct answer.*

1. The first question on a dichotomous key is: "1A. Are wings covered by an exoskeleton? vs. 1B. Are wings not covered by an exoskeleton?" Which organism is separated from the others by this question? **SC.6.L.15.1**

Ⓐ 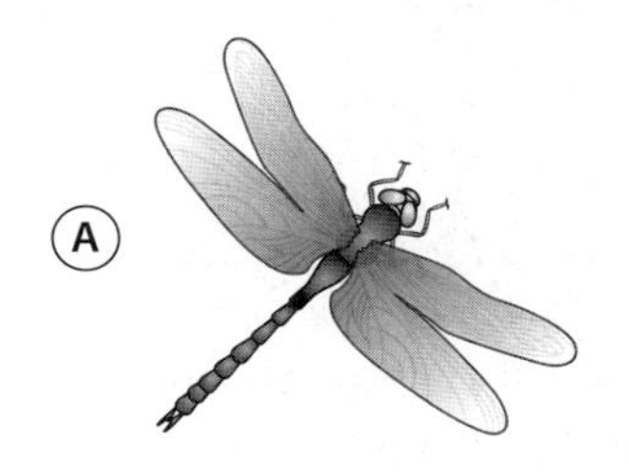

Ⓑ 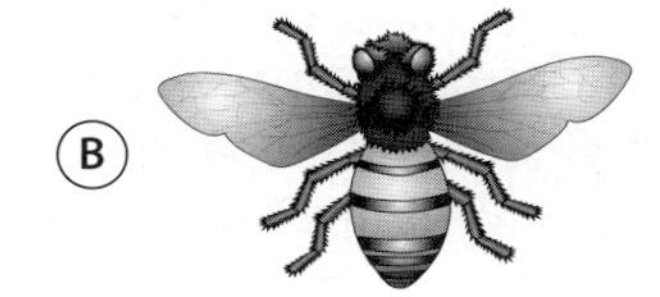

Ⓒ 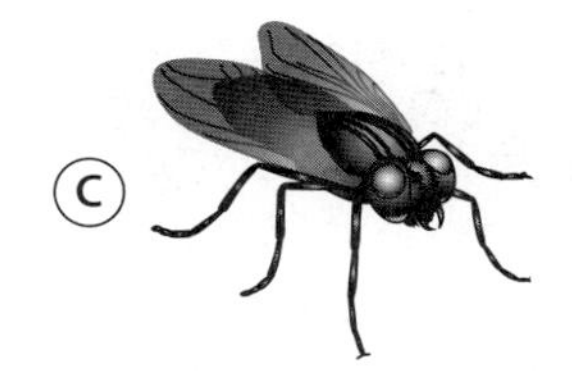

Ⓓ 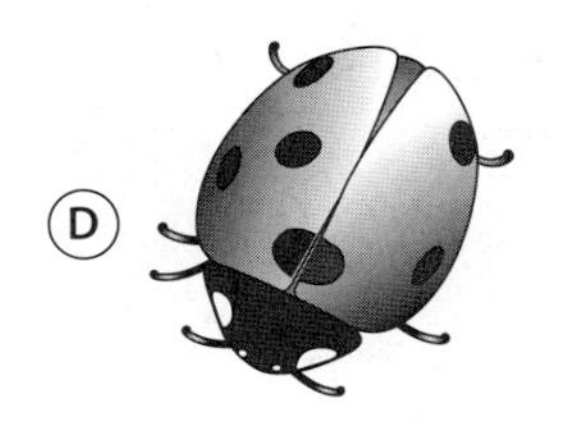

2. In which series is the order of classification correct? **SC.6.L.15.1**

Ⓕ domain, class, kingdom, species

Ⓖ domain, kingdom, class, species

Ⓗ domain, kingdom, species, class

Ⓘ domain, species, kingdom, class

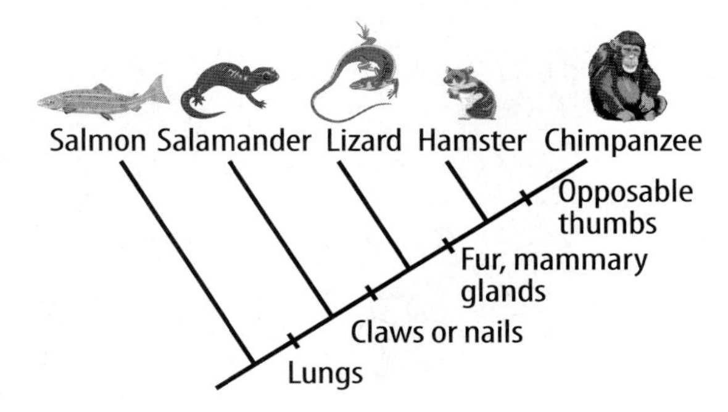

3. In the cladogram above, which pair of species is most closely related to each other? **SC.6.L.15.1**

Ⓐ hamster and salmon

Ⓑ lizard and salamander

Ⓒ salamander and chimpanzee

Ⓓ salmon and lizard

Name ______________________ Date ______________

Benchmark Mini-Assessment Chapter 9 • Lesson 3

Multiple Choice *Bubble the correct answer.*

1. Which tool would be used to examine the workings of a living unicellular organism? **SC.6.L.15.1**

(A) 

(B) 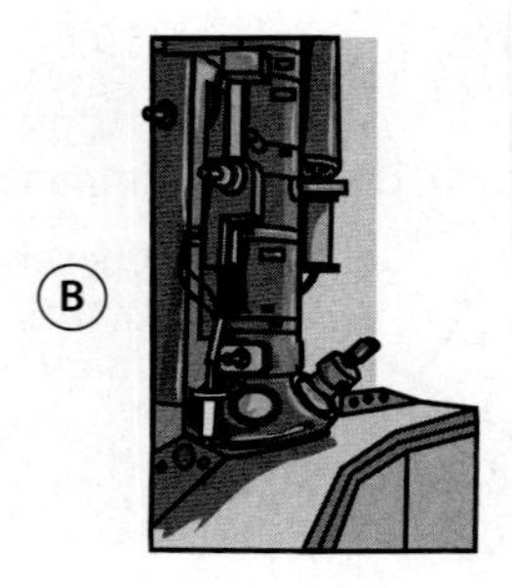

(C) 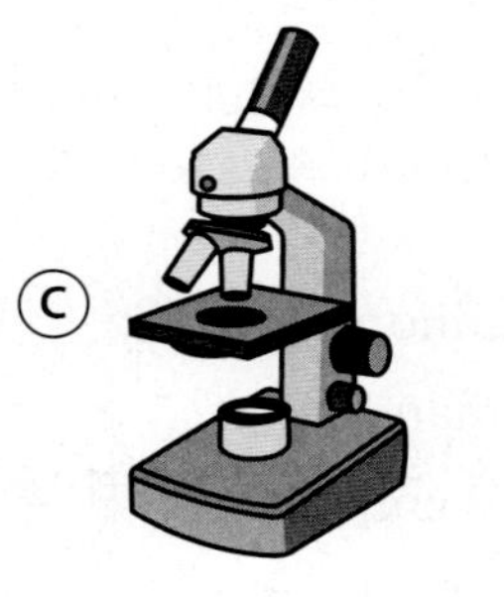

(D) 

2. Which magnifying device provides the most detail in its images? **SC.6.N.1.1**

(F) compound magnifying device

(G) electron magnifying device

(H) hand-lens magnifying device

(I) single-lens magnifying device

Microscope	Magnification	Resolution
Leeuwenhoek's invention	270X	2 micrometers
Modern light microscope	1500X	0.2 micrometers
Electron microscope	100,000+X	0.2 nanometers

3. The table above shows the different levels of magnification provided by different microscopes. About how much larger is the image from an electron microscope than an image from the first microscope invented by Anton van Leeuwenhoek? **MA.6.A.3.6**

(A) about 50 times

(B) about 100 times

(C) about 200 times

(D) about 500 times

Name ______________________ Date ______________

Notes

Name ______________________ Date ______________

Notes

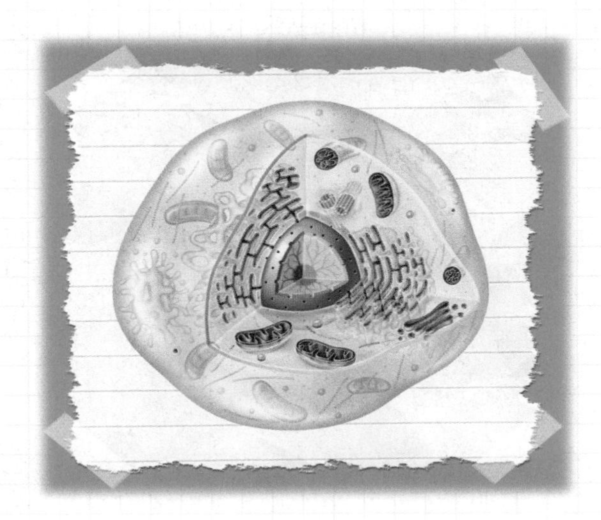

The Basic Unit of Life

The cell is called the basic unit of life. What do you think that means? Circle the answer that best matches your thinking.

A. I think it means the cell is the smallest part of matter.

B. I think it means the cell is the smallest part of mass.

C. I think it means the cell is the smallest part of volume.

D. I think it means the cell is the smallest part of mass and volume.

E. I think it means the cell is the smallest part of energy.

F. I think it means the cell is the smallest part of structure.

G. I think it means the cell is the smallest part of structure and function.

H. I think it means the cell is the smallest part of matter, structure, and function.

I. I think it means the cell is the smallest part of matter, energy, and structure.

Explain your answer. Describe your thinking about the cell as a basic unit of life.

Cell Structure and FUNCTION

FLORIDA BIG IDEAS

1 **The Practice of Science**
2 **The Characteristics of Scientific Knowledge**
14 **Organization and Development of Living Organisms**

Think About It!

How do the structures and processes of a cell enable it to survive?

You might think this unicellular organism looks like something out of a science-fiction movie. Although it looks scary, the hairlike structures in its mouth enable the organism to survive.

1. What do you think the hairlike structures do?

2. How might the shape of the hairlike structures relate to their function?

3. How do you think the structures and processes of a cell enable it to survive?

Get Ready to Read

What do you think about cells?

Before you read, decide if you agree or disagree with each of these statements. As you read this chapter, see if you change your mind about any of the statements.

	AGREE	DISAGREE
1. Nonliving things have cells.	☐	☐
2. Cells are made mostly of water.	☐	☐
3. Different organisms have cells with different structures.	☐	☐
4. All cells store genetic information in their nuclei.	☐	☐
5. Diffusion and osmosis are the same process.	☐	☐
6. Cells with large surface areas can transport more than cells with smaller surface areas.	☐	☐
7. ATP is the only form of energy found in cells.	☐	☐
8. Cellular respiration occurs only in lung cells.	☐	☐

There's More Online!
Video • Audio • Review • ⓘLab Station • WebQuest • Assessment • Concepts in Motion • Multilingual eGlossary

Lesson 1

Cells and LIFE

ESSENTIAL QUESTIONS

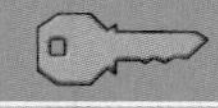 How did scientists' understanding of cells develop?

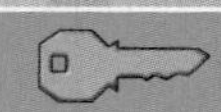 What basic substances make up a cell?

Vocabulary

cell theory p. 370
macromolecule p. 371
nucleic acid p. 372
protein p. 373
lipid p. 373
carbohydrate p. 373

Florida NGSSS

LA.6.2.2.3 The student will organize information to show understanding (e.g., representing main ideas within text through charting, mapping, paraphrasing, summarizing, or comparing/contrasting);

SC.6.L.14.2 Investigate and explain the components of the scientific theory of cells (cell theory): all organisms are composed of cells (single-celled or multi-cellular), all cells come from pre-existing cells, and cells are the basic unit of life.

SC.6.N.1.1 Define a problem from the sixth grade curriculum, use appropriate reference materials to support scientific understanding, plan and carry out scientific investigation of various types, such as systematic observations or experiments, identify variables, collect and organize data, interpret data in charts, tables, and graphics, analyze information, make predictions, and defend conclusions.

SC.6.N.1.4 Discuss, compare, and negotiate methods used, results obtained, and explanations among groups of students conducting the same investigation.

SC.6.N.2.1 Distinguish science from other activities involving thought.

SC.6.N.2.2 Explain that scientific knowledge is durable because it is open to change as new evidence or interpretations are encountered.

SC.6.N.2.3 Recognize that scientists who make contributions to scientific knowledge come from all kinds of backgrounds and possess varied talents, interests, and goals.

SC.6.N.1.1

Inquiry Launch Lab

10 minutes

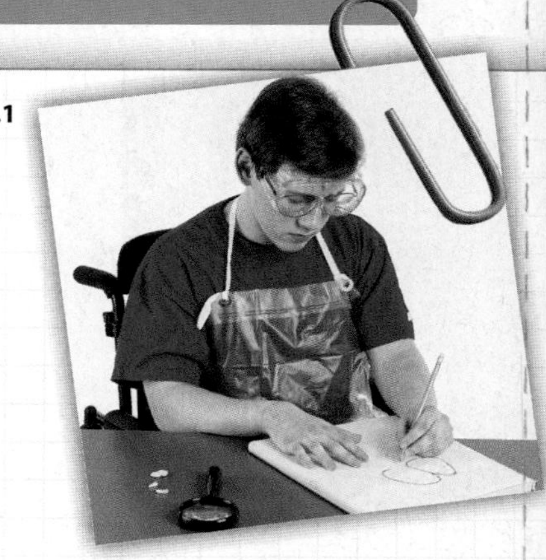

What's in a cell?

Most plants grow from seeds. A seed began as one cell, but a mature plant can be made up of millions of cells. How does a seed change and grow into a mature plant?

Procedure   

1. Read and complete a lab safety form.
2. Use a **toothpick** to gently remove the thin outer covering of a **bean seed** that has soaked overnight.
3. Open the seed with a **plastic knife,** and observe its inside with a **magnifying lens.** Draw the inside of the seed in the space below.
4. Gently remove the small, plantlike embryo, and weigh it on a **balance.** Record its mass below.
5. Gently pull a **bean seedling** from the soil. Rinse the soil from the roots. Weigh the seedling, and record the mass.

Data and Observations

Think About This

1. How did the mass of the embryo and the bean seedling differ?

2. Key Concept If a plant begins as one cell, where do all the cells come from?

Two of a Kind?

1. At first glance, the plant and animal in the photo might seem like they have nothing in common. The plant is rooted in the ground, and the iguana can move quickly. Are they more alike than they appear? How can you find out?

Understanding Cells

Have you ever looked up at the night sky and tried to find other planets in our solar system? It is hard to see them without using a telescope. This is because the other planets are millions of kilometers away. Just like we can use telescopes to see other planets, we can use microscopes to see the basic units of all living things—cells. But people didn't always know about cells. Because cells are so small, early scientists had no tools to study them. It took hundreds of years for scientists to learn about cells.

More than 300 years ago, an English scientist named Robert Hooke built a microscope. He used the microscope to look at cork, which is part of a cork oak tree's bark. What he saw looked like the openings in a honeycomb, as shown in **Figure 1.** The openings reminded him of the small rooms, called cells, where monks lived. He called the structures cells, from the Latin word *cellula* (SEL yuh luh), which means "small rooms."

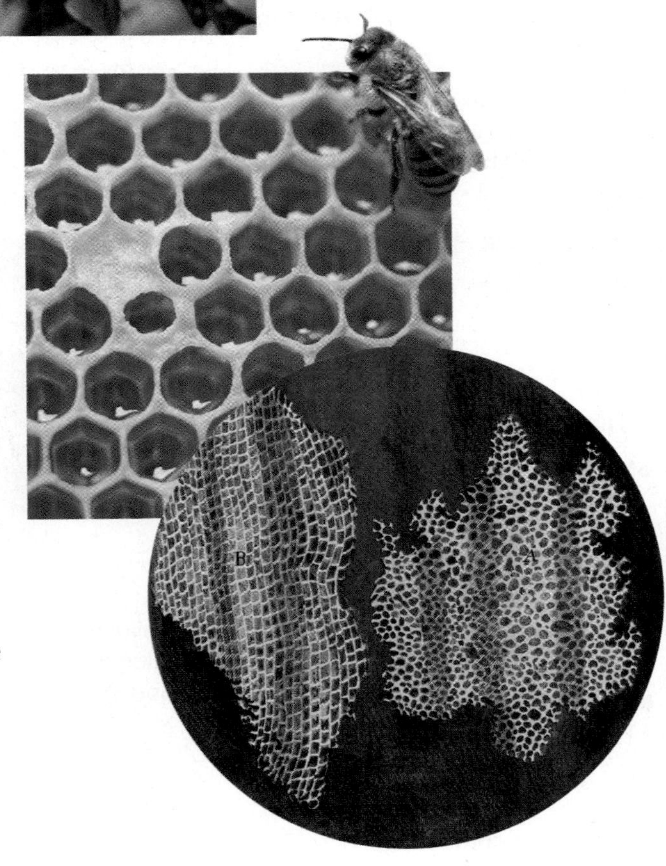

Figure 1 To Robert Hooke, the cells of cork looked like the openings in a honeycomb.

The Cell Theory

After Hooke's discovery, other scientists began making better microscopes and looking for cells in many other places, such as pond water and blood. The newer microscopes enabled scientists to see different structures inside cells. Matthias Schleiden (SHLI dun), a German scientist, used one of the new microscopes to look at plant cells. Around the same time, another German scientist, Theodor Schwann, used a microscope to study animal cells. Schleiden and Schwann realized that plant and animal cells have similar features. You'll read about many of these features in Lesson 2.

Almost two decades later, Rudolf Virchow (VUR koh), a German doctor, proposed that all cells come from preexisting cells, or cells that already exist. The observations made by Schleiden, Schwann, and Virchow were combined into one **theory.** As illustrated in **Table 1,** *the* **cell theory** *states that all living things are made of one or more cells, the cell is the smallest unit of life, and all new cells come from preexisting cells.* After the development of the cell theory, scientists raised more questions about cells. If all living things are made of cells, what are cells made of?

REVIEW VOCABULARY

theory
explanation of things or events based on scientific knowledge resulting from many observations and experiments

2. NGSSS Check **Explain** How did scientists' understanding of cells develop? SC.6.L.14.2

____

____

____

____

Table 1 Scientists developed the cell theory after studying cells with microscopes.

Active Reading **3. Summarize** Review the three principles of the cell theory by completing the table below.

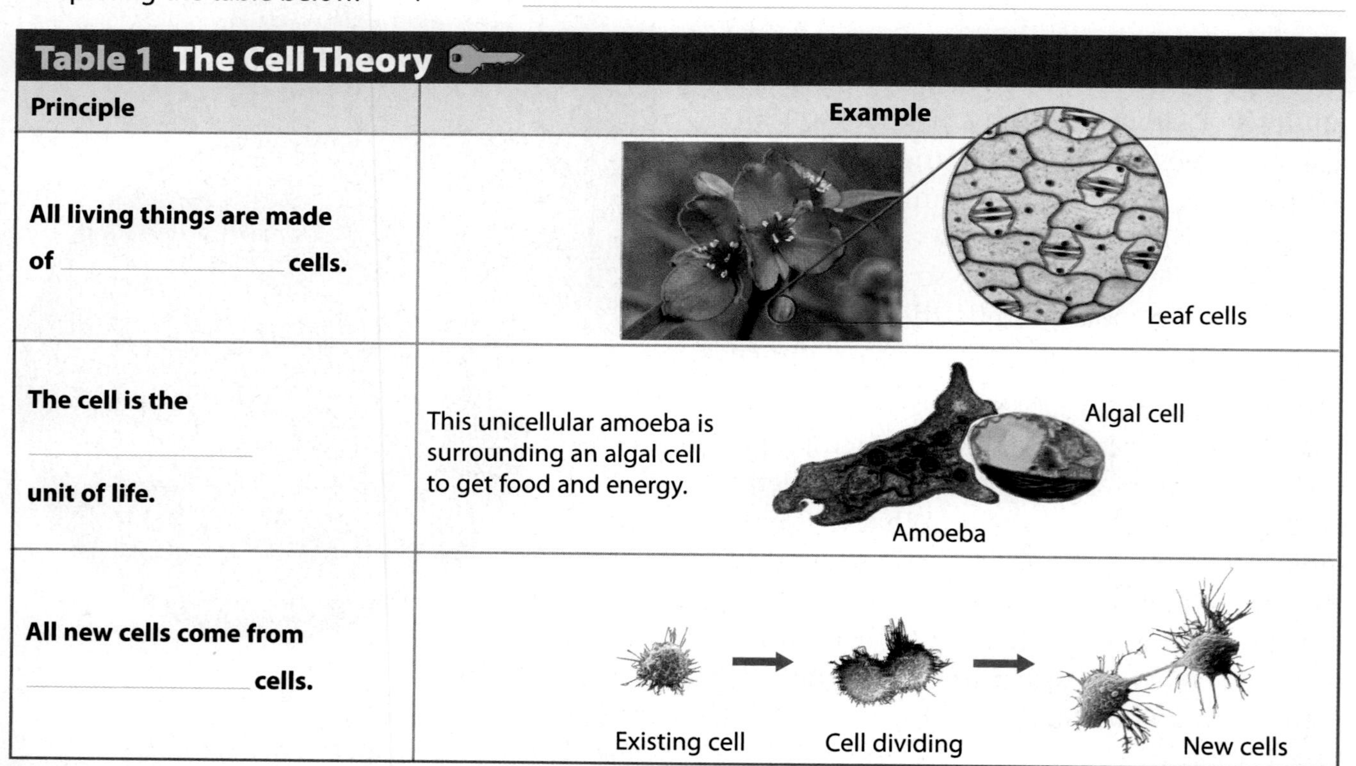

Table 1 The Cell Theory

Principle	Example
All living things are made of ________ cells.	Leaf cells
The cell is the ________ unit of life.	This unicellular amoeba is surrounding an algal cell to get food and energy. Algal cell; Amoeba
All new cells come from ________ cells.	Existing cell → Cell dividing → New cells

Basic Cell Substances

Have you ever watched a train travel down a railroad track? The locomotive pulls train cars that are hooked together. Like a train, many of the substances in cells are made of smaller parts that are joined together. *These substances, called* **macromolecules**, *form by joining many small molecules together.* As you will read later in this lesson, macromolecules have many important roles in cells. But macromolecules cannot function without one of the most important substances in cells—water.

WORD ORIGIN

macromolecule
from Greek *makro–*, means "long"; and Latin *molecula*, means "mass"

The Main Ingredient—Water

The main ingredient in any cell is water. It makes up more than 70 percent of a cell's volume and is essential for life. Why is water such an important molecule? In addition to making up a large part of the inside of cells, water also surrounds cells. The water surrounding your cells helps to insulate your body, which maintains homeostasis, or a stable internal environment.

The structure of a water molecule makes it ideal for dissolving many other substances. Substances must be in a liquid to move into and out of cells. A water molecule has two areas:

- An area that is more negative (–), called the negative end; this end can attract the positive part of another substance.
- An area that is more positive (+), called the positive end; this end can attract the negative part of another substance.

Examine **Figure 2** to see how the positive and negative ends of water molecules dissolve salt crystals.

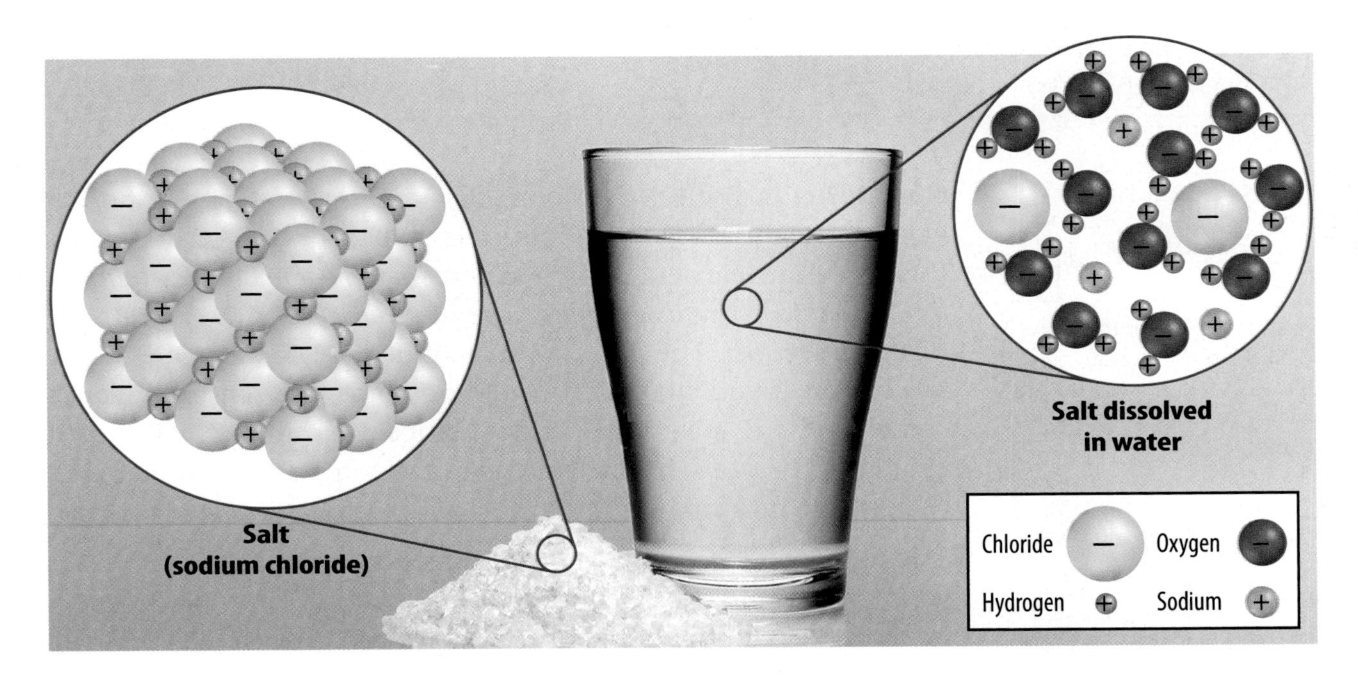

Figure 2 The positive and negative ends of a water molecule attract the positive and negative parts of another substance, similar to the way magnets are attracted to each other.

Active Reading **4. Identify** Circle the part of the salt crystal that is attracted to the oxygen in the water molecule.

Macromolecules

Although water is essential for life, all cells contain other substances that enable them to function. Recall that macromolecules are large molecules that form when smaller molecules join together. As shown in **Figure 3,** there are four types of macromolecules in cells: nucleic acids, proteins, lipids, and carbohydrates. Each type of macromolecule has unique functions in a cell. These functions range from growth and communication to movement and storage.

Active Reading **5. Identify** In **Figure 3,** circle the names of the macromolecules that provide support to a cell.

Cell Macromolecules

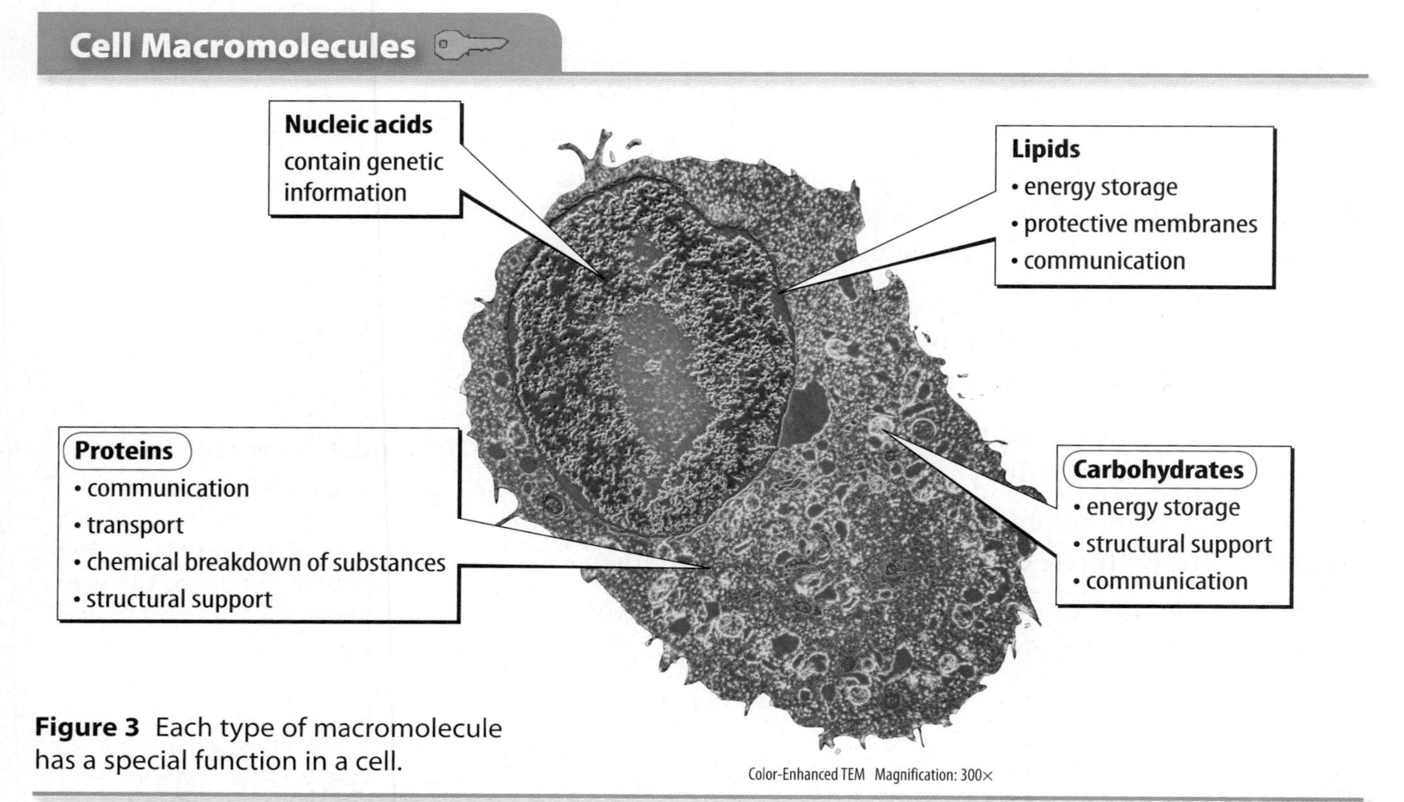

Figure 3 Each type of macromolecule has a special function in a cell.

Active Reading LA.6.2.2.3

Fold a sheet of paper to make a four-door book. Label it as shown. Use it to organize your notes on the macromolecules and their uses in a cell.

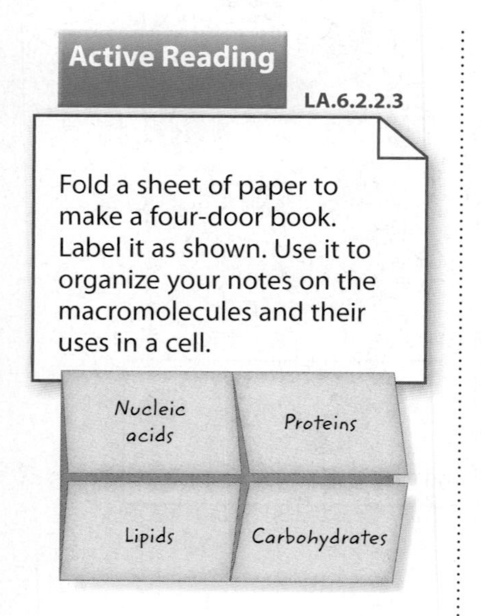

Nucleic Acids Both deoxyribonucleic (dee AHK sih ri boh noo klee ihk) acid (DNA) and ribonucleic (ri boh noo KLEE ihk) acid (RNA) are nucleic acids. **Nucleic acids** *are macromolecules that form when long chains of molecules called nucleotides (NEW klee uh tidz) join together.* The order of nucleotides in DNA and RNA is important. If you change the order of words in a sentence, you can change the meaning of the sentence. In a similar way, changing the order of nucleotides in DNA and RNA can change the genetic information in a cell.

Nucleic acids are important in cells because they contain genetic information. This information can pass from parents to offspring. DNA includes instructions for cell growth, cell reproduction, and cell processes that enable a cell to respond to its environment. DNA is used to make RNA. RNA is used to make proteins.

Proteins The macromolecules necessary for nearly everything cells do are proteins. **Proteins** *are long chains of amino acid molecules.* You just read that RNA is used to make proteins. RNA contains instructions for joining amino acids together.

Cells contain hundreds of proteins. Each protein has a unique function. Some proteins help cells communicate with each other. Other proteins transport substances around inside cells. Some proteins, such as amylase (AM uh lays) in saliva, help break down nutrients in food. Other proteins, such as keratin (KER uh tun)—a protein found in hair, horns, and feathers—provide structural support.

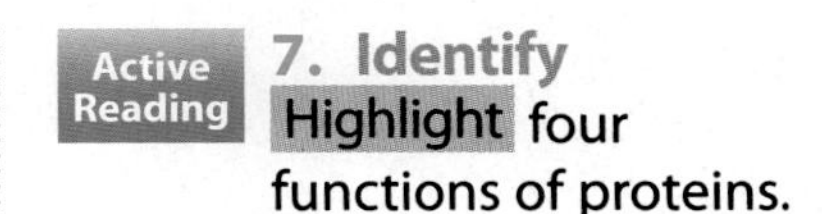

Active Reading **7. Identify** Highlight four functions of proteins.

Lipids Another group of macromolecules found in cells is lipids. *A* **lipid** *is a large macromolecule that does not dissolve in water.* Because lipids do not mix with water, they play an important role as protective barriers in cells. They are also the major part of cell membranes. Lipids play roles in energy storage and in cell communication. Examples of lipids are cholesterol (kuh LES tuh rawl), phospholipids (fahs foh LIH pids), and vitamin A.

Active Reading **6. Determine** Underline why lipids are important to cells.

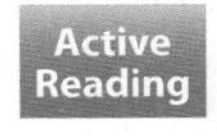

8. Recall What basic substances make up a cell?

Carbohydrates *One sugar molecule, two sugar molecules, or a long chain of sugar molecules make up* **carbohydrates** (kar boh HI drayts). Carbohydrates store energy, provide structural support, and are needed for communication between cells. Sugars and starches are carbohydrates that store energy. Fruits contain sugars. Breads and pastas are mostly starch. The energy in sugars and starches can be released quickly through chemical reactions in cells. Cellulose is a carbohydrate in the cell walls in plants that provides structural support.

Inquiry LAB STATION SC.6.N.1.4 **Try It!**

MiniLab ***How can you observe DNA?*** at connectED.mcgraw-hill.com

Apply It! After you complete the lab, answer the questions below.

1. Why do you think DNA was not visible in all of the cells you observed?

2. What other macromolecules do you think you could observe in the onion root-tip cells? Explain.

Lesson Review 1

Visual Summary

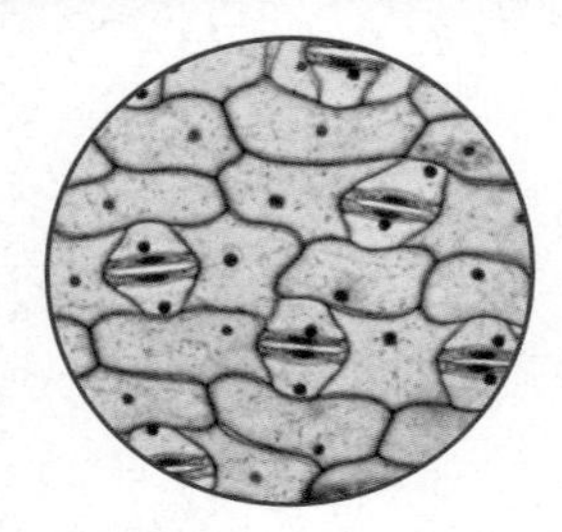

The cell theory summarizes the main principles for understanding that the cell is the basic unit of life.

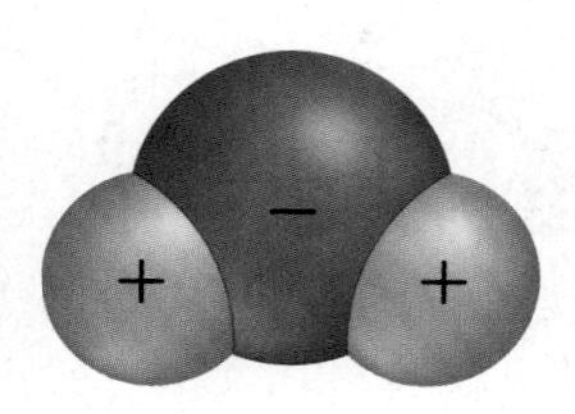

Water is the main ingredient in every cell.

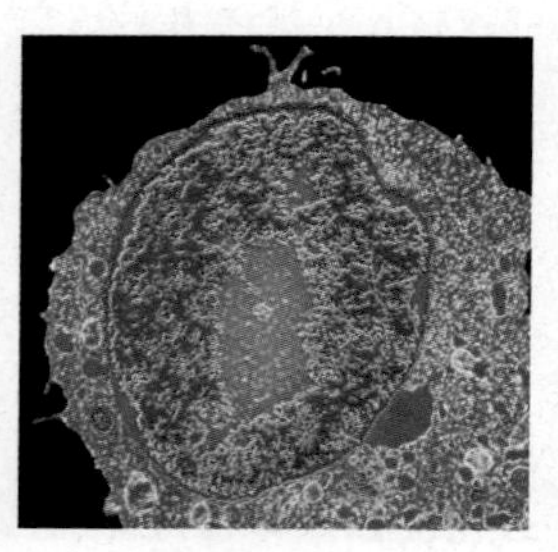

A nucleic acid, such as DNA, contains the genetic information for a cell.

Use Vocabulary

1. The ____________ ____________ states that the cell is the basic unit of all living things. SC.6.L.14.2

2. **Use the term** *nucleic acid* in a sentence.

Understand Key Concepts

3. Which macromolecule is made from amino acids?
   - (A) lipid
   - (B) protein
   - (C) carbohydrate
   - (D) nucleic acid

4. **Describe** how the invention of the microscope helped scientists understand cells.

5. **Compare** the functions of DNA and proteins in a cell.

Interpret Graphics

6. **Summarize** Fill in the graphic organizer below to summarize the main principles of the cell theory. SC.6.L.14.2

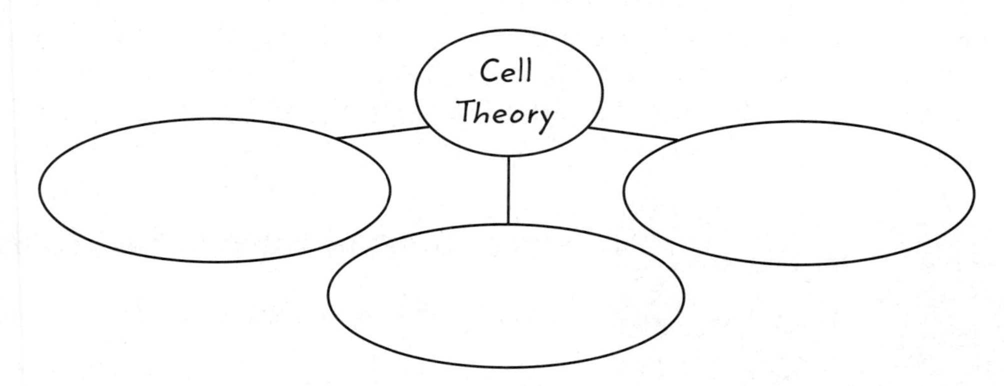

Critical Thinking

7. **Summarize** the functions of lipids in cells.

8. **Hypothesize** why carbohydrates are found in plant cell walls.

HOW IT WORKS

A Very Powerful Microscope

Using technology to look inside cells

If Robert Hooke had used an atomic force microscope (AFM), he would have observed more than just cells. He would have seen the macromolecules inside them! An AFM can scan objects that are only nanometers in size. A nanometer is one one-billionth of a meter. That's 100,000 times smaller than the width of a human hair. AFM technology has enabled scientists to better understand how cells function. It also has given them a three-dimensional look at the macromolecules that make life possible. This is how it works.

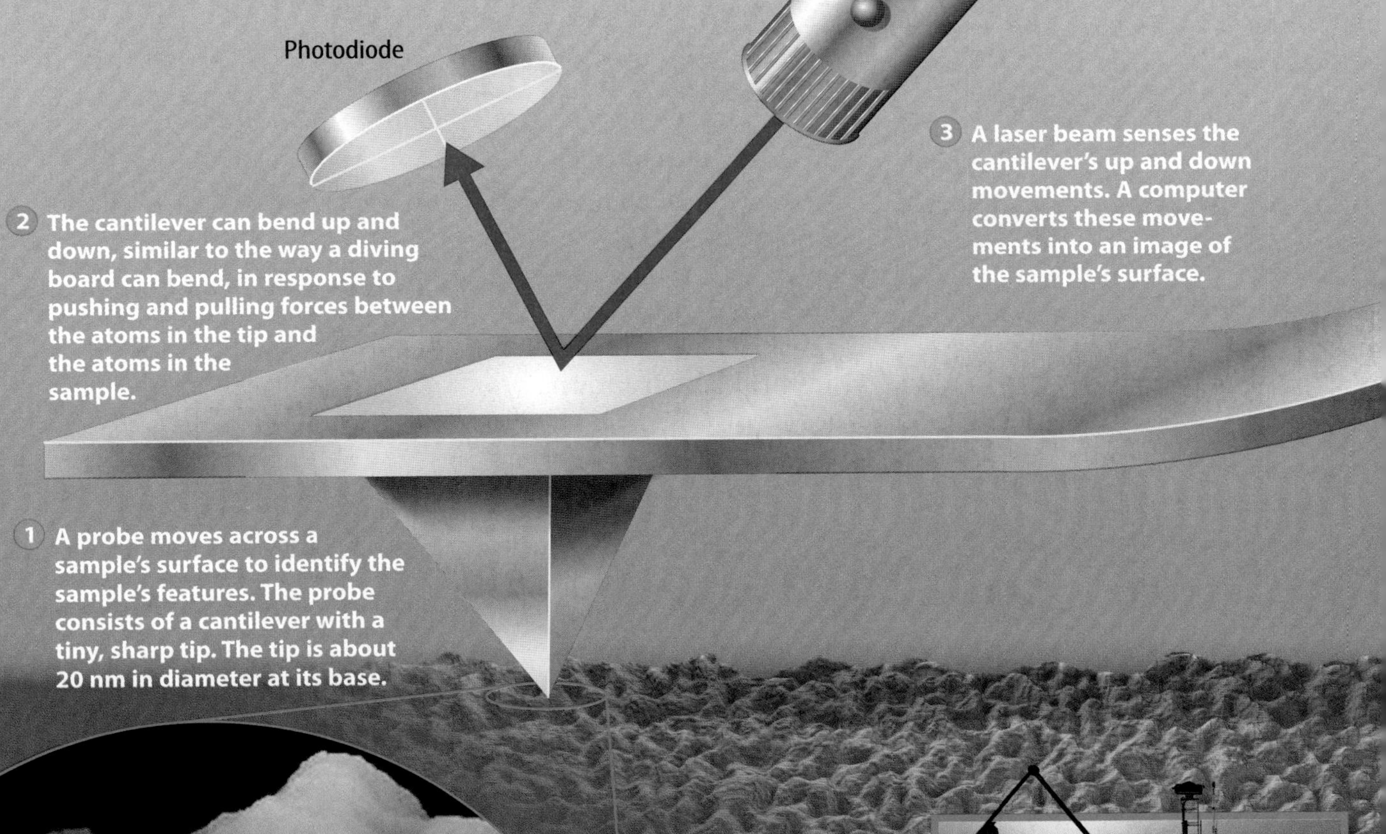

Photodiode

1 **A probe moves across a sample's surface to identify the sample's features. The probe consists of a cantilever with a tiny, sharp tip. The tip is about 20 nm in diameter at its base.**

2 **The cantilever can bend up and down, similar to the way a diving board can bend, in response to pushing and pulling forces between the atoms in the tip and the atoms in the sample.**

3 **A laser beam senses the cantilever's up and down movements. A computer converts these movements into an image of the sample's surface.**

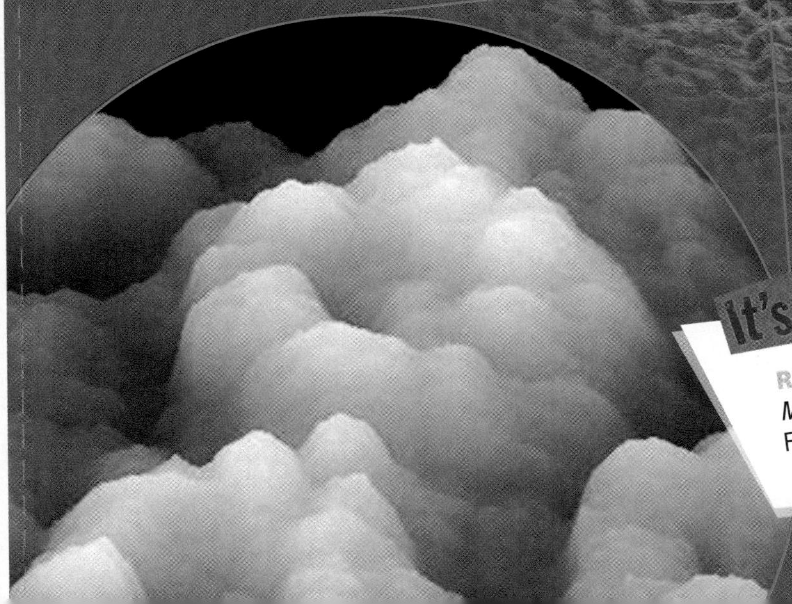

It's Your Turn

RESEARCH NASA's *Phoenix Mars Lander* included an atomic force microscope. Find out what scientists discovered on Mars with this instrument.

Lesson 2

THE CELL

ESSENTIAL QUESTIONS

 How are prokaryotic cells and eukaryotic cells similar, and how are they different?

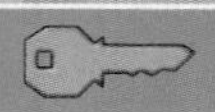 What do the structures in a cell do?

Vocabulary

cell membrane p. 378
cell wall p. 378
cytoplasm p. 379
cytoskeleton p. 379
organelle p. 380
nucleus p. 381
chloroplast p. 383

Florida NGSSS

LA.6.2.2.3 The student will organize information to show understanding (e.g., representing main ideas within text through charting, mapping, paraphrasing, summarizing, or comparing/contrasting);

SC.6.L.14.4 Compare and contrast the structure and function of major organelles of plant and animal cells, including cell wall, cell membrane, nucleus, cytoplasm, chloroplasts, mitochondria, and vacuoles.

SC.6.N.1.1 Define a problem from the sixth grade curriculum, use appropriate reference materials to support scientific understanding, plan and carry out scientific investigation of various types, such as systematic observations or experiments, identify variables, collect and organize data, interpret data in charts, tables, and graphics, analyze information, make predictions, and defend conclusions.

SC.6.N.2.1 Distinguish science from other activities involving thought.

 Launch Lab SC.6.N.1.1

10 minutes

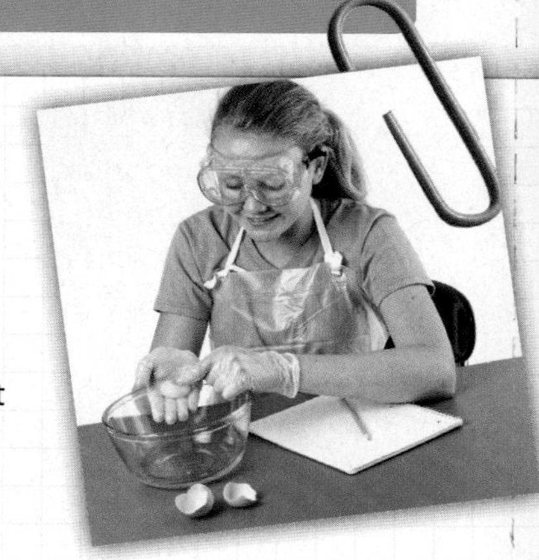

Why do eggs have shells?

Bird eggs have different structures, such as a shell, a membrane, and a yolk. Each structure has a different function that helps keep the egg safe and assists in development of the baby bird inside of it.

Procedure   

1. Read and complete a lab safety form.
2. Place an **uncooked egg** in a bowl.
3. Feel the shell, and record your observations below.
4. Crack open the egg. Pour the contents into the bowl.
5. Observe the inside of the shell and the contents of the bowl. Record your observations.

Data and Observations

Think About This

1. What do you think is the role of the eggshell?

2. Are there any structures in the bowl that have the same function as the eggshell? Explain.

3. Key Concept What does the structure of the eggshell tell you about its function?

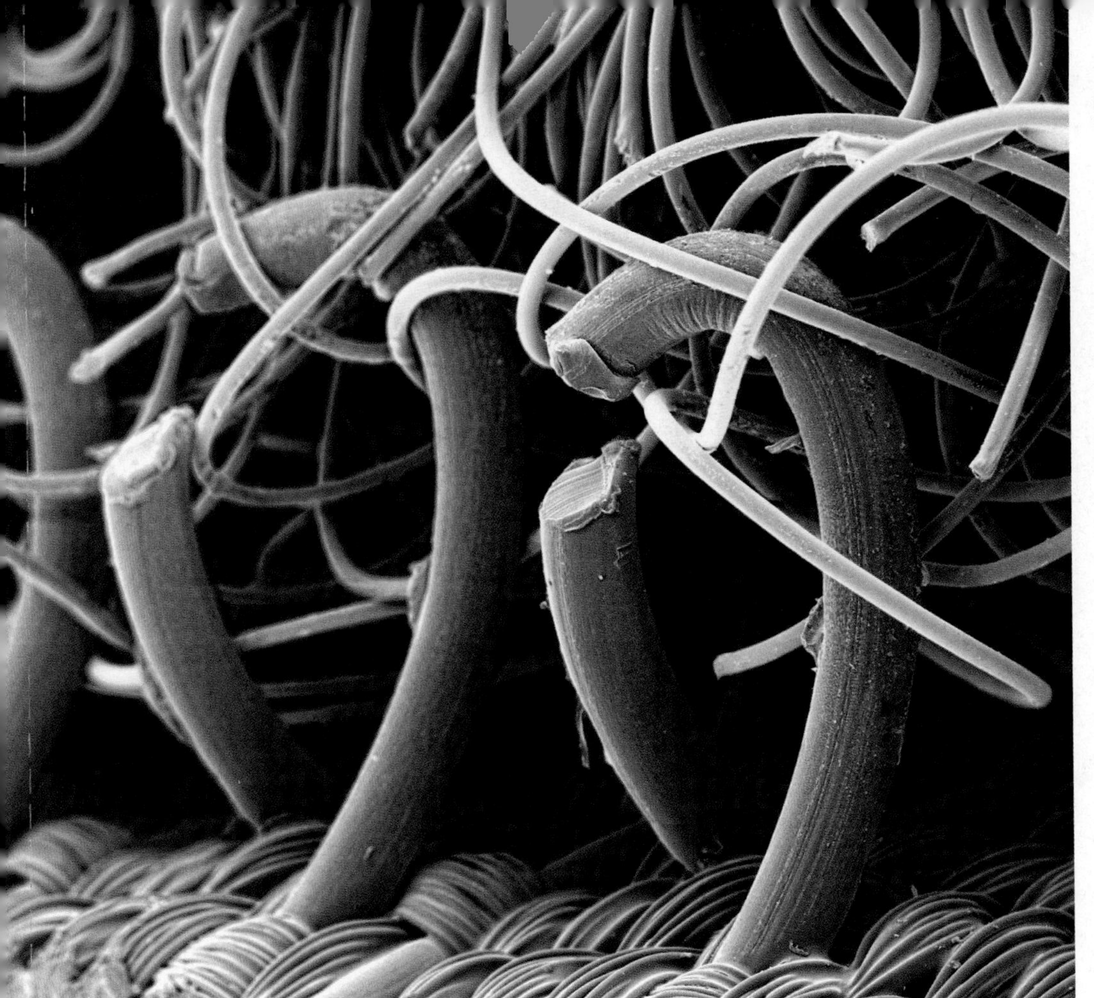

Inquiry Hooked Together?

1. What happens when one of the hooks in the photo at left goes through one of the loops? The two sides fasten together. The shapes of the hooks and loops in the hook-and-loop tape are suited to their function—to hold the two pieces together. How do you think the shape of a cell also might be suited to its function?

Cell Shape and Movement

You might recall from Lesson 1 that all living things are made up of one or more cells. As illustrated in **Figure 4,** cells come in many shapes and sizes. The size and shape of a cell relates to its job or function. For example, a human red blood cell cannot be seen without a microscope. Its small size and disk shape enable it to pass easily through the smallest blood vessels. The shape of a nerve cell enables it to send signals over long distances. Some plant cells are hollow and make up tubelike structures that carry materials throughout a plant.

The structures that make up a cell also have unique functions. Think about how the players on a football team perform different tasks to move the ball down the field. In a similar way, a cell is made of different structures that perform different functions that keep a cell alive. You will read about some of these structures in this lesson.

Figure 4 The shape of a cell relates to the function it performs.

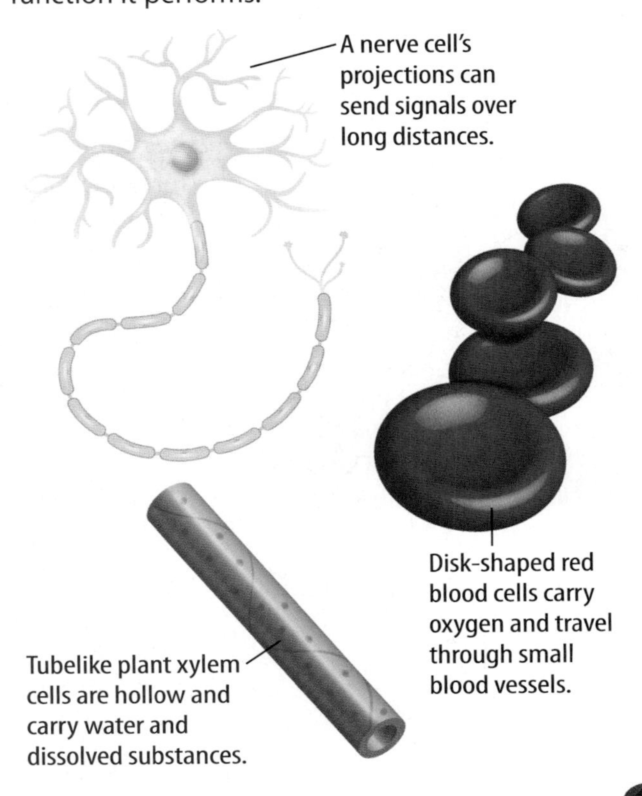

Plant Cell

Figure 5 The cell wall maintains the shape of a plant cell.

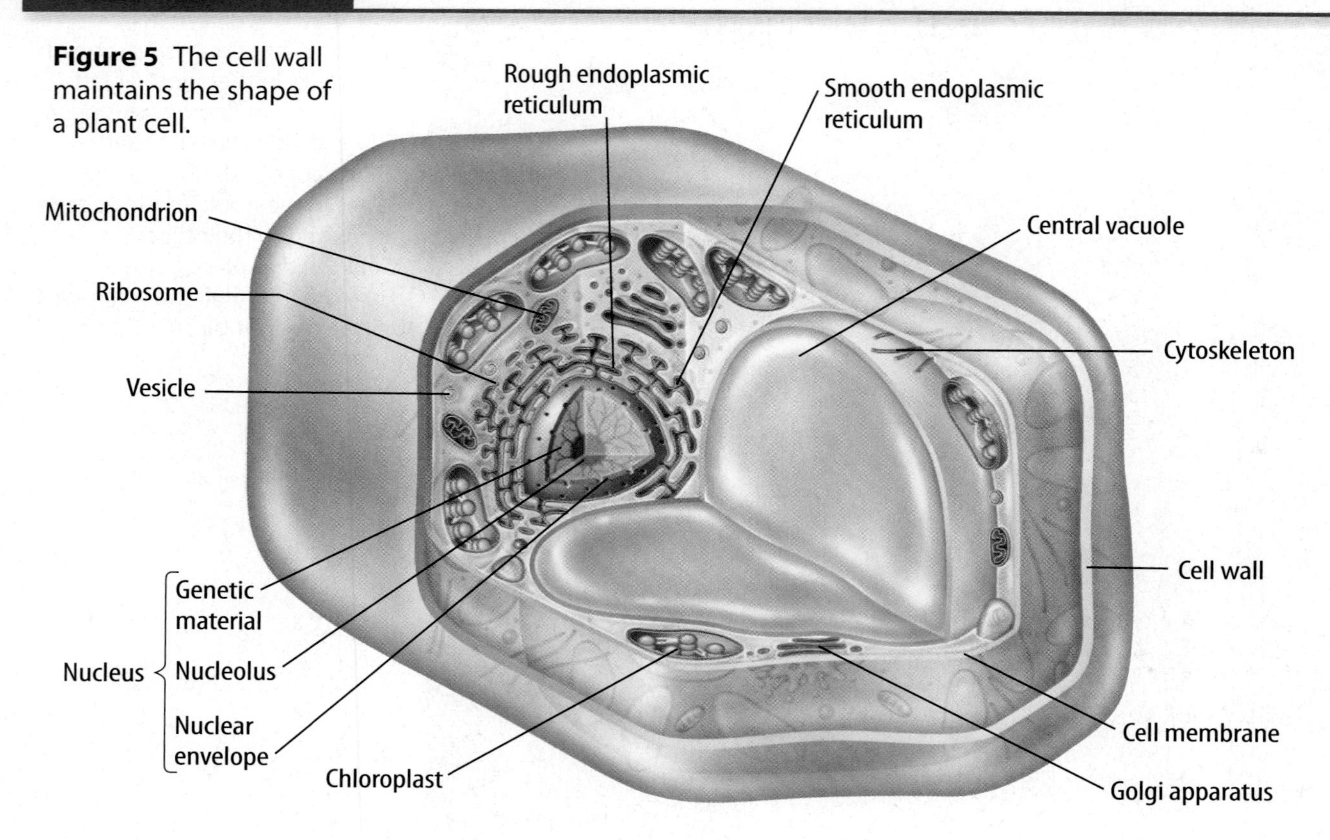

ACADEMIC VOCABULARY
function
(noun) the purpose for which something is used

Cell Membrane

Although different types of cells perform different **functions**, all cells have some structures in common. As shown in **Figure 5** and **Figure 6,** every cell is surrounded by a protective covering called a membrane. *The* **cell membrane** *is a flexible covering that protects the inside of a cell from the environment outside a cell.* Cell membranes are mostly made of two different macromolecules—proteins and a type of lipid called phospholipids. Think again about a football team. The defensive line tries to stop the other team from moving forward with the football. In a similar way, a cell membrane protects the cell from the outside environment.

Active Reading **2. Identify** Underline the two main macromolecules that make up cell membranes.

Cell Wall

Every cell has a cell membrane, but some cells are also surrounded by a structure called the cell wall. Plant cells such as the one in **Figure 5,** fungal cells, bacteria, and some types of protists have cell walls. *A* **cell wall** *is a stiff structure outside the cell membrane.* A cell wall protects a cell from attack by viruses and other harmful organisms. In some plant cells and fungal cells, a cell wall helps maintain the cell's shape and gives structural support.

Active Reading **3. Define** Describe the structure and function of a cell wall.

Animal Cell

Figure 6 The cytoskeleton maintains the shape of an animal cell.

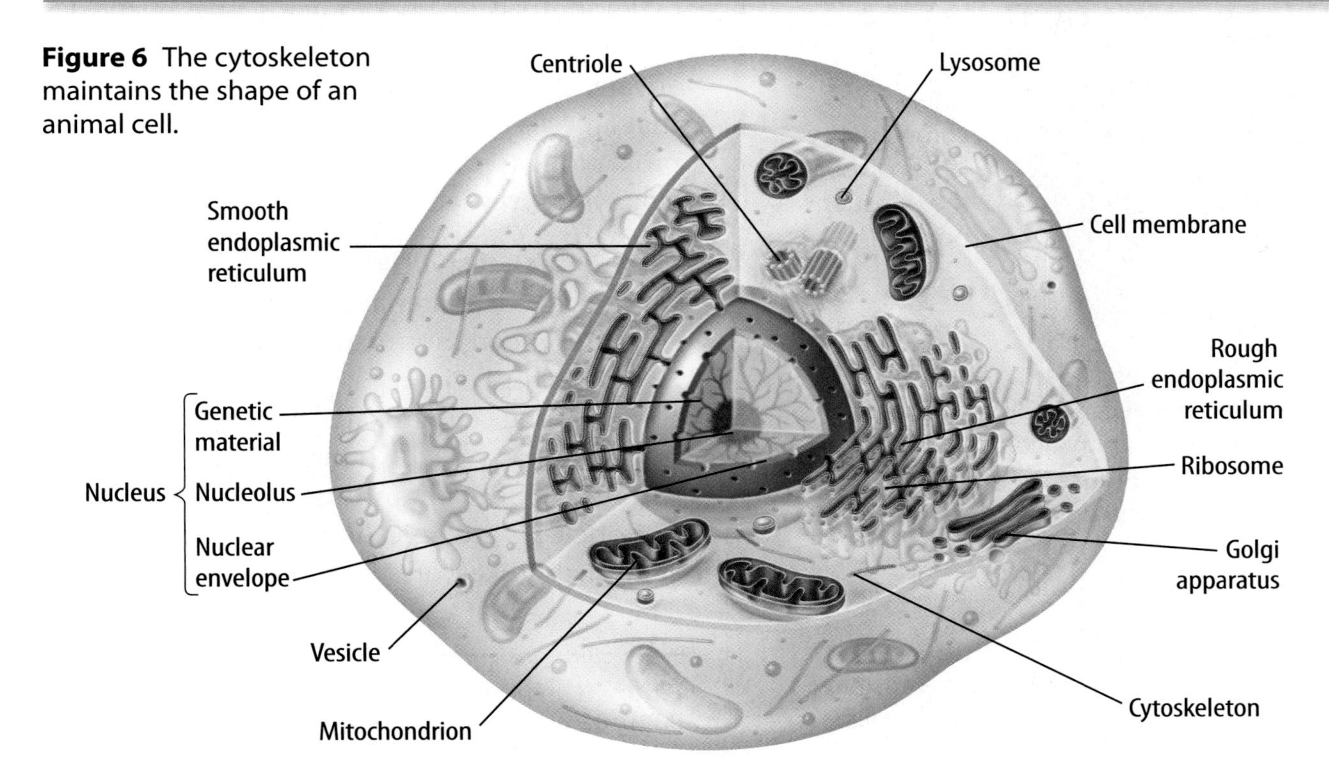

Active Reading **4. Identify** Circle the names of two parts in the plant cell in **Figure 5** that are not in the animal cell in **Figure 6.**

Cell Appendages

Arms, legs, claws, and antennae are all types of appendages. Cells can have appendages too. Cell appendages are often used for movement. Flagella (fluh JEH luh; singular, flagellum) are long, tail-like appendages that whip back and forth and move a cell. A cell can also have cilia (SIH lee uh; singular, cilium) like the ones shown in **Figure 7.** Cilia are short, hairlike structures. They can move a cell or move molecules away from a cell. A microscopic organism called a paramecium (pa ruh MEE shee um) moves around its watery environment using its cilia. The cilia in your windpipe move harmful substances away from your lungs.

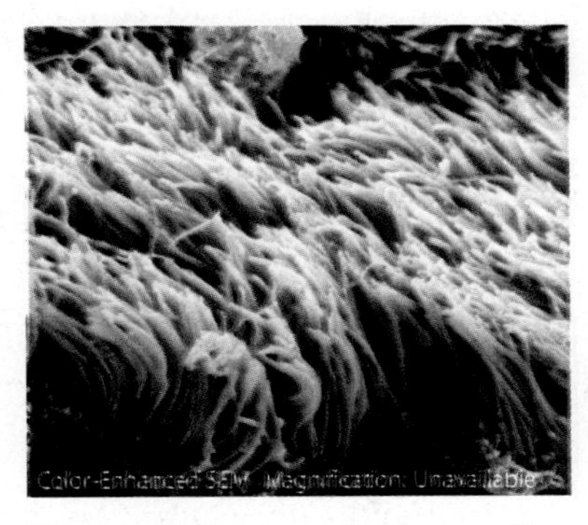

Figure 7 Lung cells have cilia that help move fluids and foreign materials.

Cytoplasm and the Cytoskeleton

In Lesson 1, you read that water is the main ingredient in a cell. Most of this water is in the **cytoplasm**, *a fluid inside a cell that contains salts and other molecules.* The cytoplasm also contains a cell's cytoskeleton. *The* **cytoskeleton** *is a network of threadlike proteins that are joined together.* The proteins form a framework inside a cell. This framework gives a cell its shape and helps it move. Cilia and flagella are made from the same proteins that make up the cytoskeleton.

WORD ORIGIN
cytoplasm
from Greek *kytos,* means "hollow vessel"; and *plasma,* means "something molded"

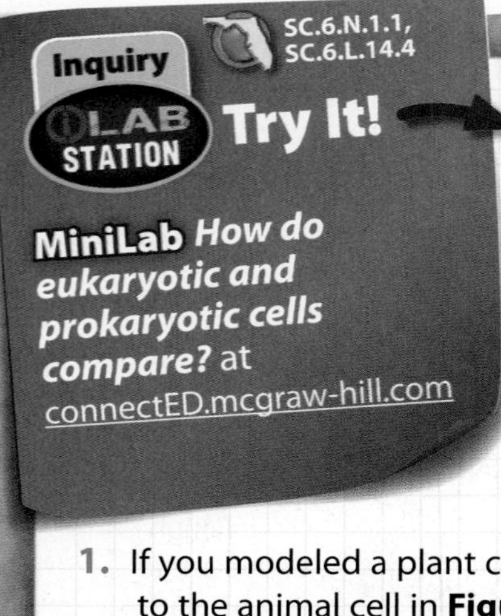

Apply It!

After you complete the lab, answer the questions below.

1. If you modeled a plant cell, compare it to the animal cell in **Figure 6.** If you modeled an animal cell, compare it to the plant cell in **Figure 5.** How do the cells compare? How do they differ?

2. Why do you think prokaryotic cells have fewer cell parts than eukaryotic cells?

5. Organize Classify cells as prokaryotic or eukaryotic by writing *P* or *E* in the right-hand column of the table below.

Characteristic	Cell Type
Cell's genetic material is surrounded by a membrane.	
Cell is usually a unicellular organism.	
Cell contains organelles.	

6. Compare and Contrast How are prokaryotic cells and eukaryotic cells similar, and how are they different?

Cell Types

Recall that the use of microscopes enabled scientists to discover cells. With more advanced microscopes, scientists discovered that all cells can be grouped into two types—prokaryotic (proh ka ree AH tihk) cells and eukaryotic (yew ker ee AH tihk) cells.

Prokaryotic Cells

The genetic material in a prokaryotic cell is not surrounded by a membrane, as shown in **Figure 8.** This is the most important feature of a prokaryotic cell. Prokaryotic cells also do not have many of the other cell parts that you will read about later in this lesson. Most prokaryotic cells are unicellular organisms and are called prokaryotes.

Figure 8 In prokaryotic cells, the genetic material floats freely in the cytoplasm.

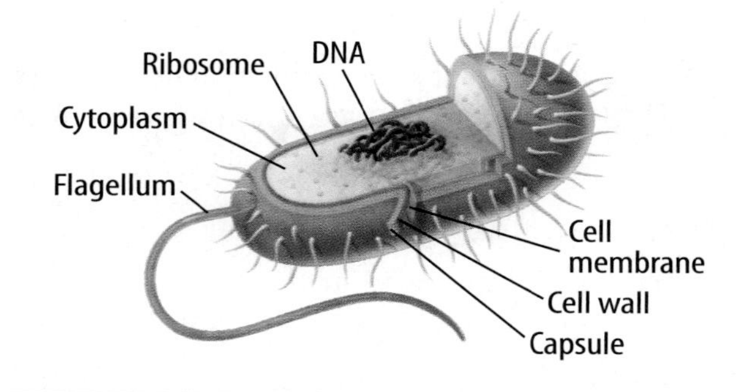

Eukaryotic Cells

Plants, animals, fungi, and protists are all made of eukaryotic cells, such as the ones shown in **Figure 5** and **Figure 6,** and are called eukaryotes. With few exceptions, each eukaryotic cell has genetic material that is surrounded by a membrane. Every eukaryotic cell has *other structures, called* **organelles,** *which have specialized functions. Most organelles are surrounded by membranes.* Eukaryotic cells are usually larger than prokaryotic cells. About ten prokaryotic cells would fit inside one eukaryotic cell.

Cell Organelles

As you have just read, organelles are eukaryotic cell structures with specific functions. Organelles enable cells to carry out different functions at the same time. For example, cells can obtain energy from food, store information, make macromolecules, and get rid of waste materials all at the same time because different organelles perform the different tasks.

Active Reading

FOLDABLES® LA.6.2.2.3

Fold a sheet of paper into a vertical half book. Use it to record information about cell organelles and their functions.

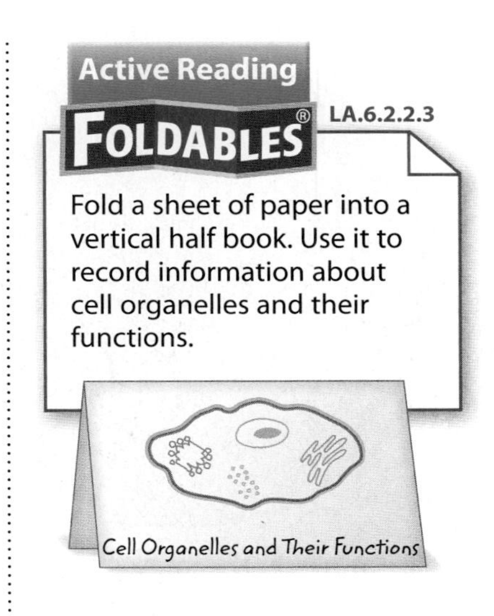

The Nucleus

The largest organelle inside most eukaryotic cells is the nucleus, shown in **Figure 9.** *The* **nucleus** *is the part of a eukaryotic cell that directs cell activities and contains genetic information stored in DNA.* DNA is organized into structures called chromosomes. The number of chromosomes in a nucleus is different for different species of organisms. For example, kangaroo cells contain six pairs of chromosomes. Most human cells contain 23 pairs of chromosomes.

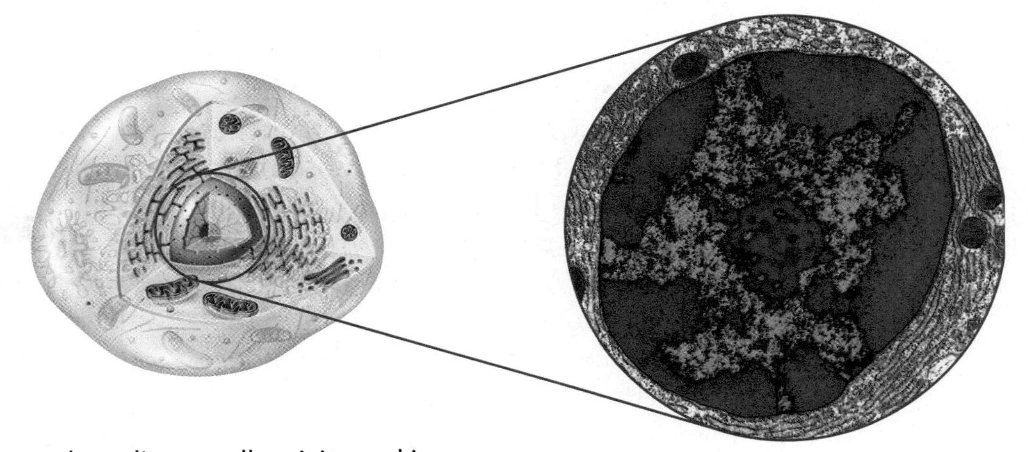

Figure 9 The nucleus directs cell activity and is surrounded by a membrane.

In addition to chromosomes, the nucleus contains proteins and an organelle called the nucleolus (new KLEE uh lus). The nucleolus is often seen as a large dark spot in the nucleus of a cell. The nucleolus makes ribosomes, organelles that are involved in the production of proteins. You will read about ribosomes later in this lesson.

Surrounding the nucleus are two membranes that form a structure called the nuclear **envelope.** The nuclear envelope contains many pores. Certain molecules, such as ribosomes and RNA, move into and out of the nucleus through these pores.

Science Use v. Common Use

envelope

Science Use an outer covering

Common Use a flat paper container for a letter

Active Reading **7. Describe** What is the nuclear envelope?

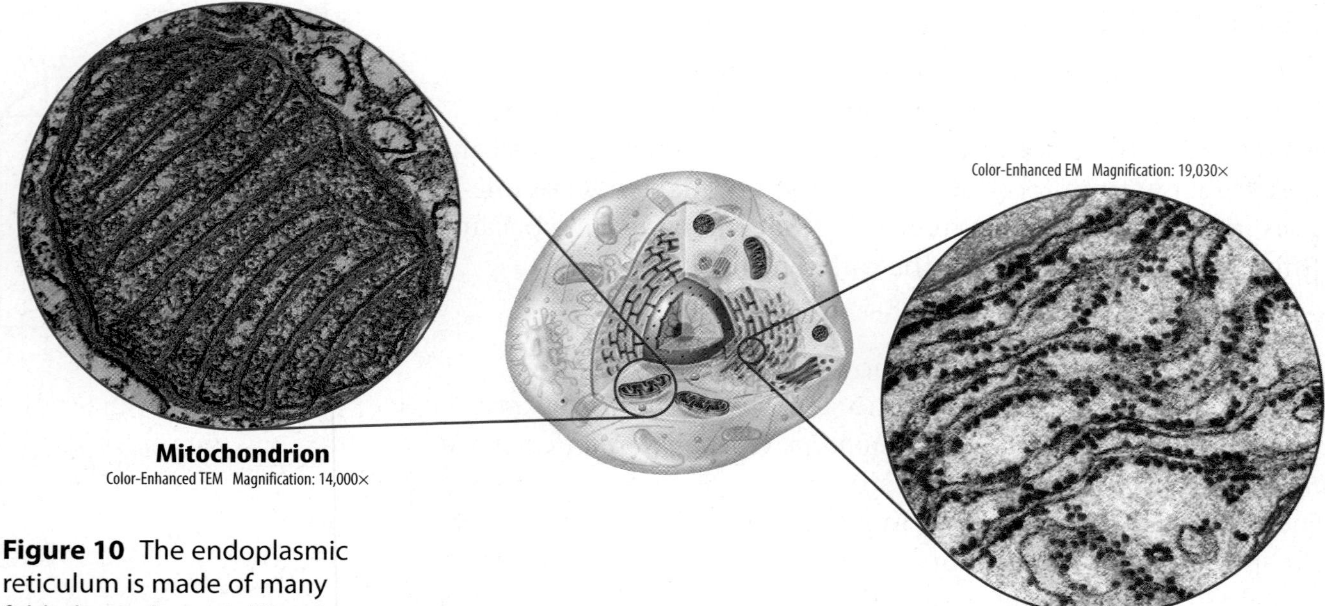

Figure 10 The endoplasmic reticulum is made of many folded membranes. Mitochondria provide a cell with usable energy.

Manufacturing Molecules

You might recall from Lesson 1 that proteins are important molecules in cells. Proteins are made on small structures called ribosomes. Unlike other cell organelles, a ribosome is not surrounded by a membrane. Ribosomes are in a cell's cytoplasm. They also can be attached to a weblike organelle called the endoplasmic reticulum (en duh PLAZ mihk • rih TIHK yuh lum), or ER. As shown in **Figure 10,** the ER spreads from the nucleus throughout most of the cytoplasm. ER with ribosomes on its surface is called rough ER. Rough ER is the site of protein production. ER without ribosomes is called smooth ER. It makes lipids such as cholesterol. Smooth ER is important because it helps remove harmful substances from a cell.

Active Reading **8. Contrast** Underline the description and function of smooth ER. Highlight the description and function of rough ER.

Processing Energy

All living things require energy in order to survive. Cells process some energy in specialized organelles. Most eukaryotic cells contain hundreds of organelles called mitochondria (mi tuh KAHN dree uh; singular, mitochondrion), shown in **Figure 10.** Some cells in a human heart can contain a thousand mitochondria.

Like the nucleus, a mitochondrion is surrounded by two membranes. Energy is released during chemical reactions that occur in the mitochondria. This energy is stored in high-energy molecules called ATP—adenosine triphosphate (uh DEH nuh seen • tri FAHS fayt). ATP is the fuel for cellular processes such as growth, cell division, and material transport.

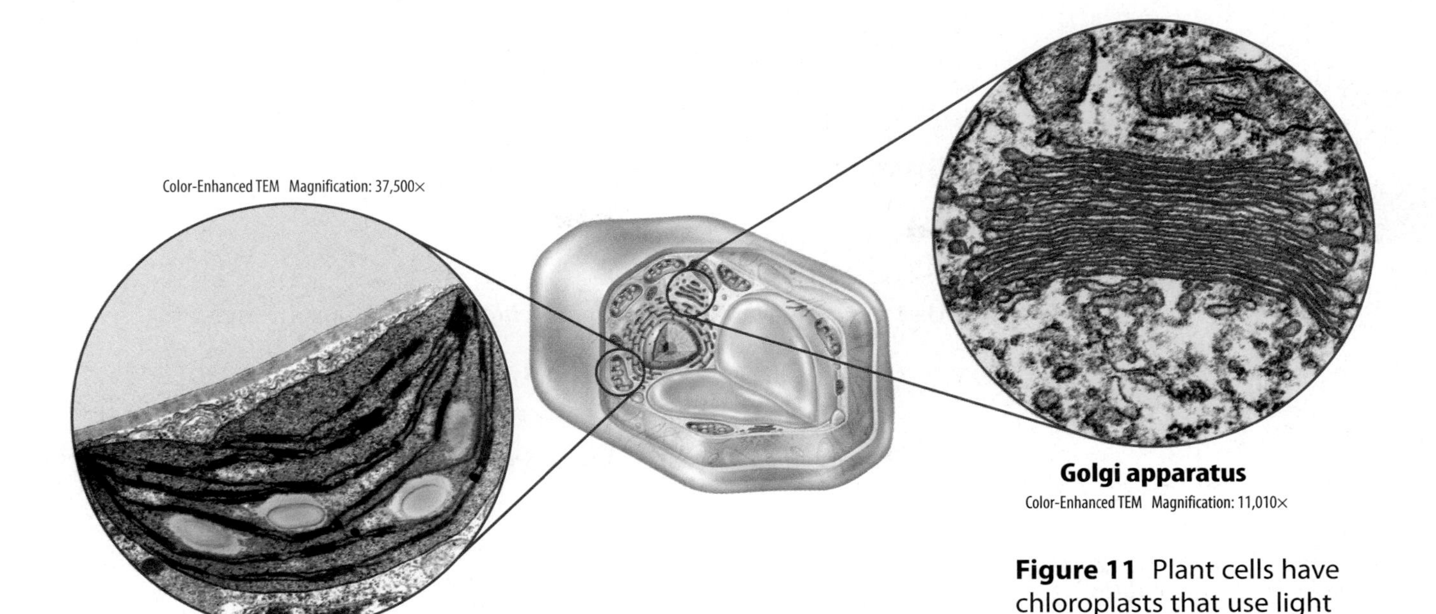

Figure 11 Plant cells have chloroplasts that use light energy and make food. The Golgi apparatus packages materials into vesicles.

Plant cells and some protists, such as algae, also contain organelles called chloroplasts (KLOR uh plasts), shown in **Figure 11.** **Chloroplasts** *are membrane-bound organelles that use light energy and make food—a sugar called glucose—from water and carbon dioxide in a process known as photosynthesis* (foh toh SIHN thuh sus). The sugar contains stored chemical energy that can be released when a cell needs it. You will read more about photosynthesis in Lesson 4.

Active Reading **9. Identify** Underline the types of cells that contain chloroplasts.

Processing, Transporting, and Storing Molecules

Near the ER is an organelle that looks like a stack of pancakes. This is the Golgi (GAWL jee) apparatus, shown in **Figure 11.** It prepares proteins for their specific jobs or functions. Then it packages the proteins into tiny, membrane-bound, ball-like structures called vesicles. Vesicles are organelles that transport substances from one area of a cell to another area of a cell. Some vesicles in an animal cell are called lysosomes. Lysosomes contain substances that help break down and recycle cellular components.

Some cells also have saclike structures called vacuoles (VA kyuh wohlz). Vacuoles are organelles that store food, water, and waste material. A typical plant cell usually has one large vacuole that stores water and other substances. Some animal cells have many small vacuoles.

Active Reading **10. Determine** Highlight the function of the Golgi apparatus.

Lesson Review 2

Visual Summary

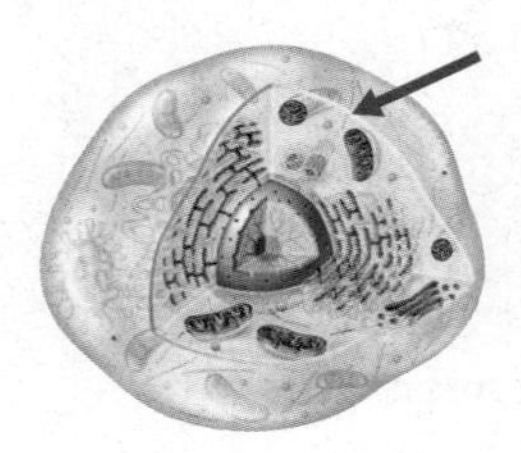

A cell is protected by a flexible covering called the cell membrane.

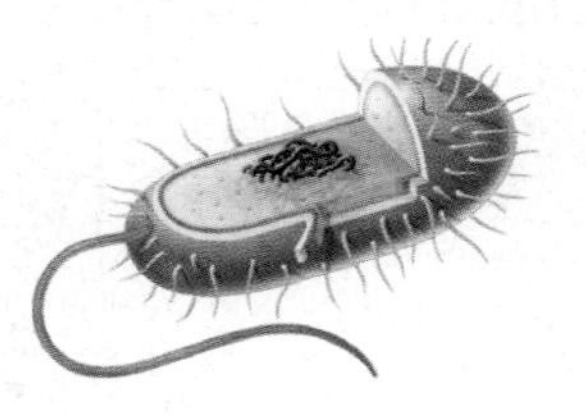

Cells can be grouped into two types—prokaryotic cells and eukaryotic cells.

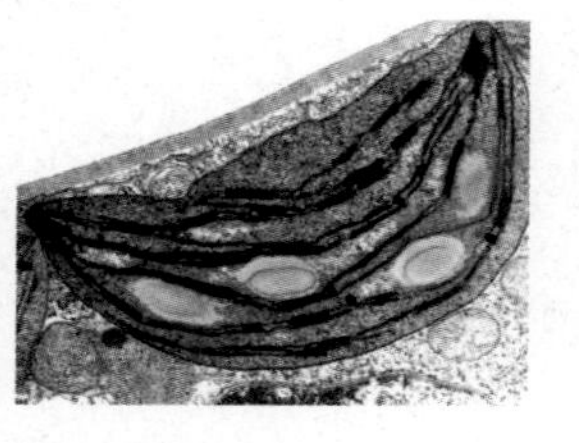

In a chloroplast, light energy is used for making sugars in a process called photosynthesis.

Skill Lab ***How are plant cells and animal cells similar and how are they different?*** at connectED.mcgraw-hill.com

Use Vocabulary

1. **Distinguish** between the cell wall and the cell membrane. SC.6.L.14.4

2. **Use the terms** *mitochondria* and *chloroplasts* in a sentence.

3. **Define** *organelle* in your own words.

Understand Key Concepts

4. Which organelle is used to store water? SC.6.L.14.4
   - (A) chloroplast
   - (B) lysosome
   - (C) nucleus
   - (D) vacuole

5. **Draw** a prokaryotic cell and label its parts.

Interpret Graphics

6. **Explain** how the structure of the cells below relates to their function.

Critical Thinking

7. **Analyze** Why are most organelles surrounded by membranes?

8. **Compare** the features of eukaryotic and prokaryotic cells.

Active Reading **Classify** information about organelles. In the right-hand column, indicate whether the organelle is in a plant cell, an animal cell, or both.

Organelle	Function	Plant, Animal, or Both?
Nucleus	1	
Ribosome	2	
Endoplasmic reticulum	3	
Mitochondria	4	
Chloroplast	5	
Golgi apparatus	6	
Vesicle	7	
Central vacuole	8	
Lysosome	9	

Lesson 3

Moving Cellular MATERIAL

ESSENTIAL QUESTIONS

How do materials enter and leave cells?

How does cell size affect the transport of materials?

Vocabulary

passive transport p. 387
diffusion p. 388
osmosis p. 388
facilitated diffusion p. 389
active transport p. 390
endocytosis p. 390
exocytosis p. 390

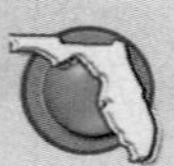

Florida NGSSS

LA.6.2.2.3 The student will organize information to show understanding (e.g., representing main ideas within text through charting, mapping, paraphrasing, summarizing, or comparing/contrasting);

MA.6.A.3.6 Construct and analyze tables, graphs, and equations to describe linear functions and other simple relations using both common language and algebraic notation.

SC.6.L.14.3 Recognize and explore how cells of all organisms undergo similar processes to maintain homeostasis, including extracting energy from food, getting rid of waste, and reproducing.

SC.6.L.14.4 Compare and contrast the structure and function of major organelles of plant and animal cells, including cell wall, cell membrane, nucleus, cytoplasm, chloroplasts, mitochondria, and vacuoles.

SC.6.N.1.1 Define a problem from the sixth grade curriculum, use appropriate reference materials to support scientific understanding, plan and carry out scientific investigation of various types, such as systematic observations or experiments, identify variables, collect and organize data, interpret data in charts, tables, and graphics, analyze information, make predictions, and defend conclusions.

SC.6.N.1.1

Launch Lab

5 minutes

What does the cell membrane do?

All cells have a membrane around the outside of the cell. The cell membrane separates the inside of a cell from the environment outside a cell. What else might a cell membrane do?

Procedure

1. Read and complete a lab safety form.
2. Place a square of **wire mesh** on top of a **beaker.**
3. Pour a small amount of **birdseed** on top of the wire mesh. Record your observations below.

Data and Observations

Think About This

1. What part of a cell does the wire mesh represent?

2. What happened when you poured birdseed on the wire mesh?

3. Key Concept How do you think the cell membrane affects materials that enter and leave a cell?

Inquiry Why the Veil?

1. A beekeeper often wears a helmet with a face-covering veil made of mesh. The openings in the mesh are large enough to let air through, yet small enough to keep bees out. In a similar way, some things must be allowed in or out of a cell, while other things must be kept in or out. How do you think the right things enter or leave a cell?

Passive Transport

Recall from Lesson 2 that membranes are the boundaries between cells and between organelles. Another important role of membranes is to control the movement of substances into and out of cells. A cell membrane is semipermeable. This means it allows only certain substances to enter or leave a cell. Substances can pass through a cell membrane by one of several different processes. The type of process depends on the physical and chemical properties of the substance passing through the membrane.

Small molecules, such as oxygen and carbon dioxide, pass through membranes by a process called passive transport. **Passive transport** *is the movement of substances through a cell membrane without using the cell's energy*. Passive transport depends on the amount of a substance on each side of a membrane. For example, suppose there are more molecules of oxygen outside a cell than inside it. Oxygen will move into that cell until the amount of oxygen is equal on both sides of the cell's membrane. Since oxygen is a small molecule, it passes through a cell membrane without using the cell's energy. The different types of passive transport are explained on the following pages.

Active Reading **2. Define** What is a semipermeable membrane?

Active Reading

FOLDABLES® LA.6.2.2.3

Fold a sheet of paper into a two-tab book. Label the tabs as shown. Use it to organize information about the different types of passive and active transport.

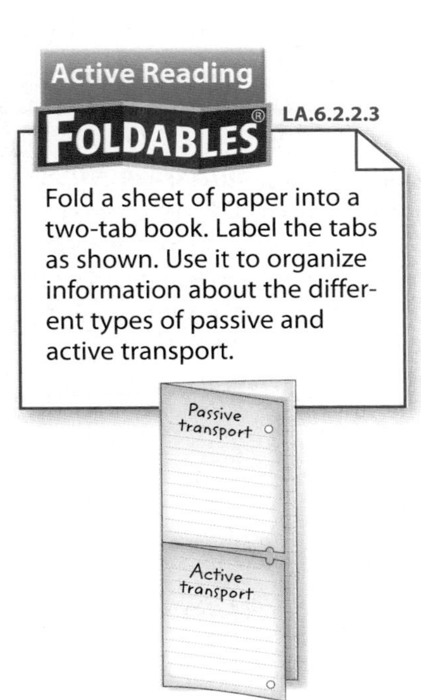

Diffusion

Word Origin

diffusion
from Latin *diffusionem,* means "scatter, pour out"

What happens when the concentration, or amount per unit of volume, of a substance is unequal on each side of a membrane? The molecules will move from the side with a higher concentration of that substance to the side with a lower concentration. **Diffusion** *is the movement of substances from an area of higher concentration to an area of lower concentration.*

Usually, diffusion continues through a membrane until the concentration of a substance is the same on both sides of the membrane. When this happens, a substance is in equilibrium. Compare the two diagrams in **Figure 12.** What happened to the red dye that was added to the water on one side of the membrane? Water and dye passed through the membrane in both directions until there were equal concentrations of water and dye on both sides of the membrane.

Figure 12 Over time, the concentration of dye on either side of the membrane becomes the same.

Diffusion

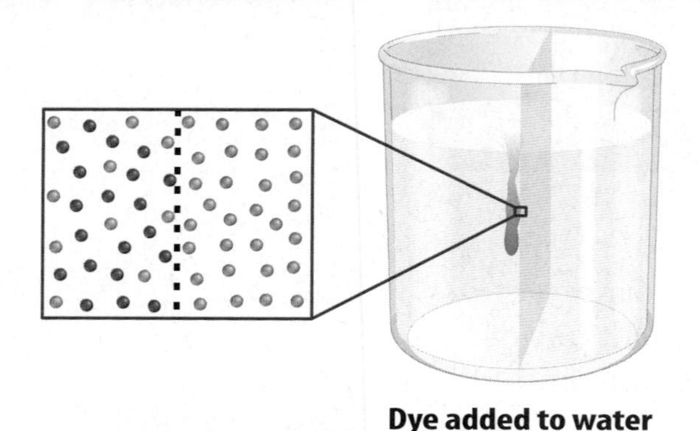

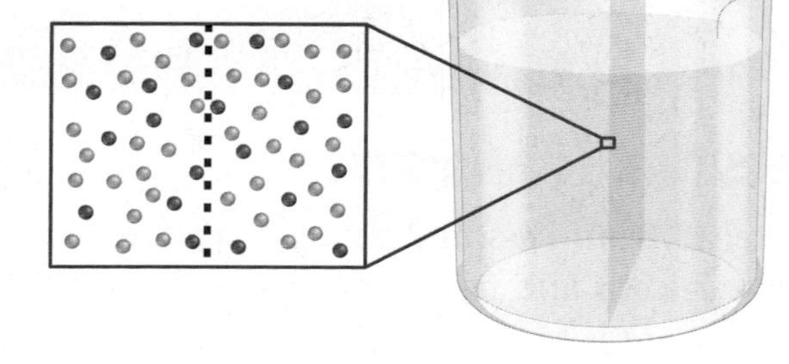

Active Reading **3. Illustrate** What would the water in the beaker on the right look like if the membrane did not let anything through? Draw your answer below.

Osmosis—The Diffusion of Water

Diffusion refers to the movement of any small molecules from higher to lower concentrations. However, **osmosis** *is the diffusion of water molecules only through a membrane.* Semipermeable cell membranes also allow water to pass through them until equilibrium occurs. For example, the amount of water stored in the vacuoles of plant cells can decrease because of osmosis. That is because the concentration of water in the air surrounding the plant is less than the concentration of water inside the vacuoles of plant cells. Water will continue to diffuse into the air until the concentrations of water inside the plant's cells and in the air are equal. If the plant is not watered to replace the lost water, it will wilt and eventually die.

Facilitated Diffusion

Some molecules are too large or are chemically unable to travel through a membrane by diffusion. *When molecules pass through a cell membrane using special proteins called transport proteins, this is* **facilitated diffusion.** Like diffusion and osmosis, facilitated diffusion does not require a cell to use energy. As shown in **Figure 13,** a cell membrane has transport proteins. The two types of transport proteins are carrier proteins and channel proteins. Carrier proteins carry large molecules, such as the sugar molecule glucose, through the cell membrane. Channel proteins form pores through the membrane. Atomic particles, such as sodium ions and potassium ions, pass through the cell membrane by channel proteins.

Active Reading **4. Describe** How do materials move through the cell membrane in facilitated diffusion?

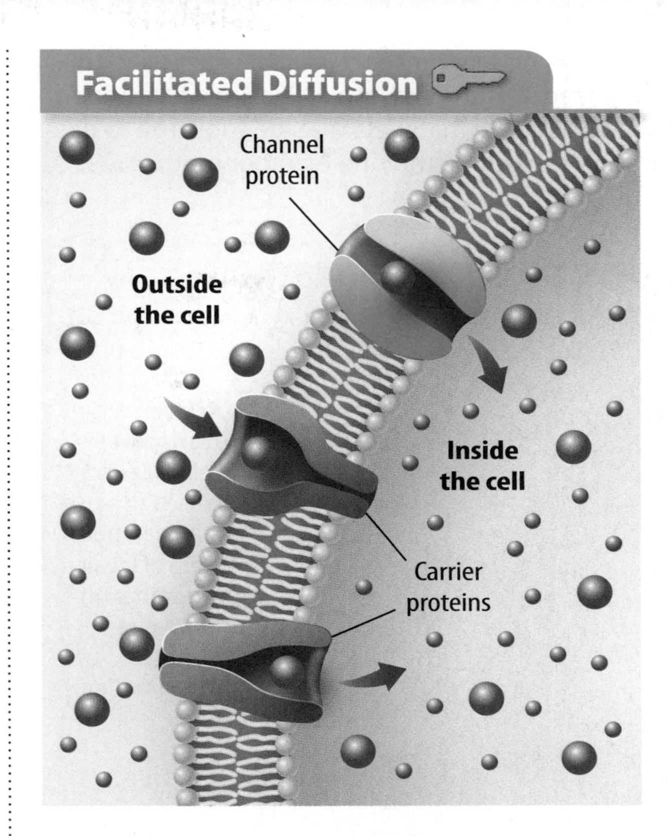

Figure 13 Transport proteins are used to move large molecules into and out of a cell.

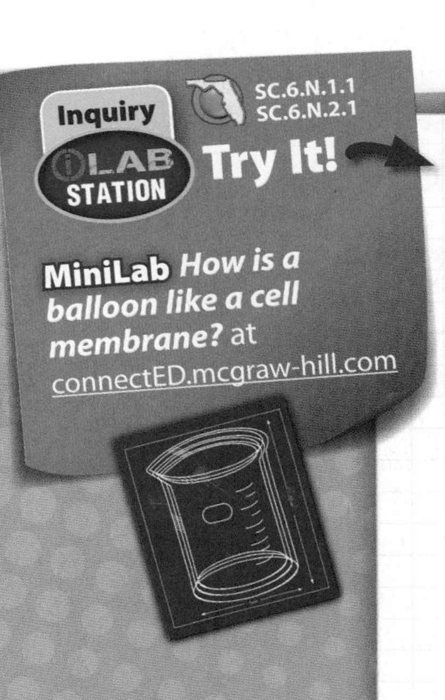

Apply It! After you complete the lab, answer these questions.

1. What type of passive transport occurred in this lab—diffusion, osmosis, or facilitated diffusion? Explain your answer.

2. What do you think the advantages of passive transport are?

Figure 14 Active transport is most often used to bring needed nutrients into a cell. Endocytosis and exocytosis move materials that are too large to pass through the cell membrane by other methods.

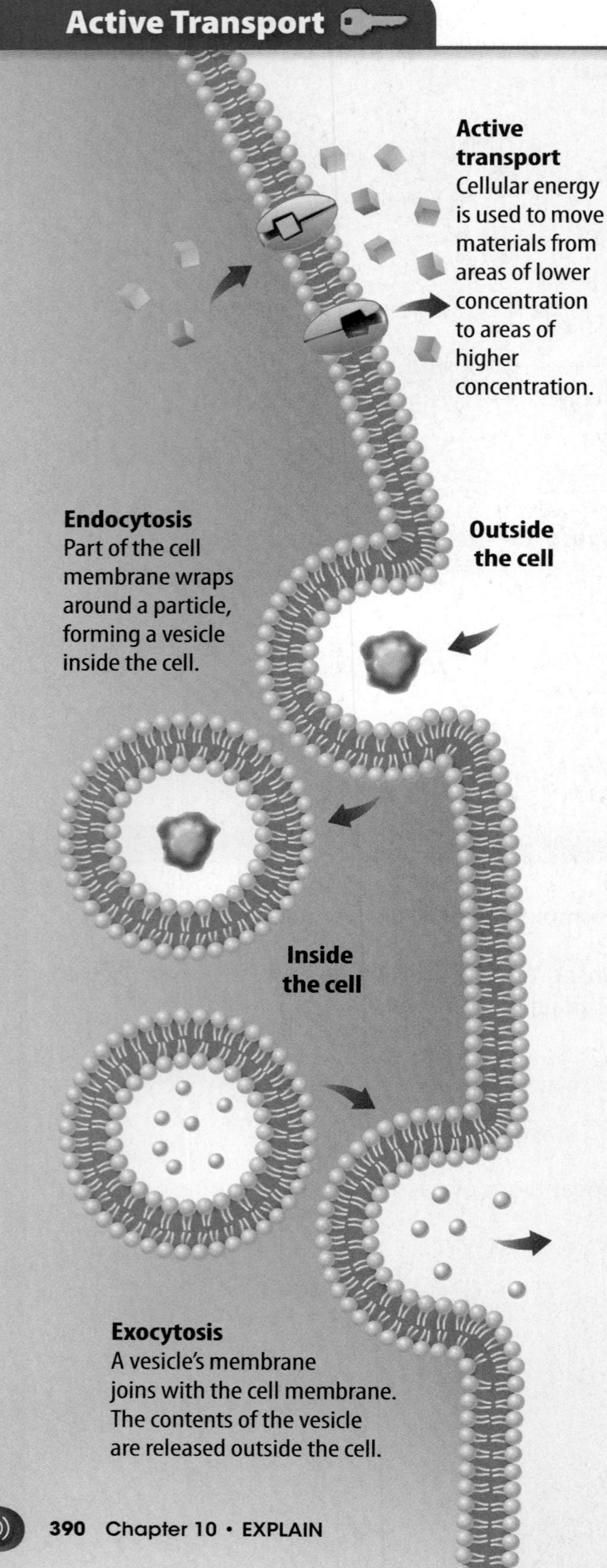

Active Transport

Sometimes when cellular materials pass through membranes it requires a cell to use energy. **Active transport** *is the movement of substances through a cell membrane only by using the cell's energy.*

Recall that passive transport is the movement of substances from areas of higher concentration to areas of lower concentration. However, substances moving by active transport move from areas of lower concentration to areas of higher concentration, as shown in **Figure 14.**

Cells can take in needed nutrients from the environment through carrier proteins by using active transport. This occurs even when concentrations of these nutrients are lower in the environment than inside the cell. Some other molecules and waste materials also leave cells by active transport.

Endocytosis and Exocytosis

Some substances are too large to enter a cell membrane by diffusion or by using a transport protein. These substances can enter a cell by another process. **Endocytosis** (en duh si TOH sus), shown in **Figure 14,** *is the process during which a cell takes in a substance by surrounding it with the cell membrane.* Some cells take in bacteria and viruses using endocytosis.

Some substances are too large to leave a cell by diffusion or by using a transport protein. These substances can leave a cell another way. **Exocytosis** (ek soh si TOH sus), shown in **Figure 14,** *is the process during which a cell's vesicles release their contents outside the cell.* Proteins and other substances are removed from a cell through this process.

5. NGSSS Check **Summarize** How do materials enter and leave cells? SC.6.L.14.3

Cell Size and Transport

Recall that the movement of nutrients, waste material, and other substances into and out of a cell is important for survival. For this movement to happen, the area of the cell membrane must be large compared to its volume. The area of the cell membrane is the cell's surface area. The volume is the amount of space inside the cell. As a cell grows, both its volume and its surface area increase. The volume of a cell increases faster than its surface area. If a cell were to keep growing, it would need large amounts of nutrients and would produce large amounts of waste material. However, the surface area of the cell's membrane would be too small to move enough nutrients and wastes through it for the cell to survive.

Active Reading **6. Analyze** How does cell size affect the transport of materials?

Math Skills MA.6.A.3.6

Use Ratios

A ratio is a comparison of two numbers, such as surface area and volume. If a cell were cube-shaped, you would calculate surface area by multiplying its length (ℓ) by its width (w) by the number of sides (6).

You would calculate the volume of the cell by multiplying its length (ℓ) by its width (w) by its height (h).

To find the surface-area-to-volume ratio of the cell, divide its surface area by its volume.

In the table below, surface-area-to-volume ratios are calculated for cells that are 1 mm, 2 mm, and 4 mm per side. Notice how the ratios change as the cell's size increases.

Surface area $= \ell \times w \times 6$

Volume $= \ell \times w \times h$

$\frac{\text{Surface area}}{\text{Volume}}$

	1 mm, 1 mm, 1 mm	2 mm, 2 mm, 2 mm	4 mm, 4 mm, 4 mm
Length	1 mm	2 mm	4 mm
Width	1 mm	2 mm	4 mm
Height	1 mm	2 mm	4 mm
Number of sides	6	6	6
Surface area (ℓ × w × no. of sides)	1 mm × 1 mm × 6 = 6 mm^2	2 mm × 2 mm × 6 = 24 mm^2	4 mm × 4 mm × 6 = 96 mm^2
Volume ($\ell \times w \times h$)	1 mm × 1 mm × 1 mm = 1 mm^3	2 mm × 2 mm × 2 mm = 8 mm^3	4 mm × 4 mm × 4 mm = 64 mm^3
Surface-area-to-volume ratio	$\frac{6\ mm^2}{1\ mm^3} = \frac{6}{1}$ or 6:1	$\frac{24\ mm^2}{8\ mm^3} = \frac{3}{1}$ or 3:1	$\frac{96\ mm^2}{64\ mm^3} = \frac{1.5}{1}$ or 1.5:1

Practice

7. What is the surface-area-to-volume ratio of a cell whose six sides are 3 mm long?

Lesson Review 3

Visual Summary

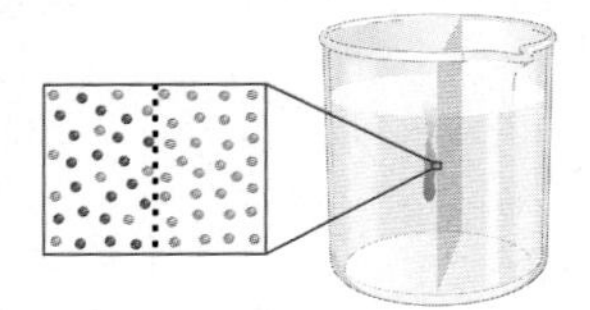

Small molecules can move from an area of higher concentration to an area of lower concentration.

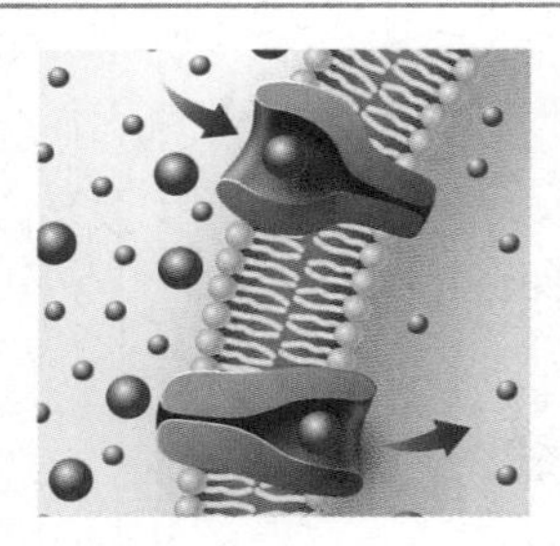

Proteins transport larger molecules through a cell membrane.

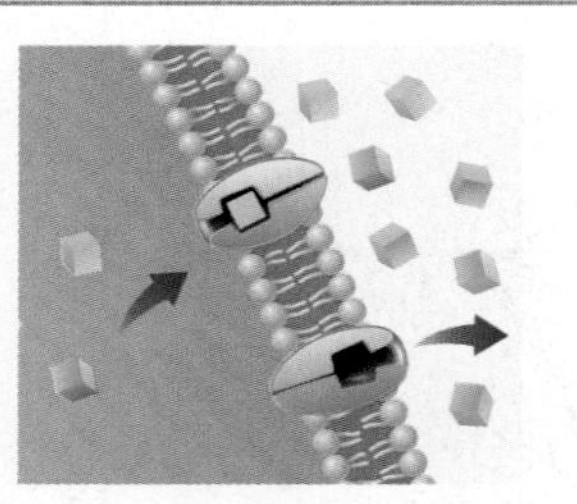

Some molecules move from areas of lower concentration to areas of higher concentration.

Use Vocabulary

1. **Distinguish** between active transport and passive transport. SC.6.L.14.3

2. The process by which vesicles move substances out of a cell is ________.

Understand Key Concepts 

3. **Summarize** the function of endocytosis.

4. What is limited by a cell's surface-area-to-volume ratio?
   - (A) cell shape
   - (B) cell size
   - (C) cell surface area
   - (D) cell volume

Interpret Graphics

5. Fill in the graphic organizer below to describe ways that cells transport substances. SC.6.L.14.3

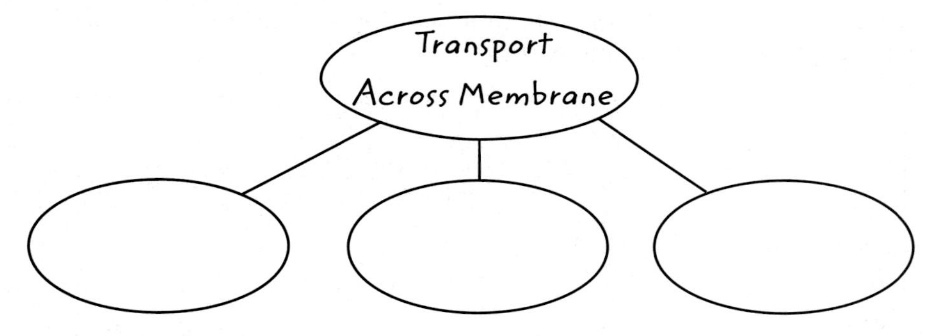

Critical Thinking

6. **Relate** the surface area of a cell to the transport of materials.

Math Skills MA.6.A.3.6

7. **Calculate** the surface-area-to-volume ratio of a cube whose sides are 6 cm long.

Explain the process of facilitated diffusion.

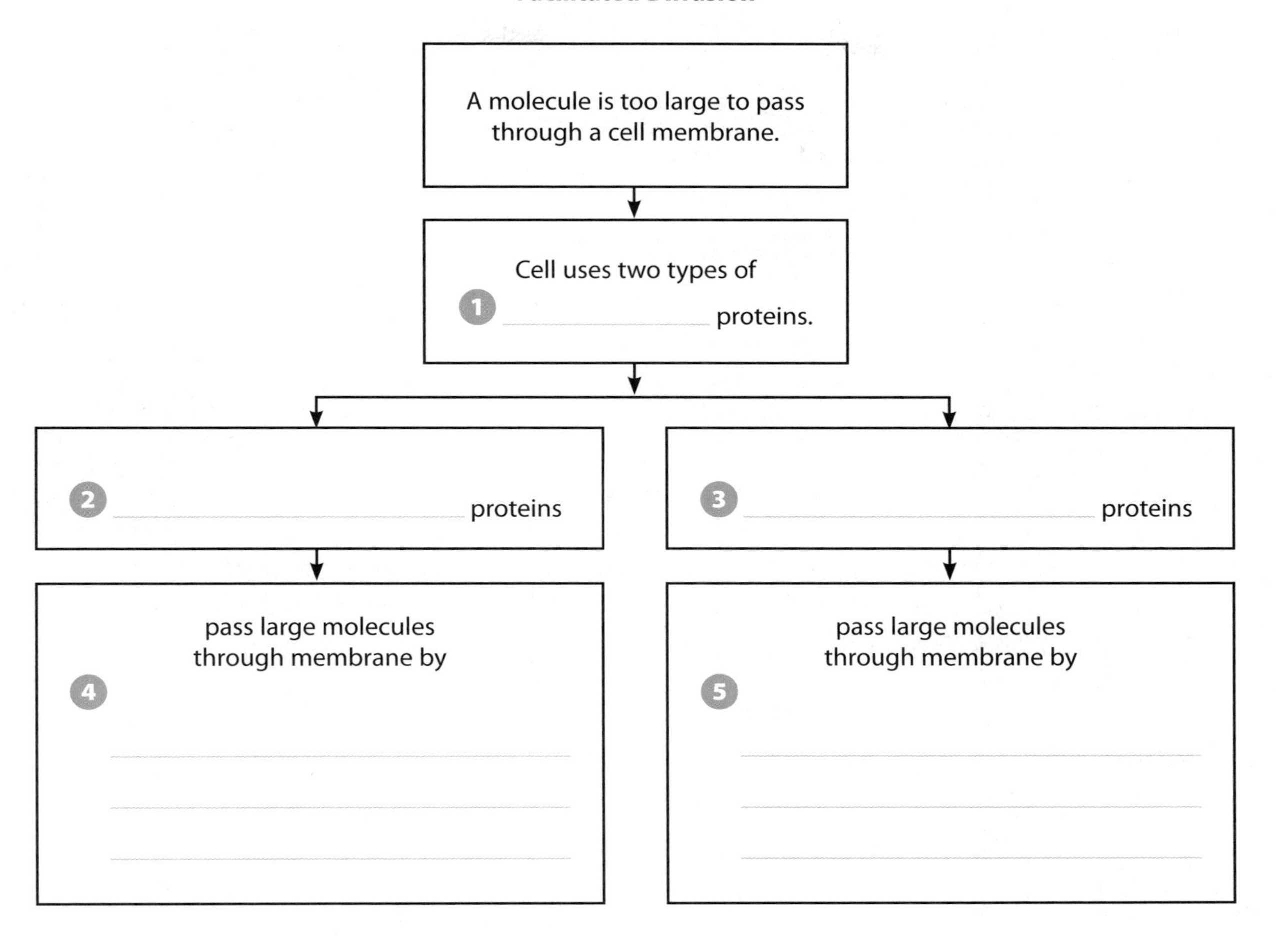

Organize information about active transport.

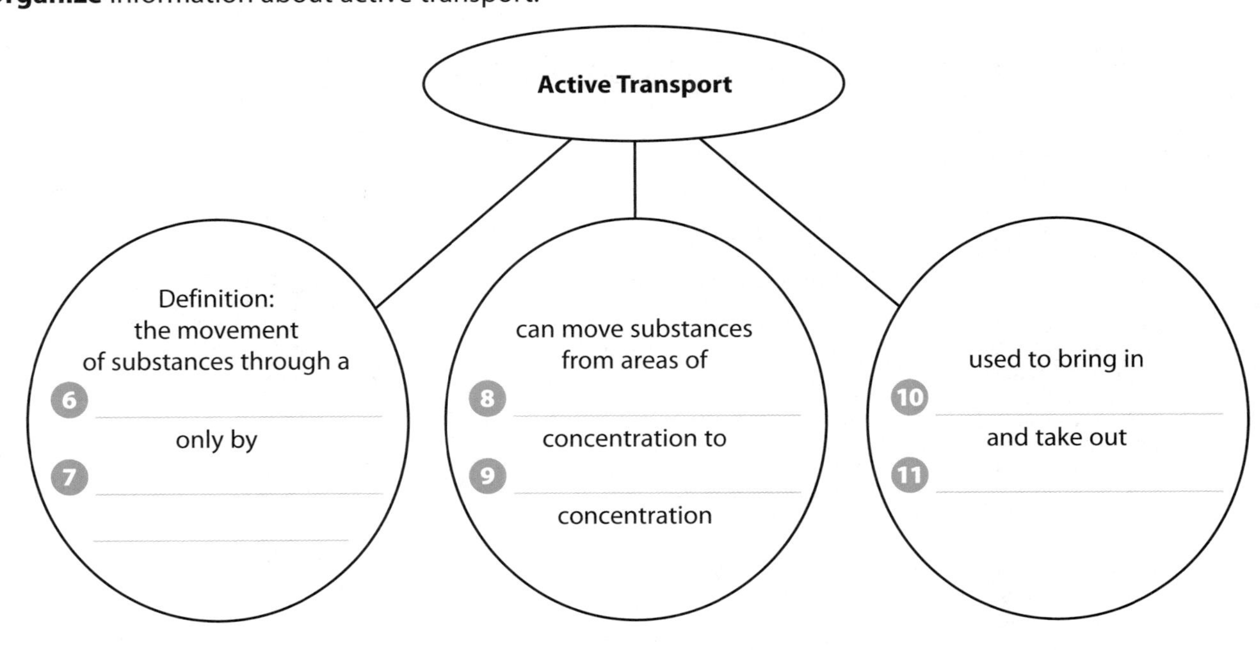

Lesson 4

Cells and ENERGY

ESSENTIAL QUESTIONS

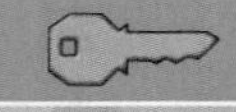 How does a cell obtain energy?

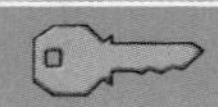 How do some cells make food molecules?

Vocabulary

cellular respiration p. 395
glycolysis p. 395
fermentation p. 396
photosynthesis p. 397

Florida NGSSS

LA.6.2.2.3 The student will organize information to show understanding (e.g., representing main ideas within text through charting, mapping, paraphrasing, summarizing, or comparing/contrasting);

MA.6.A.3.6 Construct and analyze tables, graphs, and equations to describe linear functions and other simple relations using both common language and algebraic notation.

SC.6.L.14.3 Recognize and explore how cells of all organisms undergo similar processes to maintain homeostasis, including extracting energy from food, getting rid of waste, and reproducing.

SC.6.N.1.1 Define a problem from the sixth grade curriculum, use appropriate reference materials to support scientific understanding, plan and carry out scientific investigation of various types, such as systematic observations or experiments, identify variables, collect and organize data, interpret data in charts, tables, and graphics, analyze information, make predictions, and defend conclusions.

SC.6.N.1.4 Discuss, compare, and negotiate methods used, results obtained, and explanations among groups of students conducting the same investigation.

SC.6.N.2.1 Distinguish science from other activities involving thought.

SC.6.N.1.1

Launch Lab

5 minutes

What do you exhale?

Does the air you breathe in differ from the air you breathe out?

Procedure

1. Read and complete a lab safety form.
2. Unwrap a **straw.** Use the straw to slowly blow into a small **cup** of **bromthymol blue.** Do not splash the liquid out of the cup.
3. Record any changes in the solution.

Data and Observations

Think About This

1. What changes did you observe in the solution?

2. What do you think caused the changes in the solution?

3. Key Concept Why do you think the air you inhale differs from the air you exhale?

Why are there bubbles?

1. Have you ever seen bubbles on a green plant in an aquarium? Where do you think the bubbles come from?

Cellular Respiration

When you are tired, you might eat something to give you energy. All living things, from one-celled organisms to humans, need energy to survive. Recall that cells process energy from food into the energy-storage compound ATP. **Cellular respiration** *is a series of chemical reactions that convert the energy in food molecules into a usable form of energy called ATP.* Cellular respiration is a complex process that occurs in two parts of a cell—the cytoplasm and the mitochondria.

Reactions in the Cytoplasm

The first step of cellular respiration, called glycolysis, occurs in the cytoplasm of all cells. **Glycolysis** *is a process by which glucose, a sugar, is broken down into smaller molecules.* As shown in **Figure 15,** glycolysis produces some ATP molecules. It also uses energy from other ATP molecules. You will read on the following page that more ATP is made during the second step of cellular respiration than during glycolysis.

Active Reading **2. Determine** What is produced during glycolysis?

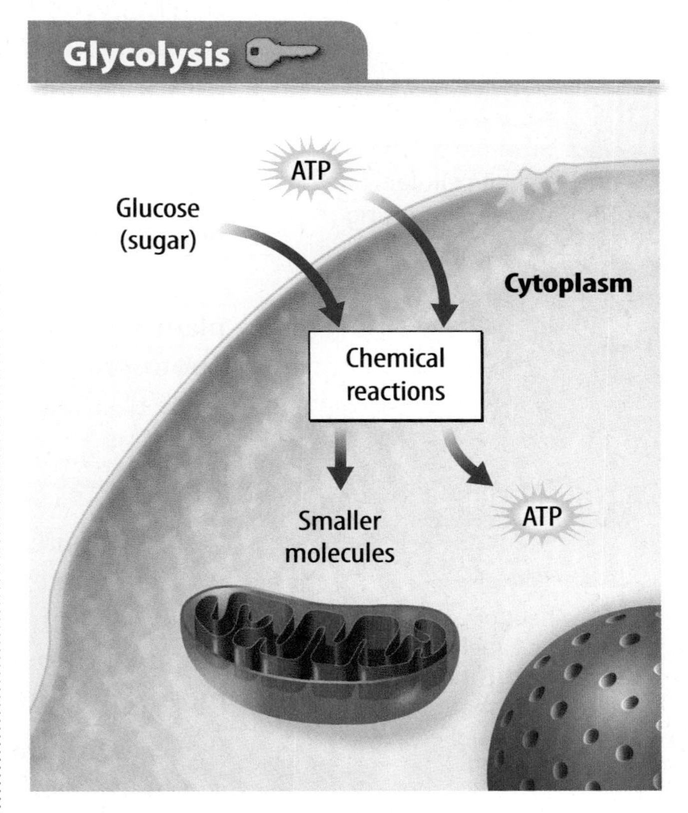

Figure 15 Glycolysis is the first step of cellular respiration.

3. Visual Check Explain Why is ATP shown twice in the figure?

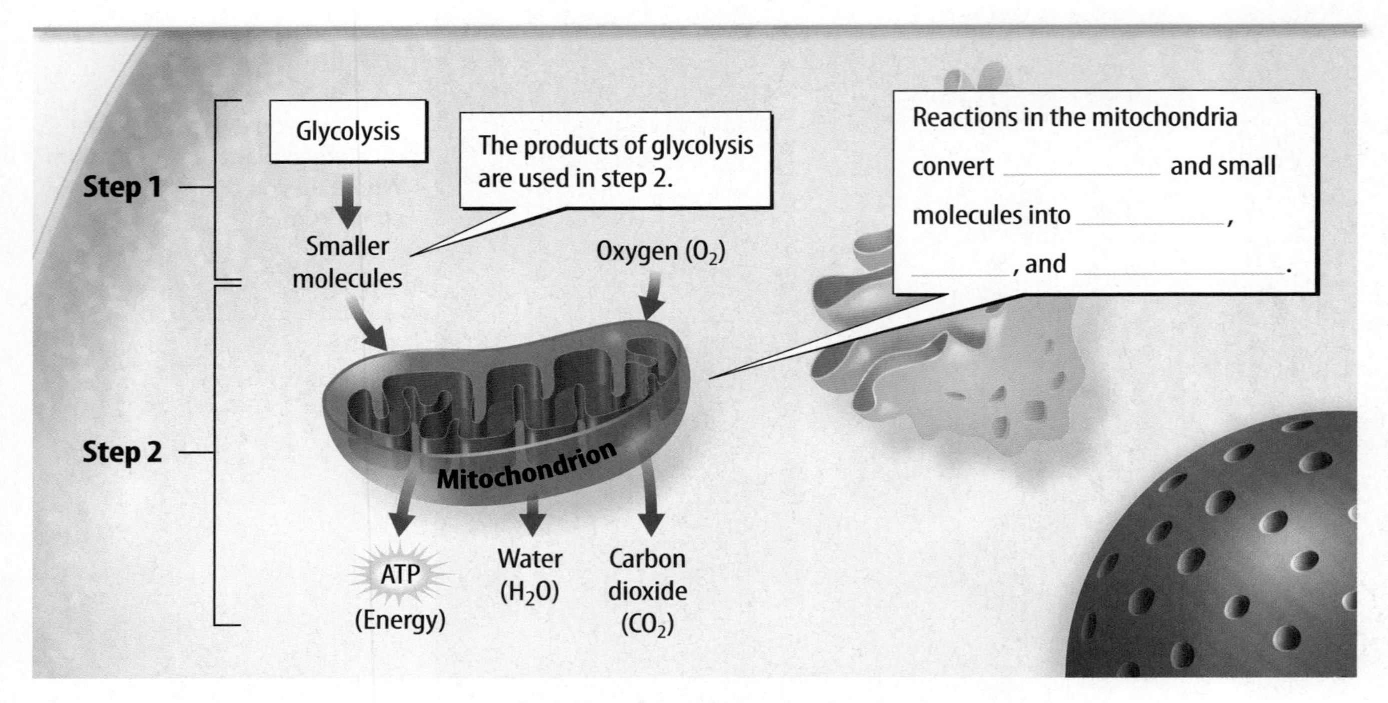

Figure 16 After glycolysis, cellular respiration continues in the mitochondria.

Active Reading **4. Apply** Summarize the second step of cellular respiration by completing the diagram.

Reactions in the Mitochondria

The second step of cellular respiration occurs in the mitochondria of eukaryotic cells, as shown in **Figure 16.** This step of cellular respiration requires oxygen. The smaller molecules made from glucose during glycolysis are broken down. Large amounts of ATP—usable energy—are produced. Cells use ATP to power all cellular processes. Two waste products—water and carbon dioxide (CO_2)—are given off during this step.

The CO_2 released by cells as a waste product is used by plants and some unicellular organisms in another process called photosynthesis. You will read more about the chemical reactions that take place during photosynthesis in this lesson.

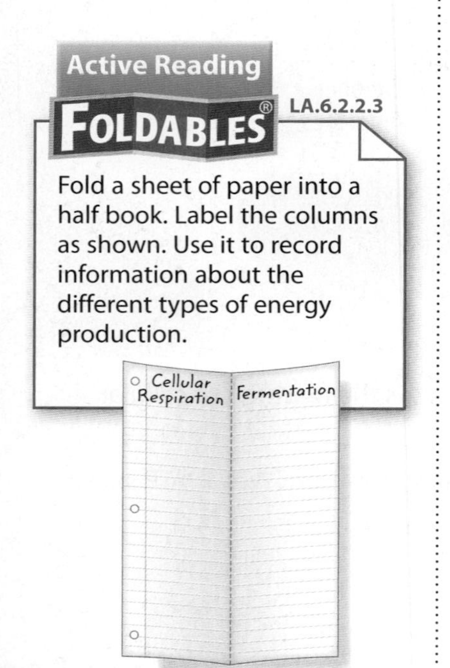

Fermentation

Have you ever felt out of breath after exercising? Sometimes when you exercise, your cells don't have enough oxygen to make ATP through cellular respiration. Then, chemical energy is obtained through a different process called fermentation. This process does not use oxygen.

Fermentation *is a reaction that eukaryotic and prokaryotic cells can use to obtain energy from food when oxygen levels are low.* Because no oxygen is used, fermentation makes less ATP than cellular respiration does. Fermentation occurs in a cell's cytoplasm, not in mitochondria.

5. NGSSS Check Explain How does a cell obtain energy? SC.6.1.14.3

Types of Fermentation

One type of fermentation occurs when glucose is converted into ATP and a waste product called lactic acid, as illustrated in **Figure 17.** Some bacteria and fungi help produce cheese, yogurt, and sour cream using lactic-acid fermentation. Muscle cells in humans and other animals can use lactic-acid fermentation and obtain energy during exercise.

Some types of bacteria and yeast make ATP through a process called alcohol fermentation. However, instead of producing lactic acid, alcohol fermentation produces an alcohol called ethanol and CO_2, also illustrated in **Figure 17.** Some types of breads are made using yeast. The CO_2 produced by yeast during alcohol fermentation makes the dough rise.

Figure 17 Your muscle cells and yeast cells can convert glucose to energy through fermentation.

Active Reading **6. Recall** Identify the waste products of the two types of fermentation by completing the figure below.

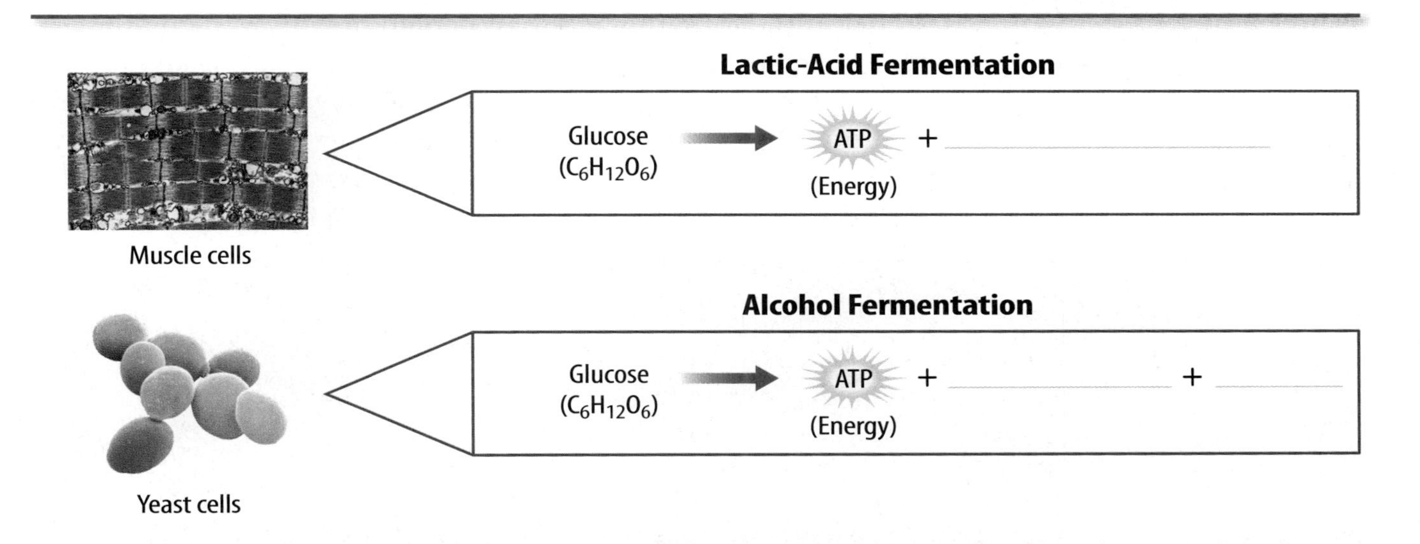

Photosynthesis

Humans and other animals convert food energy into ATP through cellular respiration. However, plants and some unicellular organisms obtain energy from light. **Photosynthesis** *is a series of chemical reactions that convert light energy, water, and CO_2 into the food-energy molecule glucose and give off oxygen.*

Active Reading **7. Recall** Underline the definition of photosynthesis.

WORD ORIGIN

photosynthesis
from Greek photo, means "light"; and synthesis, means "composition"

Lights and Pigments

Photosynthesis requires light energy. In plants, pigments such as chlorophyll absorb light energy. When chlorophyll absorbs light, it absorbs all colors except green. Green light is reflected as the green color seen in leaves. However, plants contain many pigments that reflect other colors, such as yellow and red.

Reactions in Chloroplasts

The light energy absorbed by chlorophyll and other pigments powers the chemical reactions of photosynthesis. These reactions occur in chloroplasts, the organelles in plant cells that convert light energy to chemical energy in food. During photosynthesis, light energy, water, and carbon dioxide combine and make sugars. Photosynthesis also produces oxygen that is released into the atmosphere, as shown in **Figure 18.**

Active Reading **8. Determine** Highlight the substances that some cells use to make food.

Importance of Photosynthesis

Recall that photosynthesis uses light energy and CO_2 and makes food energy and releases oxygen. This food energy is stored in the form of glucose. When an organism, such as the bird in **Figure 18,** eats plant material, such as fruit, it takes in food energy. An organism's cells use the oxygen released during photosynthesis and convert the food energy into usable energy through cellular respiration. **Figure 18** illustrates the important relationship between cellular respiration and photosynthesis.

Figure 18 The relationship between cellular respiration and photosynthesis is important for life.

Active Reading **9. Classify** Label the chemical reactions below as either *cellular respiration* or *photosynthesis.*

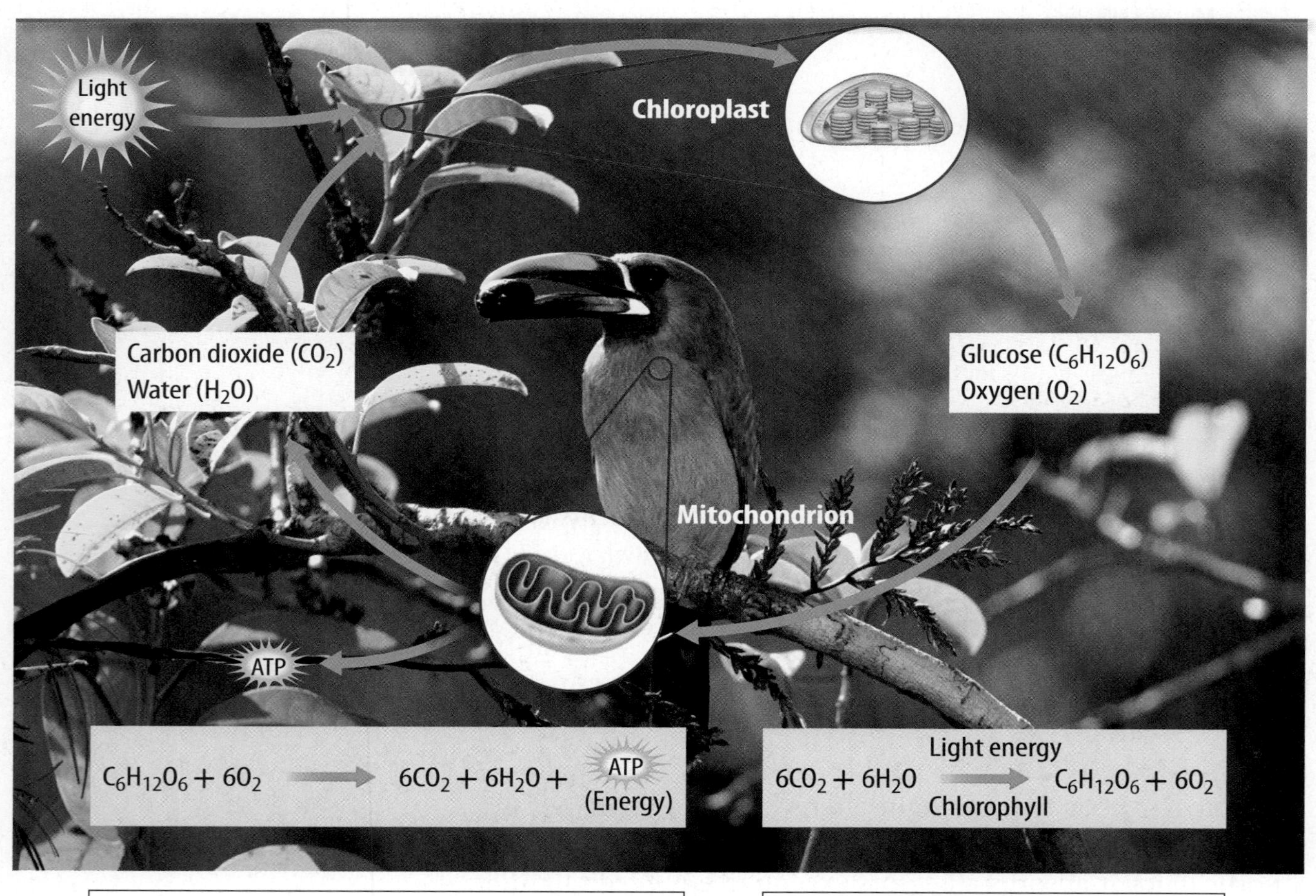

Lesson Review 4

Visual Summary

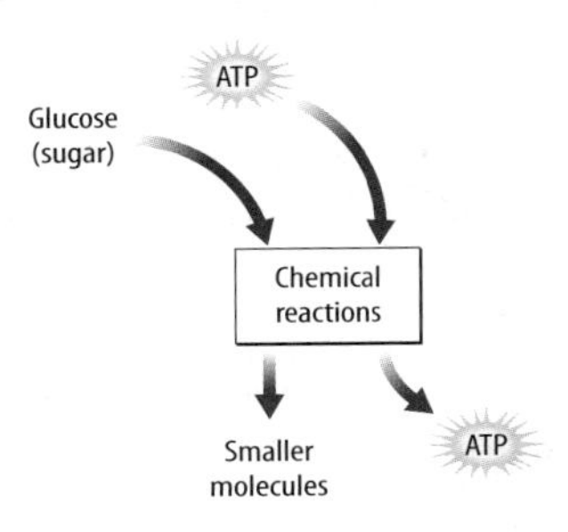

Glycolysis is the first step in cellular respiration.

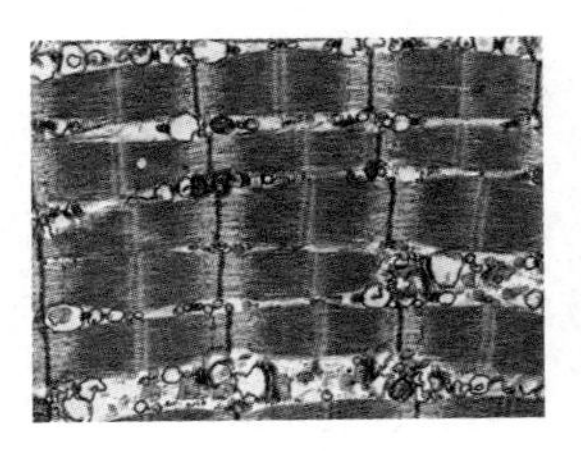

Fermentation provides cells, such as muscle cells, with energy when oxygen levels are low.

Light energy powers the chemical reactions of photosynthesis.

Use Vocabulary

1. **Define** *glycolysis* using your own words.

2. **Distinguish** between cellular respiration and fermentation. SC.6.L.14.3

3. A process used by plants to convert light energy into food energy is ______. SC.6.L.14.3

Understand Key Concepts

4. Which contains pigments that absorb light energy?
   - (A) chloroplast
   - (B) mitochondrion
   - (C) nucleus
   - (D) vacuole

5. **Relate** mitochondria to cellular respiration.

6. **Describe** the role of chlorophyll in photosynthesis.

Interpret Graphics

7. Fill in the boxes with the substances used and produced during photosynthesis.

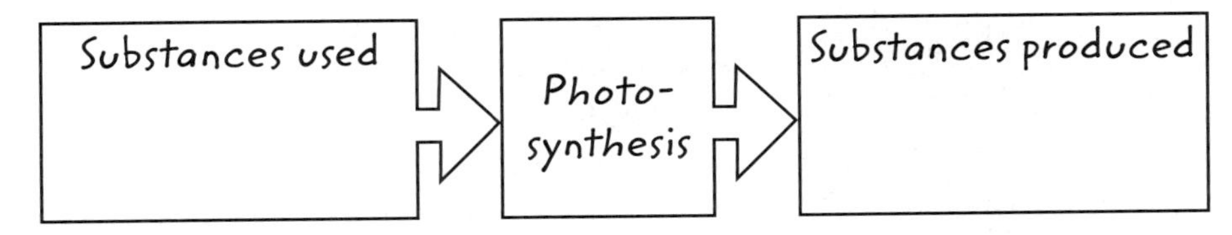

Critical Thinking

8. **Design** a concept map on a separate sheet of paper to show the relationship between cellular respiration in animals and photosynthesis in plants. SC.6.L.14.3

9. **Summarize** the roles of glucose and ATP in energy processing.

Chapter 10 Study Guide

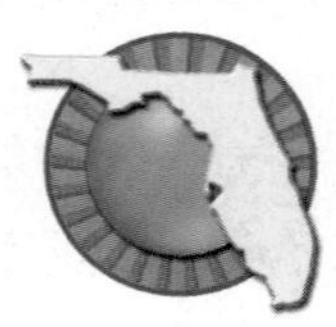

Think About It! A cell is made up of various structures that are essential for growth, reproduction, and homeostasis. They provide support and movement, process energy, and transport materials.

Key Concepts Summary

LESSON 1 Cells and Life

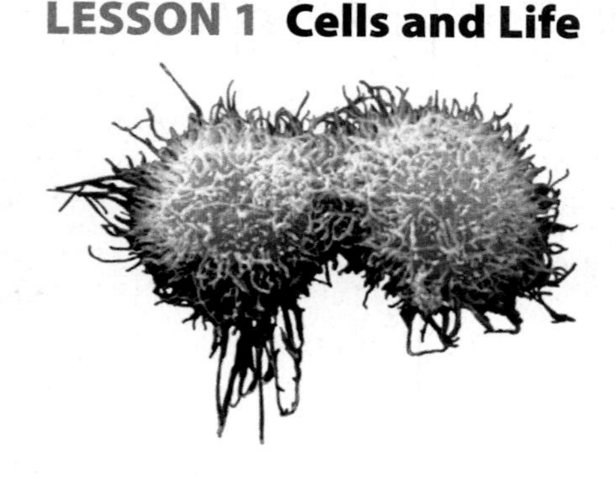

- The invention of the microscope led to discoveries about cells. In time, scientists used these discoveries to develop the **cell theory**, which explains how cells and living things are related.
- Cells are composed mainly of water, **proteins, nucleic acids, lipids**, and **carbohydrates**.

Vocabulary

cell theory p. 370
macromolecule p. 371
nucleic acid p. 372
protein p. 373
lipid p. 373
carbohydrate p. 373

LESSON 2 The Cell

- Cell structures have specific functions, such as supporting a cell, moving a cell, controlling cell activities, processing energy, and transporting molecules.
- A prokaryotic cell lacks a nucleus and other **organelles**, while a eukaryotic cell has a nucleus and other organelles.

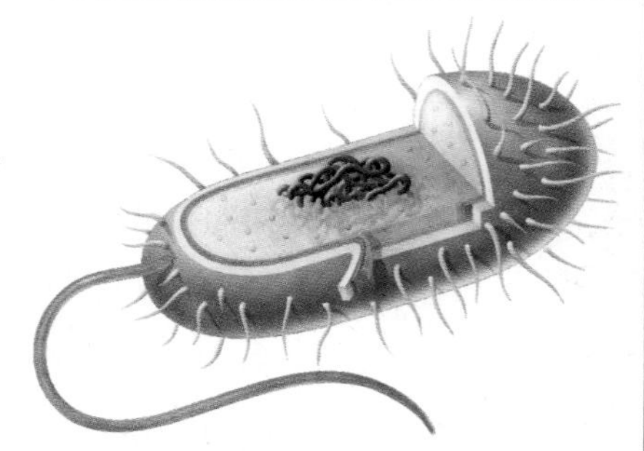

cell membrane p. 378
cell wall p. 378
cytoplasm p. 379
cytoskeleton p. 379
organelle p. 380
nucleus p. 381
chloroplast p. 383

LESSON 3 Moving Cellular Material

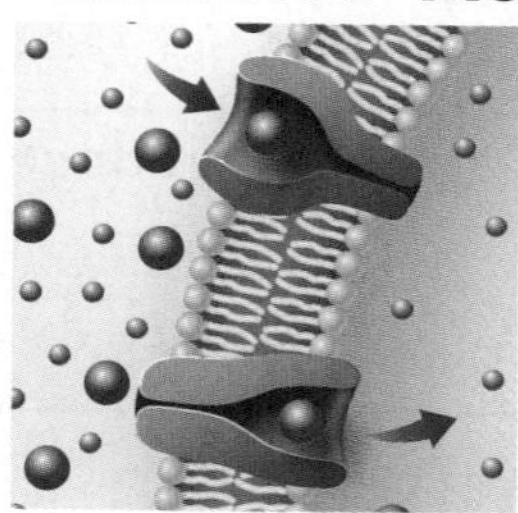

- Materials enter and leave a cell through the cell membrane using **passive transport** or **active transport, endocytosis**, and **exocytosis**.
- The ratio of surface area to volume limits the size of a cell. In a smaller cell, the high surface-area-to-volume ratio allows materials to move easily to all parts of a cell.

passive transport p. 387
diffusion p. 388
osmosis p. 388
facilitated diffusion p. 389
active transport p. 390
endocytosis p. 390
exocytosis p. 390

LESSON 4 Cells and Energy

- All living cells release energy from food molecules through **cellular respiration** and/or **fermentation**.
- Some cells make food molecules using light energy through the process of **photosynthesis**.

$$C_6H_{12}O_6 + 6O_2 \rightarrow 6CO_2 + 6H_2O + \text{ATP (Energy)}$$

Cellular respiration

$$6CO_2 + 6H_2O \xrightarrow[\text{Chlorophyll}]{\text{Light energy}} C_6H_{12}O_6 + 6O_2$$

Photosynthesis

cellular respiration p. 395
glycolysis p. 395
fermentation p. 396
photosynthesis p. 397

Active Reading

FOLDABLES® Chapter Project

Assemble your lesson Foldables as shown to make a Chapter Project. Use the project to review what you have learned in this chapter.

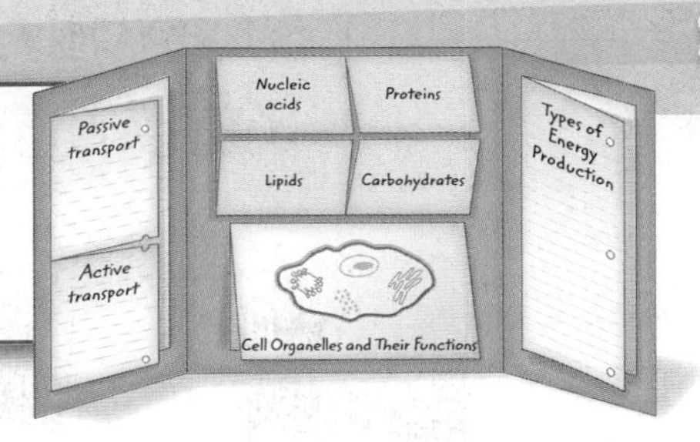

Use Vocabulary

1. Substances formed by joining smaller molecules together are called ______________.

2. The ______________ consists of proteins joined together to create fiberlike structures inside cells.

3. The movement of substances from an area of high concentration to an area of low concentration is called ______________.

4. A process that uses oxygen to convert energy from food into ATP is ______________ ______________.

Link Vocabulary and Key Concepts

Use vocabulary terms from the previous page to complete the concept map.

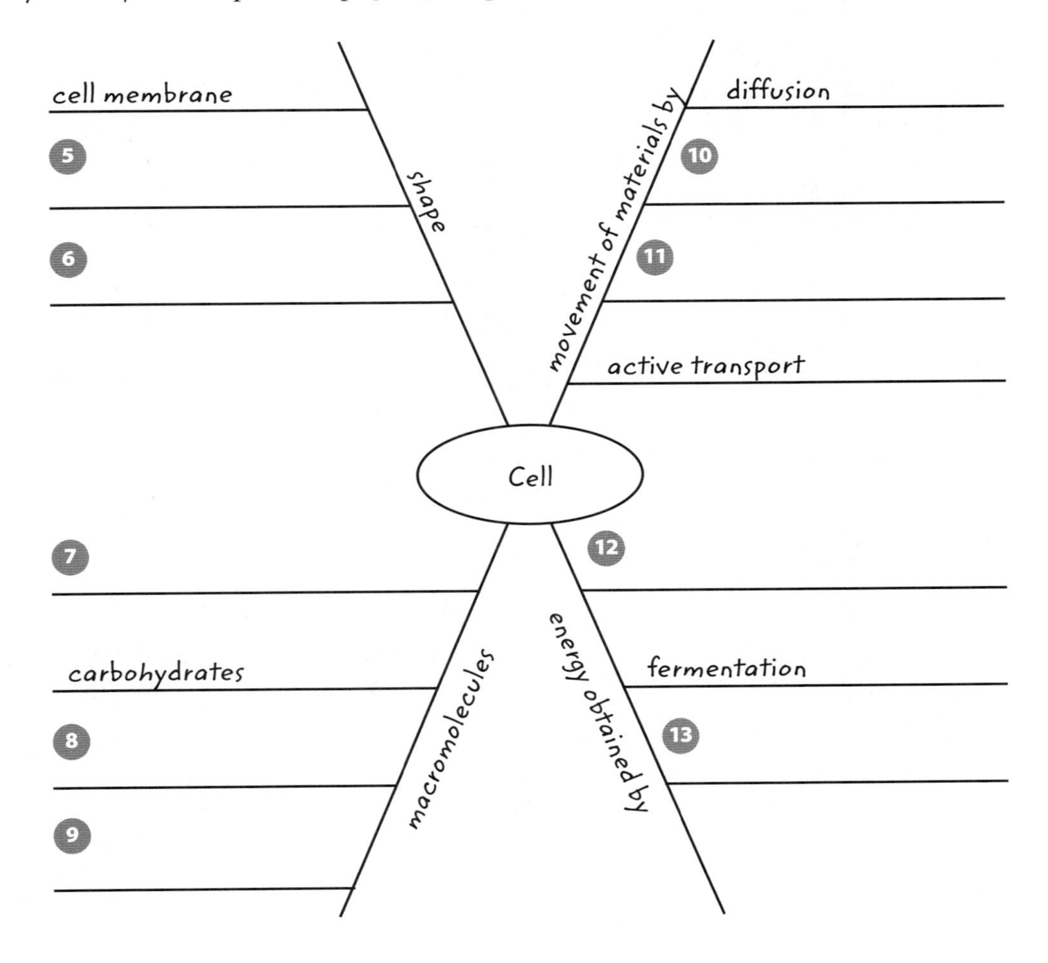

Chapter 10 Review

Fill in the correct answer choice.

Understand Key Concepts

1 Cholesterol is which type of macromolecule? SC.6.L.14.2
- Ⓐ carbohydrate
- Ⓑ lipid
- Ⓒ nucleic acid
- Ⓓ within the crust.

2 Genetic information is stored in which macromolecule? SC.6.L.14.2
- Ⓐ DNA
- Ⓑ glucose
- Ⓒ lipid
- Ⓓ starch

3 The arrow below is pointing to which cell part? SC.6.L.14.4

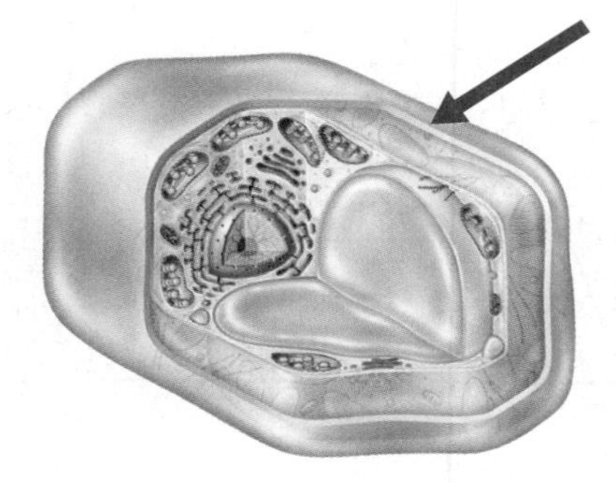

- Ⓐ chloroplast
- Ⓑ mitochondrion
- Ⓒ cell membrane
- Ⓓ cell wall

4 Which best describes vacuoles? SC.6.L.14.4
- Ⓐ lipids
- Ⓑ proteins
- Ⓒ contained in mitochondria
- Ⓓ storage compartments

5 Which is true of fermentation? SC.6.L.14.3
- Ⓐ does not generate energy
- Ⓑ does not require oxygen
- Ⓒ occurs in mitochondria
- Ⓓ produces lots of ATP

Critical Thinking

6 **Evaluate** the importance of the microscope to biology. SC.6.L.14.2

7 **Summarize** the role of water in cells. LA.6.2.2.3

8 **Infer** Why do cells need carrier proteins that transport glucose? SC.6.L.14.3

9 **Compare** the amounts of ATP generated in cellular respiration and fermentation. LA.6.2.2.3

10 **Hypothesize** how air pollution like smog affects photosynthesis. SC.6.L.14.3

11 **Compare** prokaryotes and eukaryotes by filling in the table below. SC.6.L.14.3

Structure	Prokaryote (yes or no)	Eukaryote (yes or no)
Cell membrane		
DNA		
Nucleus		
Endoplasmic reticulum		
Golgi apparatus		
Cell wall		

Writing in Science

12 **Write** On a separate sheet of paper, write a five-sentence paragraph relating the cytoskeleton to the walls of a building. Be sure to include a topic sentence and a concluding sentence in your paragraph. SC.6.L.14.4

Big Idea Review

13 How do the structures and processes of a cell enable it to survive? As an example, explain how chloroplasts help plant cells. SC.6.L.14.4

14 The photo at the beginning of the chapter shows a protozoan. What structures enable it to get food into its mouth? SC.6.L.14.4

Math Skills MA.6.A.3.6

Use Ratios

15 A rectangular solid measures 4 cm long by 2 cm wide by 2 cm high. What is the surface-area-to-volume ratio of the solid?

16 At different times during its growth, a cell has the following surface areas and volumes:

Time	Surface area (μm)	Volume (μm)
1	6	1
2	24	8
3	54	27

What happens to the surface-area-to-volume ratio as the cell grows?

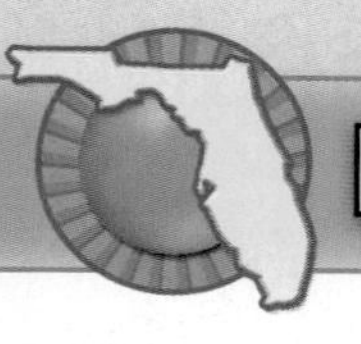

Florida NGSSS Benchmark Practice

Fill in the correct answer choice.

Multiple Choice

1 Which is NOT a principle of cell theory? SC.6.L.14.2

Ⓐ All living things are made of one or more cells.

Ⓑ The cell is the smallest unit of life.

Ⓒ All cells are exactly the same.

Ⓓ All new cells come from preexisting cells.

Use the diagram below to answer question 2.

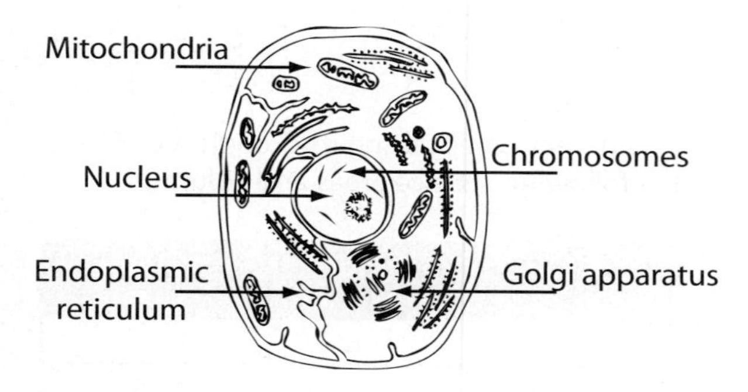

2 Where does respiration occur in the cell? SC.6.L.14.4

Ⓕ endoplasmic reticulum

Ⓖ nucleus

Ⓗ Golgi apparatus

Ⓘ mitochondria

3 Which macromolecule in a cell is responsible for storing energy, providing structural support, and is needed for communication among cells? SC.6.L.14.3, SC.6.L.14.4

Ⓐ proteins

Ⓑ lipids

Ⓒ nucleic acids

Ⓓ carbohydrates

Use the diagram below to answer questions 4 and 5.

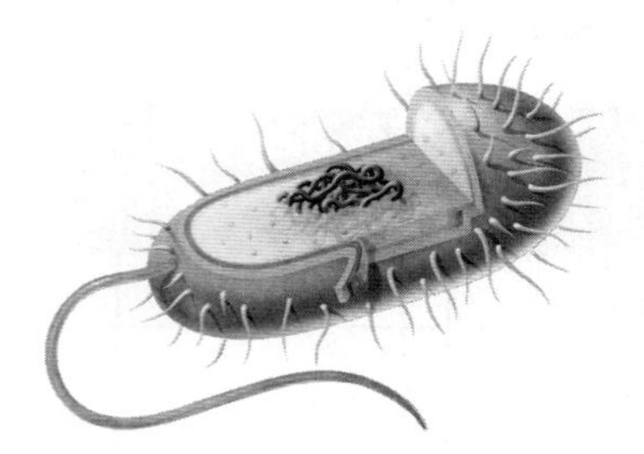

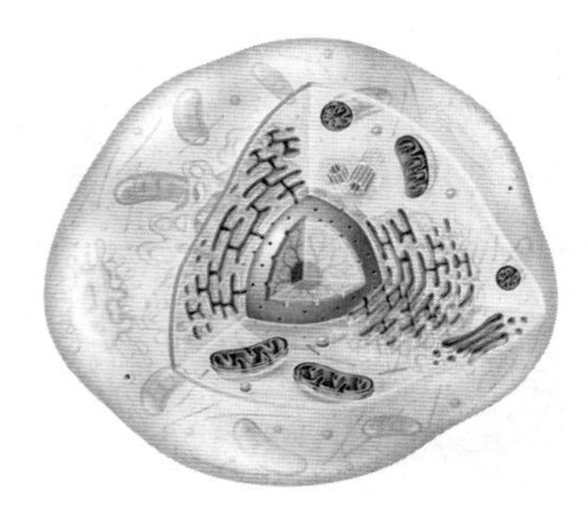

4 Which structure does the arrow point to in the eukaryotic cell? SC.6.L.14.4

Ⓐ cytoplasm

Ⓑ lysosome

Ⓒ nucleus

Ⓓ ribosome

5 Which feature does a typical prokaryotic cell have that is missing from some eukaryotic cells, like the one above? SC.6.L.14.4

Ⓕ cytoplasm

Ⓖ DNA

Ⓗ cell membrane

Ⓘ cell wall

6 Which cell organelle uses light energy to make food during photosynthesis? **SC.6.L.14.4**

Ⓐ nucleus

Ⓑ mitochondria

Ⓒ vacuole

Ⓓ chloroplast

Use the diagram below to answer question 7.

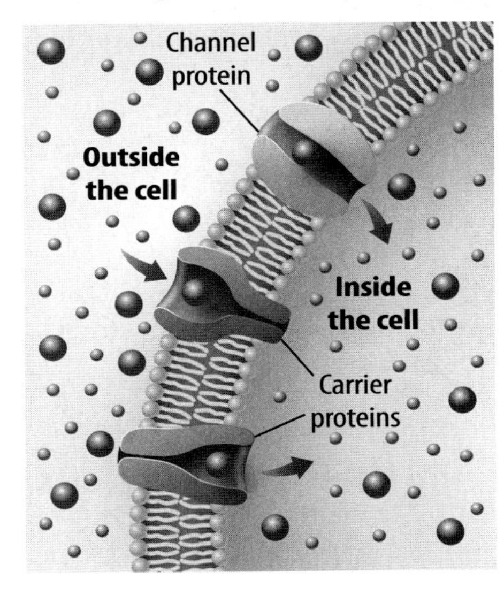

7 What process uses channel proteins and carrier proteins to control the movement of substances into and out of cells? **SC.6.L.14.3**

Ⓕ osmosis

Ⓖ facilitated diffusion

Ⓗ passive transport

Ⓘ diffusion

8 Which structure in an animal cell is a network of threadlike proteins that are joined together to provide a framework inside the cell? **SC.6.L.14.4**

Ⓕ cytoskeleton

Ⓖ cell wall

Ⓗ cytoplasm

Ⓘ cell membrane

9 What process do cells use when they need to take in nutrients from a lower concentration to a higher concentration? **SC.6.L.14.3**

Ⓐ diffusion

Ⓑ active transport

Ⓒ passive transport

Ⓓ exocytosis

10 Which organism has a cell wall? **SC.6.L.14.4**

Ⓕ mouse

Ⓖ pine tree

Ⓗ lion

Ⓘ snake

11 What is the main ingredient in any cell? **SC.6.L.14.4**

Ⓐ protein

Ⓑ nucleic acids

Ⓒ lipids

Ⓓ water

NEED EXTRA HELP?

If You Missed Question...	1	2	3	4	5	6	7	8	9	10	11
Go to Lesson...	1	2	1	2	2	2	3	2	3	2	1

Name ______________________ Date ____________

Multiple Choice *Bubble the correct answer.*

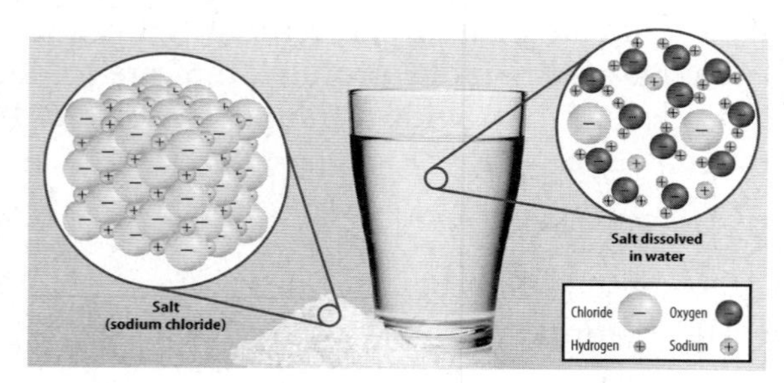

1. The picture above shows salt (NaCl) dissolved in water (H_2O). Which statement is true? **SC.6.N.1.1**

Ⓐ Chloride is attracted to the hydrogen atoms in a water molecule.

Ⓑ Sodium is attracted to the hydrogen atoms in a water molecule.

Ⓒ Neither sodium nor chloride is attracted to the atoms in a water molecule.

Ⓓ Sodium and chloride are attracted to the oxygen atoms in a water molecule.

2. Which type of macromolecule helps a cell break down food? **SC.6.L.14.2**

Ⓕ lipids

Ⓖ proteins

Ⓗ carbohydrates

Ⓘ nucleic acids

3. An amoeba can divide and form two new identical amoebas. Which macromolecules are copied and pass genetic information to the new cells? **SC.6.L.14.2**

Ⓐ lipids

Ⓑ proteins

Ⓒ carbohydrates

Ⓓ nucleic acids

4. One example of a carbohydrate that stores energy is **SC.6.L.14.2**

Ⓕ fat.

Ⓖ hair.

Ⓗ keratin.

Ⓘ sugar.

Name ______________________ Date ______________

Multiple Choice *Bubble the correct answer.*

1. Which cell belongs to a prokaryote? **SC.6.L.14.3**

Ⓐ 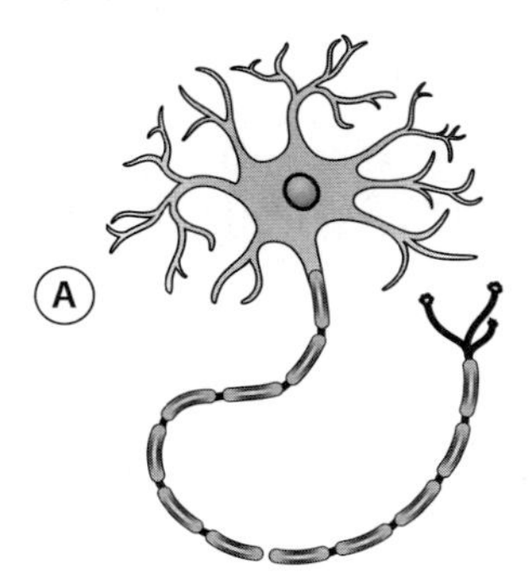

Ⓑ 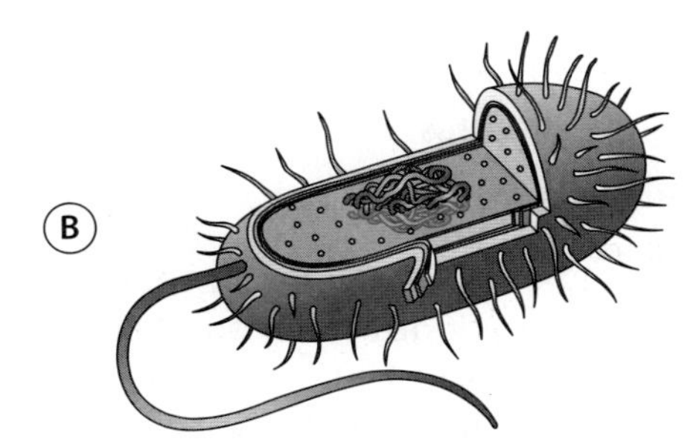

Ⓒ 

Ⓓ 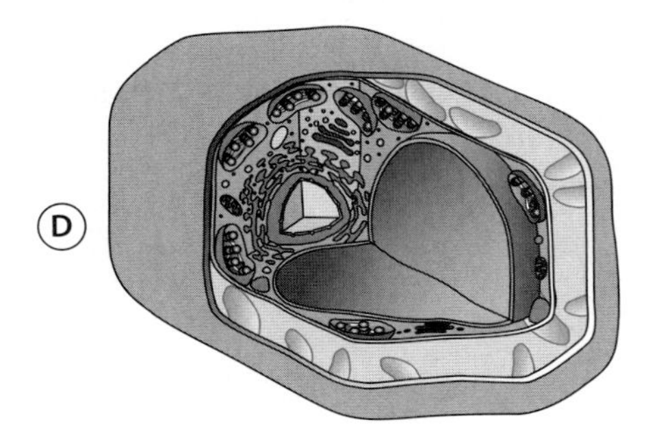

2. Your body is protected by multiple layers of skin cells. What shape would you expect a cell on the outer layer of skin to have? **SC.6.L.14.4**

Ⓕ a flat shape

Ⓖ a round shape

Ⓗ a long, branching shape

Ⓘ a short, hollow shape

3. Which cell structure could be called the packaging center of the cell? **SC.6.L.14.4**

Ⓐ lysosome

Ⓑ mitochondrion

Ⓒ endoplasmic reticulum

Ⓓ Golgi apparatus

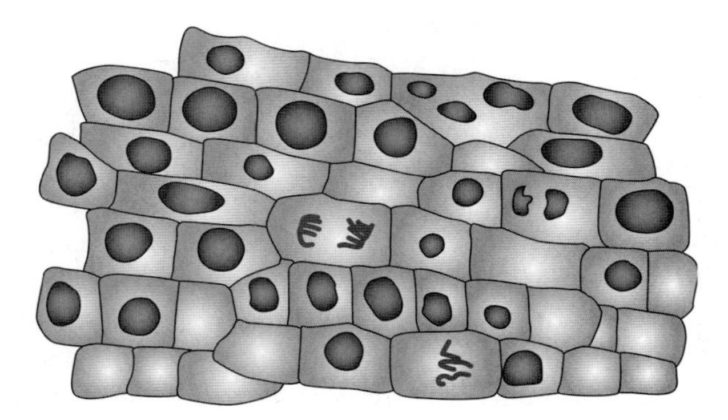

4. Rosa uses a microscope to look at a group of cells, as shown above. She sees that the cells are joined together, so she knows that they are from one organism. She also sees that all of them have cell walls. Rosa could be looking at **SC.6.L.14.3**

Ⓕ bacterial cells.

Ⓖ human cells.

Ⓗ mouse cells.

Ⓘ mushroom cells.

Name ______________________ Date ______________

Multiple Choice *Bubble the correct answer.*

Use the image below to answer questions 1 and 2.

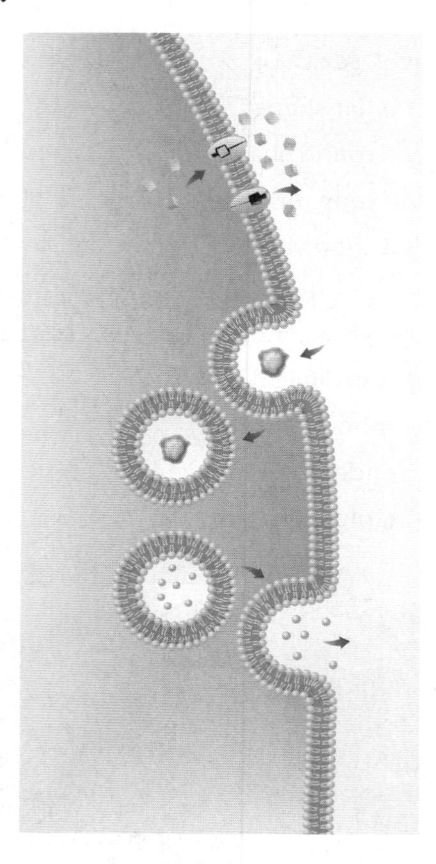

1. The diagram above shows three methods that require energy to move materials across the cell membrane. Which process do cells use to take in bacteria and viruses? **SC.6.L.14.3**

Ⓐ endocytosis

Ⓑ exocytosis

Ⓒ active transport

Ⓓ passive transport

2. Which process removes proteins and hormones from a cell? **SC.6.L.14.3**

Ⓕ endocytosis

Ⓖ exocytosis

Ⓗ active transport

Ⓘ passive transport

3. Which type of macromolecule helps move molecules into a cell through the process of facilitated diffusion? **SC.6.L.14.3**

Ⓐ lipids

Ⓑ proteins

Ⓒ carbohydrates

Ⓓ nucleic acids

4. Sasha places a semipermeable membrane in a beaker and adds water to the beaker. She then adds sugar to the water on the left side of the beaker. What will happen to the concentration of sugar on both sides of the membrane after 30 minutes? **SC.6.N.1.1**

Ⓕ All the sugar will be concentrated on the left side of the membrane.

Ⓖ All the sugar will be concentrated on the right side of the membrane.

Ⓗ Sugar concentration will be equal on both sides.

Ⓘ Sugar concentration will be high on the left side and low on the right side.

Name ______________________ Date ______________

Multiple Choice *Bubble the correct answer.*

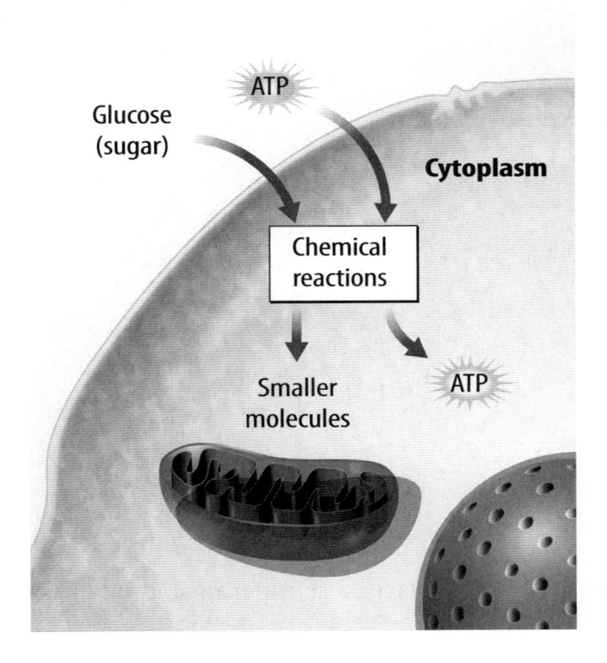

1. Which process is illustrated in the image above? **SC.6.L.14.3**

Ⓐ glycolysis

Ⓑ photosynthesis

Ⓒ alcohol fermentation

Ⓓ lactic-acid fermentation

2. Which process produces the most ATP? **SC.6.L.14.3**

Ⓕ Plant cells convert sunlight to sugar through photosynthesis.

Ⓖ Yeast cells produce ethanol through alcohol fermentation.

Ⓗ Human stomach cells convert sugar to energy through cellular respiration.

Ⓘ Human muscle cells generate energy at the end of a race through lactic-acid fermentation.

3. Which substance is a product of cellular respiration? **SC.6.L.14.3**

Ⓐ glucose

Ⓑ lactic acid

Ⓒ oxygen

Ⓓ water

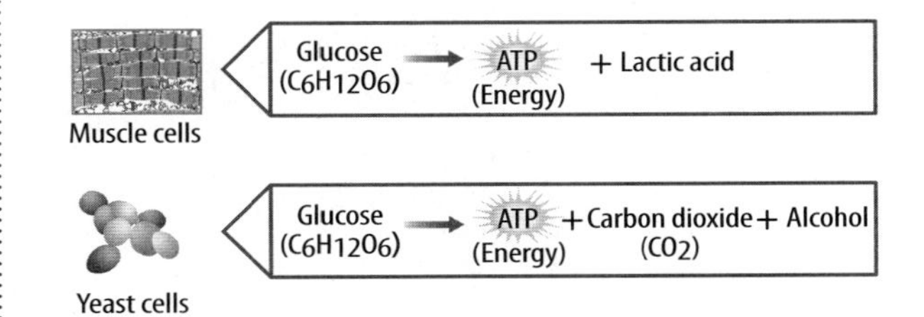

4. Which processes are shown in the image above? **SC.6.L.14.3**

Ⓕ two types of cellular respiration

Ⓖ two types of fermentation

Ⓗ two types of glycolysis

Ⓘ two types of photosynthesis

Name ______________________ Date ______________

Notes

Getting Bigger

A small baby snake grows into a larger adult snake. Choose the explanation that best describes why a small baby snake grows into a larger adult snake.

A. The baby snake's cells divide.

B. The baby snake's cells grow into longer cells.

C. The baby snake's cells grow into much larger cells.

D. The baby snake's body parts stretch out and get longer.

E. The food the baby snake eats makes it grow big and strong.

F. The baby snake's cells differentiate into different types of cells.

Explain your thinking. Describe your ideas about growth.

From a Cell to an ORGANISM

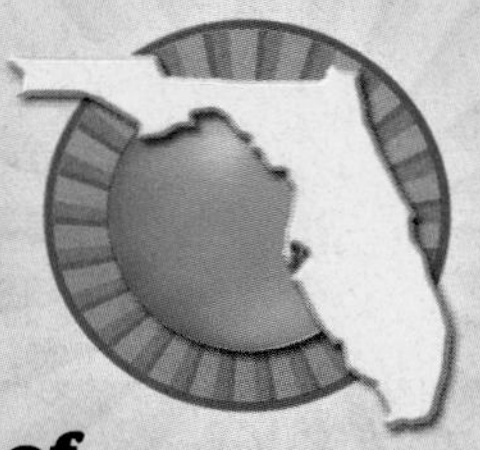

The Big Idea

FLORIDA BIG IDEAS

1 **The Practice of Science**

2 **The Characteristics of Scientific Knowledge**

14 **Organization and Development of Living Organisms**

Think About It!

How can one cell become a multicellular organism?

From the outside, a chicken egg looks like a simple oval object. But big changes are taking place inside the egg. Over several weeks, the one cell in the egg will grow and divide and become a chick.

1. How did the original cell change over time?

2. What might have happened to the chick's cells as the chick grew?

3. How can one cell become a multicellular chick?

Get Ready to Read

What do you think about organism development?

Before you read, decide if you agree or disagree with each of these statements. As you read this chapter, see if you change your mind about any of the statements.

	AGREE	DISAGREE
1 Cell division produces two identical cells.		
2 Cell division is important for growth.		
3 At the end of the cell cycle, the original cell no longer exists.		
4 Unicellular organisms do not have all the characteristics of life.		
5 All the cells in a multicellular organism are the same.		
6 Some organs work together as part of an organ system.		

There's More Online!
Video • Audio • Review • ⓘLab Station • WebQuest • Assessment • Concepts in Motion • Multilingual eGlossary

Lesson 1

The Cell Cycle and Cell DIVISION

ESSENTIAL QUESTIONS

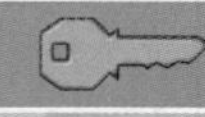
What are the phases of the cell cycle?

Why is the result of the cell cycle important?

Vocabulary

cell cycle p. 415
interphase p. 416
sister chromatid p. 418
centromere p. 418
mitosis p. 419
cytokinesis p. 419
daughter cell p. 419

Florida NGSSS

LA.6.2.2.3 The student will organize information to show understanding (e.g., representing main ideas within text through charting, mapping, paraphrasing, summarizing, or comparing/contrasting);

MA.6.A.3.6 Construct and analyze tables, graphs, and equations to describe linear functions and other simple relations using both common language and algebraic notation.

SC.6.L.14.3 Recognize and explore how cells of all organisms undergo similar processes to maintain homeostasis, including extracting energy from food, getting rid of waste, and reproducing.

SC.6.N.1.1 Define a problem from the sixth grade curriculum, use appropriate reference materials to support scientific investigation of various types, such as systematic observations or experiments, identify variables, collect and organize data, interpret data in charts, tables and graphics, analyze information, make predictions, and defend conclusions.

SC.6.N.1.5 Recognize that science involves creativity, not just in designing experiments, but also in creating explanations that fit evidence.

SC.6.N.2.1 Distinguish science from other activities involving thought.

SC.6.N.1.1

Inquiry Launch Lab

15 minutes

Why isn't your cell like mine?

All living things are made of cells. Some are made of only one cell, while others are made of trillions of cells. Where do all those cells come from?

Procedure

1. Read and complete a lab safety form.
2. Ask your team members to face away from you. Draw an animal cell on a sheet of **paper.** Include as many organelles as you can.
3. Use **scissors** to cut the cell drawing into equal halves. Fold each sheet of paper in half so the drawing cannot be seen.
4. Ask your team members to face you. Give each team member half of the cell drawing.
5. Have team members sit facing away from each other. Each person should use a **glue stick** to attach the cell half to one side of a sheet of paper. Then, each person should draw the missing cell half.
6. Compare the two new cells to your original cell.

Think About This

1. How did the new cells compare to the original cell?

2.  Key Concept What are some things that might be done in the early steps to produce two new cells that are more like the original cell?

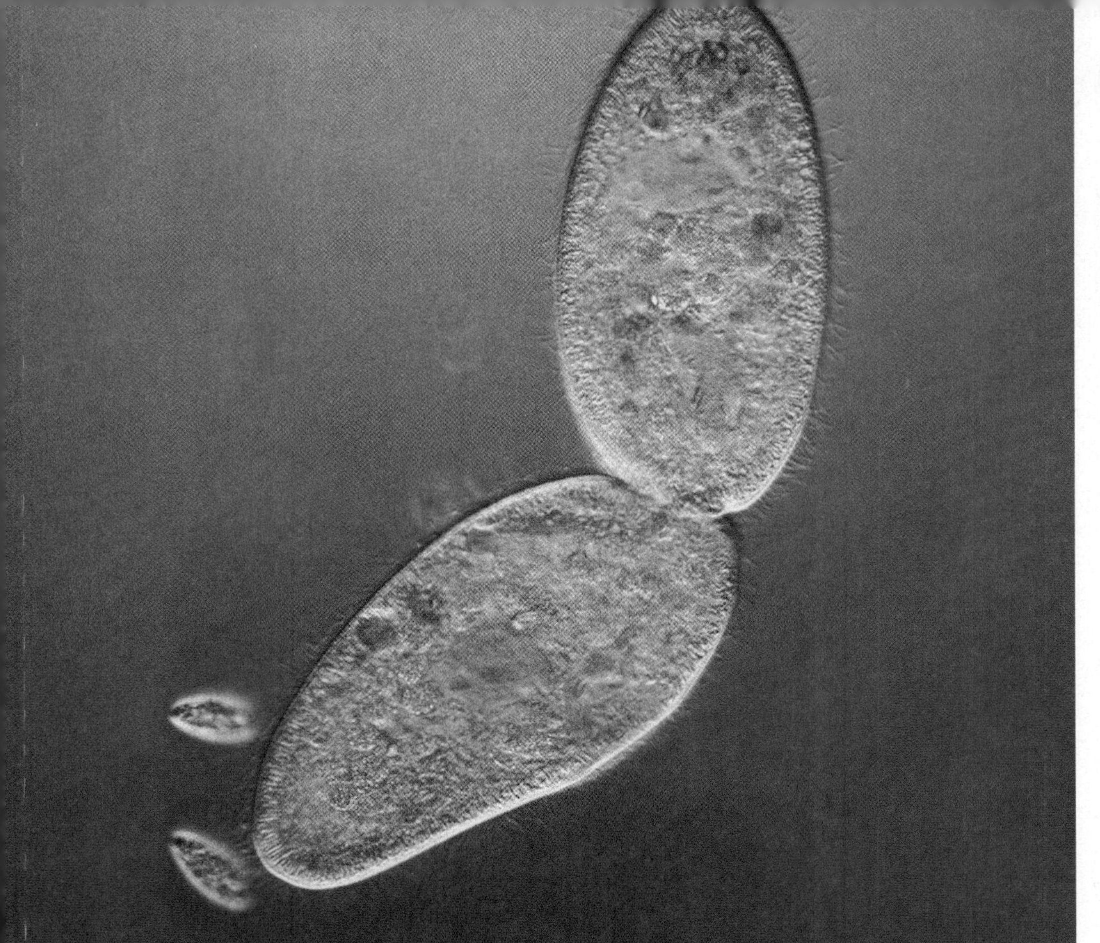

Inquiry **Time to Split?**

1. Unicellular organisms such as these reproduce when one cell divides into two new cells. The two cells are identical to each other. What do you think happened to the contents of the cell before it divided?

The Cell Cycle

No matter where you live, you have probably noticed that the weather changes in a regular pattern each year. Some areas experience four seasons—winter, spring, summer, and fall. In other parts of the world, there are only two seasons—rainy and dry. As seasons change, temperature, precipitation, and the number of hours of sunlight vary in a regular cycle.

These changes can affect the life cycles of organisms such as trees. Notice how the tree in **Figure 1** changes with the seasons. Like changing seasons or the growth of trees, cells go through cycles. *Most cells in an organism go through a cycle of growth, development, and division called the* **cell cycle**. Through the cell cycle, organisms grow, develop, replace old or damaged cells, and produce new cells.

Figure 1 This maple tree changes in response to a seasonal cycle.

2. **Visual Check** **List** What are the seasonal changes of this maple tree?

Active Reading

FOLDABLES® LA.6.2.2.3

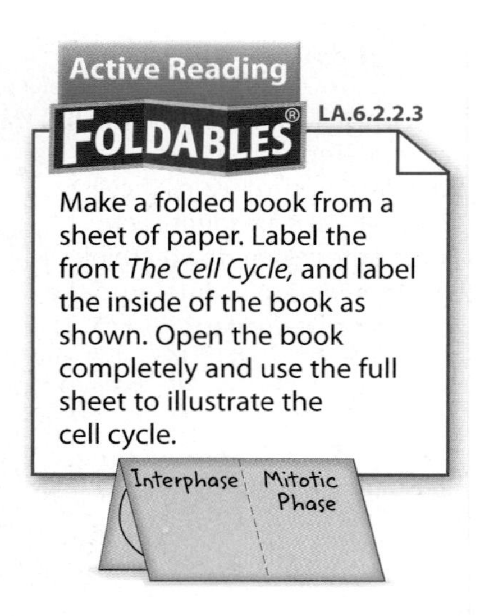

Make a folded book from a sheet of paper. Label the front *The Cell Cycle,* and label the inside of the book as shown. Open the book completely and use the full sheet to illustrate the cell cycle.

Phases of the Cell Cycle

There are two main phases in the cell cycle—interphase and the mitotic (mi TAH tihk) phase. **Interphase** *is the period during the cell cycle of a cell's growth and development.* A cell spends most of its life in interphase, as shown in **Figure 2.** During interphase, most cells go through three stages:

- rapid growth and replication, or copying, of the membrane-bound structures called organelles;
- copying of DNA, the genetic information in a cell; and
- preparation for cell division.

Interphase is followed by a shorter period of the cell cycle known as the mitotic phase. A cell reproduces during this phase. The mitotic phase has two stages, as illustrated in **Figure 2.** The nucleus divides in the first stage, and the cell's fluid, called the cytoplasm, divides in the second stage. The mitotic phase creates two new identical cells. At the end of this phase, the original cell no longer exists.

3. NGSSS Check Differentiate Underline the two main phases of the cell cycle. SC.6.L.14.3

The Cell Cycle

Figure 2 A cell spends most of its life growing and developing during interphase.

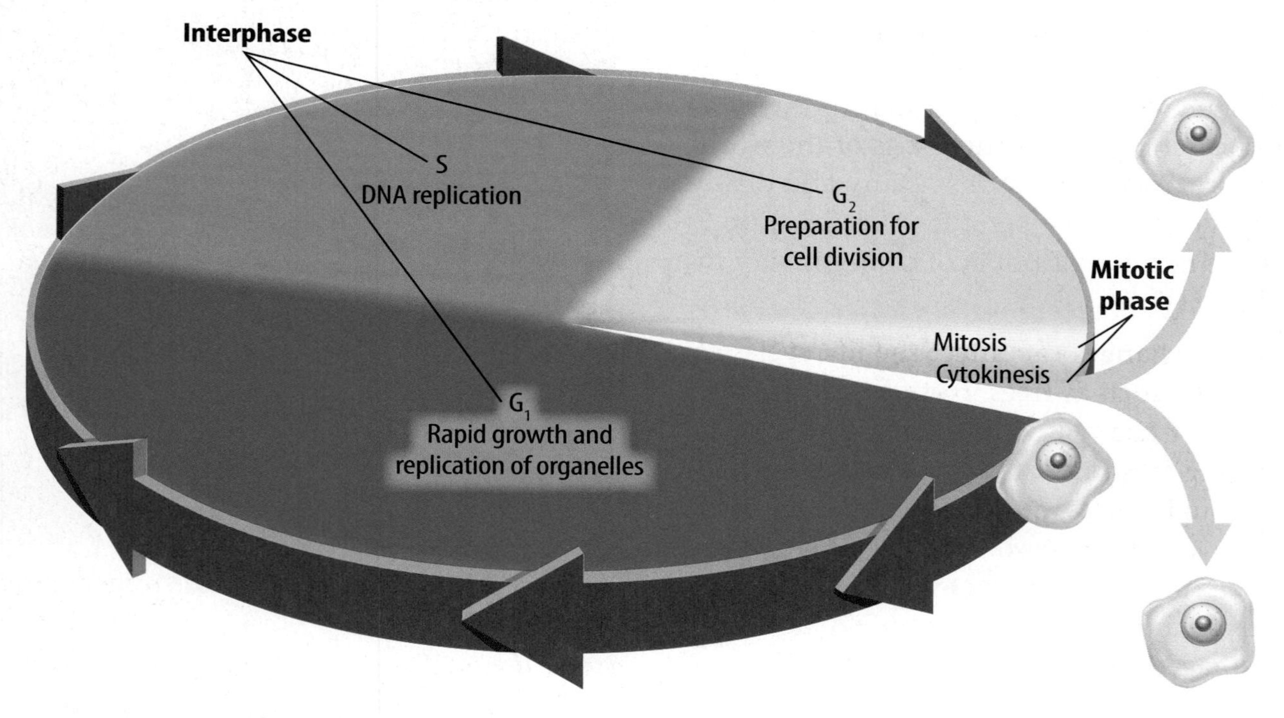

4. Visual Check Name Which stage of interphase is the longest? ______

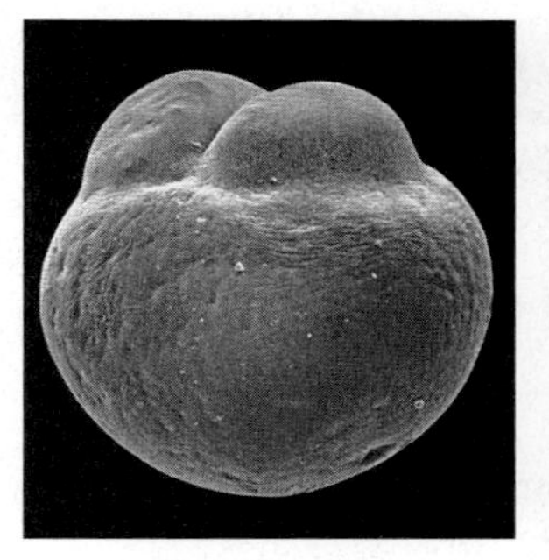
2-cell stage
SEM Magnification: 160×

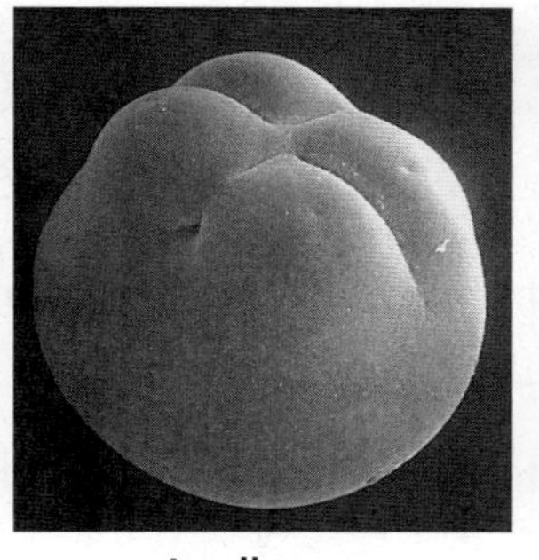
4-cell stage
SEM Magnification: 155×

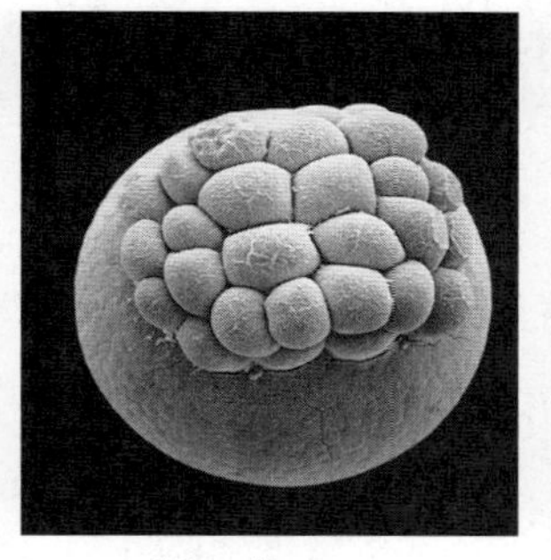
32-cell stage
SEM Magnification: 150×

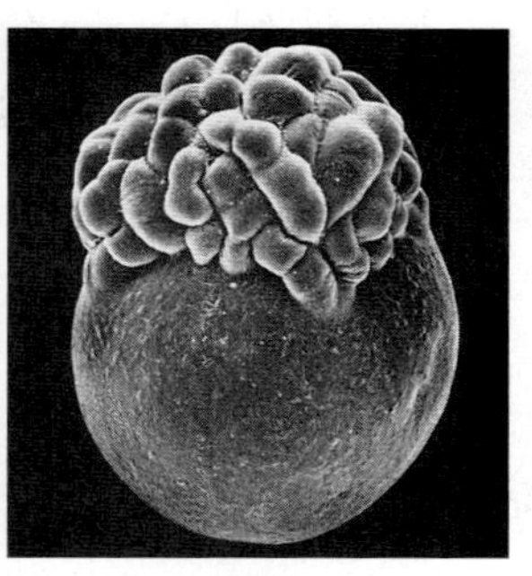
256-cell stage
SEM Magnification: 130×

Figure 3 The fertilized egg of a zebra fish divides into 256 cells in 2.5 hours.

Length of a Cell Cycle

The time it takes a cell to complete the cell cycle depends on the type of cell that is dividing. Recall that a **eukaryotic** cell has membrane-bound organelles, including a nucleus. For some eukaryotic cells, the cell cycle might last only eight minutes. For other cells, the cycle might take as long as one year. Most dividing human cells normally complete the cell cycle in about 24 hours. As illustrated in **Figure 3,** the cells of some organisms divide very quickly.

Review Vocabulary
eukaryotic
a cell with membrane-bound structures

Interphase

As you have read, interphase makes up most of the cell cycle. Newly produced cells begin interphase with a period of rapid growth—the cell gets bigger. This is followed by cellular activities such as making proteins. Next, actively dividing cells make copies of their DNA and prepare for cell division. During interphase, the DNA is called chromatin (KROH muh tun). Chromatin is long, thin strands of DNA, as shown in **Figure 4.** When scientists dye a cell in interphase, the nucleus looks like a plate of spaghetti. This is because the nucleus contains many strands of chromatin tangled together.

Figure 4 During interphase, the nuclei of an animal cell and a plant cell contain long, thin strands of DNA called chromatin.

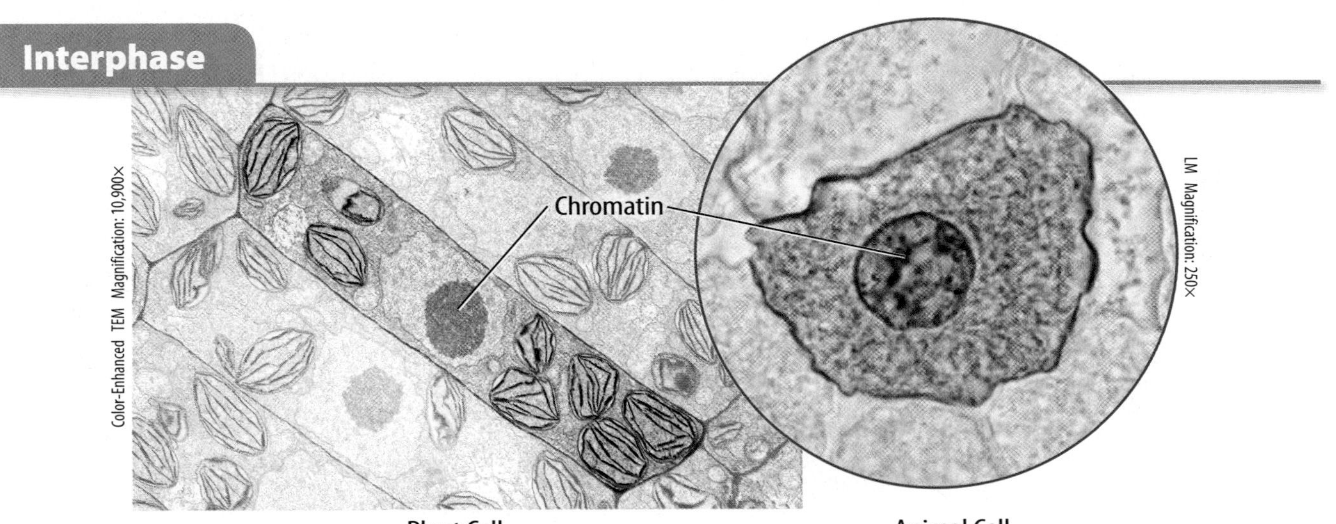

Table 1 Phases of the Cell Cycle

Phase	Stage	Description
Interphase		growth and cellular functions; organelle replication
		growth and chromosome replication; organelle replication
		growth and cellular functions; organelle replication
Mitotic phase		division of nucleus
		division of cytoplasm

Table 1 The two phases of the cell cycle can each be divided into different stages.

Active Reading **5. Discover** Fill in the missing information in the table.

Phases of Interphase

Scientists divide interphase into three stages, as shown in **Table 1.** Interphase begins with a period of rapid growth—the G_1 stage. This stage lasts longer than other stages of the cell cycle. During G_1, a cell grows and carries out its normal cell functions. For example, during G_1, cells that line your stomach make enzymes that help digest your food. Although most cells continue the cell cycle, some cells stop the cell cycle at this point. For example, mature nerve cells in your brain remain in the G_1 stage and do not divide again.

During the second stage of interphase—the S stage—a cell continues to grow and copies its DNA. There are now identical strands of DNA. These identical strands of DNA ensure that each new cell gets a copy of the original cell's genetic information. Each strand of DNA coils up and forms a chromosome. Identical chromosomes join together. The cell's DNA is now arranged as pairs of identical chromosomes. Each pair is called a duplicated chromosome. *Two identical chromosomes, called* **sister chromatids**, *make up a duplicated chromosome,* as shown in **Figure 5.** Notice that the *sister chromatids are held together by a structure called the* **centromere**.

The final stage of interphase—the G_2 stage—is another period of growth and the final preparation for mitosis. A cell uses energy copying DNA during the S stage. During G_2, the cell stores energy that will be used during the mitotic phase of the cell cycle.

Active Reading **6. Identify** Highlight what happens in the G_2 phase.

Figure 5 The coiled DNA forms a duplicated chromosome made of two sister chromatids connected at the centromere.

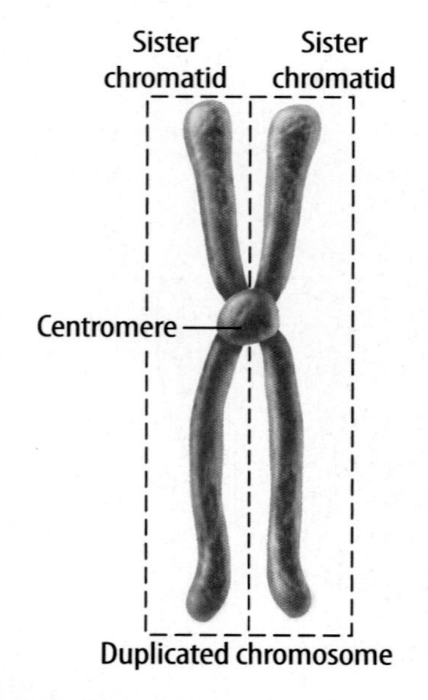

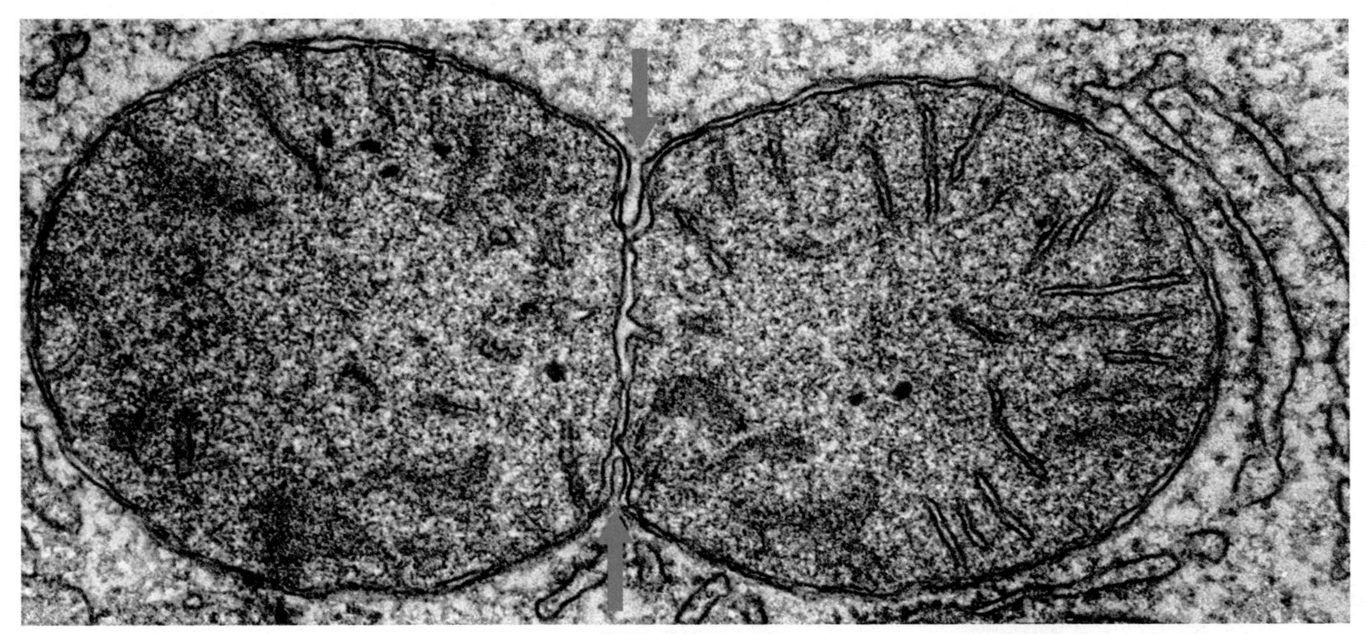

Figure 6 This mitochondrion is in the final stage of dividing.

Organelle Replication

During cell division, the organelles in a cell are distributed between the two new cells. Before a cell divides, it makes a copy of each organelle. This enables the two new cells to function properly. Some organelles, such as the energy-processing mitochondria and chloroplasts, have their own DNA. These organelles can make copies of themselves on their own, as shown in **Figure 6.** A cell produces other organelles from materials such as proteins and lipids. A cell makes these materials using the information contained in the DNA inside the nucleus. Organelles are copied during all stages of interphase.

The Mitotic Phase

The mitotic phase of the cell cycle follows interphase. It consists of two stages: mitosis (mi TOH sus) and cytokinesis (si toh kuh NEE sus). *In* **mitosis**, *the nucleus and its contents divide. In* **cytokinesis**, *the cytoplasm and its contents divide.* **Daughter cells** *are the two new cells that result from mitosis and cytokinesis.*

WORD ORIGIN

mitosis
from Greek *mitos,* means "warp thread"; and Latin *–osis,* means "process"

During mitosis, the contents of the nucleus divide, forming two identical nuclei. The sister chromatids of the duplicated chromosomes separate from each other. This gives each daughter cell the same genetic information. For example, a cell that has ten duplicated chromosomes actually has 20 chromatids. When the cell divides, each daughter cell will have ten different chromatids. Chromatids are now called chromosomes.

In cytokinesis, the cytoplasm divides and forms the two new daughter cells. Organelles that were made during interphase are divided between the daughter cells.

Phases of Mitosis

Like interphase, mitosis is a continuous process that scientists divide into different phases, as shown in **Figure 7.**

Prophase During the first phase of mitosis, called prophase, the copied chromatin coils together tightly. The coils form visible duplicated chromosomes. The nucleolus disappears, and the nuclear membrane breaks down. Structures called spindle fibers form in the cytoplasm.

Metaphase During metaphase, the spindle fibers pull and push the duplicated chromosomes to the middle of the cell. Notice in **Figure 7** that the chromosomes line up along the middle of the cell. This arrangement ensures that each new cell will receive one copy of each chromosome. It is important that new cells be identical. Incorrect genetic information in the new cell can lead to the uncontrolled division and growth of new cells, known as cancer.

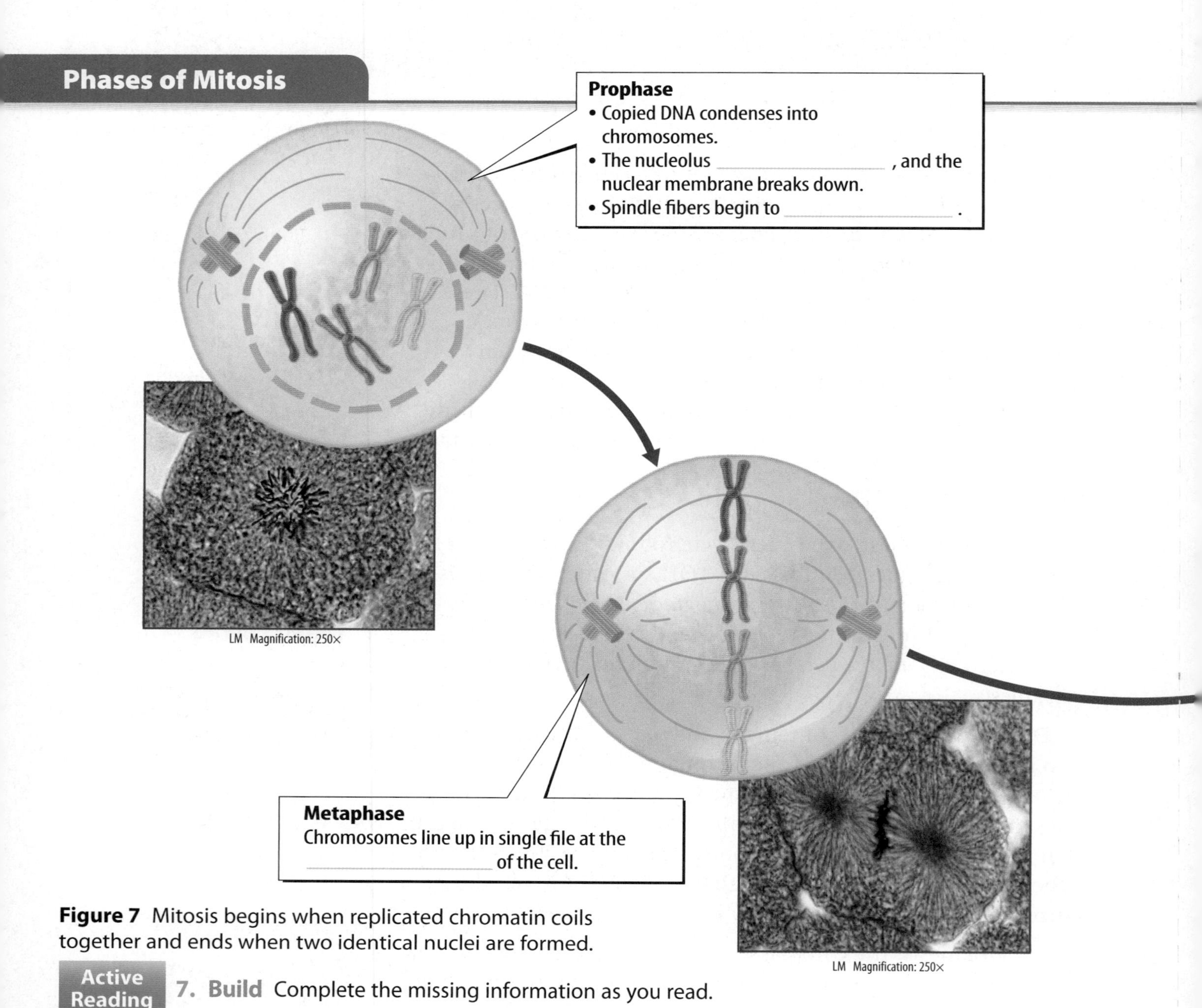

Figure 7 Mitosis begins when replicated chromatin coils together and ends when two identical nuclei are formed.

Active Reading 7. **Build** Complete the missing information as you read.

Anaphase In anaphase, the third stage of mitosis, the two sister chromatids in each chromosome separate from each other. The spindle fibers pull them in opposite directions. Once separated, the chromatids are now two identical, single-stranded chromosomes. As they move to opposite sides of a cell, the cell begins to get longer. Anaphase is complete when the two identical sets of chromosomes are at opposite ends of a cell.

Telophase During telophase, the spindle fibers begin to disappear. Also, the chromosomes begin to uncoil. A nuclear membrane forms around each set of chromosomes at either end of the cell. This forms two new identical nuclei. Telophase is the final stage of mitosis. It is often described as the reverse of prophase because many of the processes that occur during prophase are reversed during telophase.

8. State What are the phases of mitosis?

____

____

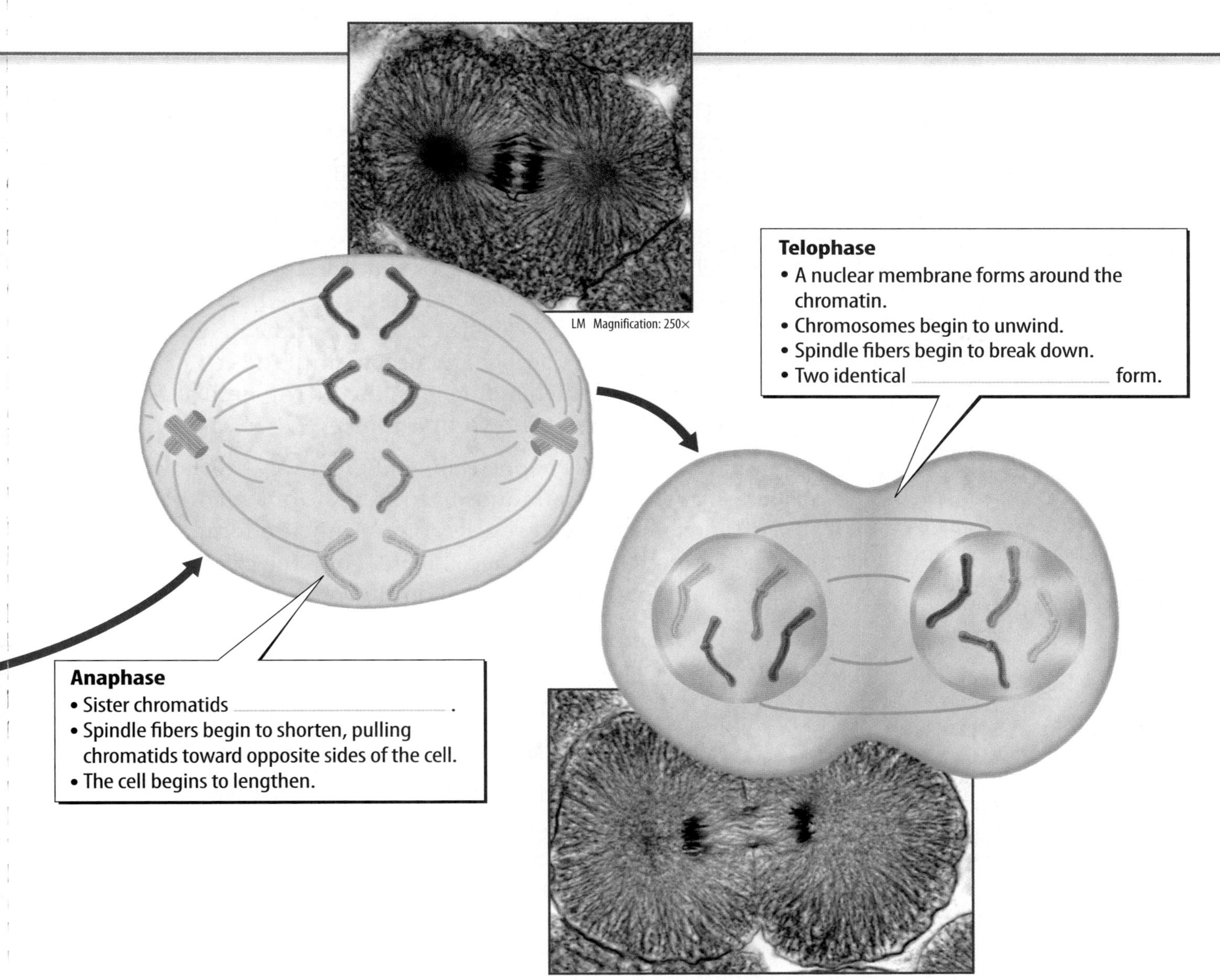

Cytokinesis

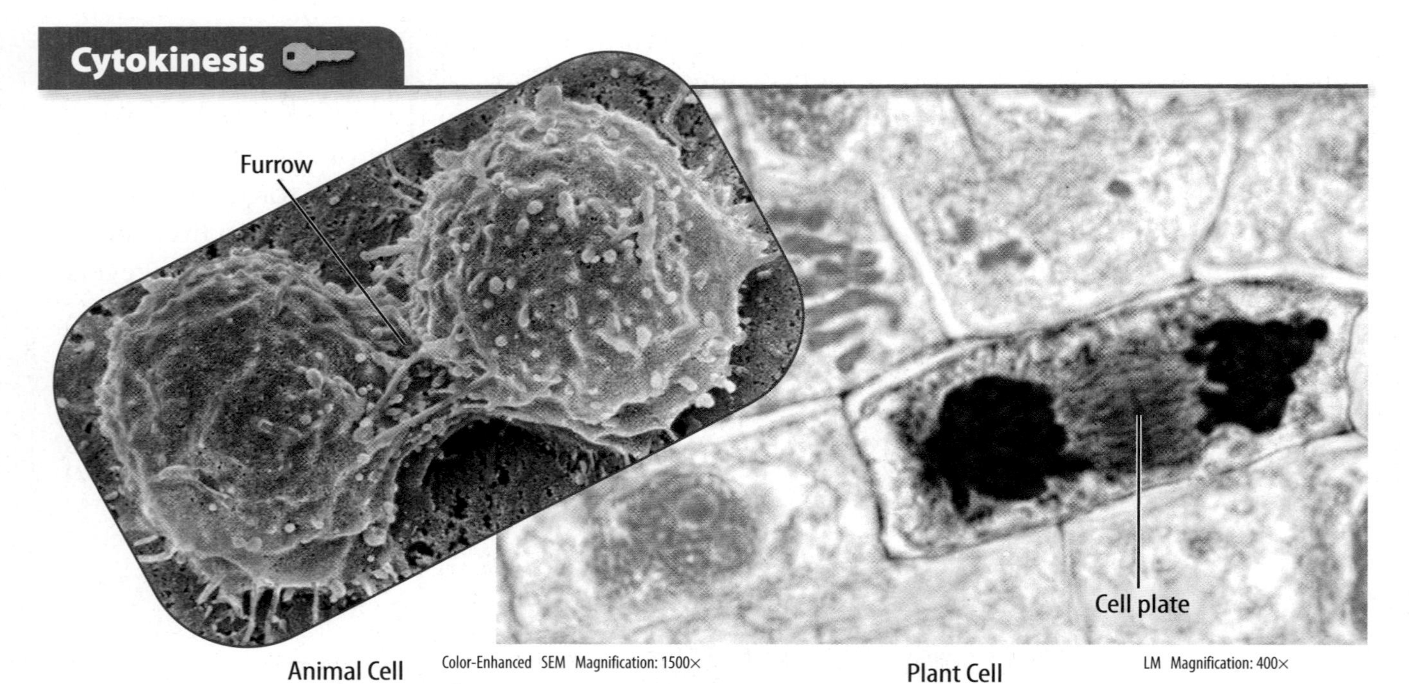

Figure 8 Cytokinesis differs in animal cells and plant cells.

Dividing the Cell's Components

Following the last phase of mitosis, a cell's cytoplasm divides in a process called cytokinesis. The specific steps of cytokinesis differ depending on the type of cell that is dividing. In animal cells, the cell membrane contracts, or squeezes together, around the middle of the cell. Fibers around the center of the cell pull together. This forms a crease, called a furrow, in the middle of the cell. The furrow gets deeper and deeper until the cell membrane comes together and divides the cell. An animal cell undergoing cytokinesis is shown in **Figure 8.**

Cytokinesis in plants happens in a different way. As shown in **Figure 8,** a new cell wall forms in the middle of a plant cell. First, organelles called vesicles join together to form a membrane-bound disk called a cell plate. Then the cell plate grows outward toward the cell wall until two new cells form.

Active Reading 9. **Compare** Highlight the process of cytokinesis in plant and animal cells.

Math Skills MA.6.A.3.6

Use Percentages

A percentage is a ratio that compares a number to 100. If the length of the entire cell cycle is 24 hours, 24 hours equals 100%. If part of the cycle takes 6.0 hours, it can be expressed as 6.0 hours/24 hours. To calculate percentage, divide and multiply by 100. Add a percent sign.

$$\frac{6.0}{24} = 0.25 \times 100 = 25\%$$

Practice

10. Interphase in human cells takes about 23 hours. If the cell cycle is 24 hours, what percentage is interphase?

Results of Cell Division

Recall that the cell cycle results in two new cells. These daughter cells are genetically identical to each other and to the original cell that no longer exists. For example, a human cell has 46 chromosomes. When that cell divides, it will produce two new cells with 46 chromosomes each. The cell cycle is important for reproduction in some organisms, growth in multicellular organisms, replacement of worn out or damaged cells, and repair of damaged tissues.

Reproduction

In some unicellular organisms, cell division is a form of asexual reproduction. For example, an organism called a paramecium often reproduces by dividing into two new daughter cells or two new paramecia. Cell division is also important in other methods of reproduction in which the offspring are identical to the parent organism.

Growth

Cell division allows multicellular organisms, such as humans, to grow and develop from one cell (a fertilized egg). In humans, cell division begins about 24 hours after fertilization and continues rapidly during the first few years of life. It is likely that during the next few years you will go through another period of rapid growth and development. This happens because cells divide and increase in number as you grow and develop.

Replacement

Even after an organism is fully grown, cell division continues. This process replaces cells that wear out or are damaged. The outermost layer of your skin is always rubbing or flaking off.
A layer of cells below the skin's surface is constantly dividing. This produces millions of new cells daily to replace the ones that are rubbed off.

Repair

Cell division is also critical for repairing damage. When a bone breaks, cell division produces new bone cells that patch the broken pieces back together.

Not all damage can be repaired, however, because not all cells continue to divide. Recall that mature nerve cells stop the cell cycle in interphase. For this reason, injuries to nerve cells often cause permanent damage.

11. NGSSS Check Judge Underline why the result of the cell cycle is important. SC.6.L.14.3

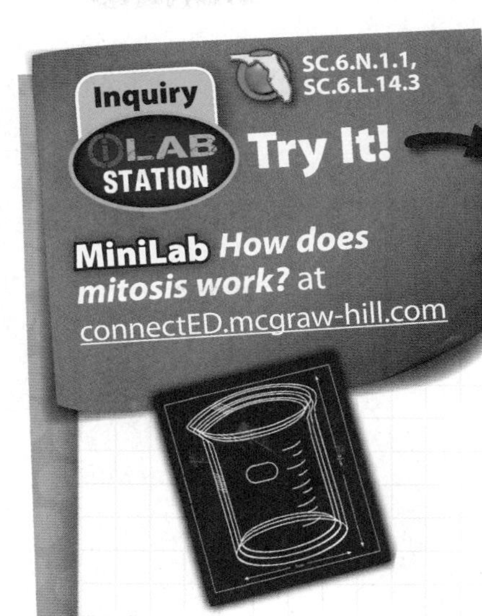

Apply It!

After you complete the lab, answer this question.

1. How is mitosis beneficial in terms of cell replacement?

Active Reading **12. Assess** Write a questsion about the main idea under each heading. Exchange quizzes with another student. Together, discuss the answers to the quizzes.

Lesson Review 1

Visual Summary

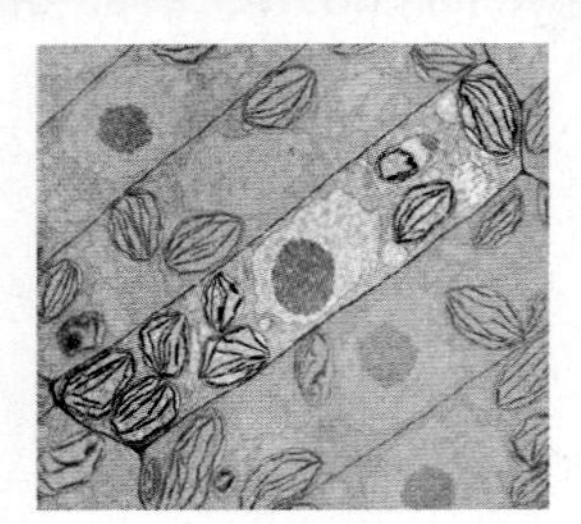

During interphase, most cells go through periods of rapid growth and replication of organelles, copying DNA, and preparation for cell division.

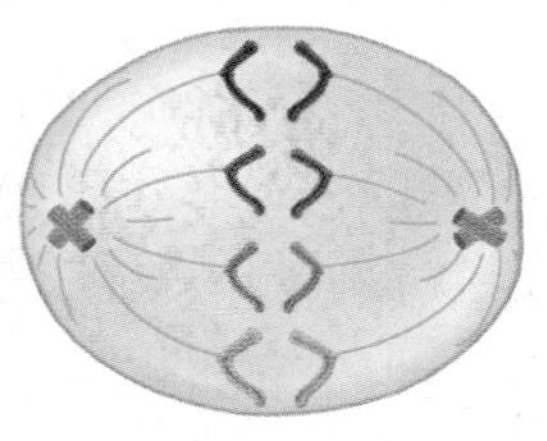

The nucleus and its contents divide during mitosis.

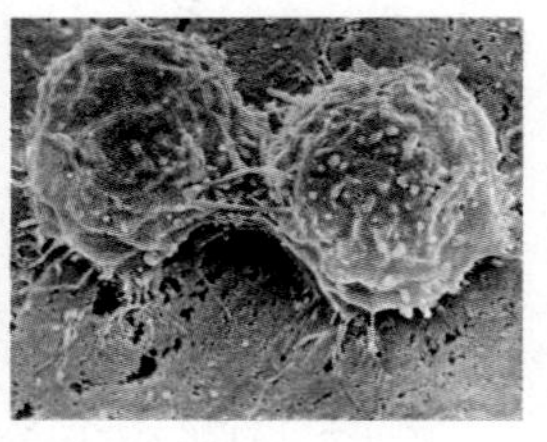

The cytoplasm and its contents divide during cytokinesis.

Use Vocabulary

1. **Distinguish** between mitosis and cytokinesis.

2. A duplicated chromosome is made of two ______.

3. **Use the term** *interphase* in a sentence.

Understand Key Concepts

4. Which is NOT part of mitosis?
   - (A) anaphase
   - (B) interphase
   - (C) prophase
   - (D) telophase

5. **Give three examples** of why the result of the cell cycle is important. SC.6.L.14.3

Interpret Graphics

6. **Organize** Fill in the graphic organizer below to show the results of cell division. LA.6.2.2.3

Critical Thinking

7. **Predict** what might happen to a cell if it were unable to divide by mitosis.

Math Skills

8. The mitotic phase of the human cell cycle takes approximately 1 hour. What percentage of the 24-hour cell cycle is the mitotic phase?

SCIENCE & SOCIETY

DNA Fingerprinting

DNA

Solving Crimes One Strand at a Time

Every cell in your body has the same DNA in its nucleus. Unless you are an identical twin, your DNA is entirely unique. Identical twins have identical DNA because they begin as one cell that divides and separates. When your cells begin mitosis, they copy their DNA. Every new cell has the same DNA as the original cells. That is why DNA can be used to identify people. Just as no two people have the same fingerprints, your DNA belongs to you alone.

Using scientific methods to solve crimes is called forensics. DNA fingerprinting is now a basic tool in forensics. Samples collected from a crime scene can be compared to millions of samples previously collected and indexed in a computer.

Every day, everywhere you go, you leave a trail of DNA. It might be in skin cells. It might be in hair or in the saliva you used to lick an envelope. If you commit a crime, you will most likely leave DNA behind. An expert crime scene investigator will know how to collect that DNA.

DNA evidence can prove innocence as well. Investigators have reexamined DNA found at old crime scenes. Imprisoned persons have been proven not guilty through DNA fingerprinting methods that were not yet available when the crime was committed.

DNA fingerprinting can also be used to identify bodies that were previously known only as a John or Jane Doe.

The Federal Bureau of Investigation (FBI) has a nationwide index of DNA samples called CODIS (Combined DNA Index System).

It's Your Turn

DISCOVER Your cells contain organelles called mitochondria. They have their own DNA, called mitochondrial DNA. Your mitochondrial DNA is identical to your mother's mitochondrial DNA. Find out how this information is used.

Lesson 2

Levels of ORGANIZATION

ESSENTIAL QUESTIONS

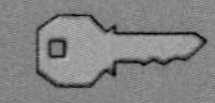
How do unicellular and multicellular organisms differ?

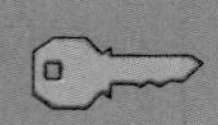
How does cell differentiation lead to the organization within a multicellular organism?

Vocabulary

cell differentiation p. 429

stem cell p. 430

tissue p. 431

organ p. 432

organ system p. 433

Florida NGSSS

LA.6.2.2.3 The student will organize information to show understanding (e.g., representing main ideas within text through charting, mapping, paraphrasing, summarizing, or comparing/contrasting);

SC.6.L.14.1 Describe and identify patterns in the hierarchical organization of organisms from atoms to molecules and cells to tissues to organs to organ systems to organisms.

SC.6.L.14.3 Recognize and explore how cells of all organisms undergo similar processes to maintain homeostasis, including extracting energy from food, getting rid of waste, and reproducing.

SC.6.N.1.1 Define a problem from the sixth grade curriculum, use appropriate reference materials to support scientific understanding, plan and carry out scientific investigation of various types, such as systematic observations or experiments, identify variables, collect and organize data, interpret data in charts, tables, and graphics, analyze information, make predictions, and defend conclusions.

Inquiry Launch Lab

SC.6.N.1.1

15 minutes

How is a system organized?

The places people live are organized in a system. Do you live in or near a city? Cities contain things such as schools and stores that enable them to function on their own. Many cities together make up another level of organization.

Procedure

1. Read and complete a lab safety form.
2. Using a **metric ruler** and **scissors,** measure and cut squares of **construction paper** that are 4 cm, 8 cm, 12 cm, 16 cm, and 20 cm on each side. Use a different color for each square.
3. Stack the squares from largest to smallest, and glue them together.
4. Cut apart the *City, Continent, Country, County,* and *State* labels your teacher gives you.
5. Use a **glue stick** to attach the *City* label to the smallest square. Sort the remaining labels from smallest to largest, and glue to the corresponding square.

Think About This

1. What is the largest level of organization a city belongs to?

2. Can any part of the system function without the others? Explain.

3. Key Concept How do you think the system used to organize where people live is similar to how your body is organized?

Inquiry Scales on Wings?

1. This butterfly has a distinctive pattern of colors on its wings. The pattern is formed by a cluster of tiny scales. In a similar way, multicellular organisms are made of many small parts working together. How do these scales work together?

Life's Organization

You might recall that all matter is made of atoms and that atoms combine and form molecules. Molecules make up cells. A large animal, such as a Komodo dragon, is not made of one cell. Instead, it is composed of trillions of cells working together. Its skin, shown in **Figure 9,** is made of many cells that are specialized for protection. The Komodo dragon has other types of cells, such as blood cells and nerve cells, that perform other functions. Cells work together in the Komodo dragon and enable it to function. In the same way, cells work together in you and in other multicellular organisms.

Recall that some organisms are made of only one cell. These unicellular organisms carry out all the activities necessary to survive, such as absorbing nutrients and getting rid of wastes. But no matter their sizes, all organisms are made of cells.

Figure 9 Skin cells are only one of the many kinds of cells that make up a Komodo dragon.

Unicellular Organisms 

Figure 10 Unicellular organisms carry out life processes within one cell.

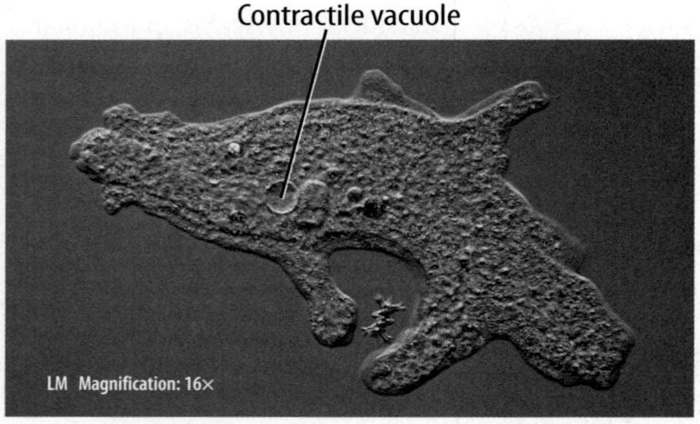

This unicellular amoeba captures a desmid for food.

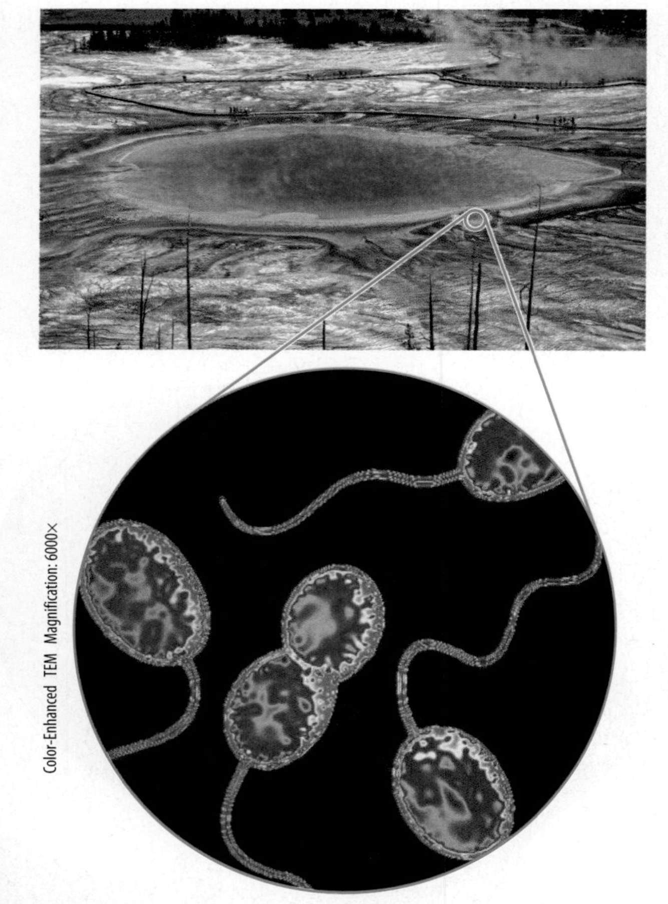

These heat-loving bacteria are often found in hot springs as shown here. They get their energy to produce food from sulfur instead of from light like plants.

Unicellular Organisms

As you read on the previous page, some organisms have only one cell. Unicellular organisms do all the things needed for their survival within that one cell. For example, the amoeba in **Figure 10** is ingesting another unicellular organism, a type of green algae called a desmid, for food. Unicellular organisms also respond to their environment, get rid of waste, grow, and even reproduce on their own. Unicellular organisms include both prokaryotes and some eukaryotes.

Prokaryotes

Recall that a cell without a membrane-bound nucleus is a prokaryotic cell. In general, prokaryotic cells are smaller than eukaryotic cells and have fewer cell structures. A unicellular organism made of one prokaryotic cell is called a prokaryote. Some prokaryotes live in groups called colonies. Some can also live in extreme environments, as shown in **Figure 10.**

Eukaryotes

You might recall that a eukaryotic cell has a nucleus surrounded by a membrane and many other specialized organelles. For example, the amoeba shown in **Figure 10** has an organelle called a contractile vacuole. It functions like a bucket that is used to bail water out of a boat. A contractile vacuole collects excess water from the amoeba's cytoplasm. Then it pumps the water out of the amoeba. This prevents the amoeba from swelling and bursting.

A unicellular organism that is made of one eukaryotic cell is called a eukaryote. There are thousands of different unicellular eukaryotes, such as algae that grow on the inside of an aquarium and the fungus that causes athlete's foot.

2. Identify (Circle) an example of a unicellular eukaryotic organism.

Multicellular Organisms

Multicellular organisms are made of many eukaryotic cells working together, like the crew on an airplane. Each member of the crew, from the pilot to the mechanic, has a specific job that is important for the plane's operation. Similarly, each type of cell in a multicellular organism has a specific job that is important to the survival of the organism.

Active Reading **3. Determine** Highlight how unicellular and multicellular organisms differ.

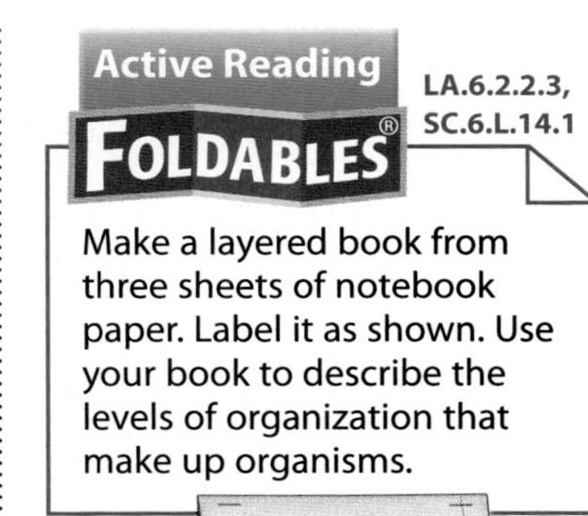

LA.6.2.2.3, SC.6.L.14.1

Make a layered book from three sheets of notebook paper. Label it as shown. Use your book to describe the levels of organization that make up organisms.

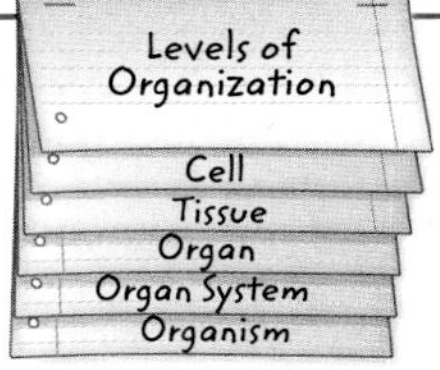

Cell Differentiation

As you read in the last lesson, all cells in a multicellular organism come from one cell—a fertilized egg. Cell division starts quickly after fertilization. The first cells made can become any type of cell, such as a muscle cell, a nerve cell, or a blood cell. *The process by which cells become different types of cells is called* **cell differentiation** (dihf uh ren shee AY shun).

You might recall that a cell's instructions are contained in its chromosomes. Also, nearly all the cells of an organism have identical sets of chromosomes. If an organism's cells have identical sets of instructions, how can cells be different? Different cell types use different parts of the instructions on the chromosomes. A few of the many different types of cells that can result from human cell differentiation are shown in **Figure 11.**

Figure 11 A fertilized egg produces cells that can differentiate into a variety of cell types.

Cell Differentiation in Eukaryotes

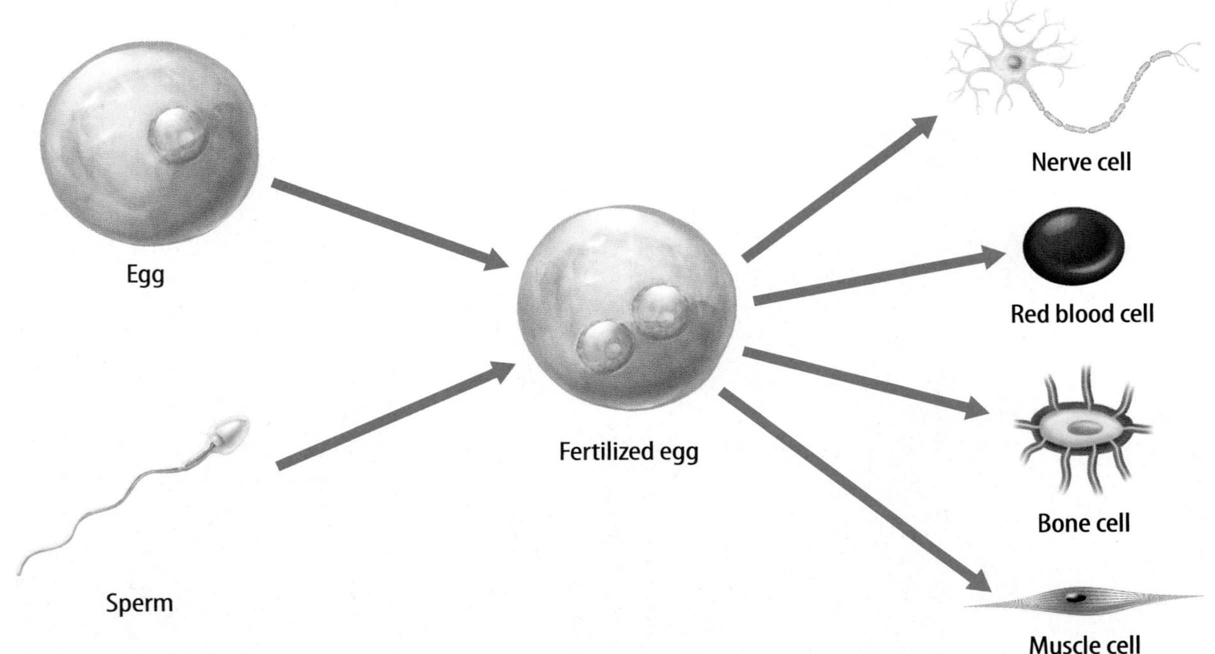

Animal Stem Cells Not all cells in a developing animal differentiate. **Stem cells** *are unspecialized cells that are able to develop into many different cell types.* There are many stem cells in embryos but fewer in adult organisms. Adult stem cells are important for the cell repair and replacement you read about in Lesson 1. For example, stem cells in your bone marrow can produce more than a dozen different types of blood cells. These replace ones that are damaged or worn out. Stem cells have also been discovered in skeletal muscles. These stem cells can produce new muscle cells when the **fibers** that make up the muscle are torn.

Science Use v. Common Use

fiber

Science Use a long muscle cell

Common Use a thread

Plant Cells Plants also have unspecialized cells similar to animal stem cells. These cells are grouped in areas of a plant called meristems (MER uh stemz). Meristems are in different areas of a plant, including the tips of roots and stems, as shown in **Figure 12.** Cell division in meristems produces different types of plant cells with specialized structures and functions, such as transporting materials, making food, storing food, or protecting the plant. These cells might become parts of stems, leaves, flowers, or roots.

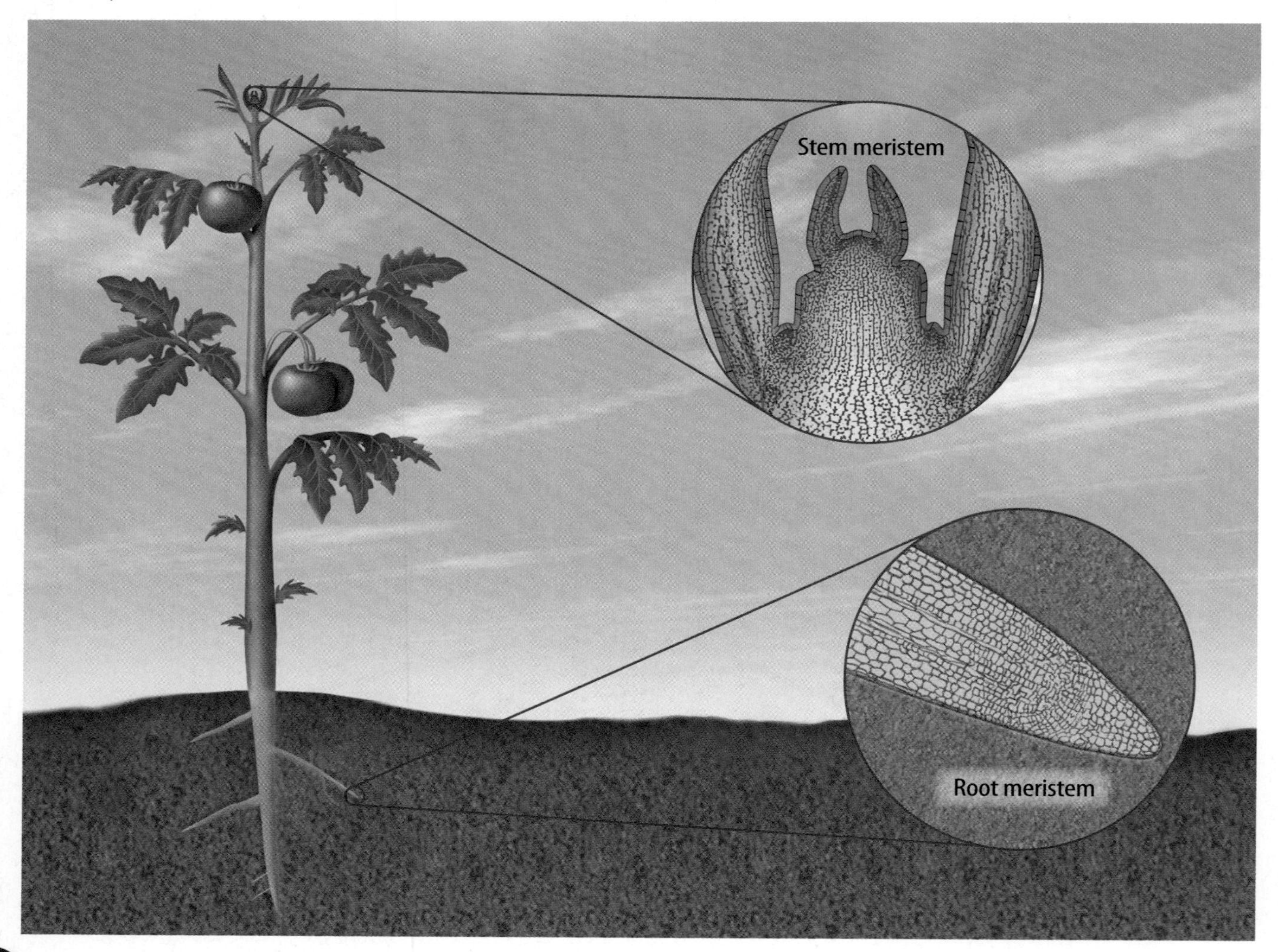

Figure 12 Plant meristems produce cells that can become part of stems, leaves, flowers, or roots.

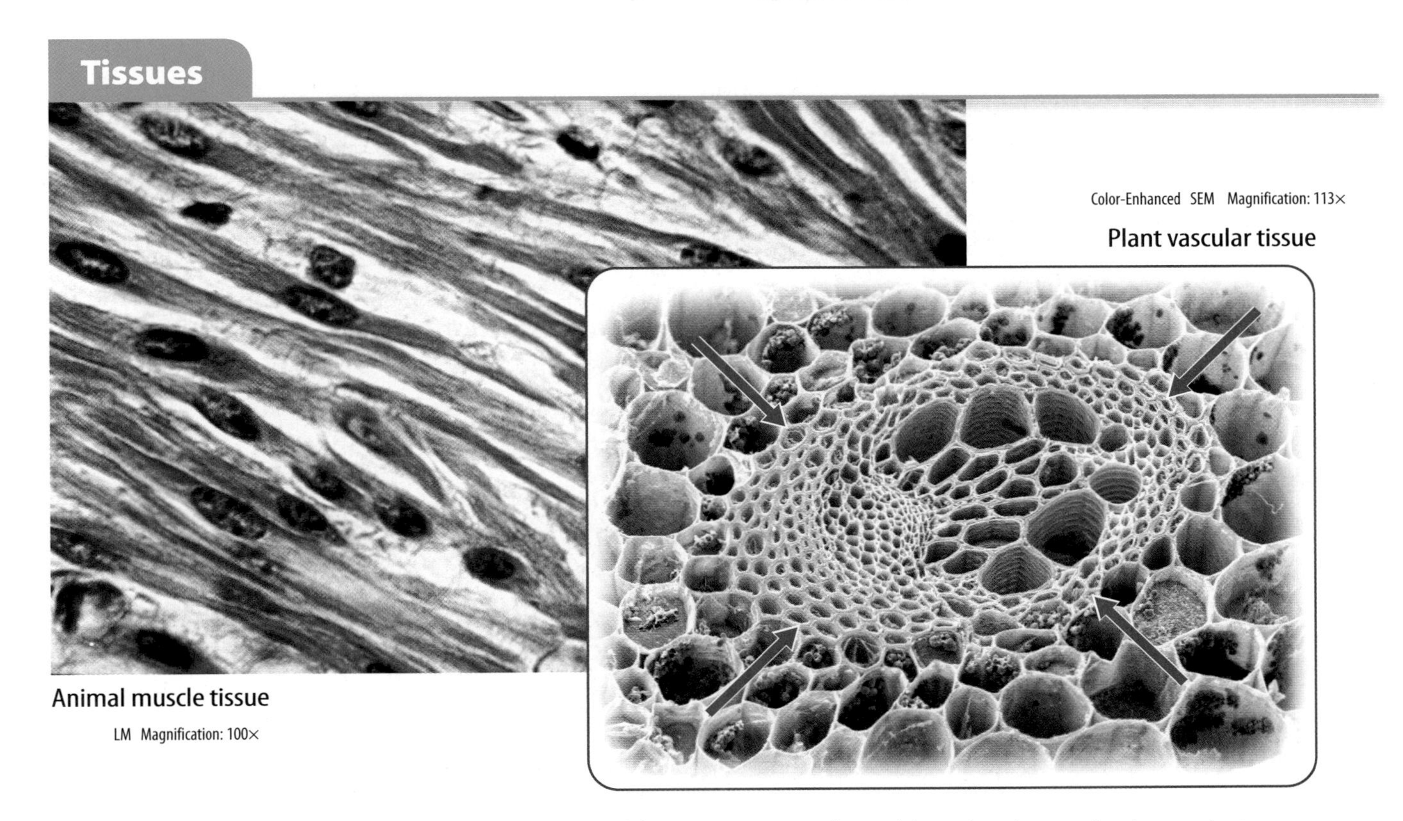

Figure 13 Similar cells work together and form tissues such as this animal muscle tissue that contracts the stomach to help digestion. Plant vascular tissue, indicated by red arrows, moves water and nutrients throughout a plant.

Tissues

In multicellular organisms, similar types of cells are organized into groups. **Tissues** *are groups of similar types of cells that work together to carry out specific tasks.* Humans, like most other animals, have four main types of tissue—muscle, connective, nervous, and epithelial (eh puh THEE lee ul). For example, the animal tissue shown in **Figure 13** is smooth muscle tissue that is part of the stomach. Muscle tissue causes movement. Connective tissue provides structure and support and often connects other types of tissue together. Nervous tissue carries messages to and from the brain. Epithelial tissue forms the protective outer layer of the skin and the lining of major organs and internal body cavities.

Plants also have different types of tissues. The three main types of plant tissue are dermal, vascular (VAS kyuh lur), and ground tissue. Dermal tissue provides protection and helps reduce water loss. Vascular tissue, shown in **Figure 13,** transports water and nutrients from one part of a plant to another. Ground tissue provides storage and support and is where photosynthesis takes place.

Active Reading **4. Contrast** Highlight the different animal and plant tissues and their descriptions.

WORD ORIGIN

tissue
from Latin *texere,* means "weave"

Organs

Complex jobs in organisms require more than one type of tissue. **Organs** *are groups of different tissues working together to perform a particular job.* For example, your stomach is an organ specialized for breaking down food. It is made of all four types of tissue: muscle, epithelial, nervous, and connective. Each type of tissue performs a specific function necessary for the stomach to work properly. Layers of muscle tissue contract and break up pieces of food, epithelial tissue lines the stomach, nervous tissue sends signals to indicate the stomach is full, and connective tissue supports the stomach wall.

Academic Vocabulary
complex
(adjective) made of two or more parts

Plants also have organs. The leaves shown in **Figure 14** are organs specialized for photosynthesis. Each leaf is made of dermal, ground, and vascular tissues. Dermal tissue covers the outer surface of a leaf. The leaf is a vital organ because it contains ground tissue that produces food for the rest of the plant. Ground tissue is where photosynthesis takes place. The ground tissue is tightly packed on the top half of a leaf. The vascular tissue moves both the food produced by photosynthesis and water throughout the leaf and the rest of the plant.

Active Reading **6. List** What are the tissues in a leaf organ?

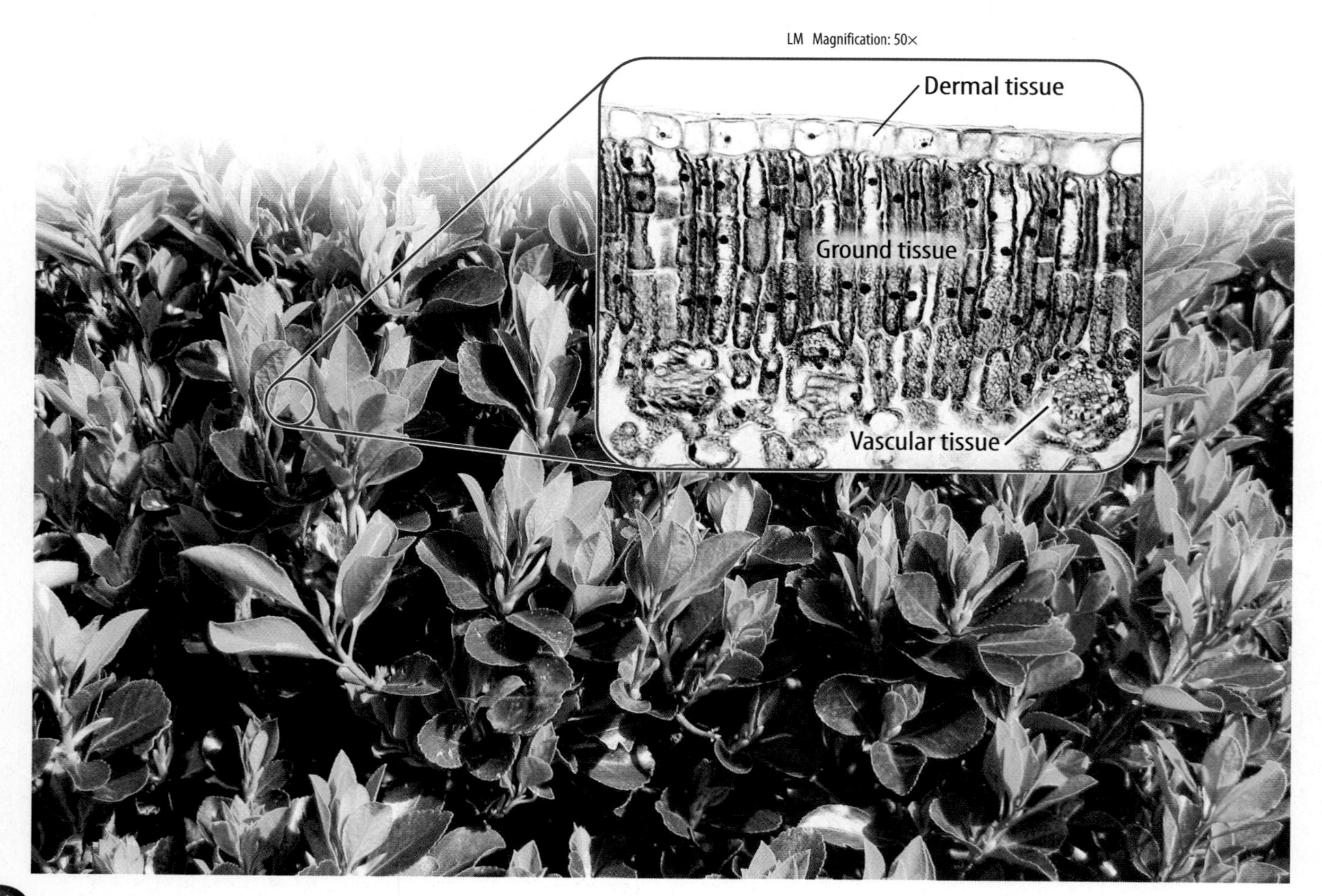

Figure 14 A plant leaf is an organ made of several different tissues.

5. Visual Check **Assess** Circle which plant tissue makes up the thinnest layer.

Organ Systems

Usually organs do not function alone. Instead, **organ systems** *are groups of different organs that work together to complete a series of tasks.* Human organ systems can be made of many different organs working together. For example, the human digestive system is made of many organs, including the stomach, the small intestine, the liver, and the large intestine. These organs and others all work together to break down food and take it into the body. Blood absorbs and transports nutrients from broken down food to cells throughout the body.

Plants have two major organ systems—the shoot system and the root system. The shoot system includes leaves, stems, and flowers. Food and water are transported throughout the plant by the shoot system. The root system anchors the plant and takes in water and nutrients.

Active Reading 7. **Name** Write down the major organ systems in plants.

____

____

Sequence the organization of cells, tissues, organs, and organ systems in a multicellular organism.

8. Cells are organized in ______________.

9. Different ______________ working together to perform a particular job are called ______________.

10. Groups of ______________ that work together to complete a series of tasks are called ______________.

11. Many ______________ working together make up a(n) ______________.

Apply It! After you complete the lab, answer this question.

1. How do the organ systems of a plant work together to support the organism?

Organisms

Multicellular organisms usually have many organ systems. These systems work together to carry out all the jobs needed for the survival of the organisms. For example, the cells in the leaves and the stems of a plant need water to live. They cannot absorb water directly. Water diffuses into the roots and is transported through the stem to the leaves by the transport system.

In the human body, there are many major organ systems. Each organ system depends on the others and cannot work alone. For example, the cells in the muscle tissue of the stomach cannot survive without oxygen. The stomach cannot get oxygen without working together with the respiratory and circulatory systems. **Figure 15** will help you review how organisms are organized.

12. NGSSS Check **Reason** How does cell differentiation lead to the organization within a multicellular organism? SC.6.L.14.1

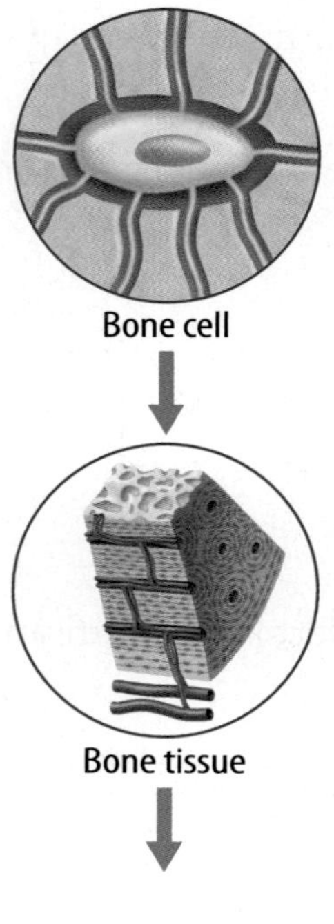

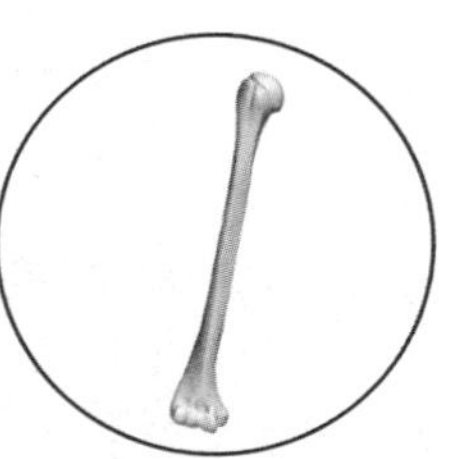

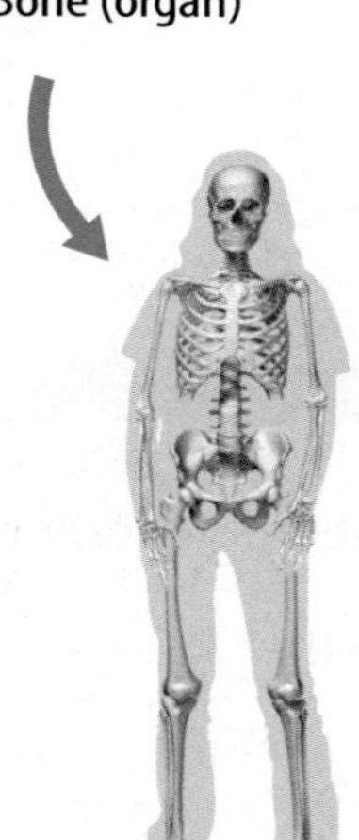

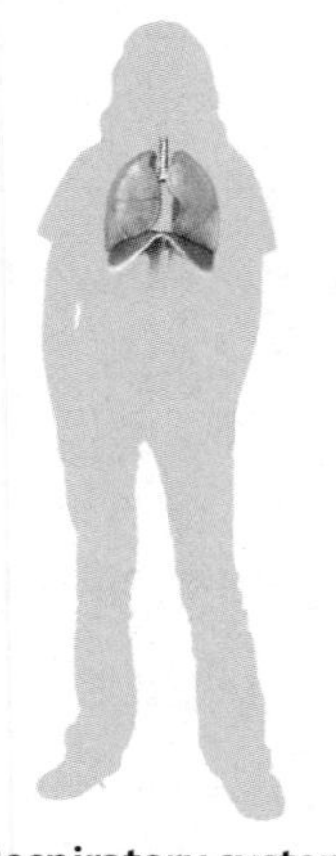

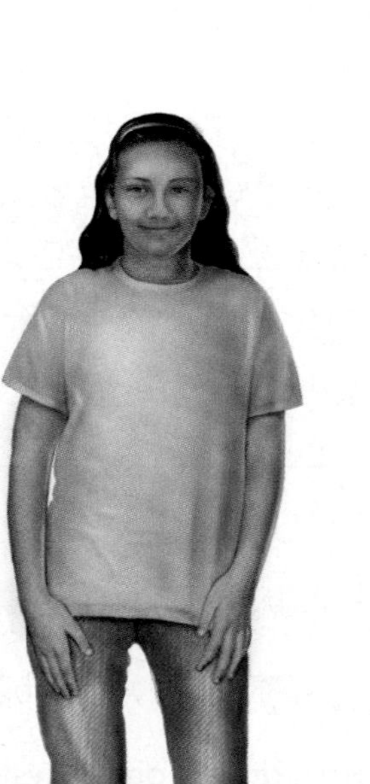

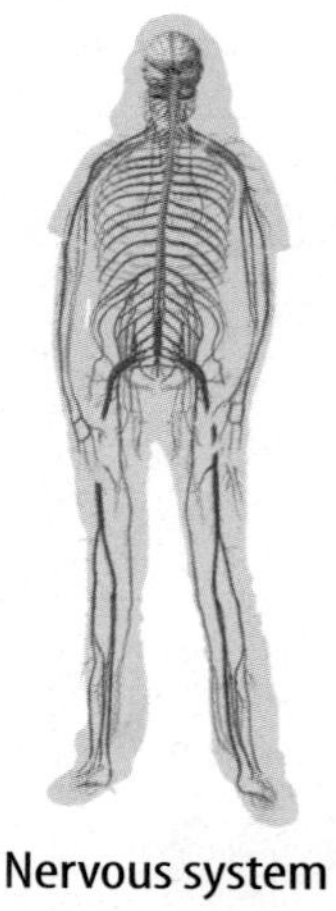

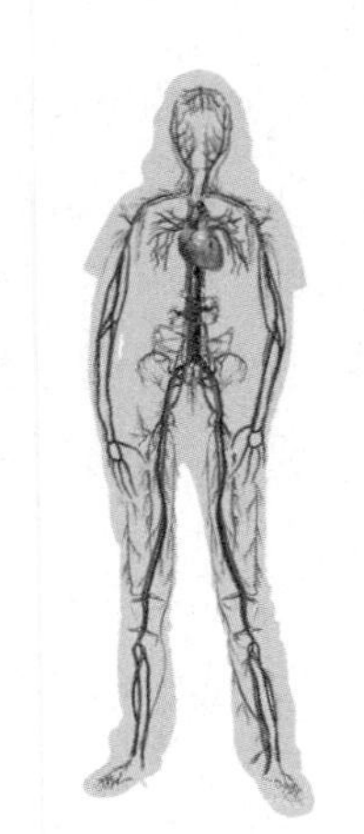

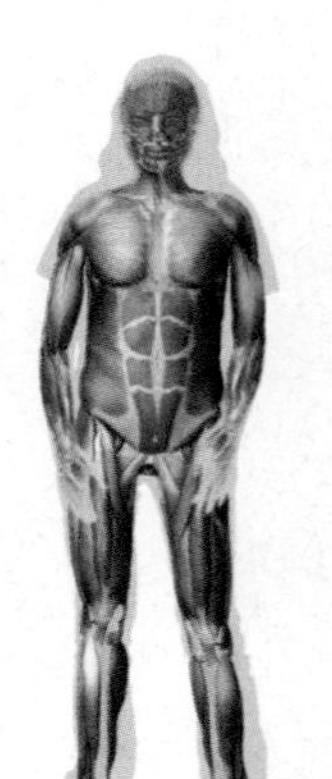

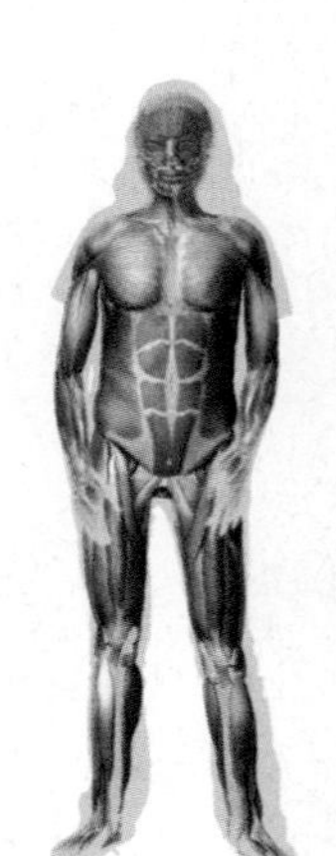

Figure 15 An organism is made of organ systems, organs, tissues, and cells that all function together and enable the organism to survive.

Lesson Review 2

Visual Summary

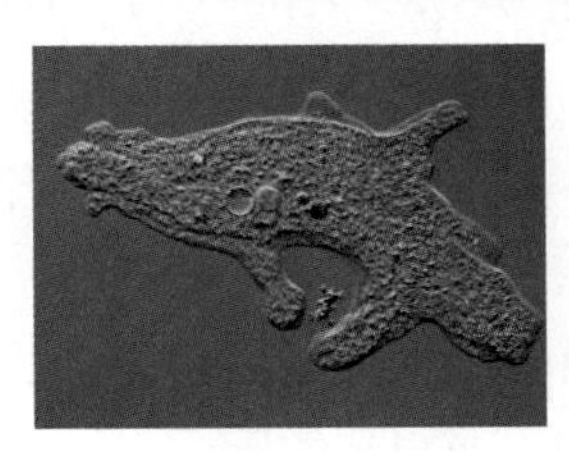

A unicellular organism carries out all the activities necessary for survival within one cell.

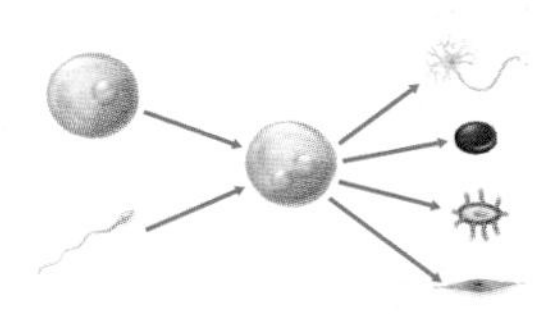

Cells become specialized in structure and function during cell differentiation.

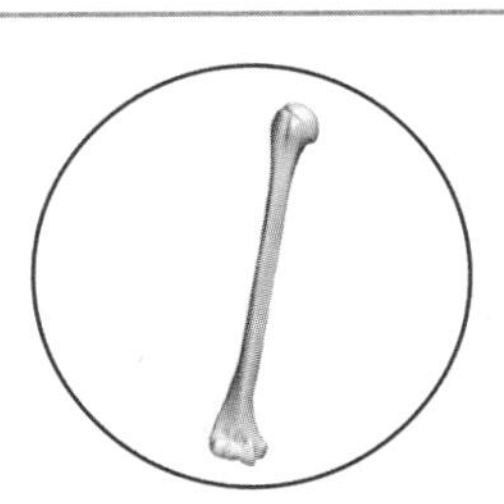

Organs are groups of different tissues that work together to perform a job.

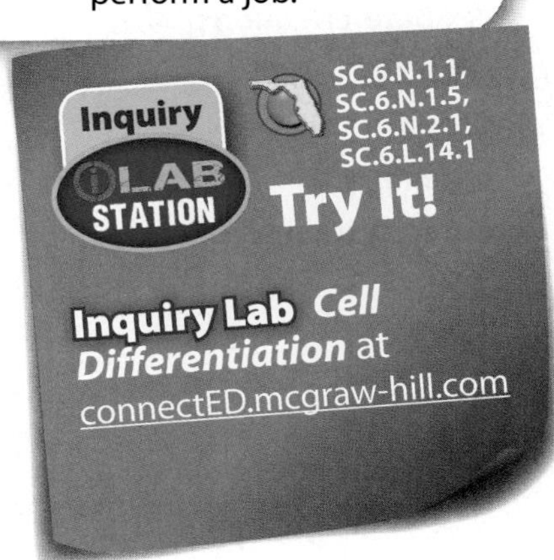

Use Vocabulary

1. **Define** *cell differentiation* in your own words.

2. **Distinguish** between an organ and an organ system. SC.6.L.14.1

Understand Key Concepts 

3. **Explain** the difference between a unicellular organism and a multicellular organism. SC.6.L.14.1

4. **Describe** how cell differentiation produces different types of cells in animals.

5. Which is the correct sequence of the levels of organization? SC.6.L.14.1
   - (A) cell, organ, tissue, organ system, organism
   - (B) organism, organ, organ system, tissue, cell
   - (C) cell, tissue, organ, organ system, organism
   - (D) tissue, organ, organism, organ system, cell

Interpret Graphics

6. **Organize** Fill in the table below to summarize the characteristics of unicellular and multicellular organisms. LA.6.2.2.3

Organism Characteristics	
Unicellular	Multicellular

Critical Thinking

7. **Predict** A mistake occurs during mitosis of a muscle stem cell. How might this affect muscle tissue?

8. **Compare** the functions of a cell to the functions of an organism, such as getting rid of wastes.

Chapter 11 Study Guide

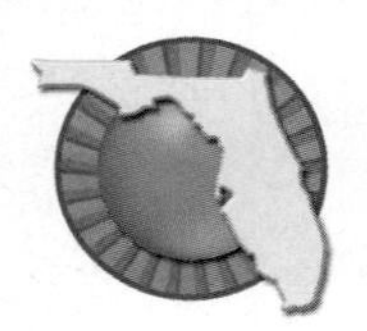

Think About It! Through various physiological functions essential for growth and reproduction, one cell can grow and develop into a multicellular organism.

Key Concepts Summary

LESSON 1 The Cell Cycle and Cell Division

- The **cell cycle** consists of two phases. During **interphase,** a cell grows and its chromosomes and organelles replicate. During the mitotic phase of the cell cycle, the nucleus divides during **mitosis,** and the cytoplasm divides during **cytokinesis.**
- The cell cycle results in two genetically identical **daughter cells.** The original parent cell no longer exists.
- The cell cycle is important for growth in multicellular organisms, reproduction in some organisms, replacement of worn-out cells, and repair of damaged cells.

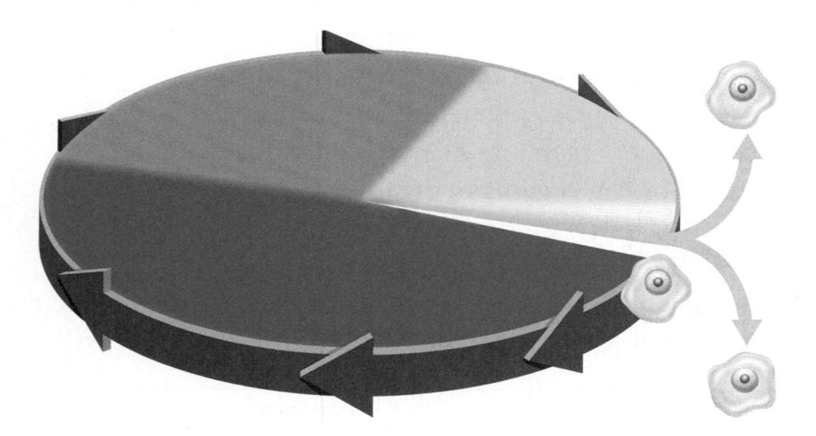

Vocabulary

cell cycle p. 415
interphase p. 416
sister chromatid p. 418
centromere p. 418
mitosis p. 419
cytokinesis p. 419
daughter cell p. 419

LESSON 2 Levels of Organization

- The one cell of a unicellular organism is able to obtain all the materials that it needs to survive.
- In a multicellular organism, cells cannot survive alone and must work together to handle the organism's needs.
- Through **cell differentiation,** cells become different types of cells with specific functions. Cell differentiation leads to the formation of **tissues, organs,** and **organ systems.**

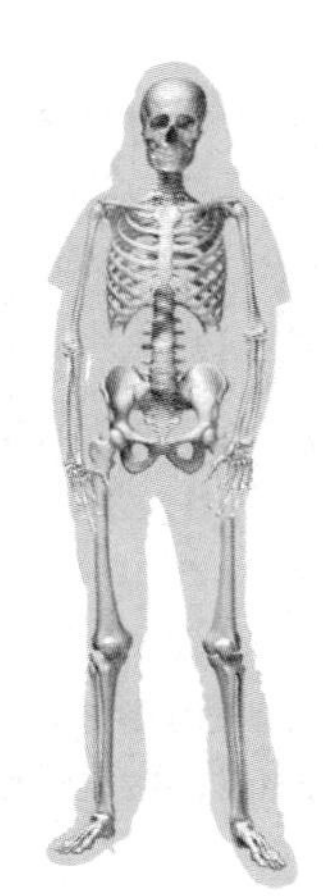

cell differentiation p. 429
stem cell p. 430
tissue p. 431
organ p. 432
organ system p. 433

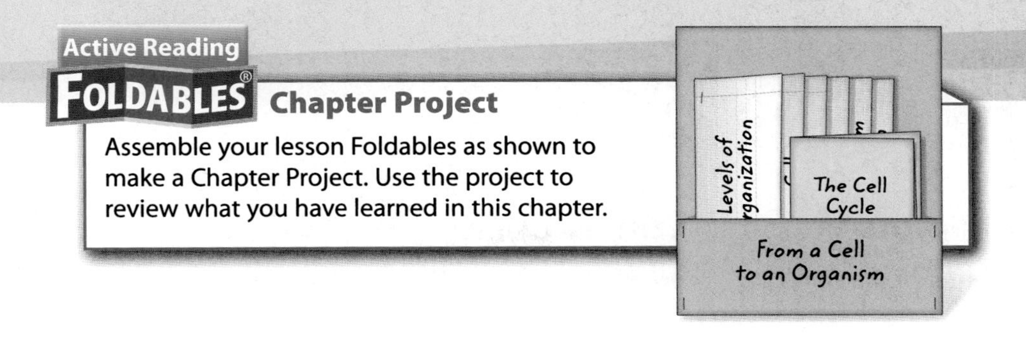

Chapter Project

Assemble your lesson Foldables as shown to make a Chapter Project. Use the project to review what you have learned in this chapter.

Use Vocabulary

1. Use the term *sister chromatids* in a sentence.

2. Define the term *centromere* in your own words.

3. The new cells formed by mitosis are called.

4. Use the term *cell differentiation* in a sentence.

5. Define the term *stem cell* in your own words.

6. Organs are groups of ______ working together to perform a specific task. SC.6.L.14.1

Link Vocabulary and Key Concepts

Use vocabulary terms from the previous page and from the chapter to complete the concept map. SC.6.L.14.3

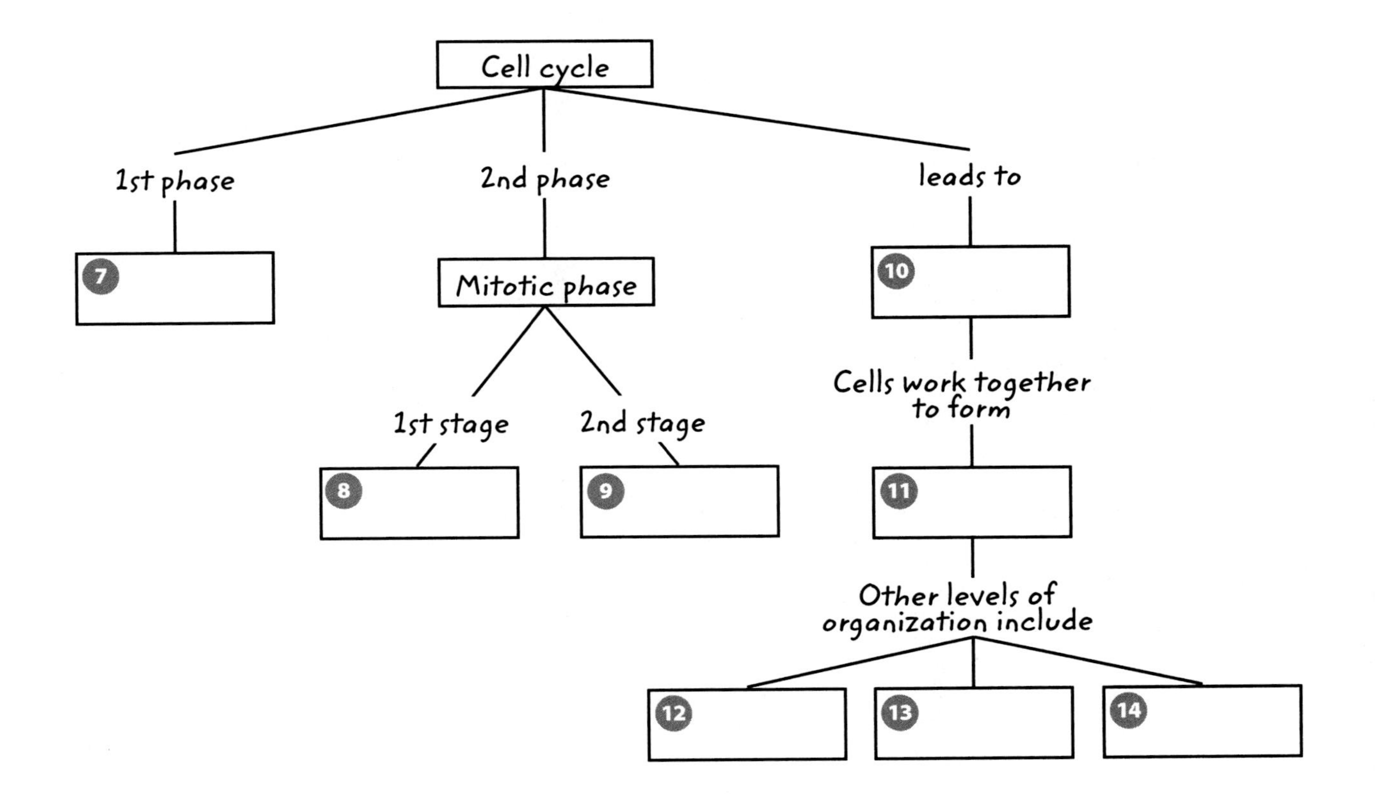

Chapter 11 Review

Fill in the correct answer choice.

Understand Key Concepts

1. Chromosomes line up in the center of the cell during which phase? **SC.6.L.14.3**
   - (A) anaphase
   - (B) metaphase
   - (C) prophase
   - (D) telophase

2. Which stage of the cell cycle precedes cytokinesis? **SC.6.L.14.3**
   - (A) G_1
   - (B) G_2
   - (C) interphase
   - (D) mitosis

Use the figure below to answer questions 3 and 4.

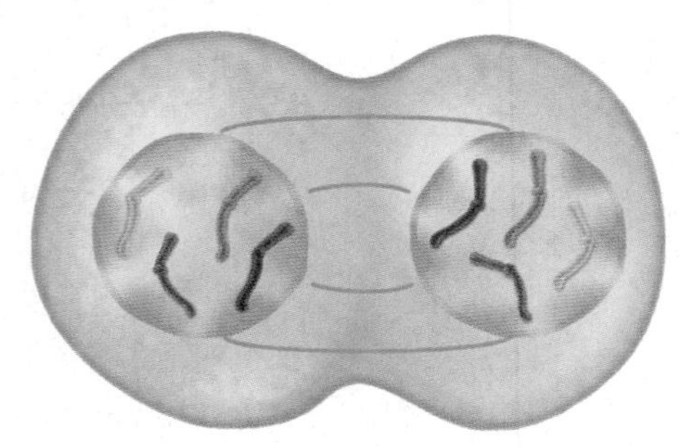

3. The figure represents which stage of mitosis? **SC.6.L.14.3**
   - (A) anaphase
   - (B) metaphase
   - (C) prophase
   - (D) telophase

4. What forms during this phase? **SC.6.L.14.3**
   - (A) centromere
   - (B) furrow
   - (C) sister chromatid
   - (D) two nuclei

5. What is the longest part of the cell cycle? **SC.6.L.14.3**
   - (A) anaphase
   - (B) cytokinesis
   - (C) interphase
   - (D) mitosis

Critical Thinking

6. **Sequence** the events that occur during the phases of mitosis. **SC.6.L.14.3**

7. **Infer** why the chromatin condenses into chromosomes before mitosis begins. **SC.6.L.14.3**

8. **Create** Use the figure below to create a cartoon that shows a duplicated chromosome separating into two sister chromatids. **SC.6.L.14.3**

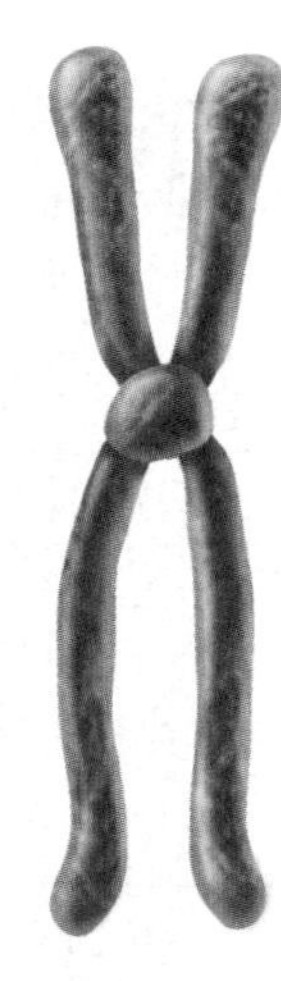

9. **Classify** a leaf as a tissue or an organ. Explain your choice. **SC.6.L.14.1**

10. **Distinguish** between a tissue and an organ. **SC.6.L.14.1**

11 **Construct** a table that lists and defines the different levels of organization. SC.6.L.14.1

12 **Summarize** the differences between unicellular organisms and multicellular organisms. SC.6.L.14.1

Writing in Science

13 **Write** a five-sentence paragraph on a seperate piece of paper describing a human organ system. Include a main idea, supporting details, and a concluding statement. LA.6.2.2.3

Big Idea Review

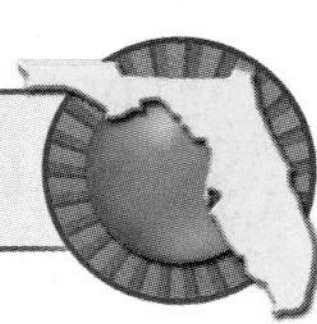

14 Why is cell division important for multicellular organisms? SC.6.L.14.1

Math Skills

MA.6.A.3.6

Use Percentages

15 During an interphase lasting 23 hours, the S stage takes an average of 8.0 hours. What percentage of interphase is taken up by the S stage?

Use the following information to answer questions 16 and 17.

During a 23-hour interphase, the G_1 stage takes 11 hours and the S stage takes 8.0 hours.

16 What percentage of interphase is taken up by the G_1 and S stages?

17 What percentage of interphase is taken up by the G_2 phase?

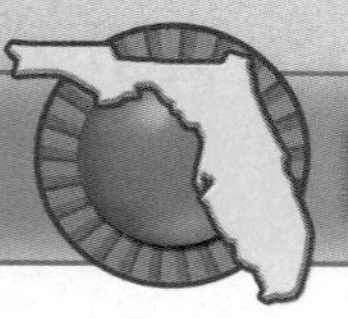

Florida NGSSS Benchmark Practice

Fill in the correct answer choice.

Multiple Choice

1 Which statement describes the number of chromosomes in a newly formed cell after mitosis? **SC.6.L.14.3**

Ⓐ They are equal to the number of the parent cell.

Ⓑ They are half the number of the parent cell.

Ⓒ They are double the number of the parent cell.

Ⓓ They are double the number of the two parent cells.

Use the diagram below to answer question 2.

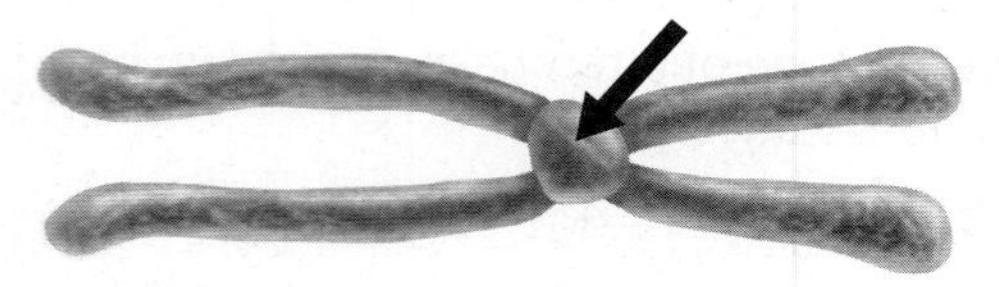

2 What is the function of the object indicated by the arrow? **SC.6.L.14.3**

Ⓕ provide energy to the cell

Ⓖ hold cytoplasm in the cell

Ⓗ hold the sister chromatids together

Ⓘ divide the organelles

3 Which is the correct sequence of organization in a multicellular organism? **SC.6.L.14.1**

Ⓐ cell → tissue → organ system → organ

Ⓑ organ system → cell → tissue → organ

Ⓒ tissue → cell → organ → organ system

Ⓓ cell → tissue → organ → organ system

4 What structures separate during anaphase? **SC.6.L.14.3**

Ⓕ centromeres

Ⓖ chromatids

Ⓗ nuclei

Ⓘ organelles

Use the diagram below to answer question 5.

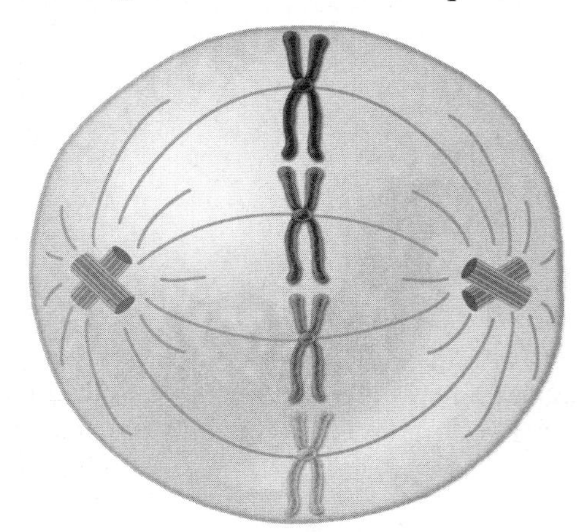

5 Which stage of mitosis does the image above represent? **SC.6.L.14.3**

Ⓐ anaphase

Ⓑ metaphase

Ⓒ prophase

Ⓓ telophase

6 Which mitotic phase occurs when a cell's cytoplasm and its contents are divided? **SC.6.L.14.3**

Ⓕ prophase

Ⓖ cytokinesis

Ⓗ interphase

Ⓘ telophase

7 Which is the most accurate description of a leaf or your stomach? **SC.6.L.14.1**

Ⓐ a cell

Ⓑ an organ

Ⓒ an organ system

Ⓓ a tissue

Use the figure below to answer question 8.

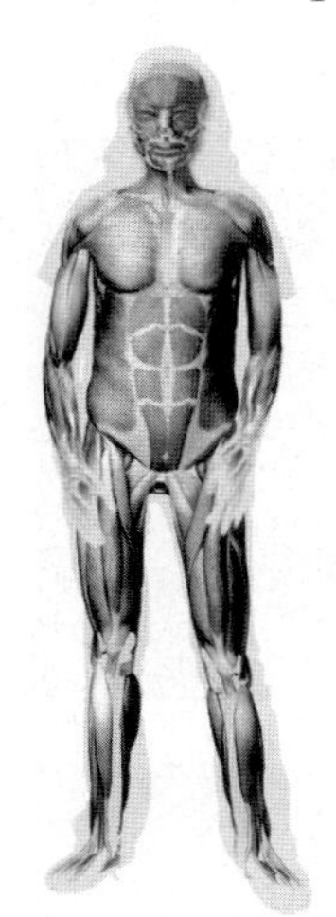

8 Which does this figure illustrate? **SC.6.L.14.1**

Ⓕ an organ

Ⓖ an organism

Ⓗ an organ system

Ⓘ a tissue

9 If a cell has 30 chromosomes at the start of mitosis, how many chromosomes will be in each new daughter cell? **SC.6.L.14.3**

Ⓐ 10

Ⓑ 15

Ⓒ 30

Ⓓ 60

10 Which are groups of similar types of cells that work together to carry out specific tasks? **SC.6.L.14.1**

Ⓕ stem cells

Ⓖ organs

Ⓗ organ systems

Ⓘ tissues

Use the figures below to answer questions 11 and 12

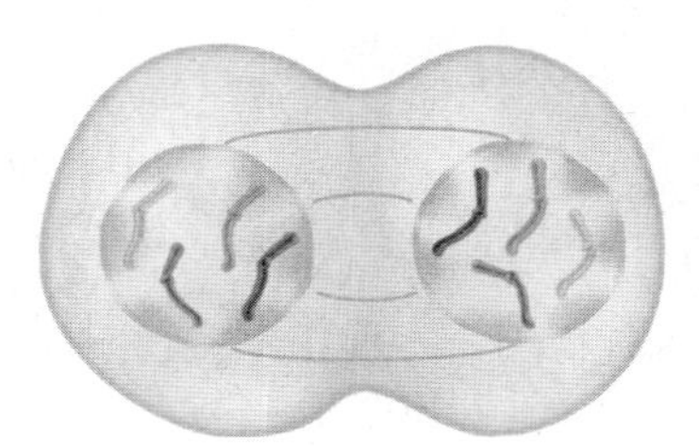

Figure A

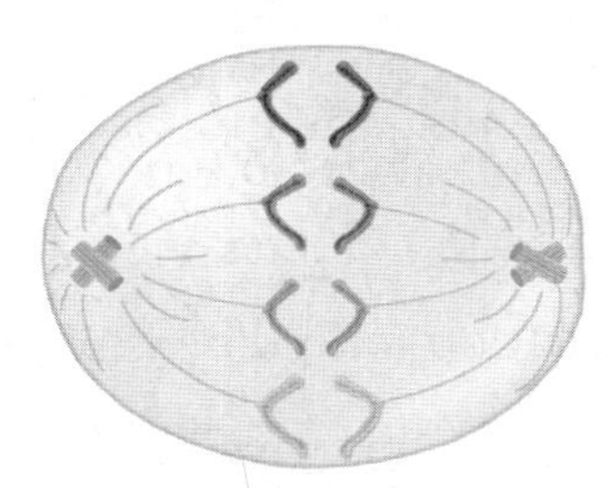

Figure B

11 In Figure A, during which stage of mitosis does a nuclear membrane form around each set of chromosomes? **SC.6.L.14.3**

Ⓐ prophase

Ⓑ metaphase

Ⓒ anaphase

Ⓓ telophase

12 What occurs in a cell undergoing anaphase, as shown in Figure B? **SC.6.L.14.3**

Ⓕ Duplicated chromosomes appear, and the nuclear membrane breaks down.

Ⓖ The sister chromatids separate.

Ⓗ The duplicated chromosomes line up in the middle of the cell.

Ⓘ A nuclear membrane develops around each set of chromosomes as they uncoil.

NEED EXTRA HELP?

If You Missed Question...	1	2	3	4	5	6	7	8	9	10	11	12
Go to Lesson...	1	1	2	1	1	1	2	2	1	2	1	1

Name ____________________ Date ____________

Multiple Choice *Bubble the correct answer.*

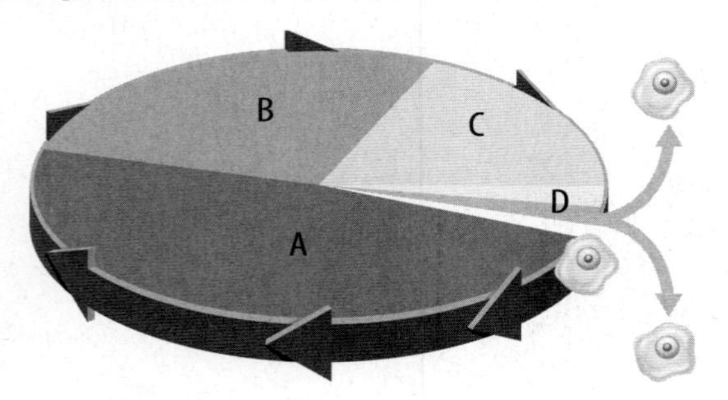

1. In the image above, which phase of a cell's life is represented by D? **SC.6.L.14.3**

Ⓐ DNA replication phase

Ⓑ mitotic phase

Ⓒ preparation for cell division phase

Ⓓ rapid growth and replication phase

2. The final phase of the cell cycle is cytokinesis. How does cytokinesis occur in plant cells? **SC.6.L.14.3**

Ⓕ A new cell wall forms in the middle of the cell.

Ⓖ The old cell wall grows inward, forming a new cell wall.

Ⓗ Fibers grow into the cell, dividing the cytoplasm and the nuclei.

Ⓘ Fibers tighten around the cell to divide the cytoplasm and nuclei.

3. Which image shows the first stage in the division of a fertilized egg? **SC.6.L.14.3**

Ⓐ 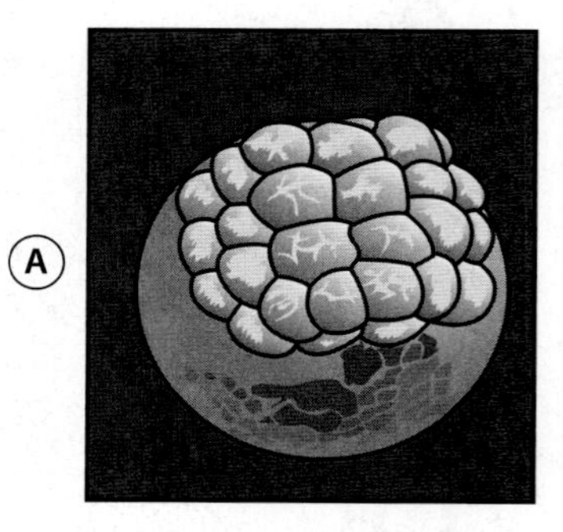

Ⓑ 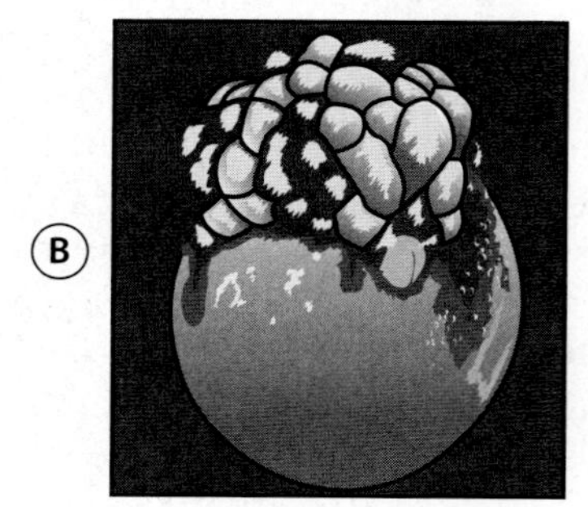

Ⓒ 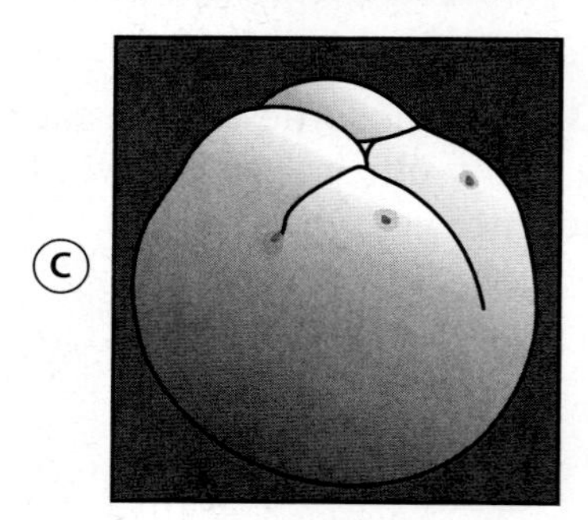

Ⓓ 

Name ______________________________ Date ______________

Multiple Choice *Bubble the correct answer.*

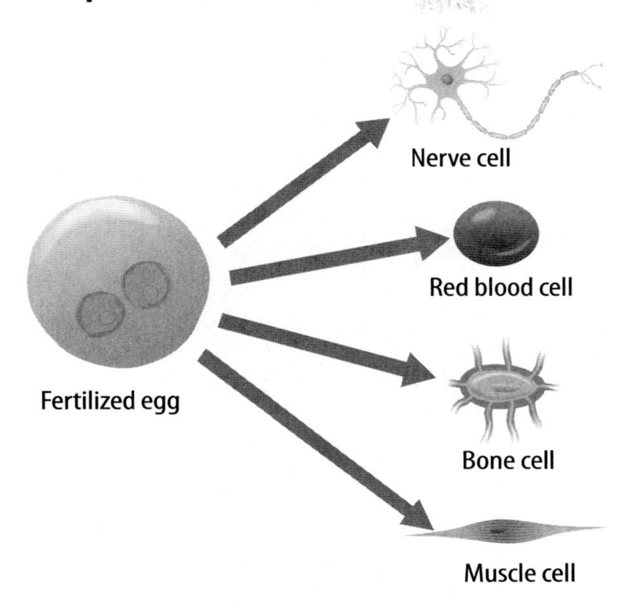

1. Which process is shown in the image above? **SC.6.L.14.3**

- (A) metaphase
- (B) photosynthesis
- (C) cell differentiation
- (D) organelle replication

2. Which of these is the most complex organization of cells in a multicellular organism? **SC.6.L.14.3**

- (F) organ
- (G) tissue
- (H) differentiated cell
- (I) organ system

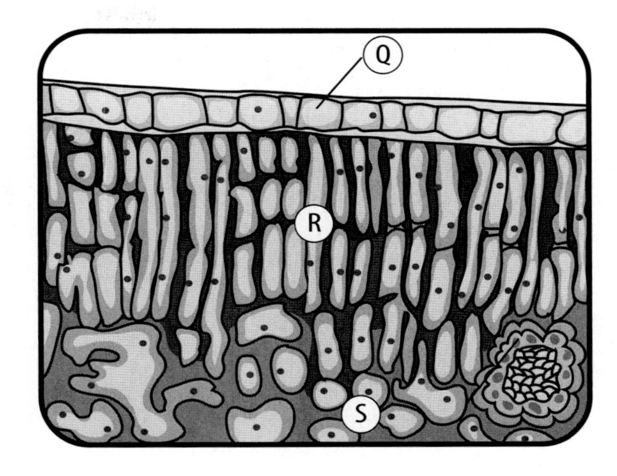

3. In the image of the plant leaf above, what do Q, R, and S represent? **SC.6.L.14.3**

- (A) bacteria
- (B) muscles
- (C) roots
- (D) tissues

Name ______________________________ Date ______________

Notes

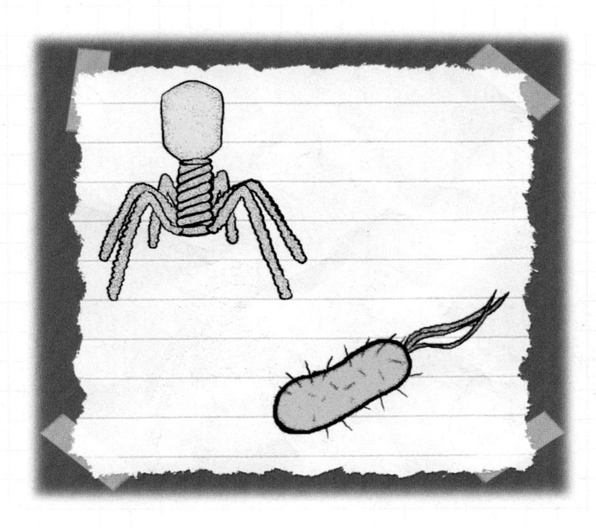

Basic Unit of Function

Three friends talked about the human body. They each agreed that the cell is the basic unit of structure in the human body. But they did not agree on the basic unit of function in the human body. This is what they said:

Sarah: I think body systems are the basic unit of function.

Todd: I think cells are the basic unit of function in the body.

Juanita: I think tissues are the basic unit of function.

Seif: I think organs are the basic unit of function.

Circle the friend you most agree with. Explain your thinking about cells.

FLORIDA
Chapter 12

Human Body Systems

FLORIDA BIG IDEAS

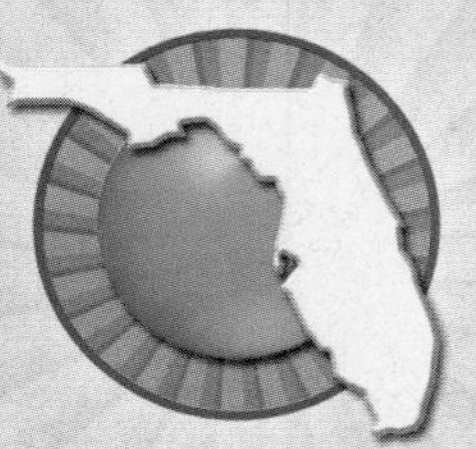

1 The Practice of Science
2 The Characteristics of Scientific Knowledge
3 The Role of Theories, Laws, Hypotheses, and Models
14 Organization and Development of Living Organisms

Think About It!

What are the functions of the human body systems?

This is a photograph of a cross section through a human body. You can see the lower part of a human arm and part of the abdomen. In the abdomen, you might be able to point out a vertebra, muscles, fat, and part of the intestine.

1. What body systems can you identify here?

2. What are the functions of the human body systems?

Get Ready to Read

What do you think about human body systems?

Before you read, decide if you agree or disagree with each of these statements. As you read this chapter, see if you change your mind about any of the statements.

	AGREE	DISAGREE
1 A human body has organ systems that carry out specific functions.		
2 The body protects itself from disease.		
3 All bones in the skeletal system are hollow.		
4 The endocrine system makes hormones.		
5 The testes produce sperm.		
6 Puberty occurs during infancy.		

There's More Online!
Video • Audio • Review • ⓘLab Station • WebQuest • Assessment • Concepts in Motion • Multilingual eGlossary

Lesson 1

Transport and DEFENSE

ESSENTIAL QUESTIONS

 How do nutrients enter and leave the body?

 How do nutrients travel through the body?

 How does the body defend itself from harmful invaders?

Vocabulary

organ system p. 449
homeostasis p. 449
nutrient p. 451
Calorie p. 451
lymphocyte p. 457
immunity p. 458

Florida NGSSS

LA.6.2.2.3 The student will organize information to show understanding (e.g., representing main ideas within text through charting, mapping, paraphrasing, summarizing, or comparing/contrasting);

MA.6.A.3.6 Construct and analyze tables, graphs, and equations to describe linear functions and other simple relations using both common language and algebraic notation.

SC.6.L.14.5 Identify and investigate the general functions of the major systems of the human body (digestive, respiratory, circulatory, reproductive, excretory, immune, nervous, and musculoskeletal) and describe ways these systems interact with each other to maintain homeostasis.

SC.6.N.1.1 Define a problem from the sixth grade curriculum, use appropriate reference materials to support scientific understanding, plan and carry out scientific investigation of various types, such as systematic observations or experiments, identify variables, collect and organize data, interpret data in charts, tables, and graphics, analyze information, make predictions, and defend conclusions.

SC.6.N.2.1 Distinguish science from other activities involving thought.

 Launch Lab

SC.6.N.1.1, SC.6.N.2.1

15 minutes

Which tool can transport water quickly?

You need to transport materials throughout your body. Each cell must receive nutrients and oxygen and get rid of wastes. What kinds of tools do you think would be most effective in moving fluids such as water quickly?

Procedure

1. Read and complete a lab safety form.
2. Choose one of the **tools** for moving water.
3. Have another student use a **stopwatch** to keep time for 30 s. Use your tool to transport as much water as you can in 30 s from the main **bowl** into a **beaker.**
4. Use a **graduated cylinder** to measure the amount of water you moved from the bowl to the beaker. Record the measurement.
5. Trade roles with your partner. Repeat steps 2 through 4.
6. Repeat step 5 until you have used all of the tools.

Data and Observations

Think About This

1. Which tool was most effective for moving water quickly? Which tool was least effective?

2. Key Concept Why do you think moving small items in fluid might be more effective than moving them all individually?

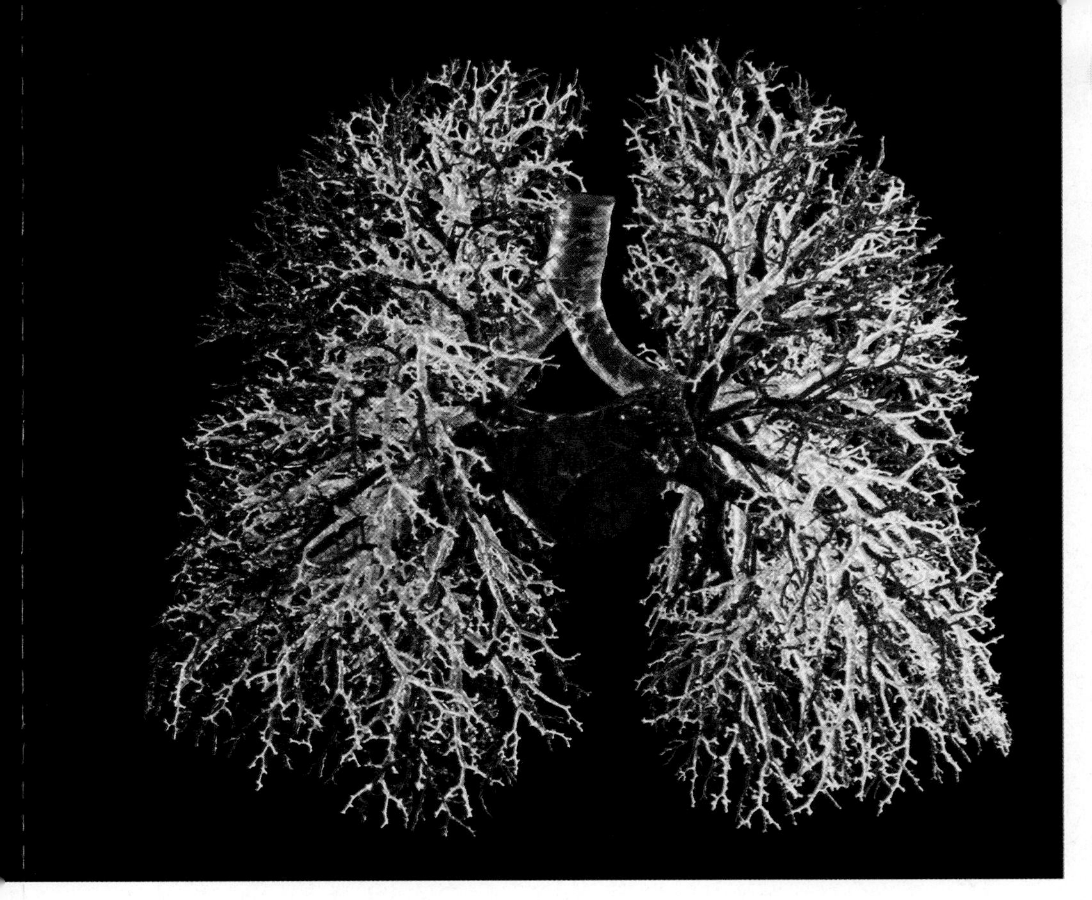

Inquiry Unusual Web?

1. This branching structure might look like a strange spider web, but it is actually a resin cast of human lungs. The yellowish tubes are large air passages, the white parts are small airways, and the blue parts are blood vessels. Why do you think the lungs need all these parts?

The Body's Organization

All organisms have different parts with special functions. Recall that cells are the basic unit of all living organisms. Organized groups of cells that work together are tissues. Groups of tissues that perform a specific function are organs. *Groups of organs that work together and perform a specific task are* **organ systems**. Organ systems provide movement, transport substances, and perform many other functions.

Organ systems work together and maintain **homeostasis** (hoh mee oh STAY sus), *or steady internal conditions when external conditions change.* Maintaining homeostasis is an important part of keeping your body healthy and safe. It occurs without any conscious thought.

The respiratory system works with the circulatory system to provide oxygen to muscles when being active, as shown in **Figure 1.** The digestive system digests food energy for the body. The excretory system helps by expelling wastes from the body. Bodies are protected from infection by the lymphatic and circulatory systems working together.

Active Reading 2. **Recall** List five examples of organ systems.

Figure 1 Sweating helps the body maintain homeostasis by releasing excess thermal energy.

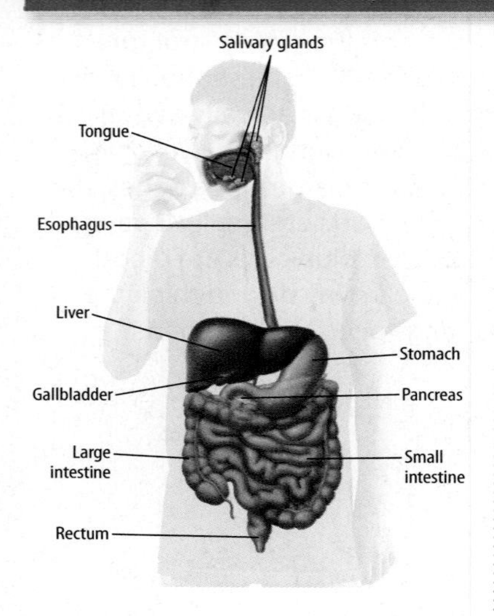

Figure 2 Food enters the digestive system through the mouth, and nutrients are absorbed by the small intestine.

Active Reading **3. Identify** Highlight the path that food takes through the digestive system.

Digestion and Elimination

Humans need food, water, and oxygen to survive. Food contains energy that is processed by the body. The process by which food is broken down is called digestion. After digestion, substances that are not used by the body are removed through the process of elimination.

The Digestive System

As shown in **Figure 2,** the digestive system is made up of several organs. Food and water enter the digestive system through the mouth.

Digestion After food enters the mouth, chewing breaks food into smaller parts. Saliva, which contains enzymes, also helps the mouth break down food. Recall that enzymes are proteins that speed up chemical reactions.

When you swallow, food, water, and other liquids move into a hollow tube called the esophagus (ih SAH fuh gus). The esophagus connects the mouth to the stomach. Digestion continues as food leaves the esophagus and enters the stomach. The stomach is a flexible, baglike organ that contains other enzymes that break down food into smaller parts that can be used by the body.

Active Reading **4. Identify** Where does food enter the body?

____

____

Absorption Next, food moves into the small intestine. By the time food gets to the small intestine, it is a soupy mixture. The small intestine is a tube that has two functions—digestion and absorption. The liver makes a substance called bile. The pancreas makes enzymes. Both bile and enzymes are used in the small intestine to break down food even more. Because the small intestine is very long, it takes food hours to move through it. During that time, particles of food and water are absorbed into the blood.

Active Reading
FOLDABLES® LA.6.2.2.3

Make three horizontal two-tab books. Label them with the body systems in this lesson, and glue them side by side to form a booklet with six tabs. Use your book to organize information about each body system in this lesson.

Elimination The large intestine, or colon (KOH lun), receives digested food that the small intestine did not absorb. The large intestine also absorbs water from the remaining waste material. Most foods are completely digested into smaller parts that can be easily absorbed by the small intestine. However, some foods travel through the entire digestive system without being digested or absorbed. For example, some types of fiber, called insoluble fiber, in vegetables and whole grains are not digested and leave the body through the rectum.

5. **NGSSS Check** **Explain** How does food enter and leave the body? SC.6.L.14.5

Nutrition

As you have read, one of the functions of the small intestine is absorption. **Nutrients** *are the parts of food used by the body to grow and survive.* There are several types of nutrients. **Proteins**, fats, carbohydrates, vitamins, and minerals are all nutrients. Nutrition labels on food, as shown in **Figure 3,** show the amount of each nutrient in that food. By looking at the labels on packaged foods, you can make sure you get the nutrients you need. Different people need different amounts of nutrients. For example, football players, swimmers, and other athletes need a lot of nutrients for energy. Pregnant women also need lots of nutrients to provide for their developing babies.

Digestion helps release energy from food. *A* **Calorie** *is the amount of energy it takes to raise the temperature of 1 kg of water by 1°C.* The body uses Calories from proteins, fats, and carbohydrates, which each contain a different amount of energy.

Figure 3 The information on a nutrition label can help you decide whether a food is healthful to eat.

Math Skills

MA.6.A.3.6

Use Proportions

A proportion is an equation of two equal ratios. You can solve a proportion for an unknown value. For example, a 50-g egg provides 70 Calories (C) of energy. How many Calories would you get from 125 g of scrambled eggs?

Write a proportion.

$$\frac{50 \text{ g}}{70 \text{ C}} = \frac{125 \text{ g}}{x}$$

Find the cross products.

$$50 \text{ g}(x) = 70 \text{ C} \times 125 \text{ g}$$

$$50 \text{ g}(x) = 8{,}750 \text{ C g}$$

Divide both sides by 50.

$$\frac{50 \text{ g}(x)}{50 \text{ g}} = \frac{8{,}750 \text{ C g}}{50 \text{ g}}$$

Simplify the equation.

$$x = 175 \text{ C}$$

Practice

6. The serving size of a large fast-food hamburger with cheese is 316 g. It contains 790 C of energy. How many Calories would you consume if you ate 100 g of the burger?

Review Vocabulary

protein
a long chain of amino acid molecules

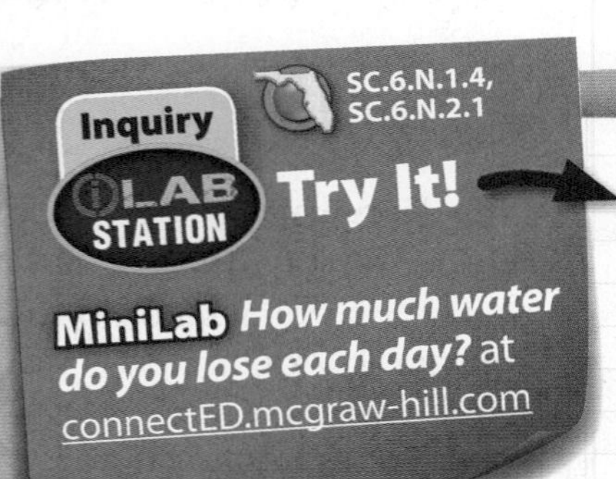

Apply It!

After you complete the lab, answer the questions below.

1. How effective is excretion through the skin?

2. Why do you think excreting excess water is important?

The Excretory System

The excretory system removes solid, liquid, and gas waste materials from the body. The lungs, skin, liver, kidneys, and bladder all are parts of the excretory system.

The lungs remove carbon dioxide (CO_2) and excess water as water vapor when you breathe out, or exhale. The skin removes water and salt when you sweat.

The liver removes wastes from the blood. As you have read, the liver also is a part of the digestive system. The digestive and excretory systems work together to break down, absorb, and remove excess material from food.

When the liver breaks down proteins, urea forms. Urea is toxic if it stays in the body. The kidneys, shown in **Figure 4,** remove urea from the body by making urine. Urine contains water, urea, and other waste chemicals. Urine leaves each kidney through a tube, called the ureter (YOO ruh tur), and is stored in a flexible sac, called the bladder. Urine is removed from the body through a tube called the urethra (yoo REE thruh).

Like the liver, the rectum is part of the excretory system and the digestive system. Food substances that are not absorbed by the small intestine are mixed with other wastes and form feces. The rectum stores feces until it moves out of the body.

Figure 4 The kidneys remove waste material from the body.

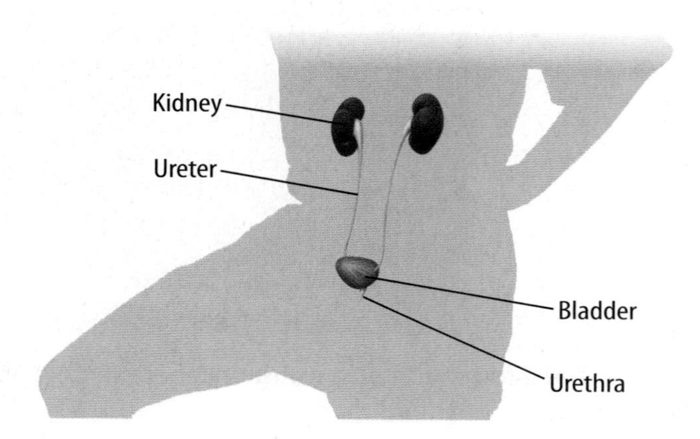

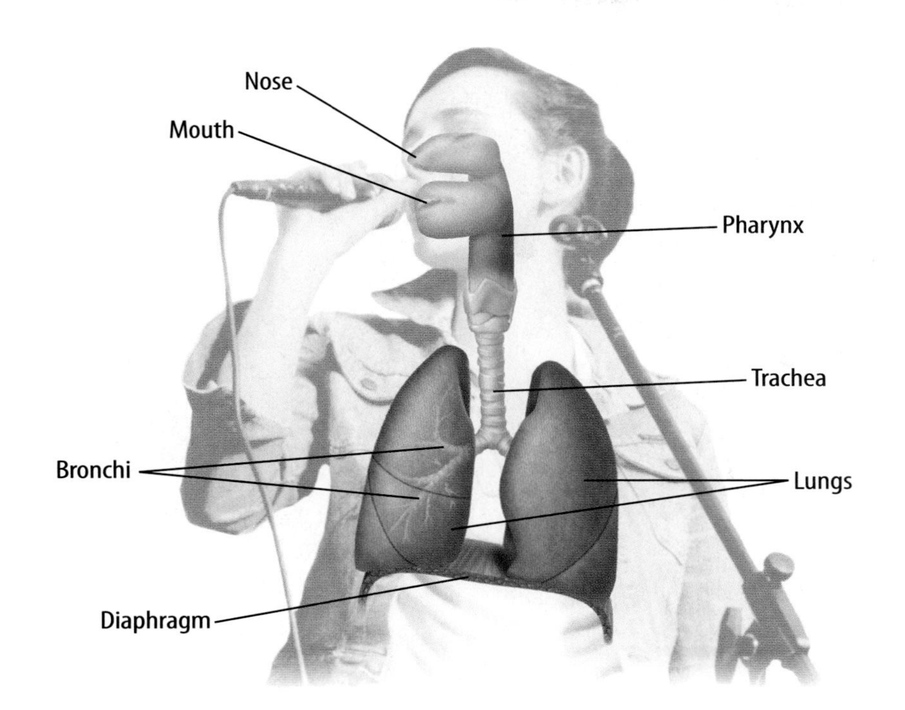

Figure 5 Air enters the respiratory system through the nose and the mouth. Oxygen enters the blood in the lungs.

Active Reading **7. Identify** Draw the path that air takes into the body.

Respiration and Circulation

You have read about how the body converts food into nutrients and how the small intestine absorbs nutrients. But how do the oxygen you breathe in and the nutrients absorbed by the small intestine get to the rest of the body? And how do waste products leave the body?

The Respiratory System

The respiratory system, shown in **Figure 5,** exchanges gases between the body and the environment. As air flows through the respiratory system, it passes through the nose and mouth, pharynx (FER ingks), trachea (TRAY kee uh), bronchi (BRAHN ki; singular, bronchus), and lungs. The parts of the respiratory system work together and supply the body with oxygen. They also rid the body of wastes, such as carbon dioxide.

Pharynx and Trachea Oxygen enters the body when you inhale, or breathe in. Carbon dioxide leaves the body when you exhale. When you inhale, air enters the nostrils and passes through the pharynx. Because the pharynx is part of the throat, it is a part of both the digestive and respiratory systems. Food goes through the pharynx to the esophagus. Air travels through the pharynx to the trachea. The trachea is also called the windpipe because it is a long, tubelike organ that connects the pharynx to the bronchi.

Active Reading **8. Describe** Which organ is part of both the digestive system and the respiratory system?

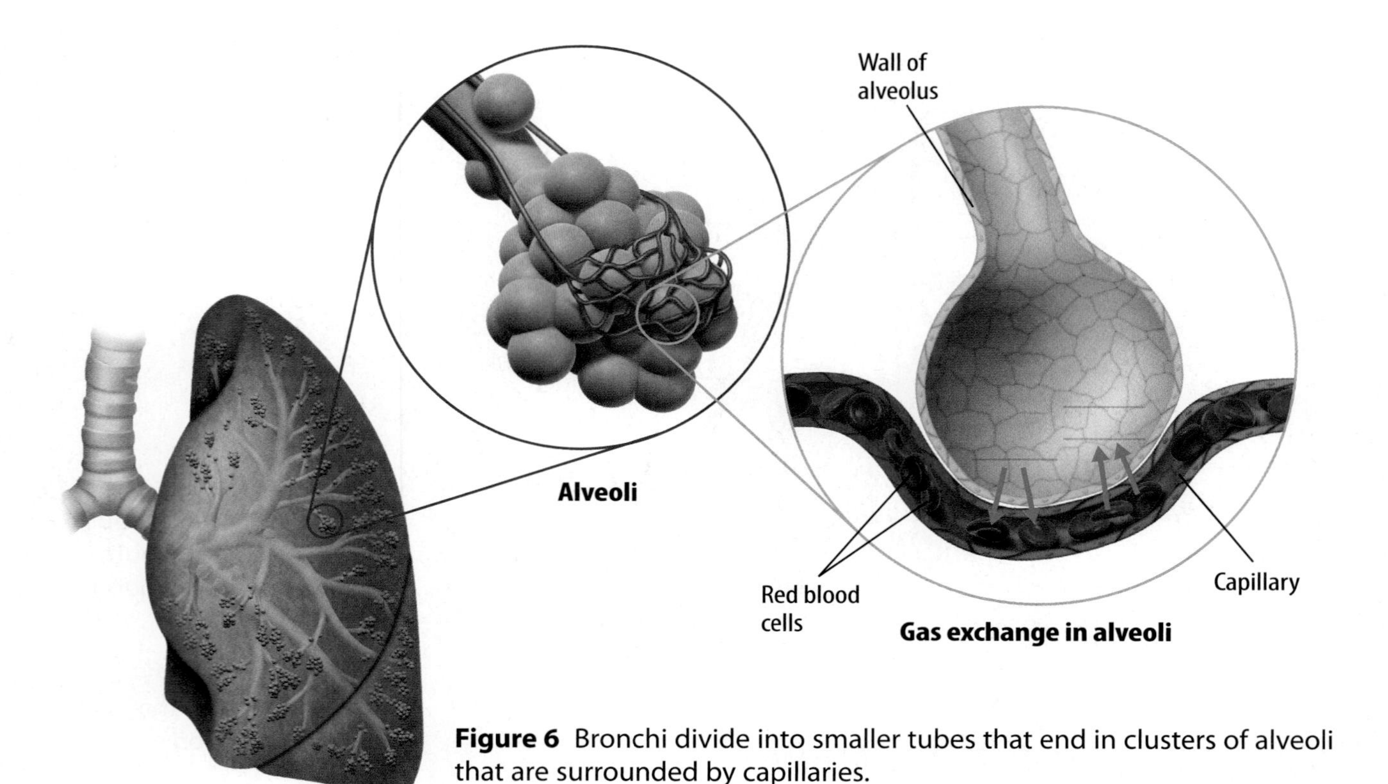

Figure 6 Bronchi divide into smaller tubes that end in clusters of alveoli that are surrounded by capillaries.

Active Reading **9. Recall** On the lines in the figure above, write in the name of the gas that enters capillaries and the gas that enters alveoli.

Bronchi and Alveoli There are two bronchi; one enters the left lung, and one enters the right lung. As shown in **Figure 6,** the bronchi divide into smaller tubes that end in tiny groups of cells that look like bunches of grapes. These groups of cells are called alveoli (al VEE uh li). Inside each lung, there are more than 100 million alveoli. The alveoli are surrounded by blood vessels called capillaries. Oxygen in the alveoli enters the capillaries. The blood inside capillaries transports oxygen to the rest of the body.

Active Reading **10. Describe** What are alveoli, and what do they do?

SCIENCE USE V. COMMON USE

vessel

Science Use a tube in the body that carries fluid such as blood

Common Use a ship

The Circulatory System

As shown in **Figure 7,** the heart, blood, and blood vessels make up the circulatory system. It transports nutrients, gases, wastes, and other substances through the body. Blood vessels transport blood to all organs in the body. Because your body uses oxygen and nutrients continually, your circulatory system transports blood between the heart, lungs, and other organs more than 1,000 times each day!

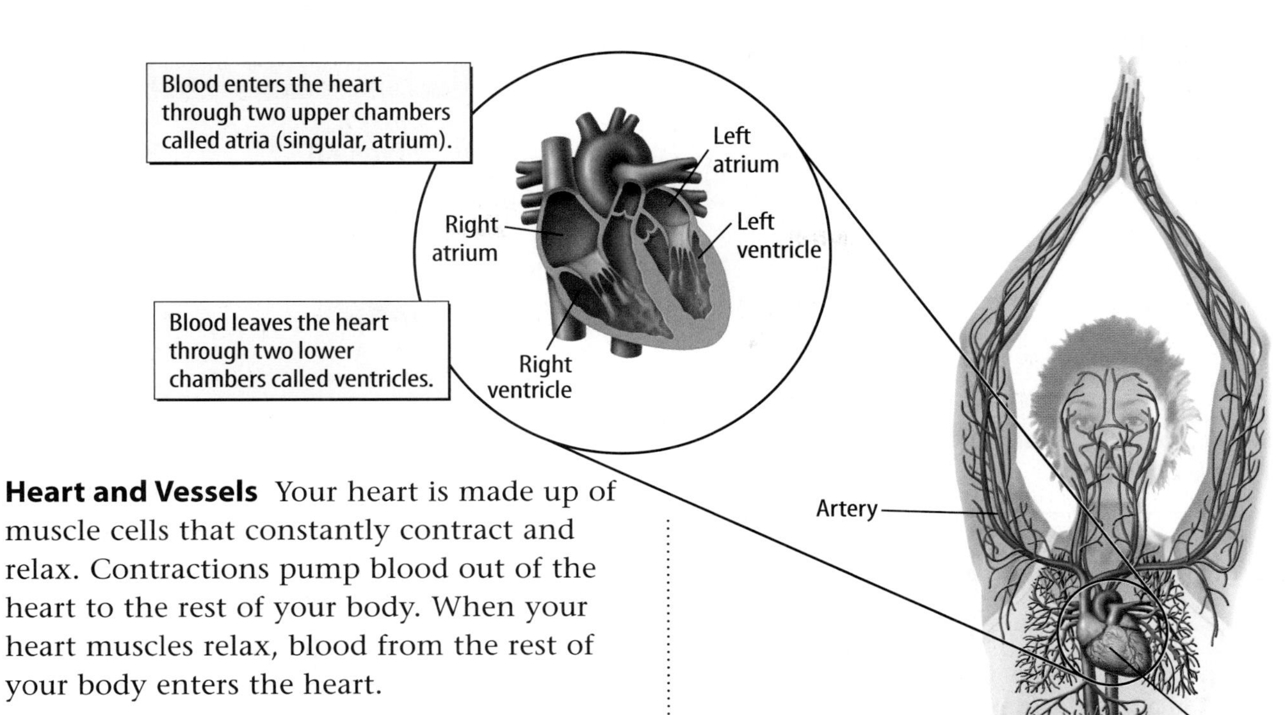

Heart and Vessels Your heart is made up of muscle cells that constantly contract and relax. Contractions pump blood out of the heart to the rest of your body. When your heart muscles relax, blood from the rest of your body enters the heart.

Blood travels through your body in tiny tubes called vessels. If all the blood vessels in your body were laid end-to-end in a single line, it would be more than 95,000 km long.

The three main types of blood vessels are arteries, veins, and capillaries. Arteries carry blood away from your heart. Usually this blood is oxygen-rich and contains nutrients, except for the blood in the pulmonary arteries that contains CO_2. Arteries are large and surrounded by muscle cells that help blood move through the vessels faster. Veins transport blood that contains CO_2 back to your heart, except for the blood in the pulmonary veins, which is oxygen-rich. Capillaries are very tiny vessels that enable oxygen, CO_2, and nutrients to move between your circulatory system and your entire body.

You just read that capillaries surround the alveoli in your lungs Capillaries also surround the small intestine, where they absorb nutrients and transport them to the rest of the body.

11. NGSSS Check **Describe** How do nutrients travel through the body? SC.6.L.14.5

Figure 7 The circulatory system transports nutrients and oxygen to all parts of the body and removes wastes, such as CO_2.

Table 1 Human Blood Types				
Blood Type	**Type A**	**Type B**	**Type AB**	**Type O**
Antigens on red blood cells	A, A, A, A, A, A	B, B, B, B, B, B	B, A, A, B, B, A	
Percentage of U.S. population with this blood type	42	10	4	44
Clumping proteins in plasma	anti-B	anti-A	none	anti-A and anti-B
Blood type(s) that can be RECEIVED in a transfusion	A or O	B or O	A, B, AB, or O	O only
This blood type can DONATE TO these blood types	A or AB	B or AB	AB only	A, B, AB, O

Table 1 The red blood cells of each blood type have different proteins on their surfaces.

12. **Visual Check**
Interpret To which blood group can type A donate?

Blood The blood that circulates through vessels has several parts. The liquid part of blood is called plasma and contains nutrients, water, and CO_2. Blood also contains red blood cells, platelets, and white blood cells. Red blood cells carry oxygen. Platelets help the body heal when you get a cut. White blood cells help the body defend itself from toxins and diseases. You will read more about white blood cells on the next page.

Everyone has red blood cells. However, different people have different proteins on the surfaces of their red blood cells, as shown in **Table 1.** Scientists classify these different red-blood-cell proteins into groups called blood types.

People with A proteins on their red blood cells have type A blood. People with B proteins on their red blood cells have type B blood. Some people have both A and B proteins on their red blood cells. They have type AB blood. People with type O blood have neither A nor B proteins on the surfaces of their red blood cells.

Medical professionals use blood types to determine which type of blood a person can receive from a blood donor. For example, because people with type O blood have no proteins on the surfaces of their red blood cells, they can receive blood only from a donor who also has type O blood.

The Lymphatic System

Have you ever had a cold and found it painful to swallow? This can happen if your tonsils swell. Tonsils are small organs on both sides of your throat. They are part of the lymphatic (lihm FA tihk) system.

The spleen, the thymus, bone marrow, and lymph nodes also are parts of the lymphatic system. The spleen stores blood for use in an emergency. The thymus, the spleen, and bone marrow make white blood cells.

Your lymphatic system has three main functions: removing excess fluid around organs, producing white blood cells, and absorbing and transporting fats. The lymphatic system helps your body maintain fluid homeostasis. About 65 percent of the human body is water. Most of this water is inside cells. Sometimes, when water, wastes, and nutrients move between capillaries and organs, not all of the fluid is taken up by the organs. When fluid builds up around organs, swelling can occur. To prevent swelling, the lymphatic system removes the fluid.

Active Reading **13. Identify** (Circle) a function of the lymphatic system.

Lymph vessels are all over your body, as shown in **Figure 8.** Fluid that travels through the lymph vessels flows into organs called lymph nodes. Humans have more than 500 lymph nodes. The lymph nodes work together and protect the body by removing toxins, wastes, and other harmful substances.

The lymphatic system makes white blood cells. They help the body defend against infection. There are many different types of white blood cells. *A* **lymphocyte** (LIHM fuh site) *is a type of white blood cell that is made in the thymus, the spleen, or the bone marrow.* Lymphocytes protect the body by traveling through the circulatory system, defending against infection.

Word Origin

lymphocyte
from Latin *lympha,* means "water"; and Greek *kytos,* means "hollow, as a cell or container"

Figure 8 Lymph vessels are throughout your body.

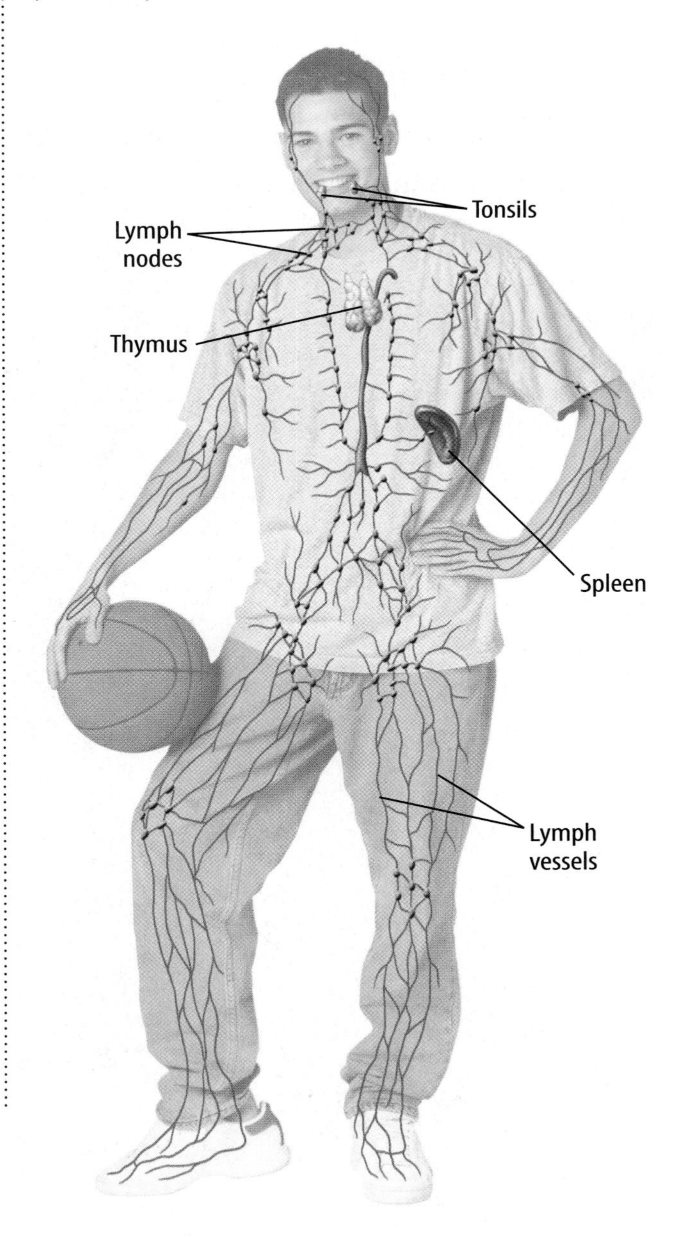

Immunity

The lymphatic system protects your body from harmful substances and infection. *The resistance to specific pathogens, or disease-causing agents, is called* **immunity.** The skeletal system produces immune cells, and the circulatory system transports them throughout the body. Immune cells include lymphocytes and other white blood cells. These cells **detect** viruses, bacteria, and other foreign substances that are not normally made in the body. The immune cells attack and destroy them, as shown in **Figure 9.**

ACADEMIC VOCABULARY

detect

(verb) to discover the presence of

If the body is exposed to the same bacterium, virus, or substance later, some immune cells remember and make proteins called antibodies. These antibodies recognize specific proteins on the harmful agent and help the body fight infection faster. Because there are many different types of bacteria and viruses, humans make billions of different types of antibodies. Each type of antibody responds to a different harmful agent.

Figure 9 Lymphocytes surround bacteria and destroy or remove them from the body.

14. **Visual Check** Recall How long did it take for the lymphocyte to completely surround the bacterium?

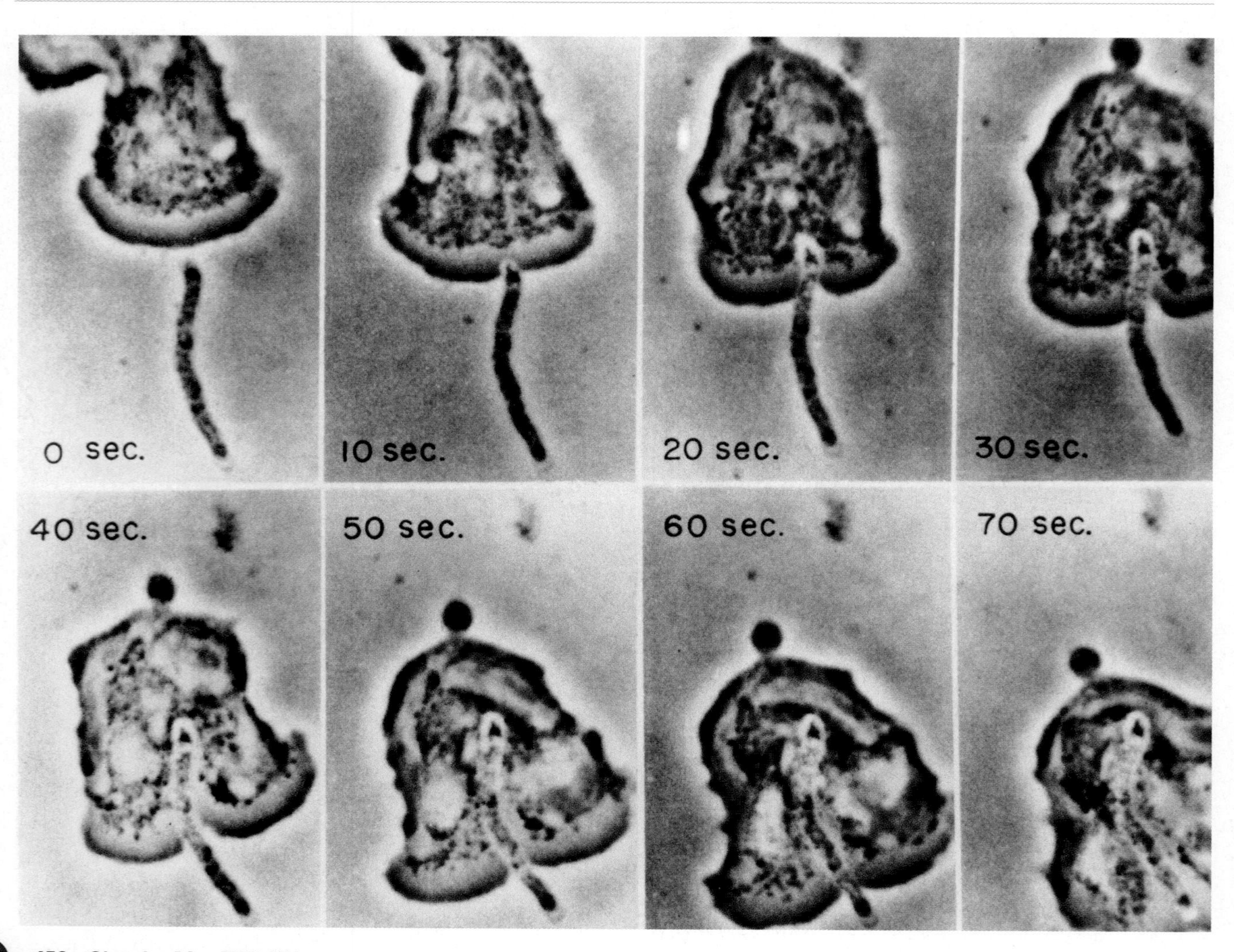

Types of Diseases

There are two main groups of diseases—infectious and noninfectious—as shown in **Table 2.** Infectious diseases are caused by pathogens, such as bacteria and viruses. Infectious diseases are usually contagious, which means they can be spread from one person to another. The flu is an example of an infectious disease. Viruses that invade organ systems of the body, such as the respiratory system, cause infectious diseases.

A noninfectious disease is caused by the environment or a genetic disorder, not a pathogen. Skin cancer, diabetes, and allergies are examples of noninfectious diseases. Noninfectious diseases are not contagious and cannot be spread from one person to another.

Lines of Defense

The human body has many ways of protecting itself from viruses, bacteria, and harmful substances. Skin and mucus (MYEW kus) are parts of the first line of defense. They prevent toxins and other substances from entering the body. Mucus is a thick, gel-like substance in the nostrils, trachea, and lungs. Mucus traps harmful substances and prevents them from entering your body.

The second line of defense is the immune response. In the immune response, white blood cells attack and destroy harmful substances, as shown in **Figure 9.**

The third line of defense protects your body against substances that have infected the body before. As you have read, immune cells make antibodies that destroy the harmful substances. Vaccines are used to help the body develop antibodies against infectious diseases. For example, many people get an influenza vaccine annually to protect them against the flu.

Table 2 Examples of Diseases

Infectious Disease	
Disease	**Pathogen**
colds	virus
AIDS	virus
strep throat	bacteria
chicken pox	virus
Noninfectious Disease	
cancer	
diabetes	
heart disease	
allergy	

Table 2 Diseases are classified into two main groups based on whether they are caused by pathogens.

Active Reading **15. Deduce** In the table, circle the name of the disease caused by bacteria and what type of disease it is.

16. NGSSS Check Expound How does the body defend itself from harmful invaders? SC.6.L.14.5

Lesson Review 1

Visual Summary

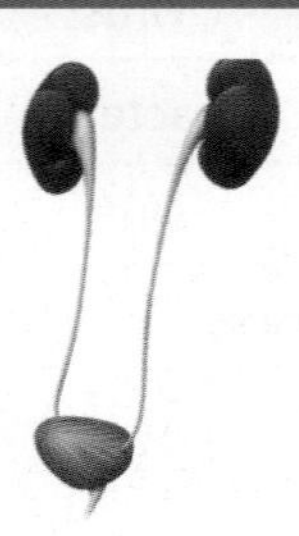

The kidneys remove liquid wastes from the body.

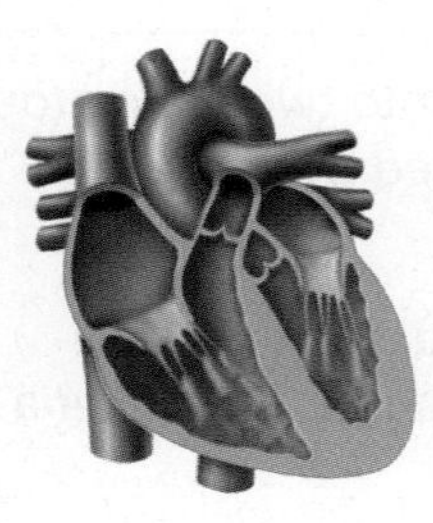

The circulatory system transports nutrients, gases, wastes, and other substances through the body.

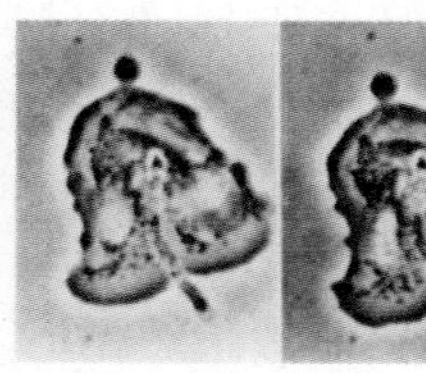

Immune cells detect and destroy viruses, bacteria, and other foreign substances.

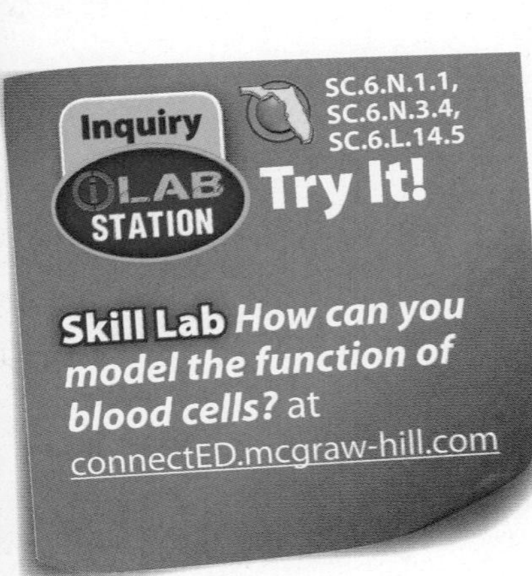

Use Vocabulary

1. **Use the term** *organ system* in a sentence. **SC.6.L.14.5**

2. **Define** *homeostasis* in your own words. **SC.6.L.14.5**

3. A(n) ________________ is a type of white blood cell. **SC.6.L.14.5**

Understand Key Concepts

4. Organs are groups of ________________ that work together.
   - (A) cells
   - (B) organisms
   - (C) systems
   - (D) tissues

5. **Examine** how the circulatory system and the respiratory system work together and move oxygen through the body. **SC.6.L.14.5**

6. **Contrast** infectious diseases and noninfectious diseases. **SC.6.L.14.5**

Interpret Graphics

7. **Summarize** Fill in the graphic organizer below to show how food travels through the digestive system. **SC.6.L.14.5, LA.6.2.2.3**

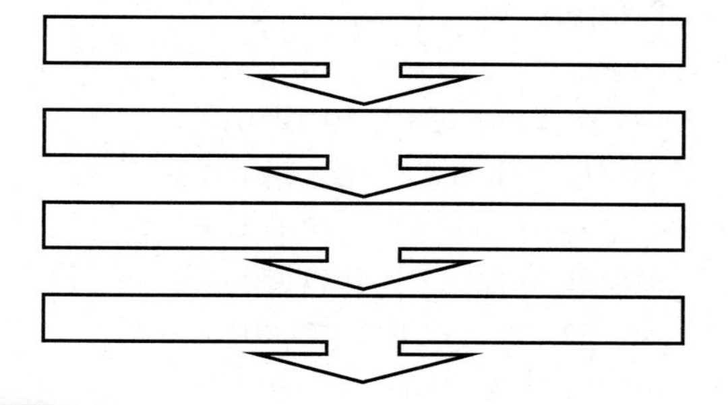

Math Skills

8. If 30.5 g of milk contains 18 C, how many Calories will you consume by drinking a glass of milk (244 g)? **MA.6.A.3.6**

1. **Order** information about the digestive system and fill in information about each part's functions. Hint: Some parts are involved during the same step and should have the same number.

Order	Part	What happens there?
	small intestine	
	esophagus	
	large intestine	
	liver	
6	rectum	
1	mouth	
	pancreas	
	stomach	

2. **Organize** information about nutrition.

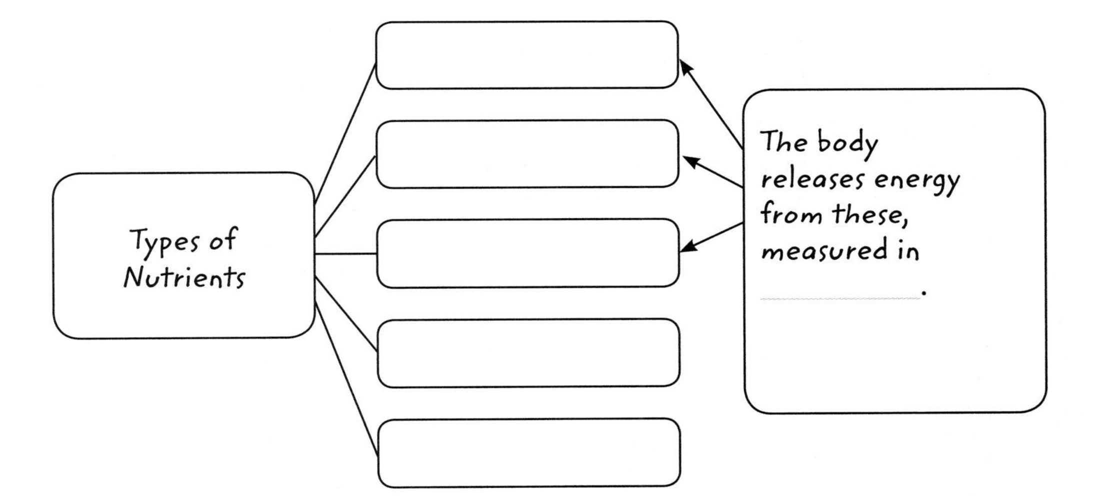

Lesson 2

Structure, Movement, AND CONTROL

ESSENTIAL QUESTIONS

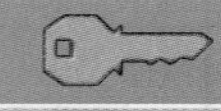 How does the body move?

 How does the body respond to changes in its environment?

Vocabulary

compact bone p. 464
spongy bone p. 464
neuron p. 466
reflex p. 467
hormone p. 469

Florida NGSSS

LA.6.2.2.3 The student will organize information to show understanding (e.g., representing main ideas within text through charting, mapping, paraphrasing, summarizing, or comparing/contrasting);

LA.6.4.2.2 The student will record information (e.g., observations, notes, lists, charts, legends) related to a topic, including visual aids to organize and record information and include a list of sources used;

SC.6.L.14.5 Identify and investigate the general functions of the major systems of the human body (digestive, respiratory, circulatory, reproductive, excretory, immune, nervous, and musculoskeletal) and describe ways these systems interact with each other to maintain homeostasis.

SC.6.N.1.1 Define a problem from the sixth grade curriculum, use appropriate reference materials to support scientific understanding, plan and carry out scientific investigation of various types, such as systematic observations or experiments, identify variables, collect and organize data, interpret data in charts, tables, and graphics, analyze information, make predictions, and defend conclusions.

SC.6.N.2.1 Distinguish science from other activities involving thought.

SC.6.N.3.4 Identify the role of models in the context of the sixth grade science benchmarks.

Launch Lab

SC.6.N.1.1, SC.6.N.2.1 SC.6.N.3.4

15 minutes

Why is the skeletal system so important?

Your skeletal system protects your body's organs, provides support, helps you move, and stores necessary minerals.

Procedure 

1. Read and complete a lab safety form.
2. Obtain one of the **disassembled human figures** and a **kit of materials.**
3. Use the materials to build a backbone for your figure. Using your backbone, connect the head and the arms to the legs of the figure.

Think About This

1. Which materials did you find helpful in creating a backbone and skeletal structure for your figure? Which were not helpful?

2. What characteristics of the "skeleton" were important as you built it? What problems would be caused by not having a skeleton?

3. Key Concept Can you make your figure move? How does having a good support structure help it to move?

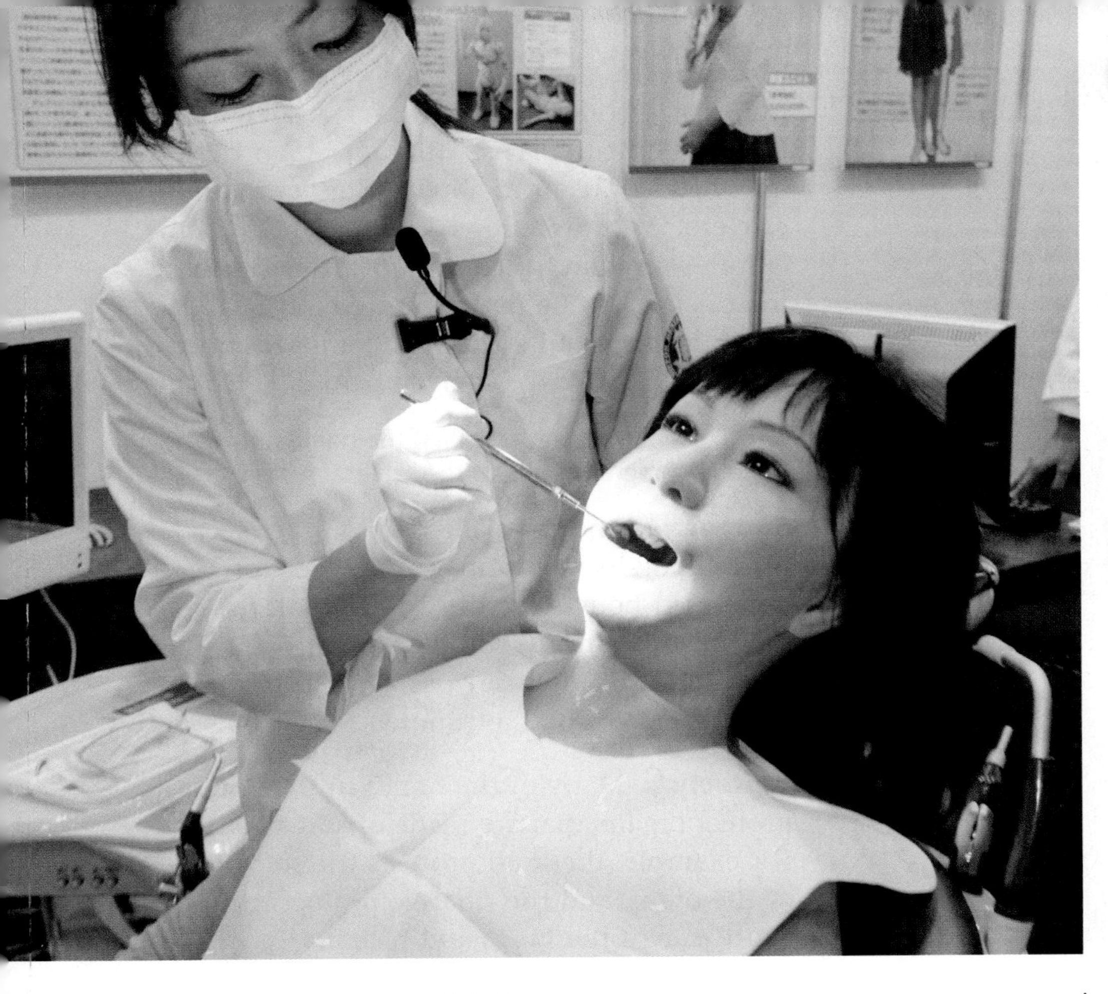

Inquiry Open wide?

1. When you have a dental checkup, you are asked to open your mouth. How are you able to open your mouth? What keeps your teeth from falling out when you chew food?

Structure and Movement

Have you ever had to open your mouth for a dental checkup as shown in the photo above? The human body can move in many different directions and perform a wide variety of tasks. It is able to do things that require many parts of the body to move, such as shooting a basketball into a hoop or swimming a lap in a pool. The human body also can remain very still, such as when posing for a picture or balancing on one leg. Many organ systems work together to accomplish these tasks and maintain homeostasis.

In this lesson, you will read more about two organ systems—the skeletal system and the muscular system—that give the body structure, help the body move, and protect other organ systems.

The Skeletal System

The skeletal system has four major jobs. It protects internal organs, provides support, helps the body move, and stores minerals. The skeletal system is mostly bones. Adults have 206 bones. Ligaments, tendons, and cartilage are also parts of the skeletal system.

Active Reading FOLDABLES® LA.6.2.2.3

Make two horizontal two-tab books. Label them as shown, and glue them side by side to form a booklet with four tabs. Glue this section to the back of the one you made in Lesson 1. Use your book to organize information about the body systems in this lesson.

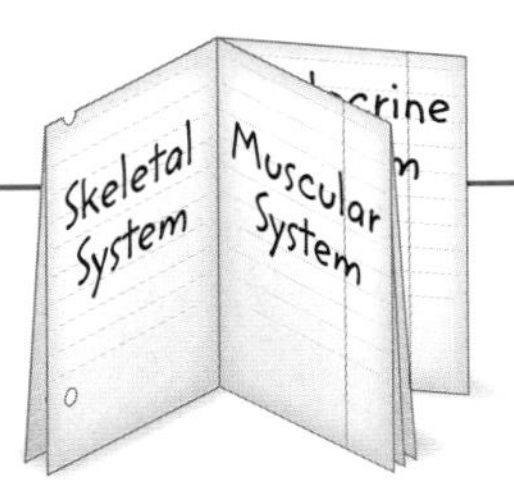

The Skeletal System

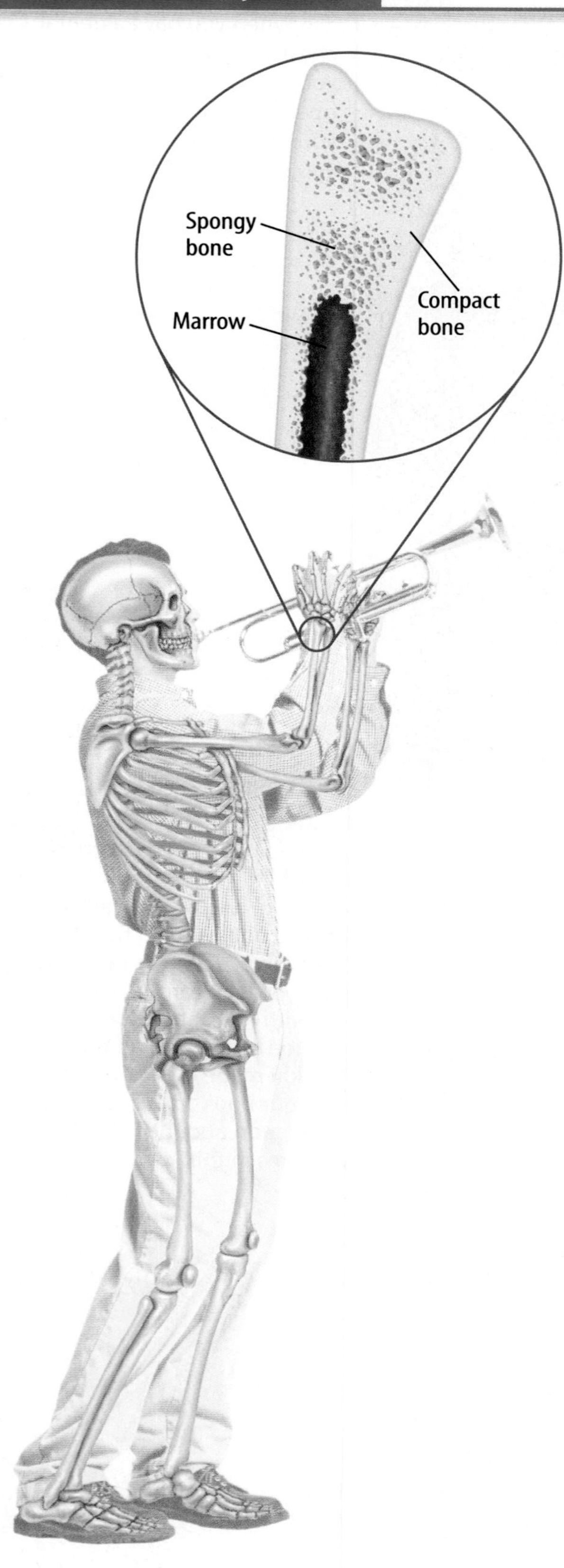

Figure 10 Bone is made up of a dense, hard exterior and a spongy interior.

Storage The skeletal system is also an important storage site for minerals such as calcium. Calcium is essential for life. It has many functions in the body. Muscles require calcium for contractions. The nervous system requires calcium for communication. Most of the calcium in the body is stored in bone. Calcium helps build stronger compact bone. Cheese and milk are good sources of calcium.

Active Reading **2. Identify** Highlight the name of the mineral stored by the skeletal system.

Support Without a skeleton, your body would look like a beanbag. Your skeleton gives your body structure and support, as shown in **Figure 10.** Your bones help you stand, sit up, and raise your arms to play an instrument, such as a trumpet.

Protection Many of the bones in the body protect organs that are made of softer tissue. For example, the skull protects the soft tissue of the brain, and the rib cage protects the soft tissue of the lungs and heart.

Movement The skeletal system helps the body move by working with the muscular system. Bones can move because they are attached to muscles. You will read more about the interaction of these two systems later in this lesson.

Bone Types Bones are organs that contain two types of tissue. **Compact bone** *is the hard outer layer of bone.* **Spongy bone** *is the interior region of bone that contains many tiny holes.* As shown in **Figure 10,** spongy bone is inside compact bone. Some bones also contain bone marrow. Recall that bone marrow is a part of the lymphatic system and makes white blood cells.

Active Reading **3. Contrast** How do the two types of bone tissue differ?

The Muscular System

You might already know that there are muscle cells in your arms and legs. But did you know that there are muscle cells in your eyes, heart, and blood vessels? Without muscle cells you would not be able to talk, write, or run.

As shown in **Figure 11,** muscle cells are everywhere in the body. Almost one-half of your body mass is muscle cells. These muscle cells make up the muscular system. By working together, they help the body move.

The muscular system is made of three different types of muscle tissue—skeletal muscle, cardiac muscle, and smooth muscle. Skeletal muscle works with the skeletal system and helps you move. Tendons connect skeletal muscles to bones. Skeletal muscle also gives you the strength to lift heavy objects.

Another type of muscle tissue is cardiac muscle. Cardiac muscle is only in the heart. It continually contracts and relaxes and moves blood throughout your body.

Smooth muscle tissue is another type of muscle tissue. Smooth muscle tissue is in organs such as the stomach and the bladder. Blood vessels also have smooth muscle tissue.

4. NGSSS Check List What systems help the body move?
SC.6.L.14.5

Figure 11 Cardiac muscle has cells with more than one nucleus. Smooth muscle cells are long, and skeletal muscle cells have dark bands.

5. Visual Check
Determine Which type of muscle is in your arms?

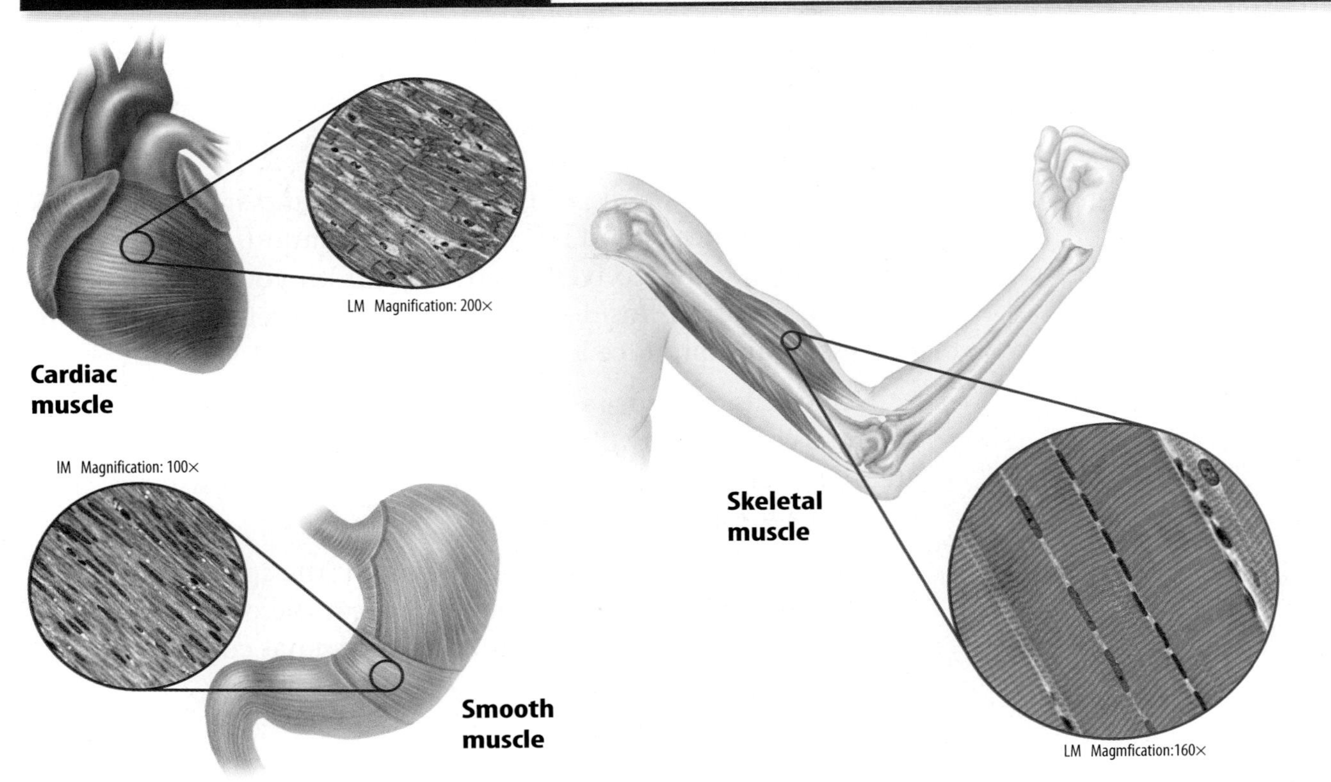

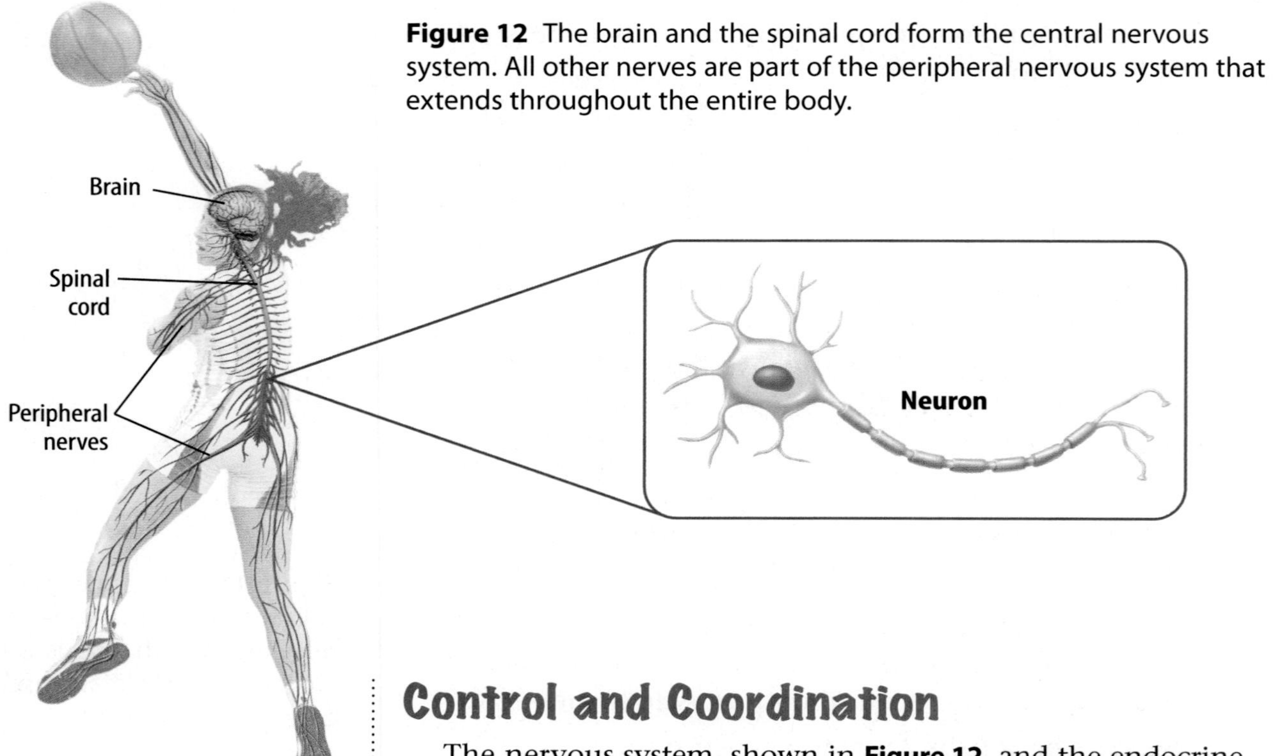

Figure 12 The brain and the spinal cord form the central nervous system. All other nerves are part of the peripheral nervous system that extends throughout the entire body.

Control and Coordination

The nervous system, shown in **Figure 12,** and the endocrine system, which you will read about later, receive and process information about your internal and external environments. These two systems control many functions, including movement, communication, and growth, by working with other systems in the body, and they help maintain homeostasis.

Active Reading **6. Recall** (Circle) the names of the parts of the central nervous system.

The Nervous System

The nervous system is a group of organs and specialized cells that detect, process, and respond to information. The nervous system constantly receives information from your external environment and from inside your body. It can receive information, process it, and produce a response in less than 1 second.

Nerve cells, or **neurons**, *are the basic units of the nervous system.* Neurons can be many different lengths. In adults, some neurons are more than 1 m long. This is about as long as the distance between a toe and the spinal cord.

WORD ORIGIN

neuron
from Greek *neuron,* means "a nerve cell with appendages"

The nervous system includes the brain, the spinal cord, and nerves. The brain and the spinal cord form the central nervous system. Nerves outside the brain and the spinal cord make up the peripheral nervous system.

Processing Information The central nervous system is protected by the skeletal system. Muscles and other organs surround the peripheral nervous system. Information enters the nervous system through neurons in the peripheral nervous system. Most of the information then is sent to the central nervous system for processing. After the central nervous system processes information, it signals the peripheral nervous system to respond.

Voluntary and Involuntary Control The body carries out many functions that depend on the nervous system. Some of these functions such as breathing and digestion are automatic, or involuntary. They do not require you to think about them to make them happen. The nervous system automatically controls these functions and maintains homeostasis.

Most of the other functions of the nervous system are not automatic. They require you to choose to make them happen. Tasks such as reading, talking, and walking are voluntary. These tasks require input, processing, and a response.

Reflexes Have you ever touched a hot pan with your hand? Touching a hot object sends a rapid signal that your hand is in pain. The signal is so fast that you do not think about moving your hand; it just happens automatically. *Automatic movements in response to a signal are called* **reflexes**. The spinal cord receives and processes reflex signals, as shown in **Figure 13.** Processing the information in the spinal cord instead of the brain helps the body respond more quickly.

Figure 13 Reflexes happen automatically.

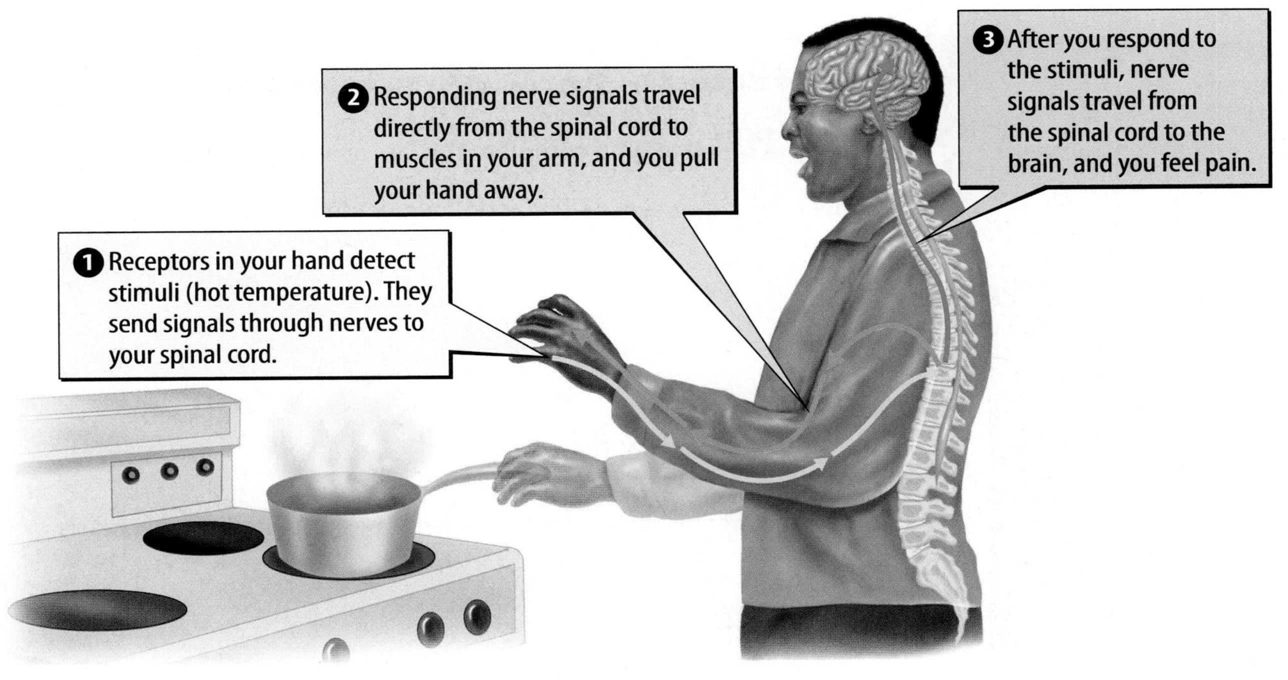

7. **Visual Check** **Recall** What detects heat when you touch a hot pan?

Figure 14 Each sense receives a different type of signal.

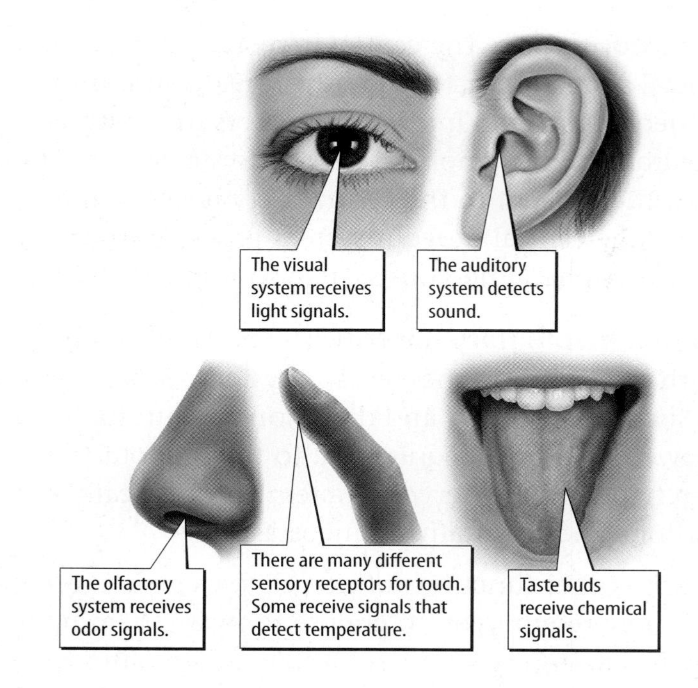

Active Reading **8. Diagram** Make a drawing showing how one type of signal in the environment is processed in the human body.

The Senses Humans detect their external environment with five senses—vision, hearing, smell, touch, and taste—as shown in **Figure 14.** Each of the five senses has specific neurons that receive signals from the environment. Information detected by the senses is sent to the spinal cord and then to the brain for processing and a response. Responses depend on the specific signal detected. Some responses cause muscles to contract and move such as when you touch a hot surface. The aroma of baking cookies might cause your mouth to produce saliva.

Apply It! After you complete the lab, answer these questions.

1. How might your sense of hearing help you maintain homeostasis?

2. In what other ways besides balance might sight affect homeostasis?

The Endocrine System

How tall were you in first grade? How tall are you now? From the time you were born until now, your body has changed. These changes are controlled by the endocrine system, shown in **Figure 15.** Like the nervous system, the endocrine system sends signals to the body. *Chemical signals released by the organs of the endocrine system are called* **hormones**. Hormones cause organ systems to carry out specific functions.

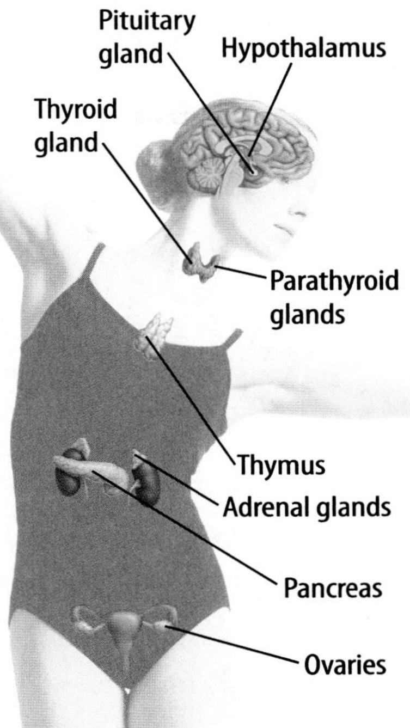

Figure 15 The endocrine system uses hormones to communicate with other organ systems.

Why does your body need two organ systems to process information? The signals sent by the nervous system travel quickly through neurons. Hormones travel in blood through blood vessels in the circulatory system. These messages travel more slowly than nerve messages. A signal sent by the nervous system can travel from your head to your toes in less than 1 s, but a hormone will take about 20 s to make the trip. Although hormones take longer to reach their target organ system, their effects usually last longer.

Active Reading **9. Explain** How do hormones help the body maintain homeostasis?

Many of the hormones made by the endocrine system work with other organ systems and maintain homeostasis. For example, parathyroid hormone works with the skeletal system and controls calcium storage. Insulin is a hormone that is released from the pancreas that signals the digestive system to control nutrient homeostasis. Other hormones, such as growth hormone, work with many organ systems to help you grow. In the next lesson, you will read about another system that the endocrine system works with.

10. NGSSS Check **Describe** How does the body respond to changes in its environment? SC.6.L.14.5

Lesson Review 2

Visual Summary

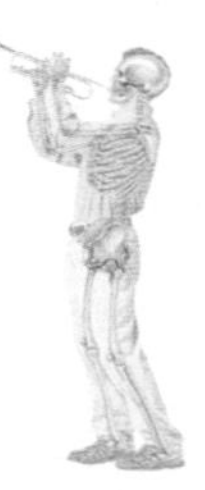

The skeletal system protects organs, provides support, helps the body move, and stores minerals.

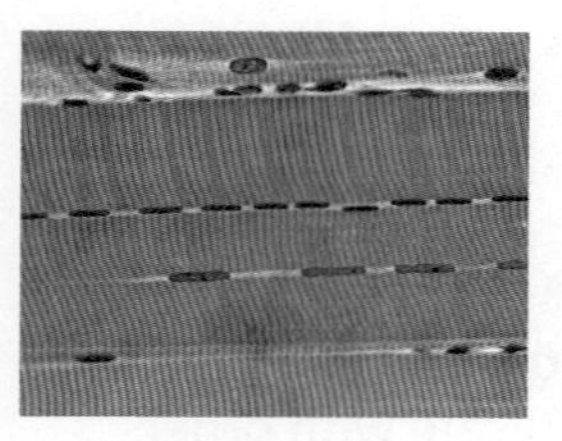

Skeletal muscle works with the skeletal system and helps you move.

Reflex signals are received by the spinal cord but are not processed by the brain. This helps the body respond quickly.

Use Vocabulary

1. **Distinguish** between compact bone and spongy bone.

2. A chemical signal that is released by the endocrine system is a(n) ______.

3. **Use the term** *neuron* in a sentence.

Understand Key Concepts

4. An automatic movement in response to a signal is called a
   - (A) hormone.
   - (B) muscle.
   - (C) neuron.
   - (D) reflex.

5. **Compare** the role of tendons in helping the skeletal system and the muscular system work together to a bridge between two cities. SC.6.L.14.5

Interpret Graphics

6. **Summarize** Fill in the graphic organizer below to show the three types of muscle tissue. LA.6.2.2.3

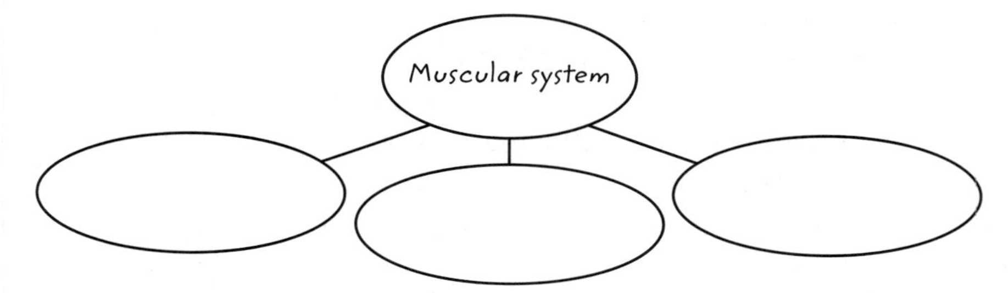

Critical Thinking

7. **Hypothesize** What would be the effect of losing one's sight on the ability to digest food? Explain your answer.

SCIENCE & SOCIETY

Bone Marrow Transplants

Why might you need new bone marrow?

Healthy blood cells are essential to overall health. Red blood cells carry oxygen throughout the body. Some white blood cells fight infections. Platelets help stop bleeding. A bone marrow transplant is sometimes necessary when a disease interferes with the body's ability to produce healthy blood cells.

Bone marrow is a tissue found inside some of the bones in your body. Healthy bone marrow contains cells that can develop into white blood cells, red blood cells, or platelets. Some diseases, such as leukemia and sickle cell disease, affect bone marrow. Replacing malfunctioning bone marrow with healthy bone marrow can help treat these diseases.

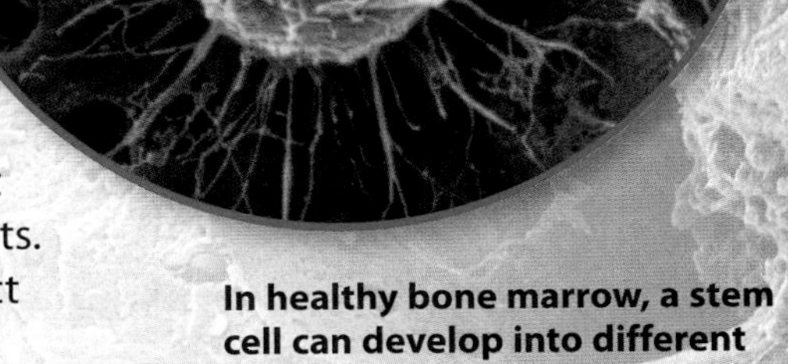

In healthy bone marrow, a stem cell can develop into different types of blood cells.

A bone marrow transplant involves several steps. The patient receiving the bone marrow must have treatments to destroy his or her unhealthy bone marrow. Healthy bone marrow must be obtained for the transplant. Sometimes, the patient's own bone marrow can be treated and used for transplant. This transplant has the greatest chance of success. Other transplants involve healthy bone marrow donated by another person. The bone marrow must be tested to ensure that it is a good match for the patient.

The bone marrow donor undergoes a procedure called harvesting. Bone marrow is taken from the donor's pelvic bone. The donor's body replaces the harvested bone marrow, so there are no long-term effects for the donor.

The donated bone marrow is introduced into the patient's bloodstream. If the transplant is successful, the new bone marrow moves into the bone cavities and begins producing healthy blood cells.

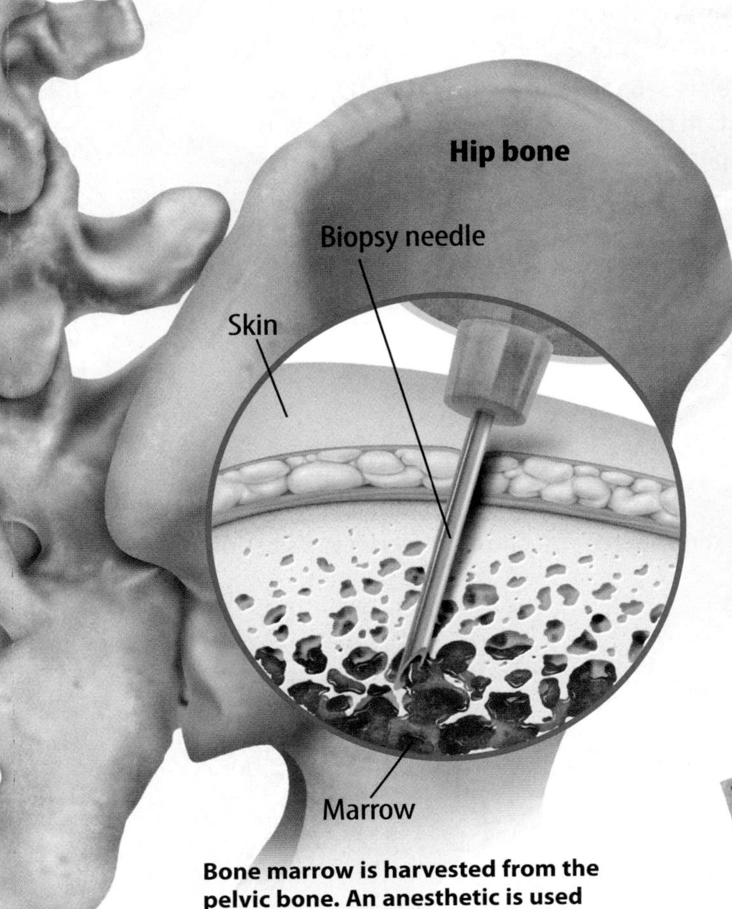

Bone marrow is harvested from the pelvic bone. An anesthetic is used to keep the donor from feeling pain during the procedure.

It's Your Turn

LA.6.4.2.2

NATIONAL MARROW DONOR PROGRAM®

RESEARCH AND REPORT Find out more about bone marrow transplants. What other diseases can be treated using a bone marrow transplant? What is the National Marrow Donor Program? Present your findings to your class.

Lesson 3

Reproduction and DEVELOPMENT

ESSENTIAL QUESTIONS

What do the male and female reproductive systems do?

How do humans grow and change?

Vocabulary

reproduction p. 473
gamete p. 473
sperm p. 473
ovum p. 473
fertilization p. 473
zygote p. 473

Florida NGSSS

LA.6.2.2.3 The student will organize information to show understanding (e.g., representing main ideas within text through charting, mapping, paraphrasing, summarizing, or comparing/contrasting);

SC.6.L.14.5 Identify and investigate the general functions of the major systems of the human body (digestive, respiratory, circulatory, reproductive, excretory, immune, nervous, and musculoskeletal) and describe ways these systems interact with each other to maintain homeostasis.

SC.6.N.1.1 Define a problem from the sixth grade curriculum, use appropriate reference materials to support scientific understanding, plan and carry out scientific investigation of various types, such as systematic observations or experiments, identify variables, collect and organize data, interpret data in charts, tables, and graphics, analyze information, make predictions, and defend conclusions.

SC.6.N.2.1 Distinguish science from other activities involving thought.

SC.6.N.3.4 Identify the role of models in the context of the sixth grade science benchmarks.

Inquiry Launch Lab

SC.6.N.3.4

15 minutes

How do the sizes of egg and sperm cells compare?

A sperm cell combines with an egg cell to create a zygote that will eventually become a fetus and then a baby. The sperm and egg cells each contribute half the genetic material to the zygote.

Procedure

1. Read and complete a lab safety form.
2. Select one of the **spheres** to use as a model of an egg cell. With a **ruler,** measure the diameter of the sphere. Record the measurement.
3. If an average sperm cell is 3–6 microns in diameter, and an average egg cell is 120–150 microns in diameter, determine the diameter of a suitable model for a sperm cell.
4. Find another sphere that is approximately the size needed to create an accurate model to represent a sperm cell. Label both of your models.

Data and Observations

Think About This

1. What were the sizes of the spheres you chose to model the sizes of the sperm and egg cells?

2. **Key Concept** How do the egg cell and sperm cells interact in reproduction? How do you think size plays a role in this interaction?

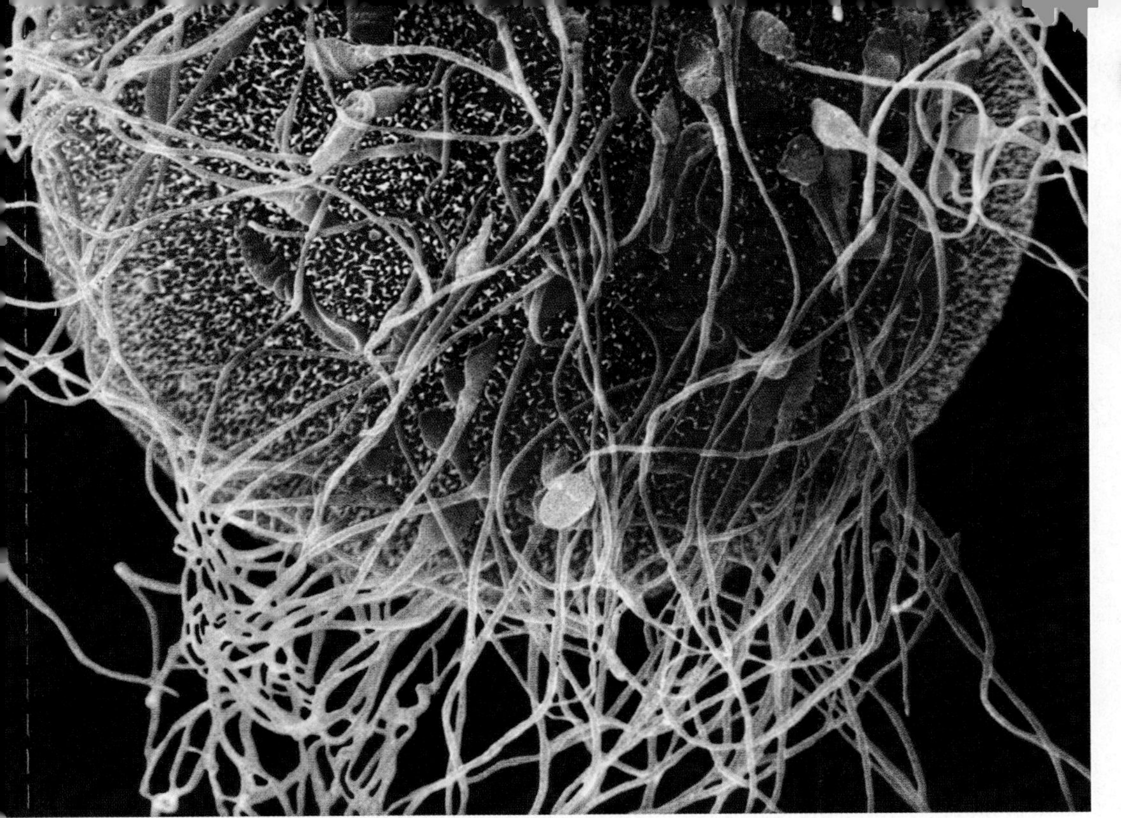

Inquiry Strands of Hair?

1. The things that look like strands of hair are sperm, the male reproductive cells. The red structure is an egg, the female reproductive cell. Why do you think there are so many sperm but only one egg?

Reproduction and Hormones

You have read how the endocrine system works with other organ systems and helps the body grow and maintain homeostasis. The endocrine system has another very important function—to ensure that humans can reproduce. Some of the organs of the endocrine system produce hormones that help humans reproduce. **Reproduction** *is the process by which new organisms are produced.* Reproduction is essential to the continuation of life on Earth.

A male and a female each have special organs for reproduction. Organs in the male reproductive system are different from those in the female reproductive system. *Human reproductive cells,* called **gametes** (GA meets), are made by the male and female reproductive systems. *Male gametes are called* **sperm**. *Female gametes are called* **ova** (OH vah; singular ovum), or eggs.

As shown in the photo at the beginning of the lesson, *a sperm joins with an egg in a reproductive process called* **fertilization**. *The cell that forms when a sperm cell fertilizes an egg cell is called a* **zygote** (ZI goht). A zygote is the first cell of a new human. It contains genetic information from both the sperm and the ovum. The zygote will grow and develop in the female's reproductive system.

Active Reading **2. Describe** How do gametes enable humans to reproduce?

Active Reading FOLDABLES® LA.6.2.2.3

Make a horizontal two-tab book. Label it as shown, and glue it side by side to the back of the booklet made in Lessons 1 and 2. Use the book to organize information about the male and female reproductive systems.

WORD ORIGIN

zygote
from Greek *zygoun,* means "to join"

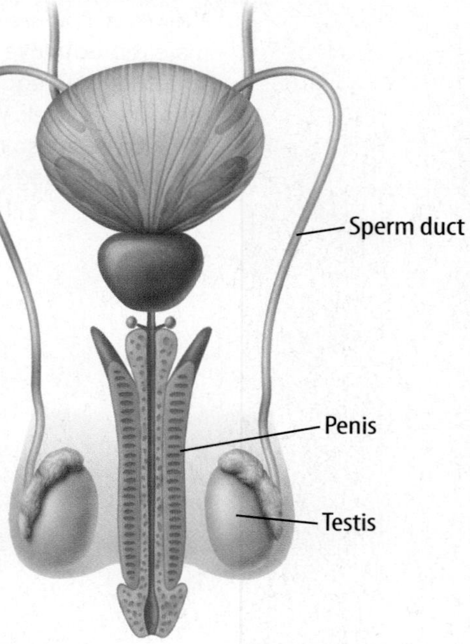

Male Reproductive System

Sperm duct
Penis
Testis

The organs of the male reproductive system produce sperm and deliver it to the female reproductive system.

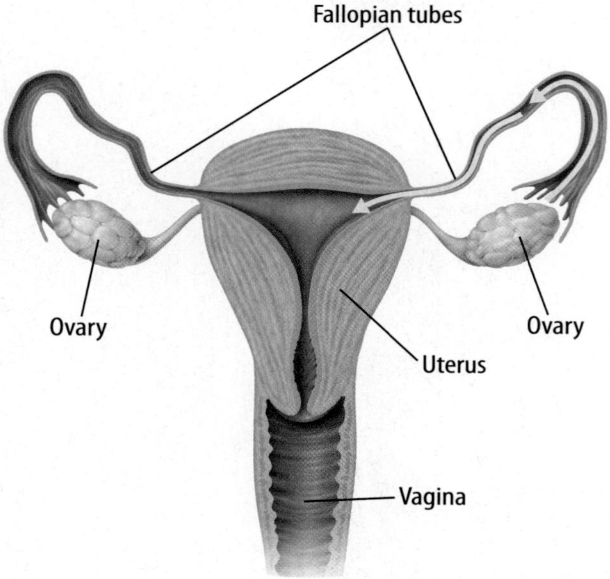

Female Reproductive System

Fallopian tubes
Ovary
Ovary
Uterus
Vagina

The female reproductive system produces eggs and provides a place for a new human to grow and develop before birth.

Figure 16 Males and females have specialized organs for reproduction.

The Male Reproductive System

The male reproductive system, shown in **Figure 16,** produces sperm and delivers it to the female reproductive system. Sperm are produced in the testes (TES teez; singular, testis). Sperm develop inside each testis and then are stored in tubes called sperm ducts. Sperm matures in the sperm ducts.

The testes also produce a hormone called testosterone. Testosterone helps sperm change from round cells to long, slender cells that can swim. Once sperm have fully developed, they can travel to the penis. The penis is a tubelike structure that delivers sperm to the female reproductive system. Sperm are transported in a fluid called semen (SEE mun). Semen contains millions of sperm and nutrients that provide the sperm with energy.

3. **NGSSS Check** **Recall** What does the male reproductive system do? **SC.6.L.14.5**

The Female Reproductive System

The female reproductive system contains two ovaries, as shown in **Figure 16.** Eggs grow and mature in the ovaries. Two hormones made by the ovaries, estrogen (ES truh jun) and progesterone (proh JES tuh rohn), help eggs mature. Once mature, eggs are released from the ovaries and enter the fallopian tubes. As shown in **Figure 16,** the fallopian tubes connect the ovaries to the uterus.

If sperm are also present in the fallopian tube, fertilization can occur as the egg enters the fallopian tube. Sperm enter the female reproductive system through the vagina, a tube-shaped organ that leads to the uterus. A fertilized egg, or zygote, can move through the fallopian tube and attach inside the uterus.

If there are no sperm in the fallopian tube, the egg will not be fertilized. Unfertilized eggs break down in the uterus and leave the body during the menstrual cycle.

 **Figure 17** During pregnancy, a unicellular zygote evelops into a fetus.

5. **Visual Check** **Identify** When is the heart fully formed?

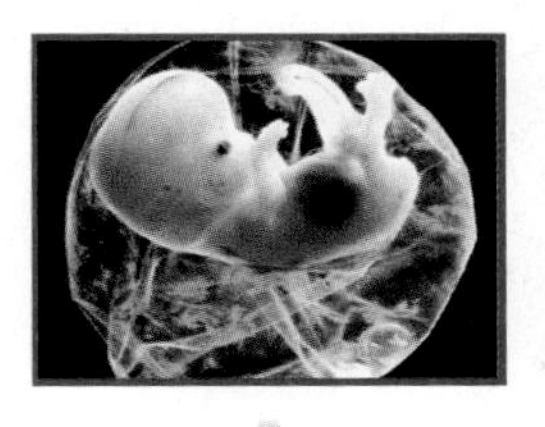

5 weeks The embryo is about 7 mm long. The heart and other organs have started to develop. The arms and legs are beginning to bud.

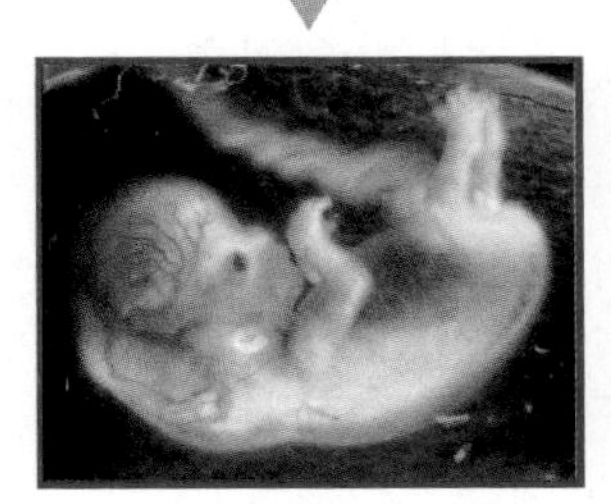

8 weeks The embryo is about 2.5 cm long. The heart is fully formed and beating, bones are beginning to harden, and nearly all muscles have appeared.

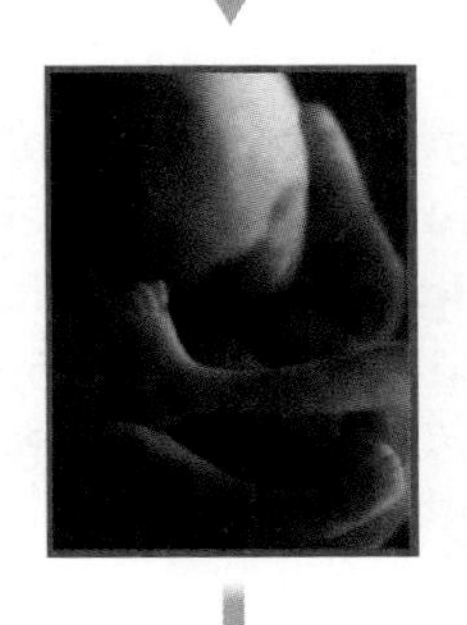

14 weeks Growth and development continue. The fetus is about 6 cm long.

16 weeks The fetus is about 15 cm long and about 140 g. The fetus can make a fist and has a range of facial expressions.

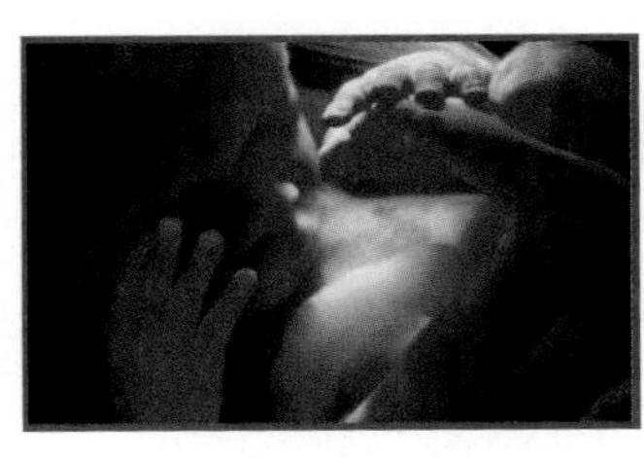

22 weeks The fetus is about 27 cm long and about 430 g. Footprints and fingerprints are forming.

The Menstrual Cycle The endocrine system controls egg maturation and release and thickening of the lining of the uterus in a process called the menstrual (MEN stroo ul) cycle. The menstrual cycle takes about 28 days and has three parts.

During the first part of the cycle, eggs grow and mature and the thickened lining of the uterus leaves the body. In the second part of the cycle, mature eggs are released from the ovaries and the lining of the uterus thickens. In the third part of the cycle, unfertilized eggs and the thickened lining break down. The lining leaves the body in the first part of the next cycle.

4. **NGSSS Check** **Restate** What does the female reproductive system do? SC.6.L.14.5

Human Development

As shown in **Figure 17,** humans develop in many stages. You have read that when a sperm fertilizes an egg, a zygote forms. The zygote develops into an embryo (EM bree oh). An embryo is a ball-shaped structure that attaches inside the uterus and continues to grow.

The embryo develops into a fetus, the last stage before birth. It takes about 38 weeks for a fertilized egg to fully develop. This developmental period is called pregnancy. During this period, the organ systems of the fetus will develop and the fetus will get larger. Pregnancy ends with birth. During birth, the endocrine system releases hormones that help the uterus push the fetus through the vagina and out of the body.

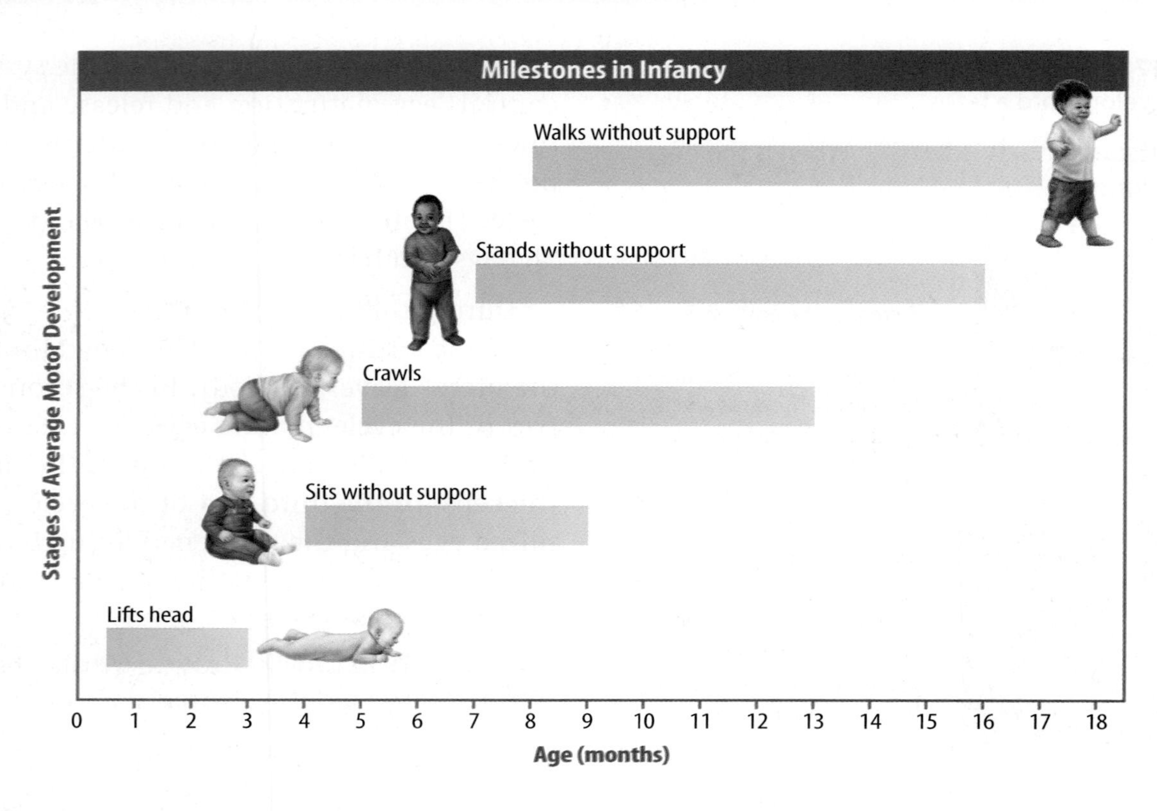

Figure 18 During infancy, a human learns to crawl and walk.

6. Visual Check **Generalize** When does an infant usually crawl?

From Birth Through Childhood

The first life stage after birth is infancy, the first 2 years of life. During infancy, the muscular and nervous systems develop and an infant begins walking, as shown in **Figure 18.** Growth and development continue in childhood, which is from about 2 years to about 12 years of age. Bones in the skeletal system grow longer and stronger, and the lymphatic system matures.

Adolescence Through Adulthood

Adolescence follows childhood. During adolescence, growth of the skeletal and muscular systems continues. Organs such as the lungs and kidneys get larger. As the endocrine system develops, the male and female reproductive systems mature. The period of time during which the reproductive system matures is called puberty.

After adolescence is adulthood, as shown in **Figure 19.** During adulthood, humans continue to change. In later adulthood, hair turns gray, wrinkles might form in the skin, and bones become weaker in a process called aging. Aging is a slow process that can last for decades.

Figure 19 Humans continue to change during adolescence and adulthood.

Active Reading **7. Describe** How do humans change during adulthood?

Lesson Review 3

Visual Summary

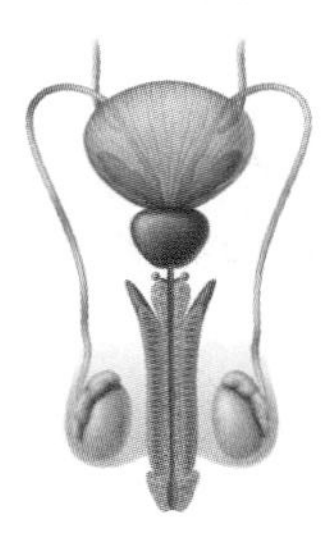

Sperm are produced in the testes and develop inside each testis in the seminiferous tubules.

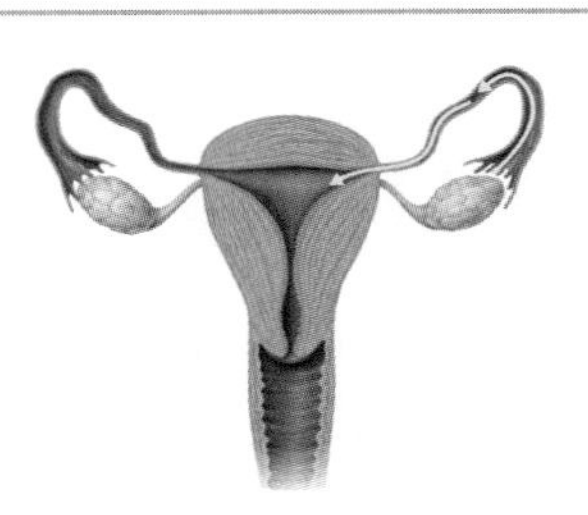

Eggs grow and mature in the ovaries.

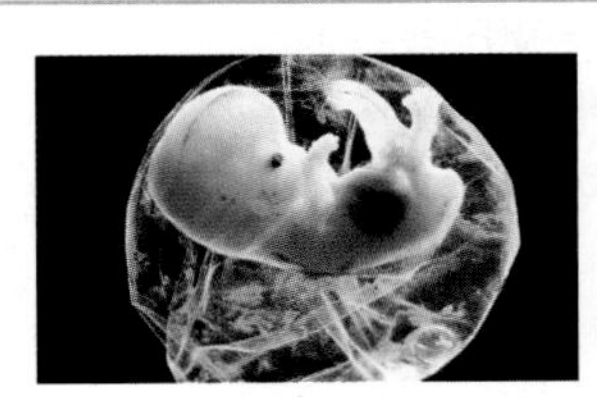

During pregnancy, a zygote develops into an embryo and then into a fetus.

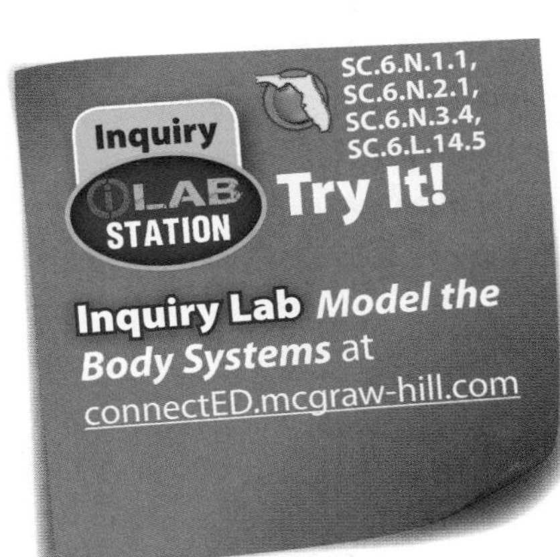

Use Vocabulary

1. Sperm and ova are types of ______________.

2. **Distinguish** between an ovum and a zygote.

3. **Define** *fertilization* in your own words. SC.6.L.14.5

Understand Key Concepts 

4. The period between birth and 2 years is called
   - (A) adolescence.
   - (B) adulthood.
   - (C) childhood.
   - (D) infancy.

5. **Compare** the functions of the ovaries and the testes. SC.6.L.14.5

6. **Distinguish** between a zygote and a fetus.

Interpret Graphics

7. **Summarize** Fill in the graphic organizer below to show the stages of life. LA.6.2.2.3

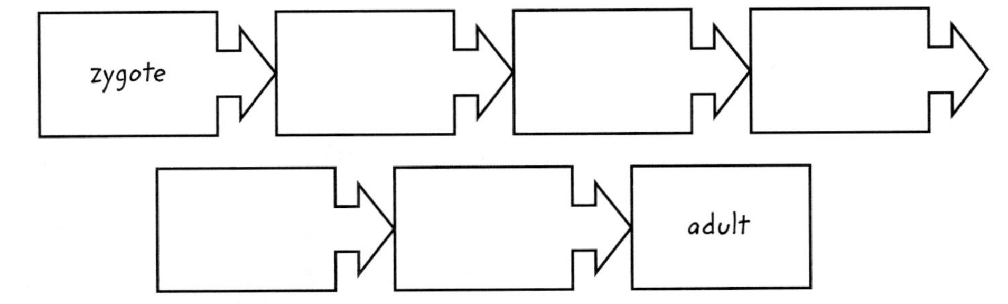

Critical Thinking

8. **Hypothesize** why development before birth takes a long time, about 38 weeks.

Chapter 12 Study Guide

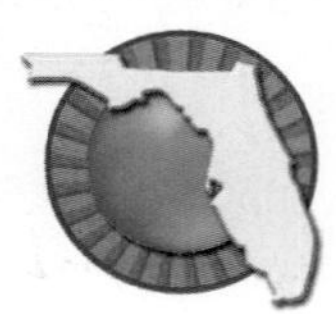

Think About It! Human body systems maintain life by various functions essential for growth, reproduction, and homeostasis, such as transporting materials and providing control.

Key Concepts Summary

LESSON 1 Transport and Defense

- Nutrients enter the body through the digestive system. Wastes and water leave the body through the excretory system. Oxygen enters the body and carbon dioxide leaves the body through the respiratory system.
- Substances such as **nutrients** and oxygen reach the body's cells through the circulatory system.
- The lymphatic system helps the body defend itself against harmful invaders.

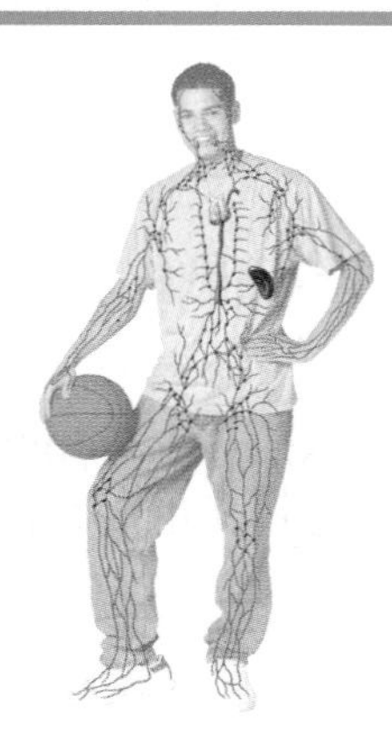

Vocabulary

organ system p. 449
homeostasis p. 449
nutrient p. 451
Calorie p. 451
lymphocyte p. 457
immunity p. 458

LESSON 2 Structure, Movement, and Control

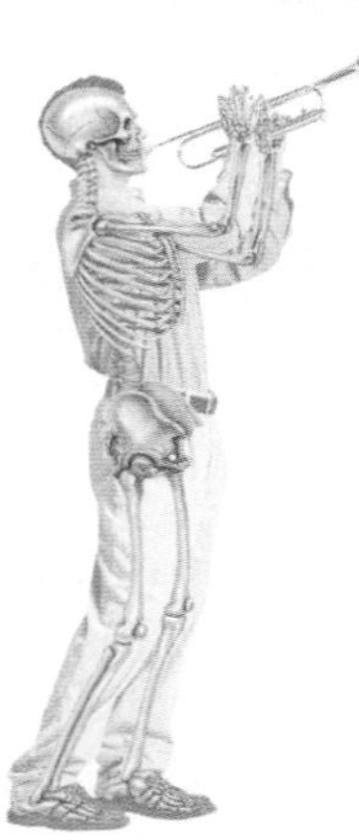

- The muscular system and the skeletal system work together and help the body move. The skeletal system provides the body with structure and protects other organ systems.
- The nervous system and the endocrine system work together and help the body respond to changes in the environment.

compact bone p. 464
spongy bone p. 464
neuron p. 466
reflex p. 467
hormone p. 469

LESSON 3 Reproduction and Development

- The male and female reproductive systems ensure survival of the human species.
- Humans develop and grow both before and after birth.

reproduction p. 473
gamete p. 473
sperm p. 473
ovum p. 473
fertilization p. 473
zygote p. 473

Use Vocabulary

1. Carbohydrates, fats, and proteins are all ______________ and contain ______________.

2. The thymus and the spleen produce white blood cells called ______________.

3. Bones in the skeletal system are made of a hard exterior called ______________.

4. Define the term *hormone* in your own words.

5. Use the terms *reproduction* and *fertilization* in a sentence.

6. Distinguish between ova and sperm.

Link Vocabulary and Key Concepts

Use vocabulary terms from the previous page to complete the concept map. SC.6.L.14.5

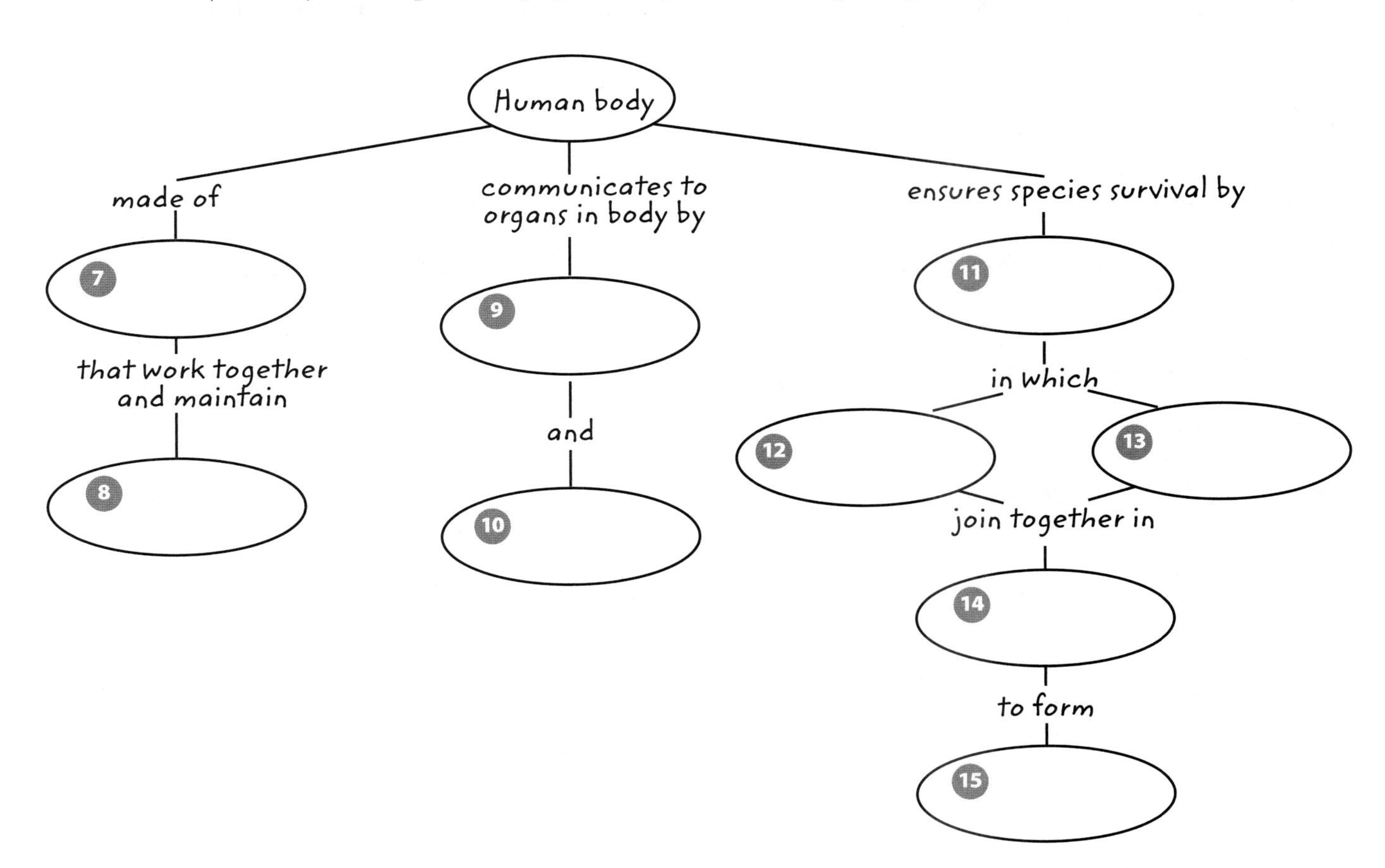

Chapter 12 Review

Fill in the correct answer choice.

Understand Key Concepts

1 Which body system removes carbon dioxide and waste? **SC.6.L.14.5**

Ⓐ circulatory
Ⓑ digestive
Ⓒ excretory
Ⓓ lymphatic

2 Which body system makes immune cells? **SC.6.L.14.5**

Ⓐ circulatory
Ⓑ digestive
Ⓒ excretory
Ⓓ lymphatic

3 Which hormone helps the cells produced in the system below mature? **SC.6.L.14.5**

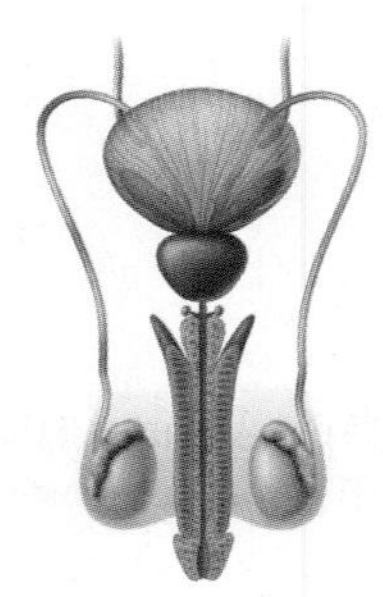

Ⓐ estrogen
Ⓑ insulin
Ⓒ progesterone
Ⓓ testosterone

4 Which are proteins that recognize specific proteins on bacteria? **SC.6.L.14.5**

Ⓐ antibodies
Ⓑ enzymes
Ⓒ nutrients
Ⓓ receptors

5 Which is NOT a type of muscle tissue? **SC.6.L.14.5**

Ⓐ cardiac
Ⓑ lymphatic
Ⓒ skeletal
Ⓓ smooth

6 Which is NOT a type of blood vessel? **SC.6.L.14.5**

Ⓐ artery
Ⓑ capillary
Ⓒ spleen
Ⓓ vein

Critical Thinking

7 **Relate** the body's organization to how homeostasis is maintained. **SC.6.L.14.5**

8 **Compare** the functions of lymphatic vessels and blood vessels. **LA.6.2.2.3**

9 **Hypothesize** how an injury to the spinal cord might affect the ability of the nervous system to sense and respond to a change in the environment. **SC.6.L.14.5**

10 **Assess** how the nervous system helps the muscular system control heart rate, digestion, and respiration. **SC.6.L.14.5**

11 **Relate** the organs of the lymphatic system to immunity. **SC.6.L.14.5**

12 **Assess** the role of the skeletal system in the storage of nutrients. SC.6.L.14.5

13 **Summarize** the role of puberty in the transition from adolescence to adulthood. LA.6.2.2.3

14 **Compare** the functions of the male and female reproductive systems. LA.6.2.2.3

Writing in Science

15 **Write** a five-sentence paragraph that distinguishes the two main types of diseases on a separate sheet of paper. Be sure to include a topic sentence and a concluding sentence in your paragraph. LA.6.2.2.3

Big Idea Review

16 How do organ systems help the body function? SC.6.L.14.5

17 What are the functions of the digestive system, the skeletal system, the muscular system, and the nervous system? SC.6.L.14.5

Math Skills

MA.6.A.3.6

Use Proportions

18 Which type of chicken in the table below has the fewest Calories per gram?

Food	Mass (g)	Calories (C)
$\frac{1}{2}$ chicken breast, baked	86	140
1 chicken leg, baked	52	112

19 A small 140-g apple and a 100-g banana each provide 70 C of energy. How many Calories would there be in a 200-g serving of fruit salad that contained equal amounts of apple and banana? [Hint: Add the values for the apple and the banana first.]

Florida NGSSS Benchmark Practice

Fill in the correct answer choice.

Multiple Choice

1 Which is NOT a waste that is removed from the body by the excretory system? **SC.6.L.14.5**

Ⓐ water

Ⓑ salt

Ⓒ blood

Ⓓ carbon dioxide

2 Which term describes organ systems that work together and maintain steady internal conditions when external conditions change? **SC.6.L.14.5**

Ⓕ tissues

Ⓖ homeostasis

Ⓗ neurons

Ⓘ adolescence

Use the diagram below to answer question 3.

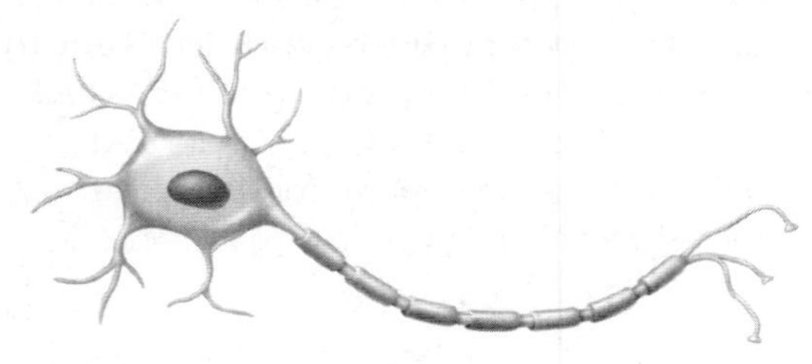

3 Which body system is made of the basic unit shown in the diagram? **SC.6.L.14.5**

Ⓐ circulatory system

Ⓑ endocrine system

Ⓒ muscular system

Ⓓ nervous system

4 Which body system provides protection from infection and toxins? **SC.6.L.14.5**

Ⓕ circulatory system

Ⓖ digestive system

Ⓗ excretory system

Ⓘ lymphatic system

5 Which two systems work together to make your body move? **SC.6.L.14.5**

Ⓐ digestive system and skeletal system

Ⓑ lymphatic system and digestive system

Ⓒ nervous system and excretory system

Ⓓ skeletal system and muscular system

Use the image below to answer question 6.

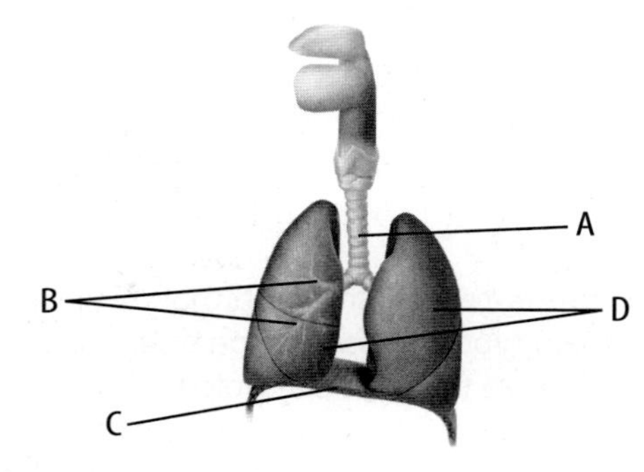

6 Which letter shows the muscle that contracts when you inhale and relaxes when you exhale? **SC.6.L.14.5**

Ⓕ A

Ⓖ B

Ⓗ C

Ⓘ D

7 Which is NOT a function of the skeletal system? **SC.6.L.14.5**

Ⓐ support
Ⓑ movement
Ⓒ protection
Ⓓ signaling

8 Which organ system works together with the nervous system to maintain homeostasis and control movement, communication, and growth? **SC.6.L.14.5**

Ⓕ excretory
Ⓖ endocrine
Ⓗ respiratory
Ⓘ muscular

Use the diagram below to answer question 9.

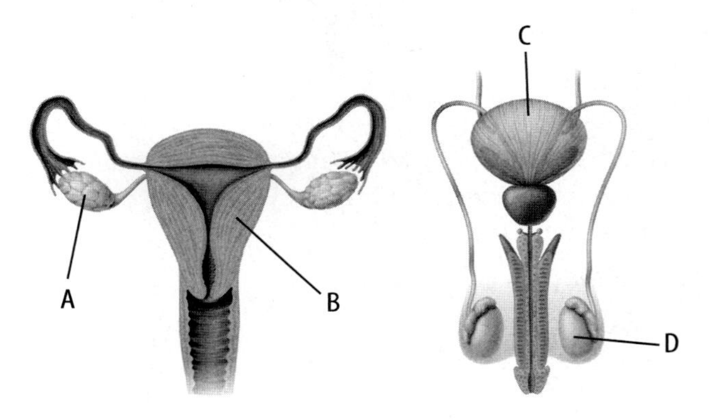

9 Which structure produces male gametes? **SC.6.L.14.5**

Ⓐ A
Ⓑ B
Ⓒ C
Ⓓ D

10 Which does NOT make up part of the circulatory system? **SC.6.L.14.5**

Ⓕ lungs
Ⓖ blood
Ⓗ heart
Ⓘ blood vessels

Use the figure below to answer question 11.

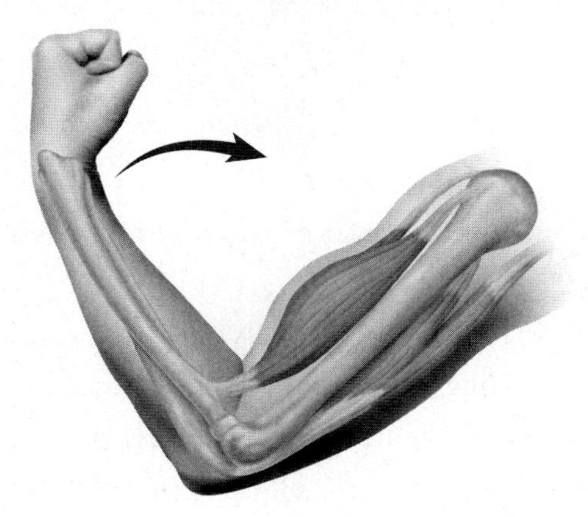

11 Which two organ systems work together in the figure above? **SC.6.L.14.5**

Ⓐ muscular and respiratory
Ⓑ skeletal and endocrine
Ⓒ muscular and skeletal
Ⓓ skeletal and excretory

12 Which is a function of the lymphatic system? **SC.6.L.14.5**

Ⓕ to transport gases
Ⓖ to protect internal organs
Ⓗ to control movement
Ⓘ to transport fats

NEED EXTRA HELP?

If You Missed Question...	1	2	3	4	5	6	7	8	9	10	11	12
Go to Lesson...	1	1	2	1	2	1	2	2	3	1	2	1

Name ______________________________ Date ______________

Multiple Choice *Bubble the correct answer.*

Use the diagram below to answer questions 1 and 2.

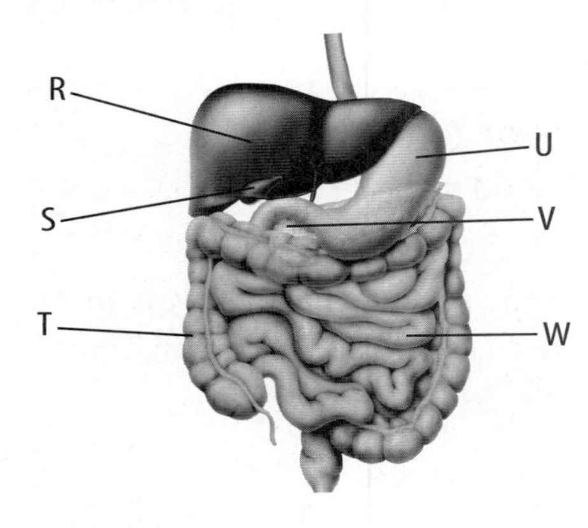

1. What is a function of the organ labeled R? **SC.6.L.14.5**

Ⓐ to absorb excess water

Ⓑ to digest food and absorb nutrients

Ⓒ to physically break down food

Ⓓ to produce bile to break down food

2. Which letter represents the organ that delivers enzymes to the small intestine? **SC.6.L.14.5**

Ⓕ R

Ⓖ S

Ⓗ U

Ⓘ V

3. Choose the correct pathway for carbon dioxide as it moves from a blood cell to the outside of the body. **SC.6.L.14.5**

Ⓐ alveolus→bronchus→trachea→pharynx→nostrils

Ⓑ bronchus→trachea→pharynx→alveolus→nostrils

Ⓒ pharynx→alveolus→trachea→bronchus→nostrils

Ⓓ trachea→pharynx→bronchus→alveolus→nostrils

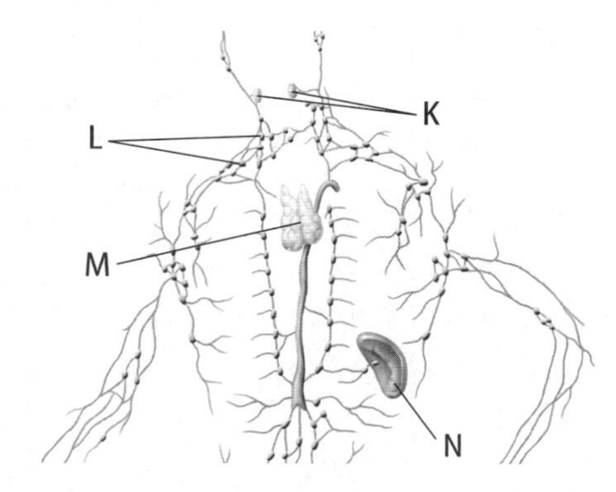

4. Which letters in the diagram above indicate the organs where lymphocytes are produced? **SC.6.L.14.5**

Ⓕ K and L

Ⓖ K and N

Ⓗ L and M

Ⓘ M and N

Name ______________________ Date ______________

Multiple Choice *Bubble the correct answer.*

1. Which statement is true about what is happening in the image above? **SC.6.L.14.5**

(A) Nerve signals from the brain are controlling the arm.

(B) Nerve signals from the spinal cord are controlling the arm.

(C) Receptors in the arm are controlling the spinal cord.

(D) Receptors in the hand are controlling the hand.

2. Which is true of cardiac muscle? **SC.6.L.14.5**

(F) It gives you strength to lift objects.

(G) It helps you digest food.

(H) It is only in the heart.

(I) It works with the skeletal system.

3. Parathyroid glands produce hormones that are essential to the proper function of the **SC.6.L.14.5**

(A) circulatory system.

(B) digestive system.

(C) respiratory system.

(D) skeletal system.

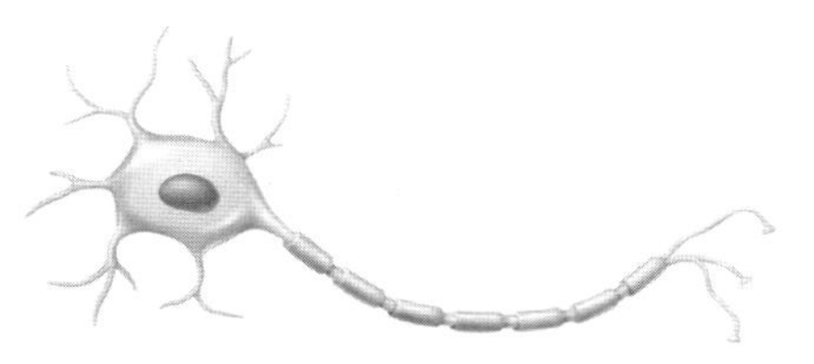

4. The cell shown above is responsible for **SC.6.L.14.5**

(F) heart contraction.

(G) hormone production.

(H) information transfer.

(I) structural support.

Name ______________________ Date ______________

Multiple Choice *Bubble the correct answer.*

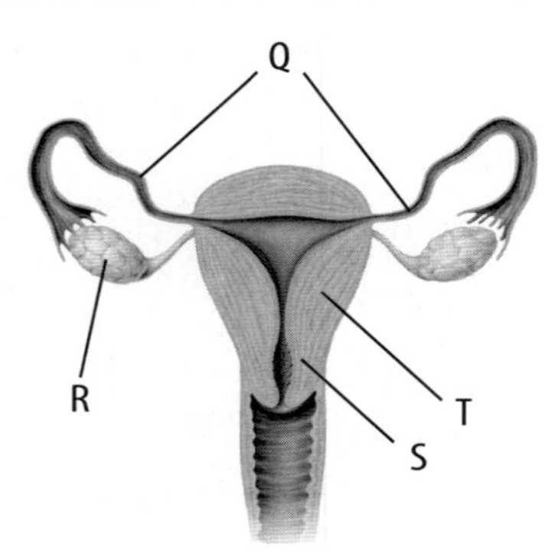

1. In the diagram above, which structure is where fertilization normally occurs? **SC.6.L.14.5**

(A) Q

(B) R

(C) S

(D) T

2. Which is NOT a part of the menstrual cycle? **SC.6.L.14.5**

(F) Eggs divide to form gametes.

(G) Eggs grow and mature.

(H) The lining of the uterus breaks down.

(I) The lining of the uterus thickens.

3. Which male reproductive structure performs a function that is similar to a function of the female ovaries? **SC.6.L.14.5**

(A) penis

(B) scrotum

(C) testes

(D) urethra

4. When would an infant normally reach the stage of development shown in the image above? **SC.6.L.14.5**

(F) between months 1 and 2

(G) between months 3 and 4

(H) between months 5 and 13

(I) between months 14 and 18

Name ______________________ Date ____________

Notes

Name ______________________________ Date ______________

Notes

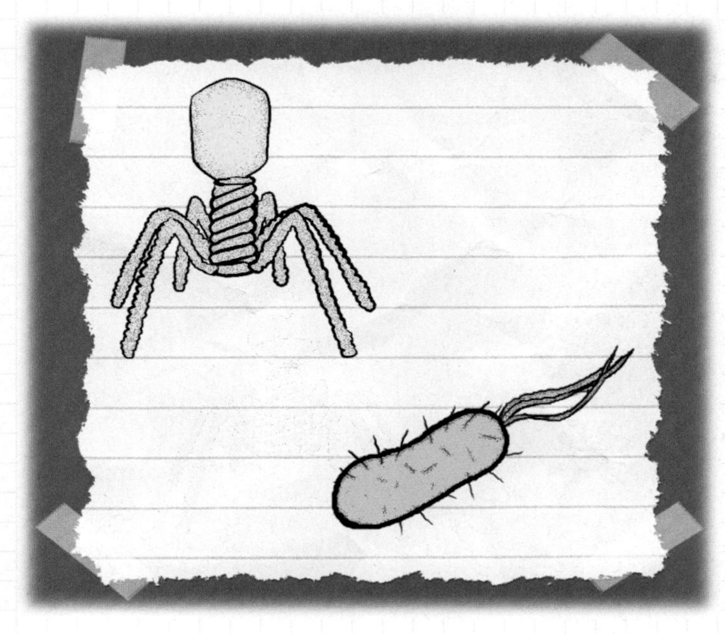

Is it an organism?

Janet wondered if bacteria and viruses are considered organisms. She asked her friends and this is what they said:

Tony: Bacteria are organisms, but viruses are not organisms.

Suze: Viruses are organisms, but bacteria are not organisms.

Lucas: Bacteria and viruses are both organisms.

Gina: Neither bacteria nor viruses are organisms.

Which friend do you agree with the most? _____________ Explain why you agree.

Bacteria and

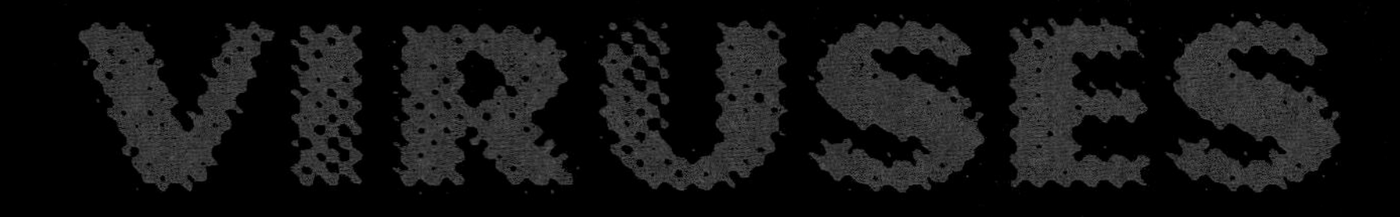

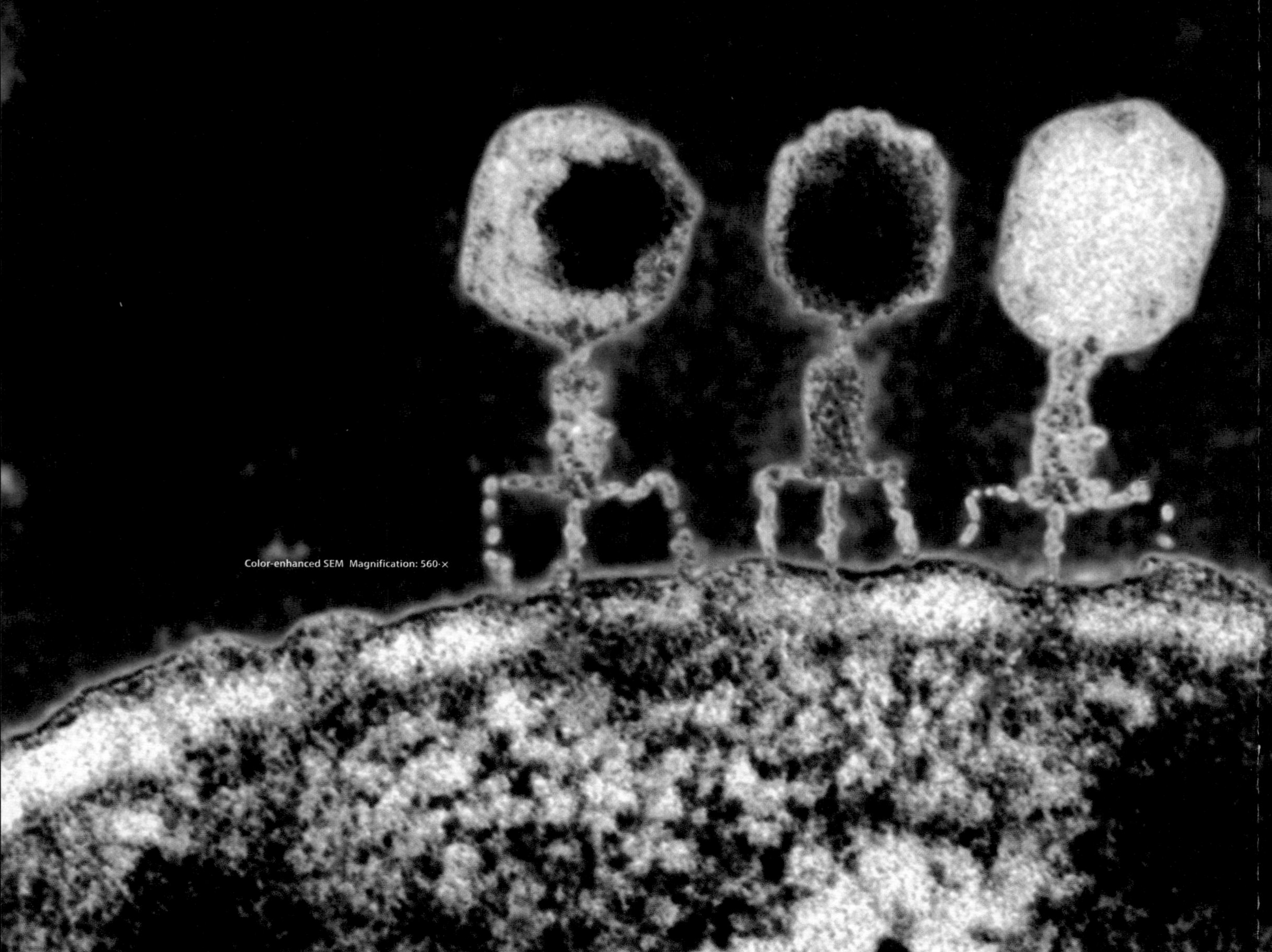

The Big Idea

FLORIDA BIG IDEAS

1 **The Practice of Science**

3 **The Role of Theories, Laws, Hypotheses, and Models**

14 **Organization and Development of Living Organisms**

Think About It!

What are bacteria and viruses and why are they important?

You might think this photo shows robots landing on another planet. Actually, this is a picture of viruses attacking a type of unicellular organism called a bacterium (plural, bacteria). Many viruses can attach to the surface of one bacterium.

1. Do you think the bacterium is harmful? Are the viruses?

2. What do you think happens after the viruses attach to the bacterium?

3. What are viruses and bacteria, and why do you think they are important?

Get Ready to Read

What do you think about bacteria and viruses?

Before you read, decide if you agree or disagree with each of these statements. As you read this chapter, see if you change your mind about any of the statements.

	AGREE	DISAGREE
1 A bacterium does not have a nucleus.	☐	☐
2 Bacteria cannot move.	☐	☐
3 All bacteria cause diseases.	☐	☐
4 Bacteria are important for making many types of food.	☐	☐
5 Viruses are the smallest living organisms.	☐	☐
6 Viruses can replicate only inside an organism.	☐	☐

There's More Online!

Video • Audio • Review • ⓘLab Station • WebQuest • Assessment • Concepts in Motion • Multilingual eGlossary

Lesson 1

What are BACTERIA?

ESSENTIAL QUESTIONS

What are bacteria?

Vocabulary
bacterium p. 493
flagellum p. 496
fission p. 496
conjugation p. 496
endospore p. 497

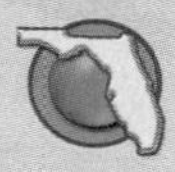

Florida NGSSS

LA.6.2.2.3 The student will organize information to show understanding (e.g., representing main ideas within text through charting, mapping, paraphrasing, summarizing, or comparing/contrasting);

LA.6.4.2.2 The student will record information (e.g., observations, notes, lists, charts, legends) related to a topic, including visual aids to organize and record information and include a list of sources used;

MA.6.A.3.6 Construct and analyze tables, graphs, and equations to describe linear functions and other simple relations using both common language and algebraic notation.

SC.6.L.14.6 Compare and contrast types of infectious agents that may infect the human body, including viruses, bacteria, fungi, and parasites.

SC.6.N.1.1 Define a problem from the sixth grade curriculum, use appropriate reference materials to support scientific understanding, plan and carry out scientific investigation of various types, such as systematic observations or experiments, identify variables, collect and organize data, interpret data in charts, tables, and graphics, analyze information, make predictions, and defend conclusions.

SC.6.N.2.1 Distinguish science from other activities involving thought.

SC.6.N.3.4 Identify the role of models in the context of the sixth grade science benchmarks.

 SC.6.N.3.4

Inquiry Launch Lab

10 minutes

How small are bacteria?

Bacteria are tiny cells that can be difficult to see, even with a microscope. You might be surprised to learn that bacteria are found all around you, including in the air, on your skin, and in your body. One way of understanding how small bacteria are is to model their size.

Procedure 

1. Read and complete a lab safety form.
2. Examine the size of a **baseball** and a **2.5-gal. bucket.** Estimate how many baseballs you think would fit inside the bucket.
3. As a class, count how many baseballs it takes to fill the bucket.

Think About This

1. How much larger is the bucket than a baseball?

2. If your skin cells were the size of the bucket and bacteria were the size of the baseballs, how many bacterial cells would fit on a skin cell?

3. Key Concept Why do you think you cannot see bacteria on your skin or on your desk?

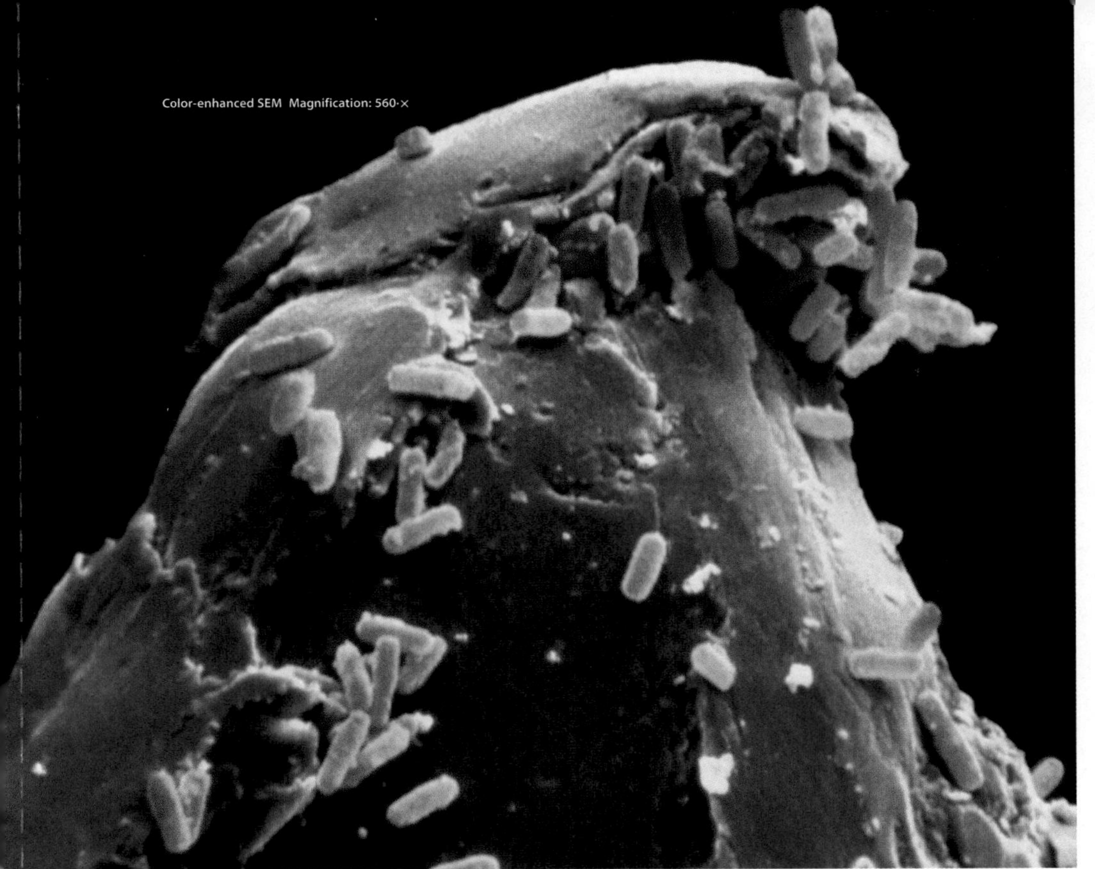

Color-enhanced SEM Magnification: 560×

How clean is this surface?

1. This photo shows a microscopic view of the point of a needle. The small orange things are bacteria. Bacteria are everywhere, even on surfaces that appear clean. Do you think bacteria are living or nonliving? Why?

Characteristics of Bacteria

Did you know that billions of tiny organisms too small to be seen surround you? These organisms, called bacteria, even live inside your body. **Bacteria** (singular, bacterium) *are microscopic prokaryotes.* You might recall that a prokaryote is a unicellular organism that does not have a nucleus or other membrane-bound organelles.

WORD ORIGIN

bacteria
from Greek *bakterion,* means "small staff"

Bacteria live in almost every habitat on Earth, including the air, glaciers, the ocean floor, and in soil. A teaspoon of soil can contain between 100 million and 1 billion bacteria. Bacteria also live in or on almost every organism, both living and dead. Hundreds of species of bacteria live on your skin. In fact, your body contains more bacterial cells than human cells! The bacteria in your body outnumber human cells by 10 to 1.

Active Reading **2. Explain** What are bacteria?

Other prokaryotes, called archaea (ar KEE uh; singular, archaean), are similar to bacteria and share many characteristics with them, including the lack of membrane-bound organelles. Archaea can live in places where few other organisms can survive, such as very warm areas or those with little oxygen. Both bacteria and archaea are important to life on Earth.

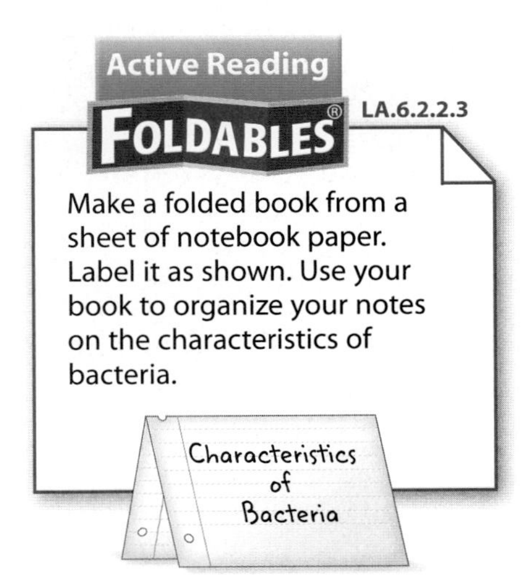

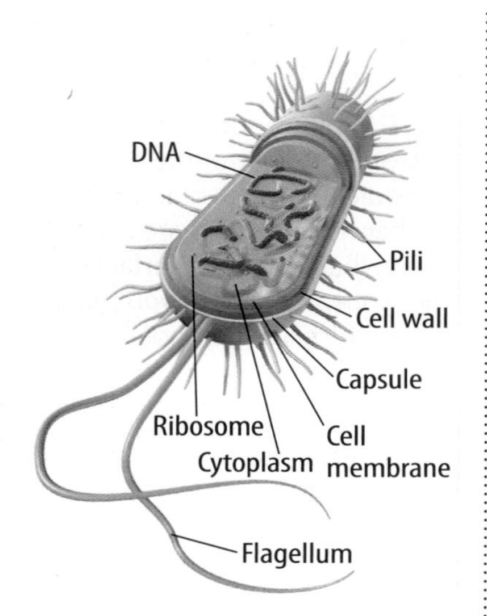

Figure 1 Bacteria have a cell membrane and contain cytoplasm.

Structure of Bacteria

A typical bacterium, such as the one shown in **Figure 1,** consists of cytoplasm and DNA surrounded by a cell membrane and a cell wall. The cytoplasm also contains ribosomes. Most bacteria have DNA that is one coiled, circular chromosome. Many bacteria also have one or more small circular pieces of DNA called plasmids that are separate from its other DNA.

Some bacteria have specialized structures that help them survive. For example, the bacterium that causes pneumonia (noo MOH nyuh), an inflammation of the lungs, has a thick covering, or capsule, around its cell wall. The capsule protects the bacterium from drying out. It also prevents white blood cells from surrounding and antibiotics from entering it. Many bacteria have capsules with hairlike structures called pili (PI li) that help the bacteria stick to surfaces.

Size and Shapes of Bacteria

Bacteria are much smaller than plant or animal cells. Bacteria are generally only 1–5 micrometers (μm) (1 m = 1 million μm) wide, while an average eukaryotic cell is 10–100 μm wide. Scientists estimate that as many as 100 bacteria could be lined up across the head of a pin. As shown in **Figure 2,** bacteria generally have one of three basic shapes.

Figure 2 Bacteria are generally shaped like a sphere, a rod, or a spiral.

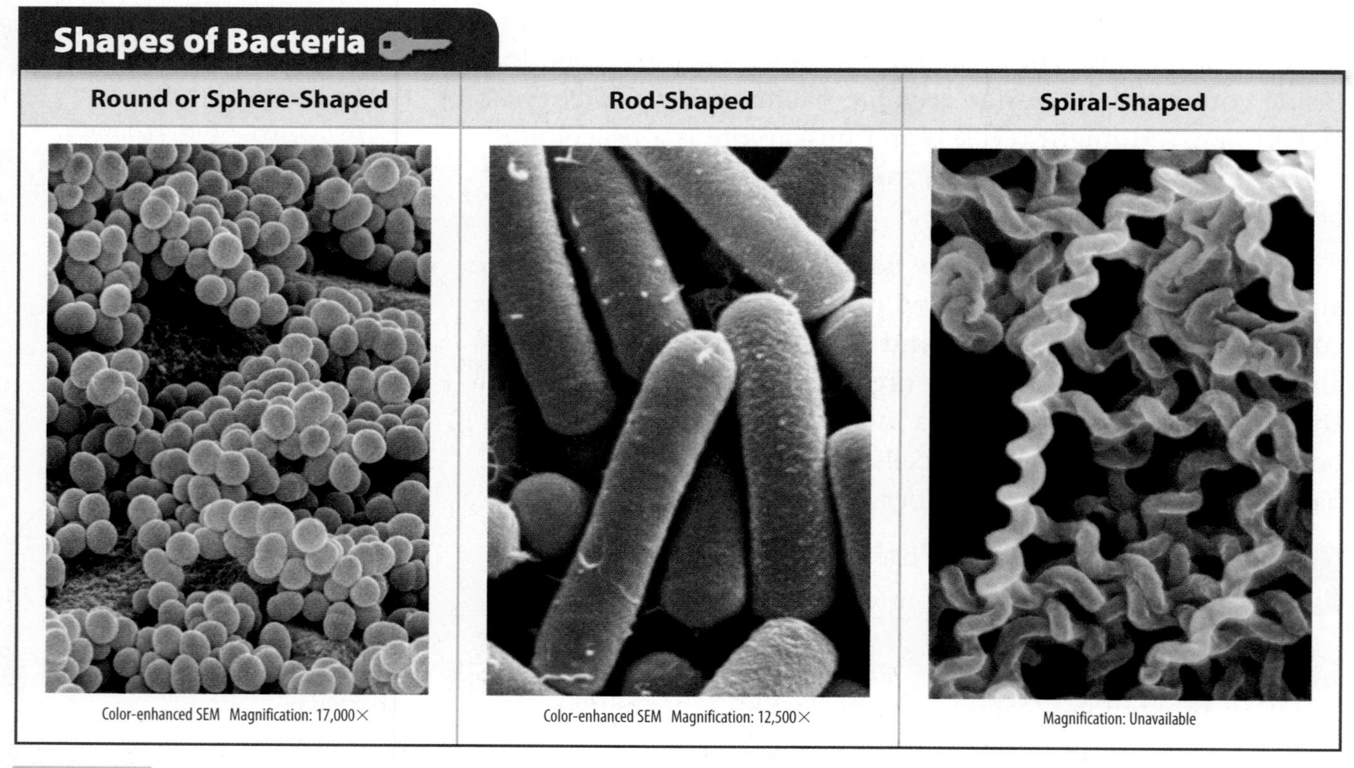

Active Reading 3. **Infer** Bacteria often form clusters. Which shape of bacteria do you think would likely form a cluster?

Apply It! After you complete the lab, answer this question.

1. How do you think Earth's nutrient cycles would be affected if bacteria cells did not have slime layers?

Obtaining Food and Energy

Bacteria live in many places. Because these environments are very different, bacteria obtain food in various ways. Some bacteria take in food and break it down and obtain energy. Many of these bacteria feed on dead organisms or organic waste, as shown in **Figure 3.** Others take in their nutrients from living hosts. For example, bacteria that cause tooth decay live in dental plaque on teeth and feed on sugars in the foods you eat and the beverages you drink.

Some bacteria make their own food. These bacteria use light energy and make food, like most plants do. These bacteria live where there is a lot of light, such as the surface of lakes and streams. Other bacteria use energy from chemical reactions and make their food. These bacteria live in places where there is no sunlight, such as the dark ocean floor.

Figure 3 This banana is rotting because bacteria are breaking it down for food.

Active Reading 4. **Restate** How do bacteria obtain food?

Most organisms, including humans, cannot survive without oxygen. However, certain bacteria do not need oxygen to survive. These bacteria are called anaerobic (a nuh ROH bihk) bacteria. Bacteria that need oxygen are called aerobic (er OH bihk) bacteria. Most bacteria in the environment are aerobic.

Active Reading 5. **Contrast** Complete the table by contrasting anaerobic bacteria with aerobic bacteria.

Aerobic Bacteria	Anaerobic Bacteria

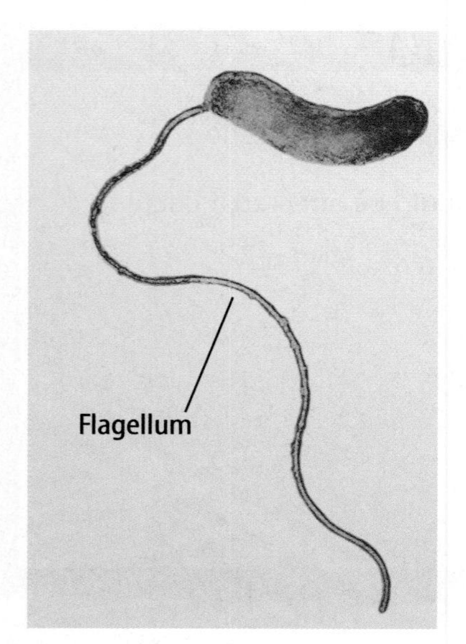

Figure 4 Some bacteria move using a flagellum.

Movement

Some bacteria are able to move around to find the resources that they need to survive. These bacteria have special structures for movement. *Many bacteria have long whiplike structures called* **flagella** (fluh JEH luh; singular, flagellum), as shown in **Figure 4.** Others twist or spiral as they move. Still other bacteria use their pili like grappling hooks or make threadlike structures that enable them to push away from a surface.

Reproduction

You might recall that organisms reproduce asexually or sexually. Bacteria reproduce asexually by fission. **Fission** *is cell division that forms two genetically identical cells.* Fission can occur quickly—as often as every 20 minutes under ideal conditions.

Bacteria produced by fission are identical to the parent cell. However, genetic variation can be increased by a process called conjugation, shown in **Figure 5.** *During* **conjugation** (kahn juh GAY shun), *two bacteria of the same species attach to each other and combine their genetic material.* DNA is transferred between the bacteria. This results in new combinations of genes, increasing genetic diversity. New organisms are not produced during conjugation, so the process is not considered reproduction.

Active Reading **6. Explain** How does conjugation increase the genetic diversity of bacteria?

Conjugation

Figure 5 Conjugation results in genetic diversity by transferring DNA between two bacterium cells.

Active Reading **7. Identify** (Circle) the structure that the donor cell uses to connect to the recipient cell.

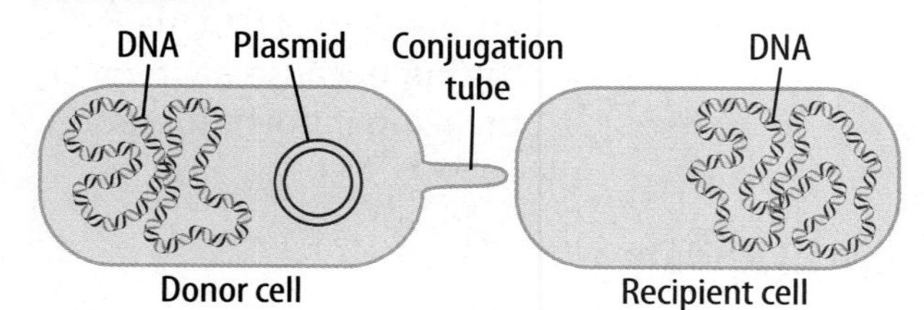

❶ The donor cell and recipient cell both have circular chromosomal DNA. The donor cell also has DNA as a plasmid. The donor cell forms a conjugation tube and connects to the recipient cell.

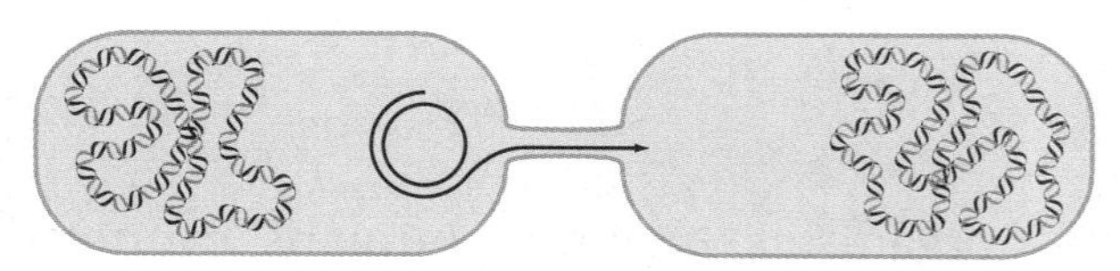

❷ The conjugation tube connects both cells. The plasmid splits in two, and one plasmid strand moves through the conjugation tube into the recipient cell.

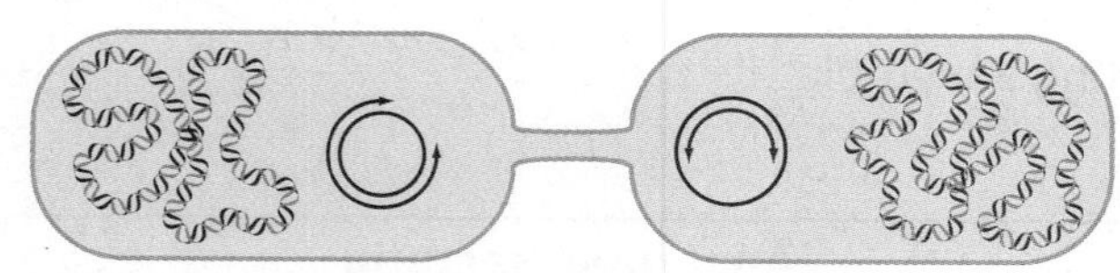

❸ The complimentary strands of the plasmids are completed in both bacteria.

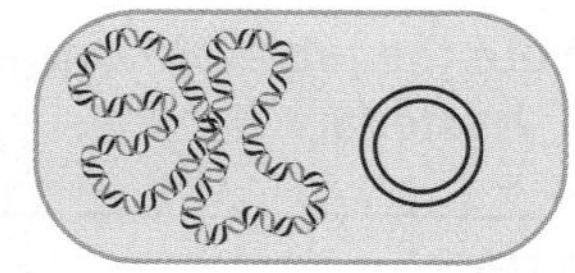
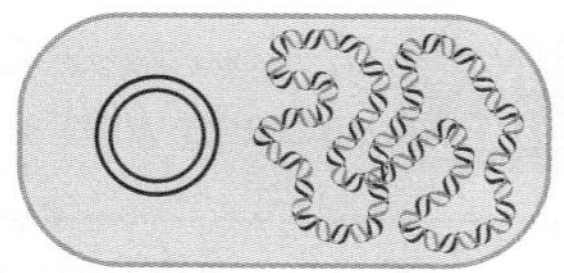

❹ With the new plasmids complete, the bacteria separate from each other. The recipient cell now contains plasmid DNA from the donor cell as well as its own chromosomal DNA.

Endospore Formation

Figure 6

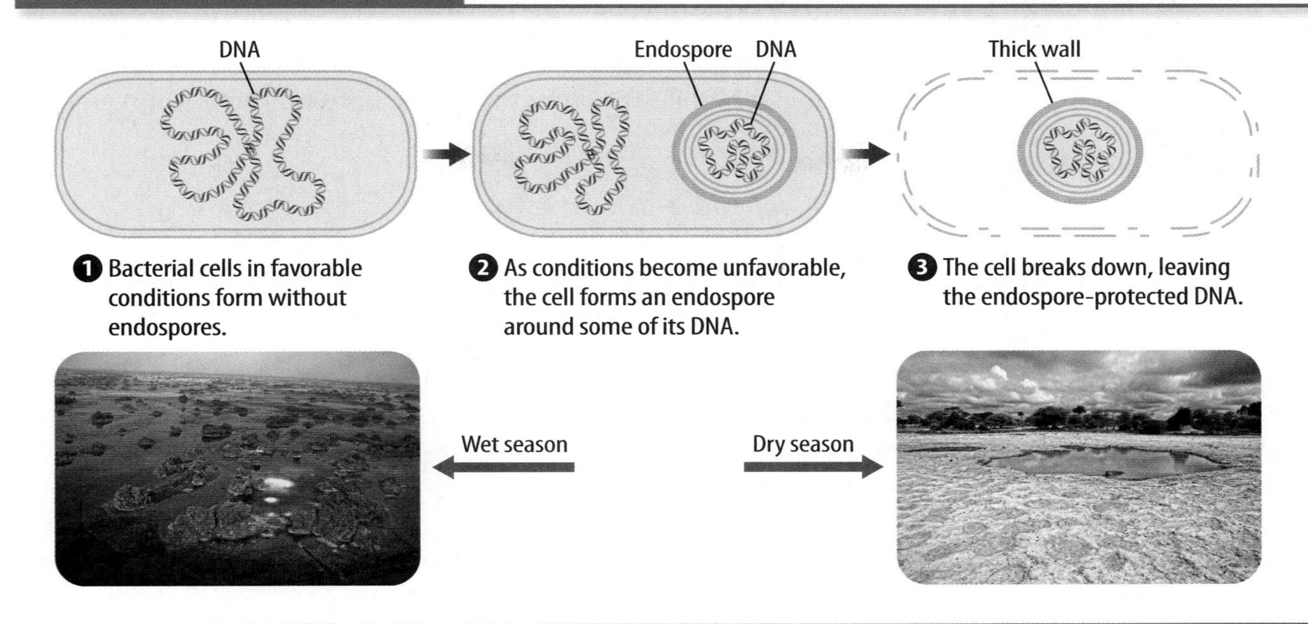

❶ Bacterial cells in favorable conditions form without endospores.

❷ As conditions become unfavorable, the cell forms an endospore around some of its DNA.

❸ The cell breaks down, leaving the endospore-protected DNA.

Active Reading 8. **Label** An endospore can protect a ______________.

Endospores

Sometimes environmental conditions are unfavorable for the survival of bacteria. In these cases, some bacteria can form endospores. *An* **endospore** (EN doh spor) *forms when a bacterium builds a thick internal wall around its chromosome and part of the cytoplasm,* as shown in **Figure 6.** An endospore can protect a bacterium from intense heat, cold, or drought. It also enables a bacterium to remain dormant for months or even centuries. The ability to form endospores enables bacteria to survive extreme conditions that would normally kill them.

Archaea

Prokaryotes called archaea were once considered bacteria. Like a bacterium, an archaean has a cell wall and no nucleus or membrane-bound organelles. Its chromosome is also circular, like those in bacteria. However, there are some important differences between archaea and bacteria. The ribosomes of archaea more closely resemble the ribosomes of eukaryotes than those of bacteria. Archaea also contain molecules in their plasma membranes that are not found in any other known organisms. Archaea often live in extreme environments, such as hot springs and salt lakes. Some scientists refer to archaea as extremophiles (ik STREE muh filez)—a term that means "those that love extremes."

Math Skills

MA.6.A.3.6

Use a Formula

Each time bacteria undergo fission, the population doubles. Use an equation to calculate how many bacteria there are:

$n = x \times 2^f$ where n is the final number of bacteria, x is the starting number of bacteria, and f is the number of times that fission occurs.

Example: 100 bacteria undergo fission 3 times.

$f = 3$, so 2^f is 2 multiplied by itself 3 times.
$(2 \times 2 \times 2 = 8)$

$n = 100 \times 8 = 800$ bacteria

Practice

9. How many bacteria would there be if 1 bacterium underwent fission 10 times?

Lesson Review 1

Visual Summary

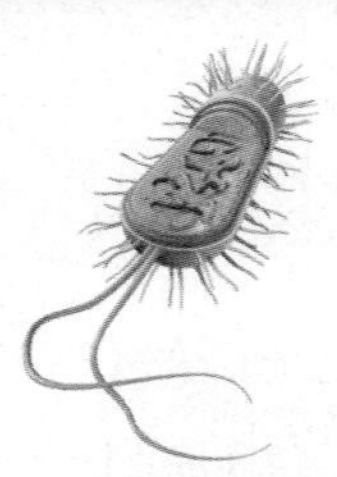

Bacteria are unicellular prokaryotes.

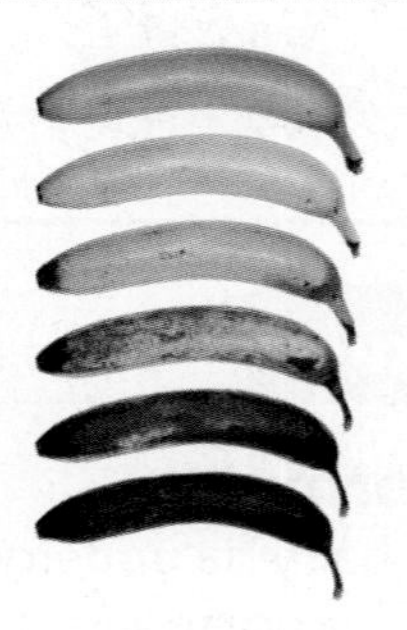

Many bacteria feed on dead organic matter.

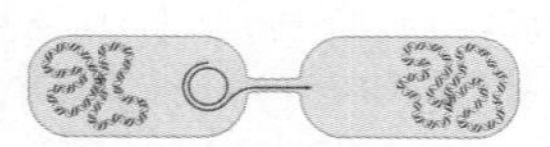

Bacteria can increase genetic diversity by sharing DNA through conjugation.

Use Vocabulary

1. The long, whiplike structure that some bacteria use for movement is a(n) ______________________________.

2. **Define** *conjugation* in your own words.

Understand Key Concepts

3. **Describe** a typical bacterium.

4. Which is NOT a common bacterium shape? SC.6.L.14.6
   - (A) rod
   - (B) sphere
   - (C) spiral
   - (D) square

Interpret Graphics

5. **Identify** Complete the table below to identify shapes of bacteria. LA.6.2.2.3

Bacterial Shapes	Illustration

Critical Thinking

6. **Analyze** how bacteria that can form endospores would have an advantage over bacteria that cannot form endospores. SC.6.L.14.6

Math Skills

7. How many bacteria would result if fission occurred 4 times with 1,000 bacteria?

Cooking Bacteria!

HOW NATURE WORKS

How Your Body Is Like Bleach

When it comes to killing germs, few things work as well as household bleach. How does bleach kill bacteria? Believe it or not, killing bacteria with bleach and boiling an egg involve similar processes.

After cooking, egg proteins become a tangled mass.

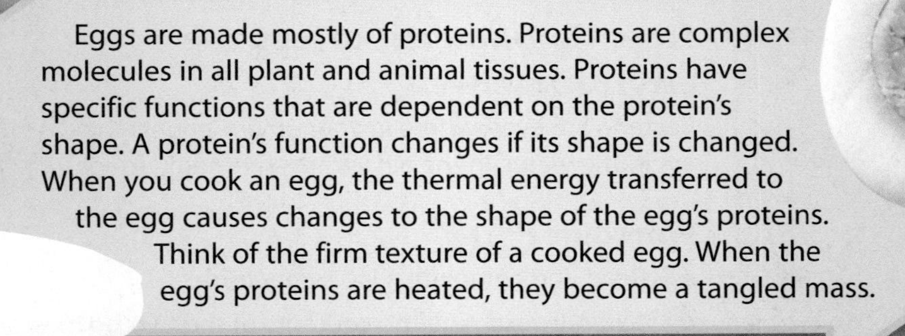

Eggs are made mostly of proteins. Proteins are complex molecules in all plant and animal tissues. Proteins have specific functions that are dependent on the protein's shape. A protein's function changes if its shape is changed. When you cook an egg, the thermal energy transferred to the egg causes changes to the shape of the egg's proteins. Think of the firm texture of a cooked egg. When the egg's proteins are heated, they become a tangled mass.

Before cooking, the proteins in eggs remain unfolded and change shape easily.

A common ingredient in bleach is also found in your body's immune cells.

Bacteria also contain proteins that change shape when exposed to heat.

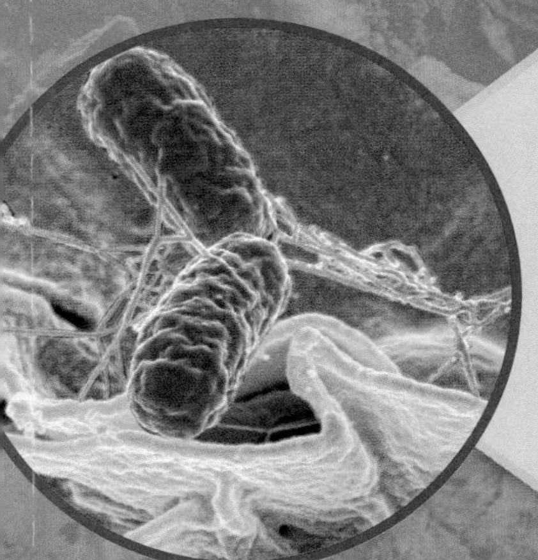

Like eggs, bacteria also contain proteins. When bacteria are exposed to high temperatures, their proteins change shape, similar to those in a boiled egg. But what is the connection with bleach? Scientists have discovered that an ingredient in bleach, hypochlorite (hi puh KLOR ite), also causes proteins to change shape. The bacterial proteins that are affected by bleach are needed for the bacterias' growth. When the shape of those proteins changes, they no longer function properly, and the bacteria die.

Scientists also know now that your body's immune cells produce hypochlorite. Your body protects itself with the same chemical you can use to clean your kitchen!

It's Your Turn

RESEARCH AND REPORT A bacterial infection often causes inflammation, or a response to tissue damage that can include swelling and pain. Research and report on what causes inflammation. LA.6.4.2.2

Lesson 2

Bacteria in NATURE

ESSENTIAL QUESTIONS

How can bacteria affect the environment?

How can bacteria affect health?

Vocabulary

decomposition p. 502
nitrogen fixation p. 502
bioremediation p. 503
pathogen p. 504
antibiotic p. 504
pasteurization p. 505

Florida NGSSS

LA.6.2.2.3 The student will organize information to show understanding (e.g., representing main ideas within text through charting, mapping, paraphrasing, summarizing, or comparing/contrasting);

SC.6.L.14.6 Compare and contrast types of infectious agents that may infect the human body, including viruses, bacteria, fungi, and parasites.

SC.6.N.1.1 Define a problem from the sixth grade curriculum, use appropriate reference materials to support scientific understanding, plan and carry out scientific investigation of various types, such as systematic observations or experiments, identify variables, collect and organize data, interpret data in charts, tables, and graphics, analyze information, make predictions, and defend conclusions.

SC.6.N.1.4 Discuss, compare, and negotiate methods used, results obtained, and explanations among groups of students conducting the same investigation.

SC.6.N.2.1 Distinguish science from other activities involving thought.

SC.6.N.1.1

Inquiry Launch Lab

10 minutes

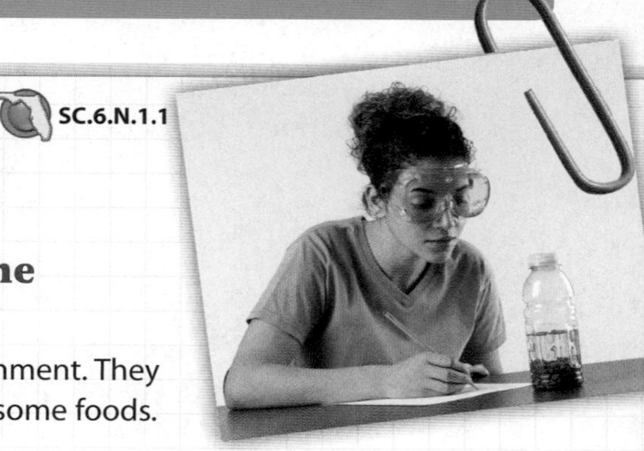

How do bacteria affect the environment?

Bacteria are everywhere in your environment. They are in the water, in the air, and even in some foods.

Procedure

1. Read and complete a lab safety form.
2. Carefully examine the contents of the two **bottles** provided by your teacher.
3. Record your observations below.

Data and Observations

Think About This

1. Compare your observations of bottle A to those of bottle B. Which one appears to have more bacteria in it? Support your answer.

2. Key Concept Based on your observations, how could bacteria affect the environment around you?

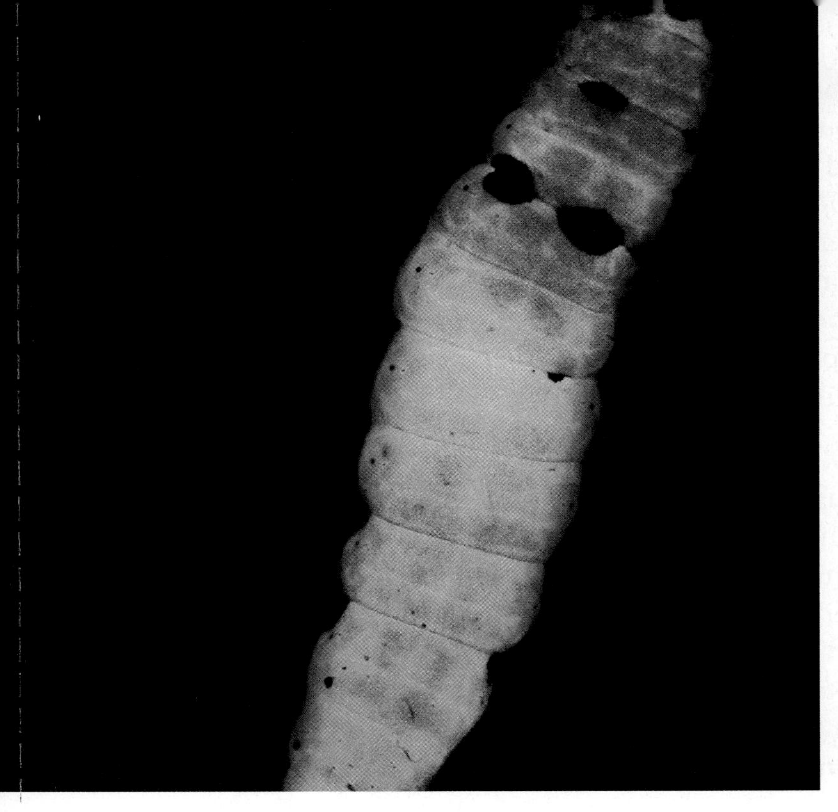

Inquiry Why does this larva glow?

1. Some bacteria have the ability to glow in the dark. The moth larva shown on this page is filled with many such bacteria. These bacteria produce toxins that can slowly kill the animal. A chemical reaction within each bacterium makes the larva's body appear to glow. Why can't you see the individual bacteria that cause the glowing of the moth larva?

Beneficial Bacteria

When you hear about bacteria, you probably think about getting sick. However, only a fraction of all bacteria cause diseases. Most bacteria are beneficial. In fact, many organisms, including humans, depend on bacteria to survive. Some types of bacterium help with digestion and other body processes. For example, one type of bacterium in your intestines makes vitamin K, which helps your blood clot properly. Several others help break down food into smaller particles. Another type of bacterium called *Lactobacillus* lives in your intestines and prevents harmful bacteria from growing.

Animals benefit from bacteria as well. Without bacteria, some organisms, such as the cow pictured in **Figure 7,** wouldn't be able to digest the plants they eat. Bacteria and other microscopic organisms live in a large section of the cow's stomach called the rumen. The bacteria help break down a substance in grass called cellulose into smaller molecules that the cow can use.

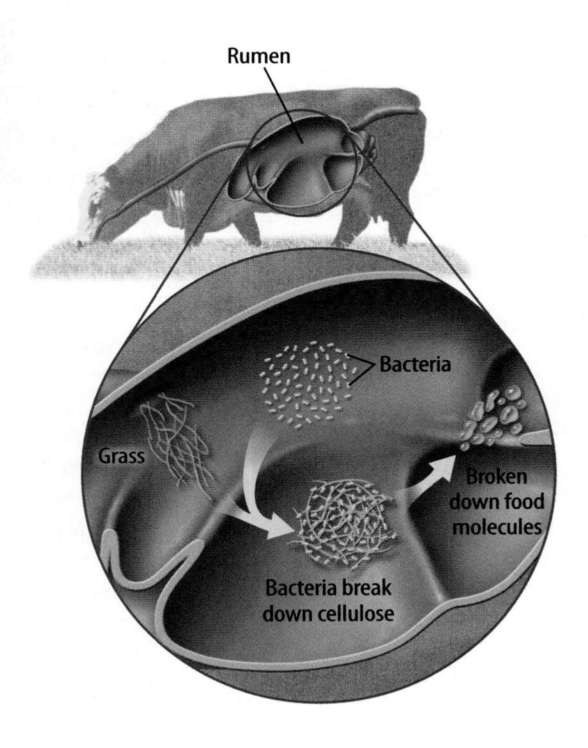

Figure 7 Cows get help digesting the cellulose in plants from the bacteria that live in their rumen—one of four stomach sections.

2. **Infer** Examine **Figure 7.** What would happen if bacteria were not present?

Active Reading 3. **Explain** What role do bacteria play in a cow's digestion?

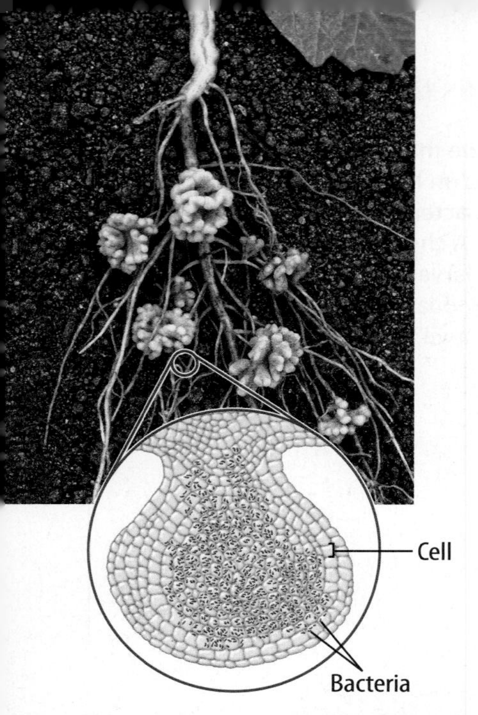

Figure 8 The roots of some plants have nodules that contain nitrogen-fixing bacteria.

Decomposition

What do you think would happen if organic waste such as food scraps and dead leaves never decayed? **Decomposition,** *the breaking down of dead organisms and organic waste,* is an important process in nature. When a tree dies, bacteria and other decomposing organisms feed on the dead organic matter. As decomposers break down the tree, they release molecules such as carbon and phosphorus into the soil that other organisms can then take in and use for life processes.

Nitrogen Fixation

Organisms use nitrogen to make proteins. Although about 78 percent of the atmosphere is nitrogen gas, it is in a form that plants and animals cannot use. Some plants can obtain nitrogen from bacteria. These plants have special structures called nodules, shown in **Figure 8,** on their roots. Bacteria in the nodules convert nitrogen from the atmosphere into a form usable to plants. **Nitrogen fixation** *is the conversion of atmospheric nitrogen into nitrogen compounds that are usable by living things.*

Active Reading **4. Explain** What is nitrogen fixation?

Active Reading **5. Explain** As you complete the lesson, provide examples of ways in which bacteria can be beneficial.

Benefit	Explanation and Example
Digestion	
Decomposition	
Nitrogen fixation	
Bioremediation	
Production of food	

Inquiry LAB STATION SC.6.L.14.3 **Try It!**

MiniLab ***Can decomposition happen without oxygen?*** at connectED.mcgraw-hill.com

Apply It! After you complete the lab, answer this question.

1. Why might decomposers be described as the ultimate recyclers even if it takes many years for them to break down a tree?

Bioremediation

Can you imagine an organism that eats pollution? Some bacteria do just that. *The use of organisms, such as bacteria, to clean up environmental pollution is called* **bioremediation** (bi oh rih mee dee AY shun). These organisms often break down harmful substances, such as sewage, into less harmful material that can be used as landfill or fertilizers.

Other kinds of bacteria can help clean up radioactive waste, such as uranium in abandoned minefields. In many cases, without using bacteria, the substances would take centuries to break down and would contaminate soils and water. Bacteria are also commonly used to clean up areas that have been contaminated by oil, such as the Gulf of Mexico. Bacteria could aid in the long-term Gulf cleanup resulting from the explosion on the *Deepwater Horizons* oil platform in 2010. These natural oil-eating bacteria will help consume the oil and other pollutants in the water as shown in **Figure 9.**

Active Reading **6. Explain** Why might using bacteria to clean up environmental spills be a good option?

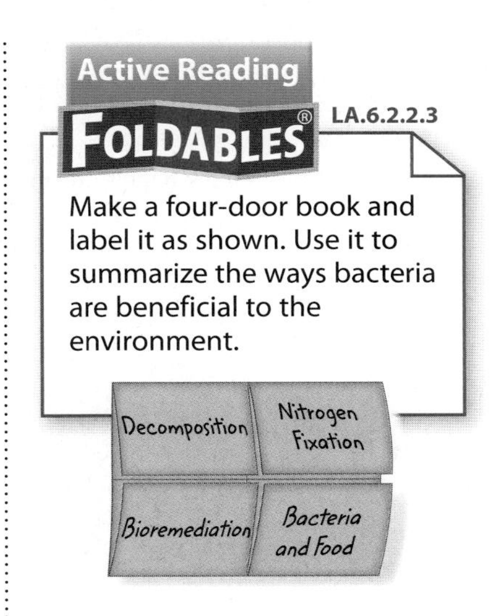

Figure 9 Some bacteria clean the environment by removing harmful pollutants from the water.

Active Reading **7. List** What events might cause oil contamination of water?

Bacteria and Food

Would you like a side of bacteria with that sandwich? If you have eaten a pickle lately, you might have had some. Some pickles are made when the sugar in cucumbers is converted into an acid by a specific type of bacteria. Pickles are just one of the many food products made with the help of bacteria. Bacteria are used to make foods such as yogurt, cheese, buttermilk, vinegar, and soy sauce. Bacteria are even used in the production of chocolate. They help break down the covering of the cocoa bean during the process of making chocolate. Bacteria are responsible for giving chocolate some of its flavor.

WORD ORIGIN
pathogen from Greek *pathos,* means "to suffer"; and *gen,* means "to produce"

Harmful Bacteria

Of the 5,000 known species of bacteria, relatively few are considered **pathogens** (PA thuh junz)—*agents that cause disease.* Some pathogens normally live in your body but cause illness only when your immune system is weakened. For example, the bacterium *Streptococcus pneumoniae* lives in the throats of most healthy people. However, it can cause pneumonia if a person's immune system is weakened. Other bacterial pathogens can enter your body through a cut, the air you breathe, or the food you eat. Once inside your body, they can reproduce and cause disease.

8. NGSSS Check Describe Give an example of one way that bacteria can be harmful to your health. SC.6.L.14.6

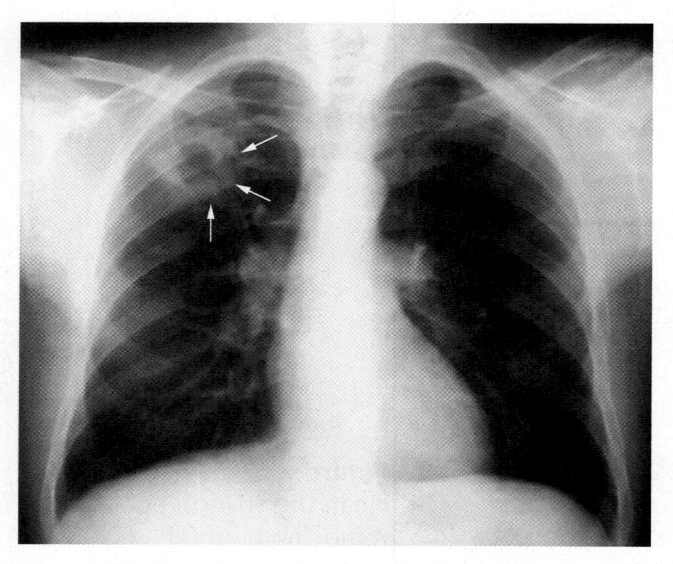

Figure 10 In an X-ray, the lungs of a person with tuberculosis may show pockets or scars where bacterial infection has begun.

9. Visual Check Infer Look at the X-ray in **Figure 10.** How do you think the bacteria that made this person sick entered his or her body?

Bacterial Diseases

Bacteria can harm your body and cause disease in one of two ways. Some bacteria make you sick by damaging tissue. For example, the disease tuberculosis, shown in **Figure 10,** is caused by a bacterium that invades lung tissue and breaks it down for food. Other bacteria cause illness by releasing toxins. For example, the bacterium *Clostridium botulinum* can grow in improperly canned foods and produce toxins. If the contaminated food is eaten, the toxins can cause food poisoning, resulting in paralyzed limbs or even death.

Treating Bacterial Diseases Most bacterial diseases in humans can be treated with antibiotics. **Antibiotics** (an ti bi AH tihks) *are medicines that stop the growth and reproduction of bacteria.* Many antibiotics work by preventing bacteria from building cell walls. Others affect ribosomes in bacteria, interrupting the production of proteins.

SCIENCE USE V. COMMON USE
resistance
Science Use the capacity of an organism to defend itself against a disease
Common Use the act of opposing something

Many types of bacteria have become **resistant** to antibiotics over time. Some diseases, such as tuberculosis, pneumonia, and meningitis, are now more difficult to treat.

Bacterial Resistance How do you think bacteria become resistant to antibiotics? This process, shown in **Figure 11,** can happen over a long or short period of time depending on how quickly the bacteria reproduce. Random mutations occur to a bacterium's DNA that enable it to survive or "resist" a specific antibiotic. If that antibiotic is used as a treatment, only the bacteria with the mutation will survive.

Over time, the resistant bacteria will reproduce and become more common. The antibiotic is no longer effective against that bacterium, and a different antibiotic must be used to fight the disease. Scientists are always working to develop more effective antibiotics to which bacteria have not developed resistance.

Active Reading **10. Describe** Underline a section in this text discussing how bacteria develop resistance to antibiotics.

Food Poisoning

All food, unless it has been treated or processed, contains bacteria. Over time these bacteria reproduce and begin breaking down the food, causing it to spoil. As you read on the previous page, eating food contaminated by some bacteria can cause food poisoning. By properly treating or processing food and killing bacteria before the food is stored or eaten, it is easier to avoid food poisoning and other illnesses.

Pasteurization (pas chuh ruh ZAY shun) *is a process of heating food to a temperature that kills most harmful bacteria.* Products such as milk, ice cream, yogurt, and fruit juice are usually pasteurized in factories before they are transported to grocery stores and sold to you. After pasteurization, foods are much safer to eat. Foods do not spoil as quickly once they have been pasteurized. Because of pasteurization, food poisoning is much less common today than it was in the past.

Active Reading **11. Explain** How does pasteurization affect human health?

___

___

How Resistance Develops

Figure 11 A population of bacteria can develop resistance to antibiotics after being exposed to them over time.

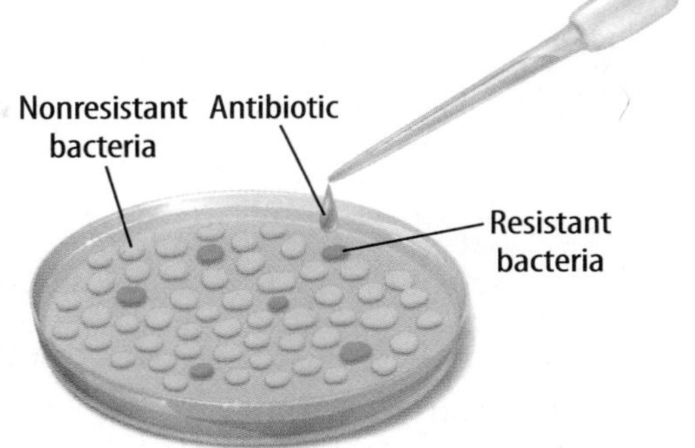

1. An antibiotic is added to a colony of bacteria. A few of the bacteria have mutations that enable them to resist the antibiotic.

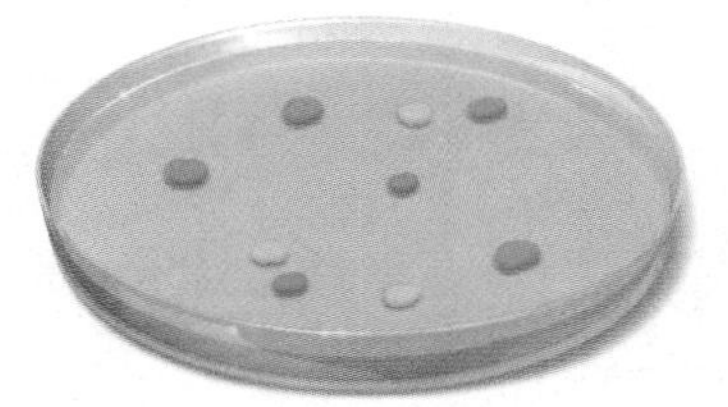

2. The antibiotic kills most of the nonresistant bacteria. The resistant bacteria survive and reproduce, creating a growing colony of bacteria.

3. Surviving bacteria are added to another plate containing more of the same antibiotic.

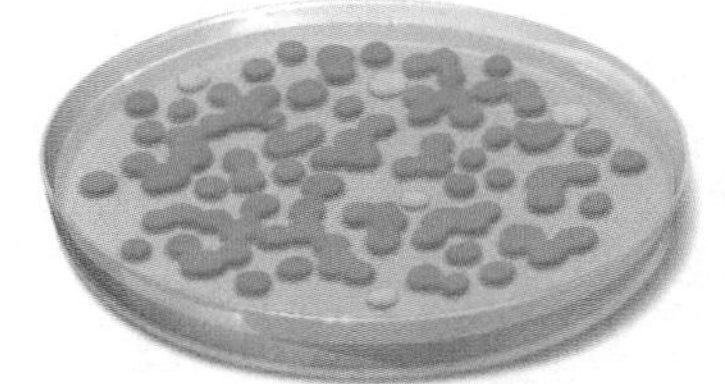

4. The antibiotic now affects only a small percentage of the bacteria. The surviving bacteria continue to reproduce. Most of the bacteria are resistant to the antibiotic.

Lesson Review 2

Visual Summary

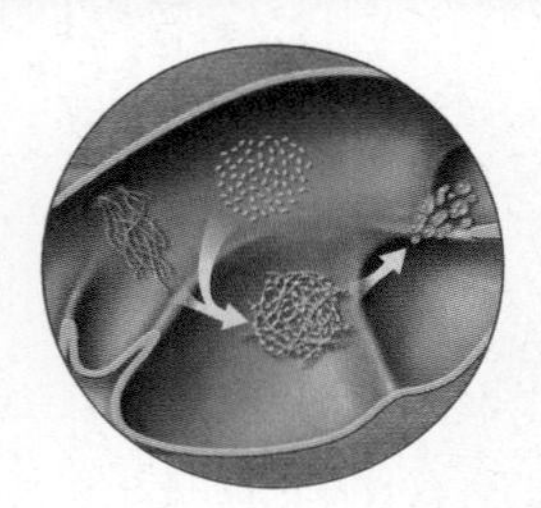

Bacteria can help some organisms, including humans and cows, digest food.

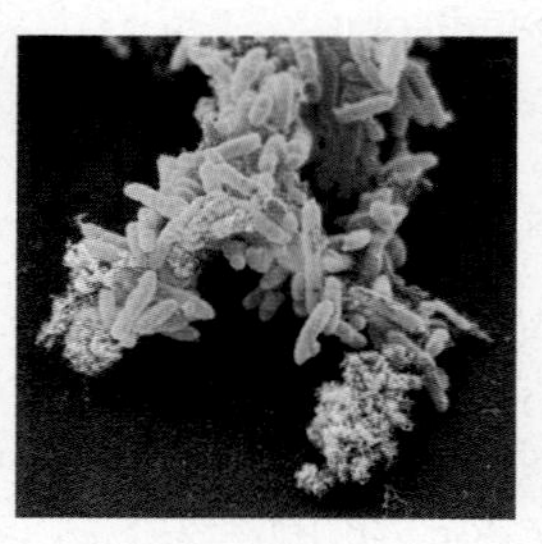

Bacteria can be used to remove harmful substances such as uranium.

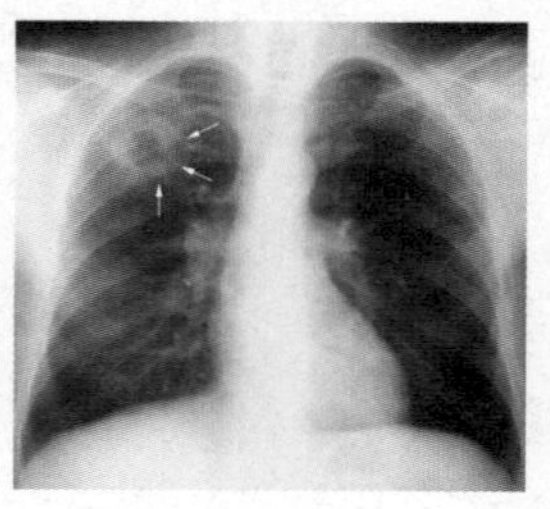

Some bacteria are pathogen, and cause diseases in humans and other organisms.

Use Vocabulary

1. **Distinguish** between an antibiotic and a pathogen.

2. **Define** *bioremediation* using your own words.

3. **Use the term** *pasteurization* in a sentence.

Understand Key Concepts

4. Which is NOT a beneficial use of bacteria? SC.6.L.14.6
   (A) bioremediation
   (B) decomposition
   (C) food poisoning
   (D) nitrogen fixation

5. **Compare** the benefits of nitrogen fixation and decomposition.

6. **Analyze** the importance of bacteria in food production.

Interpret Graphics

7. **Examine** the figure and describe what would happen if bacteria were not present.

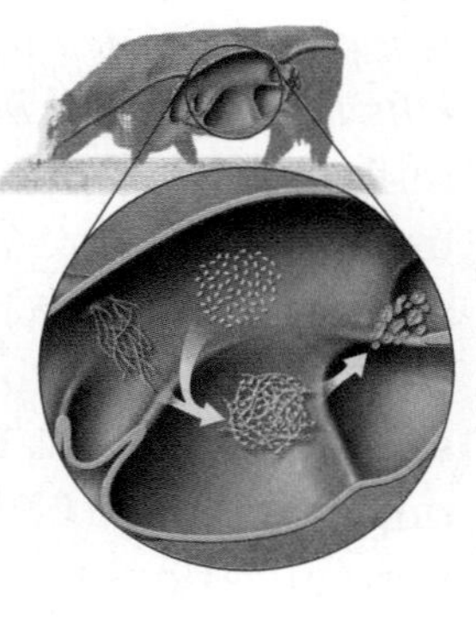

Critical Thinking

8. **Evaluate** the effect of all bacteria becoming resistant to antibiotics.

Model Complete the graphic organizer below to identify ways that bacteria can be beneficial. LA.6.2.2.3

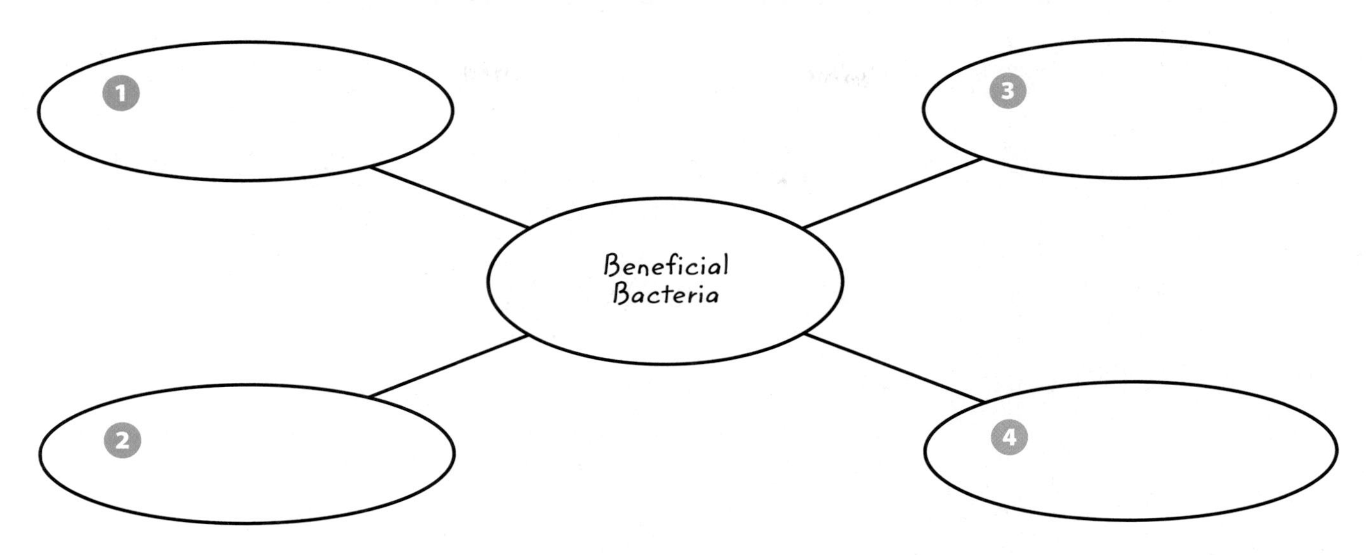

Sequence Complete the graphic organizer below to identify the development of antibiotic resistance.

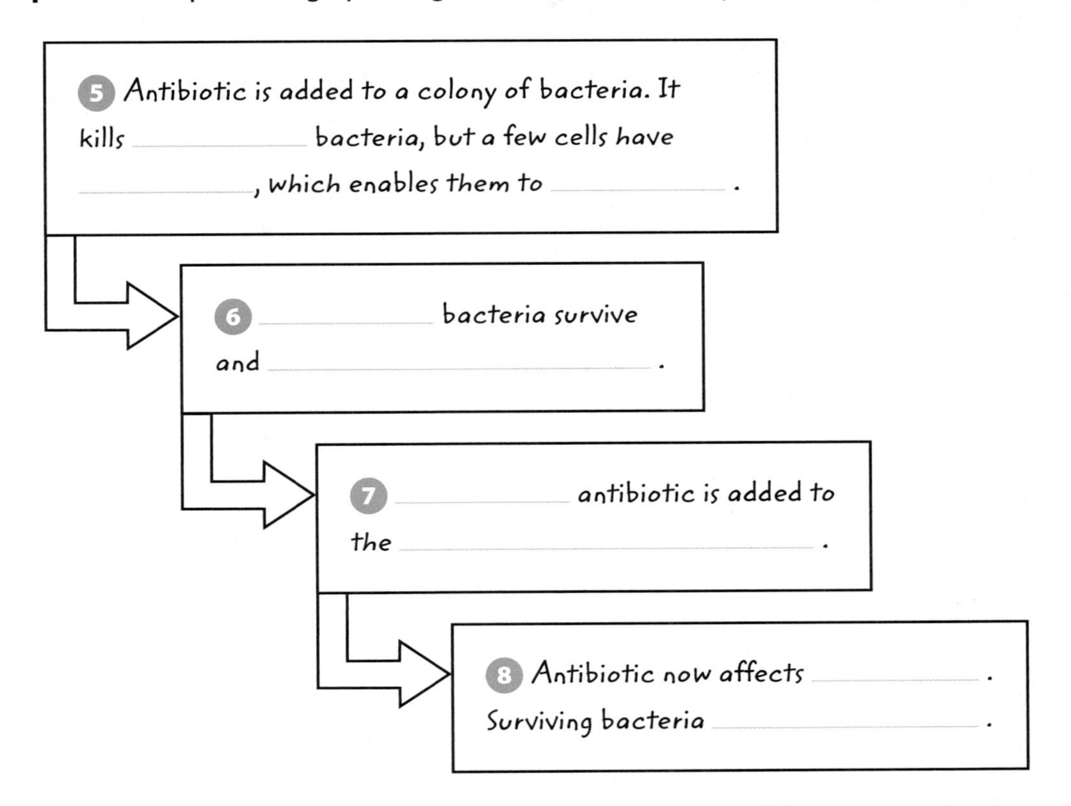

Lesson 3

What are VIRUSES?

ESSENTIAL QUESTIONS

 What are viruses?

 How do viruses affect human health?

Vocabulary

virus p. 509

antibody p. 513

vaccine p. 514

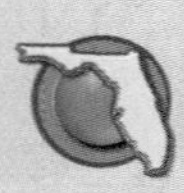

Florida NGSSS

LA.6.2.2.3 The student will organize information to show understanding (e.g., representing main ideas within text through charting, mapping, paraphrasing, summarizing, or comparing/contrasting);

SC.6.L.14.6 Compare and contrast types of infectious agents that may infect the human body, including viruses, bacteria, fungi, and parasites.

SC.6.N.1.1 Define a problem from the sixth grade curriculum, use appropriate reference materials to support scientific understanding, plan and carry out scientific investigation of various types, such as systematic observations or experiments, identify variables, collect and organize data, interpret data in charts, tables, and graphics, analyze information, make predictions, and defend conclusions.

SC.6.N.1.5 Recognize that science involves creativity, not just in designing experiments, but also in creating explanations that fit evidence.

SC.6.N.2.1 Distinguish science from other activities involving thought.

SC.6.N.3.4 Identify the role of models in the context of the sixth grade science benchmarks.

Inquiry Launch Lab

SC.6.N.1.1

10 minutes

How quickly do viruses replicate?

One characteristic that viruses share is the ability to produce many new viruses from just one virus. In this lab you can use grains of rice to model virus replication. Each grain of rice represents one virus.

Procedure  

1. Read and complete a lab safety form.
2. Estimate the number of **grains of rice** in the **fishbowl** and record this number for the first generation.
3. One student will add the contents of his or her **cup** to the fishbowl. Estimate how many viruses are now in the fishbowl and record your estimate for the second generation.
4. The rest of the class will add the contents of their cups to the fishbowl. Estimate the number of viruses and record that number of viruses for the third generation.

Data and Observations

Generation	First	Second	Third
Number of "viruses"			

Think About This

1. Recall that bacteria double every generation. How does the number of viruses produced in each generation compare with the number of bacteria produced in each generation?

2. Key Concept How could the rate at which viruses are produced affect human health?

Painted Flowers?

1. The streaking patterns on the petals of these tulips are not painted on but are caused by a virus. Tulips with these patterns are prized for their beautiful appearance. How do you think a virus could cause this flower's pattern? Do you think all viruses are harmful?

Characteristics of Viruses

Do chicken pox, mumps, measles, and polio sound familiar? You might have received shots to protect you from these diseases. You might have also received a shot to protect you from influenza, commonly known as the flu. What do these diseases have in common? They are caused by different viruses. *A* **virus** *is a strand of DNA or RNA surrounded by a layer of protein that can infect and replicate in a host cell.* If you have had a cold, you have been infected by a virus.

A virus does not have a cell wall, a nucleus, or any other organelles present in cells. The smallest viruses are between 20 and 100 times smaller than most bacteria. Recall that about 100 bacteria would fit across the head of a pin. Viruses can have different shapes, such as the crystal, cylinder, sphere, and bacteriophage (bak TIHR ee uh fayj) shapes shown in **Figure 12.**

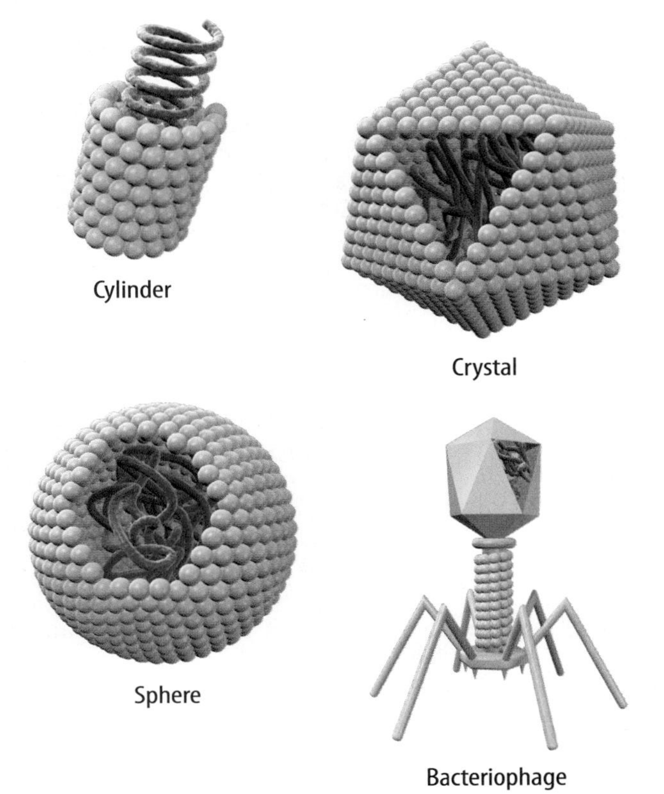

Figure 12 Viruses have a variety of shapes.

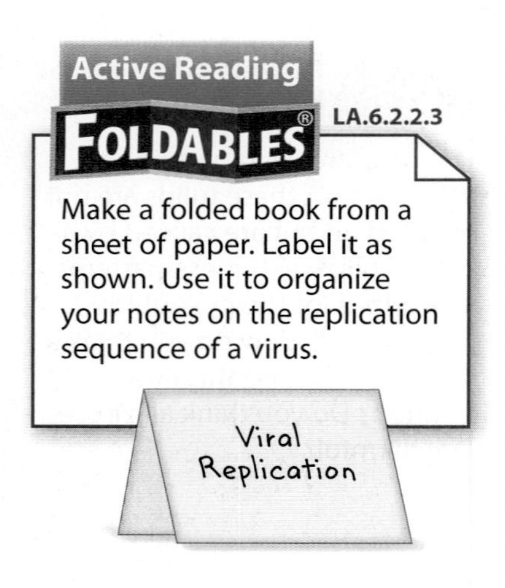

Active Reading **2. Infer** Are viruses alive? Explain why or why not.

Active Reading **3. Describe** What occurs when a virus becomes latent?

Dead or Alive?

Scientists do not consider viruses to be alive because they do not have all the characteristics of a living organism. Recall that living things are organized, respond to stimuli, use energy, grow, and reproduce. Viruses cannot do any of these things. A virus can make copies of itself in a process called replication, but it must rely on a living organism to do so.

Viruses and Organisms

Viruses must use organisms to carry on the processes that we usually associate with a living cell. Viruses have no organelles so they are not able to take in nutrients or use energy. They also cannot replicate without using the cellular parts of an organism. Viruses must be inside a cell to replicate. The living cell that a virus infects is called a host cell.

When a virus enters a cell, as shown in **Figure 13,** it can either be active or latent. Latent viruses go through an inactive stage. Their genetic material becomes part of the host cell's genetic material. For a period of time, the virus does not take over the cell to produce more viruses. In some cases, viruses have been known to be inactive for years and years. However, once it becomes active, a virus takes control of the host cell and replicates.

Figure 13 A virus infects a cell by inserting its DNA or RNA into the host cell. It then directs the host cell to make new viruses.

Active Reading **4. Model** Complete the chart below with the correct information.

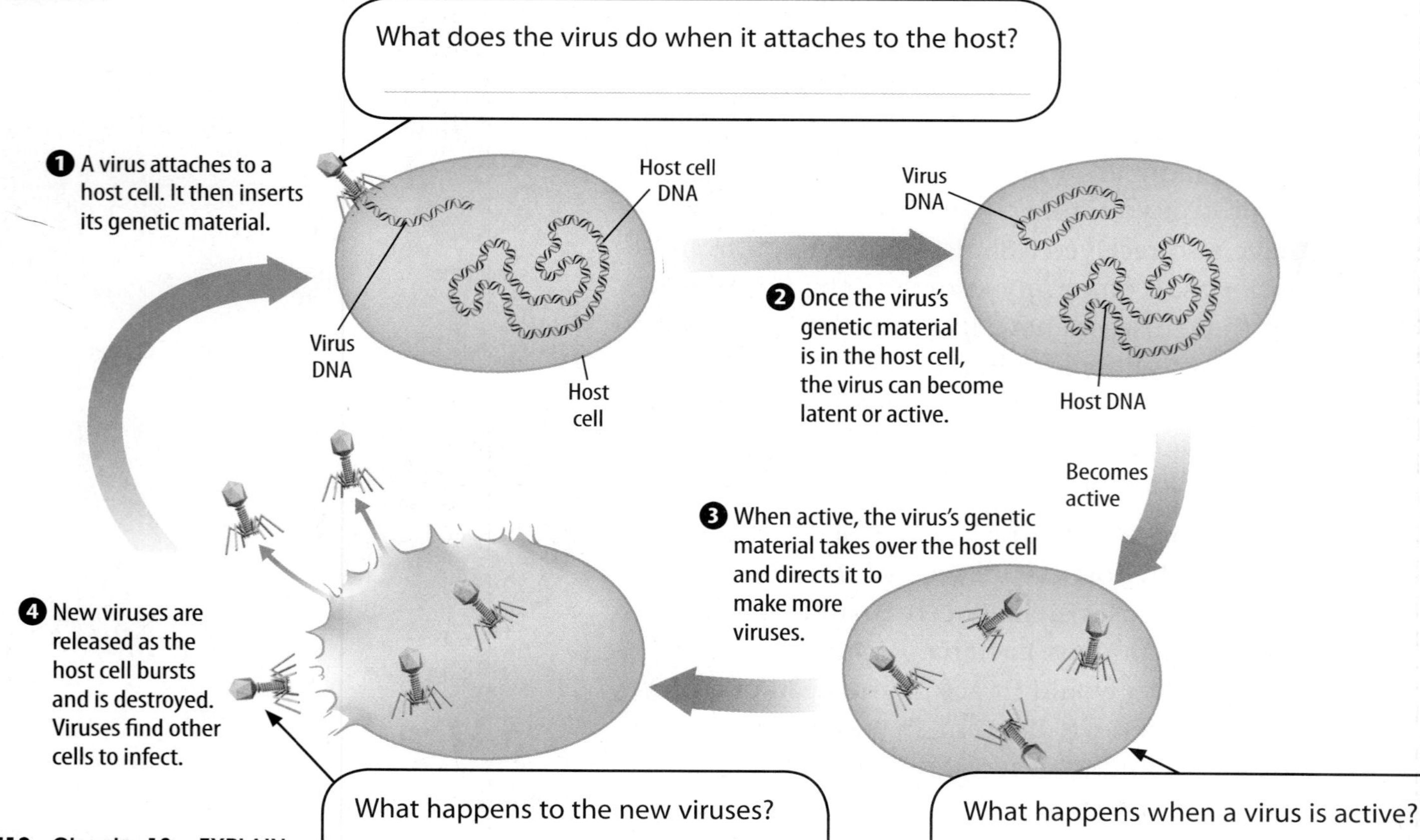

Replication

As you read earlier, a virus can make copies of itself in a process called replication, shown in **Figure 13.** A virus cannot infect every cell. A virus can only attach to a host cell with specific molecules on its cell wall or cell membrane. These molecules enable the virus to attach to the host cell. This is similar to the way that only certain electrical plugs can fit into an outlet on a wall. After a virus attaches to the host cell, its DNA or RNA enters the host cell. Once inside, the virus either starts to replicate or becomes latent, also shown in **Figure 13.** After a virus becomes active and replicates in a host cell, it destroys the host cell. Copies of the virus are then released into the host organism, where they can infect other cells.

Mutations

As viruses replicate, their DNA or RNA frequently mutates, or changes. These **mutations** enable viruses to adjust to changes in their host cells. For example, the molecules on the outside of host cells change over time to prevent viruses from attaching to the cell. As viruses mutate, they are able to produce new ways to attach to host cells. These changes happen so rapidly that it can be difficult to cure or prevent viral diseases before they mutate again.

REVIEW VOCABULARY

mutation
a change in genetic material

Active Reading **5. Describe** How does mutation enable viruses to continue causing disease?

Active Reading **6. Model** Complete the chart below with the correct information.

What happens first if the genetic material becomes latent?

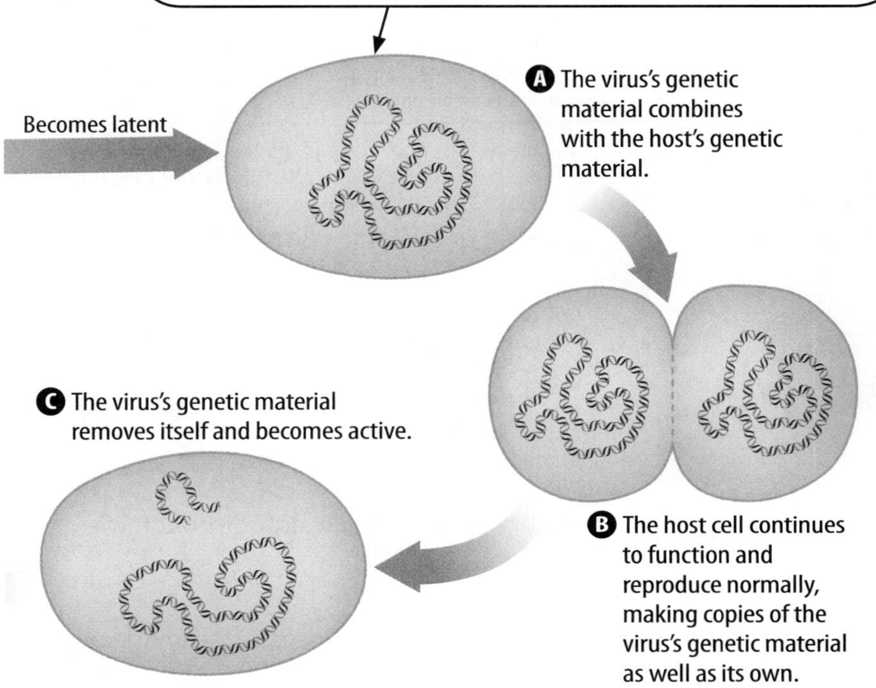

Becomes latent

A The virus's genetic material combines with the host's genetic material.

B The host cell continues to function and reproduce normally, making copies of the virus's genetic material as well as its own.

C The virus's genetic material removes itself and becomes active.

Viral Diseases

You might know that viruses cause many human diseases, such as chicken pox, influenza, some forms of pneumonia, and the common cold. But viruses also infect animals, causing diseases such as rabies and parvo. They can infect plants as well—in some cases causing millions of dollars of damage to crops. The tulips shown at the beginning of this lesson were infected with a virus that caused a streaked appearance on the petals. Most viruses attack and destroy specific cells. This destruction of cells causes the symptoms of the disease.

Some viruses cause symptoms soon after infection. Influenza viruses that cause the flu infect the cells lining your respiratory system, as shown in **Figure 14.** The viruses begin to replicate immediately. Flu symptoms, such as a runny nose and a scratchy throat, usually appear within two to three days.

Other viruses might not cause symptoms right away. These viruses are sometimes called latent viruses. Latent viruses continue replicating without damaging the host cell. HIV (human immunodeficiency virus) is one example of a latent virus that might not cause immediate symptoms.

HIV infects white blood cells, which are part of the immune system. Initially, infected cells can function normally, so an HIV-infected person might not appear sick. However, the virus can become active and destroy cells in the body's immune system, making it hard to fight other infections. It can often take a long time for symptoms to appear after infection. People infected with latent viruses might not know for many years that they have been infected.

Active Reading **7. Infer** Why is HIV considered a latent virus?

______

______

______

The Flu **Figure 14**

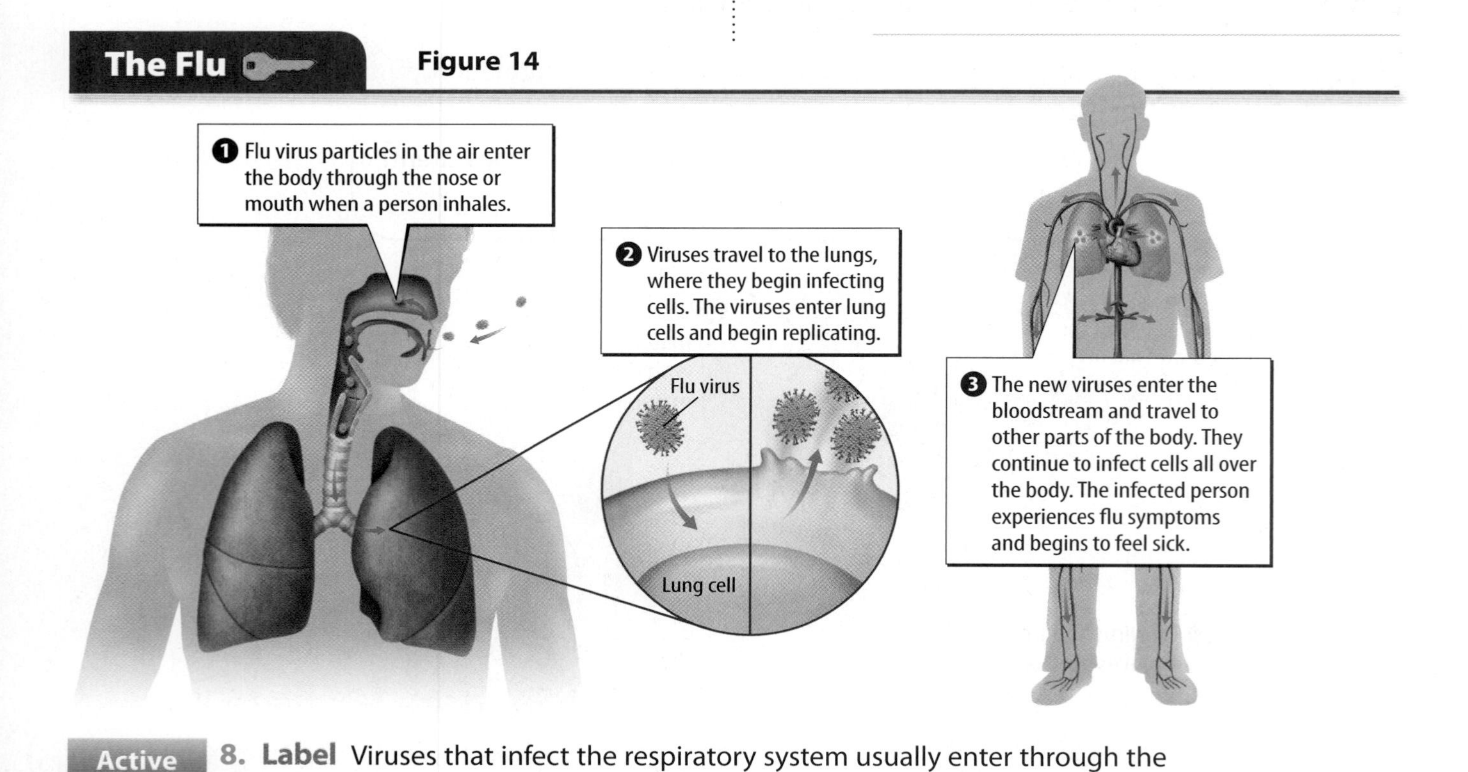

Active Reading **8. Label** Viruses that infect the respiratory system usually enter through the ______ or ______.

9. Visual Check Examine (Circle) the area where flu viruses replicate in the image above.

Treating and Preventing Viral Diseases

Since viruses are constantly changing, viral diseases can be difficult to treat. Antibiotics work only against bacteria, not viruses. Antiviral medicines can be used to treat certain viral diseases or prevent infection. These medicines prevent the virus from entering a cell or stop the virus from replicating. Antiviral medicines are specific to each virus. Like bacteria, viruses can rapidly change and become resistant to medicines.

Health officials use many methods to prevent the spread of viral diseases. One of the best ways to prevent a viral infection is to limit contact with an infected human or animal. The most important way to prevent infections is to practice good hygiene, such as washing your hands.

Immunity

Has anyone you know ever had chicken pox? Did the person get it more than once? Most people who became infected with chicken pox develop an immunity to the disease. This is an example of acquired **immunity**. When a virus infects a person, his or her body begins to make special proteins called antibodies. An **antibody** *is a protein that can attach to a pathogen and make it useless.* Antibodies bind to viruses and other pathogens and prevent them from attaching to a host cell, as shown in **Figure 15.** The antibodies also target viruses and signal the body to destroy them. These antibodies can multiply quickly if the same pathogen enters the body again, making it easier for the body to fight infection. Another type of immunity, called natural immunity, develops when a mother passes antibodies to her unborn baby.

Word Origin
immunity
from Latin *immunis,* means "exempt, free"

Antibodies

Figure 15 Antibodies bind to pathogens and prevent them from attaching to cells.

10. Visual Check **Interpret** How does the antibody prevent the virus from attaching to the host cell?

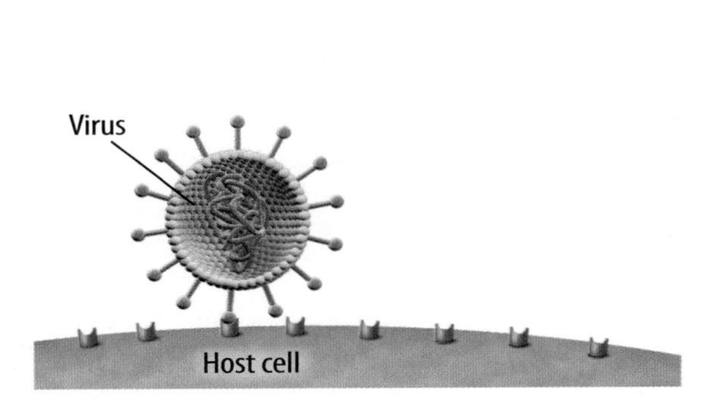

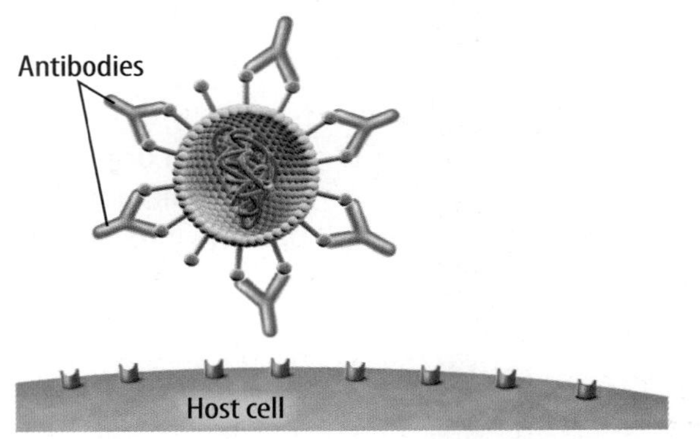

Vaccines

One way to prevent viral diseases is through vaccination. *A* **vaccine** *is a mixture containing material from one or more deactivated pathogens, such as viruses.* When an organism is given a vaccine for a viral disease, the vaccine triggers the production of antibodies. This is similar to what would happen if the organism became infected with the virus normally. However, because the vaccine contains deactivated pathogens, the organism suffers only mild symptoms or none at all. After being vaccinated against a particular pathogen, the organism will not get as sick if exposed to the pathogen again.

Vaccines can prevent diseases in animals as well as humans. For example, pet owners and farmers get annual rabies vaccinations for their animals. This protects the animals from the disease. Humans are then protected from rabies.

Research with Viruses

Scientists are researching new ways to treat and prevent viral diseases in humans, animals, and plants. Scientists are also studying the link between viruses and cancer. Viruses can cause changes in a host's DNA or RNA, resulting in the formation of tumors or abnormal growth. Because viruses can change very quickly, scientists must always be working on new ways to treat and prevent viral diseases.

You might think that all viruses are harmful. However, scientists have also found beneficial uses for viruses. Viruses may be used to treat genetic disorders and cancer using gene transfer. Scientists use viruses to insert normal genetic information into a specific cell. Scientists hope that gene transfer will eventually be able to treat genetic disorders that are caused by one gene, such as cystic fibrosis or hemophilia.

11. NGSSS Check **Analyze** How do viruses affect human health? **SC.6.L.14.6**

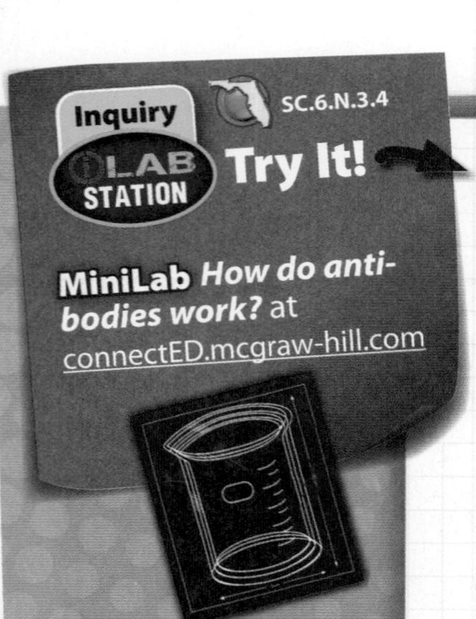

Apply It! After you complete the lab, answer these questions.

1. What is a vaccine?

2. What is a vaccine used for?

3. How is an organism's reaction to a vaccine different from the reaction that would occur if the organism became infected naturally?

Lesson Review 3

Visual Summary

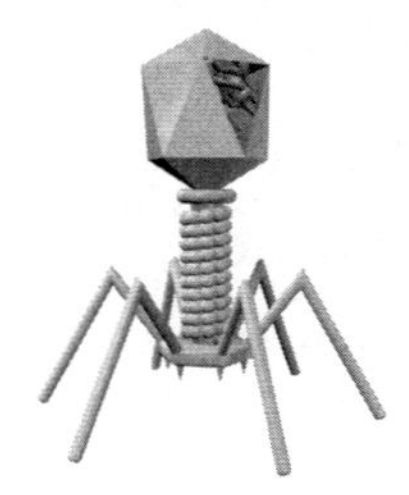

A virus is a strand of DNA or RNA surrounded by a layer of protein.

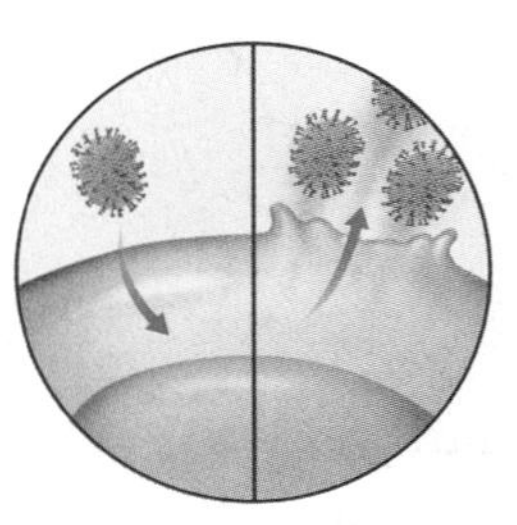

Viruses cause human diseases such as chicken pox and influenza.

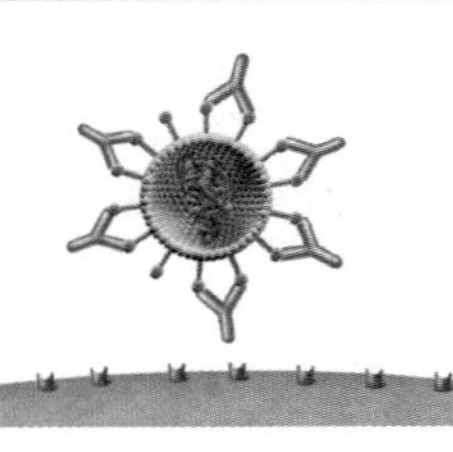

A person's body produces proteins called antibodies that prevent an infection by viruses.

Use Vocabulary

1. **List** the different virus shapes. SC.6.L.14.6

2. **Describe** in your own words how a vaccine works.

Understand Key Concepts

3. **Describe** the structure of a virus. SC.6.L.14.6

4. Which is made by the body to fight viruses?
   - (A) antibody
   - (B) bacteria
   - (C) bacteriophage
   - (D) proteins

6. **Compare** a vaccine and an antibody.

Interpret Graphics

6. **Label** the the steps that occur when a virus infects a cell. LA.6.2.2.3

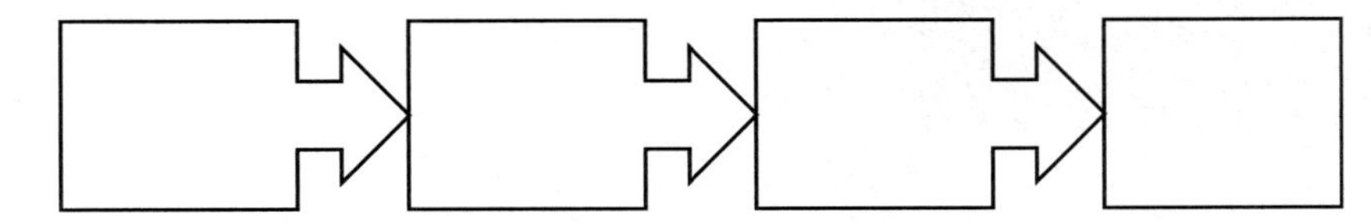

Critical Thinking

7. **Predict** the effect of preventing future mutations of the influenza virus. SC.6.L.14.6

Chapter 13 Study Guide

Think About It! Bacteria are unicellular prokaryotes, and viruses are small pieces of DNA or RNA surrounded by protein. Both bacteria and viruses may infect the human body; they can cause harmful diseases or can be useful.

Key Concepts Summary

LESSON 1 What are bacteria?

- **Bacteria** and archeans are unicellular organisms without nuclei. They have structures for movement, obtaining food, and reproduction.
- Bacteria exchange genetic information in a process called conjugation. They reproduce asexually by fission.

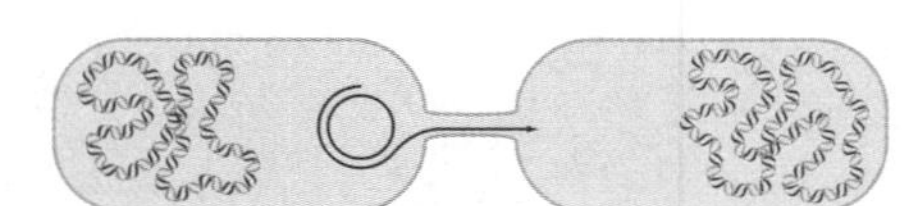

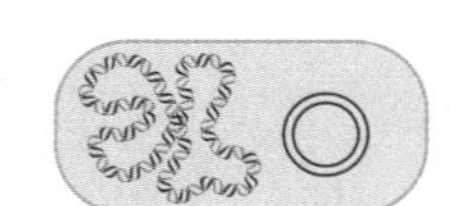

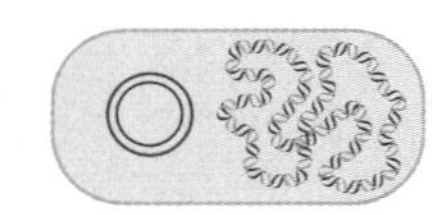

Vocabulary

bacterium p. 493
flagellum p. 496
fission p. 496
conjugation p. 496
endospore p. 497

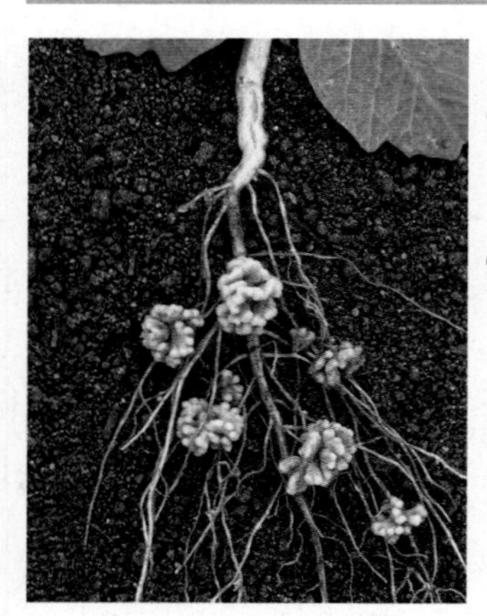

LESSON 2 Bacteria in Nature

- Bacteria decompose materials, play a role in the nitrogen cycle, clean the environment, and are used in food.
- Some bacteria cause disease, and others are used to treat it.

decomposition p. 502
nitrogen fixation p. 502
bioremediation p. 503
pathogen p. 504
antibiotic p. 504
pasteurization p. 505

LESSON 3 What are viruses?

- A **virus** is made up of DNA or RNA surrounded by a protein coat.
- Viruses can cause disease, can be made into vaccines, and are used in research.

virus p. 509
antibody p. 513
vaccine p. 514

Active Reading **FOLDABLES®** **Chapter Project**

Assemble your lesson Foldables as shown to make a Chapter Project. Use the project to review what you have learned in this chapter.

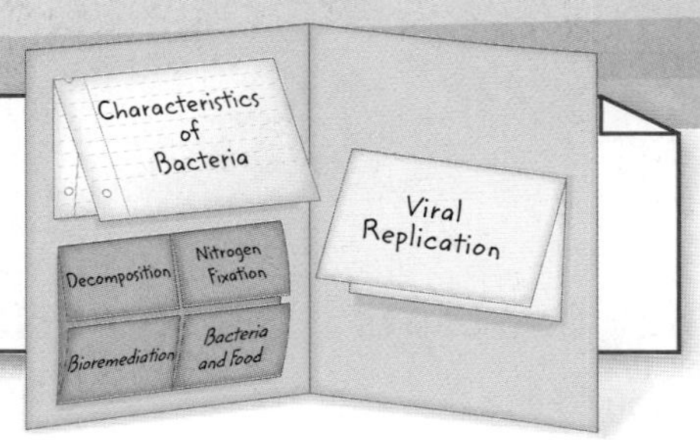

Use Vocabulary

1. Some bacteria have whiplike structures called ____________ that are used for movement.

2. Your body produces proteins called ____________ in response to infection by a virus.

3. Organisms that cause diseases are known as ____________.

4. The process of killing bacteria in a food product by heating it is called ____________.

5. Bacteria can form a(n) ____________ to survive when environmental conditions are severe.

6. A(n) ____________ is made by using pieces of deactivated viruses or dead pathogens.

Link Vocabulary and Key Concepts

Use vocabulary terms from the previous page and other terms from the chapter to complete the concept map.

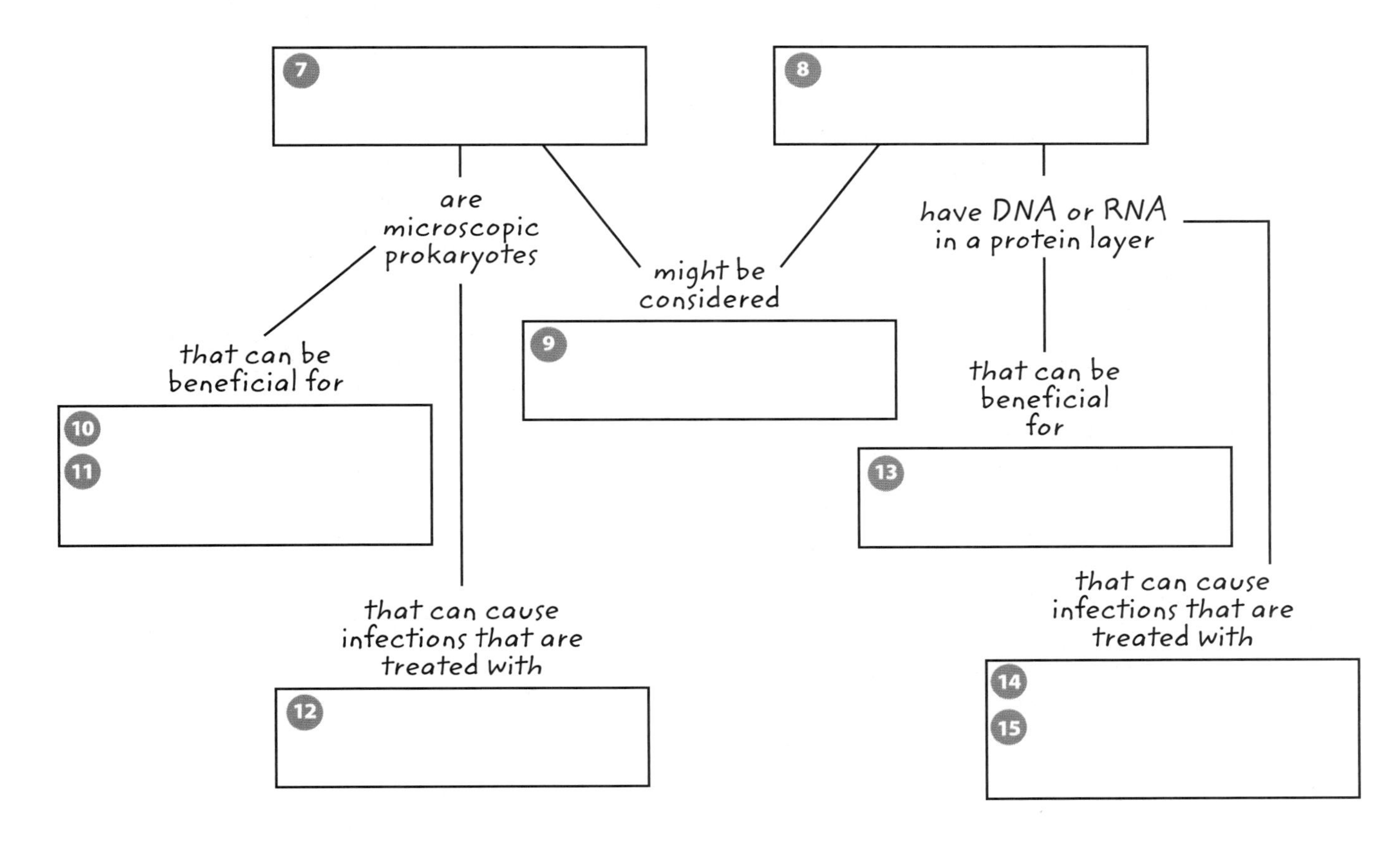

Chapter 13 Review

Understand Key Concepts

1. Which structure is NOT found in a bacterium? SC.6.L.14.4
   - (A) chromosome
   - (B) cytoplasm
   - (C) nucleus
   - (D) ribosome

2. Which structure helps a bacterium move? SC.6.L.14.4
   - (A) capsule
   - (B) endospore
   - (C) flagellum
   - (D) plasmid

3. What process is occurring in the illustration below? SC.6.L.14.6

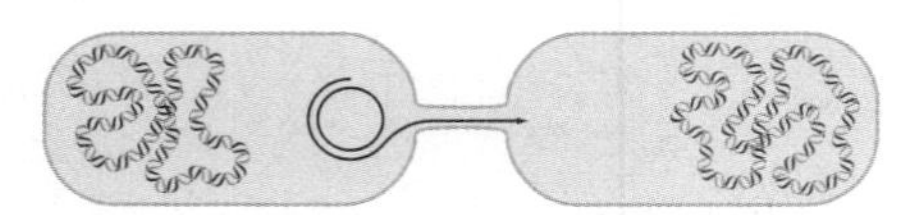

   - (A) budding
   - (B) conjugation
   - (C) fission
   - (D) replication

4. Which term describes how bacteria can be used to clean up environmental waste? SC.6.L.14.6
   - (A) bioremediation
   - (B) decomposition
   - (C) pasteurization
   - (D) nitrogen fixation

5. Which statement correctly describes pathogens? SC.6.L.14.6
   - (A) They are always bacteria.
   - (B) They are in your body only when you are sick.
   - (C) They break down dead organisms.
   - (D) They cause disease.

6. Which statement correctly describes antibiotics? SC.6.L.14.6
   - (A) They can kill any kind of bacterium.
   - (B) They help bacteria grow.
   - (C) They stop the growth and reproduction of bacteria.
   - (D) They treat all diseases.

Critical Thinking

7. **Compare and contrast** bacteria and archaea. LA.6.2.2.3

8. **Evaluate** the importance of bacterial conjugation. SC.6.L.14.6

9. **Model** the life of a bacterium that performs nitrogen fixation in the soil. SC.6.N.3.4

10. **Contrast** asexual reproduction in bacteria and replication in viruses. What are some advantages and disadvantages of each? SC.6.L.14.6

11. **Organize** the effects of bacteria on health in the table below. SC.6.L.14.6

Harmful Effects	Beneficial Effects

12. **Analyze** the importance of vaccines in preventing large outbreaks of influenza. SC.6.L.14.6

13 **Explain** what is happening in the petri dish shown below. How does this process eventually create new strains of bacteria that are resistant to antibiotics? SC.6.L.14.6

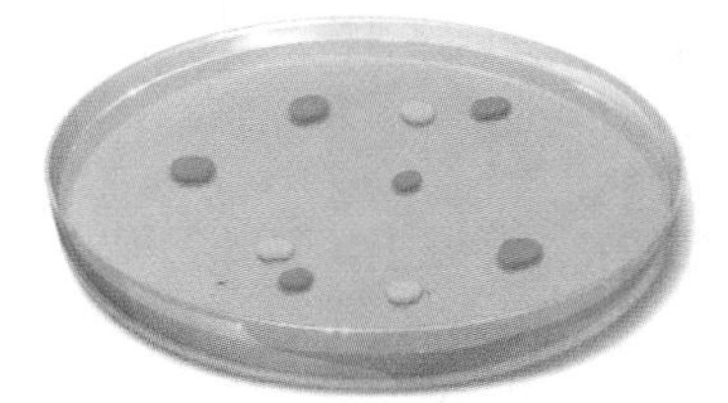

Writing in Science

14 **Summarize** On a separate piece of paper, write an argument that you could use to encourage all the families in your neighborhood to make sure their pets are vaccinated against rabies. SC.6.L.14.6

Big Idea Review

15 What are bacteria and viruses and why are they important? Include examples of how they are both beneficial and harmful to humans. SC.6.L.14.6

16 Describe what is happening in the photo below. Explain what is happening to both the bacterium and the virus. SC.6.L.14.6

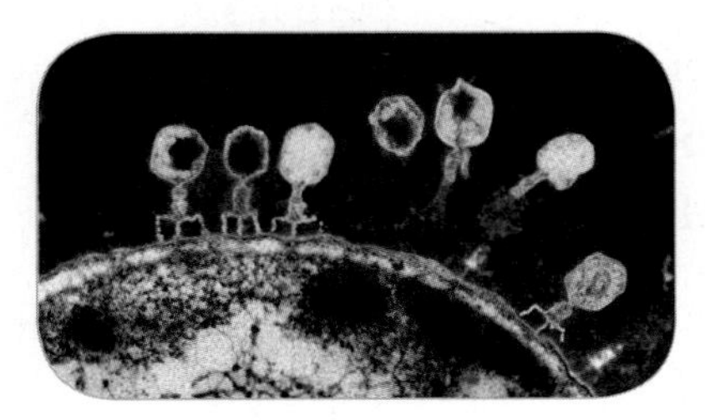

Math Skills

MA.6.A.3.6

Use a Formula

17 How many bacteria would there be if 100 bacteria underwent fission 8 times?

18 If each fission cycle takes 20 minutes, how many cycles would it take for 100 bacteria to divide into 100,000?

19 A strain of bacteria takes 30 minutes to undergo fission. Starting with 500 bacteria, how many would there be after 4 hours?

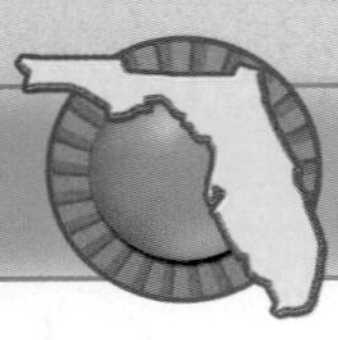

Florida NGSSS Benchmark Practice

Record your answers on the answer sheet provided by your teacher or on a sheet of paper.

Multiple Choice

1 Which is NOT a characteristic of bacteria? SC.6.L.14.6

Ⓐ They are microscopic.

Ⓑ They are unicellular.

Ⓒ They can live in many environments.

Ⓓ They have a membrane-bound nucleus.

2 Which is a characteristic of viruses? SC.6.L.14.6

Ⓕ unicellular prokaryotes

Ⓖ small pieces of DNA

Ⓗ no nucleus

Ⓘ spiral-shaped

3 Which disease is caused by bacteria? SC.6.L.14.6

Ⓐ chicken pox

Ⓑ influenza

Ⓒ tuberculosis

Ⓓ common cold

4 A chemical that harms only prokaryotic cells would affect which of the following? SC.6.L.14.6

Ⓕ viruses

Ⓖ plants

Ⓗ animals

Ⓘ bacteria

5 Bacteria and viruses have a variety of shapes. Which shape represents a virus? SC.6.L.14.6

Ⓐ bacteriophage

Ⓑ rod-shaped

Ⓒ spiral-shaped

Ⓓ round

6 How do bacteria reproduce? SC.6.L.14.6

Ⓕ conjugation

Ⓖ fission

Ⓗ mutation

Ⓘ pasteurization

Use the diagram below to answer question 7.

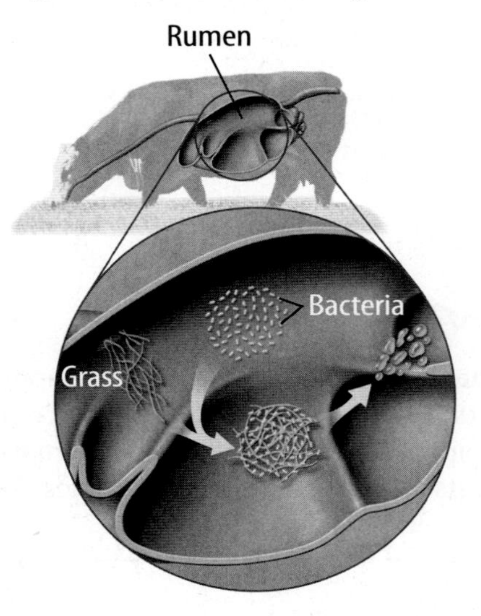

7 What role do bacteria play in the process shown above? SC.6.L.14.6

Ⓐ They break down cellulose.

Ⓑ They convert nitrogen in grass.

Ⓒ They prevent viruses from growing.

Ⓓ They remove harmful pollutants.

8 How does a virus infect the rest of a body once it replicates? SC.6.L.14.6

Ⓕ by attaching to the host cell

Ⓖ by entering the bloodstream

Ⓗ by becoming a latent virus

Ⓘ by disguising itself as a bacterium

Use the diagram below to answer question 9.

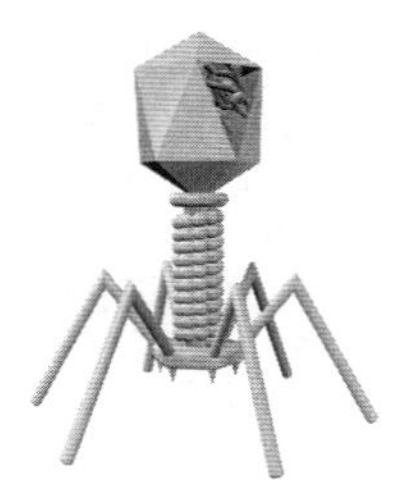

9 What is pictured in the diagram above? SC.6.L.14.6

Ⓐ an antibody

Ⓑ a bacteriophage

Ⓒ a bacterium

Ⓓ a plasmid

10 How do viruses reproduce? SC.6.L.14.6

Ⓕ conjugation

Ⓖ fission

Ⓗ mutation

Ⓘ replication

11 Which disease is NOT caused by a virus? SC.6.L.14.6

Ⓐ mumps

Ⓑ measles

Ⓒ pneumonia

Ⓓ polio

12 Which can be used to treat most bacterial diseases in humans? SC.6.L.14.6

Ⓕ antibiotics

Ⓖ bioremediation

Ⓗ pasteurization

Ⓘ pathogens

13 Which is NOT a reason viral diseases are difficult to cure or prevent? SC.6.L.14.6

Ⓐ Viruses often mutate every time they replicate.

Ⓑ Latent viruses cause immediate symptoms.

Ⓒ Antibiotics only work on bacteria.

Ⓓ Limiting contact between humans is very hard.

14 Which is a mixture that contains material from one or more deactivated pathogens, such as viruses? SC.6.L.14.6

Ⓕ antibiotic

Ⓖ endospore

Ⓗ flagellum

Ⓘ vaccine

15 What is the most important way to prevent infections? SC.6.L.14.6

Ⓐ Don't wash to build up immunity.

Ⓑ Practice good hygiene.

Ⓒ Take antibiotics even when healthy.

Ⓓ Skip all vaccination shots.

NEED EXTRA HELP?

If You Missed Question . . .	1	2	3	4	5	6	7	8	9	10	11	12	13	14	15
Go to Lesson . . .	1	3	2	1	3	1	2	3	3	3	3	2	3	3	3

Name ______________________________ Date ______________

Multiple Choice *Bubble the correct answer.*

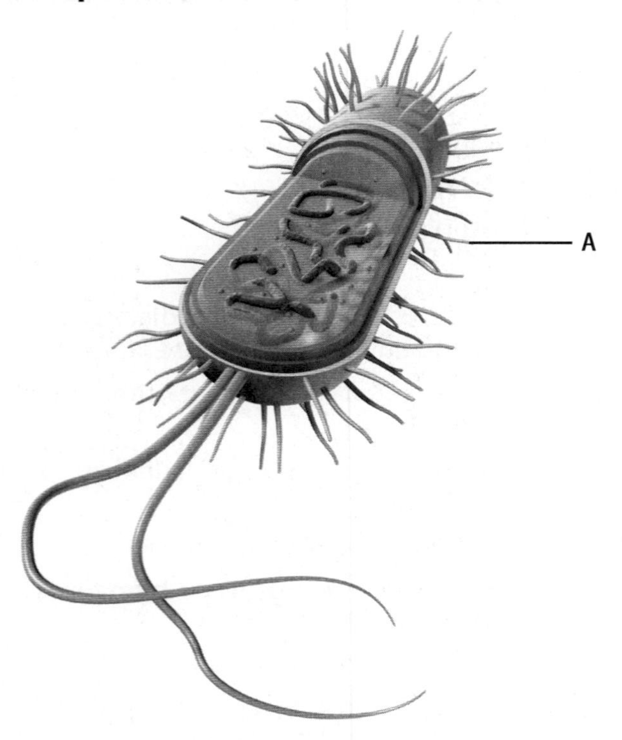

1. The diagram above shows a bacterial cell. What does structure A do? **SC.6.L.14.6**

Ⓐ It controls all processes of the bacterial cell.

Ⓑ It helps the bacterium stick to surfaces.

Ⓒ It helps the bacterial cell move.

Ⓓ It keeps the bacterium from drying out.

2 Which type of bacterium needs oxygen to live? **SC.6.L.14.6**

Ⓕ aerobic

Ⓖ anaerobic

Ⓗ archaea

Ⓘ endospore

3. Which of the following is NOT a characteristic of an archaean cell? **SC.6.L.14.6**

Ⓐ It has a cell wall.

Ⓑ It has a nucleus.

Ⓒ It has a circular strand of DNA.

Ⓓ It has an extreme environment.

4. *Lactobacillus* is a beneficial bacterium that is rod-shaped. Which image below could be *Lactobacillus*? **SC.6.L.14.6**

Ⓕ 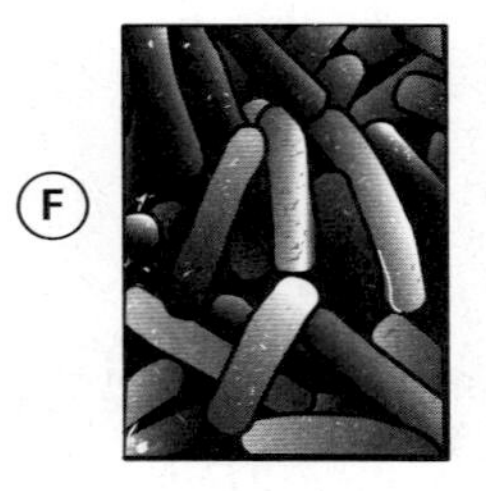

Ⓖ 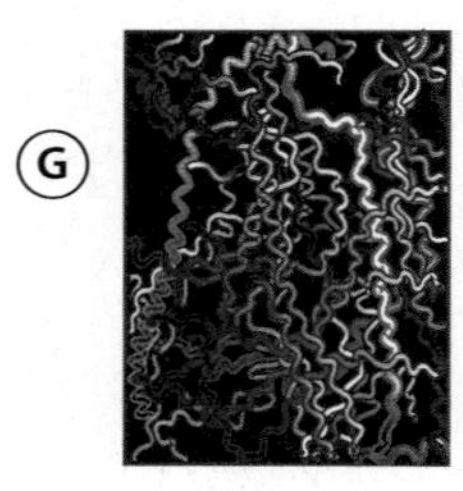

Ⓗ 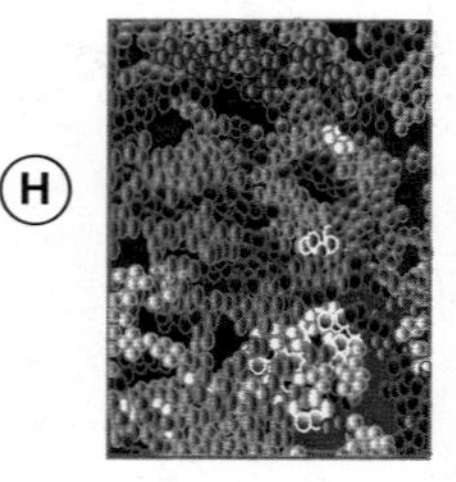

Ⓘ 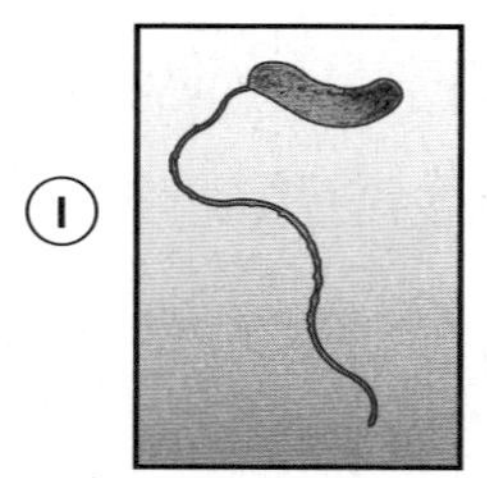

Name ______________________________ Date ______________

Multiple Choice *Bubble the correct answer.*

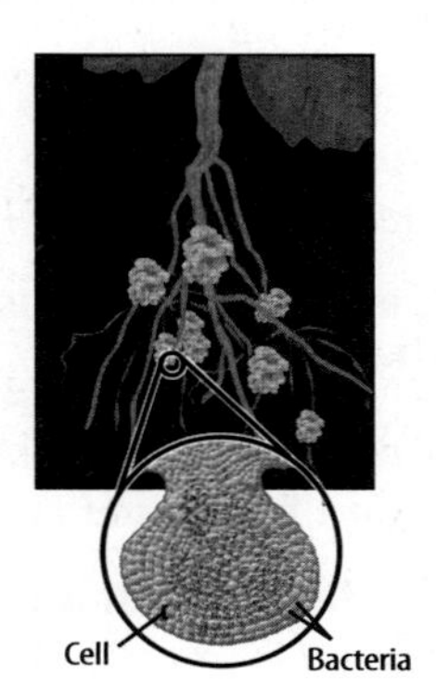

1. Examine the image above. What is the purpose of the structures in the image? **SC.6.L.14.6**

Ⓐ to break down organic matter in the soil

Ⓑ to convert nitrogen in air into a form that plants can use

Ⓒ to fight pathogens that could invade the roots

Ⓓ to steal nutrients away from the host plant

2. Leaves litter the ground in a maple forest. Bacteria break down these leaves through the process of **SC.6.L.14.6**

Ⓕ bioremediation.

Ⓖ decomposition.

Ⓗ bacterial resistance.

Ⓘ nitrogen fixation.

3. Which process is used to make dairy products safe to eat? **SC.6.L.14.6**

Ⓐ bioremediation

Ⓑ decomposition

Ⓒ pasteurization

Ⓓ pathogens

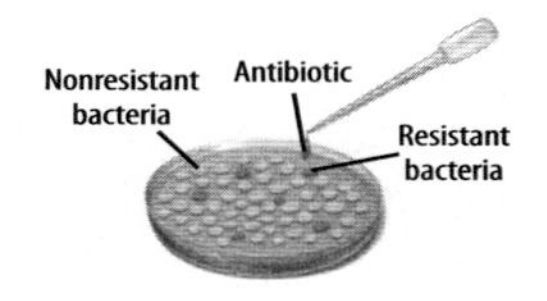

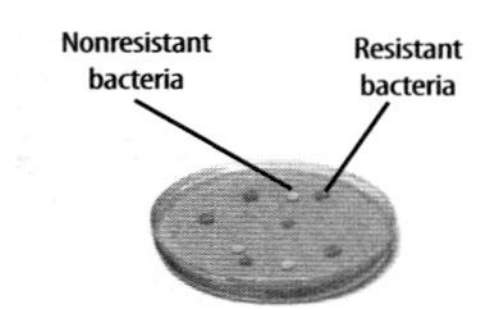

4. What is happening above in the image on the right? **SC.6.L.14.6**

Ⓕ The antibiotic has no effect on the nonresistant bacteria.

Ⓖ The antibiotic kills most of the nonresistant bacteria.

Ⓗ The antibiotic kills most of the resistant bacteria.

Ⓘ The antibiotic kills the nonresistant and resistant bacteria.

Name ______________________ Date ______________

Multiple Choice *Bubble the correct answer.*

1. Which image shows the structure of a bacteriophage? **SC.6.L.14.6**

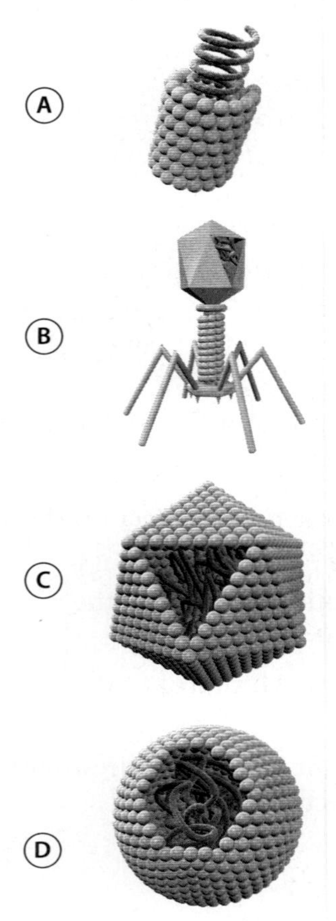

Ⓐ

Ⓑ

Ⓒ

Ⓓ

2. What happens within a dog's body when the dog is vaccinated for rabies? **SC.6.L.14.6**

Ⓕ Any rabies virus cells that are in the dog are killed.

Ⓖ Any rabies virus cells that are in the dog become latent.

Ⓗ Immune cells in the dog mutate and attack the virus in the vaccine.

Ⓘ The immune system of the dog produces antibodies to the rabies virus.

3. Which statement is true about viruses? **SC.6.L.14.6**

Ⓐ A virus can replicate on its own.

Ⓑ A virus is a living organism that responds to stimuli.

Ⓒ A virus is a strand of RNA or DNA surrounded by a layer of protein.

Ⓓ A virus is larger than a bacterium and is always the same shape.

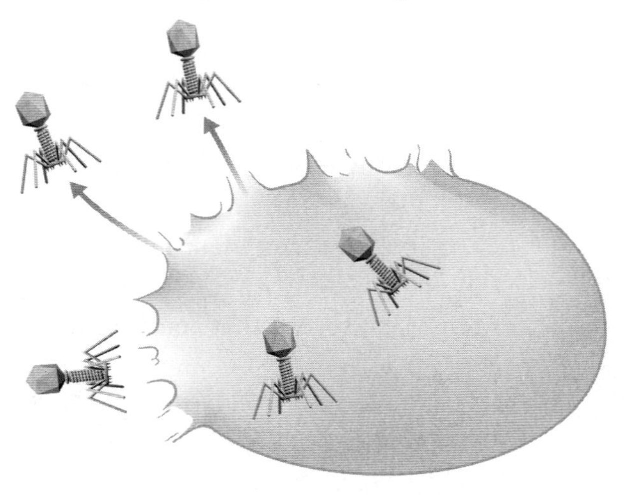

4. The image above shows one step in the process of viral replication. Which statement describes this step? **SC.6.L.14.6**

Ⓕ Latent virus DNA combines with the DNA of the cell.

Ⓖ New viruses are released into the bloodstream.

Ⓗ The virus's DNA directs the cell to make new viruses.

Ⓘ A virus can insert its DNA into a host cell.

Name ______ Date ______

Notes

Name ______________________ Date ______________

Notes

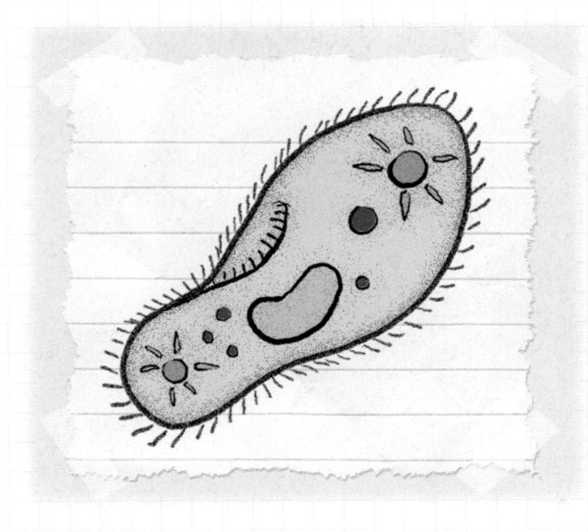

What are protists?

Protists are a diverse classification of organisms. When the students in Mrs. Applebee's science class were asked "What are protists?" they had different ideas. Here are some of their ideas:

Anna: I think they share most of their characteristics with plants.

Jordan: I think they share most of their characteristics with animals.

Ken: I think they share most of their characteristics with fungi.

Alysha: I think they share most of their characteristics with bacteria.

Joaquin: I think they share characteristics with animals, plants, and fungi.

LaVon: I don't think they share characteristics with any of the organisms you mentioned. They have their own unique characteristics.

Which student do you agree with the most? Explain why you agree.

Protists and FUNGI

The Big Idea

FLORIDA BIG IDEAS

1 **The Practice of Science**

2 **The Characteristics of Scientific Knowledge**

3 **The Role of Theories, Laws, Hypotheses, and Models**

14 **Organization and Development of Living Organisms**

Think About It! What are protists and fungi, and how do they affect an environment?

These organisms are neither plant nor animal. Protists and fungi are two groups of living things that have characteristics similar to those of plants or animals.

1. How is the organism pictured similar to a plant? An animal?

2. How might this organism benefit its environment?

Get Ready to Read What do you think about protists and fungi?

Before you read, decide if you agree or disagree with each of these statements. As you read this chapter, see if you change your mind about any of the statements.

	AGREE	DISAGREE
1 Protists are grouped together because they all look similar.	☐	☐
2 Some protists cause harm to other organisms.	☐	☐
3 Many protists make their own food.	☐	☐
4 Mushrooms and yeasts are two types of fungi.	☐	☐
5 Fungi are always helpful to plants.	☐	☐
6 Some fungi can be made into foods or medicines.	☐	☐

There's More Online!
Video • Audio • Review • ⓘLab Station • WebQuest • Assessment • Concepts in Motion • Multilingual eGlossary

Lesson 1

What are PROTISTS?

ESSENTIAL QUESTIONS

What are the different types of protists and how do they compare?

How are protists beneficial?

Vocabulary

protist p. 531
algae p. 532
diatom p. 533
protozoan p. 536
cilia p. 536
paramecium p. 536
amoeba p. 537
pseudopod p. 537

SC.6.N.1.5, SC.6.N.2.1

Launch Lab

10 minutes

How does a protist react to its environment?

Like other organisms, protists can react to their environment in many ways. One type of protist called *Euglena* has specialized structures to move, perform photosynthesis, and react to light.

Procedure

1. Read and complete a lab safety form.
2. Place a **Petri dish** containing a ***Euglena* culture** on a piece of white **paper.** Using a **hand lens,** observe the *Euglena.*
3. Carefully cut a hole the size of a dime in a piece of **aluminum foil.** Place the foil on top of the dish so that the hole is centered over the top. Shine the light from a **desk lamp** at the hole.
4. At the end of class, remove the foil and observe the *Euglena* again.

Think About This

1. Where were the *Euglena* in the dish at the beginning of class? At the end?

2. Why do you think this behavior is beneficial to *Euglena?*

3. Key Concept What structures do you think help *Euglena* react to its environment?

Florida NGSSS

LA.6.2.2.3 The student will organize information to show understanding (e.g., representing main ideas within text through charting, mapping, paraphrasing, summarizing, or comparing/contrasting);

SC.6.L.14.6 Compare and contrast types of infectious agents that may infect the human body, including viruses, bacteria, fungi, and parasites.

SC.6.N.1.5 Recognize that science involves creativity, not just in designing experiments, but also in creating explanations that fit evidence.

SC.6.N.2.1 Distinguish science from other activities involving thought.

SC.6.N.3.4 Identify the role of models in the context of the sixth grade science benchmarks.

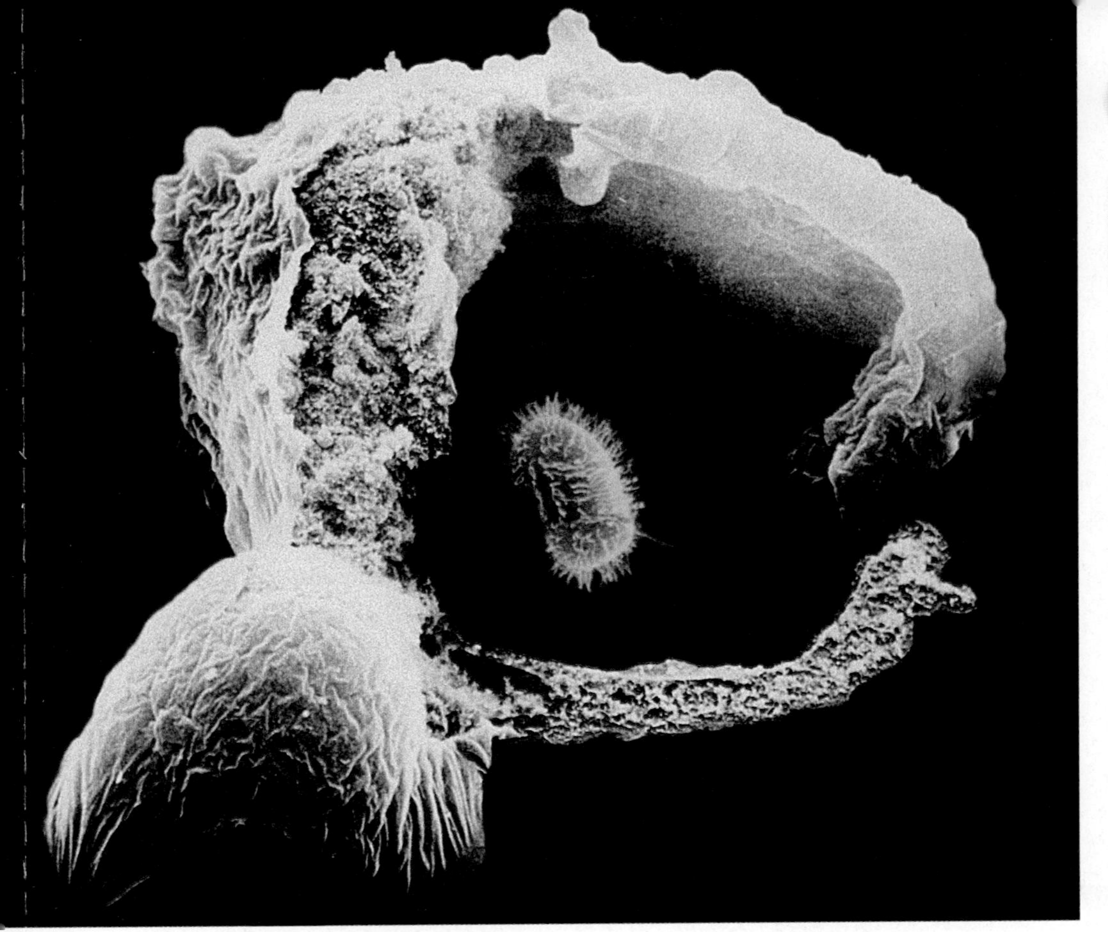

Inquiry Grabbing a Snack?

1. The protist group includes diverse organisms. What do you think the larger organism is doing in the photo? How is this organism similar to an animal?

What are protists?

When you see a living thing, one of the first questions you might have is whether it is a plant or an animal. You might recognize a dog as an animal because of its fur. You might know a flower is a plant because of its leaves. Besides appearance, organisms can also be classified by structures in their cells. A plant cell has a cell wall made of cellulose and a membrane made of flexible fats. A plant cell often contains chloroplasts, organelles that carry out photosynthesis. An animal cell also has a membrane made of flexible fats but does not contain chloroplasts or have a cell wall. These characteristics make it easy to identify both types of cells. However, some organisms, such as the protist shown in **Figure 1,** cannot be classified as easily.

A **protist** *is a member of a group of eukaryotic organisms, which have a membrane-bound nucleus.* Protists share some characteristics with plants, animals, or organisms known as fungi. However, they are not classified as any of these groups. Although protists are classified together, they are diverse and have different adaptations for movement and finding food.

Active Reading **2. Explain** What is a protist?

Figure 1 Many photosynthetic algae look like plants.

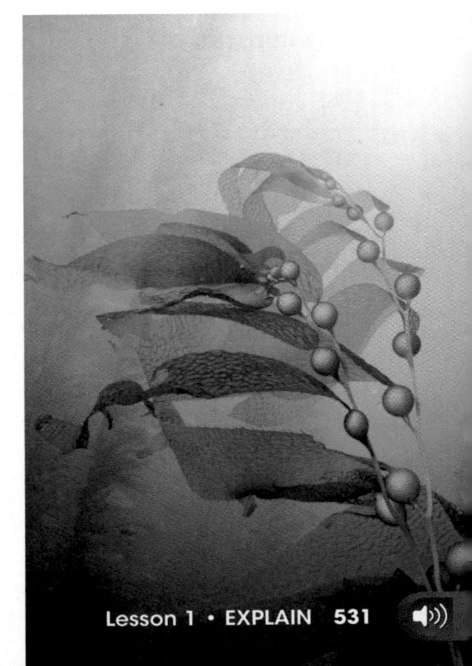

Reproduction of Protists

Most protists reproduce asexually. What does the offspring of **asexual reproduction** look like? It is an exact copy of the parent. Asexual reproduction can create new organisms quickly. However, many protists can also reproduce sexually. Offspring of sexual reproduction are genetically different from the parents. Sexual reproduction takes more time, but it creates new organisms with a variety of characteristics.

REVIEW VOCABULARY

asexual reproduction
a type of reproduction in which one parent reproduces without a sperm and an egg joining

Classification of Protists

Scientists usually classify organisms according to their similarities. However, protists are a unique and diverse classification of organisms. Typically, a protist is any eukaryote that cannot be classified as a plant, an animal, or a fungus. However, protists might look and act very much like these other types of organisms. Scientists classify protists as plantlike, animal-like, or funguslike based on which group they most resemble, as shown in **Table 1.**

Active Reading **3. Locate** Underline the three different types of protists.

Table 1 Protists Classified into One of Three Groups

Classification	Plantlike	Animal-like	Funguslike
Example	algae	paramecium	slime mold
Characteristics	• make their own food • unicellular or multicellular	• eat other organisms for food • mostly microscopic and unicellular	• break down organic matter for food • mostly multicellular

Active Reading **4. Interpret** List two types of protists that can be multicellular.

Plantlike Protists

You might have seen brown, green, or red seaweed at the beach or in an aquarium. These seaweeds are algae (AL jee; singular, alga), one type of plantlike protist. Why might they be classified as plantlike? **Algae** *are plantlike protists that produce food through photosynthesis using light energy and carbon dioxide.* Most plantlike protists, however, are much smaller than the multicellular algae shown in **Table 1.** You can't see most algae without a microscope.

Diatoms

A type of microscopic plantlike protist with a hard outer wall is a **diatom** (DI uh tahm). Diatoms are so common that if you filled a cup with water from the surface of any lake or pond, you would probably collect thousands of them. Look at the unicellular diatoms shown at the top of **Figure 2.** A diatom can resemble colored glass. In fact, the cell walls of diatoms contain a large amount of silica, the main mineral in glass.

Dinoflagellates

Can you guess how the protist in the middle of **Figure 2** moves? This organism is a dinoflagellate (di noh FLA juh lat), a unicellular plantlike protist that has flagella—whiplike parts that enable the protist to move. The flagella beat back and forth, enabling the dinoflagellate to spin and turn. Some of these protists glow in the dark because of a chemical reaction that occurs when they are disturbed.

Active Reading **5. Explain** What purpose do flagella serve?

Euglenoids

Another type of plantlike protist also uses flagella to move but has a unique structure covering its body. A euglenoid (yew GLEE noyd), shown at the bottom of **Figure 2,** is a unicellular plantlike protist with a flagellum at one end of its body. Instead of a cell wall, euglenoids have a rigid, rubbery cell coat called a pellicle (PEL ih kul). Euglenoids have eyespots that detect light and determine where to move. Euglenoids swim quickly and can creep along the surface of water when it is too shallow to swim. These protists have chloroplasts and make their own food. If there is not enough light for making food, they can absorb nutrients from decaying matter in the water. Animals such as tadpoles and small fish eat euglenoids.

Figure 2 All of these microscopic organisms are protists. The cell walls of diatoms contain silica. The dinoflagellate has two flagella that cause it to spin. The euglenoid has a flagellum and a rigid cell coat.

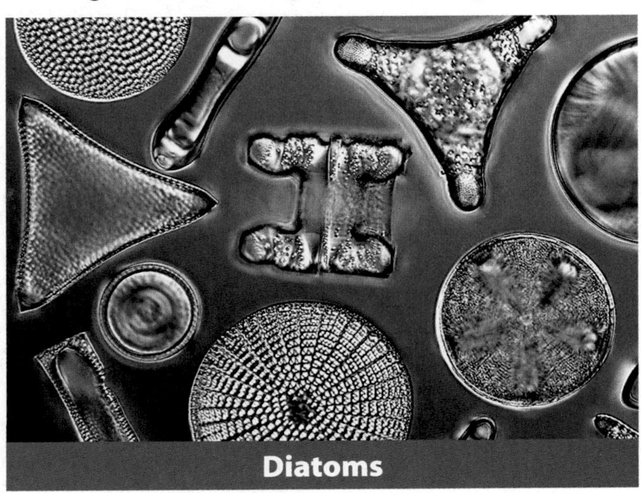

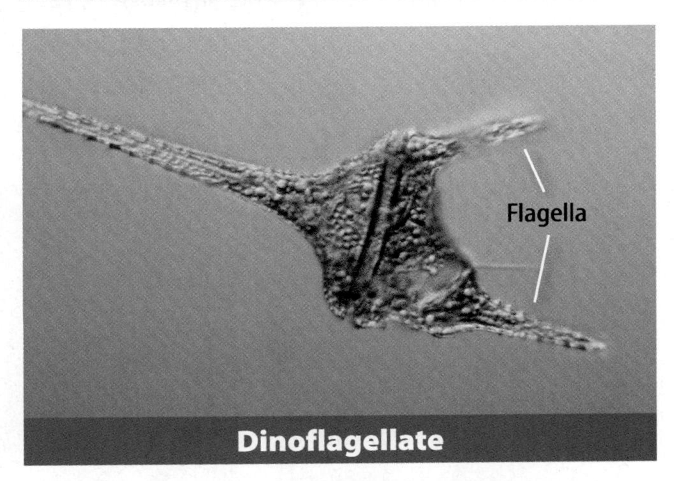

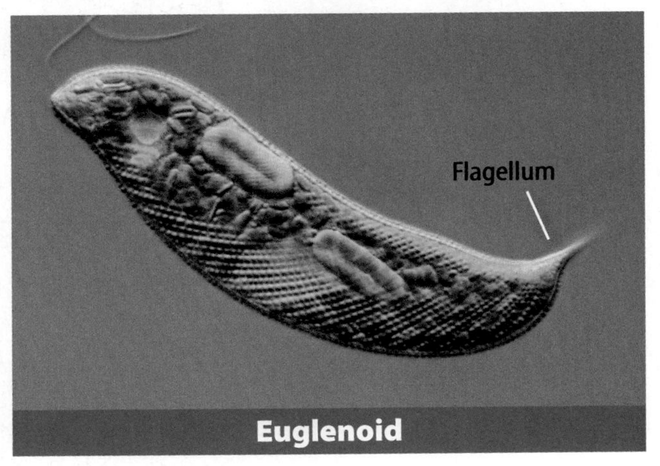

Active Reading **6. Compare** What characteristics do plantlike protists share with plants?

Active Reading **7. Describe** How do red and brown multicellular algae differ from plants?

____

____

____

Algae

Recall that algae are photosynthetic plantlike protists. Some algae are big and multicellular, like the seaweeds in **Figure 3.** Other algae are unicellular and can be seen only with a microscope. Algae are classified as red, green, or brown, depending on the pigments they contain.

Some types of red and brown algae appear similar to plants. Unlike plants, these algae do not have a complex organ system for transporting water and nutrients. Instead of roots, they have holdfasts, structures that secrete a chemical-like glue that fastens them to the rocks.

One unusual green alga is volvox. In **Figure 3** you can see that many volvox cells come together to form a larger sphere. These cells move together as one group and beat their flagella in unison. Some cells produce parts necessary for sexual reproduction. The volvox cells in the front of the group have larger eyespots that sense light for photosynthesis. Do you think volvox should be considered unicellular or multicellular?

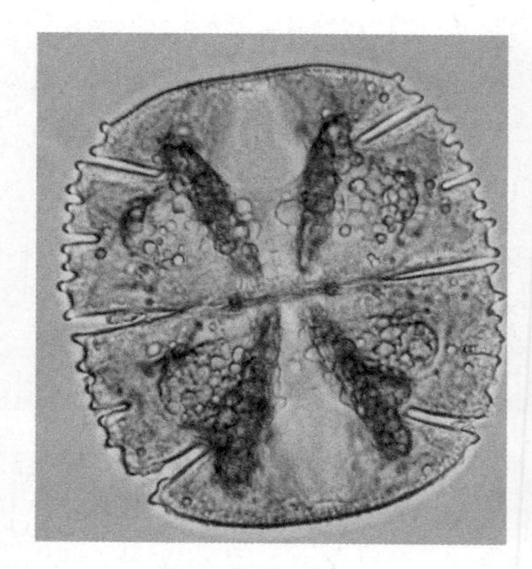

Unicellular Algae

Multicellular Algae

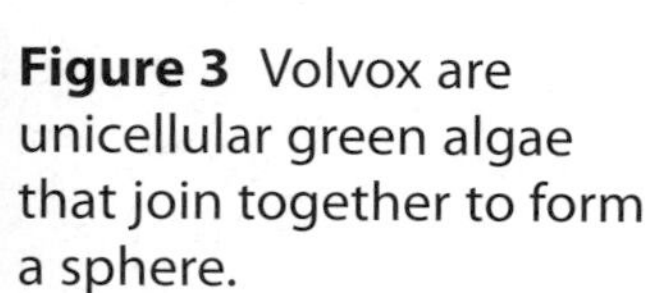

Figure 3 Volvox are unicellular green algae that join together to form a sphere.

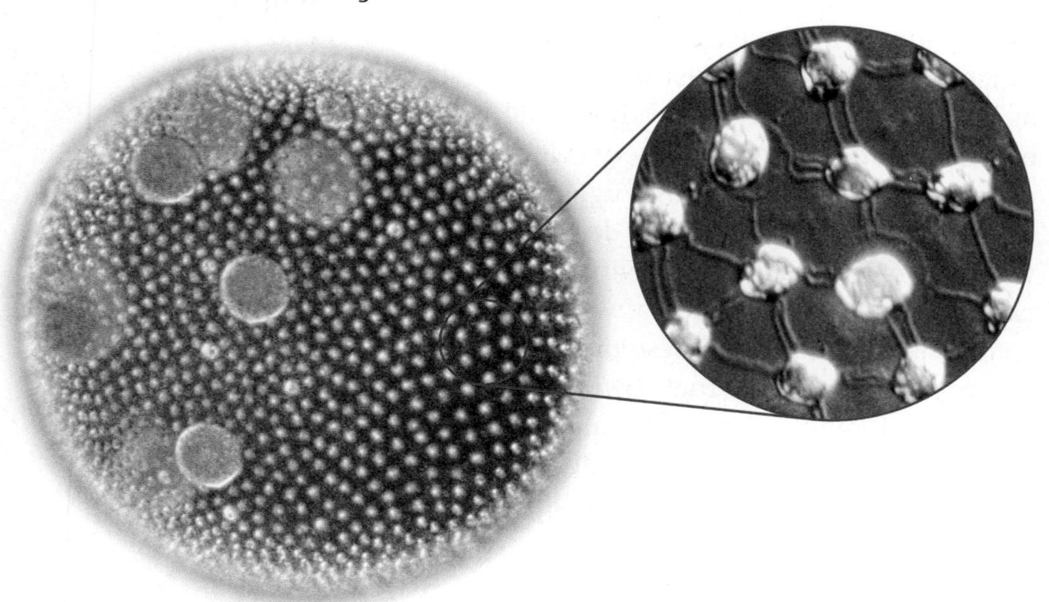

The Importance of Algae

Do you use algae in your everyday life? You might be surprised by all the materials you use that contain algae. You might be eating algae if you snack on ice cream, marshmallows, or pudding. Algae are a common ingredient in other everyday products, including toothpaste, lotions, fertilizers, and some swimming pool filters.

Algae and Ecosystems

Algae provide food for animals and animal-like protists. They also provide shelter for many aquatic organisms. In **Figure 4,** you can see that some brown algae grow tall. Thick groups of tall brown algae are called kelp forests. Sea otters and seals come to the kelp forest to eat smaller animals.

Active Reading **8. Explain** How are algae beneficial to an ecosystem?

Do you think algae ever cause problems in an ecosystem? Algae and other photosynthetic protists can help remove pollution from the water. However, this pollution can be a food source for the algae, allowing the population of algae to increase quickly. The algae produce wastes that can poison other organisms. As shown in **Figure 5,** when the number of these protists increases, the water can appear red or brown. This is called a red tide or a harmful algal bloom (HAB).

Active Reading **9. Explain** What causes a red tide?

Figure 5 *Karenia brevis* is a dinoflagellate commonly known as the Florida red tide organism. The dinoflagellate species involved in Florida HABs is red or brown in color, giving a reddish tint to the water. Florida red tides occur in the Gulf of Mexico almost every year in the late summer or early fall. A bloom typically lasts three to five months and may have a negative effect on fish, birds, and marine mammals.

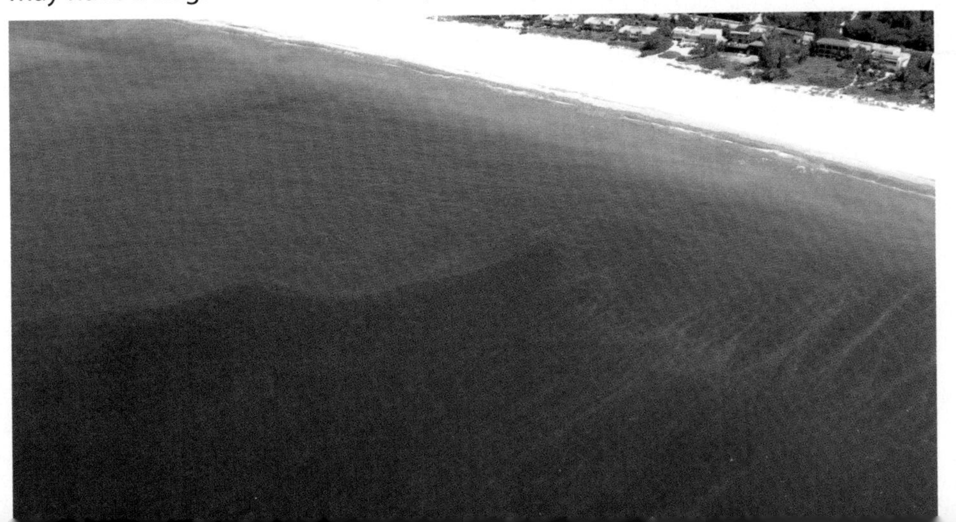

Kelp Forest

Figure 4 Brown algae can form thick kelp forests that are home to many animals and other protists.

Active Reading **10. Evaluate** Complete the chart noting the importance of algae.

Human Uses of Algae	Algae and Ecosystems

Animal-like Protists

Some protists are similar to plants, but others are more like animals. **Protozoans** (proh tuh ZOH unz) *are protists that resemble tiny animals.* Animal-like protists all share several characteristics. They do not have chloroplasts or make their own food. Protozoans are usually microscopic and all are unicellular. Most protozoans live in wet environments.

Ciliates

Cilia (SIH lee uh) *are short, hairlike structures that grow on the surface of some protists.* Protists that have these organelles are called ciliates. Cilia cover the surface of the cell. They can beat together and move the animal-like protist through the water.

Active Reading **11. Explain** What function do cilia perform?

____

____

ACADEMIC VOCABULARY
process
(noun) an event marked by gradual changes that lead toward a particular result

A common protozoan with these cilia is the **paramecium** (pa ruh MEE see um; plural, paramecia)—*a protist with cilia and two types of nuclei.* One example of a paramecium is shown in **Figure 6.** A paramecium, like most ciliates, gets its food by forcing water into a groove in its side. The groove closes and a food vacuole, or storage area, forms within the cell. The food particles are digested, and the extra water is forced back out. Ciliates reproduce asexually, but they can exchange some genetic material through a **process** called conjugation (kahn juh GAY shun). This results in more genetic variation.

Figure 6 A paramecium, like the one shown below, has two nuclei and is covered with hairlike structures called cilia.

Paramecium

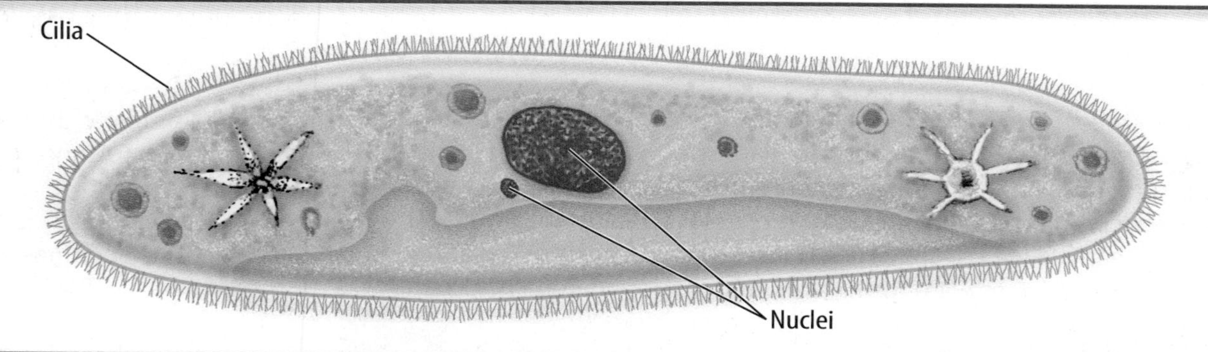

Flagellates

Recall that dinoflagellates, a type of plantlike protist, use one or more flagella to move. A type of protozoan also has one or more flagella—a flagellate. However, a flagellate does not always spin when it moves.

Flagellates eat decaying matter including plants, animals, and other protists. Many flagellates live in the digestive system of animals and absorb nutrients from food eaten by them.

Active Reading **12. Identify** Underline two different sources of food for flagellates.

Sarcodines

Animal-like protists called sarcodines (SAR kuh dinez) have no specific shape. At rest, a sarcodine resembles a random cluster of cytoplasm, or cellular material. These animal-like protists can ooze into almost any shape as they slide over mud or rocks.

An **amoeba** (uh MEE buh) *is one common sarcodine with an unusual adaptation for movement and getting nutrients.* An amoeba moves by using a **pseudopod**, *a temporary "foot" that forms as the organism pushes part of its body outward.* It moves by first stretching out a pseudopod and then oozing the rest of its body up into the pseudopod. This movement is shown in **Figure 7.**

Amoebas also use pseudopods to get nutrients. An amoeba surrounds a smaller organism or food particle with its pseudopod and then oozes around it. A food vacuole forms inside the pseudopod where the food is quickly digested. You can see an amoeba capturing its prey in the photo at the beginning of this lesson.

Some sarcodines get nutrients and energy from ingesting other organisms, while others make their own food. Some sarcodines even live in the digestive systems of humans and get nutrients and energy from the human's body.

Figure 7 An amoeba moves by extending its body to create a temporary "foot."

Amoeba Movement

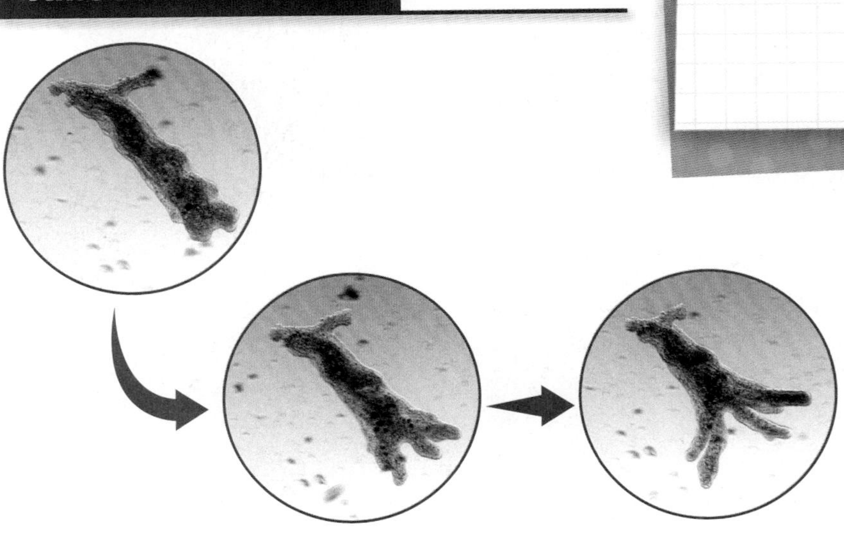

Active Reading **13. Create** Complete the spider map to identify the major characteristics of the three groups of protozoans. Record at least two characteristics of each group.

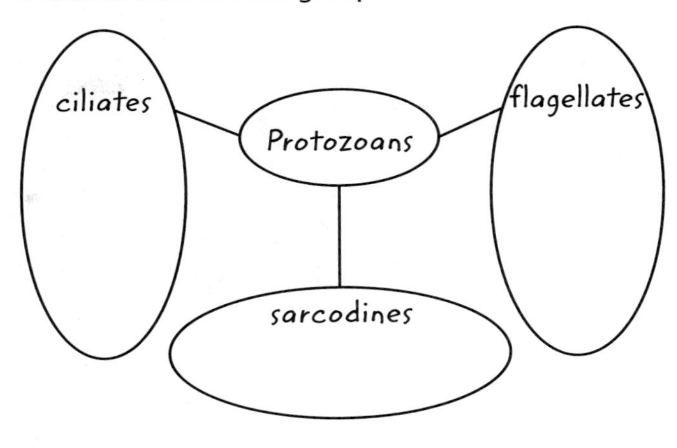

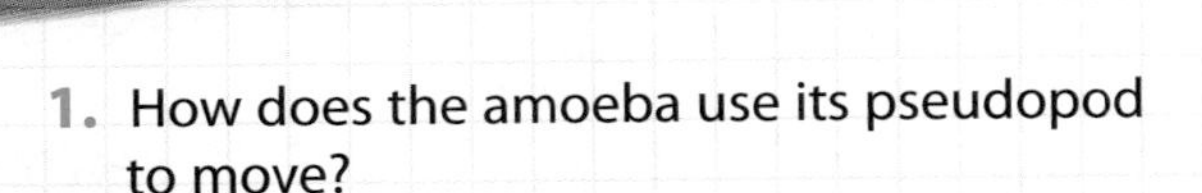

Inquiry iLAB STATION Try It! SC.6.N.3.4

MiniLab ***How can you model the movement of an amoeba?*** at connectED.mcgraw-hill.com

Apply It!

After you complete the lab, answer the questions below.

1. How does the amoeba use its pseudopod to move?

2. How do flagella differ from cilia?

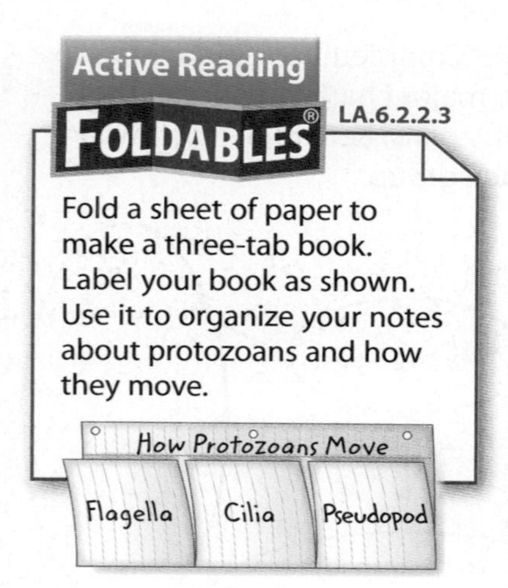

The Importance of Protozoans

Imagine living in a world without organisms that decompose other organisms. Plant material and dead animals would build up until the surface of Earth quickly became covered. Many protozoans are beneficial to an environment because they break down dead plant and animal matter. This decomposed matter is then recycled back into the environment and used by living organisms.

Some protozoans can cause disease by acting as parasites. These organisms can live inside a host organism and feed off it. Protozoan parasites are responsible for millions of human deaths every year.

One example of a disease caused by a protist is malaria. **Figure 8** illustrates how malaria develops and is spread to humans by mosquitoes. Protozoan parasites called plasmodia (singular, plasmodium) live and reproduce in red blood cells. Malaria kills more than one million people each year.

15. NGSSS Check **Infer** In what ways are protists helpful and harmful to humans? SC.6.L.14.6

____

____

____

Active Reading **14. Identify** (Circle) the stage in **Figure 8** when the parasite transfers to a healthy human.

Plasmodium Life Cycle

Figure 8 A small parasitic protozoan called plasmodium causes malaria. It is transferred among humans by mosquitoes.

1. A mosquito bites a human infected with malaria and takes in blood containing parasitic plasmodia.
2. The mosquito transfers parasites to an uninfected human when it bites him or her.
3. Parasites enter the human's liver and begin reproducing and maturing.
4. Mature parasites move from the liver and infect red blood cells, where they reproduce again.
5. The infected red blood cells burst, releasing parasites into the bloodstream.

Immature parasites

Human liver

Mature parasites

Red blood cells

Funguslike Protists

In addition to plantlike and animal-like protists, there are funguslike protists. These protists share many characteristics with fungi. However, because they differ from fungi, they are classified as protists.

Slime and Water Molds

Have you ever seen a strange organism like the one shown in **Figure 9?** These funguslike protists, called slime molds, look like they could have come from another planet. The body of the slime mold is composed of cell material and nuclei floating in a slimy mass. Most slime molds absorb nutrients from other organic matter in their environment.

Active Reading **16. Explain** Where do slime molds get their nutrients?

A Funguslike Protist

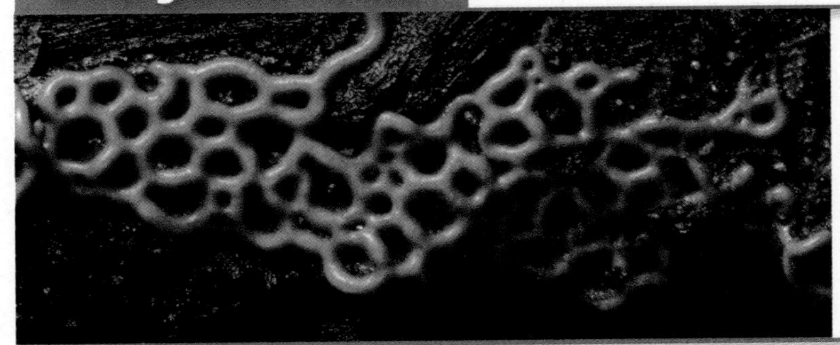

Figure 9 Slime molds come in a variety of colors and forms. These protists often live on the surfaces of plants.

A water mold is another kind of funguslike protist that lives as a parasite or feeds on dead organisms. Originally classified as fungi, water molds often cause diseases in plants.

Both slime molds and water molds reproduce sexually and asexually. The molds usually reproduce sexually when environmental conditions are harsh or unfavorable.

Active Reading **17. Analyze** Animals and animal-like protists share some characteristics. Tell how the organisms in these groups are the same. Then hypothesize why scientists have placed them in different groups.

Importance of Funguslike Protists

Funguslike protists play a valuable role in the ecosystem. They break down dead plant and animal matter, making the nutrients from these dead organisms available for living organisms. While some slime molds and water molds are beneficial, many others can be very harmful.

Many funguslike protists attack and consume living plants. The Great Irish Potato Famine resulted from damage by a funguslike protist. In 1845 this water mold destroyed more than half of Ireland's potato crop. More than one million people starved as a result.

Active Reading **18. Explain** How are funguslike protists beneficial to an environment?

Lesson Review 1

Visual Summary

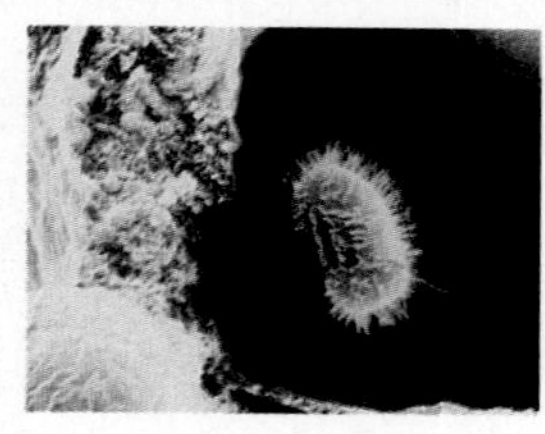

Protists are a diverse group of organisms that cannot be classified as plants, animals, or fungi.

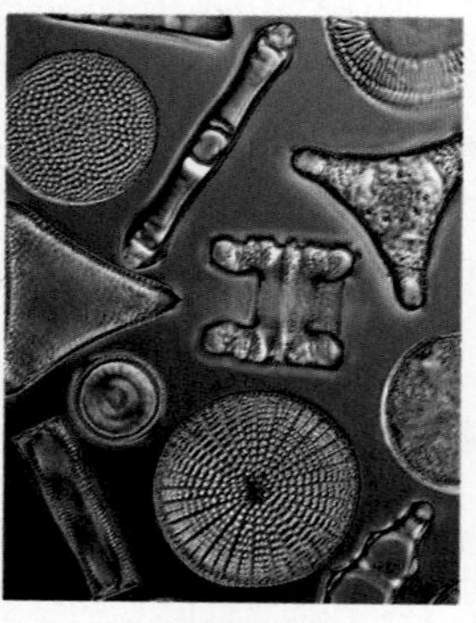

Protists are grouped according to the type of organisms they most resemble. Diatoms are one type of plantlike protist.

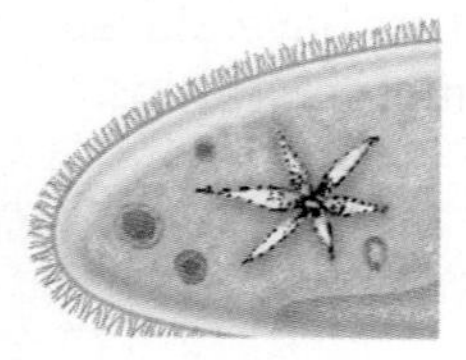

Some protists use hairlike structures called cilia to move.

Use Vocabulary

1. **Distinguish** between cilia and flagella.

2. **Define** *pseudopod* in your own words or with a drawing. SC.6.N.3.4

Understand Key Concepts

3. **List** three groups of animal-like protists and three groups of plantlike protists.

4. **Describe** one example of how protists benefit humans. SC.6.L.14.6

5. Identify which protist causes red tides. SC.6.L.14.6
   - (A) algae
   - (B) diatoms
   - (C) euglenoids
   - (D) paramecia

Interpret Graphics

6. **Identify** Fill in the graphic organizer with the three categories of protists. LA.6.2.2.3

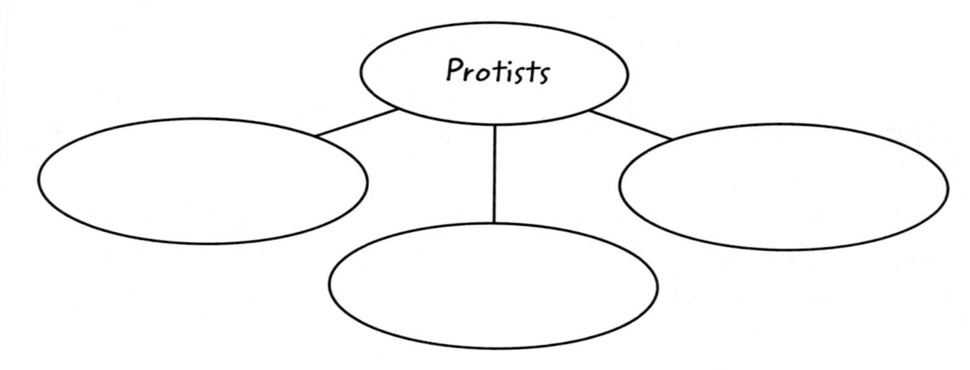

Critical Thinking

7. **Formulate** a plan for deciding how to classify a newly discovered protist.

FOCUS on FLORIDA

The Benefits of Algae

Big Benefits from Tiny Organisms

Algae are protists that can do more than just cover a pond as slimy scum. They release oxygen through photosynthesis. In fact, most of the oxygen in Earth's atmosphere comes from photosynthesis that occurs in algae, plants, and some bacteria. Algae also are food for many organisms, including humans.

Microalgae can grow outdoors in ponds or indoors under lights in photobioreactors. A photobioreactor is a tank filled with water and nutrients. Photosynthesis requires carbon dioxide. Instead of releasing carbon dioxide gas into the atmosphere, power plants can pump it into photobioreactors for microalgae to use.

A Florida-based energy company is using this technology to produce large amounts of protein-rich nutrients for food. This type of technology is beneficial because the algae are able to grow rapidly within the micro-crops without competing against other food crops for fertile land or irrigated water. The algae can be used as a protein enhancement in sports drinks, nutrition bars, nutritional supplements, and baked goods.

Processing plants using photobioreactors, such as this one, are a source of algae oil additives found in various food products.

It's Your Turn

RESEARCH Protists, including algae, are important sources of food. Research five types of organisms that depend on protists for food. Make a display of your results to share with your class. **LA.6.4.2.2**

Lesson 2

What are FUNGI?

ESSENTIAL QUESTIONS

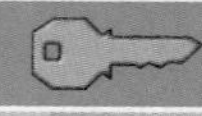 What are the different types of fungi and how do they compare?

 Why are fungi important?

 What are lichens?

Vocabulary

hyphae p. 543
mycelium p. 543
basidium p. 544
ascus p. 545
zygosporangia p. 545
mycorrhiza p. 548
lichen p. 550

Florida NGSSS

LA.6.2.2.3 The student will organize information to show understanding (e.g., representing main ideas within text through charting, mapping, paraphrasing, summarizing, or comparing/contrasting);

MA.6.A.3.6 Construct and analyze tables, graphs, and equations to describe linear functions and other simple relations using both common language and algebraic notation.

HE.6.C.1.3 Identify environmental factors that affect personal health.

SC.6.L.14.6 Compare and contrast types of infectious agents that may infect the human body, including viruses, bacteria, fungi, and parasites.

SC.6.N.1.1 Define a problem from the sixth grade curriculum, use appropriate reference materials to support scientific understanding, plan and carry out scientific investigation of various types, such as systematic observations or experiments, identify variables, collect and organize data, interpret data in charts, tables, and graphics, analyze information, make predictions, and defend conclusions.

SC.6.N.1.5 Recognize that science involves creativity, not just in designing experiments, but also in creating explanations that fit evidence.

Inquiry Launch Lab

HE.6.C.1.3

10 minutes

Is there a fungus among us?

The mold you see on food is fungi that are consuming and decomposing it. Fungi are also found as molds or mushrooms on wood, mulch, and other organic materials.

Procedure

1. Read and complete a lab safety form.
2. Examine the different **samples of fungi** your teacher provides. Use a **magnifying lens** to observe similarities and differences among the samples.
3. Record your observations in the Data and Observation section below. Include drawings of the different structures or characteristics you notice.

Data and Observations

Think About This

1. What similarities did you see among the fungi samples?
2. Why do you think your teacher had the mold samples in closed containers?
3. Key Concept In what ways do you think the fungi you observed are helpful or not helpful to people?

Up in Smoke?

1. The organism pictured is a puffball mushroom, named for the puff of material that it releases. What do you think the material is? What is the purpose of the puff of material?

Active Reading 2. **Describe** How are hyphae and mycelium related?

What are fungi?

What would you guess is the world's largest organism? A fungus in Oregon is the largest organism ever measured by scientists. It stretches almost 9 km^2. Fungi, like protists, are eukaryotes. Scientists estimate more than 1.5 million species of fungi exist.

Fungi form long, threadlike structures that grow into large tangles, usually underground. *These structures, which absorb minerals and water, are called* **hyphae** (HI fee). *The hyphae create a network called the* **mycelium** (mi SEE lee um), shown in **Figure 10.** The fruiting body of the mushroom, the part above ground, is also made of hyphae.

Fungi are heterotrophs, meaning they cannot make their own food. Some fungi are parasites, obtaining nutrients from living organisms. Fungi dissolve their food by releasing chemicals that decompose organic matter. Fungi then absorb the nutrients.

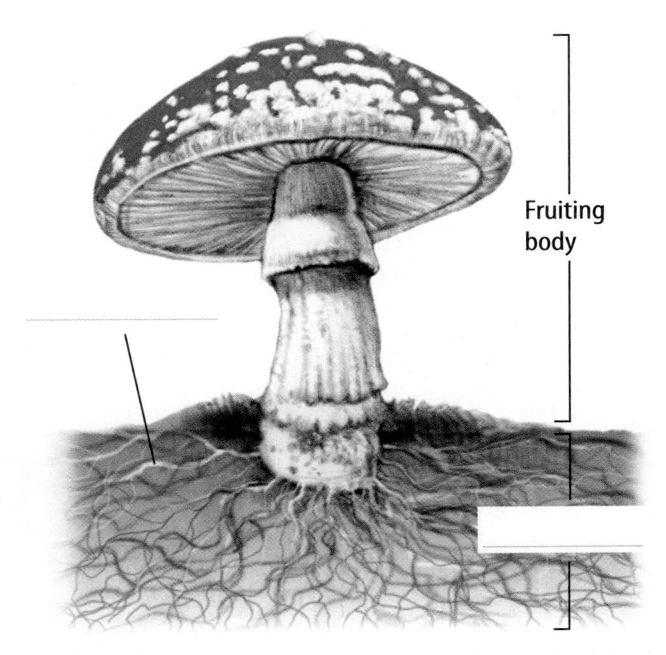

Figure 10 Mushrooms are common fungi. In the drawing, you can see mycelium, the network of hyphae. The hyphae release enzymes and absorb water and nutrients.

Active Reading 3. **Identify** In **Figure 10** above, fill in the blanks with the correct labels.

Active Reading **4. Identify** What are the four groups of fungi?

Types of Fungi

Scientists group fungi based on how they look and how they reproduce. Although fungi can reproduce sexually or asexually, almost all reproduce asexually by producing spores. Spores are small reproductive cells with a strong, protective outer covering. The spores can grow into new individuals.

The classification of fungi often changes as scientists learn more about them. Today, scientists recognize four groups of fungi: club fungi, sac fungi, zygote fungi, and imperfect fungi. As technology helps scientists understand more about fungi, the categories might change.

Club Fungi

SCIENCE USE V. COMMON USE

mushroom

Science Use a type of club fungi

Common Use the part of a fungus above the ground

When you think of fungi, you might think of a **mushroom**. Mushrooms belong to the group called club fungi. They are named for the clublike shape of their reproductive structures. However, the mushroom is just one part of the fungus. The part of the mushroom that grows above ground is a structure called a basidiocarp (bus SIH dee oh karp). Inside the basidiocarp are the **basidia** (buh SIH dee uh; singular, basidium), *reproductive structures that produce sexual spores.* Most of a club fungus is a network of hyphae that grows underground and absorbs nutrients.

Active Reading **5. State** Where does most of a club fungus grow?

Many club fungi are named for their various shapes and characteristics. Club fungi include puffballs like those at the beginning of the lesson, stinkhorns, and the bird's nest fungi shown in **Figure 11.** There is even a club fungus that glows in the dark due to a chemical reaction in its basidiocarp.

Figure 11 Club fungi, such as this bird's nest fungus, use basidiospores to reproduce.

Club Fungi

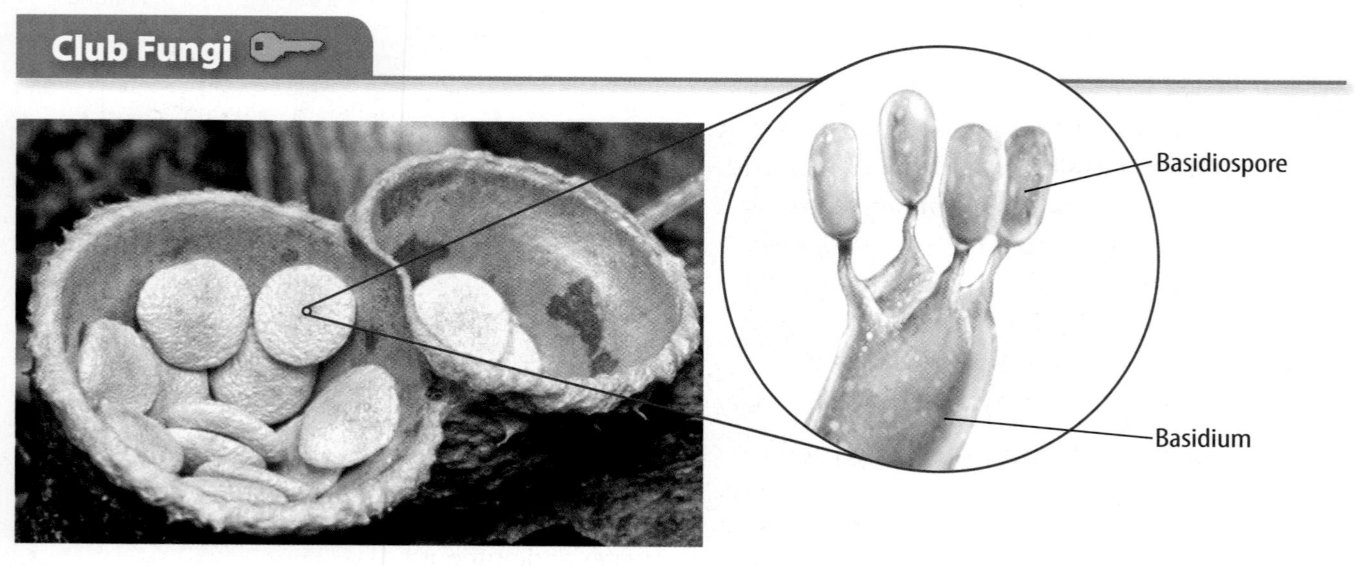

6. Visual Check **Identify** Which part of the fungus is club-shaped?

Sac Fungi

Do you know what bread and a diaper rash have in common? A type of sac fungus causes bread dough to rise. A different sac fungus is responsible for a rash that babies can develop on damp skin under their diapers. Many sac fungi cause diseases in plants and animals. Other common sac fungi, such as truffles and morels, are harvested by people for food.

Like club fungi, sac fungi are named for their reproductive structures. *The* **ascus** (AS kuhs; plural, asci) *is the reproductive structure where spores develop on sac fungi.* The ascus often looks like the bottom of a tiny bag or sack. The spores from sac fungi are called ascospores (AS kuh sporz). Sac fungi can undergo both sexual and asexual reproduction. Many yeasts are sac fungi, including the common yeast used to make bread, as shown in **Figure 12.** When the yeast is mixed with water and warmed, the yeast cells become active. They begin cellular respiration and release carbon dioxide gas. This causes the bread dough to rise.

Zygote Fungi

Another type of fungus can cause bread to develop mold. Bread mold, like the type shown in **Figure 12,** is caused by a type of fungus called a zygote fungus. You might also find zygote fungi growing in moist areas, such as a damp basement or on a bathroom shower curtain.

The hyphae of a zygote fungus grow over materials, such as bread, dissolving the material and absorbing nutrients. *Tiny stalks called* **zygosporangia** (zi guh spor AN jee uh) *form when the fungus undergoes sexual reproduction.* The zygosporangia release spores called zygospores. These zygospores then fall on other materials where new zygote fungi might grow.

7. Explain How do sac and zygote fungi differ?

____

____

____

Figure 12 Some fungi can be used to make food, but other fungi can eat the food too.

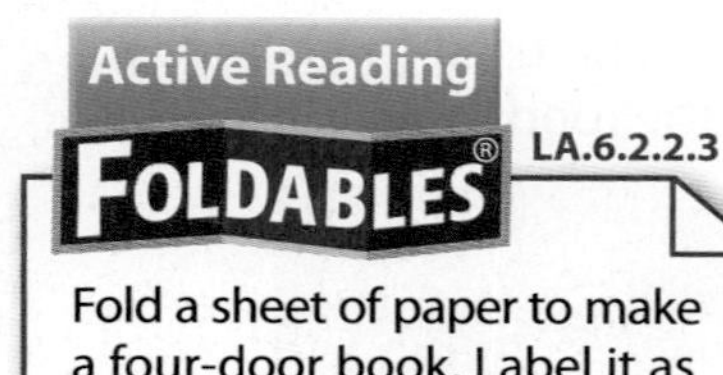

LA.6.2.2.3

Fold a sheet of paper to make a four-door book. Label it as shown. Use your book to organize information about the characteristics of the different classifications of fungi.

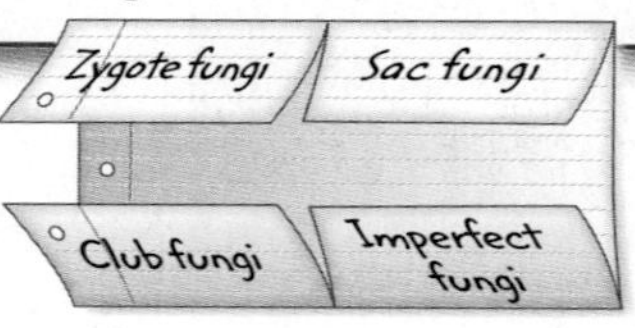

Active Reading **8. Interpret** Why are imperfect fungi classified that way?

Imperfect Fungi

How are itchy feet and blue cheese connected? They can both be caused by imperfect fungi. You might have had athlete's foot, an infection that causes flaking and itching in the skin of the feet. The imperfect fungus that causes athlete's foot grows and reproduces easily in the moist environment near a shower or in a sweaty shoe. The blue color you see in blue cheese comes from colonies of a different type of imperfect fungi. They are added to the milk or the curds during the cheesemaking process.

Imperfect fungi are named because scientists have not observed a sexual, or "perfect," reproductive stage in their life cycle. Since fungi are classified according to the shape of their reproductive structures, these fungi are left out, or labeled "imperfect." Often after a species of imperfect fungi is studied, the sexual stage is observed. The fungi is then classified as a club, sac, or zygote fungus based on these observations.

Active Reading **9. Compare** Complete the chart and compare the four groups of fungi.

Group	How They Reproduce Sexually	Examples
Club fungi		
Sac fungi		
Zygote fungi		
Imperfect fungi		

Inquiry LAB STATION SC.6.N.1.5 **Try It!**

MiniLab ***What do fungal spores look like?*** at connectED.mcgraw-hill.com

Apply It! After you complete the lab, answer this question.

1. What characteristics of club fungi give the group its name?

Fungi Products

Figure 13 Products such as bread, cheese, and medicines are made using fungi.

The Importance of Fungi

Do you like chocolate, carbonated sodas, cheese, or bread? If so, you might agree that fungi are beneficial to humans. Fungi are involved in the production of many foods and other products, as shown in **Figure 13.** Some fungi are used as a meat substitute because they are high in protein and low in cholesterol. Other fungi are used to make antibiotics.

Decomposers

Fungi help create food for people to eat, but they are also important because of the things they eat. As you read earlier, fungi are an important part of the environment because they break down dead plant and animal matter, as shown in **Figure 14.** Without fungi and other decomposers, dead plants and animals would pile up year after year. Fungi also help break down pollution, including pesticides, in soil. Without fungi to destroy it, pollution would build up in the environment.

Living things need nutrients. The nutrients available in the soil would eventually be used up if they were not replaced by decomposing plant and animal matter. Fungi help put these nutrients back into the soil for plants to use.

Active Reading **10. Identify** Underline three benefits of fungi as decomposers.

Figure 14 Fungi help decompose dead organic matter, such as this rabbit.

May 8

October 6

Math Skills MA.6.A.3.6

Using Fractions

Under certain conditions, 100 percent of the cells in fungus A reproduce in 24 hours. The number of cells of fungus A doubles once each day.

Day 1 = 10,000 cells

Day 2 = 20,000 cells

Day 3 = 40,000 cells

Day 4 = 80,000 cells

When an antibiotic is added to the fungus, the growth is reduced by 50 percent. Only half the cells reproduce each day.

Day 2 = 15,000 cells

Day 3 = 22,500 cells

Day 4 = 33,750 cells

Practice

11. Without an antibiotic, how many cells of fungus A would there be on day 6?

Fungi and Plant Roots

Plants benefit from fungi in other ways, too. Many fungi and plants grow together, helping each other. Recall that fungi take in minerals and water through the hyphae, or threadlike structures that grow on or under the surface. *The roots of the plants and the hyphae of the fungi weave together to form a structure called* **mycorrhiza** (mi kuh RI zuh; plural, micorrhizae).

Mycorrhizae can exchange molecules, as shown in **Figure 15.** As fungi break down decaying matter in the soil, they make nutrients available to the plant. They also increase water absorption by increasing the surface area of the plant's roots.

Fungi cannot photosynthesize, or make their own food using light energy. Instead, the fungi in mycorrhizae take in some of the sugars from the plant's photosynthesis. The plants benefit by receiving more nutrients and water. The fungi benefit and continue to grow by using plant sugars. Scientists suspect that most plants gain some benefit from mycorrhizae.

Active Reading **12. Explain** How do mycorrhizae benefit both the plant and the fungus?

Figure 15 The roots of this buckthorn plant and the hyphae of fungi weave together, enabling the exchange of nutrients.

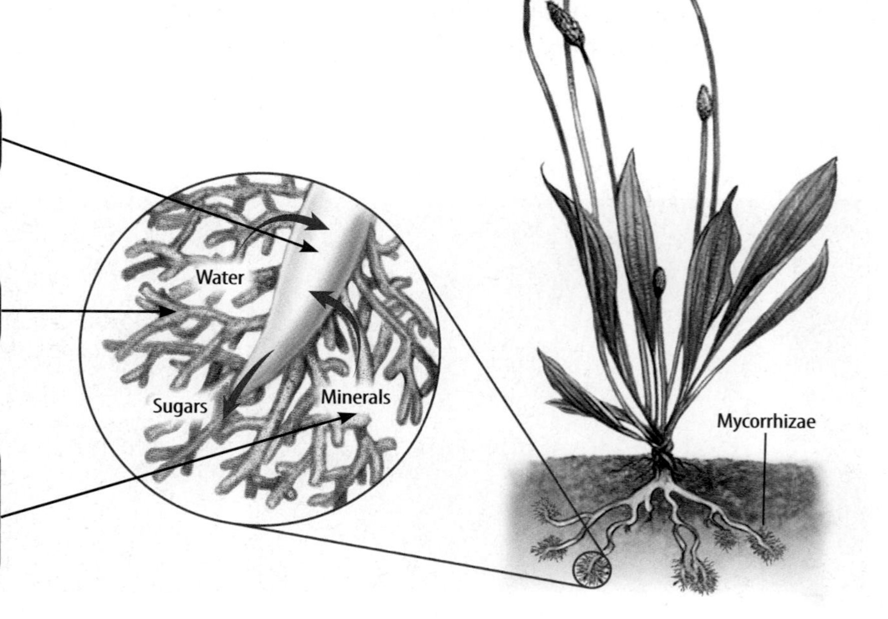

Health and Medicine

You might recall that many protists can be harmful to humans and the environment. This is true of fungi as well. A small number of people die every year after eating poisonous mushrooms or spoiled food containing harmful fungi.

You do not have to eat fungi for them to make you sick or uncomfortable. You already read that fungi cause athlete's foot rashes and diaper rashes. Some fungi cause allergies, pneumonia, and thrush. Thrush is a yeast infection that grows in the mouths of infants and people with weak immune systems.

Although fungi can cause disease, scientists also use them to make important medicines. Antibiotics, such as penicillin, are among the valuable medications made from fungi. An accident resulted in the discovery of penicillin. Alexander Fleming was studying bacteria in 1928 when spores of *Penicillium* fungus contaminated his experiment and killed the bacteria. After years of research, this fungus was used to make an antibiotic similar to the penicillin used today. **Figure 16** illustrates how penicillin affects bacterial growth.

Active Reading **13. Distinguish** Tell the difference between helpful and harmful aspects of fungi in medicine.

Harmful
Helpful

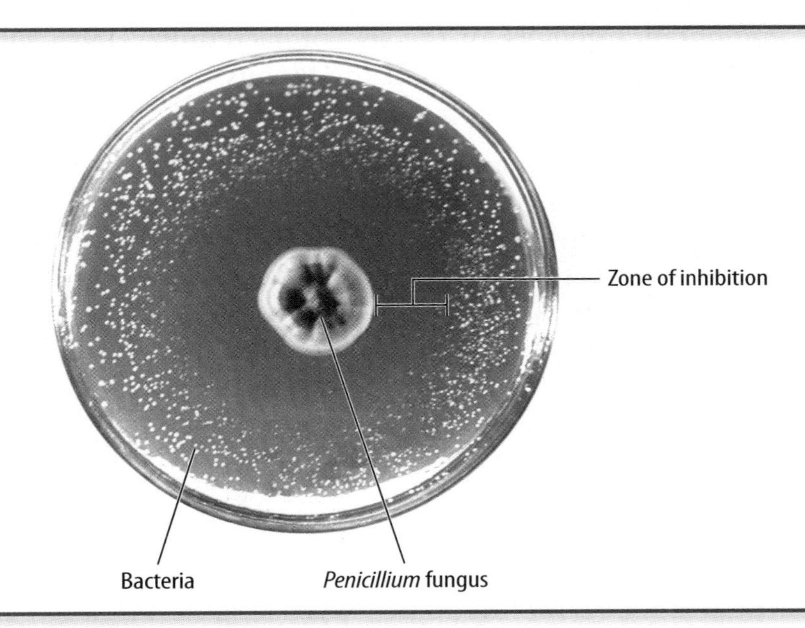

Figure 16 The *Penicillium* fungus that prevents bacteria from growing is used to make penicillin, an antibiotic medicine.

14. Visual Check
Infer How can you tell that the fungi are stopping the bacteria from growing?

Over time, some bacteria have become resistant to the many antibiotics used to fight illness. New antibiotics need to be developed to treat the same diseases. As new species of fungi are discovered and studied, scientists might find new sources of antibiotics and medicines.

Active Reading **15. Describe** Give two ways that fungi are important to humans.

What are lichens?

Word Origin

lichen
from Greek *leichen,* means "what eats around itself"

Active Reading **16. Identify** Underline which organisms usually grow together to form a lichen.

Do you recall the photo at the beginning of the chapter? The structure pictured is a lichen. *A* **lichen** (LI kun) *is a structure formed when fungi and certain other photosynthetic organisms grow together.* Usually, a lichen consists of a sac fungus or club fungus that lives in a partnership with either a green alga or a photosynthetic bacterium. The fungus's hyphae grow in a layer around the algae cells.

Green algae and photosynthetic bacteria are autotrophs, which means they can make their own food using photosynthesis. Lichens are similar to mycorrhizae because both organisms benefit from the partnership. The fungus provides water and minerals while the bacterium or alga provides the sugars and oxygen from photosynthesis.

The Importance of Lichens

Imagine living on a sunny, rocky cliff like the one in **Figure 17.** Not many organisms could live there because there is little to eat. A lichen, however, is well suited to this harsh environment. The fungus can absorb water, help break down rocks, and obtain minerals for the alga or bacterium. They can photosynthesize and make food for the fungus.

Once lichens are established in an area, it becomes a better environment for other organisms. Many animals that live in harsh conditions survive by eating lichens. Plants benefit from lichens because the fungi help break down rocks and create soil. Plants can then grow in the soil, creating a food source for other organisms in the environment.

Figure 17 Lichens are structures made of photosynthetic organisms and fungi that can live in harsh conditions.

Lichen Structure

Lesson Review 2

Visual Summary

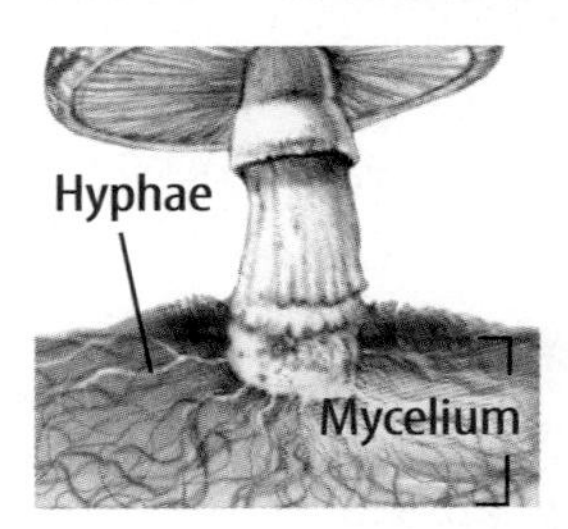

The body of a fungus is made up of threadlike hyphae that weave together to create a network of mycelium.

Club fungi produce sexual spores in the basidium.

A lichen is made of fungus and a photosynthetic bacterium or alga.

Inquiry Lab ***What does a lichen look like?*** at connectED.mcgraw-hill.com

Use Vocabulary

1. **Distinguish** between a basidium and an ascus.

2. **Identify** the structure formed between fungal hyphae and plant roots.

Understand Key Concepts 

3. **List** the four groups of fungi.

4. Which disease is caused by a fungus? SC.6.L.14.6
   - (A) athlete's foot
   - (B) influenza
   - (C) malaria
   - (D) pneumonia

Interpret Graphics

5. **Compare and Contrast** Complete the table about sac fungi and zygote fungi. LA.6.2.2.3

Sac Fungi	Zygote Fungi

Critical Thinking

6. **Support** the claim that decomposition is important for the environment.

Math Skills

7. The number of cells in fungus X doubles every 2 hours. If you begin with 10 cells, how many would be present after 24 hours?

Chapter 14 Study Guide

Think About It! Protists and fungi are diverse groups of organisms. Both may infect the human body. They are classified as neither plant nor animal and serve many functions in the ecosystem.

Key Concepts Summary

LESSON 1 What are protists?

- Scientists divide **protists** into three groups based on the type of organisms they most resemble. There are plantlike, animal-like, and funguslike protists.
- Protists are beneficial to humans in many ways. They are used to create many of the useful products you depend on. They also help decompose dead organisms and return nutrients to the environment.

Plantlike	Animal-like	Funguslike

Vocabulary

protist p. 531
algae p. 532
diatom p. 533
protozoan p. 536
cilia p. 536
paramecium p. 536
amoeba p. 537
pseudopod p. 537

LESSON 2 What are fungi?

- Scientists divide fungi into four groups, based on the type of structures they use for sexual reproduction. The four groups are club fungi, sac fungi, zygote fungi, and imperfect fungi.
- Fungi provide many foods and medicines that people use. In addition, fungi help break down dead organisms and recycle the nutrients into the environment.
- **Lichens** are structures made of a fungus and a photosynthetic organism. Both organisms work together to obtain food, water, and nutrients.

hyphae p. 543
mycelium p. 543
basidium p. 544
ascus p. 545
zygosporangia p. 545
mycorrhiza p. 548
lichen p. 550

Active Reading
FOLDABLES® Chapter Project

Assemble your lesson Foldables as shown to make a Chapter Project. Use the project to review what you have learned in this chapter.

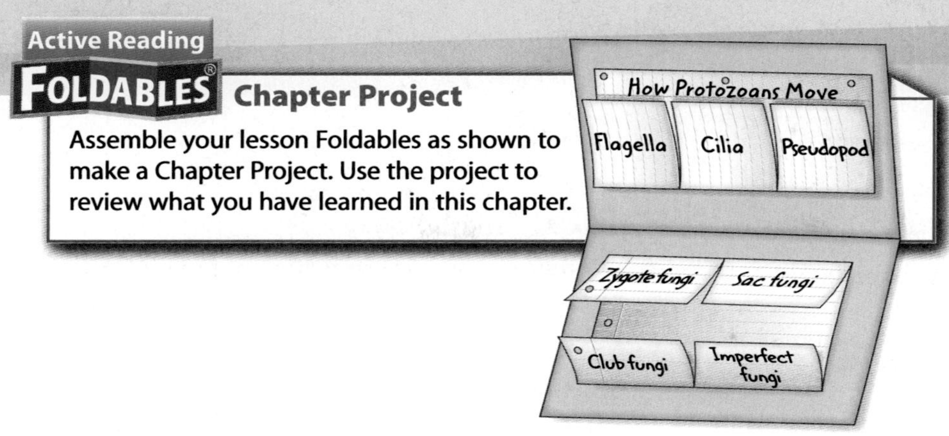

Use Vocabulary

1. A protist that resembles a tiny animal is called a(n) ______________.

2. A fungus and the roots of a plant form a structure called ______________ that benefits both organisms.

3. The ______________ is a saclike structure on a fungus that produces spores.

4. A(n) ______________ is a microscopic, plantlike protist that can resemble glass or gems.

5. Short structures that cover the outside of some protists and help them move are called ______________.

6. Fungi grow by extending threadlike body structures called ______________.

Link Vocabulary and Key Concepts

Use vocabulary terms from the previous page and other terms from this chapter to complete the concept map.

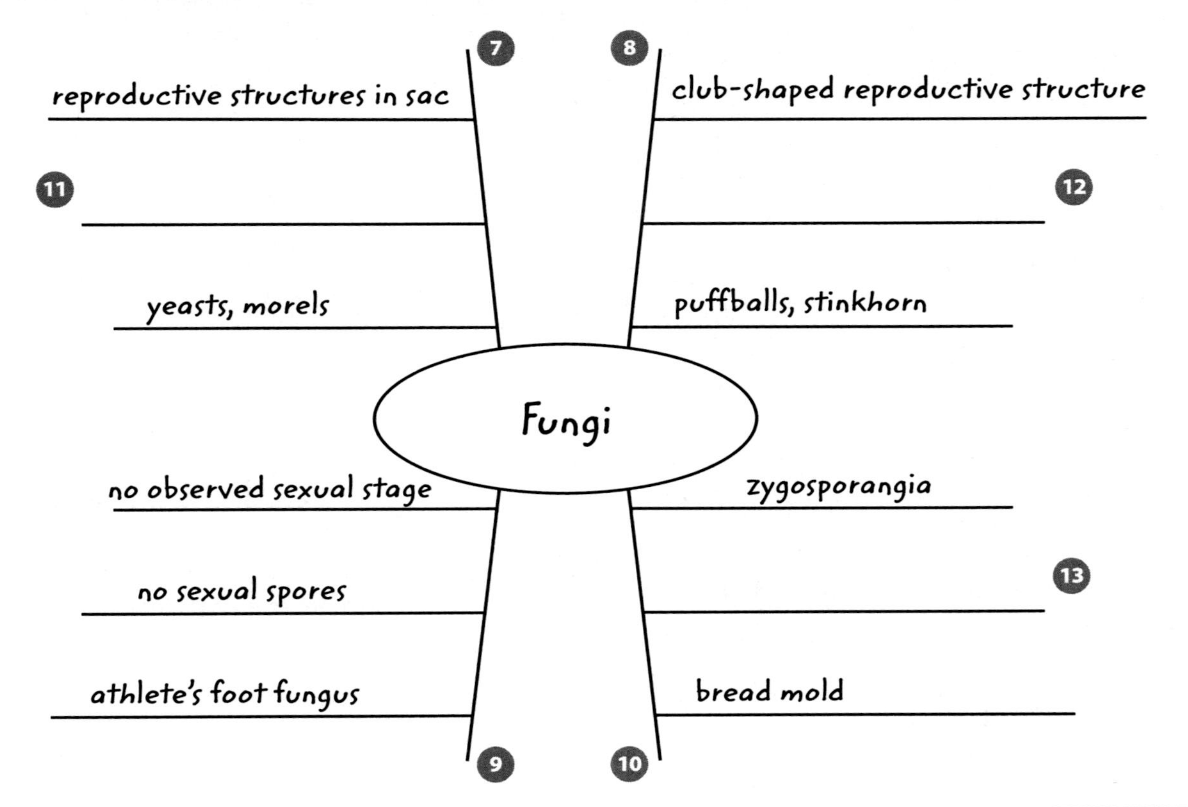

Chapter 14 Review

Fill in the correct answer choice.

Understand Key Concepts

1. Which organism causes red tides when found in large numbers? **SC.6.L.14.6**
   (A) algae
   (B) amoebas
   (C) ciliates
   (D) diatoms

2. Protists are a diverse group of organisms divided into what three categories? **SC.6.L.14.6**
   (A) animal-like, plantlike, protozoanlike
   (B) euglenoid, slime-mold, diatoms
   (C) plantlike, animal-like, and funguslike
   (D) green algae, red algae, and kelp

3. Which type of protist is commonly used in ice cream, toothpaste, soups, and body lotions? **SC.6.L.14.6**
   (A) algae
   (B) amoebas
   (C) ciliates
   (D) diatoms

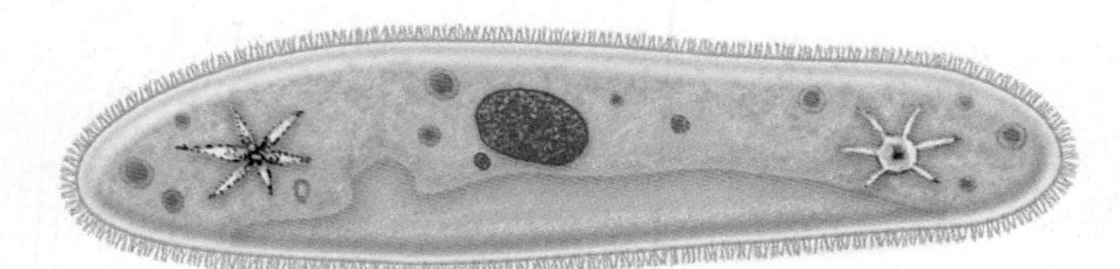

4. The organism in the figure above is a **SC.6.L.14.6**
   (A) ciliate.
   (B) diatom.
   (C) dinoflagellate.
   (D) kelp.

5. The main function of the hairlike structures surrounding the organism above is **SC.6.L.14.6**
   (A) decomposition.
   (B) movement.
   (C) photosynthesis.
   (D) reproduction.

6. What type of fungus is bread mold? **SC.6.L.14.6**
   (A) club
   (B) imperfect
   (C) sac
   (D) zygote

Critical Thinking

7. **Compare and contrast** different types of protists and fungi and how they can infect the human body. **SC.6.L.14.6**

8. **Evaluate** Imagine you are asked to justify removing kelp from an area of the ocean. Based on your knowledge of plantlike protists, what benefits or problems would you consider before you decide if the algae should be removed? **LA.6.2.2.3**

9. **Describe** Complete the table below with characteristics of the different types of animal-like protists. **LA.6.2.2.3**

	Number of nuclei	Method of eating	Method of movement
Ciliates			
Flagellates			
Sarcodines			

10. **Explain** how the movement of an amoeba differs from the movement of a dinoflagellate. **LA.6.2.2.3**

11 **List** several products you have used or seen that were made using fungi. SC.6.L.14.6

12 **Evaluate** how Alexander Fleming's experiments helped determine the importance of fungi to medicine. LA.6.2.2.3

Writing in Science

13 **Design** On a separate sheet of paper, design a brochure for a tour in which people could see several different types of lichens and fungi. What locations would be included, and which organisms would people be likely to observe? SC.6.L.14.6

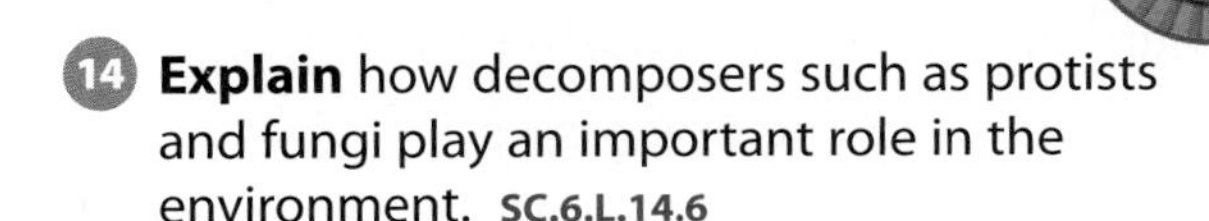

Big Idea Review

14 **Explain** how decomposers such as protists and fungi play an important role in the environment. SC.6.L.14.6

15 What is the organism shown below, and how does it affect the environment? SC.6.L.14.6

Math Skills MA.6.A.3.6

Calculating Growth

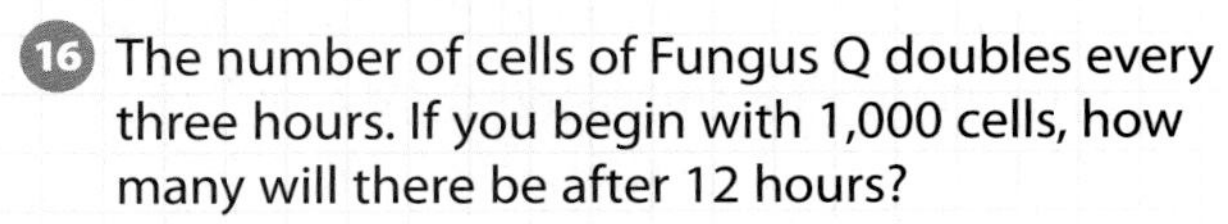

16 The number of cells of Fungus Q doubles every three hours. If you begin with 1,000 cells, how many will there be after 12 hours?

17 Scientists want to know if an antibiotic is effective in treating a fungal infection. They start with two colonies of 100 cells each. The table shows what happens during the first two days of treatment.

	Day 2 Number of Cells	Day 3 Number of Cells
Untreated fungus	400	1600
Antibiotic A	200	300

a. How long does it take the untreated fungus to double in number?

b. What effect does the antibiotic have on the growth rate of the fungus?

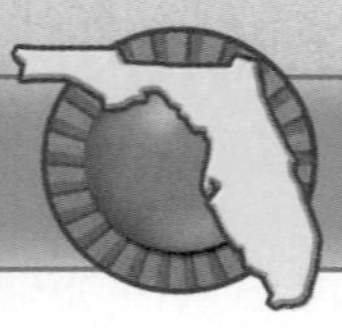

Florida NGSSS Benchmark Practice

Fill in the correct answer choice.

Multiple Choice

1 How are malaria protozoan parasites spread from human to human? **SC.6.L.14.6**

Ⓐ through contaminated drinking water

Ⓑ through contaminated food

Ⓒ through animal bites

Ⓓ through mosquito bites

Use the diagram below to answer question 2.

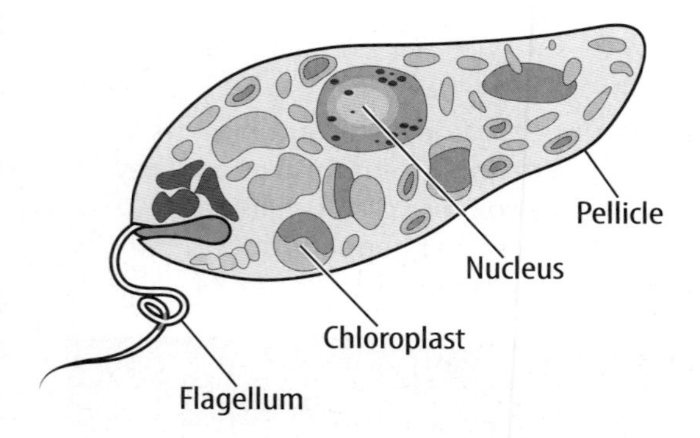

2 The euglenoid shown above does not infect humans. What structure indicates that it produces its own food? **SC.6.L.14.6**

Ⓕ chloroplast

Ⓖ flagellum

Ⓗ nucleus

Ⓘ pellicle

3 Which is an infection caused by a fungus that grows in the mouths of infants and people with weak immune systems? **SC.6.L.14.6**

Ⓐ athlete's foot

Ⓑ malaria

Ⓒ pneumonia

Ⓓ thrush

4 Which type of fungi can be poisonous to humans if eaten? **SC.6.L.14.6**

Ⓕ algae

Ⓖ paramecia

Ⓗ penicillin

Ⓘ mushrooms

5 Which animal-like protists can live in the digestive systems of humans and get nutrients and energy from the human's body? **SC.6.L.14.6**

Ⓐ sarcodines

Ⓑ dinoflagellates

Ⓒ ciliates

Ⓓ algae

6 Which small, parasitic protozoan causes malaria? **SC.6.L.14.6**

Ⓕ diatom

Ⓖ plasmodium

Ⓗ euglenoid

Ⓘ dinoflagellate

7 Where do protozoan parasites live? **SC.6.L.14.6**

Ⓐ in decomposing organisms

Ⓑ in host organisms

Ⓒ in soil

Ⓓ on leaves

8 In which human cells do plasmodia live and reproduce? **SC.6.L.14.6**

Ⓕ white blood cells

Ⓖ cancer cells

Ⓗ red blood cells

Ⓘ nerve cells

Use the diagram below to answer question 9.

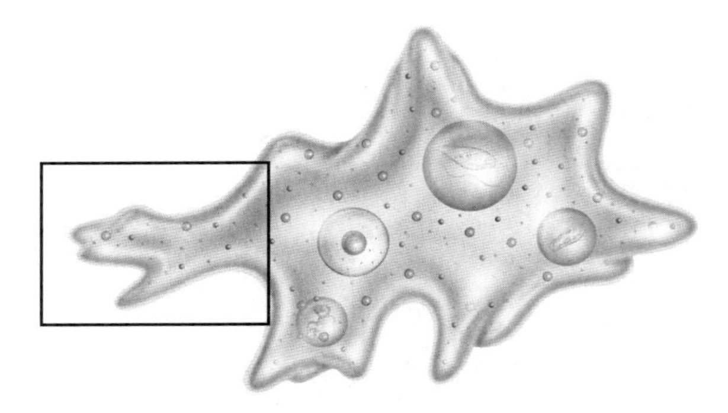

9 What is the function of the boxed area of this microscopic organism? **SC.6.L.14.6**

Ⓐ cellular respiration

Ⓑ defense

Ⓒ locomotion

Ⓓ photosynthesis

10 Which protist-caused disease is responsible for more than 1 million deaths each year? **SC.6.L.14.6**

Ⓕ thrush

Ⓖ athlete's foot

Ⓗ diaper rash

Ⓘ malaria

11 Which example of a sarcodine can live in human digestive tracts and move around using a pseudopod? **SC.6.L.14.6**

Ⓐ amoeba

Ⓑ dinoflagellate

Ⓒ diatom

Ⓓ paramecium

12 In which human organ do plasmodia spend part of their life cycle? **SC.6.L.14.6**

Ⓕ heart

Ⓖ stomach

Ⓗ liver

Ⓘ small intestine

13 Why are the fungi that cause athlete's foot labeled imperfect fungi? **SC.6.L.14.6**

Ⓐ The fungi do not show any normal characteristics of fungi.

Ⓑ Scientists classified imperfect fungi as protists.

Ⓒ The fungi reproduce in hot, dry environments.

Ⓓ Scientists have not observed a sexual reproductive stage in their life cycle.

14 Which type of fungi can cause diaper rash on babies? **SC.6.L.14.6**

Ⓕ club fungi

Ⓖ sac fungi

Ⓗ zygote fungi

Ⓘ imperfect fungi

NEED EXTRA HELP?

If You Missed Question...	1	2	3	4	5	6	7	8	9	10	11	12	13	14
Go to Lesson...	1	1	2	2	1	1	1	1	1	1	1	1	2	2

Name ______________________ Date ______________

Multiple Choice *Bubble the correct answer.*

Use the image below to answer questions 1 and 2.

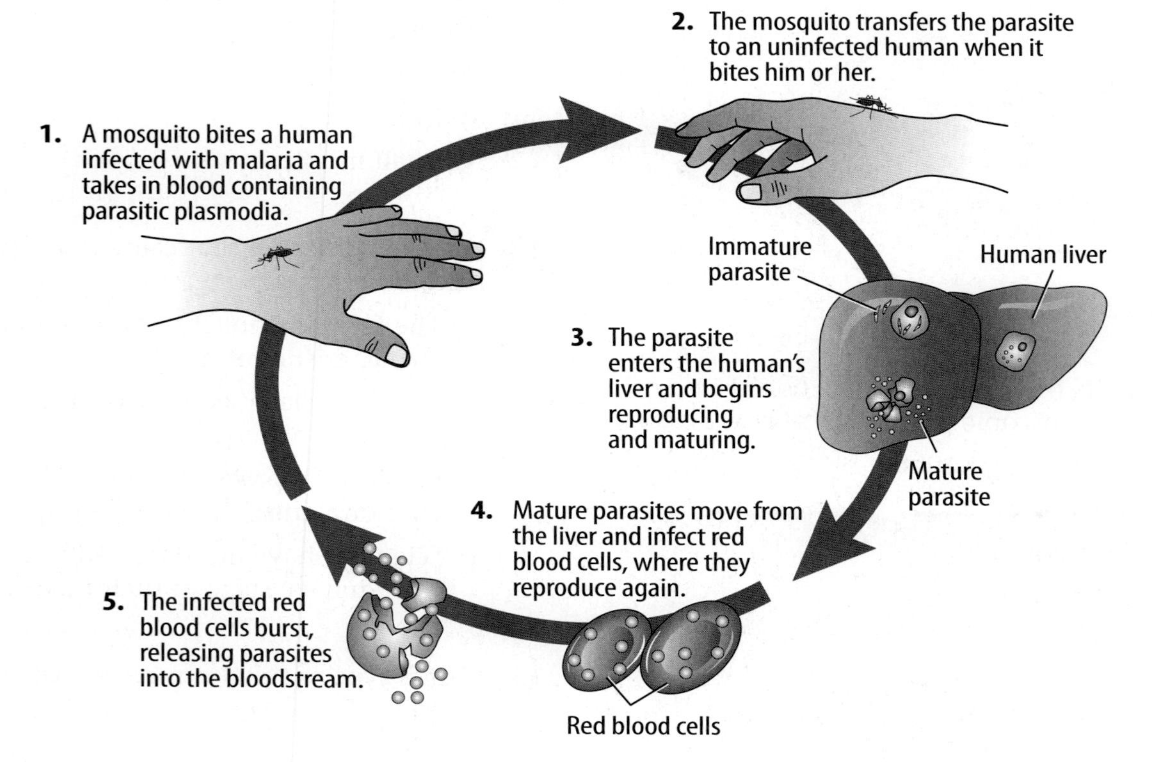

1. The parasite plasmodia causes malaria in humans. At which stage in the life cycle will the host be most affected by the symptoms of the disease? **SC.6.L.14.6**

Ⓐ Stage 2

Ⓑ Stage 3

Ⓒ Stage 4

Ⓓ Stage 5

2. At which point in the plasmodia life cycle can humans best interrupt the cycle and prevent the spread of malaria from person to person? **SC.6.L.14.6**

Ⓕ Stage 1

Ⓖ Stage 2

Ⓗ Stage 4

Ⓘ Stage 5

Name ______________________ Date ____________

Multiple Choice *Bubble the correct answer.*

Use the image below to answer questions 1 and 2.

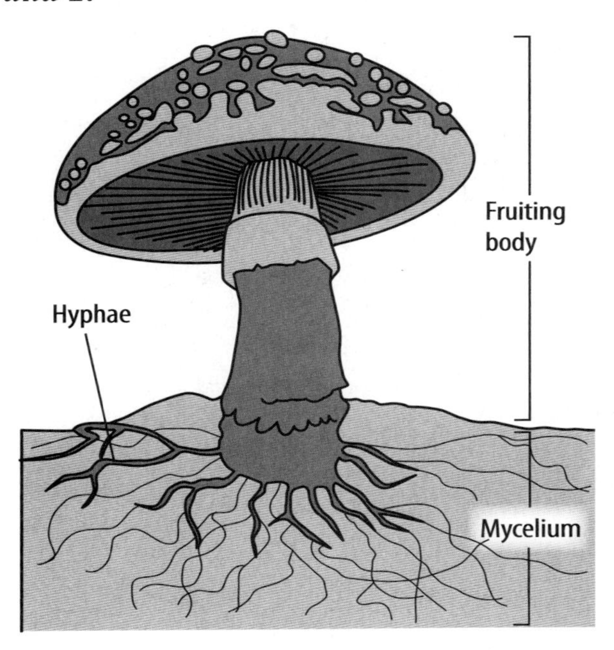

1. Which part of a typical fungus is responsible for taking in water and nutrients for the organism? **SC.6.L.14.3**

Ⓐ cap
Ⓑ gills
Ⓒ hyphae
Ⓓ mycelium

2. What type of fungus is shown above? **SC.6.L.14.6**

Ⓕ club
Ⓖ imperfect
Ⓗ sac
Ⓘ zygote

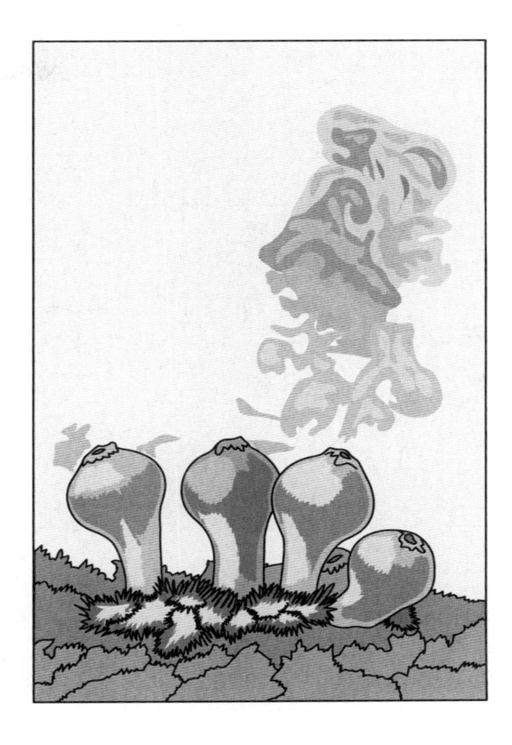

3. Look at the "smoke" rising from the puffballs above. This "smoke" is an example of **SC.6.L.14.3**

Ⓐ asexual reproduction.
Ⓑ fruiting body growth.
Ⓒ hyphae growth.
Ⓓ sexual reproduction.

4. How do fungi and other organisms sometimes interact? **SC.6.L.14.6**

Ⓕ Fungi provide minerals to animals.
Ⓖ Fungi provide sugars to plants.
Ⓗ Fungi exchange minerals for sugars with animals.
Ⓘ Fungi exchange minerals for sugars with plants.

Name ______________________ Date ____________

Notes

Name ______________________ Date ______________

Notes

Glossary/Glosario

Multilingual eGlossary

A science multilingual glossary is available on the science Web site. The glossary includes the following languages.

Arabic	Hmong	Tagalog
Bengali	Korean	Urdu
Chinese	Portuguese	Vietnamese
English	Russian	
Haitian Creole	Spanish	

Cómo usar el glosario en español:

1. Busca el término en inglés que desees encontrar.
2. El término en español, junto con la definición, se encuentran en la columna de la derecha.

Pronunciation Key

Use the following key to help you sound out words in the glossary.

a	b**a**ck (BAK)	**ew**	f**oo**d (FEWD)
ay	d**ay** (DAY)	**yoo**	p**u**re (PYOOR)
ah	f**a**ther (FAH thur)	**yew**	f**ew** (FYEW)
ow	fl**ow**er (FLOW ur)	**uh**	comm**a** (CAH muh)
ar	c**ar** (CAR)	**u (+ con)**	r**u**b (RUB)
e	l**e**ss (LES)	**sh**	**sh**elf (SHELF)
ee	l**ea**f (LEEF)	**ch**	na**t**ure (NAY chur)
ih	tr**i**p (TRIHP)	**g**	**g**ift (GIHFT)
i (i + com + e)	**i**dea (i DEE uh)	**j**	**g**em (JEM)
oh	g**o** (GOH)	**ing**	s**ing** (SING)
aw	s**o**ft (SAWFT)	**zh**	vi**si**on (VIH zhun)
or	**or**bit (OR buht)	**k**	ca**k**e (KAYK)
oy	c**oi**n (COYN)	**s**	**s**eed, **c**ent (SEED, SENT)
oo	f**oo**t (FOOT)	**z**	**z**one, rai**s**e (ZOHN, RAYZ)

English — A — Español

abrasion/air mass — **abrasión/masa de aire**

abrasion: the grinding away of rock or other surfaces as particles carried by wind, water, or ice scrape against them.

abrasión: desgaste de una roca o de otras superficies a medida que las partículas transportadas por el viento, el agua o el hielo las raspan.

acceleration: a measure of the change in velocity during a period of time.

aceleración: medida del cambio de velocidad durante un periodo de tiempo.

acid precipitation: precipitation that has a lower pH than that of normal rainwater (5.6).

precipitación ácida: precipitación que tiene un pH más bajo que el del agua de la lluvia normal (5.6).

active transport: the movement of substances through a cell membrane using the cell's energy.

transporte activo: movimiento de sustancias a través de la membrana celular usando la energía de la célula.

air mass: a large area of air that has uniform temperature, humidity, and pressure.

masa de aire: gran área de aire que tiene temperatura, humedad y presión uniformes.

air pollution: the contamination of air by harmful substances including gases and smoke.

air pressure: the pressure that a column of air exerts on the air, or a surface, below it.

air resistance: the frictional force between air and objects moving through it.

alga (plural, algae): a plantlike protist that produces food through photosynthesis using light energy and carbon dioxide.

amoeba (uh MEE buh): one common sarcodine with an unusual adaptation for movement and getting nutrients.

antibiotic (an ti bi AH tihk): a medicine that stops the growth and reproduction of bacteria.

antibody: a protein that can attach to a pathogen and makes it useless.

ascus (AS kuhs): the reproductive structure where spores develop on sac fungi.

atmosphere (AT muh sfihr): a thin layer of gases surrounding Earth.

polución del aire: contaminación del aire por sustancias dañinas, como gases y humo.

presión del aire: presión que una columna de aire ejerce sobre el aire o sobre la superficie debajo de ella.

resistencia al aire: fuerza de fricción entre el aire y los objetos que se mueven a través de él.

alga (plural, algas): protista parecida a una planta que produce el alimento por medio de la fotosíntesis, usando la energía lumínica y el dióxido de carbono.

ameba: sarcodina común con una adaptación inusual para moverse y obtener nutrientes.

antibiótico: medicina que detiene el crecimiento y reproducción de las bacterias.

anticuerpo: proteína que se adhiere a un patógeno y lo hace inútil.

ascus: estructura reproductiva donde se desarrollan las esporas en un hongo con saco.

atmósfera: capa delgada de gases que rodean la Tierra.

B

bacterium: a microscopic prokaryote.

basidium (buh SIH dee uhm): reproductive structure that produces sexual spores inside the basidiocarp.

binomial nomenclature: a naming system that gives each organism a two-word scientific name.

bioremediation (bi oh rih mee dee AY shun): the use of organisms, such as bacteria, to clean up environmental pollution.

biosphere: the parts of Earth and the surrounding atmosphere where there is life.

biota: all of the organisms that live in a region.

blizzard: a violent winter storm characterized by freezing temperatures, strong winds, and blowing snow.

bacteria: procariota microscópica.

basidio: estructura reproductiva que produce esporas sexuales en el interior de un basidiocarpo.

nomenclatura binomial: sistema de nombrar que le da a cada organismo un nombre científico de dos palabras.

biorremediación: uso de microorganismos, como bacterias, para limpiar la contaminación del medioambiente.

biosfera: partes de la Tierra y de la atmósfera que la rodea donde hay vida.

biota: todos los organismos que viven en una región.

ventisca: tormenta violenta de invierno caracterizada por temperaturas heladas, vientos fuertes, y nieve que sopla.

C

Calorie: the amount of energy it takes to raise the temperature of 1 kg of water by 1°C.

carbohydrate (kar boh HI drayt): a macromolecule made up of one or more sugar molecules, which are composed of carbon, hydrogen, and oxygen; usually the body's major source of energy.

cell: the smallest unit of life.

cell cycle: a cycle of growth, development, and division that most cells in an organism go through.

cell differentiation (dihf uh ren shee AY shun): the process by which cells become different types of cells.

cell membrane: a flexible covering that protects the inside of a cell from the environment outside the cell.

cell theory: the theory that states that all living things are made of one or more cells, the cell is the smallest unit of life, and all new cells come from preexisting cells.

cellular respiration: a series of chemical reactions that convert the energy in food molecules into a usable form of energy called ATP.

cell wall: a stiff structure outside the cell membrane that protects a cell from attack by viruses and other harmful organisms.

centromere: a structure that holds sister chromatids together.

chemical weathering: the process that changes the composition of rocks and minerals due to exposure to the environment.

chloroplast (KLOR uh plast): a membrane-bound organelle that uses light energy and makes food—a sugar called glucose—from water and carbon dioxide in a process known as photosynthesis.

caloría: cantidad de energía necesaria para aumentar la temperatura de 1 kg de agua en 1°C.

carbohidrato: macromolécula constituida de una o más moléculas de azúcar, las cuales están compuestas de carbono, hidrógeno y oxígeno; usualmente es la mayor fuente de energía del cuerpo.

célula: unidad más pequeña de vida.

ciclo celular: ciclo de crecimiento, desarrollo y división por el que pasan la mayoría de células de un organismo.

diferenciación celular: proceso por el cual las células se convierten en diferentes tipos de células.

membrana celular: cubierta flexible que protege el interior de una célula del ambiente externo de la célula.

teoría celular: teoría que establece que todos los seres vivos están constituidos de una o más células (la célula es la unidad más pequeña de vida) y que las células nuevas provienen de células preexistentes.

respiración celular: serie de reacciones químicas que convierten la energía de las moléculas de alimento en una forma de energía utilizable llamada ATP.

pared celular: estructura rígida en el exterior de la membrana celular que protege la célula del ataque de virus y otros organismos dañinos.

centrómero: estructura que mantiene unidas las cromátidas hermanas.

meteorización química: proceso que cambia la composición de las rocas y los minerales debido a la exposición al medioambiente.

cloroplasto: organelo limitado por una membrana que usa la energía lumínica para producir alimento –un azúcar llamado glucosa– del agua y del dióxido de carbono en un proceso llamado fotosíntesis.

cilia (SIH lee uh): short, hairlike structures that grow on the surface of some protists.

cilios: estructuras cortas parecidas a un cabello que crecen en la superficie de algunos protistas.

cladogram: a branched diagram that shows the relationships among organisms, including common ancestors.

cladograma: diagrama de brazos que muestra las relaciones entre los organismos, incluidos los ancestros comunes.

climate: the long-term average weather conditions that occur in a particular region.

clima: promedio a largo plazo de las condiciones del tiempo atmosférico de una región en particular.

compact bone: the hard outer layer of bone.

hueso compacto: capa externa y dura del hueso.

compound microscope: a light microscope that uses more than one lens to magnify an object.

microscopio compuesto: microscopio de luz que usa más de un lente para aumentar la imagen de un objeto.

computer model: detailed computer programs that solve a set of complex mathematical formulas.

modelo de computadora: programas de computadora que resuelven un conjunto de fórmulas matemáticas complejas.

condensation: the process by which a gas changes to a liquid.

condensación: proceso mediante el cual un gas cambia a líquido.

conduction (kuhn DUK shun): the transfer of thermal energy due to collisions between particles.

conducción: transferencia de energía térmica debido a colisiones entre partículas.

conjugation (kahn juh GAY shun): a process during which two bacteria of the same species attach to each other and combine their genetic material.

conjugación: proceso durante el cual dos bacterias de la misma especie se adhieren una a la otra y combinan sus material genético.

contact force: a push or a pull on one object by another object that is touching it.

fuerza de contacto: empuje o arrastre ejercido sobre un objeto por otro que lo está tocando.

convection: the circulation of particles within a material caused by differences in thermal energy and density; the transfer of thermal energy by the movement of particles from one part of a material to another.

convección: circulación de partículas en el interior de un material causada por diferencias en la energía térmica y la densidad; transferencia de energía térmica por el movimiento de partículas de una parte de la materia a otra.

critical thinking: comparing what you already know with information you are given in order to decide whether you agree with it.

pensamiento crítico: comparación que se hace cuando se sabe algo acerca de información nueva, y se decide si se está o no de acuerdo con ella.

cryosphere: the frozen portion of water on Earth's surface.

criosfera: la parte de agua congelada sobre la superficie de la Tierra.

cytokinesis (si toh kuh NEE sus): a process during which the cytoplasm and its contents divide.

citocinesis: proceso durante el cual el citoplasma y sus contenidos se dividen.

cytoplasm: the liquid part of a cell inside the cell membrane; contains salts and other molecules.

citoplasma: fluido en el interior de una célula que contiene sales y otras moléculas.

cytoskeleton: a network of threadlike proteins joined together that gives a cell its shape and helps it move.

citoesqueleto: red de proteínas en forma de filamentos unidos que le da forma a la célula y le ayuda a moverse.

D

daughter cells: the two new cells that result from mitosis and cytokinesis.

células hija: las dos células nuevas que resultan de la mitosis y la citocinesis.

decomposition: the breaking down of dead organisms and organic waste.

descomposición: degradación de organismos muertos y desecho orgánico.

deforestation: the removal of large areas of forests for human purposes.

deforestación: eliminación de grandes áreas de bosques con propósitos humanos.

delta: a large deposit of sediment that forms where a stream enters a large body of water.

delta: depósito grande de sedimento que se forma donde una corriente entra a un cuerpo grande de agua.

dependent variable: the factor a scientist observes or measures during an experiment.

variable dependiente: factor que el científico observa o mide durante un experimento.

deposition: the laying down or settling of eroded material.

deposición: establecimiento o asentamiento de material erosionado.

description: a spoken or written summary of observations.

descripción: resumen oral o escrito de las observaciones.

dew point: temperature at which air is fully saturated because of decreasing temperatures while holding the amount of moisture constant.

punto de rocío: temperatura en la cual el aire está completamente saturado debido a la disminución en las temperaturas aunque mantiene constante la cantidad de humedad.

diatom (DI uh tahm): a type of microscopic plantlike protist with a hard outer wall.

diatomea: tipo de protista microscópico parecido a una planta que tiene una pared externa dura.

dichotomous key: a series of descriptions arranged in pairs that lead the user to the identification of an unknown organism.

clave dicotómica: serie de descripciones organizadas en pares que dan al usuario la identificación de un organismo desconocido.

diffusion: the movement of substances from an area of higher concentration to an area of lower concentration.

difusión: movimiento de sustancias de un área de mayor concentración a un área de menor concentración.

displacement: the difference between the initial, or starting, position and the final position of an object that has moved.

desplazamiento: diferencia entre la posición inicial, o salida, y la final de un objeto que se ha movido.

distance-time graph: a graph that shows how distance and time are related.

gráfico distancia-tiempo: gráfico que muestra cómo se relacionan la distancia y el tiempo.

Doppler radar: a specialized type of radar that can detect precipitation as well as the movement of small particles, which can be used to approximate wind speed.

radar Dopler: tipo de radar especializado que detecta tanto la precipitación como el movimiento de partículas pequeñas, que se pueden usar para determinar la velocidad aproximada del viento.

drought: a period of below-average precipitation.

dune: a pile of windblown sand.

sequía: período de bajo promedio de precipitación.

duna: montón de arena que el viento transporta.

E

electric energy: energy carried by an electric current.

electron microscope: a microscope that uses a magnetic field to focus a beam of electrons through an object or onto an object's surface.

El Niño/Southern Oscillation: the combined ocean and atmospheric cycle that results in weakened trade winds across the Pacific Ocean.

endocytosis (en duh si TOH sus): the process during which a cell takes in a substance by surrounding it with the cell membrane.

endospore (EN doh spor): a thick internal wall that a bacterium builds around its chromosome and part of its cytoplasm.

energy: the ability to cause change.

erosion: the moving of weathered material, or sediment, from one location to another.

evaporation: the process of a liquid changing to a gas at the surface of the liquid.

exocytosis (ek soh si TOH sus): the process during which a cell's vesicles release their contents outside the cell.

explanation: an interpretation of observations.

energía eléctrica: energía transportada por una corriente eléctrica.

microscopio electrónico: microscopio que usa un campo magnético para enfocar un haz de electrones a través de un objeto o sobre la superficie de un objeto.

El Niño/Oscilación meridional: ciclo atmosférico y oceánico combinado que produce el debilitamiento de los vientos alisios en el Océano Pacífico.

endocitosis: proceso durante el cual una célula absorbe una sustancia rodeándola con la membrana celular.

endospora: pared interna gruesa que una bacteria produce alrededor del cromosoma y parte del citoplasma.

energía: capacidad de ocasionar cambio.

erosión: transporte de material meteorizado, o de sedimento, de un lugar a otro.

evaporación: proceso mediante el cual un líquido cambia a gas en la superficie del líquido.

exocitosis: proceso durante el cual las vesículas de una célula liberan sus contenidos fuera de la célula.

explicación: interpretación de las observaciones.

F

facilitated diffusion: the process by which molecules pass through a cell membrane using special proteins called transport proteins.

fermentation: a reaction that eukaryotic and prokaryotic cells can use to obtain energy from food when oxygen levels are low.

difusión facilitada: proceso por el cual las moléculas pasan a través de la membrana celular usando proteínas especiales, llamadas proteínas de transporte.

fermentación: reacción que las células eucarióticas y procarióticas usan para obtener energía del alimento cuando los niveles de oxígeno son bajos.

fertilization (fur tuh luh ZAY shun): a reproductive process in which a sperm joins with an egg.

fertilización: proceso reproductivo en el cual un espermatozoide se une con un óvulo.

fission: cell division that forms two genetically identical cells.

fisión: división celular que forma dos células genéticamente idénticas.

flagellum (fluh JEH lum): a long whiplike structure on many bacteria.

flagelo: estructura larga similar a un látigo que tienen muchas bacterias.

force: a push or a pull on an object.

fuerza: empuje o arrastre ejercido sobre un objeto.

friction: a contact force that resists the sliding motion of two surfaces that are touching.

fricción: fuerza que resiste el movimiento de dos superficies que están en contacto.

front: a boundary between two air masses.

frente: límite entre dos masas de aire.

G

gamete (GA meet): human reproductive cell.

gameto: célula reproductora humana.

genus (JEE nus): a group of similar species.

género: grupo de especies similares.

geosphere: the solid part of Earth.

geosfera: parte sólida de la Tierra.

glacier: a large mass of ice, formed by snow accumulation on land, that moves slowly across Earth's surface.

glaciar: masa enorme de hielo, formada por la acumulación de nieve en la tierra, que se mueve lentamente por la superficie de la Tierra.

global climate model: a set of complex equations used to predict future climates.

modelo de clima global: conjunto de ecuaciones complejas para predecir climas futuros.

global warming: an increase in the average temperature of Earth's surface

calentamiento global: incremento en la temperatura promedio de la superficie de la Tierra.

glycolysis: a process by which glucose, a sugar, is broken down into smaller molecules.

glucólisis: proceso por el cual la glucosa, un azúcar, se divide en moléculas más pequeñas.

gravity: an attractive force that exists between all objects that have mass.

gravedad: fuerza de atracción que existe entre todos los objetos que tienen masa.

greenhouse gas: a gas in the atmosphere that absorbs Earth's outgoing infrared radiation.

gas de invernadero: gas en la atmósfera que absorbe la salida de radiación infrarroja de la Tierra.

groundwater: water that is stored in cracks and pores beneath Earth's surface.

agua subterránea: agua almacenada en grietas y poros debajo de la superficie de la Tierra.

H

heat: the movement of thermal energy from a region of higher temperature to a region of lower temperature.

high-pressure system: a large body of circulating air with high pressure at its center and lower pressure outside of the system.

homeostasis (hoh mee oh STAY sus): an organism's ability to maintain steady internal conditions when outside conditions change.

horizons: layers of soil formed from the movement of the products of weathering.

hormone: a chemical signal that is produced by an endocrine gland in one part of an organism and carried in the bloodstream to another part of the organism.

humidity (hyew MIH duh tee): the amount of water vapor in the air.

hurricane: an intense tropical storm with winds exceeding 119 km/h.

hydrosphere: the system containing all Earth's water.

hyphae (HI fee): long, threadlike structures that make up the body of fungi and also form an underground structure that absorbs minerals and water.

hypothesis: a possible explanation for an observation that can be tested by scientific investigations.

calor: movimiento de energía térmica desde una región de alta temperatura a una región de baja temperatura.

sistema de alta presión: gran cuerpo de aire circulante con presión alta en el centro y presión más baja fuera del sistema.

homeostasis: capacidad de un organismo de mantener las condiciones internas estables cuando las condiciones externas cambian.

horizontes: capas de suelo formadas por el movimiento de productos meteorizados.

hormona: señal química producido por una glándula endocrina en una parte de un organismo y llevado en la corriente sanguínea a otra parte del organismo.

humedad: cantidad de vapor de agua en el aire.

huracán: tormenta tropical intensa con vientos que exceden los 119 km/h.

hidrosfera: sistema que contiene toda el agua de la Tierra.

hifas: estructuras largas en forma de filamentos que constituyen el cuerpo de los hongos y que también forman una estructura subterránea que absorbe minerales y agua.

hipótesis: explicación posible de una observación que se puede probar por medio de investigaciones científicas.

I

ice age: a period of time when a large portion of Earth's surface is covered by glaciers.

immunity: the resistance to specific pathogens, or disease-causing agents.

independent variable: the factor that is changed by the investigator to observe how it affects a dependent variable.

era del hielo: período de tiempo cuando los glaciares cubren una gran porción de la superficie de la Tierra.

inmunidad: resistencia a patógenos específicos o a agentes causantes de enfermedades.

variable independiente: factor que el investigador cambia para observar cómo afecta la variable dependiente.

inference: a logical explanation of an observation that is drawn from prior knowledge or experience.

inferencia: explicación lógica de una observación que se extrae de un conocimiento previo o experiencia.

interglacial: a warm period that occurs during an ice age.

interglacial: período tibio que ocurre durante una era del hielo.

International System of Units (SI): the internationally accepted system of measurement.

Sistema Internacional de Unidades (SI): sistema de medidas aceptado internacionalmente.

interphase: the period during the cell cycle of a cell's growth and development.

interfase: período durante el ciclo celular del crecimiento y desarrollo de una célula.

ionosphere: a region within the mesosphere and thermosphere containing ions.

ionosfera: región entre la mesosfera y la termosfera que contiene iones.

isobar: lines that connect all places on a map where pressure has the same value.

isobara: línea que conectan todos los lugares en un mapa donde la presión tiene el mismo valor.

J

jet stream: a narrow band of high winds located near the top of the troposphere.

corriente de chorro: banda angosta de vientos fuertes cerca de la parte superior de la troposfera.

K

kinetic (kuh NEH tik) energy: energy due to motion.

energía cinética: energía debida al movimiento.

L

land breeze: a wind that blows from the land to the sea due to local temperature and pressure differences.

brisa terrestre: viento que sopla desde la tierra hacia el mar debido a diferencias en la temperatura local y la presión.)

landslide: rapid, downhill movement of soil, loose rocks, and boulders.

deslizamiento de tierra: movimiento rápido del suelo, rocas sueltas y canto rodado, pendiente abajo.

law of conservation of energy: law that states that energy can be transformed from one form to another, but it cannot be created or destroyed.

ley de la conservación de la energía: ley que plantea que la energía puede transformarse de una forma a otra, pero no puede crearse ni destruirse.

lichen (LI kun): a structure formed when fungi and certain other photosynthetic organisms grow together.

líquen: estructura formada cuando crecen juntos los hongos y algunos organismos que realizan la fotosíntesis.

light microscope: a microscope that uses light and lenses to enlarge an image of an object.

microscopio de luz: microscopio que usa l uz y lentes para aumentar la imagen de un objeto.

lipid: a large macromolecule that does not dissolve in water.

lípido: macromolécula extensa que no se disuelve en agua.

loess (LUHS): a crumbly, windblown deposit of silt and clay.

loess: depósito quebradizo de limo y arcilla transportados por el viento.

longshore current: a current that flows parallel to the shoreline.

corriente costera: corriente que fluye paralela a la costa.

low-pressure system: a large body of circulating air with low pressure at its center and higher pressure outside of the system.

sistema baja presión: gran cuerpo de aire circulante con presión baja en el centro y presión más alta fuera del sistema.

lymphocyte (LIHM fuh site): a type of white blood cell that is made in the thymus, the spleen, and bone marrow.

linfocito: tipo de glóbulos blancos que se producen en el timo, el bazo y la médula del hueso.

M

macromolecule: substance that forms from joining many small molecules together.

macromolécula: Sustancia que se forma al unir muchas moléculas pequeñas.

mass wasting: the downhill movement of a large mass of rocks or soil due to gravity.

transporte en masa: movimiento cuesta debajo de gran cantidad de roca o suelo debido a la fuerza de gravedad.

meander: a broad, C-shaped curve in a stream.

meandro: curva pronunciada en forma de C en un arroyo.

mechanical energy: sum of the potential energy and the kinetic energy in a system.

energía mecánica: suma de la energía potencial y la energía cinética en un sistema.

mechanical weathering: physical processes that naturally break rocks into smaller pieces.

meteorización mecánica: proceso físico natural mediante el cual se rompe una roca en pedazos más pequeños.

microclimate: a localized climate that is different from the climate of the larger area surrounding it.

microclima: clima localizado que es diferente del clima de área más extensa que lo rodea.

mineral: a naturally occurring, inorganic solid that has a crystal structure and a definite chemical composition.

mineral: sólido inorgánico de origen natural que tiene estructura de cristal y composición química definida.

mitosis (mi TOH sus): a process during which the nucleus and its contents divide.

mitosis: proceso durante el cual el núcleo y sus contenidos se divide.

monsoon: a wind circulation pattern that changes direction with the seasons.

monsón: patrón de viento circulante que cambia de dirección con las estaciones.

moraine: a mound or ridge of unsorted sediment deposited by a glacier.

morrena: monte o colina de sedimento sin clasificar depositado por un glacial.

motion: the process of changing position.

movimiento: proceso de cambiar de posición.

multicellular: a living thing that is made up of two or more cells.

pluricelular: ser vivo formado por dos o más células.

mycelium (mi SEE lee um): an underground network of hyphae.

micelio: red subterránea de hifas.

mychorriza (mi kuh RI zuh): a structure formed when the roots of a plant and the hyphae of a fungus weave together.

micorriza: estructura formada cuando las raíces de una planta y las hifas de de un hongo se entrelazan.

neuron (NOO rahn): the basic functioning unit of the nervous system; a nerve cell.

Newton's first law of motion: law that states that if the net force acting on an object is zero, the motion of the object does not change.

Newton's second law of motion: law that states that the acceleration of an object is equal to the net force exerted on the object divided by the object's mass.

Newton's third law of motion: law that states that for every action there is an equal and opposite reaction.

nitrogen fixation (NI truh jun • fihk SAY shun): the process that changes atmospheric nitrogen into nitrogen compounds that are usable by living things.

noncontact force: a force that one object applies to another object without touching it.

nuclear energy: energy stored in and released from the nucleus of an atom.

nucleic acid: a macromolecule that forms when long chains of molecules called nucleotides join together.

nucleus: part of a eukaryotic cell that directs cell activity and contains genetic information stored in DNA.

nutrient: a part of food used by the body to grow and survive.

neurona: unidad básica de funcionamiento del sistema nervioso; célula nerviosa.

primera ley del movimiento de Newton: ley que establece que si la fuerza neta ejercida sobre un objeto es cero, el movimiento de dicho objeto no cambia.

segunda ley del movimiento de Newton: ley que establece que la aceleración de un objeto es igual a la fuerza neta que actúa sobre él divida por su masa.

tercera ley del movimiento de Newton: ley que establece que para cada acción hay una reacción igual en dirección opuesta.

fijación de nitrógeno: proceso por el cual el nitrógeno atmosférico se transforma en compuestos de nitrógeno que los seres vivos usan.

fuerza de no contacto: fuerza que un objeto puede aplicar sobre otro sin tocarlo.

energía nuclear: energía almacenada en y liberada por el núcleo de un átomo.

ácido nucléico: macromolécula que se forma cuando cadenas largas de moléculas llamadas nucleótidos se unen.

núcleo: parte de la célula eucariótica que gobierna la actividad celular y contiene la información genética almacenada en el ADN.

nutriente: parte del alimento que el cuerpo usa para crecer y vivir.

observation: the act of using one or more of your senses to gather information and take note of what occurs.

organ: a group of different tissues working together to perform a particular job.

organelle: membrane-surrounded component of a eukaryotic cell with a specialized function.

organic matter: remains of something that was once alive.

observación: acción de usar uno o más sentidos para reunir información y tomar notar de lo que ocurre.

órgano: grupo de diferentes tejidos que trabajan juntos para realizar una función específica.

organelo: componente de una célula eucariótica rodeado de una membrana con una función especializada.

materia orgánica: restos de algo que una vez estuvo vivo.

organism: something that has all the characteristics of life.

organ system: a group of organs that work together and perform a specific task.

osmosis: the diffusion of water molecules only through a membrane.

outwash: layered sediment deposited by streams of water that flow from a melting glacier.

ovum (OH vum): female reproductive cell, or gamete.

oxidation: the process that combines the element oxygen with other elements or molecules.

ozone layer: the area of the stratosphere with a high concentration of ozone.

organismo: algo que tiene todas las características de la vida.

sistema de órganos: grupo de órganos que trabajan juntos y realizar una función específica.

ósmosis: difusión de las moléculas de agua únicamente a través de una membrana.

sandur: capas de sedimentos depositados por las corrientes de agua que fluyen de un glaciar en deshielo.

óvulo: célula reproductora femenina, o gameto.

oxidación: proceso por el cual se combina el elemento oxígeno con otros elementos o moléculas.

capa de ozono: área de la estratosfera con gran concentración de ozono.

paramecium (pa ruh MEE see um): a protist with cilia and two types of nuclei.

parent material: the starting material of soil consisting of rock or sediment that is subject to weathering.

particulate (par TIH kyuh lut) matter: the mix of both solid and liquid particles in the air.

passive transport: the movement of substances through a cell membrane without using the cell's energy.

pasteurization (pas chuh ruh ZAY shun): a process of heating food or liquid to a temperature that kills most harmful bacteria.

pathogen (PA thuh jun): an agent that causes disease.

photochemical smog: air pollution that forms from the interaction between chemicals in the air and sunlight.

photosynthesis: a series of chemical reactions that convert light energy, water, and CO_2 into the food-energy molecule glucose and give off oxygen.

polar easterlies: cold winds that blow from the east to the west near the North Pole and South Pole.

Paramecio: protista con cilios y dos tipos de núcleos.

material parental: material original del suelo compuesto de roca o sedimento sujeto a meteorización.

partículas en suspensión: mezcla de partículas tanto sólidas como líquidas en el aire.

transporte pasivo: movimiento de sustancias a través de una membrana celular sin usar la energía de la célula.

pasteurización: proceso en el cual se calientan los alimentos o líquidos para matar la mayoría de bacterias dañinas.

patógeno: agente que causa enfermedad.

smog fotoquímico: polución del aire que se forma de la interacción entre los químicos en el aire y la luz solar.

fotosíntesis: serie de reacciones químicas que convierten la energía lumínica, el agua y el CO_2 en glucosa, una molécula de energía alimentaria, y libera oxígeno.

brisas polares: vientos fríos que soplan del este al oeste cerca del Polo Norte y del Polo Sur.

pores: small holes and spaces in soil.
position: an object's distance and direction from a reference point.
potential (puh TEN chul) energy: stored energy due to the interactions between objects or particles.
precipitation: water, in liquid or solid form, that falls from the atmosphere.
prediction: a statement of what will happen next in a sequence of events.
protein: a long chain of amino acid molecules; contains carbon, hydrogen, oxygen, nitrogen, and sometimes sulfur.
protist: a member of a group of eukaryotic organisms, which have a membrane-bound nucleus.
protozoan (proh tuh ZOH un): a protist that resembles a tiny animal.
pseudopod: a temporary "foot" that forms as the organism pushes part of its body outward.

poros: huecos y espacios pequeños en el suelo.
posición: distancia y dirección de un objeto según un punto de referencia.)
energía potencia: energía almacenada debido a las interacciones entre objetos o partículas.
precipitación: agua, en forma líquida o sólida, que cae de la atmósfera.
predicción: afirmación de lo que ocurrirá después en una secuencia de eventos.
proteína: larga cadena de aminoácidos; contiene carbono, hidrógeno, oxígeno, nitrógeno y, algunas veces, sulfuro.
protista: miembro de un grupo de organismos eucarióticos que tienen un núcleo limitado por una membrana.
protozoario: protista que parece un animal pequeño.
Seudópodo: "pata" temporal que se forma a medida que el organismo empuja parte del cuerpo hacia afuera.

R

radiant energy: energy carried by an electromagnetic wave.
radiation: the transfer of thermal energy by electromagnetic waves.
rain shadow: an area of low rainfall on the downwind slope of a mountain.
reference point: the starting point you use to describe the motion or the position of an object.
reflex: an automatic movement in response to a stimulus.
relative humidity: the amount of water vapor present in the air compared to the maximum amount of water vapor the air could contain at that temperature.
reproduction: the process by which new organisms are produced.
rock: a naturally occurring solid composed of minerals, rock fragments, and sometimes other materials such as organic matter.

energía radiante: energía que transporta una onda electromagnética.
radiación: transferencia de energía térmica por ondas electromagnéticas.
sombra de lluvia: área de baja precipitación en la ladera de sotavento de una montaña.
punto de referencia: punto que se escoge para describir la ubicación, o posición, de un objeto.
reflejo: movimiento automático en respuesta a un estímulo.
humedad relativa: cantidad de vapor de agua presente en el aire comparada con la cantidad máxima de vapor de agua que el aire podría contener en esa temperatura.
reproducción: proceso por el cual se producen nuevos organismos.
roca: sólido de origen natural compuesto de minerales, acumulación de fragmentos y algunas veces de otros materiales como materia orgánica.

rock cycle: the series of processes that change one type of rock into another type of rock.

ciclo geológico: series de procesos que cambian un tipo de roca en otro tipo de roca.

S

science: the investigation and exploration of natural events and of the new information that results from those investigations.

scientific law: a rule that describes a pattern in nature.

scientific theory: an explanation of observations or events that is based on knowledge gained from many observations and investigations.

sea breeze: a wind that blows from the sea to the land due to local temperature and pressure differences.

significant digits: the number of digits in a measurement that are known with a certain degree of reliability.

sister chromatids: two identical chromosomes that make up a duplicated chromosome.

soil: a mixture of weathered rock, rock fragments, decayed organic matter, water, and air.

sound energy: energy carried by sound waves.

species (SPEE sheez): a group of organisms that have similar traits and are able to produce fertile offspring.

specific heat: the amount of thermal energy (joules) needed to raise the temperature of 1 kg of material 1°C.

speed: the distance an object moves divided by the time it takes to move that distance.

speed-time graph: a graph that shows the speed of an object on the *y*-axis and time on the *x*-axis.

sperm: a male reproductive, or sex, cell; forms in a testis.

spongy bone: the interior region of bone that contains many tiny holes.

ciencia: la investigación y exploración de los eventos naturales y de la información nueva que es el resultado de estas investigaciones.

ley científica: regla que describe un patrón dado en la naturaleza.

teoría científica: explicación de observaciones o eventos con base en conocimiento obtenido de muchas observaciones e investigaciones.

brisa marina: viento que sopla del mar hacia la tierra debido a diferencias en la temperatura local y la presión.

cifras significativas: número de dígitos que se conoce con cierto grado de fiabilidad en una medida.

cromátidas hermanas: dos cromosomas idénticos que constituyen un cromosoma duplicado.

suelo: mezcla de roca meteorizada, fragmentos de rocas, materia orgánica descompuesta, agua y aire.

energía sonora: energía que transportan las ondas sonoras.

especie: grupo de organismos que tienen rasgos similares y que están en capacidad de producir crías fértiles.

calor específico: cantidad de energía térmica (julios) requerida para subir la temperatura de 1 kg de materia a 1°C.

rapidez: distancia que un objeto recorre dividida por el tiempo que éste tarda en recorrer dicha distancia.

gráfico rapidez-tiempo: gráfico que muestra la rapidez de un objeto en el eje Y y el tiempo en el eje X.

espermatozoide: célula reproductora masculina o sexual; forma en un testículo.

hueso esponjoso: región interior de un hueso que contiene muchos huecos diminutos.

stability: whether circulating air motions will be strong or weak.

estabilidad: condición en la que los movimientos del aire circulante pueden ser fuertes o débiles.

stem cell: an unspecialized cell that is able to develop into many different cell types.

célula madre: célula no especializada que tiene la capacidad de desarrollarse en diferentes tipos de células.

stratosphere (STRA tuh sfihr): the atmospheric layer directly above the troposphere.

estratosfera: capa atmosférica justo arriba de la troposfera.

surface report: a description of a set of weather measurements made on Earth's surface.

informe de superficie: descripción de un conjunto de mediciones del tiempo realizadas en la superficie de la Tierra.

talus: a pile of angular rocks and sediment from a rockfall.

talus: montón de rocas angulares y sedimentos de un derrumbe de montaña.

technology: the practical use of scientific knowledge, especially for industrial or commercial use.

tecnología: uso práctico del conocimiento científico, especialmente para uso industrial o comercial.

temperature inversion: a temperature increase as altitude increases in the troposphere.

inversión de temperatura: aumento de la temperatura en la troposfera a medida que aumenta la altitud.

thermal conductor: a material through which thermal energy flows quickly.

conductor térmico: material en el cual la energía térmica se mueve con rapidez.

thermal energy: the sum of the kinetic energy and the potential energy of the particles that make up an object.

energía térmica: suma de la energía cinética y potencial de las partículas que componen un objeto.

thermal insulator: a material through which thermal energy flows slowly.

aislante térmico: material a través del cual la energía térmica fluye con lentitud.

till: a mixture of various sizes of sediment that has been deposited by a glacier.

till: mezcla de varios tamaños de sedimento depositado por un glaciar.

tissue: a group of similar types of cells that work together to carry out specific tasks.

tejido: grupo de tipos similares de células que trabajan juntas para llevar a cabo diferentes funciones.

topography: the shape and steepness of the landscape.

topografía: forma e inclinación del paisaje.

tornado: a violent, whirling column of air in contact with the ground.

tornado: columna de aire violenta y rotativa en contacto con el suelo.

trade winds: steady winds that flow from east to west between 30°N latitude and 30°S latitude.

vientos alisios: vientos constantes que soplan del este al oeste entre 30° N de latitud y 30° S de latitud.

transpiration: the process by which plants release water vapor through their leaves.

transpiración: proceso por el cual las plantas liberan vapor de agua por medio de las hojas.

troposphere (TRO puh sfihr): the atmospheric layer closest to Earth's surface.

troposfera: capa atmosférica más cercana a la Tierra.

U

unicellular: a living thing that is made up of only one cell.

unicelular: ser vivo formado por una sola célula.

uplift: the process that moves large bodies of Earth materials to higher elevations.

levantamiento: proceso por el cual se mueven grandes cuerpos de materiales de la Tierra hacia elevaciones mayores.

upper-air report: a description of wind, temperature, and humidity conditions above Earth's surface.

informe del aire superior: descripción de las condiciones del viento, de la temperatura y de la humedad por encima de la superficie de la Tierra.

V

vaccine: a mixture containing material from one or more deactivated pathogens, such as viruses.

vacuna: mezcla que contiene material de uno o más patógenos desactivados, como los virus.

variable: any factor that can have more than one value.

variable: cualquier factor que tenga más de un valor.

velocity: the speed and the direction of a moving object.

velocidad: rapidez y dirección de un objeto en movimiento.

virus: a strand of DNA or RNA surrounded by a layer of protein that can infect and replicate in a host cell.

virus: filamento de ADN o de ARN rodeado por una capa de proteína que puede infectar una célula huésped y replicarse en ella.

W

water cycle: the series of natural processes by which water continually moves throughout the hydrosphere.

ciclo del agua: serie de procesos naturales por los que el cual el agua se mueve continuamente en toda la hidrosfera.

water vapor: water in its gaseous form.

vapor de agua: agua en forma gaseosa.

weather: the atmospheric conditions, along with short-term changes, of a certain place at a certain time.

tiempo atmosférico: condiciones atmosféricas, junto con cambios a corto plazo, de un lugar determinado a una hora determinada.

weathering: the mechanical and chemical processes that change Earth's surface over time.

meteorización: procesos mecánicos y químicos que con el paso del tiempo cambian la superficie de la Tierra.

westerlies: steady winds that flow from west to east between latitudes 30°N and 60°N, and 30°S and 60°S.

vientos del oeste: vientos constantes que soplan de oeste a este entre latitudes 30° N y 60° N, y 30° S y 60° S.

wind: the movement of air from areas of high pressure to areas of low pressure.

viento: movimiento del aire desde áreas de alta presión hasta áreas de baja presión.

work: the amount of energy used as a force moves an object over a distance.

trabajo: cantidad de energía usada como fuerza que mueve un objeto a cierta distancia.

Z

zygosporangia (zi guh spor AN jee uh): tiny stalks formed when a zygote fungus undergoes sexual reproduction.

zygote (ZI goht): the cell that forms when a sperm cell fertilizes an egg cell.

zigosporangia: tallos diminutos que se forman cuando un hongo zigoto se somete a reproducción sexual.

zigoto: célula nueva que se forma cuando un espermatozoide fertiliza un óvulo.

Index

Italic numbers = illustration/photo **Bold numbers = vocabulary term**
lab = indicates entry is used in a lab on this page

A

Abrasion
explanation of, *47,* **90,** 100
wind, *90*
Absorption
explanation of, 129
in human body, 450, *450*
Academic Vocabulary, NOS 11, 29, 50, 75, 128, 178, 218, 291, 352, **378,** 432, 458, 536
Acceleration
explanation of, *292,* **292**
negative, 292
on speed-time graphs, 300, *300*
unbalanced forces and, 308, *308,* 310
Acid precipitation, 48, 143 *lab,* 145, **145**
Acid rain, 48, 143 *lab*
Acids, 48
Active transport, *390,* **390**
Active viruses, 510
Adenosine triphosphate (ATP)
explanation of, 382, 395
glucose converted into, 397
use of, 396
Adolescence, 476, *476*
Adrenal glands, *469*
Adulthood, 476
Aerobic bacteria, 495
Aerosols, 225
Aging, 476
A-horizon, 58, *58*
Air. *See also* **Atmosphere**
on Earth, 11, *11*
movement of, 136 *lab*
stable, 133
unstable, 133
Air-breathing catfish, 339
Air circulation
explanation of, 132
global wind belts and, 138, *138*
three-cell model of, 138, *138*
Air currents
global winds and, *137,* 137–139
local winds and, 140, *140*
Air masses, 174, *174,* 174–175
Air pollution. *See also* **Pollution**
acid precipitation as, 143 *lab,* 145
explanation of, **144**
indoor, 148
monitoring of, 147
movement of, 146, *146*
particulate matter as, 146
smog as, 145, *145*
sources of, 144
temperature inversion and, 133
Air pressure
altitude and, 124, *124*
explanation of, 26, *166,* **166**
Air quality
monitoring of, 147
standards for, 147
trends in, 148, *148*
Air resistance, 306
Air temperature
explanation of, 166
pressure and, 172 *lab,* 173
water vapor and, 167, *167*
Albedo, 212
Alcohol fermentation, 397
Algae
explanation of, *532,* **532**
importance of, 535, *535,* 541
types of, 534, *534*
Alluvial fan, 81, *81*
Alpine glaciers, 97, *97*
Altitude
air pressure and, 124, *124*
temperature and, 124, *124,* 205, 206, *206*
Alveoli, 454, *454*
Amoebas
explanation of, 428, *428, 537,* **537**
Amylase, 373
Anaphase, 421, *421*
Anaerobic bacteria, 495
Anemometer, NOS 18, *NOS 18,* 166, *166*
Animalia, 342, *342*
Animal-like protists
explanation of, 532, *532*
protozoans as, 536, 538, *538*
types of, *536,* 536–537, *537*
Animals
adaptations to climate by, 210
bacteria living in, 501, *501*
as cause of mechanical weathering, *47*
Antarctica
hole in ozone layer above, 126
ice sheets in, 97
temperature in, 205
Antarctic air masses, 174
Antibiotics
explanation of, **504,** 513
from fungi, 547, 549, *549*
resistance to, 504–505, *505,* 549
Antibodies
explanation of, 458, 459, **513**
Appendages (cell), 379, *379*
Aquino, Adriana, 339
Archaea, 343, *343,* 493, 497
Arctic air masses, 174, 175
Arete, *97*
Argon, 121
Aristotle, 341
Arizona, 219
Armored catfish, 339
Arteries, 455, *455*
Ascus, 545
Asexual reproduction
explanation of, 496, **532**
in fungi, 544, 545
Ash, 121, *121*
Athlete's foot, 546
Atmosphere. *See also* **Air**
air pressure and, 124, *124*
changes in, 26–27
composition of, 121
explanation of, **13,** *30,* **119**
gases in, 13, *13*
importance of, 119
interactions in, 30
layers of, 14, *122,* 122–123
origins of, 120
solid particles in, 121
temperature and, 124
three-cell model of circulation in, 138, *138*
unstable conditions in, 133
Atomic force microscope (AFM), 375
Atoms, 351
Atria, *455*
Auditory system, 468, *468*
Auroras, 123, *123*
Autotrophs, 550
Average speed, 289, 298

B

Bacteria
aerobic, 495
affecting investigations, 507
anaerobic, 495
antibiotic resistance by, 549
archaea vs., 497
beneficial, *501,* 501–503, *502, 503*
classification of, 343, *343*
detection of, 458, *458*
endospores and, 497, *497*
in environment, 500 *lab*
explanation of, **493**
food and, 503
function of, 495
harmful, *504,* 504–505, *505*
method to kill, 499
movement of, 496, *496*
photosynthetic, 550
reproduction of, 496, *496*
resistant to antibiotics, 504–505
size and shape of, 492 *lab,* 494, *494*
structure of, 494
Balanced forces
constant motion and, 308, *308*
explanation of, 307, *307*

Barometers, *166*
Barometric pressure, 166
Basidia, **544**
Basidiocarp, 544, *544*
Bauxite, 49
Beach erosion, 89, *89*
Beakers, NOS 16, *NOS 16*
Bedrock, 56, *57,* 58
Benchmark Mini-Assessment, NOS 30, NOS 31, NOS 32, 38, 39, 68, 69, 106, 107, 108, 156, 157, 158, 159, 196, 197, 198, 236, 237, 238, 278, 279, 280, 318, 319, 320, 360, 361, 362, 406, 407, 408, 409, 442, 443, 484, 485, 486, 522, 523, 524, 558, 559
B-horizon, 58, *58*
Binoculars, NOS 18, *NOS 18*
Binomial nomenclature, **343**
Biologists, **297**
Bioremediation, **503**
Biosphere
- explanation of, **12,** *30*
- water cycle in, 25

Biota, **57**
Birth, 466, 475
Bjerknes, Jacob, 176
Bladder, 452, *452*
Blizzards, *181,* **181**
Blood, 454
Blood cells
- function of, 461, 471
- red, 456, *456,* 461
- white, *456,* 456–458, 461, 464, 471

Blood types, 456, *456*
Blood vessels, 454, 455, *455*
Blue cheese, 546
Body temperature, 335
Bone marrow, 457, 464, 471, *471*
Bone marrow transplants, 471, *471*
Bones
- calcium storage in, 464, 469
- function of, 463–464, *464*
- types of, 464

Brain
- function of, 466, *466,* 468
- skull as protection for, 464

Bromine, 126
Bronchi, 453, *453,* 454, *454*
Brown bear, *343,* 344
Bryce Canyon National Park, *80*

C

Calcium, 464, 469
Calories, **451**
Cancer research, 514
Capillaries, 455
Carbohydrates, 372, *372,* **373,** 451
Carbon dioxide
- alcohol fermentation producing, 397, *397*
- in atmosphere, 13, *13,* 119–121
- energy use and, 263
- as greenhouse gas, 224, *224*
- in human body, 452, 453, 455, *455,* 456
- infrared radiation and, 130
- in photosynthesis, 396, 398, *398*
- produced by cellular respiration, 24
- sources of, 225
- vehicle emissions of, 228

Careers in Science, 126, 221
Carlsbad Caverns (New Mexico), *87*
Carnivora, *343*
Carrier proteins, 389, *389*
Cartilage, 463
Catfish, 339
Caves, 87, 88
Cell cycle
- cytokinesis and, 422, *422*
- explanation of, 415, *415*
- interphase of, *416,* 416–418, *417, 418*
- length of, 417, *417*
- mitosis and, *420, 421*
- mitotic phase of, 416, *418,* 419, *420,* 420–422, *421, 422*
- organelle replication and, 419, *419*
- phases of, 416, *416,* 418, *418*
- results of, 422–423

Cell differentiation
- in eukaryotes, *429*
- explanation of, **429**

Cell division
- in animal cells, 422
- explanation of, 429
- in humans, 422, 423
- organelles during, 419
- in plant cells, 422
- results of, 422–423

Cell membrane
- active transport in, 390, *390*
- diffusion in, 388, 389
- explanation of, **378**
- function of, 387, 386 *lab*

Cell organelles
- explanation of, 381
- manufacturing molecules in, 382
- nucleus as, 381, *381*
- processing, transporting and storing energy in, 383, *383*
- processing energy in, 382–383

Cells
- animal stem, 430
- appendages to, 379, *379*
- communication between, 373
- composition of, 368 *lab*
- cytoplasm in, 379
- cytoskeleton in, 379
- discovery of, 369, *369*
- energy processing in, 382–383
- energy use by, 336
- eukaryotic, 380
- explanation of, **332,** 427
- features of, 370, *370*
- functions of, 378, 391, 400
- macromolecules in, *372,* 372–373
- nucleus of, 381, *381*
- passive transport in, 387
- plant, 430, *430*
- production of, 414 *lab*
- prokaryotic, 380, *380*
- proteins in, 373, 382, 383
- size and shape of, 377, *377,* 391, 393
- water in, 371, *371*

Cell theory, *370,* **370**
Cellular respiration
- explanation of, 24, 395
- glycolysis in, 395, *395*
- reactions in mitochondria in, 396, *396*
- relationship between photosynthesis and, 398, *398*

Cell wall, 378, *378*
Cementation, 29
Centromere, *418,* **418**
Channel proteins, 389
Chapter Review, NOS 28–NOS 29, 34–35, 64–65, 102–103, 152–153, 192–193, 232–233, 274–275, 314–315, 356–357, 402–403, 438–439, 480–481, 518–519, 554–555
Chemical energy, 260, 261, *261*
Chemical potential energy, 251, *251*
Chemical weathering
- acids and, 48
- explanation of, **48,** 76, *76*
- oxidation and, 49, *49*
- rate of, 50
- water and, 48, *48*

Cherrapunji, India, 219
Childhood, 476, *476*
Chlorine, 126
Chlorofluorocarbons (CFCs), 126
Chloroplasts
- explanation of, *383,* **383,** 419
- reactions in, 398

Cholesterol, 373
C-horizon, 58, *58*
Chromatics
- as chromosomes, 419
- sister, 418, *418,* 419

Chromatin
- explanation of, 417
- replicated, 420, *420*

Chromosomes
- cell division and, 418, *418*
- explanation of, 381

Cilia, 379, *379,* **536**
Ciliates, 536
Circulatory system
- blood in, 456, *456*
- function of, 454, 458
- heart and vessels in, 455, *455*

Cirque, *97*
Cladograms, *345,* **345**
Clay
- deposits of, 90
- explanation of, 45, 76
- formation of, 46
- in low-energy environments, *79*
- properties of, *55*
- resulting from mechanical weathering, 46

Clear Air Act (1970), 147
Climate. *See also* **Temperature; Weather**
- adaptations to, 210, *210*
- comparison of, 204 *lab*
- effect of albedo on, 212

effect on soil of, 56, 60
erosion and, 77, *77*
explanation of, **27,** 56, **205**
factors affecting, 205–207, 222 *lab*
methods to classify, 208, *209*
microclimates, 208
Climate change. *See also* **Global warming**
carbon dioxide in atmosphere and, 263
environmental impact of, 226
human impact on, 224–225
hurricanes and, 171
methods to predict, 227–228
methods to reduce, 228
ozone layer and, 126
regional and global, 223, *223*
sources of information on, 214, *214,* 221
Climate cycles
causes of long-term, 215, *215*
explanation of, 214
ice ages and, 214, 215
short-term, 216–219
Clostridium botulinum, 504
Clouds
cumulus, 178, *178*
effect on climate, 225, *225*
explanation of, 168
formation of, 164 *lab,* 168, 169
variables used to describe, *166, 168, 169*
water cycle and, 169
Club fungi, 544, *544*
Coastlines
climate on, 205, 207
deposition along, 88
erosion along, 80, 87
CODIS (Combined DNA Index System), 425
Cold fronts, 176, *176*
Cold waves, 219
Colon, 451
Colorado Plateau, *83*
Compact bone, *464,* **464**
Compaction, 29
Compass, NOS 18, *NOS 18*
Complex, 432
Compound microscope, *350,* **350**
Computers, NOS 17, *NOS 17*
Conclusions, NOS 7
Condensation
explanation of, *23,* **25**
in water cycle, 169
Conduction, 131, 267
Conjugation, *496,* **496,** 536
Connective tissue, 431, 432
Constant speed, 289, 298, 299
Constant velocity, *291,* **291**
Contact force
explanation of, *304,* **304**
friction as, 306, *306*
Continental climate
explanation of, *209*
vegetation in, 210
Continental crust, 19
Continental polar air masses, 175
Continental tropical air masses, 175
Contractile vacuole, *428,* **428**
explanation of, 335, *335*
Convection, 131, 132, *268,* **268**
Convection cell, 138, *138*
Convection currents, 268, *268*
Core, 19, *19*
Coriolis effect, 138, 139
Cows, 501, *501*
Creep, 95, *95*
Crust, 19, *19*
Crystallization, 28
Cumulus clouds, *168,* 178
Cystic fibrosis, 514
Cytokinesis, 419, 422, *422*
Cytoplasm
in bacteria, 494, *494*
during cytokinesis, 419, 422
explanation of, **379,** 416
Cytoskeleton, *379,* **379**

Dams, 89
Data, NOS 19
Daughter cells, 419
Death Valley, California, *90*
Deciduous trees, 210
Decomposers, 547, *547*
Decomposition
explanation of, **55, 502**
role of soil organisms in, 57
Deforestation
carbon dioxide and, 225
explanation of, 224, **225**
Deltas, *88,* **88,** 100
Density, 268
Deoxyribonucleic acid (DNA)
explanation of, 372, 381
Dependent variables, NOS 21, NOS 25
Deposition
along coastlines, 88
along streams, 88, *88*
environments for, 79
explanation of, 29, 73, 77, **79,** 100
glacial, 81, 98, *98*
in Grand Canyon, *83*
groundwater, 88
landforms created by, 81, 85
from mass wasting, 95, *95*
wind, 90, *90*
Depositional environments
explanation of, 79
low-energy, *79*
Dermal tissue, 431, 432, *432*
Desalination, 21, *21*
Design Process, NOS 19
Description, NOS 12
Detect, 458
Dew point, 167
Diaphragm, *453,* 454
Diatom, *533,* **533**
Dichotomous keys
explanation of, *344,* **344**
use of, 347
Diffusion
explanation of, *388,* **388**
facilitated, 389, *389*
Digestion, *450*
Digestive system
functions of, *450,* 450–451, 467
human, 433
nutrient homeostasis in, 469
Dinoflagellates, 533, *533*
Direction
change in, 291, *291,* 308, *308*
of motion, 292
Diseases
bacterial, *504,* 504–505, *505*
from fungi, 545, 549
protists as cause of, 538
types of, 459, *459*
viral, 509, *512,* 512–514, *513*
Displacement, *288,* **288**
Distance, 305, *305*
Distance-time graphs
comparing motion in, 298, *298*
explanation of, *296,* **296**
method to make, 297, *297*
DNA
in bacteria, 494, 496, *496*
cell cycle and, 417, *417,* 418, *418,* 425
in human cells, 425
in viruses, 509, 511
Doldrums, 139
Domains, 342, *342*
Dominate, 178
Doppler radar, 186
Downdrafts, 178
Drought
effects of, 226
explanation of, **219**
Dry climate
adaptations to, 210, *210*
explanation of, *209*
Dunes, *90,* **90**

E

Earth
curved surface of, 206
effect of orbit and tilt of axis on, 213 *lab,* 215, *215,* 216, *216*
effect of water and wind on, 84 *lab*
energy on, 129
reshaping surface of, 75
revolution of, 217, *217*
Earth science, NOS 5
Earth systems
explanation of, 11–12, 22 *lab*
interactions within, 30
Ecosystems
role of algae in, 535, *535*
role of fungus-like protists in, 539
Eggs
function of, 474
maturation of, 475
size of, 473, 472 *lab*
Elastic potential energy, 251, *251*
Electrical appliances, *253,* 260
Electric energy
changes to radiant energy, 257, *257*
explanation of, *253*
use of, 260, 261, 263

Electromagnetic waves
explanation of, *253,* 269, *269*
transfer of energy by, 128
Electron microscope, *351,* **351,** 352
El Niño, 218
El Niño/Southern Oscillation (ENSO), 218
Embryo, 475, *475*
Endocrine system
function of, 466, 469, *469*
reproduction and, 473–475
Endocytosis, 367, *390,* **390**
Endoplasmic reticulum, 382, *382*
Endospores
explanation of, **497**
formation of, *497*
Energy. *See also* **Thermal energy**
changes between forms of, 256 *lab,* 257, *257,* 258, *258*
changes caused by, 249, *249*
on Earth, 129
energy processed in, 382–383
explanation of, **249**
forms of, *253*
kinetic, 250, *250,* 255
law of conservation of, *258,* 258–259
measurement of, 251
potential, 250–251, *251,* 255
stored in carbohydrates, 373
from Sun, 128, 137
thermal, 449
use of, *260,* 260–261, *261,* 263, 336, *336, 337*
work and, 252, *252*
Envelope, 381
Environmental protection. *See also* **Land use practices**
explanation of, **50**
technological trends for, 228, *228*
Environments
climate change and, 226
depositional, **79**
high-energy, 79
low-energy, 79, *79*
Enzymes, 450
Epithelial tissue, 431, 432
Equator
solar energy and, 206
temperature near, 205
Equinox, 217, *217*
Erosion
beach, *83, 87,* 89
coastal, 80, *83,* 87, *87*
explanation of, 29, *77,* **77,** 100
glacial, 80, 97, *97*
of Grand Canyon, *83*
groundwater, 87, *87*
landforms created by, 80, *80*
from land use practices, 89, 90
from mass wasting, 95, *95*
rate of, 77
rock type and, 78
rounding and, 78, *78*
sorting and, 78, *78*
stream, 86, *86,* 92
surface, *87,* 89
water, 77, *85, 86,* 86–87
wind, 77, *85,* 90
Eskers, 81
Esophagus, 450, *450*
Ethanol, 397
Euglena, 531
Euglenoids, 533, *533*
Eukarya, 342, *342*
Eukaryotes
cell differentiation in, *429*
explanation of, 428
fungi as, 543
protists as, 531, 532
Eukaryotic cells
cellular respiration in, 396
explanation of, 380, **417,** 428
organelle in, 381–383
Evaporation
explanation of, *23,* **23**
in water cycle, 169
Excretion, 450, 451
Excretory system, 452, *452*
Exocytosis, *390,* **390**
Exosphere
explanation of, 14, *122,* 123
temperature in, 124, *124*
Experiments, NOS 21
Explanation, NOS 12
External stimuli, 334, *334*
Extremophiles, 497

F

Facilitated diffusion, 389, *389*
Facts, NOS 10
Fallopian tubes, 474, *474*
Fats, 451
Federal Bureau of Investigation (FBI), 425
Female reproductive system, 473, 474, *474,* 476
Fermentation
alcohol, 397, *397*
explanation of, **396**
types of, 397, *397*
Fetus, 475, *475*
Fiber, 430, 451
Fingerprinting, 425
Fish and Wildlife Research Institute, 339
Fission, 496
Flagella, 379, *496,* **496,** 533, 536
Flagellates, 536
Fleming, Alexander, 549
Floodplains, 89, *89*
Florida Panther, 302
Florida Solar Energy Center (FSEC), 271
Focus on Florida, 83, 135, 171, 271, 302, 339, 541
Fog, 168
Foldables, NOS 10, NOS 14, 15, 30, 33, 45, 56, 63, 81, 86, 95, 101, 123, 132, 138, 145, 151, 168, 174, 186, 191, 209, 216, 223, 231, 250, 259, 268, 273, 288, 300, 305, 313, 332, 345, 351, 355, 372, 381, 387, 396, 401, 416, 429, 437, 450, 463, 473, 479, 493, 503, 510, 517, 538, 546, 553
Food, 503
Food poisoning, 505
Force pairs, 310, *310*
Forces
balanced, 307, *307,* 308, *308*
contact, 304, *305,* 306, *306*
electric, 304
explanation of, **304**
Newton's laws of motion and, *309,* 309–310, *310*
noncontact, 304–306, *305, 306*
unbalanced, 307, *307,* 308, *308,* 310
Forensics, 425
Fossil fuels, 225, 263
Freshwater
explanation of, 12, **15,** *16*
in ground, 16, *16*
ice as, 16
in lakes and rivers, 16
Friction
air resistance and, 306
effects of, 306, *306*
explanation of, **259,** 306
law of conservation of energy and, 259, *259*
Frogs, 333, *333*
Fronts
cold, 176, *176*
explanation of, **176**
occluded, 177, *177*
stationary, 177, *177*
warm, *176,* 177
Frostbite, 181
Fujita, Ted, 179
Fungi
classification of, *342,* 544
club, 544, *544*
examination of, 542 *lab*
explanation of, 543, *543,* 552
illness related to, 549
imperfect, 546
importance of, 547, *547*
lichens and, 550, *550*
medical uses for, 549, *549*
plant roots and, 548, *548*
sac, 545, *545*
zygote, 545, *545*
Fungus-like protists
explanation of, 532, *532*
importance of, 539
slime and water molds as, 539, *539*
Furrow, 422, *422*

G

Gallbladder, *450*
Gametes, 473
Gases
in atmosphere, 13, *13*
greenhouse, **224,** 225, 228, *228,* 263
Genetic disorders, 514
Gene transfer, 514
Genus, 343
Geologists, 221
Geosphere
explanation of, **17,** *30*
interactions in, 30

materials in, *17,* 17–18, *18*
structure of, 19, *19*
water cycle in, 25
weather and, 26
G_1 interphase stage, 418
G_2 interphase stage, 418
Glacier National Park (Montana), 80, *80*
Glaciers
alpine, 97, *97*
as deposition agent, 81, 98, *98*
as erosion agent, 77, 97, *97*
explanation of, 73, **97**
ice layers in, 214
ice sheet, 97
land use practices affecting, 98
movement of, 85, 93 *lab*
sediment at edges of, 78
Glassblowers, 265
Glassware, NOS 16, *NOS 16*
Global climate model (GCM), 227
Global patterns, 137–139, 207
Global warming. *See also* **Climate change**
explanation of, **224**
hurricanes and, 171
ozone layer and, 126
Glucose, 383, 397
Glycolysis, 395, *395*
Golgi apparatus, 383, *383*
Grand Canyon, 83
Graphs
distance-time, *296,* 296–298, *297, 298*
explanation of, 295 *lab*
speed-time, *299,* 299–300, *300*
Gravel, 76
Gravitational potential energy, 251, *251, 253*
Gravity
distance and, 305, *305*
as erosion agent, 77
explanation of, 124, 303 *lab,* **305,** 308
mass and, 305, 306
mass wasting and, 94
weight and, 306
Great Irish Potato Famine, 539
Green building, 228, *228*
Greenhouse effect. *See also* **Climate change**
explanation of, 130, *130,* 224
Greenhouse gases. *See also* **Climate change**
explanation of, **224,** 263
methods to reduce, 228, *228*
sources of, 225
Greenland, 97
Green Science, 263
Grizzly bears, *343,* 344
Groins, 89, *89*
Ground tissue, 431, 432, *432*
Groundwater, 16
Groundwater
deposition by, 88
as erosion agent, 87, *87*
Growth hormone, 469
Gulf Stream, 207

H

Hail, 169. *See also* **Precipitation**
Hanging valley, *97*
Harmful algal bloom (HAB), 535
Harvesting, 471
Haze, 146
Health-care fields, 352
Hearing, 468, *468*
Heart, 454, 455, *455*
Heat. *See also* **Thermal energy**
explanation of, 128, **265**
latent, 131
thermal energy and, 265, 266, *266*
Heat waves, 219, 226
Hematite, 49
Hemophilia, 514
Heterotrophs, 543
High-energy environments, 79
High-pressure system, *173,* **173,** 174
HIV (human immunodeficiency virus), 512
Holocene Epoch, 215
Homeostasis
explanation of, **335,** 371, *449,* **449,** 457, 466, 469
importance of, 335
methods for, 335
Hoodoos, 80, *80*
Hooke, Robert, 349, 369
Horizons, *58,* **58,** 59
Hormones
function of, *469,* **469**
reproduction and, 473, 474
Horn, *97*
How It Works, 21
How Nature Works, 375, 499
Human body
circulatory system in, 454–456, *455, 456*
digestive system in, *450,* 450–451
disease in, 459, *459*
endocrine system in, 469, *469*
excretory system in, 452, *452*
fluid transport through, 448 *lab*
impact on climate change, 224–225
lymphatic system in, *457,* 457–458, *458*
muscular system in, 465, *465*
nervous system in, *466,* 466–468, *467*
organization of, 449
organ systems in, 433, 434, *434*
reproduction in, *473,* 473–475, *474, 475*
respiratory system in, *453,* 453–454, *454*
skeletal system in, 463–464, *464*
water in, 448 *lab,* 457
Human development, 475, 476, *476*
Human population growth, 228, *228*
Humidity
explanation of, 26, **166**
relative, 167
Hurricane Katrina, *89,* 171
Hurricanes, 171, 180, *180,* **180**
Hydrosphere
distribution of water in, 15–16
explanation of, **15,** *30*
interactions in, 30
water cycle in, 25
weather and, 26
Hyphae
explanation of, *543,* **543,** 548, 550
of zygote fungus, 545
Hypochlorite, 499
Hypothalamus, *469*
Hypothermia, 181
Hypothesis
explanation of, NOS 6
method to test, NOS 7, NOS 24

I

Ice, 16, 76
Ice ages, 214, 215
Ice cores
analysis of, 221
explanation of, 214, *214*
method to collect, 221
The Iceman's Last Journey (Case Study), NOS 20–NOS 27
Ice sheets
explanation of, 97
formation of, 214
in most recent ice age, 215
Ice storms, 181
Ice wedging, *47*
Ichthyologists, 339
Igneous rock
explanation of, 18, *18*
in rock cycle, 28, *28*
Immune response, 459
Immunity, 458, 513
Imperfect fungi, 546
Independent variables, NOS 21, NOS 25
Indoor air pollution, 148. *See also* **Air pollution**
Inertia, 309
Infancy, 476, *476*
Infection, 457, 458
Infectious diseases, 459, *459*
Inference
explanation of, NOS 26, *NOS 26*
Influenza, 512, *512*
Influenza vaccines, 459
Infrared radiation (IR), 128, 130
Infrared satellite images, 186, *186*
Inorganic matter, 55, *55*
Insoluble fiber, 451
Insulin, 469
Interglacials, 214, 215
Intergovernmental Panel on Climate Change (IPCC), 224
Internal stimuli, 334
International System of Units (SI)
converting between, NOS 13
explanation of, **NOS 12**
list of, *NOS 13*
prefixes for, NOS 13
Internet, NOS 17, *NOS 17*
Interphase
explanation of, *416,* **416,** 417, *417*

organelles during, 419
phases of, 418, *418*
Ionosphere, *123,* **123**
Ions, 48
Iron, 49, *49*
Isobars, 187
Isotherms, 187

J

Jet streams, 139, *139*
Joules (J), 252

K

Karenia brevis, *339*
Kelp forests, 535, *535*
Keratin, 373
Kidneys, 452, *452*
Kinetic energy
changes between potential and, 258, *258*
explanation of, **166,** 250, *250, 253*
identification of, 255
Kingdoms, *342,* **342**
Komodo dragon, 427, *427*
Köppen, Wladimir, 208

L

Lab. *See* **Launch Lab**
Lactic acid, *397,* **397**
Lactobacillus, 501
Lakes, 16
Lake Superior, *85*
Land breeze, *140,* **140**
Landforms
coastlines, 80, 87–88
comparison of, 81
deltas, 88
deposition and, 81, *81*
development of, 80, 84 *lab,* 85, 100
dunes, 80, 90
on Earth, 12
erosion and, 80, *80*
glacier, 80, 97–98
lakes, 80
mountains, 80
river, 89
weathering and, 76, *76*
Land use practices
affecting glaciers, 98
erosion from, 89, 90
mass wasting from, 96
Large intestine, *450,* 451
Latent heat, 131
Latitude, 205, 206, *206,* 216
Launch Lab, 10, 22, 44, 53, 74, 84, 93, 118, 127, 136, 143, 164, 172, 184, 204, 213, 222, 248, 256, 264, 286, 295, 303, 330, 348, 368, 376, 386, 394, 414, 426, 448, 462, 472 *lab,* 492, 500, 508, 530, 542
Lava, 18, 28
Law of conservation of energy, 258, 259, *259*
Law of gravity, 305
Leeuwenhoek, Anton van, 349, *349*
Lesson Review, NOS 11, NOS 18, NOS 27, 20, 31, 51, 61, 82, 91, 99, 125, 134, 141, 149, 170, 182, 189, 211, 220, 229, 254, 262, 270, 293, 301, 311, 338, 346, 353, 374, 384, 392, 399, 424, 435, 460, 470, 477, 498, 506, 515, 540, 551
Leukemia, 471
Lichens, 550, **550**
Life science, NOS 5
Ligaments, 463
Light, 397, 398
Light energy, *253*
Light microscope, 350, 352
Light waves, *253*
Limestone, 87
Linnaeus, Carolus, 342
Lipids, 372, *372,* **373,** 541
Liquids, 120, 271
Little Ice Age, 215
Liver, *450,* 452
Living things
characteristics of, 331, *337*
classification of, *341,* 341–345, 340 *lab, 343, 344, 345*
energy use by, 336, *336, 337*
growth and development of, 332–333, *337*
homeostasis in, 335, *335*
organization of, 331, *337*
reproduction in, 333, *337*
responses to stimuli by, 334, *337*
Loess, 90
Longshore currents, 87, 100
Low-energy environments, 79, *79*
Low-pressure system, *173,* **173**
Lungs, 452, 453, *453, 454*
Lymphatic system
function of, 457, *457,* 464
immunity and, 458, *458*
Lymph nodes, 457, *457*
Lymphocytes, *458,* **458**
Lymph vessels, 457, *457*

M

Macromolecules, 371, 372, *372*
Magma, 18, 28
Malaria, 538, *538*
Male reproductive system, 473, 474, *474,* 476
Mantle, 19, *19*
Maritime polar air masses, 175
Maritime tropical air masses, 175
Marrow, 464, *464*
Mars, 135
Mass
gravity and, 305, *305,* 306, *306*
kinetic energy and, 250, *250*
Mass wasting
deposition by, 95, *95*
erosion by, 95, *95*
examples of, *95*
explanation of, **94**
land use practices contributing to, 96
Math Skills, NOS 13, NOS 18, NOS 29, 26, 31, 35, 46, 51, 65, 96, 99, 103, 148, 149, 153, 175, 182, 193, 226, 229, 233, 260, 266, 270, 275, 289, 293, 315, 350, 353, 357, 391, 392, 403, 422, 424, 439, 451, 460, 481, 497, 498, 519, 548, 555
Matter, 248 *lab*
Mean, *NOS 15,* **NOS 15**
Meanders, 86, 100
Measurement
accuracy of, NOS 14, *NOS 14*
of energy, 252
International System of Units, NOS 12–NOS 13, *NOS 13*
Mechanical energy, *253,* 259, *259*
Mechanical weathering
causes of, 46, *47*
explanation of, **46**
rate of, 50
surface area and, 46, *46*
Median, NOS 15, *NOS 15*
Medicine, 549, *549*
Mendenhall Glacier (Alaska), *97*
Menstrual cycle, 475
Meristems, 430, *430*
Mesosphere
explanation of, 14, *122,* 123
temperature in, 124, *124*
Metal, 267
Metamorphic rock
explanation of, 18, *18*
in rock cycle, 28, *28*
Metaphase, 420, *420*
Meteorologists, 165, 186
Metersticks, NOS 16, *NOS 16*
Methane
in atmosphere, 13
as greenhouse gas, 224
infrared radiation and, 130
Microalgae, 541
Microclimates, 208
Microscopes
advancements in, 370
atomic force, 375
compound, 329, 350, *350*
development of, 349
electron, 351, *351,* 352
examining cells using, 369
light, 350, 352
use of, 352, 354
Microwaves, *253*
Mild climate, *209*
Minerals
breakdown of, 48
in human body, 451, 464
in ocean water, 15
properties of, *17,* **17**
Mini BAT. *See* **Benchmark Mini-Assessment**
Mitochondria
explanation of, 382, *382,* 419, *419*
reactions in, 396, *396*
Mitosis
explanation of, **419**
phases of, *420,* 420–421, *421*
Mitotic phase, 416, 419
Mode, NOS 15, *NOS 15*

Molecular analysis, 342
Molecules, 371, *371*
Monera, 342
Monsoon, 219
Moon, 306, *306*
Moraines, 81, **98**
Mosquitoes, 538
Motion
balanced forces and, 308, *308*
change in direction of, 292
explanation of, **288**
graphs used to describe, *296,* 296–298, *297, 298*
Newton's laws of, *309,* 309–310, *310*
ways to describe, 294
Motor vehicles
emissions of, 228
hybrid, 228
Mountains
explanation of, 27
rain shadows on, 207, *207*
weather in, 205
Mountain waves, 132, *132*
Mouth, 450, *453*
Mucus, 459
Mud flow, 94
Multicellular organisms
cell differentiation in, *429,* 429–430, *430*
explanation of, **332,** 429
growth and development of, 332, 333
organs in, 432, *432*
organ systems in, 433, 434, *434*
tissues in, 431, *431*
Muscle tissue, 431, *431,* 432
Muscular system
calcium use by, 464
function of, 465, *465*
during stages of human development, 476
Mushrooms
explanation of, **544**
as fungi, 543, *543*
poisonous, 549
Mutations, 511
Mycelium, *543,* **543**
Mycorrhiza, 548, 550

N

NASA, 126, 135
Negative acceleration, 292
Nerve cells, 377, *377*
Nervous system
calcium in, 464
function of, *466,* 466–468, *467, 468*
during stages of human development, 476
Nervous tissue, 431, 432
Neurons
explanation of, 466, 467
for senses, 468
Newton, Isaac, 309
Newton's first law of motion, 309, *309*
Newton's second law of motion, 310, *310*
Newton's third law of motion, 310, *310*
NGSSS for Science Benchmark Practice, 36–37, 66–67, 104–105, 154–155, 194–195, 234–235, 276–277, 316–317, 358–359, 404–405, 440–441, 482–483, 520–521, 556–557
NGSSS Benchmark Practice Test, FL 1–FL 9
Nitrogen, 13, *13,* 120, 121, *121*
Nitrogen fixation, *502,* **502**
Nitrous oxide, 121
Noncontact force
explanation of, **304**
gravity as, *305,* 305–306, *306*
Noninfectious diseases, 459, *459*
Nonpoint-source pollution, 144
North America, 60, *60*
North Atlantic Oscillation (NAO), 218
North Pole, 206, *206*
Nose, 453, *453*
Nuclear energy, *253*
Nuclear envelope, 381
Nuclear power plants, *253*
Nucleic acids, *372,* **372**
Nucleolus, 381, 420
Nucleus, 367, **381,** *381*
Nutrients
function of, **451,** 469
in soil, 59
transport of, 455, *455,* 456, 457
Nutrition, 451
Nutrition labels, 451, *451*

O

Observation, NOS 6, NOS 23
Occluded fronts, 177
Ocean currents, 27, 207
Oceanic crust, 19
Oceans, 15, 171
O-horizon, 58
Oil, 541
Olfactory system, 468, *468*
Opinions, NOS 10
Ores, 49
Organelles, 380–383. *See also* **Cell organelles**
explanation of, 416
membrane-bound, 417
replication of, *418,* 419
Organic matter
explanation of, **54**
in soil, 55, 57, 59
Organisms. *See also* **Living things; Multicellular organisms; Unicellular organisms**
basic components of, 427
classification of, 340 *lab, 341,* 341–345, *343, 344, 345,* 354
energy use by, 336, *336, 337*
explanation of, **331,** 354
growth and development in, 332–333, *337*
homeostasis in, 335, *337*
multicellular, *429,* 429–434, *430, 431, 432, 434*
reproduction in, 333, *337*
responses to stimuli by, 334, *337*
unicellular, 428, *428*
Organization
of living things, 427–434
of systems, 426 *lab*
Organs, 413, **432**
Organ systems
explanation of, **433, 449**
human, 433, 434, *434*
plant, 433
Osmosis, 388
Outwash, 98
Ova, 473, 474, 475
Ovaries
function of, *469,* 474, *474*
menstrual cycle and, 475
Oxidation, 49, **49**
Oxide, 49
Oxygen
in atmosphere, 13, *13,* 119–121
in human body, 453, *453,* 454, 456
in organisms, 339
from photosynthesis, 541
Ozone
in atmosphere, 121, 124, 145
explanation of, 122
ground-level, 147, 148
Ozone layer
explanation of, 14, **122**
hole in, 126

P

Page Keeley Science Probes. *See* **Science Probes**
Painted Desert (Arizona), *80,* **80**
Pancreas, *450,* 469, *469*
Paramecium, 335, *335,* 379, 532, *532, 536,* **536**
Parasites, 538, 543
Parathyroid glands, *469*
Parathyroid hormone, 469
Parent material
explanation of, *56,* **56**
formation of soil from, 59
Particulate matter, 146
Passive transport, 387
Pasteurization, 505
Pathogens, 504, 507
Pendulum, 255
Penicillium, 549, *549*
Penis, 474, *474*
Peripheral nerves, *466*
Peripheral nervous system, 466, 467
Permafrost, 210
Perspiration, 449
pH, 48, 59, *59*
Pharynx, 453, *453*
Phenomenon, 218
Phospholipids, 373, 378
Photobioreactors, 541
Photochemical smog, 145
Photosynthesis
algae and, 531, *531,* 535, 541
chloroplasts in, 398

explanation of, 120, 383, **397**
food production through, 532
function of, 548
importance of, 398, *398*
lichens and, 550, *550*
light in, 397, 398
process of, 225
Physical, NOS 5
Physical science, NOS 5
Physical weathering, *76,* **76**
Pituitary gland, *469*
Plantae, *342*
Plant cells
animal cells vs., 385
chloroplasts in, 383
shape of, *378*
Plant-like protists
algae as, 532, *532,* 535, *535*
explanation of, 532, *532*
types of, 533–534
Plants
as cause of mechanical weathering, *47*
cells in, 430, *430*
fungi and, 548, *548*
organ systems in, 433
photosynthesis and, 397–398
soil properties that support, 59
Plasma, 456
Plasmodia, 538
Plasmodium, 538, *538*
Platelets, 456, 471
Pneumonia, 494
Point source pollution, 144
Polar climate, *209*
Polar regions, 206
Pollution. *See also* **Air pollution**
acid rain as, 143 *lab*
bacteria that eat, 503
methods to reduce, 228, *228*
non-point source, 144
particulate, 146
point-source, 144
temperature inversion and, 133
Pores, **54,** *55*
Position, 286 *lab,* **287,** 288
Potential, NOS 11
Potential energy
changes between kinetic and, 258, *258*
chemical, 251, *251*
elastic, 251, *251*
explanation of, **250**
gravitational, 251, *251, 253*
identification of, 255
Precipitation. *See also* **Rain**
climate change and, 226
effect of mountains on, 27, *27*
explanation of, **25,** 26, *56,* 168, *169,* 207
types of, *169*
water cycle and, 169
Prediction, NOS 6
Pregnancy, 475, *475*
Pressure systems, 173
Processes, 29, 75, 128, 536
Prokaryotes
archaea as, 497
bacteria as, 493
explanation of, 428
Prokaryotic cells, 380, 428
Prophase, 420, *420*
Proteins
in cells, 373, 378, 381, 389
explanation of, 372, *372,* **373,** 451
in human body, 452
red-blood-cell, 456, *456*
transport, 389, *389*
Protista, *342*
Protists
animal-like, *536,* 536–538, *537, 538*
characteristics of, 530 *lab*
classification of, 532, *532*
explanation of, *531,* **531,** 552
fungus-like, 539, *539*
plant-like, 532–535, *533, 534, 535*
reproduction of, 532
Protozoans, **536,** 538, *538*
Pseudopods, *537,* **537,**

R

Radar, 186
Radiant energy
changed to thermal energy, 257, *257*
electric energy changes to, 257, *257*
explanation of, *253*
use of, 260, *260,* 261
Radiation
absorption of, 129, *129*
balance in, 130, *130*
explanation of, **128, 269**
from Sun, 14
Radio waves, 123, *123, 253*
Rain. *See also* **Precipitation**
acidic nature of, 48
adaptations to, 210
explanation of, 169
freezing, 181
Rain-shadow effect, 27, *27*
Rain shadows, *207,* **207**
Range, NOS 15, *NOS 15*
Rectum, *450*
Recycling, 228
Red blood cells, 377, *377,* 456, *456,* 461
Red tide, 535, *535*
Reference points, 288, *288*
Reflection, 129, *129*
Reflexes, *467,* **467**
Relative humidity, 167
Renewable energy, 136 *lab*
Replication, 508 *lab,* 510, *510–512,* 511
Reproduction. *See also* **Asexual reproduction; Sexual reproduction**
asexual, 532
in bacteria, 496
cell division as form of, 423
explanation of, **473**
in organisms, 333
sexual, 532
Reproductive systems
development of, 476
female, 473, 474, *474*
male, 473, 474, *474*
Reservoirs, 15
Resistant, **504,** 505, *505*
Respiration
cellular, *395,* 395–396, *396*
explanation of, 24
Respiratory system, *453,* 453–454, *454,* 467
Retaining walls, 89, *89*
Reverse osmosis, 21
Review Vocabulary, 15, 56, 120, 165, 166, 207, 351, 370, 417, 451, 511, 532
Revolution, 217
R-horizon, 58
Ribonucleic acid (RNA), 372, 373, 381
Ribosomes
of archaea, 497
in cytoplasm, 494
explanation of, 381, 382
River rapids, 86
RNA, 509, 511, 514
Rock cycle
explanation of, *28,* **28**
processes in, 28–30
Rockfall, 95, *95*
Rocks
effects of weathering on, *76*
erosion of, 77, 78, *78,* 80
explanation of, **18**
identifying change in, 44 *lab*
mass wasting of, 94, 95
oxidation of, 49
rounded, 78, *78*
surface area of, 46
types of, 18, *18*
weathering of, 45, 46, *47,* 50, 76, *76*
Rounding numbers, NOS 14
Rulers, NOS 16, *NOS 16*

S

Sac fungi, 545
Sahara, 205
Saliva, 450, 468
Salivary glands, *450*
Salt crystals, 371, *371*
Salt water, 12, 21, *21*
Sand
beach, 89
explanation of, 45, 76
formation of, 46
properties of, *55*
Sandbars, 81, *81*
Sand dunes, 90, *90*
Sandstone, *83*
Sarcodines, 537
Satellites, 186, *186*
Satellite tracking technology, 297, 302
Scanning electron microscope (SEM), 351, *351*
S cells, *417,* 418, *418*
Schleiden, Matthias, 370
Schwann, Theodor, 370
Science
applications of, NOS 4
branches of, NOS 5

explanation of, **NOS 4**
new information on, NOS 9–NOS 11
results of, NOS 8–NOS 9
Science journals, NOS 16, *NOS 16*
Science & Engineering, NOS 19
Science & Society, 425, 471
Science Use v. Common Use, 45, 98, 129, 176, 217, 257, 267, 290, 342, 381, 430, 454, 504, 544
Scientific inquiry. *See also* **The Iceman's Last Journey (Case Study)**
evaluating evidence during, NOS 10
explanation of, NOS 6–NOS 7, *NOS 7–NOS 8*
limitations of, NOS 11
use of safe practices during, NOS 11
Scientific laws, NOS 8, NOS 9
Scientific names
explanation of, 343
use of, 344
Science Probes, NOS 1, 7, 41, 71, 115, 161, 201, 245, 283, 327, 365, 411, 445, 489, 527
Scientific theories, NOS 8, NOS 9, *NOS 9*
Scientific tools
explanation of, *NOS 16,* NOS 16–NOS 17, *NOS 17*
types of general, *NOS 16,* NOS 16–NOS 17, *NOS 17*
used by Earth scientists, NOS 18, *NOS 18*
Sea arches, 87
Sea breeze, *140,* **140**
Sea caves, 87
Seasons
explanation of, 216
monsoon winds and, 219, *219*
Sea stacks, 87
Sea turtles, 297, *297, 298*
Sediment
distribution of, *57*
explanation of, 45, 76
in Grand Canyon, *83*
groins to trap, 89, *89*
in high-energy environments, 79
layers of, 79
from longshore currents, 87, *87*
shape and size of, 74 *lab,* 78
in soil, 57, *57*
sorting of, 78, *78*
transported by streams, 86
Sedimentary rock
explanation of, 18, *18*
in rock cycle, 28
Semen, 474
Senses, 468
Sexual reproduction
in fungi, 545
in humans, 473, 474, *474,* 476
offspring of, 532
Shale, 78, *79*
Shells, 376 *lab*
Shindell, Drew, 126
Sickle cell disease, 471
Sight, 468, *468*
Significant digits, *NOS 14,* **NOS 14**
Silica, 533, *533*
Silt
causes of, 45
deposits of, 90
explanation of, 76
in low-energy environments, *79*
properties of, *55*
Siltstone, 78
S interphase stage, 418
Sister chromatids, 418, *418,* 419
Skeletal system
calcium storage in, 464, 469
function of, 458, 462 *lab,* 463–464, *464*
protection of central nervous system by, 467
Skepticism, NOS 10
Skin, 452, 459
Skulls, 464
Sleet, *169. See also* **Precipitation**
Slime molds, 532, *532,* 539, *539*
Slump, 95, *95*
Small intestine, *450,* 451
Smell, 468, *468*
Smog, 145
Snow. *See also* **Precipitation**
albedo of, 212
as driving hazard, 181
explanation of, 169
Sodium chloride, 48
Soil
biota in, 57
effect of weathering on, 46
explanation of, **54,** 62
formation of, 56–57, *57*
horizons of, 58, *58,* 62
inorganic matter in, 55, *55*
observation of, 53 *lab*
organic matter in, 55
properties of, 59, *59*
types and locations of, 60, *60*
Solar energy
climate cycles and, 215, 216, *216*
electricity and, 271
reflected back into atmosphere, 212
temperatures and, 205, 206, *206*
Solstice, 217, *217*
Sound energy, *253,* 260
South Pole, 206, *206*
Space exploration, NOS 8
Species, 343
Specific heat, 207
Speed
average, 289, 298
change in, 291, 298, 300, 308, *308*
constant, 289, 298, 299
explanation of, *289,* **289,** 290
kinetic energy and, 250, *250*
negative acceleration and, 292
Speed-time graphs, 299, *299,* **299,** 300, *300*
Sperm, 472 *lab,* **473**
Sperm duct, 474, *474*
Spinal cord, *466,* 466–468
Spindle fibers, 420
Spindler, Konrad, NOS 22
Spitzer Space Telescope, NOS 8
Spleen, 457, *457*
Spongy bone, *464,* **464**
Spores
explanation of, 544
in sac fungi, 545
Stability, 132, 132–133
Stalactites, 88
Stalagmites, 88
Stationary fronts, 177
Station model, 187, *187*
Steidinger, Karen, *339*
Stem cells, 430
Stimuli, 334, *334*
Stomach, 450, *450*
Stratosphere
explanation of, 14, *122,* **122**
ozone in, 124, 126, 145
Streak plate, NOS 18, *NOS 18*
Streams
deposition along, 88
as erosion agent, 86, *86*
stages in development of, *86*
water erosion and deposition along, 88, *88,* 92
Streptococcus pneumoniae, 504
Study Guide, NOS 28–NOS 29, 32–33, 62–63, 100–101, 150–151, 190–191, 230–231, 272–273, 312–313, 354–355, 400–401, 436–437, 478–479, 516–517, 552–553
Subsoil, *57*
Sulfur dioxide, 121
Sun
energy from, 23, 128, *253,* 269, 336
exposure, 181
nuclear energy released by, *253*
radiation from, 14
reflection of rays from, 212
Sunlight, 137
Surface area, 46, *46*
Surface erosion, 89, *89*
Surface reports, 185
Swamps, 79, *79*
Sweat, 449
Systematics, 342
Systems, 426 *lab*

T

Table salt, 48
Tadpoles, 332, *332, 333*
Taste, 468, *468*
Technology, NOS 8
Telophase, 421, *421*
Temperature. *See also* **Climate; Weather**
adaptations to, 210
of air, 166, 167, *167,* 172 *lab*
altitude and, 124, *124*
body, 335
effect of greenhouse gases on, 224
effect of increasing, 226–227
in rocks, 30
thermal energy and change in, 265, 266, 271
trends in, 223, *223*
variations in, 205, 206, *206*

Temperature inversion
explanation of, **133**
pollution and, 146, *146*
Tendons, 463
Testes, 474, *474*
Testosterone, 474
Theory, 370
Thermal conductors, 267
Thermal energy. *See also* **Heat**
conduction and, 267, *267*
convection and, 268, *268*
effects of, 210
explanation of, 127 *lab*, *253*
friction and, 259, *259*
heat and, 265, *265,* 266
movement of, 264 *lab*
in ocean water, 207
production of, 128
radiant energy changed to, 257, *257*
radiation and, 269, *269*
released by human body, 449
transfer of, 131, *131*
use of, 260, 261
Thermal equilibrium, 266
Thermal insulators, *267,* **267**
Thermometers, NOS 17, *NOS 17*
Thermosphere
explanation of, *122,* 123
temperature in, 124, *124*
Thompson, Lonnie, 221
Throat, 453
Thrush, 549
Thunderstorms
explanation of, 133, 178, *178, 179*
safety precautions during, 181
Thymus, 457, *457, 469*
Thyroid gland, *469*
Till, 98
Tissues
animal, 431, *431,* 432
explanation of, **431**
plant, 431, *431, 432*
Tongue, *450*
Tonsils, 457, *457*
Topography, 57, *57,* **57,** 76
Topsoil, 90
Tornado Alley, 179
Tornadoes, 179, *179*
Touch, 468, *468*
Toxins, 459
Trace gases, 13, *13*
Trachea, 453, *453*
Trade winds, 139, 218
Transmission electron microscope (TEM), 351, *351*
Transpiration, 24
Transport
active, 390, *390*
passive, 387
Transport membranes, 389, *389,* 393
Triglycerides, 541
Triple-beam balance, NOS 17, *NOS 17*
Tropical climate, *209*
Tropical cyclones, 180
Tropics, 137
Troposphere
explanation of, 14, *122,* **122**
jet streams in, 139
temperature in, 124, *124*
Tuberculosis, 504, *504*
Typhoons, 180

U

Ultraviolet (UV) light, 128
Unbalanced forces
acceleration and, 308, *308,* 310
explanation of, 307, *307*
Unicellular organisms
explanation of, **332,** 428, *428*
growth and development of, 332
homeostasis in, 335, *335*
reproduction in, *333*
United States, *60*
Updrafts, 178, 179, *179*
Uplift, *28,* **28**
Upper-air reports, 185
Upwelling, 218
Uranium, 503, *503*
Urban areas
ozone in, 147
population growth projections for, 228
temperatures in, 205, *209*
Urban heat island, 208, *208*
Urea, 452
Ureter, 452, *452*
Urethra, 452, *452*
Urine, 452
U.S. National Weather Service, 181
U-shaped valley, 97, *97*
Uterus, 474, *474,* 475

V

Vaccines, 459, **514**
Vacuoles, 383
Vagina, 474, *474,* 475
Variables, NOS 21, 165
Vascular tissue, 431, 432, *432*
Vegetation
adaptations to climate by, 210
classifying climate by native, 208
in mountains, 207
Veins, 455, *455*
Velocity, 290, 291, 292
Ventricles, *455*
Vesicles, 383
Virchow, Rudolf, 370
Viruses
beneficial, 514
detection of, 458
explanation of, **509,** 511, 512, *512*
immunity to, 513, *513*
latent, 510
medications to treat, 513
mutations in, 511
organisms and, 510
replication of, 508 *lab,* 510, *510–511,* 511
research with, 514
shapes of, 509, *509*
types of, 509, 512
vaccines to prevent, 514
Visible light satellite images, 186, *186*
Vision, 468, *468*
Vitamin A, 373
Vitamins, 451
Volcanic eruptions, 121, *121*
Volvox, 534, *534*

W

Warm fronts, *176,* 177
Waste energy, 261
Water
in cells, 371, *371,* 388
changes on Earth from, 84 *lab*
diffusion of, 388
on Earth, 12, *12,* 15, *15*
as erosion agent, 77, *85, 86,* 86–87
in human body, 448 *lab,* 457
running downhill, 57
salt, 12, 21, *21*
specific heat of, 207
as weathering agent, 48, *48,* 76
Water bodies, 205, 207–208
Water cycle
condensation process in, 25
evaporation process in, *23, 24*
explanation of, 23, *23, 24, 169,* **169**
precipitation process in, 25
transpiration and respiration process in, 24
Waterfalls, 86
Water molds, 539
Water vapor
air temperature and, 167, *167*
in atmosphere, 13, 121
explanation of, **120**
as greenhouse gas, 224
in human body, 452
infrared radiation and, 130
water cycle and, 169
Weather. *See also* **Climate; Temperature**
cycles of, 214–219
explanation of, 14, **26,** 45, **165,** 205
life cycle and, 415
patterns in, 27
reasons for change in, 183
safety precautions during severe, 181
terms to describe, 26, *26*
types of severe, *178,* 178–181, *179, 180, 181*
variables related to, 165–169, *166, 167, 168, 169*
water cycle and, 169, *169*
Weather forecasts
explanation of, 185
satellite and radar images for, 186
station models used for, 187
surface reports as, 185
understanding, 184 *lab*
upper-air reports as, 185
use of technology in, 186, 188, *188*
Weathering
agents of, 76
chemical, *48,* 48–49, 62
climate and, 56

explanation of, 29, 43, **45,** 62, *76,* **76**
of Grand Canyon, *83*
mechanical, 46, *47,* 52, 62
modeling of, 52
rate of, 50, *50,* 56, 76, *76,* 78
Weather maps, 186–187, *187*
Weather patterns, 173–181
Weight, 306
White blood cells
function of, 456, *456,* 461
immunity and, 457, 458, 471
production of, 464
Whittaker, Robert H., 342
Wind
air pollution and, 146
changes on Earth from, 84 *lab*
deposition from, 90, *90*
as erosion agent, 77, *85,* 90
explanation of, 26, **137,** 166
global, *137,* 137–139, *138*
local, 140, *140*
measurement of, 166, *166*
monsoon, 219
pressure systems and, 173
as renewable energy source, 136 *lab*
trade, 139
as weathering agent, 76
Wind belts
explanation of, 137, *138*
global, 138–139
Windchill, 181
Windpipe. *See* **Trachea**
Wind vane, NOS 18, *NOS 18*
Winter storms, 181
Word Origin, NOS 5, 13, 25, 54, 58, 90, 98, 119, 131, 146, 168, 180, 187, 209, 215, 224, 249, 259, 268, 287, 296, 306, 309, 335, 343, 352, 371, 379, 388, 397, 419, 431, 457, 466, 473, 493, 504, 513, 533, 550
Work, 252, *252*
Writing In Science, NOS 29, 65, 103, 153, 193, 233, 275, 315, 357, 403, 439, 481, 519

X

***x*-axis, 296**

Y

***y*-axis, 296**
Yeast, 545
Yellowstone National Park, *86*

Z

zygosporangia, *545,* **545**
zygospores, 545
zygote, 473, 475, *475*
zygote fungi, 545

Credits

Photo Credits

NOS 2–3 Ashley Cooper/Woodfall Wild Images/Photoshot; **NOS 4** Thomas Del Brase/Getty Images; **NOS 5** (t)Roger Ressmeyer/CORBIS, (c)Daniel J. Cox/CORBIS, (b)Medicimage/PhotoLibrary; **NOS 6** Chris Howes/Wild Places Photography/Alamy, (c)john angerson/Alamy; **NOS 8** (b)NASA/JPL-Caltech/Harvard-Smithsonian CfA; **NOS 9** DETLEV VAN RAVENSWAAY/SCIENCE PHOTO LIBRARY; **NOS 10** Sigrid Olsson/PhotoAlto/CORBIS; **NOS 11** Bob Daemmrich/PhotoEdit; **NOS 12** CORBIS; **NOS 14** Matt Meadows; **NOS 15** Pixtal/SuperStock; **NOS 16** (t)Matt Meadows, (c)By Ian Miles-Flashpoint Pictures/Alamy, (b)Aaron Graubart/Getty Images; **NOS 17** (t)MICHAEL DALTON/FUNDAMENTAL PHOTOS/SCIENCE PHOTO LIBRARY, (c)The McGraw-Hill Companies, (b)Steve Cole/Getty Images; **NOS 18** (tl)CORBIS, (tr)Paul Rapson/Photo Researchers, Inc., (bl)The McGraw-Hill Companies, Inc., (br) Mark Steinmetz; **NOS 19** (b)Panoramic Images/Getty images, (inset)Tony Linck//Time Life Pictures/Getty Images, (bkgd)Image Source/Getty Images; **NOS 20** Landespolizeikommando für Tirol/Austria; **NOS 21 - NOS 23** (l) South Tyrol Museum of Archaeology, Italy (www.iceman.it); **NOS 24** (inset) Klaus Oeggl, AP Images; **8–9** StockTrek/Getty Images; **11** (t)Philippe Bourseiller/Getty Images, (b)CORBIS; **12** (l)Chris Close/Getty Images, (tr) Pixtal/age footstock, (br)Joseph Sohm-Visions of America/Getty Images; **13** Darrell Gulin/Getty Images; **17** (l)José Manuel Sanchis Calvete/CORBIS. (r) Mark A. Schneider/Photo Researchers, Inc.; **18** (t)Albert Copley/Visuals Unlimited/Alamy, (c)Dr. Parvinder Sethi/The McGraw-Hill Companies, (b) Andreas Einsiedel/Dorling Kindersley; **19** Hutchings Photography/Digital Light Source; **20** (t)Pixtal/age footstock, (c)CORBIS; **20** (b)Chris Close/Getty Images; **22 23** Hutchings Photography/Digital Light Source; **23** Oxford Scientific/Photolibrary/Getty Images; **28** Danita Delimont/Getty Images; **30** (tl)Brand X Pictures/PunchStock, (tr)Gary Vestal/Getty Images, (c)NASA, (bl)CORBIS/SuperStock, (br)age fotostock/SuperStock, Creatas/PunchStock; **32** Pixtal/age fotostock; **34** Hutchings Photography/Digital Light Source; **37** (r)The Visible Human Project, U.S. National Library of Medicine, (l)Dr. Richard Kessel & Dr. Gene Shih/Visuals Unlimited/Getty Images; **42–43** Robert Harding Picture Library Ltd/Alamy; **44** Hutchings Photography/Digital Light Source; **45** Masterfile Royalty-Free; **46** Hutchings Photography/Digital Light Source; **47** (t to b)Richard Hamilton Smith/CORBIS, (2)Altrendo nature/Getty Images, (3)Peter Arnold, Inc./Alamy, (4) Guy Edwardes/Getty Images; **48** (l)Worldwide Picture Library/Alamy, (r)Dr. Marli Miller/Getty Images; **49** MARTIN LAND/SCIENCE PHOTO LIBRARY; **50** Paul Stutzman/National Institute of Standards and Technology; **53** Hutchings Photography/Digital Light Source; **54** Lester Lefkowitz/CORBIS; **58** Matthew Ward/Getty Images; **72–73** Beate Muenter/PhotoLibrary; **74** Hutchings Photography/Digital Light Source; **[075** Alex Gore/Alamy; **76** (b) Image Source; **77** (l)Medioimages/Photodisc/Getty Images, (r)Photodisc/SuperStock; **78** (tl)Pixoi Ltd/Alamy, (cl)Stephen Reynolds, (bl)Photo by Stan Celestian, (bc)Photo by Stan Celestian, (br)Photo by Stan Celestian; **79** (t) Creatas/PunchStock, (b)Comstock/PunchStock; **80** (t)Joseph Szkodzinski/Getty Images, (c)CORBIS, (b)Adam Jones/Getty Images; **81** (l)Cameron Davidson**,** (r)Dr. Marli Miller/Getty Images; **83** (tr)ASSOCIATED PRESS, (br) Peter Arnold, Inc./Alamy, (bkgd)Vstock; **84** (tl)Harald Sund/Photographer's Choice/Getty Images, (tc)Steve Hamblin/CORBIS, (tr)Image Source/Getty Images, (bl)Michael Melford/Getty Images, (bc)Robert Glusic/Photodisc/Getty Images, (br)CORBIS, (br)Image Source/Getty Images; **85** (t) Jupiterimages, (r)imagebroker/Alamy; **86** (t)Michael Melford/Getty Images, (bl)Digital Archive Japan/Alamy, (bc)Connie Coleman/Photographer's Choice/Getty Images; **87** (b)imagebroker/Alamy; **88** CORBIS; **89** (t)Tony Hopewell/Getty Images; **89** (b)AP Photo/Tallahassee Democrat, Phil Sears; **90** (t)Robert Glusic/Photodisc/Getty Images, (b)Patrick Lynch/Alamy; **92** (l) Paul Franklin/Alamy, (r)NASA, (b)Art Wolfe; **93** Hutchings Photography/Digital Light Source; **94** Getty Images; **95** (l)Dr. Marli Miller/Getty Images, (c)Getty Images, (r)Dr. Marli Miller/Getty Images; **96** Getty Images; **100** (t) Karl Johaentges/Getty Images, (c)Dirk Anschutz 2007/Getty Images, (b) DEA/G.SIOEN/Getty Images; **106 107** Hutchings Photography/Digital Light Source; **116–117** Daniel H. Bailey/Alamy; **118** Hutchings Photography/Digital Light Source; **119** CORBIS; **121** (t)PhotoLink/Getty Images, (b)C. Sherburne/PhotoLink/Getty Images; **123** Per Breiehagen/Getty Images; **124** CORBIS; **126** (t)American Museum of Natural History, (b)Pedro Guzman, NASA; **126** NASA/JPL/Texas A&M/Cornell; **127** Hutchings Photography/Digital Light Source; **128** John King/Alamy; **129** Eric James/Alamy; **132** James Brunker/Alamy; **136** Hutchings Photography/Digital Light Source; **137** Lester Lefkowitz/Getty Images; **139** CORBIS; **144** (t) Reuters/CORBIS, (b)C. Sherburne/PhotoLink/Getty Images; **145** MICHAEL S. YAMASHITA/National Geographic Image Collection; **148** (t)Digital Vision Ltd./SuperStock, (c)Masterfile, (b)Creatas/PictureQuest; **162–163** Chris Livingston/Stringer/Getty Images; **164** Hutchings Photography/Digital Light Source; **165** Peter de Clercq/Alamy; **166** (l)Jan Tadeusz/Alamy, (r) matthias engelien/Alamy; **168** (l)WIN-Initiative/Getty Images, (r) MIMOTITO/Getty Images, (r)age fotostock/SuperStock; **171** (inset)NASA/Jeff Schmaltz, MODIS Land Rapid Response Team, (bkgd)Jocelyn Augustino/FEMA, Stocktrek/age footstock, imac/Alamy; **172** Hutchings Photography/Digital Light Source; **173** StockTrek/Getty Images; **178** (l)Amazon-Images/Alamy, (c)Roger Coulam/Alamy, (r)mediacolor's/Alamy; **179** Eric Nguyen/CORBIS; **181** AP Photo/Dick Blume, Syracuse Newspapers; **183** Royalty-Free/CORBIS; **184** Hutchings Photography/Digital Light Source; **185** (t)Seth Resnick/Getty Images, (b)Gene Rhoden/Visuals Unlimited; **186** National Oceanic and Atmospheric Administration (NOAA); **188** Dennis MacDonald/Alamy; **202–203** Ashley Cooper/CORBIS; **205** Hemis.fr/SuperStock; **208** (tl) Rolf Hicker/age footstock, (tc)Brand X Pictures/PunchStock, (tr)Jeremy Woodhouse/Getty Images, (bl)Andoni Canela/age footstock, (br)Steve Cole/Getty Images; **210** age fotostock/SuperStock; **211** (b)Digital Vision/Getty Images; **213** Hutchings Photography/Digital Light Source; **214** J. A. Kraulis/Masterfile, (b)Nick Cobbing/Alamy; **221** (t)American Museum of Natural History, (b)American Museum of Natural History; **222** Hutchings Photography/Digital Light Source; **223** Jupiterimages/Getty Images; **225** Chris Cheadle/Getty Images; **226** Steve McCutcheon/Visuals Unlimited, Inc.; **228** Bruce Harber/age fotostock; **246–247** David R. Frazier Photolibrary, Inc. / Alamy; **248** Hutchings Photography/Digital Light Source; **249** (l) Thinkstock/CORBIS, (c)Sharon Dominick/Getty Images, (r)Dimitri Vervitsiotis/Getty Images; **249** Edward Kinsman / Photo Researchers, Inc.; **251** (t)Hutchings Photography/Digital Light Source**,** (b)Image Source/CORBIS; **253** (t to b)David Madison/Getty Images, (2)Andrew Lambert Photography / Photo Researchers, Inc., (3)Design Pics / age footstock, (4) VStock / Alamy, (5)Matthias Kulka/zefa/CORBIS, (6)Goodshoot/PunchStock; **256** Hutchings Photography/Digital Light Source; **257** Alaska Stock LLC / Alamy; **259** Hutchings Photography/ Digital Light Source; **260** Peter Cade/Getty Images; **261** (b)Lorcan/Getty Images; **263** (t)Ivy Close Images / Alamy, (b)Lawrence Manning/CORBIS, (bkgd) CORBIS; **264** Hutchings Photography/Digital Light Source; **265** (t)Reuters/CORBIS, (bl)Rob Walls / Alamy, (br)age fotostock / SuperStock; **266** Photolink / Getty Images; **267** Fancy/Veer/CORBIS; **268** Matt Meadows; **269** Daisy Rae/Jupiter Images; **284–285** Chris Milliman/Aurora Photos/CORBIS; **286 287** Hutchings Photography/Digital Light Source; **287** Paul Springett 09 / Alamy; **296** (inset)SAM YEH/AFP/Getty Images, (bkgd)James L. Amos / Peter Arnold Inc.; **303** Hutchings Photography/Digital Light Source; **304** (b)Hideki Yoshihara/age footstock, (bl)Yellow Dog Productions Inc./Getty Images, (t)Tim O'Sullivan / Imagestate; **305** Image Source/Getty Images; **306** Zhang Jie/ChinaFotoPress/Getty Images; **307** (t)Hutchings Photography/Digital Light Source, (br)Stefan Wackerhagen/age fotostock; **309** TRL Ltd./Photo Researchers; **310** (t)John Giustina/Getty Images, (b)Image Source/Getty Images; **328–329** Vaughn Fleming/Garden Picture Library/Photolibrary; **330** Hutchings Photography/Digital Light Source; **331** (t)Angela Wyant/Getty Images, (b)Mark Gibson; **332** (l)BRUCE COLEMAN, INC./Alamy,

Credits

(r)Gary Meszaros/Photo Researchers; **333** (l)Joe McDonald/CORBIS, (t)Gary Gaugler/The Medical File/The Medical File/Peter Arnold, Inc., (t)Mark Smith/Photo Researchers, (r)BRUCE COLEMAN INC./Alamy; **334** John Kaprielian/Photo Researchers; **335** Michael Abbey/Visuals Unlimited; **337** (t)Digital Vision/PunchStock, (tc)Robert Clay/Alamy, (c)John Kaprielian/Photo Researchers, (c)Splinter Images/Alamy, (b)Lee Rentz, (bc)liquidlibrary/PictureQuest; **339** (inset)D. Finn/American Museum of Natural History, (bkgd)Reinaldo Minillo/Getty Images; **340** Hutchings Photography/Digital Light Source; **341** (t)Spencer Grant/Photolibrary, (b)PhotoLink/Getty Images; **343** Shin Yoshino/Minden Pictures/Getty Images; **344** Mark Steinmetz; **347** Florida Fish and Wildlife Conservation Commission's Fish and Wildlife Research Institute; **348** Hutchings Photography/Digital Light Source; **349** (t)Eye of Science/Photo Researchers, (c) Dr Jeremy Burgess/Photo Researchers, (b)Imagestate/Photolibrary; **350** (l)Steve Gschmeissner/Photo Researchers, (r)JGI/Getty Images; **351** (l)Stephen Schauer/Getty Images, (cl)ISM/Phototake, (cr)A. Syred/Photo Researchers, (r)A. Syred/Photo Researchers, (b)ISM/Phototake; **354** age fotostock/SuperStock; **356** The McGraw-Hill Companies; **366–367** Dr. Dennis Kunkel/Visuals Unlimited/Getty Images; **368** Hutchings Photography/Digital Light Source; **369** (t)Tui De Roy/Minden Pictures, (c)Bon Appetit/Alamy, (b)Omikron/Photo Researchers; **370** (l)Biophoto Associates/Photo Researchers, (tl)Tim Fitzharris/Minden Pictures/Getty Images, (tr)James M. Bell/Photo Researchers (b)Dr. Richard Kessel & Dr. Gene Shih/Visuals Unlimited/Getty Images; **371** FoodCollection/SuperStock; **372** Dr. Gopal Murti/Photo Researchers; **375** (l)VICTOR SHAHIN, PROF. DR. H.OBERLEITHNER, UNIVERSITY HOSPITAL OF MUENSTER/Photo Researchers, (r)NASA-JPL; **376** Hutchings Photography/Digital Light Source; **377** Eye of Science/Photo Researchers; **379** SPL/Photo Researchers; **381** Dr. Donald Fawcett/Visuals Unlimited/Getty Images; **382** (l)Dr. Donald Fawcett/Visuals Unlimited/Getty Images, (r)Dennis Kunkel / Phototake; **383** (l)Dr. R. Howard Berg/Visuals Unlimited/Getty Images, (r)Dennis Kunkel / Phototake; **386** Macmillan/McGraw-Hill; **387** LIU JIN/AFP/Getty Images; **394** Hutchings Photography/Digital Light Source; **395** Colin Milkins/Photolibrary; **397** (t)Biology Media/Photo Researchers, (b)Andrew Syred/Photo Researchers; **398** Michael & Patricia Fogden/Minden Pictures; **412** supershoot/Alamy; **412–413** Robert Pickett/CORBIS; **414** Hutchings Photography/Digital Light Source; **415** (t) Michael Abbey/Photo Researchers, (b)Bill Brooks/Alamy; **417** (t)Dr. Richard Kessel & Dr. Gene Shih/Visuals Unlimited, (bl)Ed Reschke/Peter Arnold, Inc., br)Biophoto Associates/Photo Researchers; **419** Don W. Fawcett/Photo Researchers; **420 421** Ed Reschke/Peter Arnold, Inc.; **424** (bl)P.M. Motta & D. Palermo/Photo Researchers; **425** (l)Alex Wilson/Marshall University/AP Images, (r)Bananastock/Alamy; **426** Hutchings Photography/Digital Light Source; **427** (t)Biosphoto/Bringard Denis/Peter Arnold, Inc., (c)Steve Gschmeissner/Photo Researchers, (b)Cyril Ruoso/ JH Editorial/Minden Pictures, (inset)Biosphoto/Lopez Georges/Peter Arnold, Inc.; **428** (t)Wim van Egmond/Visuals Unlimited/Getty Images, (c)Jeff Vanuga/CORBIS, (b) Alfred Pasieka/Photo Researchers; **431** (l)Biophoto Associates/Photo Researchers, (r)Andrew Syred/Photo Researchers; **432** (inset)Ed Reschke/Peter Arnold, Inc., (bkgd)Michael Drane/Alamy; **434** StockTrek/Getty Images; **446–447** The Visible Human Project, U.S. National Library of Medicine; **448** Hutchings Photography/Digital Light Source; **449** (t)Ralph Hutchings/Visuals Unlimited/Getty Images, (b)John Terence Turner/Alamy; **450** Hutchings Photography/Digital Light Source; **451** Jill Braaten/The McGraw-Hill Companies; **453** Holger Winkler/zefa/CORBIS; **455** Image Source/Jupiter Images; **457** C Squared Studios/Getty Images; **458** Omikron/Photo Researchers; **462** Hutchings Photography/Digital Light Source; **463** YOSHIKAZU TSUNO/AFP/Getty Images; **464**)CMCD/Getty Images; **465** (l)Dr, Gladden Willis/Visuals Unlimited/Getty Images, (r)Innerspace Imaging/Photo Researchers; **466** Doug Pensinger/Getty Images; **469** Mark Andersen/Getty Images; **471** (t)Andrew Paul Leonard/Photo Researchers, (b)National Marrow Donor Program, (bkgd)Steve Gschmeissner/Photo Researchers; **472** Hutchings Photography/Digital Light Source; **473** David M. Phillips/Photo Researchers; **475** (tl)Dr G. Moscoso/Photo Researchers, (tc)Anatomical Travelogue/Photo Researchers, (tr)Anatomical Travelogue/Photo Researchers, (bl)Biophoto Associates/Photo Researchers, (br)Neil Bromhall/Photolibrary; **476** Digital Vision; **490–491** Eye of Science/Photo Researchers; **492** Hutchings Photography/Digital Light Source; **493** Dr. Tony Brain/Photo Researchers; **494** (l)Dr. David M. Phillips/Visuals Unlimited/Getty Images, (c)Medical-on-Line/Alamy, (r)Visuals Unlimited/CORBIS; **495** Stefan Sollfors/Getty Images; **496** Science VU/Visuals Unlimited/Getty Images; **497** (l)Richard Du Toit/Minden Pictures, (r)Beverly Joubert/National Geographic/Getty Images; **498** Stefan Sollfors/Getty Images; **499** (tl) Photodisc/Getty Images, (tr)Michael Rosenfeld/Getty Images, (c)Rich Legg/iStockphoto, (bl)Science Source/Photo Researchers, (br)Hercules Robinson/Alamy, (bkgd)Lawrence Lawry/Getty Images; **500** (t)Hutchings Photography/Digital Light Source; **501** Dante Fenolio/Photo Researchers; **502** Wally Eberhart/Visuals Unlimited/Getty Images; **503** Spencer Platt/Getty Images; **504** Zephyr/Photo Researchers; **506** (c)Eye of Science/Photo Researchers, (b)Zephyr/Photo Researchers; **509** Juliette Wade/Gap Photo/Visuals Unlimited; **516** Wally Eberhart/Visuals Unlimited/Getty Images; **519** Eye of Science/Photo Researchers; **522** (r)Manfred Kage/Peter Arnold, Inc., (t to b)Medical-on-Line/Alamy, (2)Visuals Unlimited/CORBIS, (3)Dr. David M. Phillips/Visuals Unlimited/Getty Images, (4)Science VU/Visuals Unlimited/Getty Images; **523** Wally Eberhart/Visuals Unlimited/Getty Images; **528–529** Ed Reschke/Peter Arnold, Inc.; **530** Hutchings Photography/Digital Light Source; **531** (l)Steve Gschmeissner/Photo Researchers, (t)Visuals Unlimited/CORBIS, (b)Flip Nicklin/Minden Pictures/Getty Images; **532** (l)Flip Nicklin/Minden Pictures/Getty Images, (c)Michael Abbey/Photo Researchers, (r)Matt Meadows/Peter Arnold, Inc.; **533** (t) Manfred Kage/Peter Arnold, Inc., (c)Wim van Egmond/Visuals Unlimited/Getty Images; (b)Wim van Egmond/Visuals Unlimited/Getty Images; **534** (tl)Melba Photo Agency/PunchStock, (tr)WILDLIFE/Peter Arnold, Inc., (bl)M. I. Walker/Photo Researchers, (br)blickwinkel/Alamy; **535** (t)Images&Stories/Alamy; **535** (b)Bill Bachman/Photo Researchers; **537** Michael Abbey/Photo Researchers; **539** Mark Steinmetz; **541** (l)Ashley Cooper/Alamy, (r) JUPITERIMAGES/Brand X /Alamy, (bkgd)Kaz Chiba/Getty Images, Global Warming Images / Alamy; **542** Hutchings Photography/Digital Light Source; **543** Felix Labhardt/Getty Images; **544** Mark Steinmetz; **545** Hutchings Photography/Digital Light Source, (inset)Gregory G. Dimijian/Photo Researchers; **547** www.offwell.info, (t)Hutchings Photography/Digital Light Source, **549** C. James Webb / Phototake; **550** David Robertson/stockscotland.

Name ______________________ Date ______________ Class __________

Science Benchmark Practice Test

Multiple Choice *Bubble the correct answer.*

1 A scientist collected sediment from four different locations along the St. Johns River. The pictures below show each sample. They differ in how much they have been sorted and eroded. Which sample has likely been moved around the most by wind or water? **SC.6.E.6.1**

Ⓐ

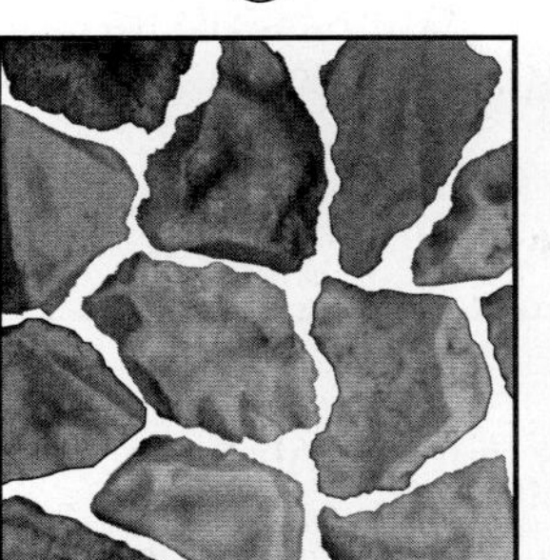

Ⓑ

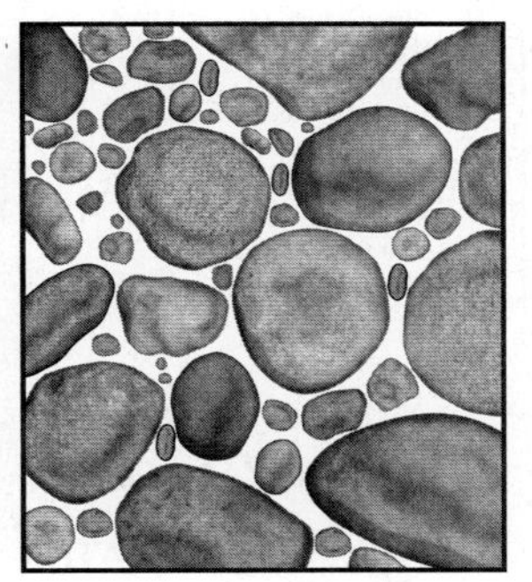

Ⓒ

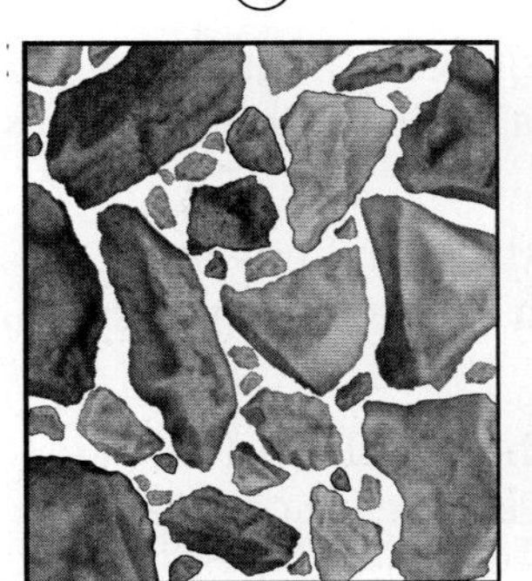

Ⓓ

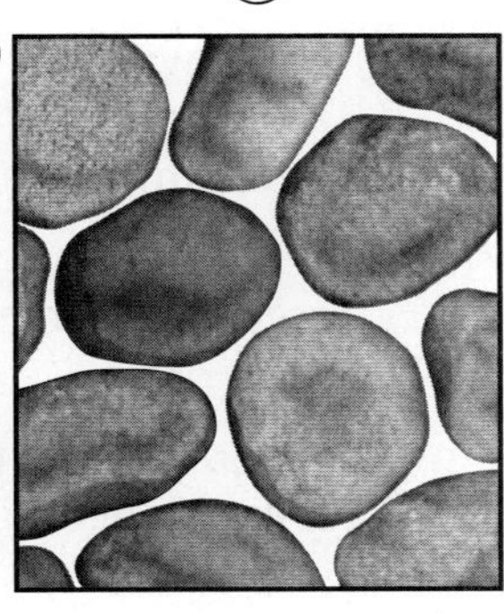

2 In northern Florida, the Suwannee River empties into the Gulf of Mexico. At the mouth of the river, there is a flat, fan-shaped region of land called a delta. Which process created the delta? **SC.6.E.6.2**

Ⓕ lava flow

Ⓖ glacial erosion

Ⓗ rock compression

Ⓘ sediment deposition

3 Three ways that heat transfers through Earth systems are radiation, conduction, and convection. Radiation from the Sun heats Earth's land. The warm land heats air near it by conduction. Which situation involves convection? **SC.6.E.7.1**

Ⓐ An air mass just above a warm ocean current absorbs thermal energy from the warmer water.

Ⓑ Heated air is less dense than cooler air above it, so it rises, warming the air at higher altitudes.

Ⓒ The bright surfaces of clouds reflect sunlight and will be heating the air above them.

Ⓓ Water changing from one phase to another exchanges thermal energy with the surrounding air.

NGSSS for Science Benchmark Practice continued

4 More than half of Earth's freshwater is frozen year-round as snow and ice. It is on mountaintops, in polar icecaps, and in glaciers. Climate change resulting in global warming could decrease the amount of ice on Earth. What could happen as a result of this decrease? **SC.6.E.7.4**

Ⓕ The drinkable freshwater available for human use could increase.

Ⓖ The ocean levels could rise due to an increase in liquid water worldwide.

Ⓗ The total amount of water, including freshwater and salt water, on Earth could increase.

Ⓘ The total amount of water, including freshwater and salt water, on Earth could decrease.

5 During the evening news, a meteorologist used the map below to describe different kinds of fronts.

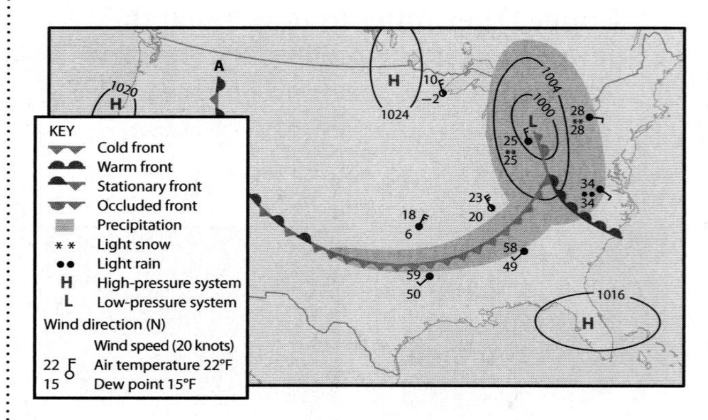

Which statement is true of this type of map? **SC.6.E.7.6**

Ⓐ It shows what the weather is in an area by giving current atmospheric conditions.

Ⓑ It shows what the climate is in an area by giving a range of temperatures over time.

Ⓒ It shows what the climate is in an area by giving a range of air pressures over time.

Ⓓ It shows how the weather in an area relates to climate by comparing conditions.

NGSSS for Science Benchmark Practice continued

6 A recommendation on the police website for the University of West Florida is to stay away from large metal objects and unplug electrical cords during severe thunderstorms. What does this safety tip prevent? **SC.6.E.7.8**

- (F) a surge of electricity from a lightning strike that is traveling through the wires
- (G) short circuits in electrical cords when heavy rains from storms cause flooding
- (H) injuries resulting when hail and powerful winds break windows and tip over heavy metal objects
- (I) injuries from heavy metal objects falling over because of violent vibrations from loud thunder

7 When Carlita goes sailing, she wears sunscreen to protect herself from the ultraviolet rays that reach Earth's surface. The table below shows the atmosphere's five main layers.

Layer	Altitude Range
Exosphere	500 km and higher
Thermosphere	85 to 500 km
Mesosphere	50 to 85 km
Stratosphere	15 to 50 km
Troposphere	0 to 15 km

Which layer contains a high concentration of ozone gas that absorbs much of the Sun's ultraviolet rays? **SC.6.E.7.9**

- (A) exosphere
- (B) stratosphere
- (C) thermosphere
- (D) troposphere

Name ______________________ Date ______________ Class __________

NGSSS for Science Benchmark Practice continued

8 As Bernie swings back and forth on a swing at a regular rate, energy transformations take place. At the lowest point in his swing, he moves the fastest. As he swings upward, he slows down. At the highest point in his swing, he momentarily slows to a stop. Which statement must be true? **SC.6.P.11.1**

- (F) He always has the same amounts of kinetic and potential energy.
- (G) He has the least kinetic energy at the lowest point in his swing.
- (H) He has the most potential energy at the highest point in his swing.
- (I) His total kinetic energy and potential energy changes as he swings.

9 Sodium chloride, also known as table salt, is an ionic compound made up of positive sodium ions and negative chloride ions. The diagram below models how sodium chloride dissolves in water.

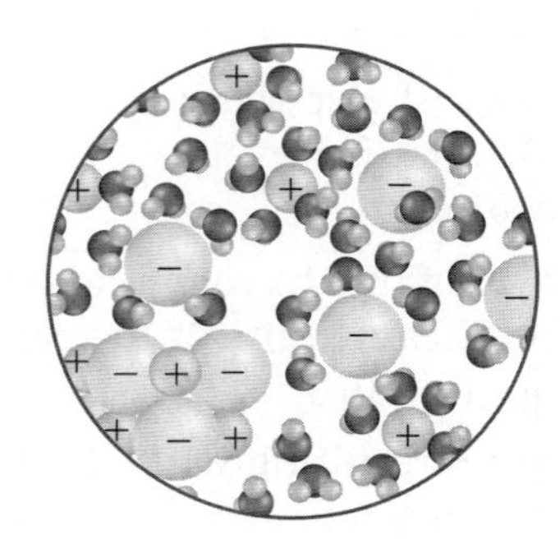

What type of force attracts the positive ends of the water molecules to the negative chloride ions? **SC.6.P.13.1**

- (A) contact force
- (B) electrical force
- (C) gravitational force
- (D) magnetic force

10 There is a greater force of gravity between objects that have a lot of mass than between objects that have very little mass. What other factor affects the gravity between two objects? **SC.6.P.13.2**

(F) the shape of the objects
(G) the volume of the objects
(H) the distance between objects
(I) the temperature of objects

11 Imagine that a play's stage crew must move a 30-kg piece of scenery on and off the stage. As they move the piece of scenery off the stage at a constant speed, the forces shown in the diagram below suddenly affect it.

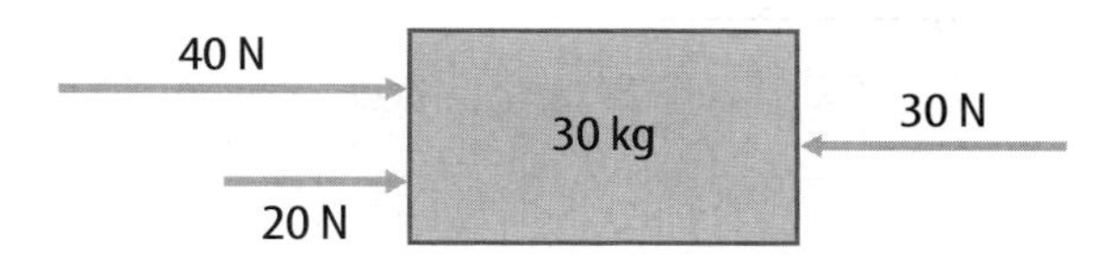

Which statement best describes what happens to the piece of scenery and why? **SC.6.P.13.3**

(A) Its motion stops because the forces acting on it are balanced.
(B) Its motion changes because the forces acting on it are balanced.
(C) Its motion changes because the forces acting on it are unbalanced.
(D) Its motion doesn't change because the forces acting on it are unbalanced.

12 The cell theory says that all cells come from other cells and that all living things are made of cells. According to the cell theory, what is the smallest unit of life? **SC.6.L.14.2**

(F) atom
(G) cell
(H) molecule
(I) organism

13 The illustration below shows a part of the human body.

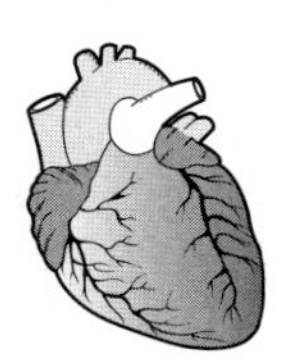

Which level of organization does this body part best represent? **SC.6.L.14.1**

(A) organ
(B) organism
(C) organ system
(D) tissue system

NGSSS for Science Benchmark Practice continued

14 Cellular respiration converts the energy in food molecules into a form of energy that a cell can use. This process begins with glycolysis, which occurs in the cytoplasm. Which organelle completes cellular respiration? **SC.6.L.14.4**

Ⓕ chloroplast

Ⓖ mitochondrion

Ⓗ nucleus

Ⓘ vacuole

15 Both single-celled organisms and the cells of multicellular organisms go through the process shown below.

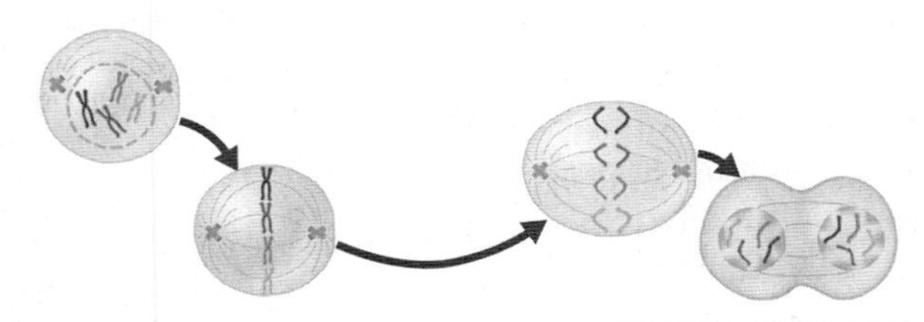

For a multicellular organism, this process can result in growth. For a single-celled organism, what is the result of this process? **SC.6.L.14.3**

Ⓐ growth

Ⓑ reproduction

Ⓒ wound healing

Ⓓ energy production

NGSSS for Science Benchmark Practice continued

16 All living things can be classified into one of three domains and then into one of six kingdoms. The table below shows which kingdoms are in each domain.

Domains	Kingdoms
Bacteria	Bacteria
Archaea	Archaea
Eukarya	Protista
	Fungi
	Plantae
	Animalia

If a new single-celled organism with a nucleus were discovered, into which domains might it be classified? **SC.6.L.15.1**

Ⓕ Archaea or Bacteria

Ⓖ Archaea or Eukarya

Ⓗ Bacteria or Eukarya

Ⓘ Bacteria, Archaea, or Eukarya

17 Sofia's 10-year study of a bird species has shown that its members make nests of twigs and dried grasses that measure 5 cm to 7 cm in diameter. This year, these birds built nests that were almost all over 7 cm in diameter. Over half of the nests observed contained bits of metal or silver-colored wrappers. What will Sofia need to do to develop an explanation that fits her observations? **SC.6.N.1.5**

Ⓐ change results

Ⓑ think creatively

Ⓒ determine where she made an error

Ⓓ make new observations more easily explained

18 Francisco has been trying different combinations of dyes for his sculpture. Tyfara is conducting an experiment that compares how long certain dyes will last under different conditions and on different materials. What makes Tyfara's work with dyes different from Francisco's work with dyes? **SC.6.N.2.1**

Ⓕ Tyfara's work is based on facts.

Ⓖ Tyfara's work must be reproducible.

Ⓗ Tyfara's work cannot have mistakes or errors.

Ⓘ Tyfara's work is easier than Francisco's work.

19 An ulcer is a type of sore that occurs in a person's stomach. For many years, it was thought that several factors, including stress, caused ulcers. It is now thought that a bacterial infection, not stress, causes ulcers. Which could have led to this change in scientific understanding? **SC.6.N.2.2**

Ⓐ new types of ulcers

Ⓑ new viral outbreaks

Ⓒ new opinions on ulcers

Ⓓ new evidence about ulcers

NGSSS for Science Benchmark Practice continued

20 Some scientists work outdoors, observing organisms or habitats. Other scientists work indoors, in a laboratory or office. Almost all scientists collaborate with and present their work to other scientists. Which best describes the characteristics of a scientist? **SC.6.N.2.3**

- (F) Scientists have a variety of backgrounds, talents, interests, and goals.
- (G) Scientists have a variety of goals, but being famous is number one.
- (H) Scientists have a variety of interests, but all must be good at math.
- (I) Scientists have a variety of talents, but all like working in a laboratory.

21 Karim's theory is that his math teacher gives pop quizzes on Mondays. Amal's theory is that their teacher gives pop quizzes on Fridays. How are scientific theories different from Karim's and Amal's theories? **SC.6.N.3.1**

- (A) Only one individual can make a scientific theory about a topic.
- (B) Scientific theories must be voted on and win the majority vote.
- (C) Scientific theories are supported by evidence and are widely accepted.
- (D) There are usually more than two scientific theories about the same topic.

22 Scientific laws describe patterns in nature. Which is an example of a scientific law? **SC.6.N.3.3**

- (F) For every action, there is an equal but opposite reaction.
- (G) It is illegal to do research on human beings without permission.
- (H) Within park boundaries, owners of dogs without leashes may be fined.
- (I) If fruit flies prefer apples to bananas, then they will land more on apple slices.

23 A scientific law is different from a societal law. Which areas of science are governed by scientific law? **SC.6.N.3.2**

- (A) gravity and conservation of energy
- (B) lab safety and experimental procedure
- (C) plate tectonics and natural selection
- (D) the use of hypotheses and inferences

NGSSS for Science Benchmark Practice continued

24 Natalia is studying the processes of the water cycle that are shown in the illustration below.

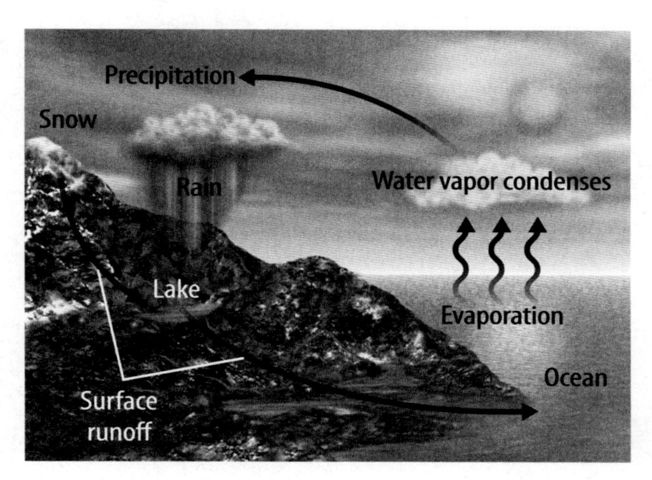

Natalia made a model of the water cycle. She put water, rocks, soil, and two small plants in a jar, sealed it, and placed it in sunlight. What advantage does studying a model have over studying the actual water cycle? **SC.6.N.3.4**

Ⓕ The model increases complexity of the cycle.

Ⓖ The model shows the processes more slowly.

Ⓗ The model can be measured much more accurately.

Ⓘ The model can be observed from the classroom more easily.

25 Although bacteria and viruses differ in many ways, they share some similarities. What is one similarity between a bacterium and a virus? **SC.6.L.14.6**

Ⓐ Both can be performed by mitosis.

Ⓑ Both can be destroyed by antibiotics.

Ⓒ Both can fix nitrogen in plant roots.

Ⓓ Both can cause illness in the human body.

PERIODIC TABLE OF THE ELEMENTS

Element	Hydrogen
Atomic number	1
Symbol	H
Atomic mass	1.01
State of matter	

Gas
Liquid
Solid
Synthetic

	1	2	3	4	5	6	7	8	9
1	Hydrogen 1 **H** 1.01								
2	Lithium 3 **Li** 6.94	Beryllium 4 **Be** 9.01							
3	Sodium 11 **Na** 22.99	Magnesium 12 **Mg** 24.31							
4	Potassium 19 **K** 39.10	Calcium 20 **Ca** 40.08	Scandium 21 **Sc** 44.96	Titanium 22 **Ti** 47.87	Vanadium 23 **V** 50.94	Chromium 24 **Cr** 52.00	Manganese 25 **Mn** 54.94	Iron 26 **Fe** 55.85	Cobalt 27 **Co** 58.93
5	Rubidium 37 **Rb** 85.47	Strontium 38 **Sr** 87.62	Yttrium 39 **Y** 88.91	Zirconium 40 **Zr** 91.22	Niobium 41 **Nb** 92.91	Molybdenum 42 **Mo** 95.96	Technetium 43 **Tc** (98)	Ruthenium 44 **Ru** 101.07	Rhodium 45 **Rh** 102.91
6	Cesium 55 **Cs** 132.91	Barium 56 **Ba** 137.33	Lanthanum 57 **La** 138.91	Hafnium 72 **Hf** 178.49	Tantalum 73 **Ta** 180.95	Tungsten 74 **W** 183.84	Rhenium 75 **Re** 186.21	Osmium 76 **Os** 190.23	Iridium 77 **Ir** 192.22
7	Francium 87 **Fr** (223)	Radium 88 **Ra** (226)	Actinium 89 **Ac** (227)	Rutherfordium 104 **Rf** (267)	Dubnium 105 **Db** (268)	Seaborgium 106 **Sg** (271)	Bohrium 107 **Bh** (272)	Hassium 108 **Hs** (270)	Meitnerium 109 **Mt** (276)

A column in the periodic table is called a **group.**

A row in the periodic table is called a **period.**

The number in parentheses is the mass number of the longest lived isotope for that element.

Lanthanide series	Cerium 58 **Ce** 140.12	Praseodymium 59 **Pr** 140.91	Neodymium 60 **Nd** 144.24	Promethium 61 **Pm** (145)	Samarium 62 **Sm** 150.36	Europium 63 **Eu** 151.96
Actinide series	Thorium 90 **Th** 232.04	Protactinium 91 **Pa** 231.04	Uranium 92 **U** 238.03	Neptunium 93 **Np** (237)	Plutonium 94 **Pu** (244)	Americium 95 **Am** (243)

Metal
Metalloid
Nonmetal
Recently discovered

10	11	12	13	14	15	16	17	18
								Helium 2 **He** 4.00
			Boron 5 **B** 10.81	Carbon 6 **C** 12.01	Nitrogen 7 **N** 14.01	Oxygen 8 **O** 16.00	Fluorine 9 **F** 19.00	Neon 10 **Ne** 20.18
			Aluminum 13 **Al** 26.98	Silicon 14 **Si** 28.09	Phosphorus 15 **P** 30.97	Sulfur 16 **S** 32.07	Chlorine 17 **Cl** 35.45	Argon 18 **Ar** 39.95
Nickel 28 **Ni** 58.69	Copper 29 **Cu** 63.55	Zinc 30 **Zn** 65.38	Gallium 31 **Ga** 69.72	Germanium 32 **Ge** 72.64	Arsenic 33 **As** 74.92	Selenium 34 **Se** 78.96	Bromine 35 **Br** 79.90	Krypton 36 **Kr** 83.80
Palladium 46 **Pd** 106.42	Silver 47 **Ag** 107.87	Cadmium 48 **Cd** 112.41	Indium 49 **In** 114.82	Tin 50 **Sn** 118.71	Antimony 51 **Sb** 121.76	Tellurium 52 **Te** 127.60	Iodine 53 **I** 126.90	Xenon 54 **Xe** 131.29
Platinum 78 **Pt** 195.08	Gold 79 **Au** 196.97	Mercury 80 **Hg** 200.59	Thallium 81 **Tl** 204.38	Lead 82 **Pb** 207.20	Bismuth 83 **Bi** 208.98	Polonium 84 **Po** (209)	Astatine 85 **At** (210)	Radon 86 **Rn** (222)
Darmstadtium 110 **Ds** (281)	Roentgenium 111 **Rg** (280)	Copernicium 112 **Cn** (285)	Ununtrium * 113 **Uut** (284)	Ununquadium * 114 **Uuq** (289)	Ununpentium * 115 **Uup** (288)	Ununhexium * 116 **Uuh** (293)		Ununoctium * 118 **Uuo** (294)

* The names and symbols for elements 113-116 and 118 are temporary. Final names will be selected when the elements' discoveries are verified.

Gadolinium 64 **Gd** 157.25	Terbium 65 **Tb** 158.93	Dysprosium 66 **Dy** 162.50	Holmium 67 **Ho** 164.93	Erbium 68 **Er** 167.26	Thulium 69 **Tm** 168.93	Ytterbium 70 **Yb** 173.05	Lutetium 71 **Lu** 174.97
Curium 96 **Cm** (247)	Berkelium 97 **Bk** (247)	Californium 98 **Cf** (251)	Einsteinium 99 **Es** (252)	Fermium 100 **Fm** (257)	Mendelevium 101 **Md** (258)	Nobelium 102 **No** (259)	Lawrencium 103 **Lr** (262)